Cylindrical coordinates

$$\nabla\Phi = \mathbf{a}_r \frac{\partial \Phi}{\partial r} + \mathbf{a}_\phi \frac{1}{r}\frac{\partial \Phi}{\partial \phi} + \mathbf{a}_z \frac{\partial \Phi}{\partial z}$$

$$\nabla \cdot \mathbf{A} = \frac{1}{r}\frac{\partial}{\partial r}(rA_r) + \frac{1}{r}\frac{\partial A_\phi}{\partial \phi} + \frac{\partial A_z}{\partial z}$$

$$\nabla \times \mathbf{A} = \mathbf{a}_r\left(\frac{1}{r}\frac{\partial A_z}{\partial \phi} - \frac{\partial A_\phi}{\partial z}\right) + \mathbf{a}_\phi\left(\frac{\partial A_r}{\partial z} - \frac{\partial A_z}{\partial r}\right) + \mathbf{a}_z\left[\frac{1}{r}\frac{\partial (rA_\phi)}{\partial r} - \frac{1}{r}\frac{\partial A_r}{\partial \phi}\right]$$

$$\nabla^2\Phi = \frac{1}{r}\frac{\partial}{\partial r}\left(r\frac{\partial \Phi}{\partial r}\right) + \frac{1}{r^2}\frac{\partial^2 \Phi}{\partial \phi^2} + \frac{\partial^2 \Phi}{\partial z^2}$$

Spherical coordinates

$$\nabla\Phi = \mathbf{a}_r \frac{\partial \Phi}{\partial r} + \mathbf{a}_\theta \frac{1}{r}\frac{\partial \Phi}{\partial \theta} + \mathbf{a}_\phi \frac{1}{r\sin\theta}\frac{\partial \Phi}{\partial \phi}$$

$$\nabla \cdot \mathbf{A} = \frac{1}{r^2}\frac{\partial}{\partial r}(r^2 A_r) + \frac{1}{r\sin\theta}\frac{\partial}{\partial \theta}(\sin\theta A_\theta) + \frac{1}{r\sin\theta}\frac{\partial A_\phi}{\partial \phi}$$

$$\nabla \times \mathbf{A} = \mathbf{a}_r\left\{\frac{1}{r\sin\theta}\left[\frac{\partial}{\partial \theta}(A_\phi \sin\theta) - \frac{\partial A_\theta}{\partial \phi}\right]\right\} + \mathbf{a}_\theta\left[\frac{1}{r\sin\theta}\frac{\partial A_r}{\partial \phi} - \frac{1}{r}\frac{\partial}{\partial r}(rA_\phi)\right] + \mathbf{a}_\phi\left\{\frac{1}{r}\left[\frac{\partial}{\partial r}(rA_\theta) - \frac{\partial A_r}{\partial \theta}\right]\right\}$$

$$\nabla^2\Phi = \frac{1}{r^2}\frac{\partial}{\partial r}\left(r^2\frac{\partial \Phi}{\partial r}\right) + \frac{1}{r^2\sin\theta}\frac{\partial}{\partial \theta}\left(\sin\theta\frac{\partial \Phi}{\partial \theta}\right) + \frac{1}{r^2\sin^2\theta}\frac{\partial^2 \Phi}{\partial \phi^2}$$

INTRODUCTION TO ELECTROMAGNETIC FIELDS

McGRAW-HILL SERIES IN ELECTRICAL ENGINEERING

Consulting Editor
Stephen W. Director, *Carnegie-Mellon University*

CIRCUITS AND SYSTEMS
COMMUNICATIONS AND SIGNAL PROCESSING
CONTROL THEORY
ELECTRONICS AND ELECTRONIC CIRCUITS
POWER AND ENERGY
ELECTROMAGNETICS
COMPUTER ENGINEERING
INTRODUCTORY
RADAR AND ANTENNAS
VLSI

Previous Consulting Editors

Ronald N. Bracewell, Colin Cherry, James F. Gibbons, Willis W. Harman, Hubert Heffner, Edward W. Herold, John G. Linvill, Simon Ramo, Ronald A. Rohrer, Anthony E. Siegman, Charles Susskind, Frederick E. Terman, John G. Truxal, Ernst Weber, and John R. Whinnery

ELECTROMAGNETICS

Consulting Editor
Stephen W. Director, *Carnegie-Mellon University*

Chen THEORY OF ELECTROMAGNETIC WAVES: A Coordinate-Free Approach
Dearhold and McSpadden ELECTROMAGNETIC WAVE PROPAGATION
Goodman INTRODUCTION TO FOURIER OPTICS
Hayt ENGINEERING ELECTROMAGNETICS
Kraus ELECTROMAGNETICS
Paul and Nasar INTRODUCTION TO ELECTROMAGNETIC FIELDS
Plonus APPLIED ELECTROMAGNETICS

INTRODUCTION TO ELECTROMAGNETIC FIELDS

SECOND EDITION

Clayton R. Paul
Syed A. Nasar

Department of Electrical Engineering
University of Kentucky

McGraw-Hill Book Company

New York St. Louis San Francisco Auckland Bogotá Hamburg
London Madrid Mexico Milan Montreal New Delhi
Panama Paris São Paulo Singapore Sydney Tokyo Toronto

INTRODUCTION TO ELECTROMAGNETIC FIELDS

34567890 DOCDOC 894321098

ISBN 0-07-045908-8

This book was set in Times Roman by Composition House Limited. The editors were Sanjeev Rao and J. W. Maisel; the designer was Rafael Hernandez; the production supervisor was Leroy A. Young. The drawings were done by J & R Services, Inc. R. R. Donnelley & Sons Company was printer and binder.

Library of Congress Cataloging-in-Publication Data

Paul, Clayton R.
Introduction to electromagnetic fields.
(McGraw-Hill series in electrical engineering. Electromagnetics)
Includes bibliographical references and index.
1. Electromagnetic fields. I. Nasar, S.A. II. Title. III. Series.
QC665.E4P38 1987 537 86-18545
ISBN 0-07-045908-8
ISBN 0-07-045909-6 (solutions manual)

ABOUT THE AUTHORS

Clayton R. Paul is a professor of electrical engineering at the University of Kentucky in Lexington. He received the bachelor of science degree from the Citadel, the master of science degree from Georgia Institute of Technology and the doctor of philosophy degree from Purdue University, all in electrical engineering.

He has been a member of the faculty of the Department of Electrical Engineering at the University of Kentucky since 1971. He has published numerous articles in professional journals concerning his research into interference effects of electromagnetic phenomena in electronic systems.

His primary academic interests are the modeling and prediction of crosstalk in cables and printed circuit boards.

Syed A. Nasar received the Ph.D. degree in electrical engineering from the University of California, Berkeley.

He is a professor and director of graduate studies in electrical engineering at the University of Kentucky, Lexington; the author or coauthor of 16 other books and over 100 journal papers; and the chief editor of *Electric Machines and Power Systems*—an international monthly. He has been involved in teaching, research, and consulting in electrical engineering for 30 years.

He is a Fellow of IEEE, a Fellow of the IEE (London), and a member of Sigma Xi and Eta Kappa Nu.

To
the memory of my father,
Oscar Paul,
November 8, 1896–November 27, 1974.
By the conduct of his life,
he taught compassion and fairness.

Clayton R. Paul

To
Professor Robert M. Saunders,
who made me aware
of the immensity that lies
beyond Maxwell's equations,
with gratitude and best regards.

Syed A. Nasar

CONTENTS

CHAPTER 5

Maxwell's Equations 215

CHAPTER 6

Propagation of Uniform Plane Waves 277

CHAPTER 10
Equations Governing Potential Functions 571

APPENDIX A
Vector Identities and Vector Operations 643

APPENDIX B
Faraday's Law for Moving Contours 649

APPENDIX C
The Smith Chart 669

Index 735

PREFACE

This text is intended for use as an introduction to the subject of electromagnetic fields at the undergraduate level of an electrical engineering curriculum. As a prerequisite, we assume that the reader has the standard background in calculus, differential equations, and elementary physics.

The subject of electromagnetic fields is perhaps the most fundamentally important topic in electrical engineering. Therefore, the student should be motivated by the material and its presentation to engage in a serious study of the topic. Since the quantities of interest are functions of not only time but also spatial parameters, the material is inherently more difficult for the beginning student than, for example, electric circuit theory. In order to solve most electromagnetic fields problems, the student must be able to visualize and understand the meaning of the governing equations. Consequently, an important aspect of the presentation of this material is a clear explanation of the fundamental principles and concepts. Our intent in writing this text is to make the subject matter interesting and motivating and to present the important concepts with a minimum of unnecessary detail so that the reader can distinguish the "forest from the trees."

This text grew out of a need for an introductory electromagnetic fields text which bridges the gap between the existing texts which cover static fields in considerable detail but do not give sufficient coverage of time-varying fields and those that cover both topics but with considerably more detail and sophistication than is required at this level of instruction. We presume that the student has been introduced to the basic static field concepts such as Coulomb's law and Gauss' law through the standard elementary physics courses. Consequently, the discussion of static field concepts is minimized so that the more important topics of time-varying fields and the engineering applications (uniform plane waves, transmission lines, waveguides, and antennas) can be covered in sufficient depth.

Chapter 1 provides an introduction and motivational survey. In revising Chapter 1 we have included more illustrative examples of applications of electromagnetic field theory. Chapter 2 presents all of the necessary vector algebra and vector calculus tools and concepts. Much of the material in Chapter 2 is review material, which can be covered rapidly. In this edition we have included a discussion of generalized, orthogonal coordinate systems in order to provide a more unified basis for discussion of the specific coordinate systems. Chapter 3 contains the static electric field concepts (Coulomb's law, electric field, Gauss' law, potential, energy, capacitance, resistance, and mechanical forces). A section on the concept and calculation of resistance for arbitrary structures has been added in this second edition. This follows the section on the calculation of capacitance for arbitrary structures present in the first edition. A section on power dissipation has also been included in this edition. Chapter 4 presents the static magnetic field concepts (the Biot-Savart law, Ampère's law, energy, inductance, mechanical forces, and magnetic circuits). This chapter has been substantially rewritten to emphasize the duality between the electric and magnetic fields. Relatively brief discussions of material properties are included in Chapters 3 and 4. In many institutions, these static field topics are covered to some degree in the elementary physics courses. In this case, Chapters 3 and 4 may serve as review material.

Chapter 5 begins the discussion of time-varying field concepts by introducing and discussing Maxwell's equations, the boundary conditions, Poynting vector, and the important sinusoidal, steady-state solution technique. This chapter is essentially unchanged from the first edition. Chapter 6 begins the discussion of the applications and implications of Maxwell's equations from an engineering standpoint. The concept of electromagnetic waves is discussed in considerable detail so that this fundamental concept will be firmly understood. Reflection and transmission of uniform plane waves as well as polarization of these waves are also discussed. We have added a section on group velocity which was not present in the first edition. A major addition in this second edition is the discussion of oblique incidence of uniform plane waves on plane material boundaries. The topics of plane wave propagation in arbitrary space directions, Snell's laws, Brewster angle, and the critical angle have been included along with the general developments on oblique incidence.

Chapter 7 contains a discussion of wave propagation on transmission lines. Both transient and sinusoidal, steady-state behavior are discussed. The emphasis is on fundamental principles, and the Smith chart and its applications are discussed in detail in Appendix C. We have included in this second edition a discussion of the per-unit-length resistance calculation for the standard transmission line structures which was not present in the first edition. Chapter 8 presents a discussion of rectangular waveguides, which, although brief, covers the essential points. In this second edition, we have added sections on attenuation in waveguides, and cavity resonators. The intent again was to highlight the essential concepts and keep the discussion brief.

The topic of antennas is covered in Chapter 9 in somewhat more detail than is customary in a text aimed at this level. The elemental, electric (Hertzian dipole), the elemental, magnetic (loop) dipole, and the long, linear dipole are discussed in a somewhat standard manner as is the topic of linear arrays. Antenna directivity and gain are also discussed. Coupling between two antennas is considered, and the important concepts of reciprocity with regard to impedance and pattern for an antenna in either a transmitting or receiving mode are derived. The Friis transmission equation is also derived. A section on the effect of reflections from an imperfect ground has been added to this second edition.

Chapter 10 contains the traditional techniques for solution of static field problems for which simple, closed-form solutions are not obtainable. Discussions of Laplace's equation in cylindrical and spherical coordinates are added in this revised version, and the section on image methods has been expanded. Solution techniques for Laplace's and Poisson's equations, as well as numerical methods (finite-difference and method of moments) and analog and graphical methods, are discussed. Ordinarily, in other texts, this chapter is placed after the material covered in Chapter 4. We have chosen to include this chapter as the last one in the text so that the reader is led to an early consideration of the important topics of time-varying fields after a brief review of static field concepts in Chapters 3 and 4. With this organization, the reader has a proper appreciation of the hierarchy of importance of the material in each chapter.

Appendix A summarizes various vector identities, vector calculus operations, and transformations between coordinate systems. Appendix B contains a discussion of Faraday's law for moving contours which is unchanged from the first edition. In Appendix C, the discussion of the Smith chart is essentially the same as in the first edition. We have, however, added sections on double stub tuners, quarter-wave transformers, broadband matching and pads and use of the chart for lossy lines.

Most of the additions to the first edition were suggested by those who have used the text. We are grateful to those who have provided their comments for improvement.

Over half the end-of-chapter problems are new in this edition. Answers to selected problems are given at the end of the text, as was requested by most of the users of the first edition.

The text should be suitable for either a one-semester or a two-semester sequence in electromagnetic fields. In a one-semester course, it would be appropriate and possible to review Chapter 2 (vectors) and cover Chapters 5, 6, 7, 8, and 9, and Appendix C (Smith charts). A two-semester sequence would be a more leisurely coverage of the entire text. Chapters 2, 3, 4, 10, 5, and possibly a portion of Chapter 6 may be covered in detail in the first semester, and the remainder of Chapter 6 and Chapters 7, 8, 9 and Appendixes B and C may be covered in the second semester.

The question of what is the most appropriate and effective way of presenting electromagnetic fields at this level is difficult to answer. We have chosen to blend the more traditional approach of discussing static fields first with the attitude of minimizing that discussion in order to get to the topic of time-varying fields as soon as possible. This approach seems to have the pedagogical advantage of discussing the more easily understood static field concepts first before delving into the inherently more difficult time-varying field concepts. In line with the attitude that it is essential that the reader visualize and understand the basic concepts in electromagnetic fields in order to begin to master the subject matter, we have tried to simplify the notation and minimize the mathematical details where possible. Elective and graduate courses will be able to delve more deeply into the details once these basic concepts are firmly understood.

The very capable typing of this maunscript and its various modifications by Mrs. Vickie L. Brann is gratefully acknowledged.

We would also like to thank those individuals who took the time to review this manuscript: Kenneth A. Connor, Rensselaer Polytechnic Institute; R. J. Garbacz, Ohio State University; Bhag S. Guru, GMI Engineering and Management Institute; P. R. Herczfeld, Drexel University; Richard Kwor, University of Notre Dame; Michael Steer, North Carolina State University; and Clayborne D. Taylor, Mississippi State University.

Clayton R. Paul
Syed A. Nasar

INTRODUCTION TO ELECTROMAGNETIC FIELDS

CHAPTER 1

Introduction

1.1 Brief Historical Development

The basic concepts of modern electromagnetic field theory have evolved over many years. In fact, an awareness of magnetism appears to be as old as recorded history. The discovery of the polarities of lodestone by Pierre de Maricourt dates to around 1269, and from that time through the early seventeenth century the progress in the study of magnetism was rather slow.[1] During the seventeenth century, however, there was a considerable revival of interest and there were several notable contributions by several scientists toward understanding magnetism. A. Kirchner demonstrated that the two poles of a magnet have equal strength, and Newton attempted to formulate the law for a bar magnet. But the correct inverse square law was postulated by John Michell in 1750 and reconfirmed by Coulomb at a later date. In 1785 Coulomb also demonstrated the law of electric force between charged bodies.† Coulomb's inverse square law (or simply Coulomb's law) may be said to be the starting point of modern electromagnetic field theory. Subsequent landmarks in the development of electromagnetic field theory include the derivations of Laplace's equation in 1782, Poisson's equation in 1813, and Gauss' divergence theorem in the same year. These developments essentially belong to the general topic of electrostatics.

We know that motion of charges constitutes electric current. Experiments with electric current could be performed only after the invention of the battery by Volta in 1800. Having a source of continuous current available, Oersted, in 1820, was able to demonstrate the production of magnetic fields by electric currents. Oersted's discovery prompted others to investigate the relationships

† It is claimed that the inverse square law of electric force was established in 1773 by Cavendish, who did not publicize his findings.[1]

between electric currents and magnetic fields. In 1820, Ampère announced a discovery relating to forces between electric current-carrying conductors and magnets and the mutual attraction (repulsion) of two electric currents. These experiments led to the formulation of Ampère's law. During 1820, Biot and Savart repeated Oersted's experiment to determine a law of force governing the forces between current-carrying conductors and gave us the Biot-Savart law. These developments belong to the general topic of magnetostatics.

During the period of Oersted and Ampère, Faraday was also experimenting on the interaction between current-carrying conductors and magnetic fields, and he developed an electric motor in 1821. Furthermore, Faraday's experiments on developing induced currents by changing the magnetism (or magnetic field) led to the law of electromagnetic induction in 1831.† Faraday also proposed the concept of magnetic lines of force. Thus, the foundation of all electromagnetic phenomena was laid. In 1864, Maxwell proposed "A Dynamical Theory of the Electromagnetic Field" and thus unified the experimental researches of over a century through a set of equations known as Maxwell's equations.[2] These equations were later verified experimentally by Hertz in 1887. It is generally accepted that all macroscopic electromagnetic phenomena are governed by Maxwell's equations.

In the following chapters we will study the basic laws governing electromagnetic fields. Before we take up the details of the analyses, we will look briefly at the range of applications of electromagnetic field theory.

1.2 Some Applications of Electromagnetic Field Theory

It was mentioned previously that Maxwell's equations govern all macroscopic electromagnetic phenomena. Therefore, it is not practicable to list here a large number of applications of electromagnetic field theory. Rather, the general nature of problems that are best handled by field theory will be pointed out, and the broad range of applications of electromagnetic field theory will be identified.

The term *field* is a concept used to describe a distribution of some quantity throughout a region of space. For instance, the electric field is measured by the force on a unit charge of electricity, and the magnetic field is measured by the force on a magnetic dipole. (Electric and magnetic fields will be defined precisely in Chaps. 3 and 4, respectively.) Thus, we notice that fields are three-dimensional spatial phenomena, and the mathematical formulation of field phenomena is always in terms of distributed parameters (in contrast to the lumped-parameter description of electric circuits). We might say that distributed-parameter field phenomena are given by partial differential equations and that lumped-parameter circuit behavior is expressed by ordinary differential equations.

† Joseph Henry of Albany, New York, is said to have made this discovery earlier than Faraday.

Indeed, Maxwell's equations are a set of partial differential equations, as we shall see in Chap. 5. Strictly speaking circuits are approximate analogs for field relationships, and are used for the sake of convenience. A simple example of this approximation is shown in Fig. 1.1. Figure 1.1*a* shows a solid cylinder of length l, area of cross section A, and made of a conducting material of conductivity σ. If we assume a uniform current-density distribution over the conductor cross section, its resistance is $R = l/\sigma A$, and the circuit analog then becomes as shown in Fig. 1.1*b*. The Ohm's law relationship for the solid conductor and its circuit analog are also shown in Fig. 1.1. As we shall see in later chapters, the other two common circuit elements, inductance and capacitance, are also approximate analogs corresponding respectively to magnetic and electric field in a given region.

From the preceding discussion, it is clear that the formulation in terms of fields implies an exact description of the electromagnetic phenomenon occurring in a specified region. There is a danger in using this approach to solve every problem in electrical engineering. Circuit concepts and other valid approximations are convenient to use in numerous situations. Electromagnetic theory must be applied where the approximations leading to the analogs break down. For instance, the radiation of electromagnetic waves from an antenna can only be described by Maxwell's equations. In Fig. 1.1*a*, if the distribution of the current is nonuniform over the conductor cross section, as would be the case for an alternating current of high frequency, the determination of the resistance involves an application of the field equations. The criteria for the use of the field equations are: (1) that the problem at hand is a distributed-parameter problem, (2) that the resulting equations can be solved without unreasonable difficulty,

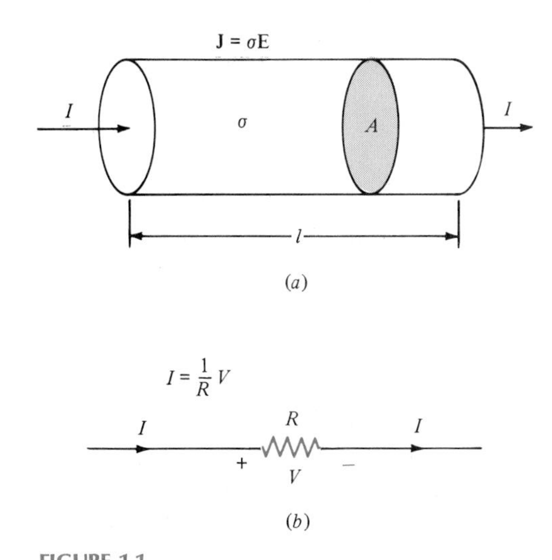

FIGURE 1.1
(*a*) A solid conductor showing the field relationship $\mathbf{J} = \sigma\mathbf{E}$, known as Ohm's law. (*b*) A circuit analog.

and (3) that the approximations that could lead to simpler solutions are not valid. Although we will see numerous examples of applications of the field equations throughout the book, the following list briefly reviews some applications that require a knowledge of electromagnetic fields.

1 *Bioelectromagnetics:* There are two major aspects of biology that require a thorough understanding of electromagnetic field theory. One relates to the modeling of biological organs and phenomena. For instance, the modeling of the heart, the modeling of the mechanism of nerve propagation, and the electrical activity of the brain are based on field theory. Reference 3 gives a good perspective of some electrophysiological systems.

   The second aspect of biology requiring a knowledge of fields theory relates to human exposure to electric and magnetic fields. The biological effects of electromagnetic fields constitute a very interesting study; Refs. 4 through 6 summarize some recent results.

   We also find the application of electric and magnetic fields to medicine in the general areas of diagnosis and treatment. Examples of the use of fields for diagnosis include imaging regions of the human body and investigating the nervous system. In treatment, electric and magnetic fields are used to aid the healing of bone fractures and soft tissue injuries[7,8]

2 *High-intensity magnetic fields:* High-intensity magnetic fields of the order of several tesla, for continuous duty, as well as for pulsed duty, have applications in magnetohydrodynamics, nuclear research, and plasma and fusion physics.[9,10,11] Proper design of magnets, including superconducting magnets, for these applications certainly requires a thorough understanding of field theory. References 9, 10, and 11 give several interesting examples of the design and application of high-field magnets.

3 *Applications to electric machines:* The application of electromagnetic field theory to electric machines ranges from the parameter determination of the machine to the evaluation of its performance characteristics. Although almost every kind of electric machine has been analyzed by the field theory, this area is among the lesser known for the usefulness of Maxwell's equations. As an example, the induction motor has been analyzed on the basis of the Poynting vector, the Lorentz force equation, the concept of wave impedance, and the diffusion equation. Similarly, other types of electric machines have been studied through the aid of field theory. References 12 and 13 give numerous examples of the application of electromagnetic field theory to electric machines.

4 *Traditional engineering applications:* Our study of electromagnetic fields will concern the more common engineering applications of the theory.

Perhaps the more common engineering applications of Maxwell's equations are in the study of microwave devices, antennas, and transmission lines. Examples of such applications are given in the following chapters. Consider a pair of parallel conductors shown in Fig. 1.2*a*, which we refer to as a transmission line and will study in Chap. 7. A pulse voltage source is attached to the load with this pair of conductors. We will find that the pulse will not appear instantaneously at the load but will require a certain amount of time to propagate from one end of the line to the other. For example, the propagation time for a pair of conductors of length 30 cm is approximately 1 ns (10^{-9} s). Although this seems to be an insignificant length of time, it becomes very significant in modern, high-speed digital computers, which transfer data at speeds of over 100 million bits per second. Lumped-parameter circuit theory may be used to *approximate* this phenomenon to some degree. Electromagnetic field theory predicts this result in an immediate, thorough, and transparent way.

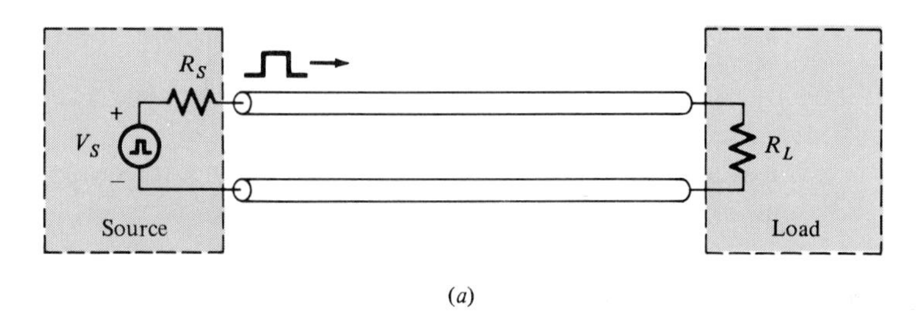

(*a*)

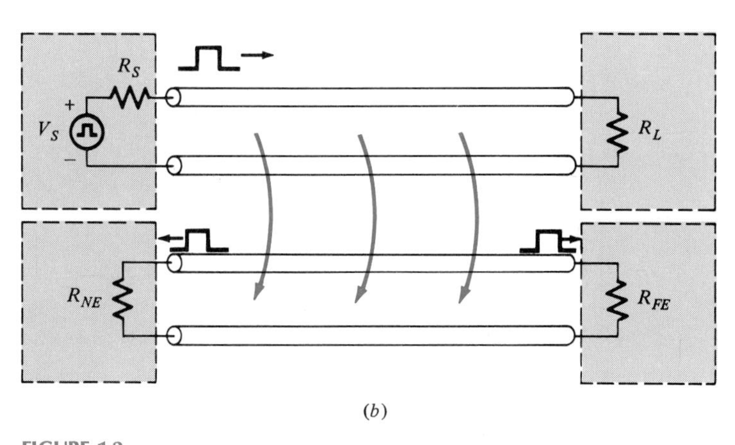

(*b*)

FIGURE 1.2

Transmission lines: (*a*) propagation delay; (*b*) crosstalk.

As another example, consider a pair of wires located in close proximity to each other (Fig. 1.2*b*). A signal being carried by one pair of wires generates electric and magnetic fields, as we will see. These fields interact with the neighboring pair of wires and induce signals at the terminations. This is referred to as crosstalk, or noise. Lumped-parameter circuit theory can be used to approximate this result, but electromagnetic field theory provides exact results and makes the underlying physical phenomena clear.

Present-day transmissions of voice and data require very large transmission bandwidths, given the increasing volume of information required to be transmitted. One way of accomplishing this is to transmit the information by attaching it to a high-frequency carrier. The higher the carrier frequency, the larger the effective bandwidth of the transmission. Using two conductors (such as in Fig. 1.2*a*) to transmit information becomes infeasible at carrier frequencies above approximately 500 MHz owing to the excessive losses of the line. This has led to the use of single-conductor hollow pipes, referred to as waveguides, which will be studied in Chap. 8. Lumped-parameter circuit theory does not permit signal transmission via single conductors, yet waveguides work very effectively in the gigahertz (10^9 Hz) frequency range. Single dielectric fibers perform a similar function at optical frequencies.

Finally, we will study signal transmission via antennas in Chap. 9. Consider the monopole antenna shown in Fig. 1.3. Lumped-parameter circuit theory would predict that no current would flow along the antenna since there is no complete circuit to yield a return path. There is, in fact, a "return path" by means of displacement current, such as occurs between the plates of a capacitor. We will find that waves are radiated from these antennas and may be used for the transfer of information.

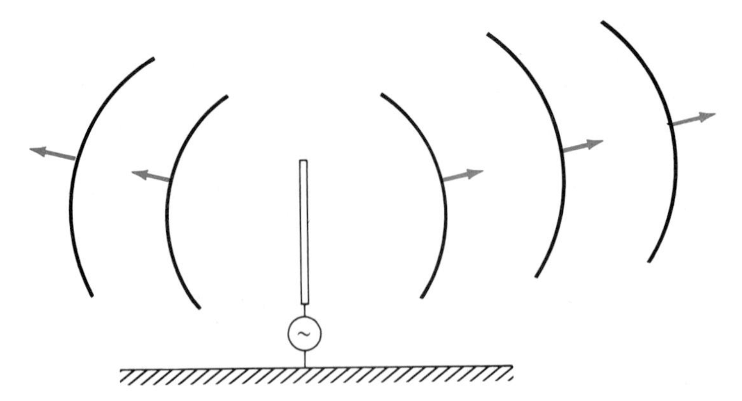

FIGURE 1.3
The monopole antenna.

It is certainly not true that all problems involving electromagnetic phenomena must be handled with the field theory approach that we will study. Many problems are adequately approximated with lumped-parameter circuit theory. As in all other areas of engineering, judgment is required in obtaining the simplest (yet satisfactory) model of a particular problem.

1.3 Units

The International System of Units (SI) is now widely used in scientific investigations and will be used exclusively in this book. A particular set of units, therefore, will not be referred to as being consistent with the International System since this will be implicit in our future discussions.

In the International System there are several basic units from which many others are obtained as certain combinations of these basic units. The basic SI units of mass (m), length (l), and time (t) are the kilogram, meter, and second, respectively. The other fundamental units are temperature (T) in degrees Kelvin (K), luminous intensity in candelas (cd), and electric current in amperes (A). The *ampere* is defined as that current which, when it flows in two parallel, straight, infinitely long conductors of negligible cross section, produces a force of 2×10^{-7} newtons per meter (N/m) when the conductors are one meter apart in a vacuum. *Charge* is defined as the time integral of current, so that the unit of charge is the ampere-second (As), which is defined as the coulomb (C). Other derived units in terms of the basic units are listed in Table 1.1. The important powers-of-10 multipliers are listed in Table 1.2.

The equations that we will obtain presume that the variables are given in a self-consistent set of units (preferably SI). The various proportionality factors in these equations will be given in SI units, and thus the equation variables must also be in SI units. Nevertheless, we may encounter the use of non-SI units in some situations. Several common non-SI units and their equivalence in SI units are given in Table 1.3. A simple way of converting between the units of one system and those of another that avoids mistakes is illustrated by the following. Suppose we wish to convert 100 miles to kilometers. Simply multiply 100 mi by unity ratios between the two systems; for example, 1 mi = 5280 ft, 1 ft = 12 in, 1 in = 2.54 cm, 100 cm = 1 m. Cancellation of the unit names in this conversion avoids the improper multiplication (division) of a unity ratio when division (multiplication) should be used:

$$100\,\cancel{\text{mi}} \times \frac{5280\,\cancel{\text{ft}}}{1\,\cancel{\text{mi}}} \times \frac{12\,\cancel{\text{in}}}{1\,\cancel{\text{ft}}} \times \frac{2.54\,\cancel{\text{cm}}}{1\,\cancel{\text{in}}} \times \frac{1\,\cancel{\text{m}}}{100\,\cancel{\text{cm}}} \times \frac{1\text{ km}}{1000\,\cancel{\text{m}}} = 160.93\text{ km}$$

These practical aspects of proper unit usage will be illustrated in the chapter problems.

TABLE 1.1

The International System of Units (SI) (abbreviated)

Name	*Typical Symbol*	*Unit*	*Unit Abbreviation*	*In Terms of Other Units*
Length	$l, x\ d, \ldots$	meter	m	
Mass	m	kilogram	kg	
Time	t	second	s	
Current	i, I	ampere	A	
Charge	q, Q	coulomb	C	As
Frequency	f	hertz	Hz	s^{-1}
Force	f, F	newton	N	$kg \cdot m/s^2$
Work energy	w, W	joule	J	N·m
Power	p, P	watt	W	J/s
Voltage	v, V	volt	V	W/A = N·m/C
Electric field intensity	E	volt/meter		V/m
Electric flux	ψ_e	coulomb	C	
Electric flux density	D	coulomb/square meter		C/m^2
Capacitance	C	farad	F	C/V = As/V
Magnetic flux	ψ_m	weber	Wb	V·s
Magnetic field intensity	H	ampere/meter		A/m
Magnetic flux density	B	tesla	T	$Wb/m^2 = V \cdot s/m^2$
Inductance	L	henry	H	Wb/A = V·s/A
Resistance	R	ohm	Ω	V/A
Conductance	G	mho or siemen	℧ or S	A/V
Conductivity	σ	mho/meter or siemen/meter	℧/m or S/m	
Free-space permittivity	ϵ_0	farad/meter	F/m	
Free-space permeability	μ_0	henry/meter	H/m	

TABLE 1.2

Powers of Ten

Power of Ten	*Prefix*	*Symbol*
10^{12}	tera	T
10^{9}	giga	G
10^{6}	mega	M
10^{3}	kilo	k
10^{2}	hecto	h
10	deka	da
10^{-1}	deci	d
10^{-2}	centi	c
10^{-3}	milli	m
10^{-6}	micro	μ
10^{-9}	nano	n
10^{-12}	pico	p
10^{-15}	femto	f
10^{-18}	atto	a

TABLE 1.3

Equivalence of Units

Unit 1	=	*Unit 2*
Inch		2.54 cm
Foot		12 in
Mile		5280 ft
Hour		60 min
Minute		60 s
Pound		0.45359 kg
Horsepower (electrical)		746 W
Foot pound		1.3558 W
Pound-force		4.4482 N
Foot pound-force		1.3558 J
Erg		10^{-7} J
Dyne		10^{-5} N
Gauss		10^{-4} T (Wb/m^2)
Oersted		79.577 A/m
Maxwell		10^{-8} Wb
Gilbert		0.7957 ampere-turn (At)

References

1. R. S. Elliot, *Electromagnetics*, McGraw-Hill, New York, 1966.
2. J. C. Maxwell, *A Treatise on Electricity and Magnetism*, Dover, New York, 1954.
3. R. Plonsey, *Bioelectric Phenomena*, McGraw-Hill, New York, 1969.
4. K. Marha et al., *Electromagnetic Fields and the Life Environment*, San Francisco Press, San Francisco, 1971.
5. A. R. Sheppard and M. Eisenbud, *Biological Effects of Electric and Magnetic Fields of Extremely Low Frequency*, New York University Press, New York, 1977.
6. W. T. Norris, et al., "People in Alternating Electric and Magnetic Fields near Electric Power Equipment," *Electronics and Power*, vol. 31, no. 2, 1985, pp. 137–141.
7. A. T. Barker and I. L. Freeston, "Medical Applications of Electric and Magnetic Fields." *Electronics and Power*, vol. 31, no. 10, 1985, pp. 757–760.
8. M. Rowbottom and C. Susskind, *Electricity and Medicine: History of Their Interaction*, San Francisco Press, San Francisco, 1984.
9. H. Kolm et al. (eds), *High Magnetic Fields*, MIT Press, Cambridge, 1962.
10. D. B. Montgomery, *Solenoid Magnet Design*, Wiley-Interscience, New York, 1969.
11. H. Knoepfel, *Pulsed High Magnetic Fields*, North-Holland, Amsterdam, 1970.
12. S. A. Nasar, "Electromagnetic Theory of Electrical Machines," *Proc. IEE*, vol. 111, 1964, pp. 1125–1131.
13. B. Hague, *The Principles of Electromagnetism Applied to Electrical Machines*, Dover, New York, 1962.

CHAPTER 2

Vector Analysis

The electromagnetic field is the result of the presence and movement of electric charges. We will find that these charges exert forces on each other and that the forces possess both a magnitude and a direction of effect that may be conveniently represented as a vector quantity. Thus, in order to compute these forces and analyze the resulting electromagnetic field, it is essential that we have a thorough understanding of the manipulation of vector quantities. This chapter is intended to introduce these concepts and mathematical operations.

Vector analysis may be divided into two general categories—vector algebra and vector calculus. Vector algebra concerns algebraic operations on vectors, such as addition, subtraction, and multiplication. In order to provide a means of facilitating our numerical calculations, this chapter will introduce certain coordinate systems. For reasons which will soon become apparent, we will only consider orthogonal coordinate systems, in which the three surfaces used to define the coordinate system are orthogonal to each other. Moreover, we will concentrate on the most important and commonly used orthogonal coordinate systems—rectangular (cartesian), cylindrical, and spherical.

Vector calculus concerns differential and integral operations involving vector functions. Electromagnetic fields are functions of position in certain regions of space. At various points within this region the field may possess only a magnitude (*scalar field*), whereas in other cases the field may possess both a magnitude and a direction (*vector field*). The various vector calculus operations allow us to state concisely the fundamental laws governing these electromagnetic fields and to perform calculations with these field quantities.

Perhaps the most important reasons for beginning our study of electromagnetic fields with the analysis of vectors are the following. As indicated previously, the electromagnetic field *quantities* are conveniently described in terms of vectors; the fundamental laws governing these basic vector field quantities can be concisely stated using the vector calculus concepts presented in this chapter. In addition to performing numerical calculations for various electromagnetic field problems, it is also important to obtain a *qualitative* understanding of these laws and their implications. To obtain this insight, it is vitally important that the reader have a thorough understanding of the mathematical concepts of this chapter. Furthermore, the study of the subject of vector analysis also has numerous other applications in mechanics, fluid flow, heat flow, etc., and thus need not (and should not) be viewed as an end in itself. Consequently, a serious study of this chapter will be of considerable future benefit to the reader.

2.1 Vectors and Scalars

To illustrate the concept of a vector, suppose that we are attempting to move an object along a surface by pulling it with a rope. The force F that is being exerted on the object is directed along the rope. Thus, we might pictorially represent this force as a directed line segment of length F pointing in the direction in which the force is being exerted: along the rope. The length of this line segment is proportional to the force F, and the direction of this line segment is denoted by an arrow. If the rope is not aligned parallel to the surface but is at an angle θ with the surface, then only a portion of the total force, $F \cos \theta$, is utilized in moving the object along the surface. We might denote this component of the total force as another directed line segment of length $F \cos \theta$ parallel to the surface.

The preceding illustrates the concept of a vector. A *vector* is a directed line segment that contains two items of information: a direction and a magnitude of some quantity. On the other hand, a scalar quantity possesses no directional information. Thus, a number (real or complex) is a scalar. Vectors will be denoted in boldface type, for example, **A**.

Consider the two vectors **A** and **B** shown in Fig. 2.1*a*. In order to add these two vectors, we translate (move without changing the length or direction) vector **B** to the tip of vector **A**, as shown in Fig. 2.1*b*. The graphical result is the sum $\mathbf{A} + \mathbf{B}$. Obviously, we could have translated **A** to the tip of **B** (as shown in Fig. 2.1*b*) and obtained $\mathbf{B} + \mathbf{A}$. Clearly then,

$$\mathbf{A} + \mathbf{B} = \mathbf{B} + \mathbf{A} \tag{1}$$

and the order of addition (and translation to do so) of the vectors is immaterial.

We may also multiply a vector by a scalar k, as shown in Fig. 2.1*c*. Multiplication of a vector by a scalar k simply changes the length of the vector (its magnitude) by a factor of k. Note that the scalar k can be positive or negative. If the scalar is negative, we simply reverse the direction of the vector and scale its length by the magnitude of the scalar. Now it is easy to see how we may define the subtraction of two vectors $\mathbf{A} - \mathbf{B}$: simply add **A** to $k\mathbf{B}$ with $k = -1$, as in Fig. 2.1*d*.

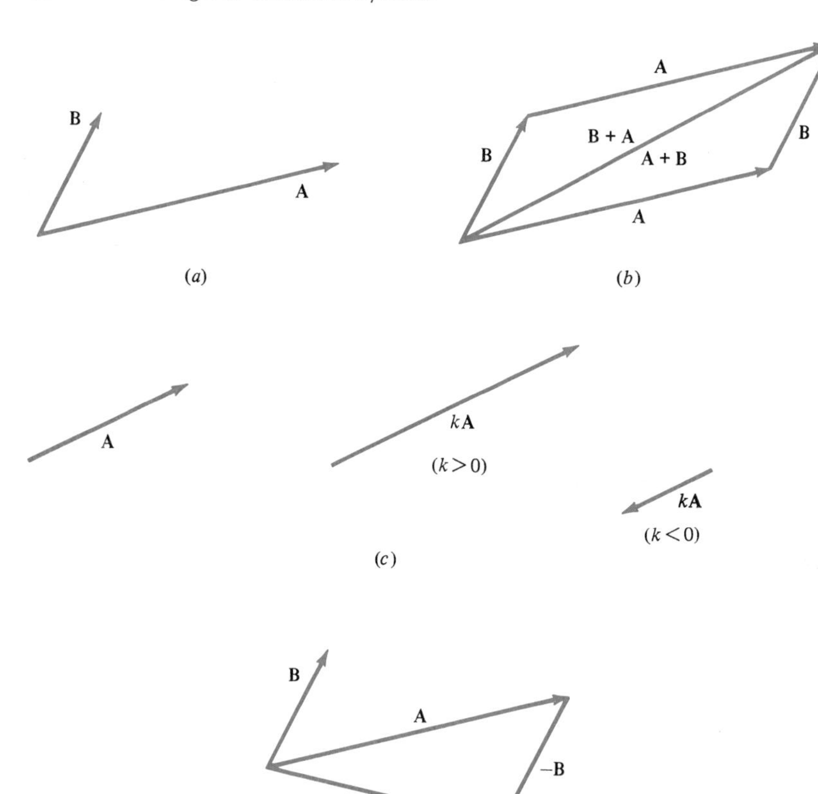

FIGURE 2.1
Vector addition, subtraction, and multiplication by a scalar constant.

These ideas can be extended to handle the addition of more than two vectors. For example, it is clear that vectors **A**, **B**, and **C** may be added in any order so that $\mathbf{A} + \mathbf{B} + \mathbf{C} = \mathbf{A} + (\mathbf{B} + \mathbf{C}) = (\mathbf{A} + \mathbf{B}) + \mathbf{C} = (\mathbf{A} + \mathbf{C}) + \mathbf{B}$.

Equality of two vectors follows logically from these ideas. We say that two vectors **A** and **B** are equal if the magnitudes of the vectors are the same and both vectors are pointing in the same direction.

2.2 Orthogonal Coordinate Systems

So far, we have discussed the concept of vectors only in general terms. It is quite easy simply to draw a picture to show, for example, the addition of two vectors (as in Fig. 2.1*b*). If, however, we are to utilize these vector concepts fully in understanding electromagnetic fields, we must understand coordinate systems. The various coordinate systems that we will study in this chapter provide the important link between merely visualizing the solution to a problem and being able to compute numerical results.

Coordinate systems enable us to locate a point in three-dimensional space as the intersection of three surfaces. These coordinate surfaces are defined by $u_1 = \text{constant}$, $u_2 = \text{constant}$, and $u_3 = \text{constant}$. For reasons that will soon become apparent, it is desirable to choose these surfaces to be orthogonal to each other. This is referred to as an orthogonal coordinate system. An example is shown in Fig. 2.2. The location of point P is given by the intersection of the three surfaces as the triple $P[u_1, u_2, u_3]$ in terms of the specific coordinates of the intersection.

In order to define a vector $\mathbf{A}$ at point P, we introduce unit vectors $\mathbf{a}_{u1}$, $\mathbf{a}_{u2}$, and $\mathbf{a}_{u3}$, as shown in Fig. 2.2. These unit vectors are defined orthogonal to the appropriate surface at P. For example, $\mathbf{a}_{u2}$ is orthogonal to the surface $u_2 = \text{constant}$ at P. For this assumed orthogonal coordinate system, the unit vectors are also orthogonal. Thus, we may decompose $\mathbf{A}$ into components that are tangent to these unit vectors:

$$\mathbf{A} = A_{u1}\mathbf{a}_{u1} + A_{u2}\mathbf{a}_{u2} + A_{u3}\mathbf{a}_{u3} \tag{2}$$

The magnitude, or length, of this vector is

$$\begin{aligned} |\mathbf{A}| &= A \\ &= \sqrt{A_{u1}^2 + A_{u2}^2 + A_{u3}^2} \end{aligned} \tag{3}$$

since the components are orthogonal. A unit vector in the direction of $\mathbf{A}$ can be obtained as

$$\mathbf{a}_A = \frac{\mathbf{A}}{|\mathbf{A}|} \tag{4}$$

Consider two vectors $\mathbf{A}$ and $\mathbf{B}$ that are defined at a point P:

$$\mathbf{A} = A_{u1}\mathbf{a}_{u1} + A_{u2}\mathbf{a}_{u2} + A_{u3}\mathbf{a}_{u3} \tag{5a}$$

$$\mathbf{B} = B_{u1}\mathbf{a}_{u1} + B_{u2}\mathbf{a}_{u2} + B_{u3}\mathbf{a}_{u3} \tag{5b}$$

Since the corresponding components are parallel, the addition of these two vectors is quite simple:

$$\mathbf{A} + \mathbf{B} = (A_{u1} + B_{u1})\mathbf{a}_{u1} + (A_{u2} + B_{u2})\mathbf{a}_{u2} + (A_{u3} + B_{u3})\mathbf{a}_{u3} \tag{6}$$

Quite often we will be interested in differential path lengths, surface areas, and volumes in these coordinate systems. Some (or all) of these coordinates may

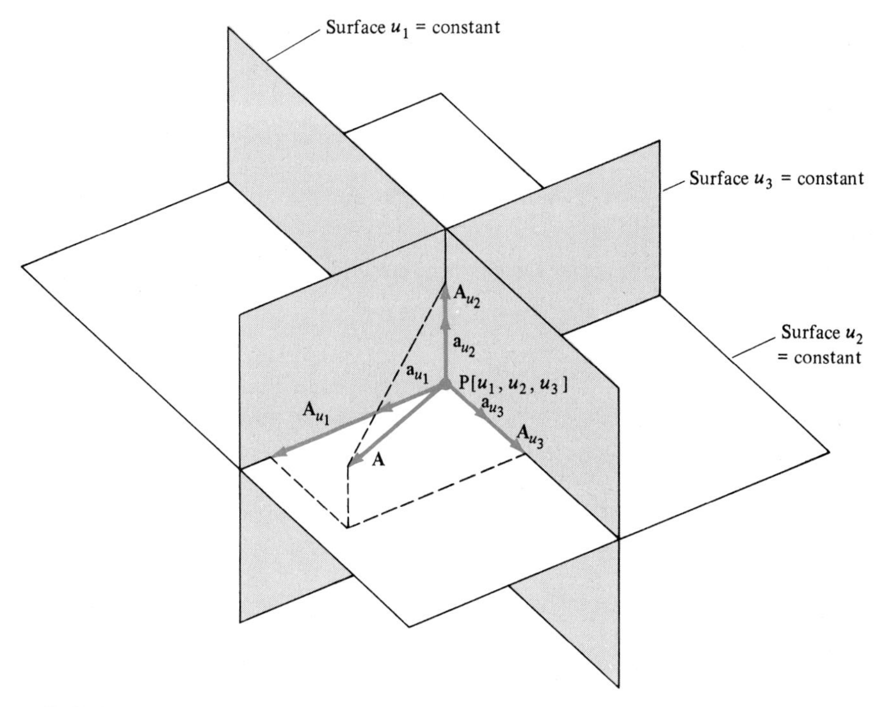

FIGURE 2.2

Orthogonal surfaces defining an orthogonal coordinate system.

not be lengths. Thus, we need a conversion coefficient to relate a differential change in a coordinate variable, du_i, to a differential change in length along that coordinate surface, dl_i. These are referred to as *metric coefficients* and are denoted for the coordinate surface as h_i. Thus

$$dl_i = h_i\, du_i \tag{7}$$

A vector representing a differential change in path in some arbitrary direction becomes

$$\begin{aligned} d\mathbf{l} &= dl_1 \mathbf{a}_{u1} + dl_2 \mathbf{a}_{u2} + dl_3 \mathbf{a}_{u3} \\ &= (h_1\, du_1)\mathbf{a}_{u1} + (h_2\, du_2)\mathbf{a}_{u2} + (h_3\, du_3)\mathbf{a}_{u3} \end{aligned} \tag{8}$$

The length of this differential change becomes

$$\begin{aligned} |d\mathbf{l}| &= dl \\ &= \sqrt{(h_1\, du_1)^2 + (h_2\, du_2)^2 + (h_3\, du_3)^2} \end{aligned} \tag{9}$$

If the surfaces are incremented by the differential amounts du_1, du_2, and du_3 in directions $\mathbf{a}_{u1}$, $\mathbf{a}_{u2}$, and $\mathbf{a}_{u3}$, a differential volume is formed:

$$\begin{aligned} dv &= (dl_1)(dl_2)(dl_3) \\ &= h_1 h_2 h_3 \, du_1 \, du_2 \, du_3 \end{aligned} \tag{10}$$

Similarly, the differential surface areas become

$$\begin{aligned} ds_1 &= dl_2 \, dl_3 \\ &= (h_2 h_3 \, du_2 \, du_3) \end{aligned} \tag{11a}$$

$$\begin{aligned} ds_2 &= dl_1 \, dl_3 \\ &= (h_1 h_3 \, du_1 \, du_3) \end{aligned} \tag{11b}$$

$$\begin{aligned} ds_3 &= dl_1 \, dl_2 \\ &= (h_1 h_2 \, du_1 \, du_2) \end{aligned} \tag{11c}$$

We will find that our major interest in differential surface areas involves the normal to these surfaces. Anticipating this need, we define

$$d\mathbf{s}_1 = (h_2 h_3 \, du_2 \, du_3)\mathbf{a}_{u1} \tag{12a}$$

$$d\mathbf{s}_2 = (h_1 h_3 \, du_1 \, du_3)\mathbf{a}_{u2} \tag{12b}$$

$$d\mathbf{s}_3 = (h_1 h_2 \, du_1 \, du_2)\mathbf{a}_{u3} \tag{12c}$$

A large number of choices of coordinate surfaces may be used to define a particular coordinate system (planes, cylinders, spheres, etc.); however, the electromagnetic field laws are independent of the coordinate system used to describe the problem. In this text we will find it necessary to use only three orthogonal coordinate systems—rectangular (cartesian), cylindrical, and spherical. The physical geometry of the particular problem will dictate the most appropriate choice of coordinate system to simplify the calculations. For example, if we are considering the flow of electric charge through a cylindrical wire, the use of a cylindrical coordinate system will facilitate the calculations since the boundaries of the problem (the cylindrical wire) match one of the surfaces of the coordinate system (a cylinder). Choosing the best coordinate system for a particular problem is an important first step in obtaining a simple solution.

2.3 The Rectangular (Cartesian) Coordinate System†

The rectangular, or cartesian, coordinate system is perhaps the most common orthogonal coordinate system. Three coordinate axes are designated as x, y, and z. A point $P_1[x_1, y_1, z_1]$ is located at the intersection of three orthogonal planes, $x = x_1$, $y = y_1$, $z = z_1$, as shown in Fig. 2.3. The unit vectors at this point are $\mathbf{a}_x$, $\mathbf{a}_y$, and $\mathbf{a}_z$, which are orthogonal to the appropriate planes. Another point $P_2[x_2, y_2, z_2]$ is located at the intersection of three planes $x = x_2$, $y = y_2$, and $z = z_2$. Note that the unit vectors at P_2 are parallel to the corresponding unit vectors at P_1. A vector at P_1 can be written as

$$\mathbf{A} = A_x\mathbf{a}_x + A_y\mathbf{a}_y + A_z\mathbf{a}_z \tag{13}$$

Another vector, **B**, at that point can be added to **A** to yield

$$\mathbf{A} + \mathbf{B} = (A_x + B_x)\mathbf{a}_x + (A_y + B_y)\mathbf{a}_y + (A_z + B_z)\mathbf{a}_z \tag{14}$$

Differential arc lengths, surface areas, and volumes can be defined by incrementing the planes by dx, dy, and dz, as shown in Fig. 2.4. Note that the metric coefficients h_x, h_y, and h_z are all unity since the coordinate surfaces represent length. Thus, the differential arc length is

$$d\mathbf{l} = dx\,\mathbf{a}_x + dy\,\mathbf{a}_y + dz\,\mathbf{a}_z \tag{15}$$

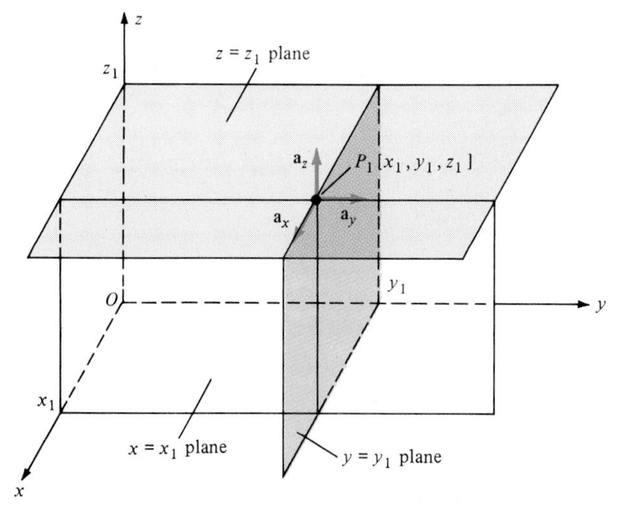

FIGURE 2.3

The rectangular (cartesian) coordinate system.

† The term *rectangular* is more properly used with reference to two-dimensional coordinate systems. However, it has become common to use that term for three-dimensional coordinate systems also, as we will do.

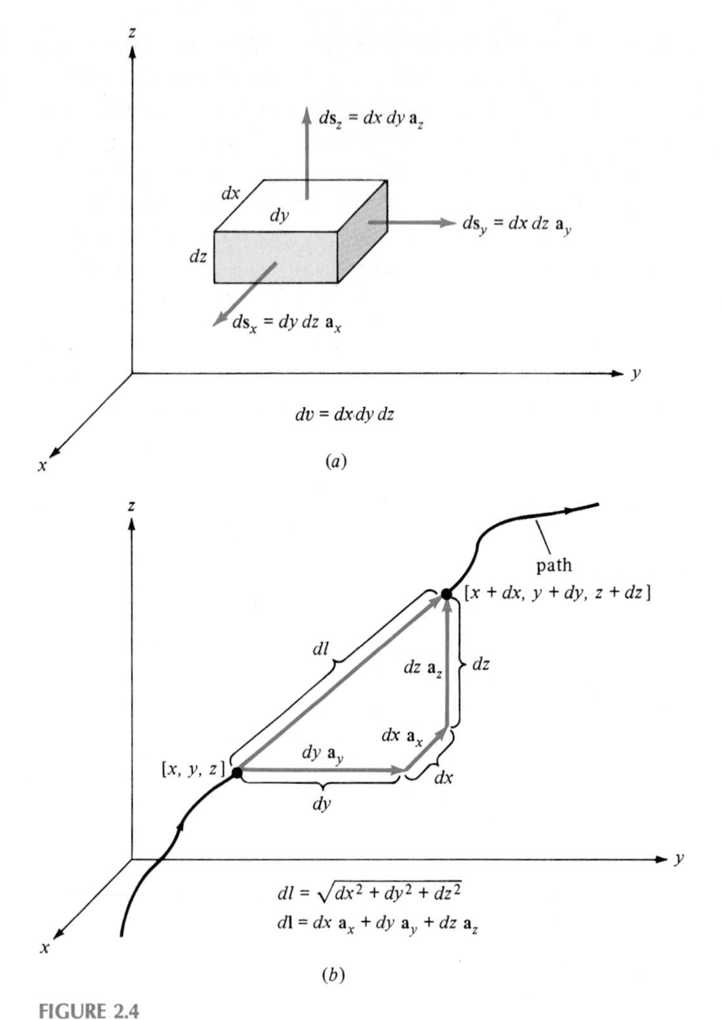

FIGURE 2.4
Differential arc lengths, surface areas, and volume elements in a rectangular coordinate system.

The differential surface elements are

$$d\mathbf{s}_x = dy\, dz\, \mathbf{a}_x \tag{16a}$$

$$d\mathbf{s}_y = dx\, dz\, \mathbf{a}_y \tag{16b}$$

$$d\mathbf{s}_z = dx\, dy\, \mathbf{a}_z \tag{16c}$$

and a differential volume element becomes

$$dv = dx\, dy\, dz \tag{17}$$

In the next two sections we will investigate two additional orthogonal coordinate systems: cylindrical and spherical. These additional coordinate systems are considered so that we may choose the coordinate system that best fits a particular problem. As will become apparent, choosing one coordinate system to describe a particular electromagnetic field problem might mean that we obtain the solution only after a great deal of effort; but choosing another to describe the same problem might make the solution process almost trivial. Consequently, since the answer to the problem will be the same no matter which coordinate system we choose, it is important to make a judicious choice.

EXAMPLE 2.1 Consider the two points in a rectangular coordinate system $P_1[3, 1, 3]$ and $P_2[1, 3, 2]$, as shown in Fig. 2.5. Determine the distance between the two points.

Solution The length of a vector between the two points is

$$|\mathbf{D}| = \sqrt{(1-3)^2 + (3-1)^2 + (2-3)^2}$$
$$= 3$$

where a vector $\mathbf{D}$ directed from P_1 to P_2 is

$$\mathbf{D} = (1-3)\mathbf{a}_x + (3-1)\mathbf{a}_y + (2-3)\mathbf{a}_z$$
$$= -2\mathbf{a}_x + 2\mathbf{a}_y - \mathbf{a}_z$$

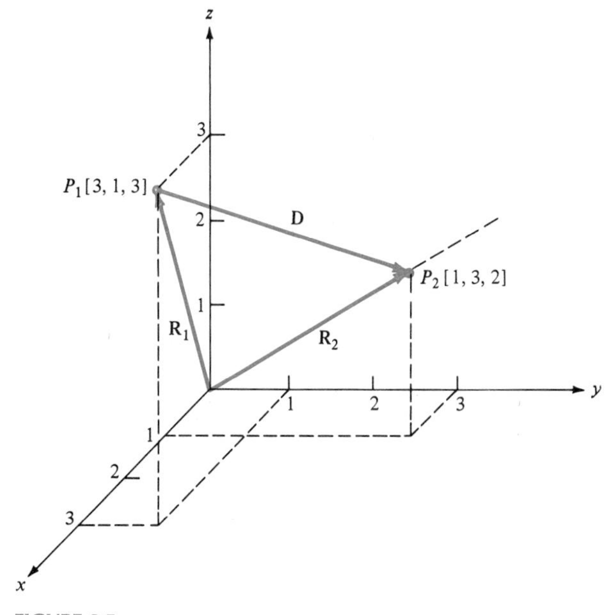

FIGURE 2.5
Example 2.1.

Also, **D** may be found as the difference of two vectors $\mathbf{R}_1$ and $\mathbf{R}_2$, each of which is defined from the origin to the point:

$$\mathbf{R}_1 = 3\mathbf{a}_x + 1\mathbf{a}_y + 3\mathbf{a}_z$$

$$\mathbf{R}_2 = 1\mathbf{a}_x + 3\mathbf{a}_y + 2\mathbf{a}_z$$

$$\begin{aligned}\mathbf{D} &= \mathbf{R}_2 - \mathbf{R}_1 \\ &= -2\mathbf{a}_x + 2\mathbf{a}_y - \mathbf{a}_z\end{aligned}$$

A unit vector in the direction of **D** becomes

$$\begin{aligned}\mathbf{a}_D &= \frac{\mathbf{D}}{|\mathbf{D}|} \\ &= -\tfrac{2}{3}\mathbf{a}_x + \tfrac{2}{3}\mathbf{a}_y - \tfrac{1}{3}\mathbf{a}_z\end{aligned}$$

The distance between the two points can also be obtained by

$$|\mathbf{D}| = \int_{P_1}^{P_2} dl$$

In order to evaluate this integral, we need to write

$$dl = \sqrt{dx^2 + dy^2 + dz^2}$$

in terms of one variable, say x. To this end we write the equations of the line joining the two points in terms of x:

$$y = -x + 4$$

$$z = \tfrac{1}{2}x + \tfrac{3}{2}$$

so that

$$dy = -dx$$

$$dz = \tfrac{1}{2}dx$$

and

$$\begin{aligned}dl &= dx\sqrt{1 + \left(\frac{dy}{dx}\right)^2 + \left(\frac{dz}{dx}\right)^2} \\ &= dx\sqrt{1 + (-1)^2 + (\tfrac{1}{2})^2} \\ &= \tfrac{3}{2}dx\end{aligned}$$

Thus

$$|\mathbf{D}| = \int_{x=1}^{3} \tfrac{3}{2} dx$$
$$= 3$$

2.4 The Circular, Cylindrical Coordinate System

The surfaces used to define a circular, cylindrical coordinate system (referred to in the future simply as the cylindrical coordinate system) consist of a plane of constant z, a cylinder of radius r with the z axis as its axis, and a half plane perpendicular to the xy plane and at an angle ϕ with respect to the xz plane, as shown in Fig. 2.6. A point $P_1[r_1, \phi_1, z_1]$ is located at the intersection of these for $r_1 = \text{constant}$, $\phi_1 = \text{constant}$, and $z_1 = \text{constant}$. Note that ϕ is measured from the xz plane and that the units of ϕ are radians.

Unit vectors $\mathbf{a}_r$, $\mathbf{a}_\phi$, and $\mathbf{a}_z$ are orthogonal to these surfaces at P_1. The unit vector $\mathbf{a}_z$ is parallel to the z axis and directed toward increasing z. The unit vector $\mathbf{a}_r$ lies in a plane that is parallel to the xy plane and is normal to the

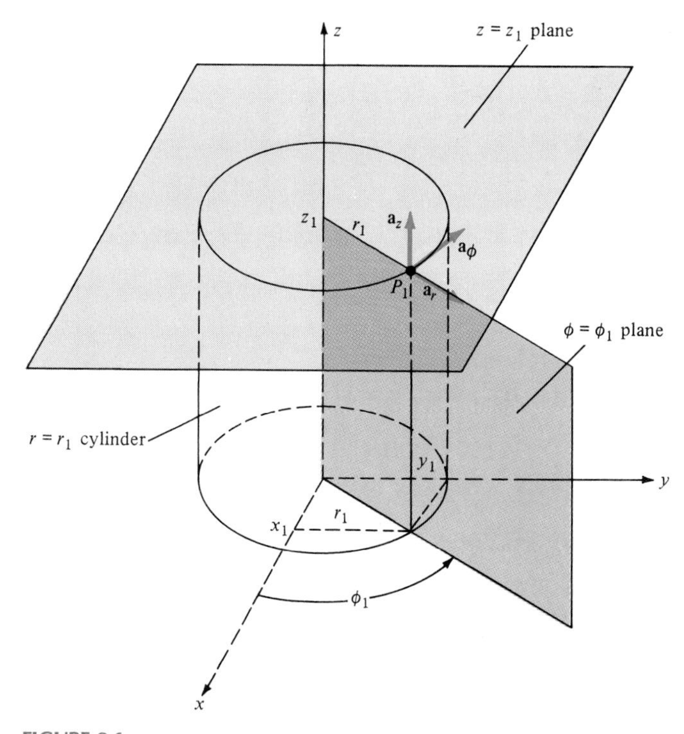

FIGURE 2.6

The cylindrical coordinate system.

surface of the cylinder at point P. Unit vector $\mathbf{a}_\phi$ is also in a plane parallel to the xy plane but is tangent to the cylinder and pointing in the direction of increasing ϕ at point P. Note that these three unit vectors are orthogonal. Also observe that the directions of the unit vectors $\mathbf{a}_r$ and $\mathbf{a}_\phi$ will vary from point to point, as opposed to a rectangular coordinate system in which all three unit vectors have fixed directions once the orientations of the x, y, and z coordinate axes are chosen.

A vector $\mathbf{A}$ at point P can be written as

$$\mathbf{A} = A_r\mathbf{a}_r + A_\phi\mathbf{a}_\phi + A_z\mathbf{a}_z \tag{18}$$

Another vector $\mathbf{B}$ defined at this point can be added to $\mathbf{A}$ by adding components, resulting in

$$\mathbf{A} + \mathbf{B} = (A_r + B_r)\mathbf{a}_r + (A_\phi + B_\phi)\mathbf{a}_\phi + (A_z + B_z)\mathbf{a}_z \tag{19}$$

This is a result of the fact that the corresponding unit vectors are parallel. It would not be possible to add two vectors that are defined at two different points since the corresponding unit vectors $\mathbf{a}_r$ and $\mathbf{a}_\phi$ for the two vectors would not be parallel.

Note that two of the three coordinates, z and r, are lengths; thus, the metric coefficients are unity: $h_z = h_r = 1$. However, the ϕ coordinate is an angle. A change in length in the ϕ direction on this cylinder of radius r is given by $dl_\phi = r\,d\phi$, and thus the metric coefficient h_ϕ is $h_\phi = r$. Therefore,

$$\boxed{h_z = 1 \qquad h_r = 1 \qquad h_\phi = r.} \tag{20}$$

Differential arc lengths, surfaces, and volumes are formed by incrementing the surfaces by dr, $d\phi$, and dz, as shown in Fig. 2.7. Thus, the differential arc length is

$$d\mathbf{l} = dr\,\mathbf{a}_r + r\,d\phi\,\mathbf{a}_\phi + dz\,\mathbf{a}_z \tag{21}$$

The differential surfaces become

$$d\mathbf{s}_r = r\,d\phi\,dz\,\mathbf{a}_r \tag{22a}$$

$$d\mathbf{s}_\phi = dr\,dz\,\mathbf{a}_\phi \tag{22b}$$

$$d\mathbf{s}_z = r\,d\phi\,dr\,\mathbf{a}_z \tag{22c}$$

and the differential volume element is

$$dv = r\,d\phi\,dr\,dz \tag{23}$$

Converting the coordinates of a point P in a rectangular coordinate system $[x_P, y_P, z_P]$, to cylindrical coordinates $[r_P, \phi_P, z_P]$ is easily accomplished. For

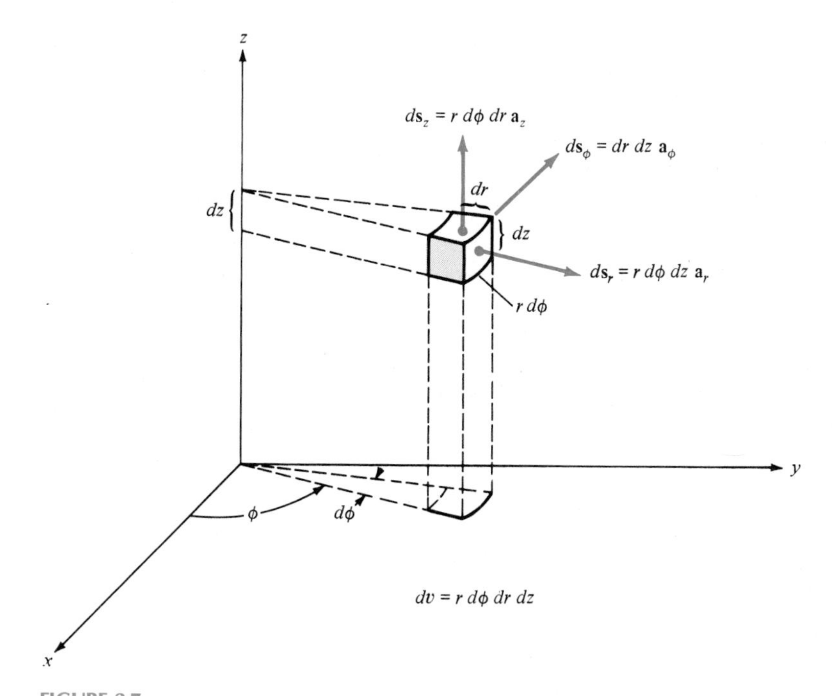

FIGURE 2.7
Differential arc lengths, surface areas, and volume elements in a cylindrical coordinate system.

example, a view of the cylindrical coordinate system in the xy plane is shown in Fig. 2.8. The z_P coordinate is, of course, the same in both coordinate systems, but the r_P and ϕ_P coordinates in the cylindrical system are related to the x_P and y_P coordinates in the rectangular system by

$$r_P = \sqrt{x_P^2 + y_P^2} \tag{24a}$$

$$\phi_P = \tan^{-1} \frac{y_P}{x_P} \tag{24b}$$

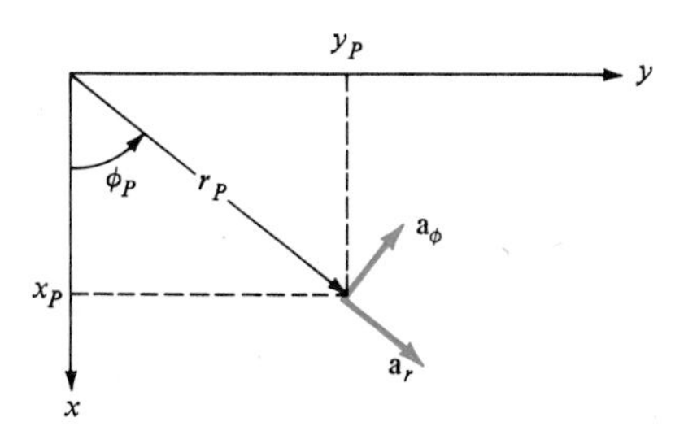

FIGURE 2.8
Conversion of cylindrical coordinates to rectangular coordinates, and vice versa.

Conversely, given the cylindrical coordinates r_P and ϕ_P, we can obtain the corresponding rectangular system coordinates as

$$x_P = r_P \cos \phi_P \tag{25a}$$

$$y_P = r_P \sin \phi_P \tag{25b}$$

A further discussion of the conversion between coordinate systems is given in Appendix A.

We would expect to use this coordinate system to describe problems having some form of cylindrical symmetry. For example, if we were interested in describing the flow of some fluid in a pipe, our natural impulse would be (or should be) to use a cylindrical coordinate system and not a rectangular one. In setting up the cylindrical coordinate system, we would also be careful to align the axis of the pipe with the z coordinate axis; otherwise, the cylindrical symmetry of the problem would not be used to its best advantage.

EXAMPLE 2.2 Determine, by direct integration, the volume enclosed by a cylinder of radius R and length L. Also determine the surface area of the cylinder.

Solution To obtain the volume, we align the cylinder with the z axis extending from $z = 0$ to $z = L$ and integrate

$$\begin{aligned} V &= \int dv \\ &= \int_{r=0}^{R} \int_{\phi=0}^{2\pi} \int_{z=0}^{L} r\, d\phi\, dr\, dz \\ &= \int_{r=0}^{R} \int_{\phi=0}^{2\pi} L\, r\, d\phi\, dr \\ &= \int_{r=0}^{R} 2\pi L\, r\, dr \\ &= \pi R^2 L \end{aligned}$$

To obtain the total surface area, we integrate over the surface of the cylinder and over the two end caps:

$$\begin{aligned} S &= \int_{\text{cylinder}} ds_r + \int_{\text{top cap}} ds_z + \int_{\text{bottom cap}} ds_z \\ &= \int_{z=0}^{L} \int_{\phi=0}^{2\pi} R\, d\phi\, dz + 2 \int_{r=0}^{R} \int_{\phi=0}^{2\pi} r\, d\phi\, dr \\ &= 2\pi R L + 2\pi R^2 \end{aligned}$$

2.5 The Spherical Coordinate System

The surfaces used to define a spherical coordinate system, as shown in Fig. 2.9, consist of (1) a sphere of radius r centered at the origin; (2) a right circular cone with its apex at the origin, its axis as the z axis, and having a half angle θ; and (3) a half plane perpendicular to the xy plane and making an angle ϕ with the xz plane.† A point $P_1[r_1, \theta_1, \phi_1]$ is located at the intersection of the plane r_1 = constant, θ_1 = constant, and ϕ_1 = constant. Note that ϕ is measured from the xz plane and that θ is measured from the z axis. The units of ϕ and θ are radians.

Unit vectors $\mathbf{a}_r$, $\mathbf{a}_\theta$, and $\mathbf{a}_\phi$ are orthogonal to these surfaces at P_1. Unit vector $\mathbf{a}_r$ is directed from the center of the sphere to the point. The unit vector $\mathbf{a}_\theta$ is tangent to the sphere and oriented in the direction of increasing θ. Similarly, the unit vector $\mathbf{a}_\phi$ is also tangent to the sphere but is oriented in the direction of increasing ϕ. As in the cylindrical coordinate system, the directions of these unit vectors will depend on the point at which they are defined.

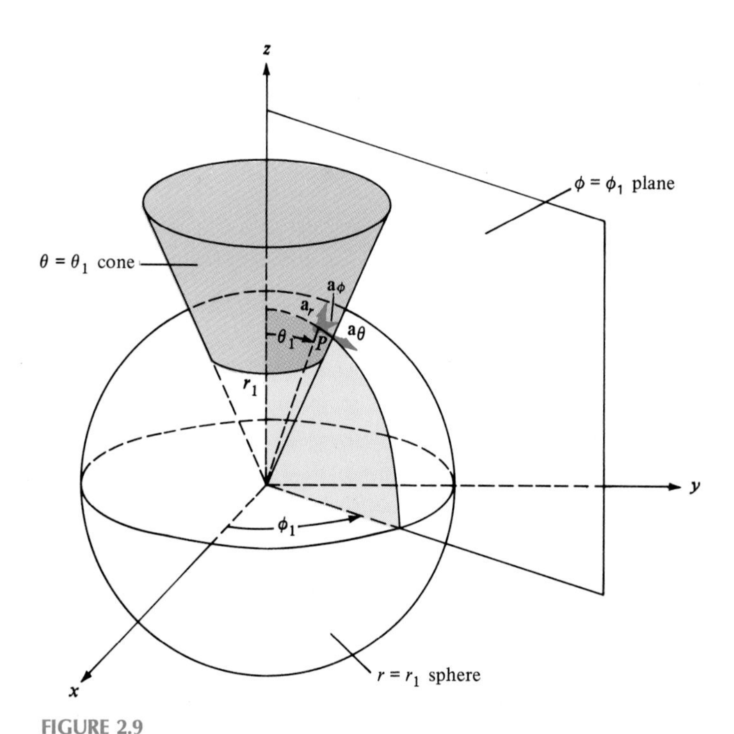

FIGURE 2.9
The spherical coordinate system.

† The symbols r and ϕ will be used in both the cylindrical and spherical coordinate systems. Clearly, r in the cylindrical system is different from r in the spherical system, but the appropriate coordinate system should be clear from the context of the discussion.

Note that all three unit vectors are mutually orthogonal. Therefore, we can express a vector **A** (defined at point P) in terms of its spherical coordinates as

$$\mathbf{A} = A_r \mathbf{a}_r + A_\theta \mathbf{a}_\theta + A_\phi \mathbf{a}_\phi \tag{26}$$

Again, we may add two vectors **A** and **B**, both of which are defined at the same point P by adding corresponding components:

$$\mathbf{A} + \mathbf{B} = (A_r + B_r)\mathbf{a}_r + (A_\theta + B_\theta)\mathbf{a}_\theta + (A_\phi + B_\phi)\mathbf{a}_\phi \tag{27}$$

since the corresponding unit vectors for **A** and **B** are parallel.

Note that the metric coefficient h_z is unity since the z coordinate is a length. However, the θ and ϕ coordinates are angles and require metric coefficients in order to convert differential changes in the coordinates to lengths. These coefficients are

$$h_z = 1 \qquad h_\theta = r \qquad h_\phi = r \sin\theta \tag{28}$$

Differential arc lengths, surfaces, and volumes are formed by incrementing the surfaces by dr, $d\theta$, and $d\phi$, as shown in Fig. 2.10. Thus, the differential arc length is

$$d\mathbf{l} = dr\ \mathbf{a}_r + r\ d\theta\ \mathbf{a}_\theta + r \sin\theta\ d\phi\ \mathbf{a}_\phi \tag{29}$$

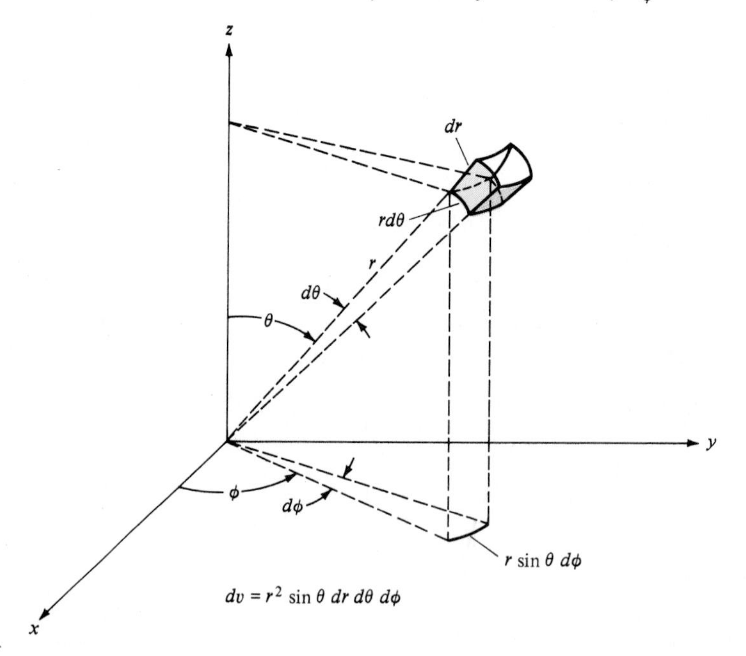

FIGURE 2.10
Differential arc lengths, surface areas, and volume elements in a spherical coordinate system.

The differential surfaces become

$$d\mathbf{s}_r = r^2 \sin\theta \, d\phi \, d\theta \, \mathbf{a}_r \tag{30a}$$

$$d\mathbf{s}_\phi = r \, d\theta \, dr \, \mathbf{a}_\phi \tag{30b}$$

$$d\mathbf{s}_\theta = r \sin\theta \, dr \, d\phi \, \mathbf{a}_\theta \tag{30c}$$

and the differential volume element is

$$dv = r^2 \sin\theta \, dr \, d\theta \, d\phi \tag{31}$$

Converting the coordinates of a point from either rectangular or cylindrical coordinates to spherical coordinates, and vice versa, is easily accomplished from Fig. 2.10. For example,

$$x = r \sin\theta \cos\phi \tag{32a}$$

$$y = r \sin\theta \sin\phi \tag{32b}$$

$$z = r \cos\theta \tag{32c}$$

and

$$r = \sqrt{x^2 + y^2 + z^2} \tag{33a}$$

$$\theta = \tan^{-1} \frac{\sqrt{x^2 + y^2}}{z} \tag{33b}$$

$$\phi = \tan^{-1} \frac{y}{x} \tag{33c}$$

A discussion of the conversion of a vector expressed in one coordinate system to another coordinate system is given in Appendix A.

EXAMPLE 2.3 Show by direct integration that the surface area of a sphere of radius R is $4\pi R^2$ and that the enclosed volume is $\frac{4}{3}\pi R^3$.

Solution To find the surface area, we integrate

$$\begin{aligned} S &= \int ds_r \\ &= \int_{\phi=0}^{2\pi} \int_{\theta=0}^{\pi} r^2 \sin\theta \, d\theta \, d\phi \\ &= R^2 \int_{\phi=0}^{2\pi} \int_{\theta=0}^{\pi} \sin\theta \, d\theta \, d\phi \\ &= 4\pi R^2 \end{aligned}$$

where we have substituted $r = R$. The volume is similarly obtained as

$$\begin{aligned} V &= \int dv \\ &= \int_{r=0}^{R} \int_{\phi=0}^{2\pi} \int_{\theta=0}^{\pi} r^2 \sin\theta \, dr \, d\theta \, d\phi \\ &= \frac{4\pi R^3}{3} \end{aligned}$$

2.6 Products of Vectors

Obviously, we cannot simply "multiply" two vectors as if they were scalars. Therefore, we must define some rules for obtaining the product of two vectors. The two rules for the product of two vectors (dot product and cross product) defined below are the most common and will be sufficient throughout our study of electromagnetic fields. The origin of these two definitions of vector products lies in their appearance in the equations governing electromagnetic fields, as we shall soon see. They also appear in other subjects requiring vector descriptions, such as mechanics, fluid flow, and heat flow.

2.6.1 Dot Product

The *first product rule* is called the dot product. The *dot product* of two vectors **A** and **B** is the product of the magnitudes of **A** and **B** and the cosine of the smallest angle between them, as shown in Fig. 2.11:

$$\mathbf{A} \cdot \mathbf{B} = |\mathbf{A}||\mathbf{B}| \cos\theta_{AB} \tag{34}$$

(read "**A** dot **B**"). Obviously, it makes no difference in which order **A** and **B** appear in the dot product, in other words,

$$\mathbf{A} \cdot \mathbf{B} = \mathbf{B} \cdot \mathbf{A} \tag{35}$$

Note that the result of the dot product of two vectors is a scalar and not a vector. Therefore, this product rule will be called a *scalar product*.

The dot product can also be interpreted as the product of the magnitude of **A**, $|\mathbf{A}|$, and the magnitude of the projection of **B** onto **A**. It can also be interpreted as the product of the magnitude of **B**, $|\mathbf{B}|$, and the magnitude of the projection of **A** onto **B**, as shown in Fig. 2.11. Clearly, from this interpretation the dot product of a vector with itself yields the square of the magnitude of the vector:

$$\mathbf{A} \cdot \mathbf{A} = |\mathbf{A}|^2 \tag{36}$$

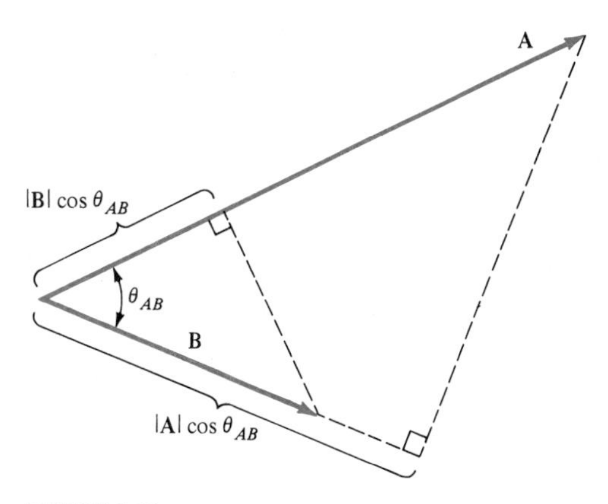

FIGURE 2.11
Vector projections and the dot product $\mathbf{A}\cdot\mathbf{B} = |\mathbf{A}||\mathbf{B}|\cos\theta_{AB}$.

Therefore, regardless of the coordinate system,

$$|\mathbf{A}| = \sqrt{\mathbf{A}\cdot\mathbf{A}} \tag{37}$$

The angle between two vectors (the smallest angle) can be computed with the dot product as

$$\theta_{AB} = \cos^{-1}\frac{\mathbf{A}\cdot\mathbf{B}}{|\mathbf{A}||\mathbf{B}|} \tag{38}$$

Note that if **A** and **B** are perpendicular, $\theta_{AB} = 90°$ and $\mathbf{A}\cdot\mathbf{B} = 0$. This property will be useful in determining whether two vectors are mutually perpendicular when it may not be obvious.

The dot product of a vector **A** and the sum of two vectors **B** and **C** is

$$\mathbf{A}\cdot(\mathbf{B}+\mathbf{C}) = \mathbf{A}\cdot\mathbf{B} + \mathbf{A}\cdot\mathbf{C} \tag{39}$$

This is easy to see if we think of the dot product of two vectors in terms of the projections of one on the other. In Fig. 2.12 we see that the projection of $(\mathbf{B}+\mathbf{C})$ on **A** is the same as the sum of the projections of **B** on **A** and of **C** on **A**. From this, we observe that

$$\begin{aligned}
(\mathbf{A}_1 + \mathbf{B}_1 + \mathbf{C}_1 + \cdots)&\cdot(\mathbf{A}_2 + \mathbf{B}_2 + \mathbf{C}_2 + \cdots) \\
&= (\mathbf{A}_1 + \mathbf{B}_1 + \mathbf{C}_1 + \cdots)\cdot\mathbf{A}_2 + (\mathbf{A}_1 + \mathbf{B}_1 + \mathbf{C}_1 + \cdots)\cdot\mathbf{B}_2 + \cdots \\
&= \mathbf{A}_1\cdot\mathbf{A}_2 + \mathbf{B}_1\cdot\mathbf{A}_2 + \mathbf{C}_1\cdot\mathbf{A}_2 + \cdots \\
&\quad + \mathbf{A}_1\cdot\mathbf{B}_2 + \mathbf{B}_1\cdot\mathbf{B}_2 + \mathbf{C}_1\cdot\mathbf{B}_2 + \cdots \\
&\quad + \mathbf{A}_1\cdot\mathbf{C}_2 + \mathbf{B}_1\cdot\mathbf{C}_2 + \mathbf{C}_1\cdot\mathbf{C}_2\cdots
\end{aligned} \tag{40}$$

and the dot product of the sums of vectors is performed in a manner very similar to the multiplication of sums of scalars.

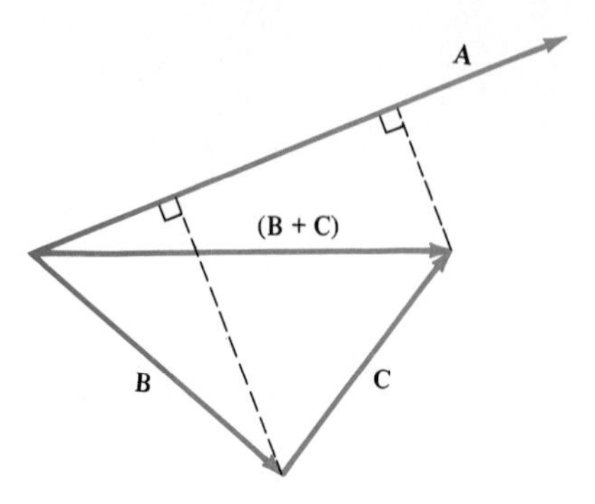

FIGURE 2.12
Illustration of $\mathbf{A}\cdot(\mathbf{B}+\mathbf{C}) = \mathbf{A}\cdot\mathbf{B} + \mathbf{A}\cdot\mathbf{C}$.

For general orthogonal coordinate systems,

$$\mathbf{a}_{u1}\cdot\mathbf{a}_{u1} = \mathbf{a}_{u2}\cdot\mathbf{a}_{u2} = \mathbf{a}_{u3}\cdot\mathbf{a}_{u3} = 1 \tag{41}$$

whereas

$$\mathbf{a}_{u1}\cdot\mathbf{a}_{u2} = \mathbf{a}_{u1}\cdot\mathbf{a}_{u3} = \mathbf{a}_{u2}\cdot\mathbf{a}_{u3} = 0 \tag{42}$$

since these unit vectors are mutually orthogonal. Therefore, the dot product of two vectors becomes

$$\begin{aligned}\mathbf{A}\cdot\mathbf{B} &= (A_{u1}\mathbf{a}_{u1} + A_{u2}\mathbf{a}_{u2} + A_{u3}\mathbf{a}_{u3})\cdot(B_{u1}\mathbf{a}_{u1} + B_{u2}\mathbf{a}_{u2} + B_{u3}\mathbf{a}_{u3}) \\ &= A_{u1}B_{u1} + A_{u2}B_{u2} + A_{u3}B_{u3}\end{aligned} \tag{43}$$

Thus, we obtain the dot product in a rectangular coordinate system as

$$\mathbf{A}\cdot\mathbf{B} = A_xB_x + A_yB_y + A_zB_z \tag{44}$$

In a cylindrical coordinate system we obtain

$$\mathbf{A}\cdot\mathbf{B} = A_rB_r + A_\phi B_\phi + A_zB_z \tag{45}$$

Similarly, in a spherical coordinate system we obtain

$$\mathbf{A}\cdot\mathbf{B} = A_rB_r + A_\theta B_\theta + A_\phi B_\phi \tag{46}$$

There are some pitfalls in blindly applying these results. For example, it makes no sense to write $\mathbf{A}\cdot\mathbf{B}\cdot\mathbf{C}$ since, for example, $\mathbf{A}\cdot\mathbf{B}$ yields a scalar that cannot be "dotted with" the vector **C**. Throughout our work in applying vector analysis to electromagnetic field problems, we should constantly be asking ourselves whether a vector operation that we intend to perform makes sense.

2.6.2 Cross Product

The second type of product of two vectors is called the *cross product*. It is classified as a vector product since the cross product of two vectors yields another vector as the result. The vector cross product (or simply the cross product) of two vectors **A** and **B** is defined as

$$\mathbf{A} \times \mathbf{B} = |\mathbf{A}||\mathbf{B}| \sin \theta_{AB} \, \mathbf{a}_n \tag{47}$$

(read "**A** cross **B**") where $\mathbf{a}_n$ is a unit vector normal to the plane that contains **A** and **B**, as shown in Fig. 2.13*a*. The angle θ_{AB} between **A** and **B** is the smallest angle between these two vectors. The direction of the unit normal vector $\mathbf{a}_n$ is

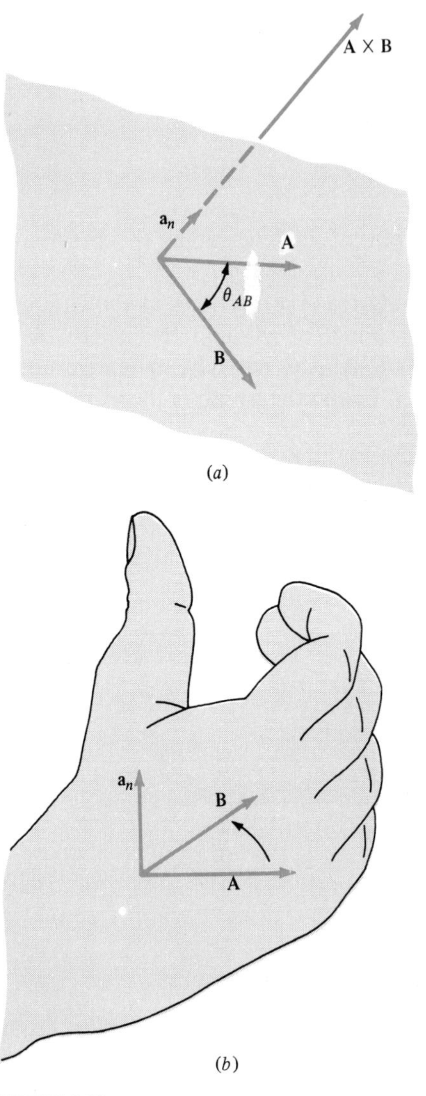

FIGURE 2.13
The vector cross product.

defined according to the so-called right-hand rule. With this rule, the direction of $\mathbf{a}_n$ is the direction that a right-hand screw would advance if turned in the direction from **A** to **B**, as illustrated in Fig. 2.13*b*.

The angle θ_{AB} between two vectors **A** and **B** may be obtained from the cross product as

$$\theta_{AB} = \sin^{-1} \frac{|\mathbf{A} \times \mathbf{B}|}{|\mathbf{A}||\mathbf{B}|} \tag{48}$$

Note that the cross product of two parallel vectors is zero. This property of the cross product is often used to determine whether two vectors are parallel.

The magnitudes of $\mathbf{A} \times \mathbf{B}$ and $\mathbf{B} \times \mathbf{A}$ are the same, but the directions of the resultant vectors will be opposite each other. In other words,

$$\mathbf{A} \times \mathbf{B} = -\mathbf{B} \times \mathbf{A} \tag{49}$$

and the order of this multiplication is important in determining the direction of the resultant vector.

The cross product can be thought of in terms of projections, as shown in Fig. 2.14. The magnitude of $\mathbf{A} \times \mathbf{B}$ is the product of $|\mathbf{A}|$ and the length of the projection of **B** onto a line perpendicular to **A**. Alternatively, the magnitude of $\mathbf{A} \times \mathbf{B}$ can be thought of as the product of $|\mathbf{B}|$ and the length of the projection of **A** onto a line perpendicular to **B**. Thinking of the cross product in this fashion, we can see in Fig. 2.15 that

$$\mathbf{A} \times (\mathbf{B} + \mathbf{C}) = \mathbf{A} \times \mathbf{B} + \mathbf{A} \times \mathbf{C} \tag{50}$$

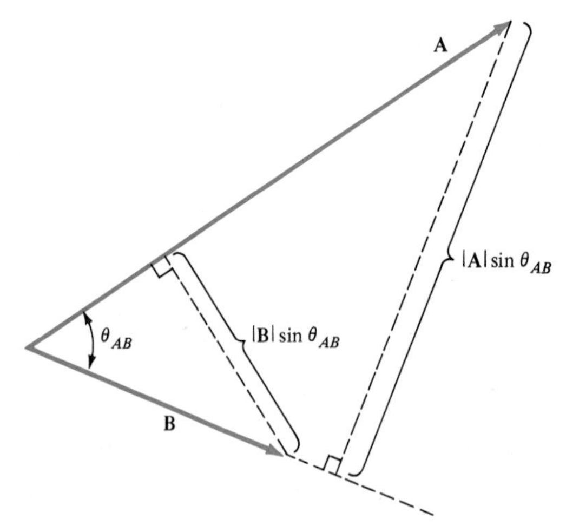

FIGURE 2.14
Illustration of $|\mathbf{A} \times \mathbf{B}| = |\mathbf{B} \times \mathbf{A}|$.

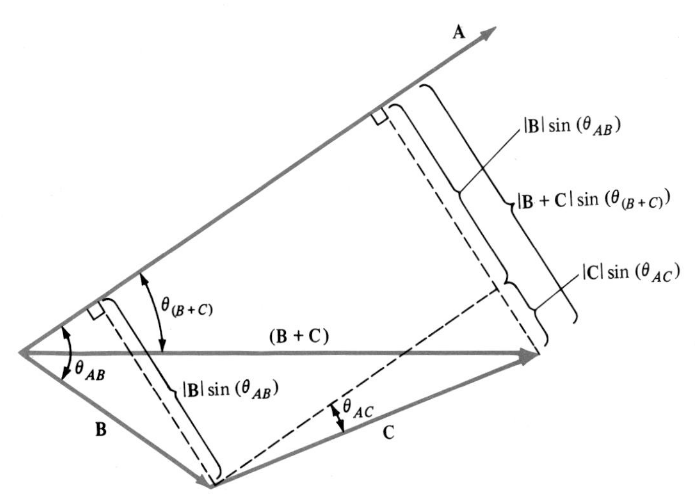

FIGURE 2.15
Illustration of $\mathbf{A} \times (\mathbf{B} + \mathbf{C}) = \mathbf{A} \times \mathbf{B} + \mathbf{A} \times \mathbf{C}$.

From this result, it follows that

$$
\begin{aligned}
(A_1 + \mathbf{B}_1 + \mathbf{C}_1 + \cdots) \times (\mathbf{A}_2 + \mathbf{B}_2 + \mathbf{C}_2 + \cdots) \\
= \mathbf{A}_1 \times \mathbf{A}_2 + \mathbf{A}_1 \times \mathbf{B}_2 + \mathbf{A}_1 \times \mathbf{C}_2 + \cdots \\
+ \mathbf{B}_1 \times \mathbf{A}_2 + \mathbf{B}_1 \times \mathbf{B}_2 + \mathbf{B}_1 \times \mathbf{C}_2 + \cdots \\
+ \mathbf{C}_1 \times \mathbf{A}_2 + \mathbf{C}_1 \times \mathbf{B}_2 + \mathbf{C}_1 \times \mathbf{C}_2 + \cdots
\end{aligned} \tag{51}
$$

In general orthogonal coordinate systems we find that

$$
\begin{aligned}
&\mathbf{a}_{u1} \times \mathbf{a}_{u1} = \mathbf{a}_{u2} \times \mathbf{a}_{u2} = \mathbf{a}_{u3} \times \mathbf{a}_{u3} = 0 \\
&\mathbf{a}_{u1} \times \mathbf{a}_{u2} = \mathbf{a}_{u3} \qquad \mathbf{a}_{u2} \times \mathbf{a}_{u1} = -\mathbf{a}_{u3} \\
&\mathbf{a}_{u1} \times \mathbf{a}_{u3} = -\mathbf{a}_{u2} \qquad \mathbf{a}_{u3} \times \mathbf{a}_{u1} = \mathbf{a}_{u2} \\
&\mathbf{a}_{u2} \times \mathbf{a}_{u3} = \mathbf{a}_{u1} \qquad \mathbf{a}_{u3} \times \mathbf{a}_{u2} = -\mathbf{a}_{u1}
\end{aligned} \tag{52}
$$

Orthogonal coordinate systems having these properties are said to be right-hand coordinate systems with the sequence $u_1 \to u_2 \to u_3$, in that crossing $\mathbf{a}_{u1}$ into $\mathbf{a}_{u2}$ yields $\mathbf{a}_{u3}$.

With the results in Eqs. (51) and (52), we may obtain the cross product of two vectors expressed in any orthogonal coordinate system as

$$
\begin{aligned}
\mathbf{A} \times \mathbf{B} &= (A_{u1}\mathbf{a}_{u1} + A_{u2}\mathbf{a}_{u2} + A_{u3}\mathbf{a}_{u3}) \times (B_{u1}\mathbf{a}_{u1} + B_{u2}\mathbf{a}_{u2} + B_{u3}\mathbf{a}_{u3}) \\
&= (A_{u2}B_{u3} - A_{u3}B_{u2})\mathbf{a}_{u1} \\
&\quad + (A_{u3}B_{u1} - A_{u1}B_{u3})\mathbf{a}_{u2} \\
&\quad + (A_{u1}B_{u2} - A_{u2}B_{u1})\mathbf{a}_{u3}
\end{aligned} \tag{53}
$$

This result for the cross product may appear difficult to remember. However, there is a simple memory aid that enables us to write the result immediately. Recall that the coordinate system is cyclic in the order $u_1 \to u_2 \to u_3 \to u_1 \to u_2 \to u_3 \to u_1 \cdots$, or

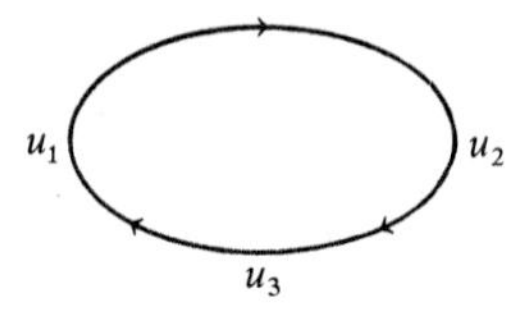

Each of the components of the cross product is of the form $\mathbf{a}_a(A_b B_c - A_c B_b)$. Once we obtain the proper subscripts for the first portion of this component, $A_b B_c$, the second portion, $A_c B_b$, is determined. If we are finding the u_3 component of $\mathbf{A} \times \mathbf{B}$, the next coordinate in our cyclic sequence is u_1, followed by u_2; therefore, we assign $a = u_3$, $b = u_1$, $c = u_2$. If we are finding the u_2 component of $\mathbf{A} \times \mathbf{B}$, the next coordinate in the sequence is u_3, followed by u_1; therefore, for the u_2 component of the cross product, we assign $a = u_2$, $b = u_3$, $c = u_1$.

An alternative way of finding the components of the cross product is by use of the following determinant rule:

$$\mathbf{A} \times \mathbf{B} = \begin{vmatrix} \mathbf{a}_{u1} & \mathbf{a}_{u2} & \mathbf{a}_{u3} \\ A_{u1} & A_{u2} & A_{u3} \\ B_{u1} & B_{u2} & B_{u3} \end{vmatrix} \tag{54}$$

Note that the first row contains the symbols for the unit vectors and that the second and third rows contain the components of A and B, respectively. The ordering of the entries in each row is u_1 then u_2 then u_3. If we evaluate this determinant, pretending for this evaluation that the unit vectors are treated like scalars, we obtain the correct result. Equation (54) is simply a mnemonic rule. Equation (53) is the fundamental result.

In the rectangular coordinate system $u_1 = x$, $u_2 = y$, and $u_3 = z$. Therefore,

$$\mathbf{A} \times \mathbf{B} = (A_y B_z - A_z B_y)\mathbf{a}_x + (A_z B_x - A_x B_z)\mathbf{a}_y + (A_x B_y - A_y B_x)\mathbf{a}_z \tag{55}$$

Note that with the above simple memory aid it is easy to write any of the three components of the cross product without writing the entire vector. Use of (54) is unnecessary.

The evaluation of the cross product when the two vectors are expressed in a cylindrical or a spherical coordinate system is identical to the evaluation in a rectangular coordinate system if the two vectors are again defined at the same point so that the corresponding unit vectors are parallel. Again, this is due to the fact that the three unit vectors in each of these coordinate systems form a mutually orthogonal set. They are also ordered cyclically as $r \to \phi \to z \to r \to$

$\phi \rightarrow \cdots$ for a cylindrical system and $r \rightarrow \theta \rightarrow \phi \rightarrow r \rightarrow \theta \rightarrow \cdots$ for a spherical system. The result for a cylindrical system is

$$\mathbf{A} \times \mathbf{B} = (A_\phi B_z - A_z B_\phi)\mathbf{a}_r + (A_z B_r - A_r B_z)\mathbf{a}_\phi + (A_r B_\phi - A_\phi B_r)\mathbf{a}_z \quad (56)$$

and for a spherical coordinate system

$$\mathbf{A} \times \mathbf{B} = (A_\theta B_\phi - A_\phi B_\theta)\mathbf{a}_r + (A_\phi B_r - A_r B_\phi)\mathbf{a}_\theta + (A_r B_\theta - A_\theta B_r)\mathbf{a}_\phi \quad (57)$$

Note that each of these results can be found quite easily by using the above memory aid and remembering the cyclic ordering of the coordinate surfaces.

Again we must be careful to avoid pitfalls in blindly applying these results. For example,

$$\mathbf{A} \cdot (\mathbf{B} \times \mathbf{C}) \neq (\mathbf{A} \cdot \mathbf{B}) \times (\mathbf{A} \cdot \mathbf{C})$$

This, of course, should be obvious since both $\mathbf{A} \cdot \mathbf{B}$ and $\mathbf{A} \cdot \mathbf{C}$ yield scalars and the cross product of two scalars is meaningless. However, the operation $\mathbf{A} \cdot (\mathbf{B} \times \mathbf{C})$ does not result in any problems of inconsistency, such as attempting to obtain the dot product of a vector and a scalar. [We certainly could not compute $(\mathbf{A} \cdot \mathbf{B}) \times \mathbf{C}$ since $\mathbf{A} \cdot \mathbf{B}$ yields a scalar.]

EXAMPLE 2.4 Suppose

$$\mathbf{A} = 1\mathbf{a}_x + 2\mathbf{a}_y + 3\mathbf{a}_z$$

$$\mathbf{B} = 4\mathbf{a}_x + 5\mathbf{a}_y + 6\mathbf{a}_z$$

$$\mathbf{C} = 7\mathbf{a}_x + 8\mathbf{a}_y + 9\mathbf{a}_z$$

Find the result of $\mathbf{A} \cdot (\mathbf{B} \times \mathbf{C})$.

Solution The cross product $\mathbf{B} \times \mathbf{C}$ yields

$$\mathbf{B} \times \mathbf{C} = -3\mathbf{a}_x + 6\mathbf{a}_y - 3\mathbf{a}_z$$

The dot product of $\mathbf{A}$ with $\mathbf{B} \times \mathbf{C}$ yields

$$\mathbf{A} \cdot (\mathbf{B} \times \mathbf{C}) = 0$$

On the other hand, we may prove the identity (see Appendix A)

$$\begin{aligned}\mathbf{A} \cdot (\mathbf{B} \times \mathbf{C}) &= \mathbf{B} \cdot (\mathbf{C} \times \mathbf{A}) \\ &= \mathbf{C} \cdot (\mathbf{A} \times \mathbf{B})\end{aligned}$$

so that

$$\mathbf{C} \times \mathbf{A} = 6\mathbf{a}_x - 12\mathbf{a}_y + 6\mathbf{a}_z$$

and

$$\mathbf{B} \cdot (\mathbf{C} \times \mathbf{A}) = 0$$

Also, we can show that

$$\mathbf{C}\cdot(\mathbf{A}\times\mathbf{B}) = 0$$

From the results of this problem we see that **A**, **B**, and **C** are contained in a common plane.

EXAMPLE 2.5 Determine the (smallest) angle between the two vectors

$$\mathbf{A} = 1\mathbf{a}_x - 3\mathbf{a}_y + 2\mathbf{a}_z$$

$$\mathbf{B} = -3\mathbf{a}_x + 4\mathbf{a}_y - \mathbf{a}_z$$

and determine a unit vector perpendicular to the plane containing **A** and **B**.

Solution The angle between the two vectors can be found by using either the dot product or the cross product. Using the dot product, we obtain

$$\cos\theta_{AB} = \frac{\mathbf{A}\cdot\mathbf{B}}{|\mathbf{A}|\,|\mathbf{B}|}$$

Forming

$$\mathbf{A}\cdot\mathbf{B} = -3 - 12 - 2$$
$$= -17$$

$$|\mathbf{A}| = \sqrt{\mathbf{A}\cdot\mathbf{A}}$$
$$= \sqrt{1 + 9 + 4}$$
$$= \sqrt{14}$$
$$= 3.74$$

$$|\mathbf{B}| = \sqrt{\mathbf{B}\cdot\mathbf{B}}$$
$$= 5.1$$

we obtain

$$\cos\theta_{AB} = \frac{-17}{3.74\times 5.1}$$
$$= -0.89$$

so that

$$\theta_{AB} = 153^\circ$$

From the cross product we may also obtain

$$\sin\theta_{AB} = \frac{|\mathbf{A} \times \mathbf{B}|}{|\mathbf{A}|\,|\mathbf{B}|}$$

Forming

$$\mathbf{A} \times \mathbf{B} = -5\mathbf{a}_x - 5\mathbf{a}_y - 5\mathbf{a}_z$$

$$|\mathbf{A} \times \mathbf{B}| = 8.66$$

we obtain

$$\sin\theta_{AB} = \frac{8.66}{3.74 \times 5.1} = 0.45$$

so that

$$\theta_{AB} = 27^\circ \quad \text{or} \quad 153^\circ$$

From the dot product, the ambiguity in angle is resolved. There are two unit vectors perpendicular to the plane containing **A** and **B**. From (47)

$$\mathbf{a}_n = \frac{\mathbf{A} \times \mathbf{B}}{|\mathbf{A}||\mathbf{B}|\sin\theta_{AB}} = \frac{-5\mathbf{a}_x - 5\mathbf{a}_y - 5\mathbf{a}_z}{3.74 \times 5.1 \times 0.45} = -0.58\mathbf{a}_x - 0.58\mathbf{a}_y - 0.58\mathbf{a}_z$$

or

$$\mathbf{a}_n = 0.58\mathbf{a}_x + 0.58\mathbf{a}_y + 0.58\mathbf{a}_z$$

obtained as the negative of the first unit vector.

2.7 Fields

One of the most important concepts in the subject of electromagnetic fields is that of a field. There are two types of fields that we will encounter in our study: scalar fields and vector fields. The concept of a scalar field is simpler and will be discussed first.

2.7.1 Scalar Fields

Rather than trying to define a scalar field mathematically, it is sufficient to illustrate the concept with an example. The temperature distribution in a room is perhaps the most common example of a scalar field. Consider Fig. 2.16, which shows the lines, or contours, of constant temperature. For example, the temperature is T_1 at all points on the contour labeled as T_1. (Only selected contours are shown. To determine those contours which are not shown, we would interpolate between those shown and assume that there is a smooth and uniform variation in the contour change.) This temperature distribution is called a *field* in the sense that the temperature will have values at various times and positions in the room (region). To denote this dependence of the temperature on positions and x, y, and z and on time t, we write symbolically $T(x, y, z, t)$.

This is the essential concept of a *scalar field.* The quantity of interest, temperature in this example, is a scalar quantity since it has no "direction of effect" associated with it and can take on a multitude of values that depend on the location of the point of interest (the field point) and on the time of observation.

2.7.2 Vector Fields

The basic difference between a scalar field and a *vector field* is that the quantity of interest in a vector field is assumed to have a directional property as well as a magnitude at points in the region. A common example of a vector field is the flow of a fluid in a constricted pipe (Fig. 2.17). The flow of the fluid at

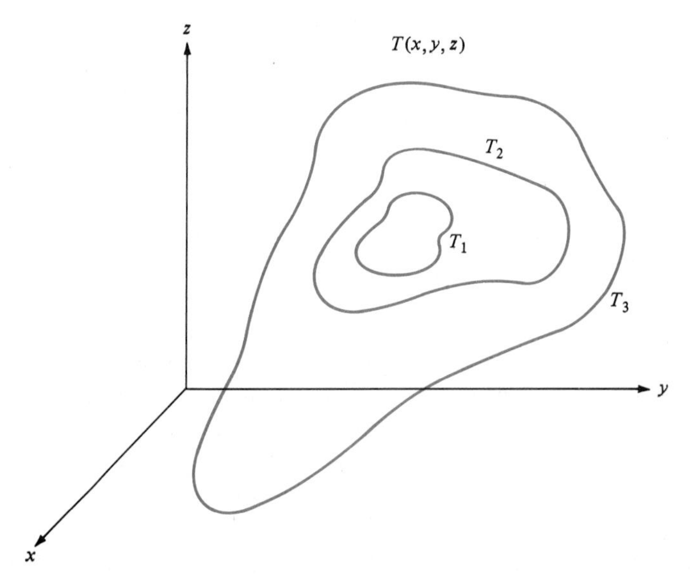

FIGURE 2.16

A scalar field (temperature distribution in a room).

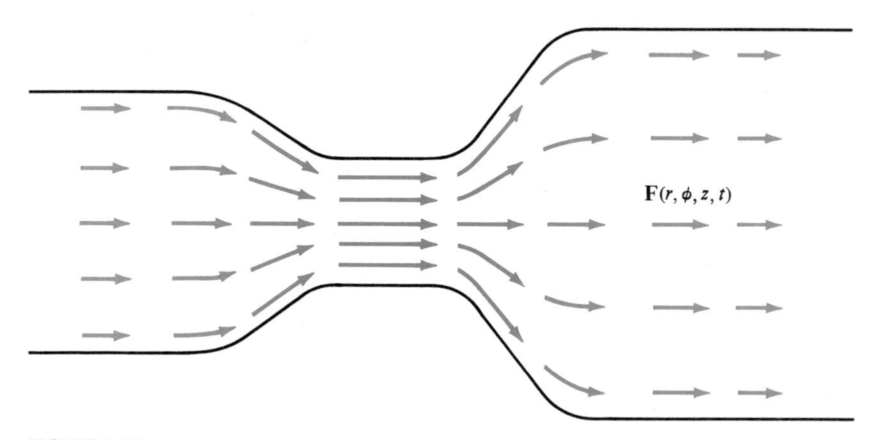

FIGURE 2.17
A vector field (flow of a fluid in a constricted pipe).

various points in the pipe would have a direction associated with it as well as a magnitude (rate of flow). We may denote the flow of the fluid at various points in the pipe with vectors. The relative lengths of the vectors show the relative rates of flow, and the directions of the vectors show the direction of flow at these points.† We could denote the dependence of this vector field on position and time as $\mathbf{F}(x, y, z, t)$. Or, if we did not wish to be partial to a rectangular coordinate system, we could denote this vector field as $\mathbf{F}(r, \phi, z, t)$ in a cylindrical coordinate system. Obviously, either coordinate system may be used. Proper choice of the coordinate system used to describe a problem, however, will simplify the problem's solution in many cases.

Therefore, both scalar and vector fields contain information about the value of a quantity (such as the temperature or rate of flow of a fluid) at points of space and (perhaps) of time in some region. The vector field also has information about the direction of effect of the quantity of interest, whereas the scalar field does not.

2.8 The Gradient of a Scalar Field

Having considered what may be appropriately classified as vector algebra concepts, we now embark in the remainder of this chapter on a consideration of some vector calculus concepts. The gradient operation that we will consider in this section is a vector calculus concept and is related to the space rate of change of some scalar field. For example, suppose we wish to take a cross-country hike. If we are fortunate to have a topographical map showing contours of constant

† A more conventional method of depicting a vector field is with flux lines. The flux lines are directed in the direction of the field vectors, and their densities indicate the magnitudes of the field vectors. This will be discussed in later chapters.

elevation, we can plan our route to avoid climbing steep grades. For example, consider the contour map shown in Fig. 2.18. Contours of constant elevation that are close together indicate a steep climb (or descent): a rapid change in elevation for a small change in horizontal distance. To minimize our effort on the hike, we could plan the route so as to avoid areas where the contour lines are closest together. For example, if we plan to climb from the base of the mountain depicted in Fig. 2.18 to the top, we might choose the path P_1. Path P_2 would be the shortest distance to the top, but the climb would be more strenuous than along path P_1.

In the example of a topographical map, we have been discussing a scalar field $EL(x, y, z)$, where EL is a scalar function that gives the elevation of the terrain at various points. Clearly, if we knew the functional form of EL rather than having it depicted as a map, we should be able to determine mathematically not only the path of minimum effort but also the rate of change of elevation with horizontal distance at points along this path. There are numerous examples of scalar electromagnetic fields for which we would also like to determine similar properties. These properties are adequately determined with the gradient operation.

Consider a scalar field $f(u_1, u_2, u_3)$ defined in a generalized orthogonal coordinate system. The scalar field is constant over surface s_1, *i.e.*, $f(u_1, u_2, u_3) = c$, as shown in Fig. 2.19. An example would be the surfaces of constant temperature in a room (Fig. 2.16). Consider another surface, s_2, over which f is constant and is displaced by an infinitesimal distance from s_1. This surface is defined by $f = c + df$. Suppose we move a distance dl between two points on these surfaces, P_1 and P_2, as shown in Fig. 2.19. The differential change in f is

$$\begin{aligned} df &= \frac{\partial f}{\partial u_1} du_1 + \frac{\partial f}{\partial u_2} du_2 + \frac{\partial f}{\partial u_3} du_3 \\ &= \frac{\partial f}{\partial l_1} dl_1 + \frac{\partial f}{\partial l_2} dl_2 + \frac{\partial f}{\partial l_3} dl_3 \end{aligned} \tag{58}$$

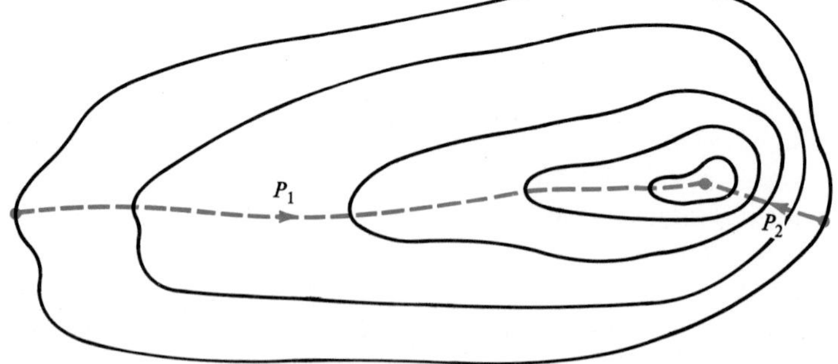

FIGURE 2.18

Topographical map. Path P_2 is the shortest distance to the top of the mountain, but path P_1 avoids steep climbs.

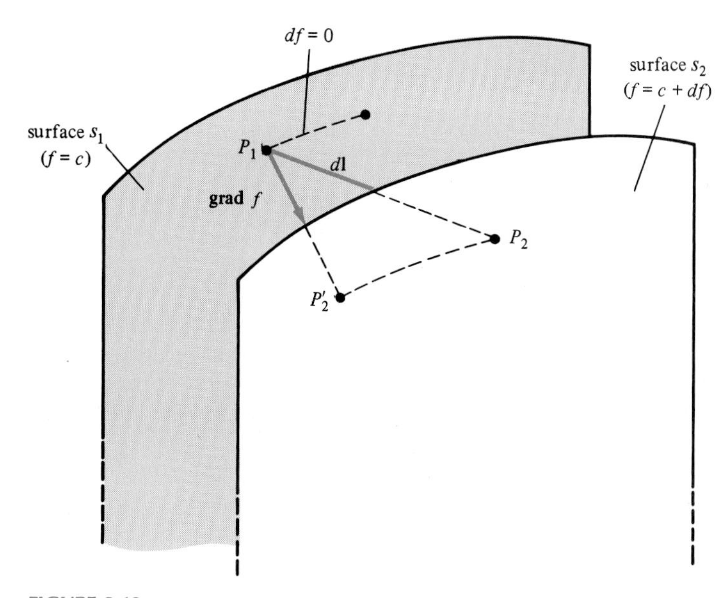

FIGURE 2.19
Illustration of the gradient vector.

The vector distance between these two points is

$$\mathbf{dl} = dl_1\mathbf{a}_{u1} + dl_2\mathbf{a}_{u2} + dl_3\mathbf{a}_{u3} \tag{59}$$

so that df can be written as

$$df = \left(\frac{\partial f}{\partial l_1}\mathbf{a}_{u1} + \frac{\partial f}{\partial l_2}\mathbf{a}_{u2} + \frac{\partial f}{\partial l_3}\mathbf{a}_{u3}\right)\cdot \mathbf{dl} \tag{60}$$

as a simple calculation will show. The item in parentheses will be called the *gradient of f* and will be given the symbol

$$\mathbf{grad}\, f = \frac{\partial f}{\partial l_1}\mathbf{a}_{u1} + \frac{\partial f}{\partial l_2}\mathbf{a}_{u2} + \frac{\partial f}{\partial l_3}\mathbf{a}_{u3} \tag{61}$$

so that we may write

$$df = \mathbf{grad}\, f \cdot \mathbf{dl} \tag{62}$$

As a shorthand notation, we will denote

$$\mathbf{grad}\, f = \nabla f \tag{63}$$

where ∇ is pronounced "del." Substituting the metric coefficients $dl_i = h_i\, du_i$, we obtain

$$\nabla f = \frac{1}{h_1}\frac{\partial f}{\partial u_1}\mathbf{a}_{u1} + \frac{1}{h_2}\frac{\partial f}{\partial u_2}\mathbf{a}_{u2} + \frac{1}{h_3}\frac{\partial f}{\partial u_3}\mathbf{a}_{u3} \tag{64}$$

From the above we may write

$$\begin{aligned} df &= \nabla f \cdot d\mathbf{l} \\ &= |\nabla f|\, dl \cos\theta \end{aligned} \tag{65}$$

where θ is the angle between ∇f and the length vector $d\mathbf{l}$, which is along the chosen path.

Several important properties of the gradient become apparent from Fig. 2.19 and the above development. Movement between two points on the same surface (for example, P_1 on s_1) results in no change in f: $df = 0$. From Eq. (65), this implies that **grad** f is orthogonal to that surface. Thus, *at a point on a surface for which f = constant, the gradient vector* **grad** *f is orthogonal to that surface.* Now consider the case where P_1 and P_2 are on different constant-f surfaces. Movement between P_1 and P_2 will yield a change in f given by (65). The space rate of change in f is given by

$$\frac{df}{dl} = |\nabla f| \cos\theta \tag{66a}$$

The *maximum space rate of change* in f will occur when the path $d\mathbf{l}$ is chosen such that $\theta = 0$; i.e., the path is in the direction of the gradient, or orthogonal to s_1 at P_1. This maximum space rate of change in f is then given by

$$\left.\frac{df}{dl}\right|_{\max} = |\nabla f| \tag{66b}$$

Therefore, *the gradient vector denotes both the direction and the magnitude of the maximum space rate of change of the scalar field f.*

In a rectangular coordinate system the metric coefficients are all unity: $h_x = h_y = h_z = 1$. Thus

$$\nabla f(x, y, z) = \frac{\partial f}{\partial x}\mathbf{a}_x + \frac{\partial f}{\partial y}\mathbf{a}_y + \frac{\partial f}{\partial z}\mathbf{a}_z \tag{67}$$

In this coordinate system we may define the del operator as

$$\nabla = \frac{\partial}{\partial x}\mathbf{a}_x + \frac{\partial}{\partial y}\mathbf{a}_y + \frac{\partial}{\partial z}\mathbf{a}_z \tag{68}$$

In a cylindrical coordinate system the metric coefficients are $h_r = 1$, $h_\phi = r$, and $h_z = 1$, so that

$$\nabla f(r, \phi, z) = \frac{\partial f}{\partial r}\mathbf{a}_r + \frac{1}{r}\frac{\partial f}{\partial \phi}\mathbf{a}_\phi + \frac{\partial f}{\partial z}\mathbf{a}_z \tag{69}$$

In a spherical coordinate system $h_r = 1$, $h_\theta = r$, $h_\phi = r \sin\theta$, and

$$\nabla f(r, \theta, \phi) = \frac{\partial f}{\partial r}\mathbf{a}_r + \frac{1}{r}\frac{\partial f}{\partial \theta}\mathbf{a}_\theta + \frac{1}{r\sin\theta}\frac{\partial f}{\partial \phi}\mathbf{a}_\phi \tag{70}$$

A definition of the del operator in cylindrical and spherical coordinate systems similar to (68) as

$$\nabla = \frac{1}{h_1}\frac{\partial}{\partial u_1}\mathbf{a}_{u1} + \frac{1}{h_2}\frac{\partial}{\partial u_2}\mathbf{a}_{u2} + \frac{1}{h_3}\frac{\partial}{\partial u_3}\mathbf{a}_{u3} \tag{71}$$

would result in the correct answers given in (69) and (70). In certain other vector operations discussed in later sections, however, this interpretation will not yield the correct result for cylindrical and spherical coordinate systems—but for rectangular coordinate systems the correct result will be obtained by using the del operator in (68). Thus, we will reserve the del operator only for use in rectangular coordinate systems. In other coordinate systems it will only serve a symbolic role.

EXAMPLE 2.6 Determine the gradient of the following scalar fields:

(a) $f(x, y, z) = xy^2 + 2z$
(b) $f(r, \phi, z) = 2r \sin\phi$
(c) $f(r, \theta, \phi) = 2\theta + r^2$

Solution

(a) $\nabla f = \frac{\partial f}{\partial x}\mathbf{a}_x + \frac{\partial f}{\partial y}\mathbf{a}_y + \frac{\partial f}{\partial z}\mathbf{a}_z = y^2\mathbf{a}_x + 2xy\mathbf{a}_y + 2\mathbf{a}_z$

(b) $\nabla f = \frac{\partial f}{\partial r}\mathbf{a}_r + \frac{1}{r}\frac{\partial f}{\partial \phi}\mathbf{a}_\phi + \frac{\partial f}{\partial z}\mathbf{a}_z = 2\sin\phi\,\mathbf{a}_r + 2\cos\phi\,\mathbf{a}_\phi$

(c) $\nabla f = \frac{\partial f}{\partial r}\mathbf{a}_r + \frac{1}{r}\frac{\partial f}{\partial \theta}\mathbf{a}_\theta + \frac{1}{r\sin\theta}\frac{\partial f}{\partial \phi}\mathbf{a}_\phi = 2r\mathbf{a}_r + \frac{2}{r}\mathbf{a}_\theta$

EXAMPLE 2.7 Show that the gradient of the scalar field $f(x, y, z) = x + y$ is normal to lines of constant f.

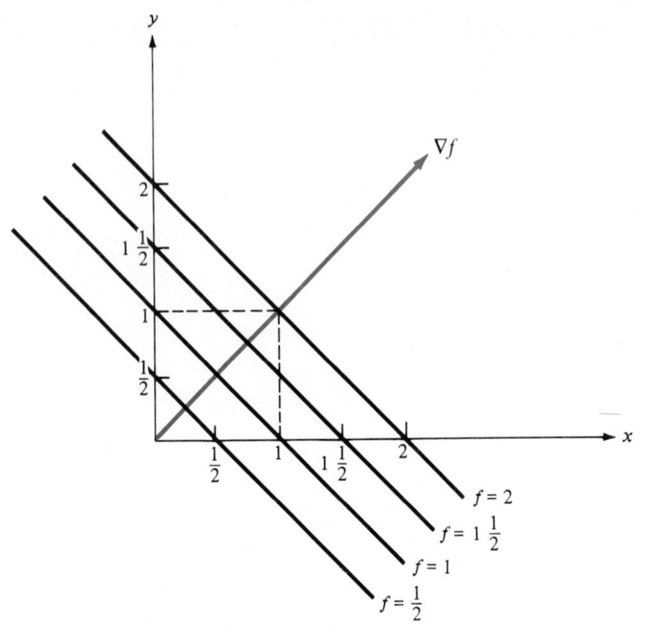

FIGURE 2.20
Example 2.7. The gradient.

Solution The gradient of this function is $\nabla f = \mathbf{a}_x + \mathbf{a}_y$. A plot of f in the xy plane is given in Fig. 2.20. The equations of lines (planes) of constant f are $f = x + y$. Note that ∇f is normal to these lines of constant f. A more direct proof can be obtained by utilizing the property of the dot product in which the dot product of two vectors that are mutually perpendicular is zero. If we write the contours of constant f as a vector

$$\mathbf{V} = k(\mathbf{a}_x - \mathbf{a}_y)$$

we find

$$(\nabla f) \cdot \mathbf{V} = 0$$

2.9 The Line Integral of a Vector Field

The gradient operation defined in the previous section served a useful role in that it allowed us to determine the maximum space rate of change of a scalar field at points in a region. The line integral to be discussed in this section serves a similar purpose.

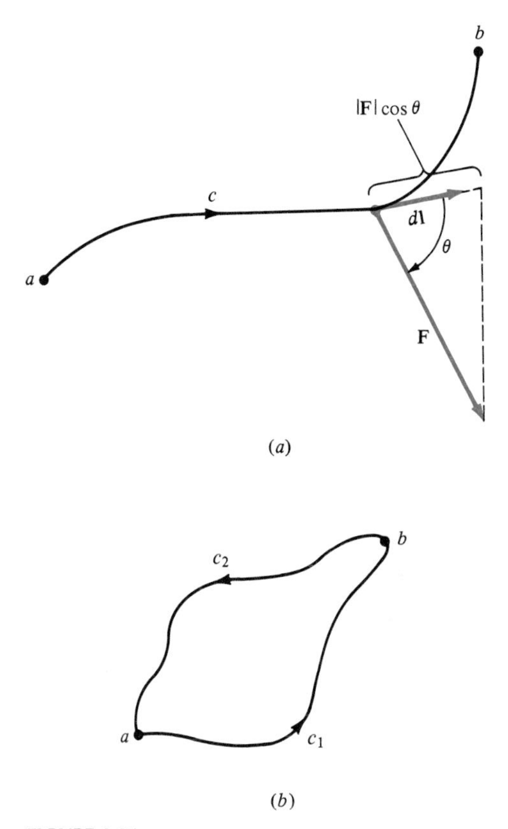

FIGURE 2.21
The line integral of a vector field along a contour.

In computing the work done in moving an object along some path or contour, we simply add the incremental expenditures of energy along the contour. For example, consider the contour c shown in Fig. 2.21a. At each point along the contour, we obtain the component of force directed along the contour $|\mathbf{F}| \cos \theta$, and the work required to move along the contour from a to b is

$$W = \int_a^b |\mathbf{F}| \cos \theta \, dl = \int_a^b \mathbf{F} \cdot d\mathbf{l} \tag{72}$$

Note that θ is a function of position along the contour as is $d\mathbf{l}$, which is tangent to the contour. The integral in (72) is called the *line integral* of the vector field $\mathbf{F}$.

The evaluation of line integrals in orthogonal coordinate systems is straightforward. For example, in a generalized orthogonal coordinate system

$$\int_a^b \mathbf{F}(u_1, u_2, u_3) \cdot d\mathbf{l} = \int_{l_1} F_{u1} \, dl_1 + \int_{l_2} F_{u2} \, dl_2 + \int_{l_3} F_{u3} \, dl_3 \tag{73}$$

Substituting the metric coefficients, $dl_i = h_i\, du_i$, yields

$$\int_a^b \mathbf{F}(u_1, u_2, u_3)\cdot d\mathbf{l} = \int_{u_1} h_1 F_{u1}\, du_1 + \int_{u_2} h_2 F_{u2}\, du_2 + \int_{u_3} h_3 F_{u3}\, du_3 \quad (74)$$

Each of the above integrals requires integration with respect to the individual coordinates. Thus, the components F_{u1}, F_{u2}, and F_{u3} must be converted to a function of only one coordinate variable: u_1, u_2, and u_3, respectively. This conversion will be made using the equation(s) of the path of integration. In rectangular, cylindrical, and spherical coordinate systems these become

$$\int_a^b \mathbf{F}(x, y, z)\cdot d\mathbf{l} = \int_{x_a}^{x_b} F_x\, dx + \int_{y_a}^{y_b} F_y\, dy + \int_{z_a}^{z_b} F_z\, dz \quad (75)$$

$$\int_a^b \mathbf{F}(r, \phi, z)\cdot d\mathbf{l} = \int_{r_a}^{r_b} F_r\, dr + \int_{\phi_a}^{\phi_b} rF_\phi\, d\phi + \int_{z_a}^{z_b} F_z\, dz \quad (76)$$

$$\int_a^b \mathbf{F}(r, \theta, \phi)\cdot d\mathbf{l} = \int_{r_a}^{r_b} F_r\, dr + \int_{\theta_a}^{\theta_b} rF_\theta\, d\theta + \int_{\phi_a}^{\phi_b} r \sin\theta\, F_\phi\, d\phi \quad (77)$$

EXAMPLE 2.8 Evaluate the line integral of the vector field

$$\mathbf{F}(x, y, z) = (x + y)\mathbf{a}_x - x\mathbf{a}_y + z\mathbf{a}_z$$

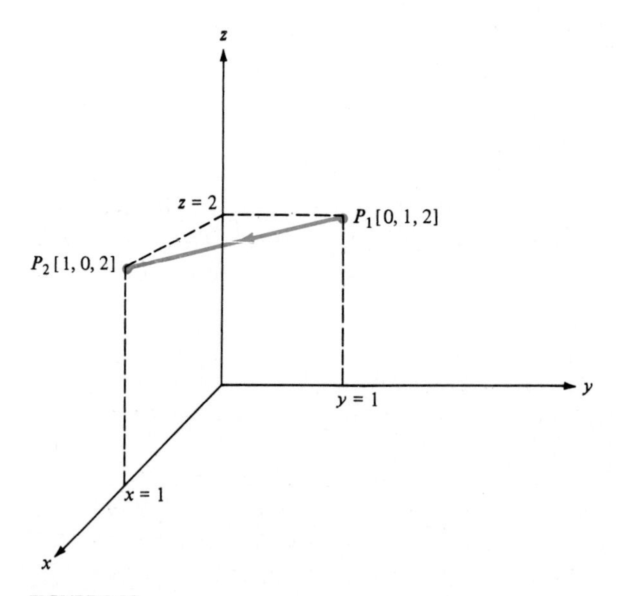

FIGURE 2.22
Example 2.8. The line integral.

along a path consisting of a straight line from $P_1[0, 1, 2]$ to $P_2[1, 0, 2]$, as shown in Fig. 2.22.

Solution The integral becomes

$$\int_{P_1}^{P_2} \mathbf{F} \cdot d\mathbf{l} = \int_{x=0}^{1} (x + y)dx - \int_{y=1}^{0} x\, dy + \int_{z=2}^{2} z\, dz$$

The integrands of the first and second integrals must be converted to functions of x and y, respectively. The equation of the path can be determined as

$$y = 1 - x$$

Substituting this relation yields

$$\begin{aligned} \int_{P_1}^{P_2} \mathbf{F} \cdot d\mathbf{l} &= \int_{x=0}^{1} dx - \int_{y=1}^{0} (1 - y)dy \\ &= x\Big|_0^1 - \left(y - \frac{y^2}{2}\right)\Big|_1^0 \\ &= \tfrac{3}{2} \end{aligned}$$

Now suppose we have a scalar field $f(u_1, u_2, u_3)$. If we integrate the incremental change in f along this contour, we obtain

$$\int_a^b df = f_b - f_a = \int_a^b \nabla f \cdot d\mathbf{l} \tag{78}$$

where f_b and f_a are the values of f at b and a, respectively. In (78) we have substituted the previously determined relation $df = \nabla f \cdot d\mathbf{l}$. Note that since the integral of df depends only on the value of f at the endpoints of the contour, the result in (78) is independent of the path taken from a to b. If we choose a contour c_1 from a to b and a return contour c_2 from b to a, as shown in Fig. 2.21b, we find that

$$\oint_c df = \int_{c_1} df + \int_{c_2} df = \oint_c \nabla f \cdot d\mathbf{l} = 0 \tag{79}$$

where the integration around the closed contour c consisting of contour c_1 and contour c_2 is denoted by a circle on the integral sign.

We thus find that the line integral of the gradient of a scalar field around a closed path or contour is identically zero. Actually, this result follows directly from the definition of the gradient function. More important, we observe that if a vector field $\mathbf{F}$ can be written as the gradient of some scalar field f such as

$$\mathbf{F} = \nabla f \tag{80}$$

then (79) shows that

$$\oint_c \mathbf{F} \cdot d\mathbf{l} = \oint_c \nabla f \cdot d\mathbf{l} = 0 \tag{81}$$

regardless of the contour c. Such a vector field $\mathbf{F}$ is said to be *conservative.* Similarly, we find that if a vector field $\mathbf{F}$ can be written as the gradient of a scalar field f, as in (80), then the line integral of $\mathbf{F}$ between two points is independent of the path taken between these two points since $\mathbf{F} = \nabla f$ and the line integral of ∇f depends only on the value of f at the endpoints of the path:

$$\int_{a,c_1}^{b} \mathbf{F} \cdot d\mathbf{l} = \int_{a,c_2}^{b} \mathbf{F} \cdot d\mathbf{l} \tag{82}$$

The reader should be cautioned that (82) is necessarily true only if $\mathbf{F}$ represents a conservative field such that $\mathbf{F}$ can be written as the gradient of a scalar field f.

EXAMPLE 2.9 Consider the scalar field

$$f(x, y, z) = 2xy + 3$$

If the vector field $\mathbf{F}$ is given by

$$\mathbf{F} = \nabla f = 2y\mathbf{a}_x + 2x\mathbf{a}_y$$

verify that $\oint_c \mathbf{F} \cdot d\mathbf{l} = 0$ for the closed contour consisting of paths c_1, c_2, and c_3 shown in Fig. 2.23.

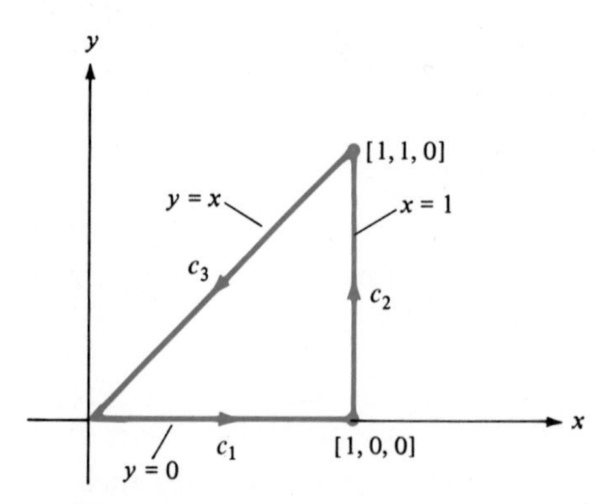

FIGURE 2.23

Example 2.9. A conservative field.

Solution For the given field,

$$\mathbf{F}\cdot d\mathbf{l} = 2y\,dx + 2x\,dy$$

and

Path c_1
$$\int_{[0,0,0]}^{[1,0,0]} \mathbf{F}\cdot d\mathbf{l} = \int_{x=0}^{x=1} 2\cancel{y}^{\,0}\,dx = 0$$

Path c_2
$$\int_{[1,0,0]}^{[1,1,0]} \mathbf{F}\cdot d\mathbf{l} = \int_{y=0}^{y=1} 2\cancel{x}^{\,1}\,dy = 2$$

Path c_3
$$\int_{[1,1,0]}^{[0,0,0]} \mathbf{F}\cdot d\mathbf{l} = \int_{x=1}^{0} 2\cancel{y}^{\,y=x}\,dx + \int_{y=1}^{0} 2\cancel{x}^{\,x=y}\,dy$$
$$= x^2\Big|_1^0 + y^2\Big|_1^0 = -2$$

Clearly

$$\oint_c \mathbf{F}\cdot d\mathbf{l} = 0$$

As in Example 2.8, note that along the various paths certain constraints among the x and y variables exist and must be substituted into each integrand. For example, along path c_1 there is no variation with y, so that only the dx component of $\mathbf{F}\cdot d\mathbf{l}$ need be integrated from $x = 0$ to $x = 1$. However, the path constraint $y = 0$ must be substituted into the integrand. Similarly, along path c_3 we must integrate both components of $\mathbf{F}\cdot d\mathbf{l}$. The dx component is integrated from $x = 1$ to $x = 0$, but the path constraint $y = x$ must be substituted for y in the integrand. Similarly, along c_3 the dy component is integrated from $y = 1$ to $y = 0$, and the path constraint $x = y$ must be substituted into the integrand.

2.10 The Divergence of a Vector Field

We will find it very useful in our study of electromagnetic fields to think in terms of the net outflow, or flux, of a vector field through some surface. Consider a vector field in a region of space shown in Fig. 2.24. The *flux* of the vector field $\mathbf{F}$ through some surface s is defined as

$$\psi = \int_s |\mathbf{F}|\cos\theta\,ds = \int_s \mathbf{F}\cdot d\mathbf{s} \tag{83}$$

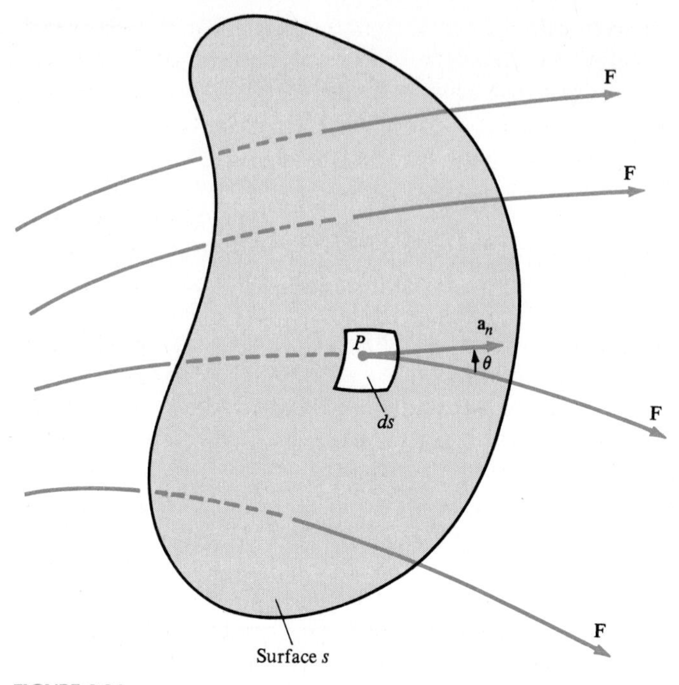

FIGURE 2.24
The flux of a vector field through an open surface s.

where ds is a differential surface area at a point P on the surface and θ is the angle between the vector $\mathbf{F}$ and a unit normal $\mathbf{a}_n$ to the surface at that point. The vector surface area $d\mathbf{s}$ is defined as

$$d\mathbf{s} = ds\,\mathbf{a}_n \tag{84}$$

Note that at the point P on the surface the vector field $\mathbf{F}$ can be divided into a component normal to the surface, $|\mathbf{F}|\cos\theta$, and a component tangent to the surface, $|\mathbf{F}|\sin\theta$. The component of $\mathbf{F}$ that is tangent to the surface contributes nothing to the net flux, or flow, of the vector through the surface, thus, we obtain (83).

Note that there are two possibilities for the direction for $d\mathbf{s}$ for an open surface: normal to the surface, but there are two sides to the surface. This ambiguity is removed when we consider a closed surface and define $d\mathbf{s}$ as pointing out of the surface. If the surface s is closed, we denote the net outward flux of the vector from the surface as

$$\psi = \oint_s \mathbf{F} \cdot d\mathbf{s} \tag{85}$$

where the small circle is used on the integral sign to denote a closed surface and $d\mathbf{s}$ is defined as in (84), with $\mathbf{a}_n$ pointing out of the closed surface. Surface integrals are, in general, difficult to evaluate. Fortunately, the problems we will encounter are such that the surfaces over which integration is required conform to the surfaces used to define the coordinate systems. In this case, the surface integrals are simple to evaluate.

The net outward flux of a vector field from a closed surface provides a measure of the magnitude of the *sources* or *sinks* of the field that may be enclosed by the surface. (Lines of flux originate at sources and terminate at sinks.) For example, suppose we associate N_i flux or vector field lines emanating from a source of strength S_i. Similarly, we may associate N_i flux or vector field lines directed toward a sink of strength S_i. If we determine the *net* outward flux of the vector through the surface s, we have determined whether a net source or sink exists within s. The magnitude or strength of the source or sink is found simply by examining the flux of the vector field through the surface. If ψ in (85) is positive, there is a net source of flux in s (the combined strengths of the sources exceeds that of the sinks). Similarly, if ψ in (85) is negative, the strength of the sinks in s exceeds the strength of the sources in s. If ψ in (85) is zero, then either there are no sources or sinks in s or the strengths of the sources in s equal the strengths of the sinks. Thus, the magnitude and sign of ψ relate the flux of a vector through a surface to the *net* sources or sinks of that vector field within the volume enclosed by s.

Clearly the net outward flux of a vector field from some enclosed surface gives an indication of the presence of sources or sinks within the surface. Of particular interest is the localization of these sources. For example, if we shrink the surface and associated enclosed volume to zero, we have an indication of the presence of sources or sinks at some point (enclosed by the now infinitesimal volume). This operation gives the number of flux lines "diverging" from the point and will be associated with the divergence of the vector field $\mathbf{F}$ at the point. This *divergence* of $\mathbf{F}$ is abbreviated as div $\mathbf{F}$ and defined as

$$\text{div } \mathbf{F} = \lim_{\Delta v \to 0} \frac{\oint_s \mathbf{F} \cdot d\mathbf{s}}{\Delta v} \tag{86}$$

Thus, div $\mathbf{F}$ is the net outward flux of the vector field $\mathbf{F}$ per unit volume as the volume shrinks to zero.

Computation of div $\mathbf{F}$ from the definition in (86) is obviously undesirable. We may obtain a simpler expression by considering a differential volume. The result can be derived for general orthogonal coordinate systems and the result specialized to the rectangular, cylindrical, and spherical coordinate systems by using the appropriate metric coefficients. However, it is more illustrative to do this for the rectangular coordinate system and then to state the general result.

Consider a differential volume $\Delta v = \Delta x \, \Delta y \, \Delta z$ in a rectangular coordinate system formed about some point $P_0(x_0, y_0, z_0)$, as shown in Fig. 2.25. In order

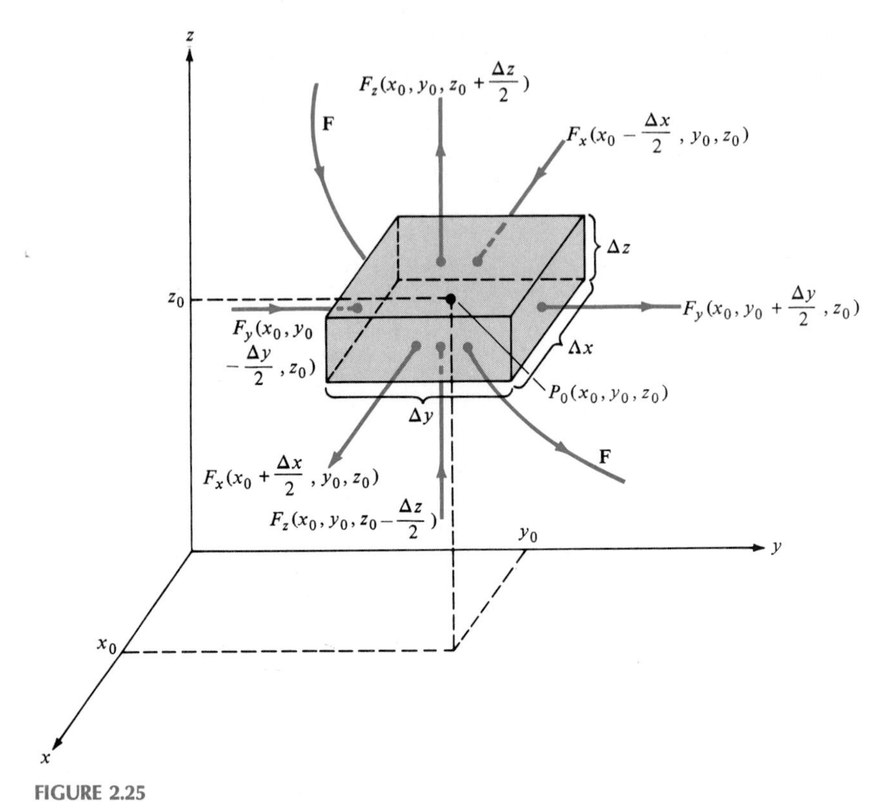

FIGURE 2.25

A differential volume element for computing the divergence of the vector field **F** in a rectangular coordinate system.

to compute the net flux of some vector field $\mathbf{F}(x, y, z)$ out of this volume, we must determine the net flux out of each of the six surfaces:

$$\oint_s \mathbf{F} \cdot d\mathbf{s} = \int_{\text{front face}} \mathbf{F} \cdot d\mathbf{s} + \int_{\text{back face}} \mathbf{F} \cdot d\mathbf{s} + \int_{\text{left face}} \mathbf{F} \cdot d\mathbf{s} + \int_{\text{right face}} \mathbf{F} \cdot d\mathbf{s}$$
$$+ \int_{\text{bottom face}} \mathbf{F} \cdot d\mathbf{s}$$
$$+ \int_{\text{top face}} \mathbf{F} \cdot d\mathbf{s} \tag{87}$$

where the normal to each of these surfaces is directed out of the volume. We will evaluate the contributions in the x direction (over the front and back faces), and the corresponding results for the other integrals in (87) will be obvious.

Over the front and back faces only the x component of **F**, F_x, provides a contribution since F_y and F_z are tangent to these surfaces.

Over the front face

$$F_x\left(x_0 + \frac{\Delta x}{2}, y_0, z_0\right) \simeq F_x(x_0, y_0, z_0) + \frac{\Delta x}{2} \frac{\partial F_x}{\partial x}\bigg|_{x_0, y_0, z_0} \tag{88}$$

Thus

$$\begin{aligned}\int_{\text{front face}} \mathbf{F} \cdot d\mathbf{s} &= F_x\left(x_0 + \frac{\Delta x}{2}, y_0, z_0\right)\Delta y\, \Delta z \\ &\simeq F_x(x_0, y_0, z_0)\Delta y\, \Delta z + \frac{\Delta x\, \Delta y\, \Delta z}{2} \frac{\partial F_x}{\partial x}\bigg|_{x_0, y_0, z_0}\end{aligned} \tag{89}$$

Over the back face

$$F_x\left(x_0 - \frac{\Delta x}{2}, y_0, z_0\right) \simeq F_x(x_0, y_0, z_0) - \frac{\Delta x}{2} \frac{\partial F_x}{\partial x}\bigg|_{x_0, y_0, z_0} \tag{90}$$

But over this back face, F_x is directed *into* the volume so that

$$\begin{aligned}\int_{\text{back face}} \mathbf{F} \cdot d\mathbf{s} &= -F_x\left(x_0 - \frac{\Delta x}{2}, y_0, z_0\right)\Delta y\, \Delta z \\ &= -F_x(x_0, y_0, z_0)\Delta y\, \Delta z + \frac{\Delta x\, \Delta y\, \Delta z}{2} \frac{\partial F_x}{\partial x}\bigg|_{x_0, y_0, z_0}\end{aligned} \tag{91}$$

Thus, the net flux leaving Δv in the x direction is the sum of (89) and (91):

$$\int_{\text{front face}} \mathbf{F} \cdot d\mathbf{s} + \int_{\text{back face}} \mathbf{F} \cdot d\mathbf{s} = \Delta x\, \Delta y\, \Delta z \frac{\partial F_x}{\partial x}\bigg|_{x_0, y_0, z_0} \tag{92}$$

Similarly,

$$\int_{\text{right face}} \mathbf{F} \cdot d\mathbf{s} + \int_{\text{left face}} \mathbf{F} \cdot d\mathbf{s} = \Delta x\, \Delta y\, \Delta z \frac{\partial F_y}{\partial y}\bigg|_{x_0, y_0, z_0} \tag{93}$$

$$\int_{\text{top face}} \mathbf{F} \cdot d\mathbf{s} + \int_{\text{bottom face}} \mathbf{F} \cdot d\mathbf{s} = \Delta x\, \Delta y\, \Delta z \frac{\partial F_z}{\partial z}\bigg|_{x_0, y_0, z_0} \tag{94}$$

The net flux leaving Δv is the sum of (92), (93), and (94):

$$\oint_S \mathbf{F} \cdot d\mathbf{s} = \left(\frac{\partial F_x}{\partial x} + \frac{\partial F_y}{\partial y} + \frac{\partial F_z}{\partial z}\right)\bigg|_{x_0, y_0, z_0} \Delta x\, \Delta y\, \Delta z \tag{95}$$

Dividing (95) by $\Delta v = \Delta x\, \Delta y\, \Delta z$ and shrinking the volume yields

$$\begin{aligned} \operatorname{div} \mathbf{F} &= \lim_{\Delta v \to 0} \frac{\oint_s \mathbf{F} \cdot d\mathbf{s}}{\Delta v} \\ &= \frac{\partial F_x}{\partial x} + \frac{\partial F_y}{\partial y} + \frac{\partial F_z}{\partial z} \end{aligned} \tag{96}$$

which is to be evaluated at the point of interest.

With the del operator defined only for rectangular coordinate systems in (68), we may write

$$\operatorname{div} \mathbf{F} = \nabla \cdot \mathbf{F} \tag{97}$$

as can be directly verified. For other coordinate systems, (97) will serve only a symbolic role.

In general orthogonal coordinate systems it can be shown that

$$\nabla \cdot \mathbf{F} = \frac{1}{h_1 h_2 h_3} \left[\frac{\partial (h_2 h_3 F_1)}{\partial u_1} + \frac{\partial (h_1 h_3 F_2)}{\partial u_2} + \frac{\partial (h_1 h_2 F_3)}{\partial u_3} \right] \tag{98}$$

Thus, in cylindrical coordinate systems $(u_1, u_2, u_3) = (r, \phi, z)$, so that $h_1 = 1$, $h_2 = r$, and $h_3 = 1$

$$\nabla \cdot \mathbf{F}(r, \phi, z) = \frac{1}{r} \frac{\partial}{\partial r} (r F_r) + \frac{1}{r} \frac{\partial F_\phi}{\partial \phi} + \frac{\partial F_z}{\partial z} \qquad \text{cylindrical coordinates} \tag{99}$$

For a spherical coordinate system $(u_1, u_2, u_3) = (r, \theta, \phi)$, the metric coefficients are $h_1 = 1$, $h_2 = r$, $h_3 = r \sin \theta$, so that

$$\nabla \cdot \mathbf{F}(r, \theta, \phi) = \frac{1}{r^2} \frac{\partial}{\partial r} (r^2 F_r) + \frac{1}{r \sin \theta} \frac{\partial (F_\theta \sin \theta)}{\partial \theta} + \frac{1}{r \sin \theta} \frac{\partial F_\phi}{\partial \phi} \tag{100}$$

EXAMPLE 2.10 Consider a vector field in a rectangular coordinate system defined by

$$\mathbf{F}(x, y, z) = \mathbf{a}_y$$

Determine the flux of the vector field out of a closed cylinder of length 2 m and radius 2 m centered on the z axis and extending from $z = 0$ to $z = 2$.

Solution A sketch of the problem is given in Fig. 2.26. Note that since the field is constant in direction and does not vary with position, it is clear that the

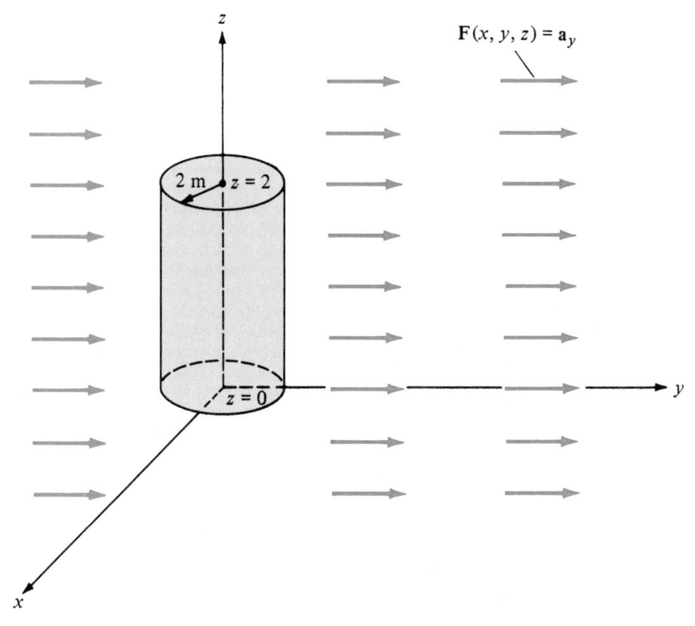

FIGURE 2.26

Example 2.10. The flux through a closed surface and its relation to the divergence of the field.

net flux out of the cylinder is zero; that is, the same amount of flux leaves the cylinder as enters it. In order to check this observation, we compute

$$\oint_s \mathbf{F} \cdot d\mathbf{s}$$

over the surface of the closed cylinder. First, because of the surface shape, it would be advantageous to convert the vector field to a cylindrical coordinate system. (See Appendix A.) The result is

$$\mathbf{F} = \sin\phi \ \mathbf{a}_r + \cos\phi \ \mathbf{a}_\phi$$

The net flux out of the ends of the cylinder is zero owing to the direction of the field. Over the sides we obtain

$$d\mathbf{s} = r \ d\phi \ dz \ \mathbf{a}_r + dr \ dz \ \mathbf{a}_\phi + r \ d\phi \ dr \ \mathbf{a}_z$$

so that

$$\mathbf{F} \cdot d\mathbf{s} = \cancel{r}^{\,2} \sin\phi \ d\phi \ dz + \cos\phi \ \cancel{dr}^{\,0} \ dz$$

and

$$\oint \mathbf{F} \cdot d\mathbf{s} = \int_{\phi=0}^{2\pi} \int_{z=0}^{2} 2 \sin \phi \, d\phi \, dz$$

$$= \int_{\phi=0}^{2\pi} 4 \sin \phi \, d\phi$$

$$= -4 \cos \phi \Big|_{0}^{2\pi}$$

$$= 0$$

Alternatively, taking the divergence of the field, we obtain

$$\nabla \cdot \mathbf{F} = \frac{1}{r} \frac{\partial}{\partial r} (r \sin \phi) + \frac{1}{r} \frac{\partial}{\partial \phi} \cos \phi$$

$$= 0$$

Thus, the divergence of the field at all points is zero and the result is reasonable.

2.11 The Divergence Theorem

In computing the net flux of some vector field **F** out of a closed surface s, we simply add (integrate) the components of **F** normal to the surface (and pointing outward) over the surface. An alternative method for determining this net flux is given by the divergence theorem:

$$\oint_s \mathbf{F} \cdot d\mathbf{s} = \int_v \nabla \cdot \mathbf{F} \, dv \tag{101}$$

In words, the divergence theorem states that the integral of the normal component of a vector field over a closed surface yields the same result as the integral of the divergence of the vector field throughout the volume enclosed by the surface. The important point to note about the divergence theorem in (101) is that we may perform either a surface integration over some closed surface s or a volume integration throughout the volume v enclosed by s. In choosing to evaluate the volume integral, we must first find $\nabla \cdot \mathbf{F}$ at all points within the volume. Remembering that $\nabla \cdot \mathbf{F}$ is the net outward flux of **F** per unit volume from a differentially small volume, we see that the right-hand side of (101) requires that we sum the net outward flux from all of the differential volumes enclosed by s:

$$\int_v \nabla \cdot \mathbf{F} \, dv = \lim_{\Delta v_i \to 0} \sum_{i=1}^{N} (\nabla \cdot \mathbf{F})_i \, \Delta v_i \tag{102}$$

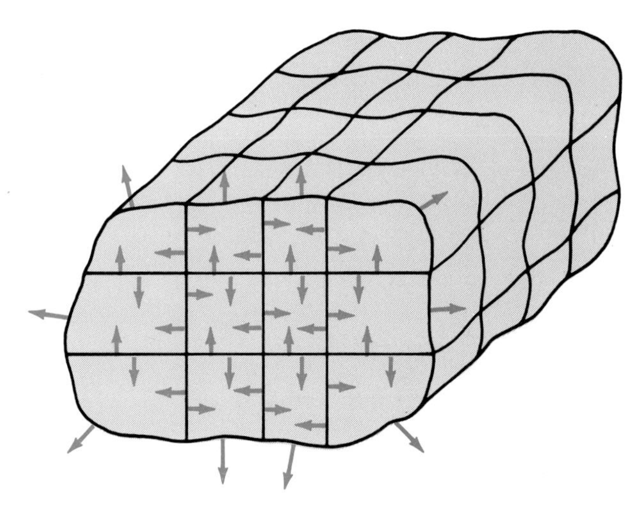

FIGURE 2.27
Proof of the divergence theorem for a general vector field.

This interpretation allows a simple, heuristic proof of the divergence theorem. To show this, consider a cross-sectional view of the volume shown in Fig. 2.27, in which the volume v is divided into differential cells. Note that $\int_v \nabla \cdot \mathbf{F}\, dv$ requires that we add the net flux emanating from each of the cells. [See (102).] At the boundary between two cells, the net flux leaving one cell cancels the net flux leaving the other cell. Therefore, in evaluating $\int_v \nabla \cdot \mathbf{F}\, dv$ over cells that do not have a side on the surface s, we obtain the result of zero. A nonzero result is obtained only for cells that have a side on the surface s, and for these cells, we obtain a nonzero result only over the side of the cell that is on the surface s. Thus, it should be clear that the divergence theorem is valid for any vector field.

EXAMPLE 2.11 As an example, consider a general vector field given by $\mathbf{A} = 2xy\mathbf{a}_x + 3\mathbf{a}_y + z^2y\mathbf{a}_z$. Suppose the surface is a cube with sides of unity area, as shown in Fig. 2.28. Verify the divergence theorem for this vector field and surface.

Solution The divergence of $\mathbf{A}$ is $\nabla \cdot \mathbf{A} = 2y + 2zy = 2y(z + 1)$, and a differential volume element is $dx\, dy\, dz$. Integrating over volume v, we obtain

$$\int_v \nabla \cdot \mathbf{A}\, dv = \int_{z=0}^{1} \int_{y=0}^{1} \int_{x=0}^{1} 2y(z + 1)dx\, dy\, dz = \tfrac{3}{2}$$

The integral of the normal component of $\mathbf{A}$ over the surface of this volume is obtained by integrating those components of $\mathbf{A}$ which are normal to the surfaces

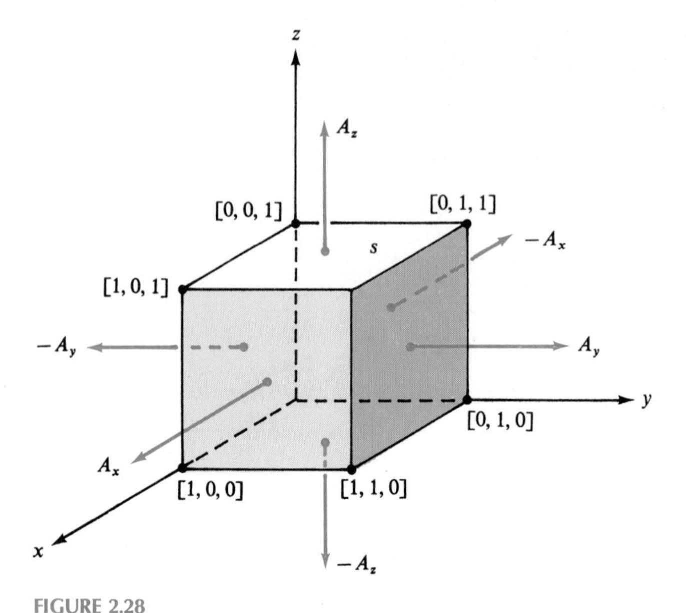

FIGURE 2.28
Example 2.11. The divergence theorem.

of this cube, as was done in obtaining the equation for divergence in the previous section. From Fig. 2.28 we observe that

$$\begin{aligned}\oint_s \mathbf{A}\cdot d\mathbf{s} &= \int_{x=0}^{1}\int_{y=0}^{1} A_z\Big|_{z=1}\,dx\,dy + \int_{x=0}^{1}\int_{y=0}^{1}(-A_z)\Big|_{z=0}\,dx\,dy\\ &+ \int_{z=0}^{1}\int_{x=0}^{1} A_y\Big|_{y=1}\,dx\,dz + \int_{z=0}^{1}\int_{x=0}^{1}(-A_y)\Big|_{y=0}\,dx\,dz\\ &+ \int_{y=0}^{1}\int_{z=0}^{1} A_x\Big|_{x=1}\,dz\,dy + \int_{y=0}^{1}\int_{z=0}^{1}(-A_x)\Big|_{x=0}\,dz\,dy = \tfrac{3}{2}\end{aligned}$$

2.12 The Curl of a Vector Field

The divergence of a vector field considered in the previous two sections gives an indication of the net outward flow, or flux, of the field from some closed surface (or a point). The next vector calculus operation that we shall consider gives the net circulation of a vector field around some closed contour (or about a point).

Consider Fig. 2.29, which shows a region in space such as a pipe in which a fluid flows. The vector field **F** represents the direction and rate of flow of the fluid. Suppose we insert a small paddle wheel into the fluid (as shown), with the axis of the paddle wheel directed into the page. If the flow rates are the same at all points on a cross section of the pipe (as shown in Fig. 2.29*a*), there will be no rotation of the paddle wheel and we say that the vector field **F** has no circulation

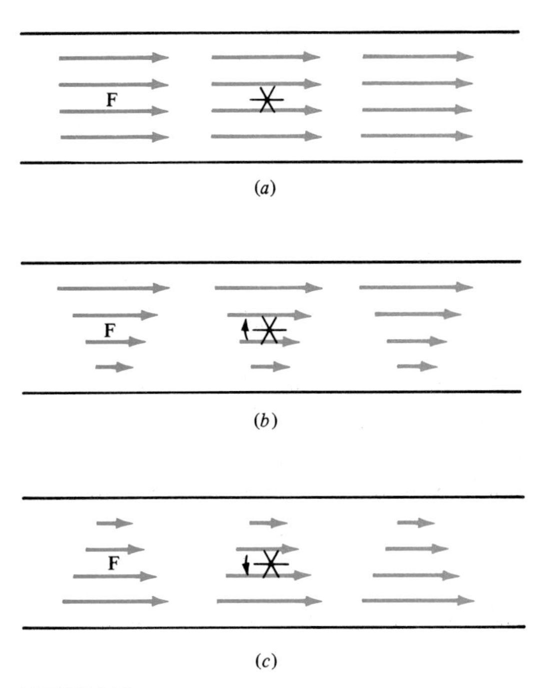

FIGURE 2.29
The curl of a vector field.

in a plane parallel to the page. In Fig. 2.29*b* the flow rate is greater at the top of the pipe than at the bottom, resulting in a net rotation of the paddle wheel in the clockwise direction. The vector field is said to have a *circulation* in the plane of the page. The similar case of counterclockwise circulation is shown in Fig. 2.29*c*.

The net circulation of a vector field **F** about some closed contour c is logically defined as the line integral of **F** along c :

$$C = \oint_c \mathbf{F} \cdot d\mathbf{l} \tag{103}$$

The circulation about a point can be obtained by shrinking the contour c. This leads to the definition of the *curl* (or circulation) of a vector field, curl **F**, about a point as the circulation of **F** per unit surface (enclosed by c) as the contour and surface shrink to zero:

$$\text{curl } \mathbf{F} \cdot \mathbf{a}_n = \lim_{\Delta s \to 0} \frac{\oint_c \mathbf{F} \cdot d\mathbf{l}}{\Delta s} \tag{104}$$

Equation (104) gives the component of the curl that is in the direction of the unit normal to Δs, $\mathbf{a}_n$. The direction of the contour c and the unit normal to $\mathbf{a}_n$ are related by the right-hand rule.

In the paddle-wheel illustration, the circulation of the fluid is manifested as a rotation of the axis of the wheel that is perpendicular to the plane containing the paddle wheel. This aspect is accounted for by the unit normal to the surface s, $\mathbf{a}_n$, in (104). Note that in the paddle-wheel problem there are three important and independent orientations of the paddle wheel: (1) in the plane shown; (2) perpendicular to the plane, with the axis of the paddle wheel aligned with the axis of the pipe; and (3) perpendicular to the plane, with the axis of the paddle wheel pointing up or down. Each of these three orientations gives a measure of the circulation in three orthogonal planes. The net circulation of the field will be the vector result of the three circulations in these orthogonal planes. The determination of this net circulation of a vector field will occur in numerous instances in our investigation of electromagnetic fields.

Again, as with the definition of divergence of a vector field, the definition of the curl of a vector field given in (104) is not convenient from a computational standpoint. We will now derive a more convenient form. First we obtain curl **F** in a rectangular coordinate system. The corresponding results for cylindrical and spherical coordinate systems may be derived in a similar manner.

We will obtain the z component of curl **F**, and the remaining x and y components will then become obvious. We choose the infinitesimal area $ABCD$ in the xy plane with sides Δx and Δy centered at $P_0[x_0, y_0, z_0]$, as shown in Fig. 2.30. The circulation around contour c_{xy} will be composed of four contributions:

$$\oint_{c_{xy}} \mathbf{F}\cdot d\mathbf{l} = \int_{AB} \mathbf{F}\cdot d\mathbf{l} + \int_{BC} \mathbf{F}\cdot d\mathbf{l} + \int_{CD} \mathbf{F}\cdot d\mathbf{l} + \int_{DA} \mathbf{F}\cdot d\mathbf{l} \tag{105}$$

Along AB, only the x component of F provides a contribution, and along this part of the contour

$$F_x\left(x_0, y_0 - \frac{\Delta y}{2}, z_0\right) \simeq F_x(x_0, y_0, z_0) - \frac{\Delta y}{2}\frac{\partial F_x}{\partial y}\bigg|_{x_0, y_0, z_0} \tag{106}$$

Similarly, along CD

$$F_x\left(x_0, y_0 + \frac{\Delta y}{2}, z_0\right) \simeq F_x(x_0, y_0, z_0) + \frac{\Delta y}{2}\frac{\partial F_x}{\partial y}\bigg|_{x_0, y_0, z_0} \tag{107}$$

Along DA, only the y component of F provides a contribution, and along this part of the contour

$$F_y\left(x_0 - \frac{\Delta x}{2}, y_0, z_0\right) \simeq F_y(x_0, y_0, z_0) - \frac{\Delta x}{2}\frac{\partial F_y}{\partial x}\bigg|_{x_0, y_0, z_0} \tag{108}$$

Similarly, along BC

$$F_y\left(x_0 + \frac{\Delta x}{2}, y_0, z_0\right) \simeq F_y(x_0, y_0, z_0) + \frac{\Delta x}{2}\frac{\partial F_y}{\partial x}\bigg|_{x_0, y_0, z_0} \tag{109}$$

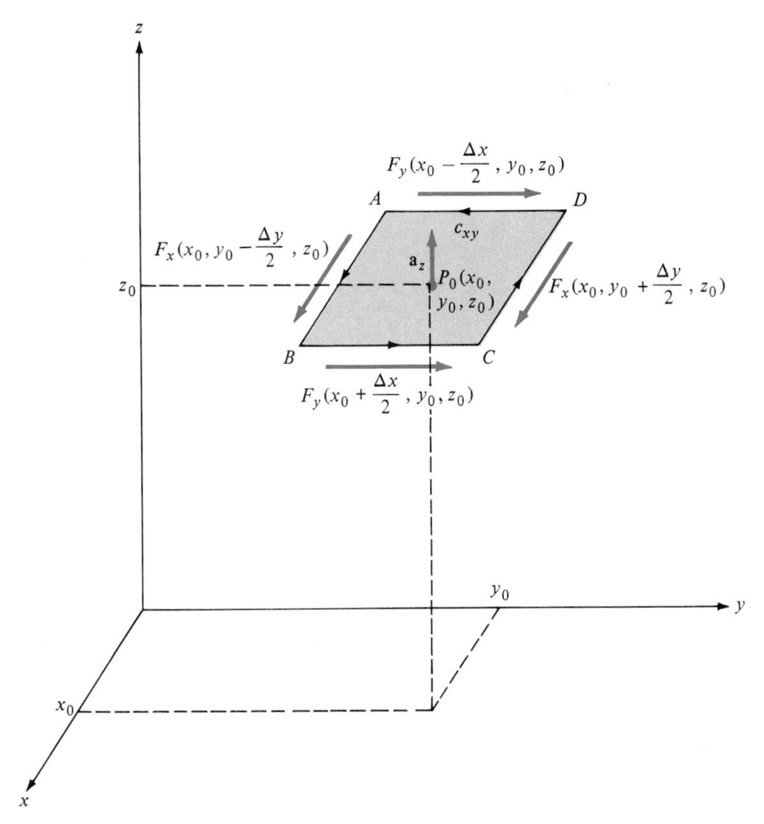

FIGURE 2.30
Determination of the z component of the curl of a vector field $\mathbf{F}(x, y, z)$ in a rectangular coordinate system.

Noting the directions of the contours and the directions of the tangential components, we obtain the net circulation as

$$\oint_{c_{xy}} \mathbf{F}\cdot d\mathbf{l} = F_x\left(x_0, y_0 - \frac{\Delta y}{2}, z_0\right)\Delta x - F_x\left(x_0, y_0 + \frac{\Delta y}{2}, z_0\right)\Delta x$$
$$+ F_y\left(x_0 + \frac{\Delta x}{2}, y_0, z_0\right)\Delta y - F_y\left(x_0 - \frac{\Delta x}{2}, y_0, z_0\right)\Delta y \quad (110)$$

Substituting the above results yields

$$\oint_{c_{xy}} \mathbf{F}\cdot d\mathbf{l} = \left(\frac{\partial F_y}{\partial x} - \frac{\partial F_x}{\partial y}\right)\Bigg|_{x_0, y_0, z_0} \Delta x\, \Delta y \quad (111)$$

Dividing (111) by the area of this surface, $\Delta s_z = \Delta x\, \Delta y$, we obtain from (104)

$$(\textbf{curl F})_z = \frac{\partial F_y}{\partial x} - \frac{\partial F_x}{\partial y} \quad (112)$$

which is the z component of curl $\mathbf{F}$. Proceeding in a similar fashion, we can write the x and y components as

$$(\text{curl } \mathbf{F})_x = \frac{\partial F_z}{\partial y} - \frac{\partial F_y}{\partial z} \tag{113}$$

$$(\text{curl } \mathbf{F})_y = \frac{\partial F_x}{\partial z} - \frac{\partial F_z}{\partial x} \tag{114}$$

Combining (112), (113), and (114) gives

$$\text{curl } \mathbf{F} = \left(\frac{\partial F_z}{\partial y} - \frac{\partial F_y}{\partial z}\right)\mathbf{a}_x + \left(\frac{\partial F_x}{\partial z} - \frac{\partial F_z}{\partial x}\right)\mathbf{a}_y + \left(\frac{\partial F_y}{\partial x} - \frac{\partial F_x}{\partial y}\right)\mathbf{a}_z \tag{115}$$

which may also be written as

$$\text{curl } \mathbf{F} = \nabla \times \mathbf{F} \tag{116}$$

where ∇ is the del operator, again given for rectangular coordinate systems by

$$\nabla = \mathbf{a}_x \frac{\partial}{\partial x} + \mathbf{a}_y \frac{\partial}{\partial y} + \mathbf{a}_z \frac{\partial}{\partial z} \tag{117}$$

Note that the expression for curl $\mathbf{F}$ in rectangular coordinates given in (115) and (116) can be easily obtained by expanding $\nabla \times \mathbf{F}$ as

$$\nabla \times \mathbf{F} = \begin{vmatrix} \mathbf{a}_x & \mathbf{a}_y & \mathbf{a}_z \\ \dfrac{\partial}{\partial x} & \dfrac{\partial}{\partial y} & \dfrac{\partial}{\partial z} \\ F_x & F_y & F_z \end{vmatrix} \tag{118}$$

However, there is no need to perform the expansion of (118) if we use the simple memory aid developed for the cross product of two vectors in Sec. 2.6.2. Each of the components of curl $\mathbf{F}$ can be written as

$$(\nabla \times \mathbf{F})_a = \mathbf{a}_a\left(\frac{\partial F_c}{\partial b} - \frac{\partial F_b}{\partial c}\right) \tag{119}$$

With the cyclic ordering of the axis labels as $x \to y \to z \to x \to y \to z \to x \cdots$, we may easily form the individual components of curl $\mathbf{F}$. For example, if we wish to find the y component, we set $a = y$ in (119). The next coordinate in the cyclic coordinate sequence is z, followed by x. Therefore, we set $b = z$ and $c = x$, resulting in

$$(\nabla \times \mathbf{F})_y = \mathbf{a}_y\left(\frac{\partial F_x}{\partial z} - \frac{\partial F_z}{\partial x}\right) \tag{120}$$

Equation (115) provides a convenient means for calculating the curl of a vector field in rectangular coordinates. For general orthogonal coordinate systems, a similar but more tedious development gives

$$\nabla \times \mathbf{F} = \frac{1}{h_2 h_3}\left[\frac{\partial(h_3 F_3)}{\partial u_2} - \frac{\partial(h_2 F_2)}{\partial u_3}\right]\mathbf{a}_{u1} + \frac{1}{h_1 h_3}\left[\frac{\partial(h_1 F_1)}{\partial u_3} - \frac{\partial(h_3 F_3)}{\partial u_1}\right]\mathbf{a}_{u2} + \frac{1}{h_1 h_2}\left[\frac{\partial(h_2 F_2)}{\partial u_1} - \frac{\partial(h_1 F_1)}{\partial u_2}\right]\mathbf{a}_{u3} \tag{121}$$

This may be easily remembered with the memory aid developed in Sec. 2.6.2 or computed from

$$\nabla \times \mathbf{F} = \frac{1}{h_1 h_2 h_3}\begin{vmatrix} h_1\mathbf{a}_1 & h_2\mathbf{a}_2 & h_3\mathbf{a}_3 \\ \dfrac{\partial}{\partial u_1} & \dfrac{\partial}{\partial u_2} & \dfrac{\partial}{\partial u_3} \\ h_1 F_1 & h_2 F_2 & h_3 F_3 \end{vmatrix} \tag{122}$$

For a cylindrical coordinate system, $(u_1, u_2, u_3) = (r, \phi, z)$ and $h_1 = 1$, $h_2 = r$, and $h_3 = 1$. Thus

$$\textbf{curl } \mathbf{F}(r, \phi, z) = \left(\frac{1}{r}\frac{\partial F_z}{\partial \phi} - \frac{\partial F_\phi}{\partial z}\right)\mathbf{a}_r + \left(\frac{\partial F_r}{\partial z} - \frac{\partial F_z}{\partial r}\right)\mathbf{a}_\phi + \frac{1}{r}\left[\frac{\partial}{\partial r}(rF_\phi) - \frac{\partial F_r}{\partial \phi}\right]\mathbf{a}_z \tag{123}$$

For spherical coordinates, $(u_1, u_2, u_3) = (r, \theta, \phi)$ and $h_1 = 1$, $h_2 = r$, and $h_3 = r \sin\theta$. Thus

$$\textbf{curl } \mathbf{F}(r, \theta, \phi) = \frac{1}{r\sin\theta}\left[\frac{\partial}{\partial\theta}(F_\phi \sin\theta) - \frac{\partial F_\theta}{\partial \phi}\right]\mathbf{a}_r + \frac{1}{r}\left[\frac{1}{\sin\theta}\frac{\partial F_r}{\partial \phi} - \frac{\partial}{\partial r}(rF_\phi)\right]\mathbf{a}_\theta + \frac{1}{r}\left[\frac{\partial}{\partial r}(rF_\theta) - \frac{\partial F_r}{\partial \theta}\right]\mathbf{a}_\phi \tag{124}$$

For these coordinate systems, we will also symbolically write **curl F** $= \nabla \times \mathbf{F}$, with the understanding that in other than rectangular coordinates the del operator is not defined; thus, $\nabla \times \mathbf{F}$ is merely a shorthand notation for **curl F**.

EXAMPLE 2.12 Consider the vector field

$$\mathbf{F}(x, y, z) = z\mathbf{a}_y$$

Determine the net circulation of this field per unit surface.

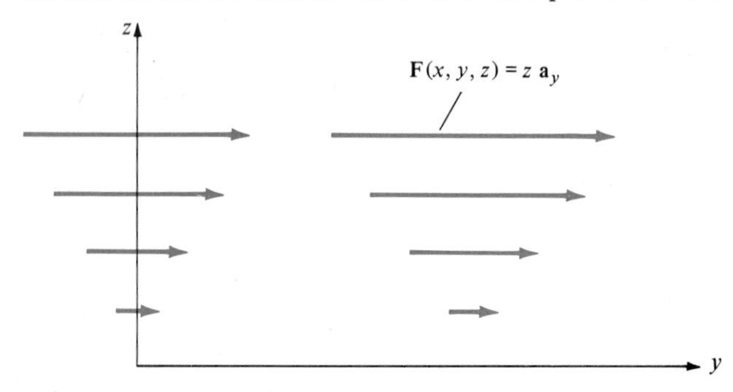

FIGURE 2.31
Example 2.12. The curl of a vector field.

Solution The vector field is sketched in Fig. 2.31. From this sketch it is clear that there is a net circulation only in the yz plane and in the clockwise direction. To confirm this, we write the curl with our convenient mnemonic rule as

$$\begin{aligned}\nabla \times \mathbf{F} &= \left(\frac{\partial F_z}{\partial y}^{\nearrow 0} - \frac{\partial F_y}{\partial z}\right)\mathbf{a}_x + \left(\frac{\partial F_x}{\partial z}^{\nearrow 0} - \frac{\partial F_z}{\partial x}^{\nearrow 0}\right)\mathbf{a}_y + \left(\frac{\partial F_y}{\partial x}^{\nearrow 0} - \frac{\partial F_x}{\partial y}^{\nearrow 0}\right)\mathbf{a}_z \\ &= -\frac{\partial F_y}{\partial z}\mathbf{a}_x \\ &= -1\mathbf{a}_x\end{aligned}$$

Thus, the curl is in the negative x direction, symbolizing (by the right-hand rule) circulation in the clockwise direction in the yz plane, obvious from Fig. 2.31.

2.13 Stokes' Theorem

Our interpretation of the curl of a vector field as the net circulation per unit surface provides a simple, heuristic proof of Stokes' theorem which is expressed as:

$$\int_s (\nabla \times \mathbf{F}) \cdot d\mathbf{s} = \oint_c \mathbf{F} \cdot d\mathbf{l} \tag{125}$$

In words, Stokes' theorem states that the net flux of $\nabla \times \mathbf{F}$ over some *open surface* s is equal to the line integral of $\mathbf{F}$ along the closed contour c that bounds s. Since s is an open surface, we must be careful to define, in unambiguous terms, the direction of $d\mathbf{s}$, i.e., the normal to the surface. This is defined in (125) by the right-hand rule. If we place the fingers of our right hand in the direction of the contour c (the direction of c may be chosen arbitrarily), then the direction the thumb will point is the direction of $d\mathbf{s}$.

To give a simple heuristic proof of Stokes' theorem, let us consider the surface s shown in Fig. 2.32. The shape of the surface s is chosen to be somewhat general in order to illustrate the generality of Stokes' theorem. Note the direction of c and the resulting direction of the normal to the surface. Let the surface be divided into infinitesimal surface areas. For each of these areas, we have

$$(\nabla \times \mathbf{F})_i \cdot \Delta s_i = \oint_{\Delta c_i} \mathbf{F} \cdot d\mathbf{l} \tag{126}$$

where Δc_i is the incremental contour around the ith incremental element of the surface Δs_i. Now observe that if we add the contribution of all the areas, then the line integrals in (126) at the common boundaries of two adjacent surface areas cancel. The only place where these do not cancel is at the boundary of a surface area that is not adjacent to the boundary of another surface area, namely, along contour c. Figure 2.32 shows how areas 1 to 4 combine to give the line integral around the outer boundaries. If we sum (126) over the entire surface and take the limit as $\Delta s_i \to 0$, we obtain (125).

In the above derivation the open surface s is shown to be of some general shape—not necessarily flat. This is an important aspect of Stokes' theorem: The contour c bounds the open surface s, and the shape of s is arbitrary. This concept may be compared to inflating a balloon. The contour c represents the mouth of the balloon. As we inflate the balloon, the surface of the balloon changes shape but the contour c around the mouth remains the same. The evaluation of (125)

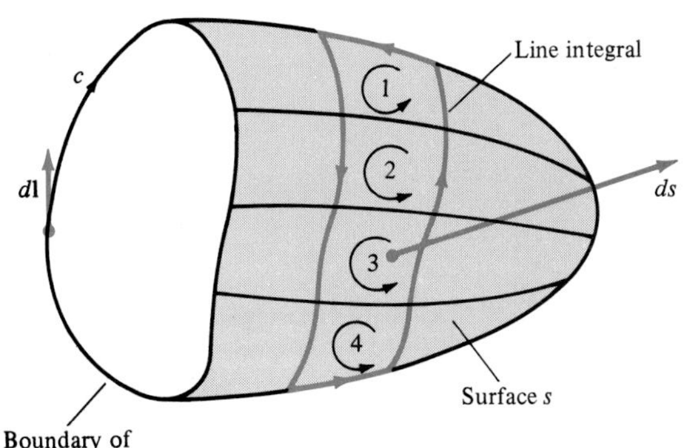

FIGURE 2.32
Proof of Stokes' theorem for a general vector field.

gives identical results for all these balloon surface areas so long as c remains the same for these different surfaces: an interesting result.

EXAMPLE 2.13 Consider the vector field $\mathbf{F} = \mathbf{a}_x + zy^2\mathbf{a}_y$. Verify Stokes' theorem for this vector field and the flat surface in the yz plane bounded by [0, 0, 0], [0, 1, 0], [0, 1, 1], and [0, 0, 1], as shown in Fig. 2.33. Choose the contour c in the clockwise direction, as shown.

Solution First we evaluate the line integral around c. The dot product $\mathbf{F}\cdot d\mathbf{l}$ along c is

$$\mathbf{F}\cdot d\mathbf{l} = (\mathbf{a}_x + zy^2\mathbf{a}_y)\cdot(dx\,\mathbf{a}_x + dy\,\mathbf{a}_y + dz\,\mathbf{a}_z) = dx + zy^2\,dy$$

Along the portions of the closed contour c denoted by c_1, c_2, c_3, and c_4 we have

$$\begin{aligned}
\oint_c \mathbf{F}\cdot d\mathbf{l} &= \int_{c_1}\mathbf{F}\cdot d\mathbf{l} + \int_{c_2}\mathbf{F}\cdot d\mathbf{l} + \int_{c_3}\mathbf{F}\cdot d\mathbf{l} + \int_{c_4}\mathbf{F}\cdot d\mathbf{l} \\
&= \int_{z=0}^{z=1}(\cancelto{0}{dx} + \cancelto{0}{zy^2}\,\cancelto{0}{dy}) + \int_{y=0}^{y=1}(\cancelto{0}{dx} + \cancelto{1}{z}\,y^2\,dy) \\
&\quad + \int_{z=1}^{z=0}(\cancelto{0}{dx} + \cancelto{0}{zy^2}\,\cancelto{0}{dy}) + \int_{y=1}^{y=0}(\cancelto{0}{dx} + \cancelto{0}{z}\,y^2\,dy) \\
&= 0 + \left.\frac{y^3}{3}\right|_0^1 + 0 + 0 \\
&= \frac{1}{3}
\end{aligned}$$

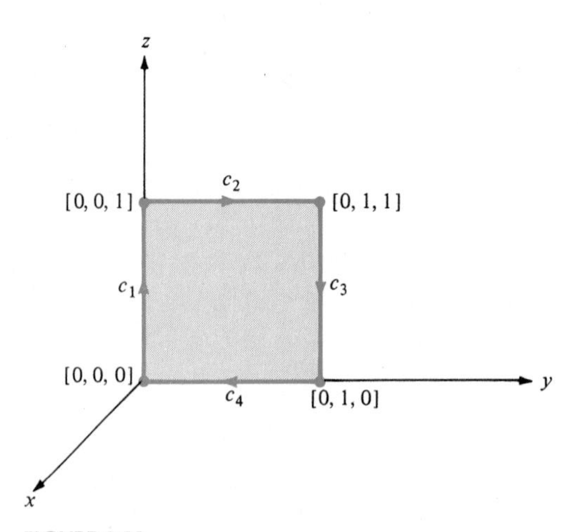

FIGURE 2.33
Example 2.13. Stokes' theorem.

Evaluating the surface integral over the (arbitrarily chosen as flat) surface s bounded by c, we obtain the following. Note that because of the arbitrarily chosen direction of c, the normal to the differential surface is directed in the negative x direction and

$$d\mathbf{s} = -dy\, dz\, \mathbf{a}_x$$

Forming $\nabla \times \mathbf{F}$, we obtain

$$\begin{aligned}\nabla \times \mathbf{F} &= \left(\frac{\partial F_z}{\partial y} - \frac{\partial F_y}{\partial z}\right)\mathbf{a}_x + \left(\frac{\partial F_x}{\partial z} - \frac{\partial F_z}{\partial x}\right)\mathbf{a}_y + \left(\frac{\partial F_y}{\partial x} - \frac{\partial F_x}{\partial y}\right)\mathbf{a}_z \\ &= -y^2\mathbf{a}_x\end{aligned}$$

and

$$\begin{aligned}\int_s (\nabla \times \mathbf{F}) \cdot d\mathbf{s} &= \int_{y=0}^{y=1} \int_{z=0}^{z=1} y^2\, dy\, dz \\ &= \frac{y^3}{3}\bigg|_0^1 \\ &= \frac{1}{3}\end{aligned}$$

2.14 Two Important Vector Identities

We will have numerous occasions to call on the following two important vector identities. The reader should commit these to memory.

The first vector identity is

$$\nabla \times \nabla f = 0 \tag{127}$$

or the curl of the gradient of some scalar field is identically zero. This may be proved rather directly by expanding (127) in rectangular coordinates. A more illustrative proof is to use Stokes' theorem. Integrating (127) over some open surface s and applying Stokes' theorem to the result, we obtain

$$\int_s [\nabla \times (\nabla f)] \cdot d\mathbf{s} = \oint_c (\nabla f) \cdot d\mathbf{l} = 0 \tag{128}$$

where c is the contour bounding s. This result is zero since the line integral of the gradient of a scalar field around some closed contour is identically zero, as was shown previously. Since the surface s and associated contour c are arbitrary, we conclude that (127) must be true.

Conversely, (127) shows that if a vector field $\mathbf{F}$ has zero curl, $\nabla \times \mathbf{F} = 0$, then $\mathbf{F}$ may be expressed, to within a scalar constant, as the gradient of a scalar field. We will have numerous occasions throughout our study of electromagnetic fields to utilize this observation. Such a field is said to be irrotational, or conservative.

The second important vector identity is

$$\nabla \cdot (\nabla \times \mathbf{F}) = 0 \tag{129}$$

or the divergence of the curl of a vector field is identically zero. Again, this may be proved in a direct manner by expanding (129) in rectangular coordinates. A more illustrative proof is obtained by integrating $\nabla \times \mathbf{F}$ over some closed surface s and applying the divergence theorem, resulting in

$$\oint_s (\nabla \times \mathbf{F}) \cdot d\mathbf{s} = \int_v \nabla \cdot (\nabla \times \mathbf{F}) dv \tag{130}$$

The closed surface s can be partitioned into two adjoining open surfaces s_1 and s_2, as shown in Fig. 2.34. Contours c_1 and c_2 bound these surfaces, and their directions are related to the unit normals to the surfaces by the right-hand rule. Equation (130) can be separated as

$$\int_{s_1} (\nabla \times \mathbf{F}) \cdot d\mathbf{s} + \int_{s_2} (\nabla \times \mathbf{F}) \cdot d\mathbf{s} = \oint_{c_1} \mathbf{F} \cdot d\mathbf{l} + \oint_{c_2} \mathbf{F} \cdot d\mathbf{l} \tag{131}$$

Note, however, that because the normals to the two open surfaces must both be defined out of the closed surface s, the adjacent contours c_1 and c_2 must be opposing each other. Thus, the right-hand side of (131) evaluates to zero and we obtain the desired result.

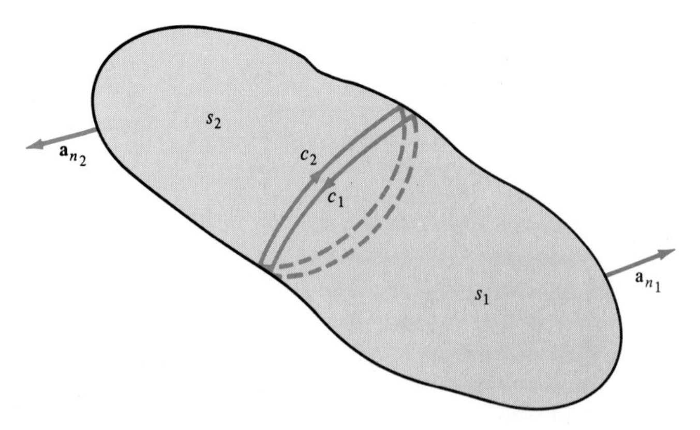

FIGURE 2.34

Illustration of the proof that $\nabla \cdot \nabla \times \mathbf{F} = 0$.

If a vector field has zero divergence, $\nabla \cdot \mathbf{F} = 0$, it is said to be divergenceless or solenoidal. From (129) we may define $\mathbf{F}$ to be the curl of some other vector field. We will also have numerous occasions to use this observation.

Numerous other vector identities are summarized in Appendix A. The above two, however, are the most important ones. Interestingly, if all combinations of gradient, divergence, and curl are taken two at a time, we obtain

(1)	$\nabla \times (\nabla f) = 0$	
(2)	$\nabla \times (\nabla \cdot \mathbf{F})$	Meaningless
(3)	$\nabla \times (\nabla \times \mathbf{F})$	(Appendix A)
(4)	$\nabla \cdot (\nabla f)$	(Appendix A)
(5)	$\nabla \cdot (\nabla \cdot \mathbf{F})$	Meaningless
(6)	$\nabla \cdot (\nabla \times \mathbf{F}) = 0$	
(7)	$\nabla (\nabla f)$	Meaningless
(8)	$\nabla (\nabla \cdot \mathbf{F})$	
(9)	$\nabla (\nabla \times \mathbf{F})$	Meaningless

We find that operations (2), (5), (7), and (9) are meaningless. Operations (1) and (6) are zero, as we have seen. Operations (3) and (4) have very useful identities and are given in Appendix A. The remaining operations in (8) and in (3) and (4) are related in one very important identity:

$$\nabla^2 \mathbf{F} = \nabla(\nabla \cdot \mathbf{F}) - \nabla \times \nabla \times \mathbf{F} \tag{132}$$

Expanding the right-hand side of (132) in rectangular coordinates, we find that $\nabla^2 \mathbf{F}$ is given by

$$\nabla^2 \mathbf{F} = \nabla^2 F_x \mathbf{a}_x + \nabla^2 F_y \mathbf{a}_y + \nabla^2 F_z \mathbf{a}_z \tag{133}$$

and $\nabla^2 F_x$, $\nabla^2 F_y$, and $\nabla^2 F_z$ are determined from

$$\nabla^2 f = \nabla \cdot \nabla f = \frac{\partial^2 f}{\partial x^2} + \frac{\partial^2 f}{\partial y^2} + \frac{\partial^2 f}{\partial z^2} \tag{134}$$

$\nabla^2 \mathbf{F}$ is called the *vector laplacian*; $\nabla^2 f$ is called the *scalar laplacian*, or simply the *laplacian.*

2.15 Summary

In this chapter we have considered the fundamental concepts of vector algebra and vector calculus required in our study of electromagnetic fields. We first considered the three primary orthogonal coordinate systems—rectangular, cylindrical, and spherical. The vector algebra operations of addition and subtraction of two vectors and the multiplication of a vector by a scalar were obtained in terms of the vector components. In addition, the dot product and cross product of two vectors were defined. Differential path lengths, surface areas, and volumes were determined for each coordinate system.

Scalar and vector fields were considered, and the important vector calculus operations on these fields—gradient, line integral, divergence, and curl—were defined. The del (∇) operator provides a simple way of remembering the gradient [$\nabla f(x, y, z)$], divergence [$\nabla \cdot \mathbf{F}(x, y, z)$], and curl [$\nabla \times \mathbf{F}(x, y, z)$] operations in rectangular coordinates. No such simplification results for cylindrical or spherical coordinate systems, and these results are given in Appendix A. Finally, two important vector identities that will be used extensively were proved. A summary of the vector algebra and vector calculus operations in all three coordinate systems is given on the inside of the front and back covers of this text.

Problems

2-1 An airplane having a total weight of 3500 lb is in a steady climb, as shown in Fig. P2.1. The flight path of the airplane is at an angle of 10° to the horizon, and the longitudinal axis of the airplane is at an angle of 5° to the flight path. (The angle of

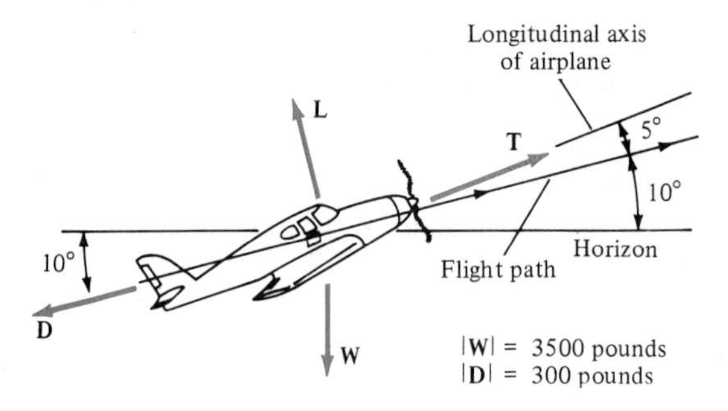

FIGURE P2.1

attack is 5°.) The thrust vector **T** is directed along the longitudinal axis of the airplane, and the drag vector **D** is directed along the flight path. The lift vector **L** is assumed to be perpendicular to the flight path. If the total drag is 300 lb, determine the required thrust of the engine, $|\mathbf{T}|$, and lift of the wings, $|\mathbf{L}|$.

2-2 A vector $\mathbf{A}(x, y, z)$ is defined in a rectangular coordinate system as being directed from [0, − 1, 3] to [5, 1, − 2]. Determine

- **(a)** an expression for **A**,
- **(b)** the distance between the two points, and
- **(c)** a unit vector pointing in the direction of **A**.

2-3 Determine the distance between two points P_1 and P_2 when these are expressed

- **(a)** in a cylindrical coordinate system as $P_1 = [2, 3\pi/2, 1]$ and $P_2 = [3, \pi, 0]$ and
- **(b)** in a spherical coordinate system as $P_1 = [1, \pi/2, 3\pi/2]$ and $P_2 = [2, \pi/2, \pi/4]$.

2-4 Three vectors **A**, **B**, and **C** are given in a rectangular coordinate system as

$$\mathbf{A} = 2\mathbf{a}_x + 3\mathbf{a}_y - \mathbf{a}_z$$

$$\mathbf{B} = \mathbf{a}_x + \mathbf{a}_y - 2\mathbf{a}_z$$

$$\mathbf{C} = 3\mathbf{a}_x - \mathbf{a}_y + \mathbf{a}_z$$

Compute:

(a) $\mathbf{A} + \mathbf{B}$
(b) $\mathbf{B} - \mathbf{C}$
(c) $\mathbf{A} + 3\mathbf{B} - 2\mathbf{C}$
(d) $|\mathbf{A}|$
(e) $\mathbf{a}_B$
(f) $\mathbf{A} \cdot \mathbf{B}$
(g) $\mathbf{B} \cdot \mathbf{A}$
(h) $\mathbf{B} \times \mathbf{C}$
(i) $\mathbf{C} \times \mathbf{B}$
(j) $\mathbf{A} \cdot \mathbf{B} \times \mathbf{C}$

2-5 If $\mathbf{A} = \mathbf{a}_x + 2\mathbf{a}_y - 3\mathbf{a}_z$ and $\mathbf{B} = 2\mathbf{a}_x - \mathbf{a}_y + \mathbf{a}_z$, determine

(a) the magnitude of the projection or component of **B** (scalar) on **A**,
(b) the angle (smallest) between **A** and **B**,
(c) the vector projection of **A** onto **B**, and
(d) a unit vector perpendicular to the plane containing **A** and **B**.

2-6 Given two vectors

$$\mathbf{A} = y\mathbf{a}_x + 3x\mathbf{a}_y - \mathbf{a}_z$$

$$\mathbf{B} = 2\mathbf{a}_y - xy\mathbf{a}_z$$

determine at $x = 1$, $y = 2$, and $z = 4$:

(a) $\mathbf{A} + \mathbf{B}$
(b) $\mathbf{A} \cdot \mathbf{B}$
(c) $\mathbf{B} \times \mathbf{A}$
(d) the smallest angle between them
(e) $\mathbf{a}_x \cdot \mathbf{A}$ (interpret this result in words)
(f) $\mathbf{a}_y \times \mathbf{B}$ (interpret this result in words)

2-7 If two vectors are expressed in cylindrical coordinates as

$$\mathbf{A} = 2\mathbf{a}_r + \pi\mathbf{a}_\phi + \mathbf{a}_z$$

$$\mathbf{B} = -\mathbf{a}_r + \frac{3\pi}{2}\mathbf{a}_\phi - 2\mathbf{a}_z$$

compute:

(a) $\mathbf{A} + \mathbf{B}$
(b) $\mathbf{A} \cdot \mathbf{B}$
(c) $\mathbf{A} \times \mathbf{B}$
(d) $\mathbf{a}_B$
(e) $|\mathbf{A}|$
(f) the smallest angle between **A** and **B**
(g) a unit vector perpendicular to the plane containing **A** and **B**

2-8 Repeat Prob. 2-7 when **A** and **B** are given in spherical coordinates as

$$\mathbf{A} = 2\mathbf{a}_r + \pi\mathbf{a}_\theta - \frac{\pi}{2}\mathbf{a}_\phi$$

$$\mathbf{B} = \mathbf{a}_r - \frac{\pi}{3}\mathbf{a}_\theta$$

2-9 If two vectors are given by

$$\mathbf{A} = \mathbf{a}_x + 2\mathbf{a}_y - \mathbf{a}_z$$

$$\mathbf{B} = \alpha\mathbf{a}_x + \mathbf{a}_y + 3\mathbf{a}_z$$

determine α such that the two vectors are perpendicular.

2-10 If two vectors are given by

$$\mathbf{A} = \mathbf{a}_r + \pi\mathbf{a}_\phi + 3\mathbf{a}_z$$

$$\mathbf{B} = \alpha\mathbf{a}_r + \beta\mathbf{a}_\phi - 6\mathbf{a}_z$$

determine α and β such that the two vectors are parallel.

2-11 Determine two unit normals to a plane described by $2x + y - z + 1 = 0$ at $[1, 3, 6]$.

2-12 Sketch the following scalar fields by showing contours of constant value of the field for $f = 0, 1, 2$:

(a) $f(x, y, z) = x + y$
(b) $f(x, y, z) = x^2 + y^2$
(c) $f(x, y, z) = x + y + z$
(d) $f(r, \phi, z) = r$ [Compare with **(b)**. Are they equivalent?]
(e) $f(r, \theta, \phi) = \phi/\pi$
(f) $f(r, \phi, z) = r - z$
(g) $f(r, \theta, \phi) = r - \theta/\pi$

2-13 Sketch the following vector fields:

(a) $\mathbf{F}(x, y, z) = x\mathbf{a}_x$
(b) $\mathbf{F}(x, y, z) = y\mathbf{a}_x$
(c) $\mathbf{F}(r, \phi, z) = z\mathbf{a}_r$
(d) $\mathbf{F}(r, \theta, \phi) = \mathbf{a}_\phi$
(e) $\mathbf{F}(r, \theta, \phi) = r\mathbf{a}_\theta$
(f) $\mathbf{F}(x, y, z) = x\mathbf{a}_x + y\mathbf{a}_y + z\mathbf{a}_z$
(g) $\mathbf{F}(x, y, z) = x\mathbf{a}_x - y\mathbf{a}_y$
(h) $\mathbf{F}(x, y, z) = -x\mathbf{a}_x - y\mathbf{a}_y$

2-14 Determine the volume V of a region defined in a cylindrical coordinate system as $1\text{ m} \le r \le 2\text{ m}$, $0 \le \phi \le \pi/3$ rad, and $0 \le z \le 1$ m by integrating $V = \int_v dv$. Check your result without performing the integral.

2-15 Determine the area S of a surface defined in a spherical coordinate system as $r = 2$ m and $\pi/4$ rad $\le \theta \le \pi/3$ rad by integrating $S = \int_s ds$.

2-16 Determine the gradient of the following scalar fields:

(a) $f(x, y, z) = 5x + 10xz - xy + 6$
(b) $f(r, \phi, z) = 2 \sin \phi - rz + 4$
(c) $f(r, \theta, \phi) = 2r \cos \theta - 5\phi + 2$

2-17 A certain scalar field is given by $V = -Q \cos \theta / r^2 (r \neq 0)$, where Q is a constant. Find the gradient of this field. The gradient denotes a vector field. For a given r, at what value of θ are the r and θ components of this vector field equal? Observe that this field (compared to the given scalar field) decays very rapidly with distance.

2-18 The temperature in a room is described as a scalar field in degrees Celsius (dimensions in meters) as

$$T(x, y, z) = 10x + 20y$$

(a) Sketch the surface of constant temperature for $T = 10°\text{C}$, $T = 20°\text{C}$, and $T = 5°\text{C}$;
(b) determine a unit vector in the direction of maximum temperature change [show that this is reasonable from the sketch in **(a)**]; and,
(c) determine this maximum temperature change with distance.

2-19 Determine the rate of change of the scalar field $f(x, y, z) = xy + 2z^2$ at [1, 1, 1] in the direction of the vector $\mathbf{a}_x - 2\mathbf{a}_y + \mathbf{a}_z$.

2-20 If the force exerted on an object is given by

$$\mathbf{F}(x, y, z) = 2x\mathbf{a}_x + 3z\mathbf{a}_y - 4\mathbf{a}_z \quad \text{N}$$

determine the work required to move the object in a straight line

(a) from [0, 0, 1 m] to [0, 0, − 3 m],
(b) from [1 m, 1 m, 0] to [0, 1 m, 0], and
(c) from [1 m, 1 m, 1 m] to [0, 0, 1 m].

2-21 Evaluate the line integral of

$$\mathbf{F} = x\mathbf{a}_x + 2xy\mathbf{a}_y - y\mathbf{a}_z$$

from [1, – 1, 0] to [0, 0, 0] along paths consisting of

(a) a straight line between the two points, and
(b) a two-segment path with segment 1 from [1, – 1, 0] to [1, 0, 0] and segment 2 from [1, 0, 0] to [0, 0, 0].

2-22 Evaluate the line integral of the vector field $\mathbf{F} = \mathbf{a}_x + 2\mathbf{a}_y + \mathbf{a}_z$ along a circular path of unity radius from [1, 0, 1] to [0, 1, 1].

2-23 Show that $\oint_c \mathbf{F} \cdot d\mathbf{l} = 0$ for the vector field

$$\mathbf{F} = (2 - y)\mathbf{a}_x + (z - x)\mathbf{a}_y + y\mathbf{a}_z$$

for any closed contour c.

2-24 If a scalar field is given by $f(x, y, z) = 2x + yz - xy$, evaluate the line integral of $\mathbf{F} = \nabla f$ from [1, – 1, 1] to [0, 0, 0] along paths consisting of

(a) a straight line between the two points, and
(b) a three-segment path with segment 1 from [1, – 1, 1] to [1, – 1, 0], segment 2 from [1, – 1, 0] to [1, 0, 0], and segment 3 from [1, 0, 0] to [0, 0, 0].

(Both paths should yield the same result. Why?)

2-25 Show that if a vector field is expressed in spherical coordinates as

$$\mathbf{F} = \frac{K}{r^2}\mathbf{a}_r \qquad \text{then} \qquad \oint_c \mathbf{F} \cdot d\mathbf{l} = 0$$

for any closed contour c.

2-26 Determine the net flux of the vector field

$$\mathbf{F}(x, y, z) = 2x^2 y\mathbf{a}_x + z\mathbf{a}_y + y\mathbf{a}_z$$

leaving the closed surfaces

(a) of a cube defined by [0, 0, 0], [0, 1, 0], [0, 1, 1], [0, 0, 1], [1, 0, 0], [1, 0, 1], [1, 1, 0], [1, 1, 1]; and
(b) of a cube defined by [1, 0, 0], [1, 1, 0], [1, 0, 1], [1, 1, 1], [2, 0, 0], [2, 0, 1], [2, 1, 0], and [2, 1, 1].

Check your result with the divergence theorem.

2-27 Water flows through a cylindrical pipe of radius R. The flow vector over the cross section is given by $\mathbf{F} = (R - r/1 + r)\mathbf{a}_z$ (in a cylindrical coordinate system). Show that the amount of water entering any arbitrary closed surface is the same as the amount leaving the surface.

2-28 Determine the net flux of the vector field

$$\mathbf{F}(r, \phi, z) = r\mathbf{a}_r + \mathbf{a}_\phi + z\mathbf{a}_z$$

leaving a cylindrical closed surface defined by $r = 1$, $0 \leq \phi \leq \pi$, and $0 \leq z \leq 1$. Verify your result with the divergence theorem.

2-29 Compute the divergence of the following vector fields:

(a) $\mathbf{F} = yz\mathbf{a}_x + zy\mathbf{a}_y + xz\mathbf{a}_z$
(b) $\mathbf{F} = r\mathbf{a}_r + z \sin \phi\mathbf{a}_\phi + 2\mathbf{a}_z$
(c) $\mathbf{F} = 2\mathbf{a}_r + r \cos \theta\mathbf{a}_\theta + r\mathbf{a}_\phi$

2-30 Compute the divergence of the following vector fields, and show by sketching the fields that the results are reasonable:

(a) $\mathbf{F}(x, y, z) = \mathbf{a}_y$
(b) $\mathbf{F}(x, y, z) = y\mathbf{a}_y$
(c) $\mathbf{F}(x, y, z) = x\mathbf{a}_y$
(d) $\mathbf{F}(x, y, z) = x\mathbf{a}_x - y\mathbf{a}_y$

2-31 Verify the divergence theorem for a vector field expressed in rectangular coordinates as $\mathbf{F} = x\mathbf{a}_x - 2y\mathbf{a}_z$ when the closed surface is that of a box with corners at [2, 1, 0], [2, 1, 1], [2, 0, 1], [2, 0, 0], [0, 1, 0], [0, 1, 1], [0, 0, 0], and [0, 0, 1].

2-32 Compute the curl of the following vector fields:

(a) $\mathbf{F} = xy\mathbf{a}_x + 2yz\mathbf{a}_y - \mathbf{a}_z$
(b) $\mathbf{F} = 2\mathbf{a}_r + \sin \phi\mathbf{a}_\phi - z\mathbf{a}_z$
(c) $\mathbf{F} = r\mathbf{a}_r + \mathbf{a}_\theta + \sin \theta\mathbf{a}_\phi$

2-33 The flow vector for a fluid flowing in a cylindrical pipe of unit inner radius is given by $\mathbf{F} = [(1 - r)/(1 + r)]\mathbf{a}_z$. Determine the curl of $\mathbf{F}$ at the axis and at the inner surface of the pipe. Sketch the profile of the curl over the pipe's cross section.

2-34 Verify Stokes' theorem for a flat rectangular surface in the xy plane bounded by [0, 0, 0], [1, 0, 0], [1, 1, 0], and [0. 1, 0] when the vector field is

(a) $\mathbf{F} = 2\mathbf{a}_x + \mathbf{a}_y$
(b) $\mathbf{F} = 2xy\mathbf{a}_x - y\mathbf{a}_z$

2-35 Verify Stokes' theorem for the vector field

$$\mathbf{F}(r, \theta, \phi) = \mathbf{a}_r + \mathbf{a}_\theta + \mathbf{a}_\phi$$

where the surface is defined by $r = 2$, $0 \leq \phi \leq \pi/2$, and $0 \leq \theta \leq \pi/2$.

2-36 Verify the following vector identities by direct expansion in a coordinate system of your choice:

(a) $\nabla \times \nabla f = 0$
(b) $\nabla \cdot \nabla \times \mathbf{F} = 0$

2-37 If a vector field is obtained from a scalar field $f(r, \phi, z) = r \sin \phi$ as $\mathbf{F} = \nabla f$, evaluate the line integral between the origin and $x = 0$, $y = 1$, and $z = 1$ along two paths

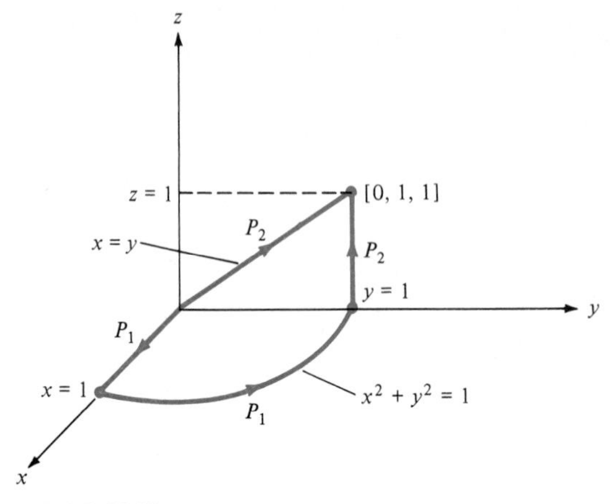

FIGURE P2.37

shown in Fig. P2.37. From these results can you determine whether your answers are incorrect? Why?

2-38 Evaluate the line integral of the vector field

$$\mathbf{F}(x, y, z) = y\mathbf{a}_x + (x + z)\mathbf{a}_y + 3yz\mathbf{a}_z$$

along straight-line paths between the following points:

$$[1, 0, 0] \rightarrow [0, 1, 0] \rightarrow [0, 1, 1] \rightarrow [0, 0, 1]$$

2-39 Evaluate the line integral of the vector field

$$\mathbf{F}(x, y, z) = xy\mathbf{a}_x + 2\mathbf{a}_y$$

along the closed triangular contour extending via straight lines from [0, 0, 0] to [1, 0, 0] to [0, 1, 0] and back to [0, 0, 0]. Evaluate

$$\int_s (\nabla \times \mathbf{F}) \cdot d\mathbf{s}$$

over the surface bounded by this contour. Do the results agree: Why should they?

2-40 Verify the divergence theorem for the vector field

$$\mathbf{F}(r, \theta, \phi) = r\mathbf{a}_r$$

over a closed surface that is a quadrant of a sphere defined by $r = 1, 0 \leqq \theta \leqq \pi/2$.

2-41 Verify Stokes' theorem for the vector field $\mathbf{V}(r, \theta, \phi) = \cos\theta\mathbf{a}_r$ by using a hemisphere of unit radius defined by $r = 1, 0 \leqq \phi \leqq 2\pi$, and $0 \leqq \theta \leqq \pi/2$.

CHAPTER 3

Electrostatic Fields

In this chapter we will study the fields produced by stationary distributions of electric charges.† We will find that these stationary charge distributions produce only one of the fundamental quantities constituting a general electromagnetic field: the electric field (which is a vector field). In Chap. 4 we will find that charges which are moving at a *constant rate* (such as a dc electric current) produce another of the fundamental quantities constituting a general electromagnetic field: the magnetic field (also a vector field). In Chap. 5 we will see that when charges are moving but their velocities are not constant, their movement produces the general electromagnetic field that involves both the electric field and the magnetic field. For this case, the two field quantities cannot be separated, and each field quantity affects the other.

3.1 Charge

Charge is the fundamental physical quantity involved in the study of electromagnetic fields. Charges constitute the source of all electromagnetic fields, so the concept of charge is basic to this study. The unit of charge is the coulomb (C), and the smallest known quantity of charge is associated with an electron. An electron has a charge of approximately 1.6×10^{-19} C, and the charge associated with an electron is classified as a negative charge. Positive charges are associated with the protons in the nuclei of atoms.

Although charge at a microscopic level has been mentioned, we will not consider such a detailed view but will, instead, be interested in macroscopic

† An electromagnetic field will often be referred to simply as a field.

(relative to atomic dimensions) distributions of charge—such as the flow of charge (electric current) in a conductor. We will study various forms of charge distributions: point charge distributions, line charge distributions, surface charge distributions, and volume charge distributions. Each of these charge distributions will be useful in describing certain physical problems. Examples of these various charge distributions are shown in Fig. 3.1. We will use the symbol ρ with subscripts to denote the line, surface, and volume charge distributions; the line, surface, and volume charge distributions will be denoted by ρ_l, ρ_s, and ρ_v, respectively.

The point charges in Fig. 3.1*a* are assumed to exist (be concentrated) at isolated points in space. Although we will not be concerned with a microscopic view of charge, the point charge is quite useful in illustrating many of the concepts involved in our study. For this purpose, the concept of a point charge has a useful role.

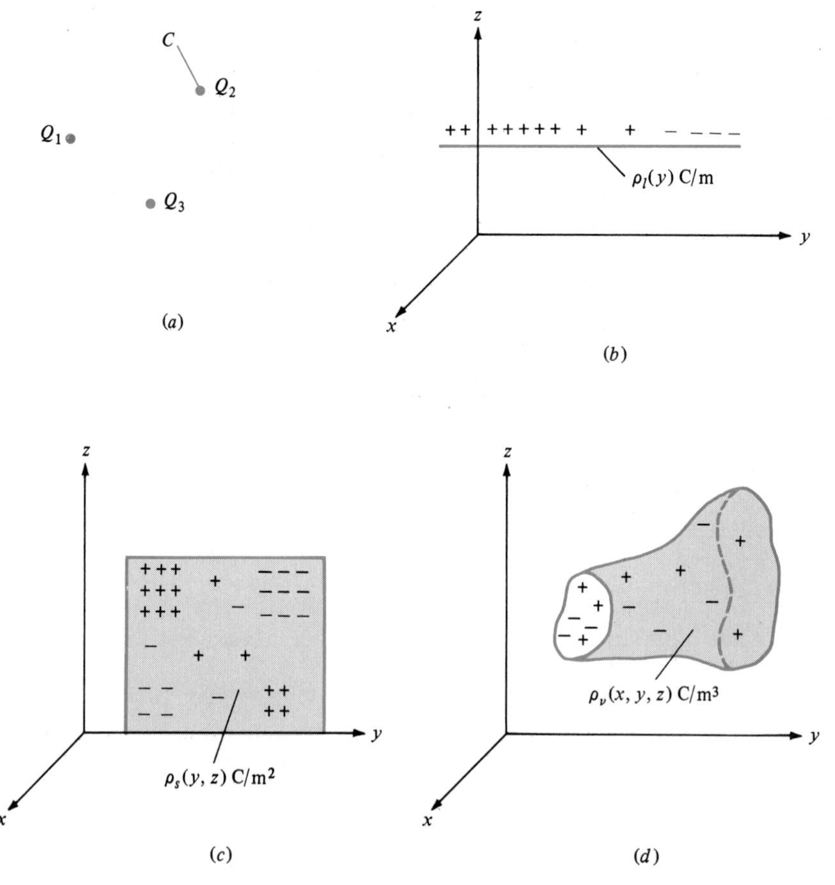

FIGURE 3.1
Forms of charge distributions.

The line charge distribution shown in Fig. 3.1*b* is essentially a distribution of charge in only one dimension and is shown as being parallel to the y axis of a rectangular coordinate system. The orientation of the coordinate system in Fig. 3.1*b* was chosen so that the distribution is a function of only one coordinate variable, y; therefore, we may denote this distribution as $\rho_l(y)$, and the units are coulombs per meter of length along the line axis (C/m). For example, if $\rho_l(y) = 2y + 1$ C/m, then the distribution of charge at $y = 0$ is 1 C/m and at $y = -5$ is -9 C/m. Note that the charge at any specific point on the line is zero.

EXAMPLE 3.1 Determine the total charge contained in the line charge distribution $\rho_l(x, y, z) = 2x + 3y - 4z$ C/m extending from [2, 1, 5] to [4, 3, 6].

Solution The total charge is the sum of the differential contributions, $dQ = \rho_l\, dl$, and is given by

$$Q = \int_{[2,1,5]}^{[4,3,6]} \rho_l\, dl$$

A differential length of the line is found from

$$dl = \sqrt{dx^2 + dy^2 + dz^2} = \sqrt{1 + \left(\frac{dy}{dx}\right)^2 + \left(\frac{dz}{dx}\right)^2}\, dx$$

Substituting the equations of the line (or contour) $y = x - 1$ and $z = \frac{1}{2}x + 4$ into the equation for the charge distribution and the equation for dl, we obtain $\rho_l = 3x - 19$ and $dl = \frac{3}{2}dx$. The total charge contained in the line is therefore

$$Q = \int_{x=2}^{4} (3x - 19)(\tfrac{3}{2})\, dx = -30 \text{ C}$$

The units of the various charge densities ρ_l, ρ_s, and ρ_v are coulombs per meter (C/m), coulombs per square meter (C/m^2), and coulombs per cubic meter (C/m^3), respectively. In each case the charge contained in any region of the charge distribution is the sum of the differential contributions:

$$Q = \int_l \rho_l\, dl \quad \text{C} \tag{1}$$

$$Q = \int_s \rho_s\, ds \quad \text{C} \tag{2}$$

$$Q = \int_v \rho_v\, dv \quad \text{C} \tag{3}$$

If any of these charge distributions ρ_l, ρ_s, or ρ_v is constant over the respective region (for example, independent of x, y, and z), it is said to be a *uniform charge distribution.* Note that the volume integral in Eq. (3) also applies, symbolically, to the line and surface integrals in (1) and (2). In the case of a line charge distribution, the volume v in (3) is taken only over the line containing the line charge distribution, and similarly for the surface charge distribution. We will therefore avoid writing three separate types of integrals—with the understanding that the volume integrals apply to surface and line integrals also, in the sense that the intent is to add differential contributions.

EXAMPLE 3.2 A circular disk of radius R has a surface charge density that increases linearly away from the center, the constant of proportionality being k. Determine the total charge on the disk.

Solution From Fig. 3.2, an elemental annular ring at a distance r from the center of the disk has a surface area

$$ds = 2\pi r \, dr$$

The distribution within this ring is

$$\rho_s = kr$$

Hence, from Eq. (2) the total charge on the disk is given by

$$Q = \int_0^R 2\pi k r^2 \, dr = \tfrac{2}{3}\pi k R^3 \qquad \text{C}$$

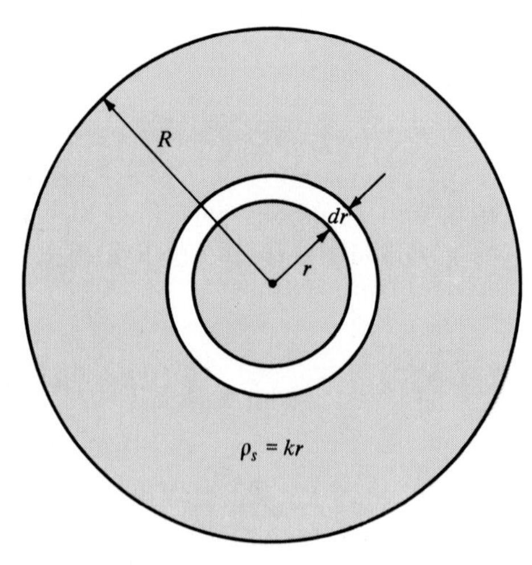

FIGURE 3.2
Example 3.2

EXAMPLE 3.3 An electron beam may be approximated by a right circular cylinder of radius R that contains a volume charge density $\rho_v = k/(c + r^2)$ C/m^3, where k and c are constants. Evaluate the total charge per unit length of the beam.

Solution Proceeding as in Example 3.2, the elemental volume of a unit-long cylindrical shell of inner radius r and outer radius $r + dr$ is

$$dv = 2\pi r\, dr$$

Hence, from (3) the total charge contained in the beam is given by

$$\begin{aligned} Q &= \int_0^R \frac{k2\pi r\, dr}{c + r^2} \\ &= \pi k \int_0^R \frac{2r\, dr}{c + r^2} \\ &= \pi k[\ln(c + r^2)]_0^R \\ &= \pi k \ln \frac{c + R^2}{c} \qquad \text{C} \end{aligned}$$

3.2 Coulomb's Law

In 1785 a French army engineer, Colonel Charles Coulomb, constructed a delicate torsional balance to measure the forces that charges exert on each other. The result of his experiments is an equation relating the magnitude of the force exerted on a point charge q by another charge Q when the point charges are separated a distance R in free space. This result is called *Coulomb's law* and is given by

$$F_Q = \frac{qQ}{4\pi\epsilon_0 R^2} \qquad \text{N} \tag{4}$$

This equation shows that the force between the charges varies directly as the product of their magnitudes and inversely as the square of the distance between them. The SI unit of the force is the newton when the charges are in coulombs and the distance between them, R, is in meters.

The constant ϵ_0 that appears in the denominator of Coulomb's law is known as the *permittivity of free space* and is given by (in farads per meter)†

$$\epsilon_0 = 8.854 \times 10^{-12} \simeq \frac{1}{36\pi} \times 10^{-9} \qquad \text{F/m} \tag{5}$$

Actually, the units of ϵ_0 are coulombs squared per newton-meter squared ($C^2/N \cdot m^2$) according to (4). In Sec. 3.9 the unit of capacitance will be defined as the farad, which has the units of coulombs squared per newton-meter ($C^2/N \cdot m$). Thus, the units of ϵ_0 are the same as those of capacitance per distance or length.

The direction of the force exerted on q by Q is the direction that q would move if it were free to move and Q were fixed in space. This results from our knowledge that charges of like sign repel each other and that charges of unlike sign attract each other. We could put this idea into a vector form by writing

$$\mathbf{F}_Q = \frac{qQ}{4\pi\epsilon_0 R^2}\mathbf{a}_R \qquad \text{N} \tag{6}$$

The unit vector $\mathbf{a}_R$ is directed along the line joining the two charges and points in the direction from Q to q, as shown in Fig. 3.3*a*. Charges exert forces on each other in a mutual fashion; thus, we could also discuss the force exerted on Q by q, $\mathbf{F}_q$, in which case the unit vector $\mathbf{a}_R$ is directed from q to Q.

The unit vector $\mathbf{a}_R$ can be easily determined in a rectangular coordinate system. For example, if q is located at $[x, y, z]$ and Q is located at $[X, Y, Z]$, then a vector pointing in the direction from Q to q can be constructed as

$$\mathbf{R} = (x - X)\mathbf{a}_x + (y - Y)\mathbf{a}_y + (z - Z)\mathbf{a}_z \tag{7a}$$

and the unit vector $\mathbf{a}_R$ can be obtained by normalizing R to a length of unity:

$$\mathbf{a}_R = \frac{\mathbf{R}}{R} \tag{7b}$$

where

$$R = |\mathbf{R}| = \sqrt{(x - X)^2 + (y - Y)^2 + (z - Z)^2} \tag{7c}$$

---

† Coulomb observed that the force is proportional to the products of the charge values and inversely proportional to the square of the separation distance R:

$$F_Q = k\frac{qQ}{R^2}$$

The constant of proportionality, k, depends upon the system of units used. In the International System of Units, the proportionality constant k is equal to $1/4\pi\epsilon_0$.

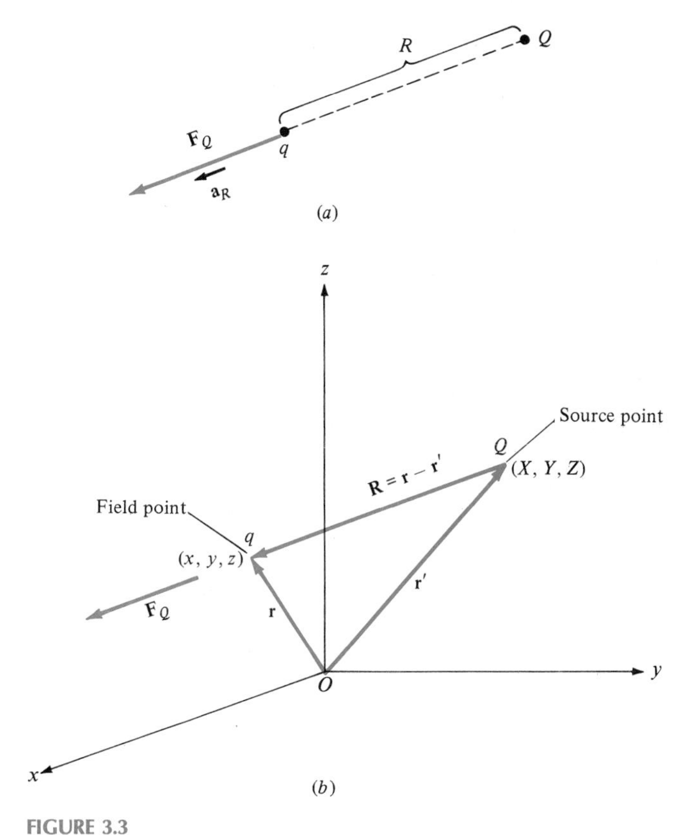

FIGURE 3.3
Coulomb's law. (*a*) Force between two point charges. (*b*) Vector diagram illustrating vectors $\mathbf{r}'$ to the source point, $\mathbf{r}$ to the field point and the distance vector $\mathbf{R} = \mathbf{r} - \mathbf{r}'$.

Let us now refer to Fig. 3.3*b*, which shows the location of the charges q and Q and the vector $\mathbf{R}$ with reference to the origin 0. If we draw two vectors $\mathbf{r}$ and $\mathbf{r}'$ from 0 to locate q and Q, respectively, then from Fig. 3.3*b* we have $\mathbf{R} = \mathbf{r} - \mathbf{r}'$ and $R = |\mathbf{r} - \mathbf{r}'|$. Hence, from (7*b*) we have

$$\mathbf{a}_R = \frac{\mathbf{r} - \mathbf{r}'}{|\mathbf{r} - \mathbf{r}'|} \tag{8}$$

Consequently, (6) may be written in a more versatile form as

$$\mathbf{F}_Q = \frac{qQ}{4\pi\epsilon_0} \frac{(\mathbf{r} - \mathbf{r}')}{|\mathbf{r} - \mathbf{r}'|^3} \tag{9}$$

Whereas (6) and (9) are equivalent, one may prefer to use one form rather than the other. The location of q will be referred to as the *field point*, whereas the location of Q will be referred to as the *source point*.

EXAMPLE 3.4 Consider a pair of point charges in free space. Charge q is located at [2, 4, 5] m and has a charge of $-300\ \mu$C. Charge Q is located at [1, 1, 3] m and has a charge of 10 μC. Determine the force exerted on q.

Solution The magnitude of the force exerted on q by Q is

$$F_Q = -\frac{27}{R^2} \quad \text{N}$$

The length vector between the two charges and directed from Q to q is $\mathbf{R} = \mathbf{a}_x + 3\mathbf{a}_y + 2\mathbf{a}_z$ and the length of this vector becomes $R = 3.74$ m. Consequently, the magnitude of the force exerted by Q on q is -1.93 N.

The unit vector $\mathbf{a}_R$ can be obtained as

$$\mathbf{a}_R = 0.27\mathbf{a}_x + 0.8\mathbf{a}_y + 0.53\mathbf{a}_z$$

and the expression for the vector force is

$$\mathbf{F}_Q = -1.93\mathbf{a}_R = -0.52\mathbf{a}_x - 1.55\mathbf{a}_y - 1.03\mathbf{a}_z \quad \text{N}$$

Suppose that we now wish to find the force exerted on charge q by several charges, $Q_1, Q_2, \ldots, Q_N$. It should be clear that we may superimpose the effects of each of these charges on q as shown in Fig. 3.4*a*. In doing so, we cannot simply add the magnitudes of the forces exerted on q by the other charges since the force exerted on q by one of the charges (for example, Q_1) is not in the same direction as the force exerted on q by any of the other charges (for example, Q_2). Consequently, we must perform a vector addition of the individual forces exerted by each charge on q.

EXAMPLE 3.5 Consider the problem in Example 3.4, but denote Q in that example as Q_1 and add an additional point charge Q_2 located at [1, −2, 6] m and having a charge of 50 μC. Compute the force exerted on q.

Solution The individual force exerted on q by this charge (in the absence of Q_1) is

$$\mathbf{F}_{Q2} = -0.58\mathbf{a}_x - 3.46\mathbf{a}_y + 0.58\mathbf{a}_z \quad \text{N}$$

The net force exerted on q by Q_1 and Q_2 is the vector sum

$$\mathbf{F}_{Q1+Q2} = \mathbf{F}_{Q1} + \mathbf{F}_{Q2} = -1.10\mathbf{a}_x - 5.01\mathbf{a}_y - 0.45\mathbf{a}_z \quad \text{N}$$

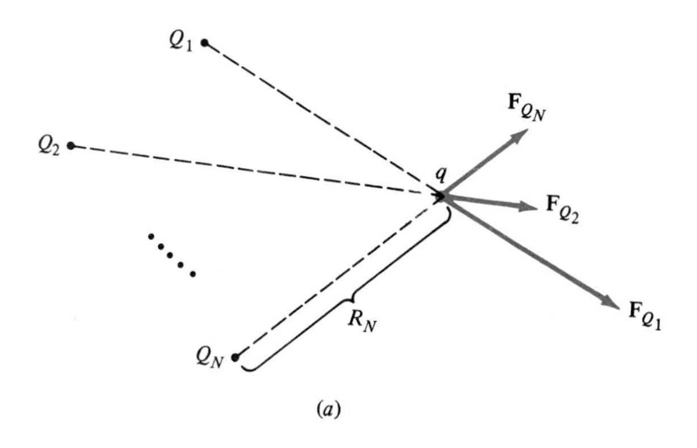

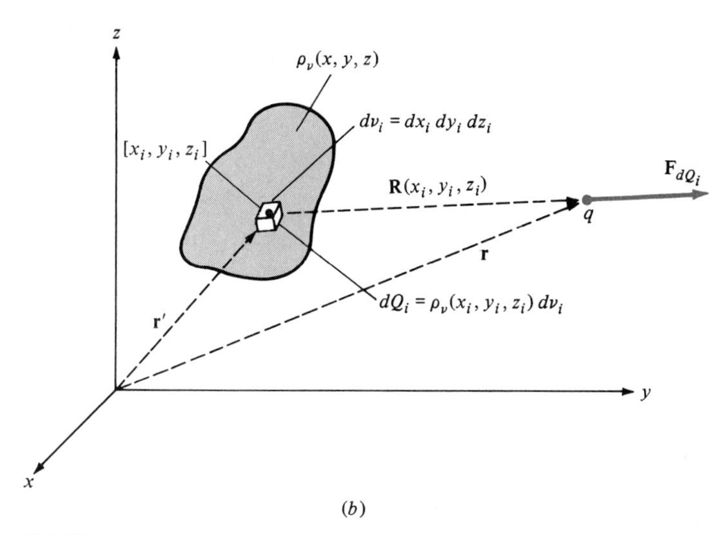

FIGURE 3.4
Coulomb's law: (*a*) groups of point charges; (*b*) charge distributions.

This superposition of the forces on charge q due to several point charges Q_1, $Q_2, \ldots, Q_N$ can be written as

$$\mathbf{F}_{Q1+Q2+\cdots+QN} = \frac{qQ_1}{4\pi\epsilon_0 R_1^2}\mathbf{a}_{R1} + \frac{qQ_2}{4\pi\epsilon_0 R_2^2}\mathbf{a}_{R2} + \cdots + \frac{qQ_N}{4\pi\epsilon_0 R_N^2}\mathbf{a}_{RN}$$

$$= \sum_{i=1}^{N} \frac{qQ_i}{4\pi\epsilon_0 R_i^2}\mathbf{a}_{Ri} \tag{10}$$

where R_i is the distance between q and Q_i, and $\mathbf{a}_{Ri}$ is the unit vector pointing from Q_i to q.

The form of Coulomb's law given in (10) can also be used to find the force exerted on some point charge q due to a line, surface, or volume distribution of charge. In each of these cases, the summation sign must be replaced by an integral since a smooth distribution of charge is being considered. For example, consider the volume charge distribution $\rho_v(x, y, z)$ shown in Fig. 3.4*b*. It is desired to find the force exerted on a point charge q by this volume charge distribution. If we subdivide the volume charge distribution into differential blocks of charge, dQ_i, we may add (vectorially) the contributions to the total force exerted on q due to each of the blocks of charge that comprise the volume charge distribution by considering each block to be a point charge. At a point $[x_i, y_i, z_i]$ within the charge distribution, we form a differential volume $dv_i = dx_i\, dy_i\, dz_i$. The total charge contained within this differential volume is $dQ_i = \rho_v(x_i, y_i, z_i)dv_i$. The net force exerted on q by the entire volume charge distribution is the vector sum of the contributions from all differential blocks of charge within this distribution; that is,

$$\mathbf{F} = \int_v \frac{q\rho_v}{4\pi\epsilon_0 R^2}\,\mathbf{a}_R\,dv = \int_v \frac{q\rho_v}{4\pi\epsilon_0}\frac{(\mathbf{r}-\mathbf{r}')}{|\mathbf{r}-\mathbf{r}'|^3}\,dv \qquad \text{N} \tag{11}$$

Note that the subscript i has been removed from R_i since R is now the distance between q and each differential element of charge within the entire volume. Also note that the unit vector $\mathbf{a}_R$ is not fixed in space but will vary from differential element to differential element. Therefore, R and $\mathbf{a}_R$ will be functions of the variables of integration, for example, x, y, and z. The alternative form is also given in (11); the vector $\mathbf{r}'$ is again said to be directed from the origin to the source point (the differential element of change), whereas the vector $\mathbf{r}$ is said to be directed from the origin to the field point (the location of q). In this case, vector $\mathbf{r}'$ varies whereas vector $\mathbf{r}$ is fixed. The result in (11) also applies in a similar fashion to line and surface charge distributions in the sense that differential contributions are to be obtained.

EXAMPLE 3.6 Consider a line charge of length L having a uniform charge distribution of ρ_l C/m. Determine an expression for the force exerted on a point charge q located a distance d from the center of the line charge.

Solution We locate a rectangular coordinate system at the center of the line charge, as shown in Fig. 3.5. The contribution to the force from a differential element of charge $dQ_i = \rho_i\, dl_i$ at $[0, 0, z_i]$ is

$$dF_i = \frac{q\rho_l\, dl_i}{4\pi\epsilon_0 R_i^2}\cos\alpha$$

Thus, the total force exerted on q is

$$\mathbf{F}_y = 2\int_{z=0}^{L/2} \frac{q\rho_l}{4\pi\epsilon_0 R_i^2}\cos\alpha\, dz\, \mathbf{a}_y$$

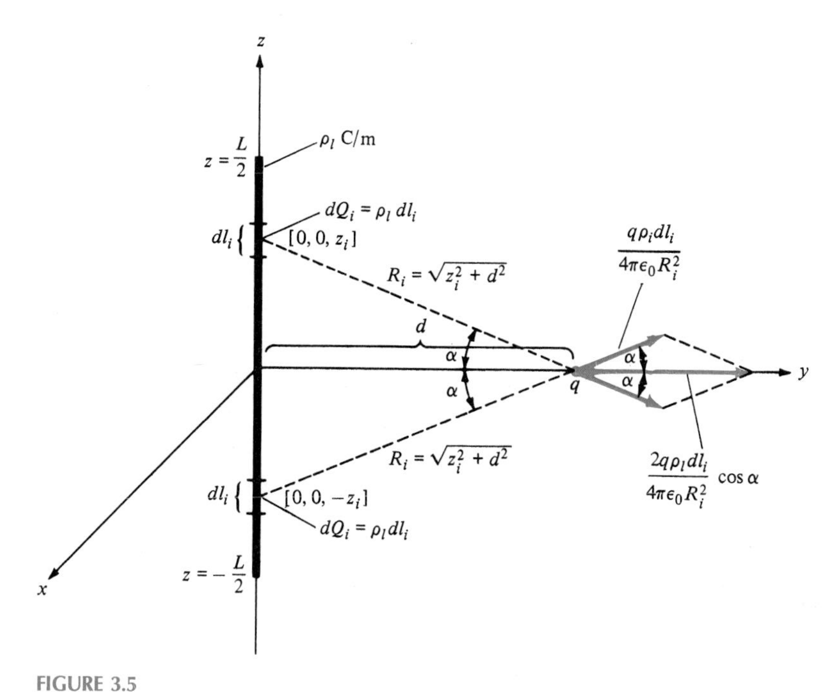

FIGURE 3.5
Example 3.6. Resultant force exerted on a point charge by a line charge distribution.

Since the charge distribution is uniform with $R_i^2 = z_i^2 + d^2$ and $\cos \alpha = d/R_i$, the result is

$$\mathbf{F}_y = \frac{q\rho_l}{2\pi\epsilon_0} \frac{L}{d\sqrt{L^2 + 4d^2}} \mathbf{a}_y \qquad \text{N}$$

The result for an infinite line charge can be obtained by allowing $L \to \infty$, or

$$\mathbf{F}_y = \frac{q\rho_l}{2\pi\epsilon_0 d} \mathbf{a}_y \qquad \text{N} \qquad L \to \infty$$

Note that the force exerted on a point charge by an infinite line charge having a uniform distribution varies inversely as the distance away from the line.

3.3 Electric Field Intensity

A point charge or a distribution of charge exerts a force on any other charge; thus, we may visualize a force field in the vicinity of these charges. Of particular interest is the determination of the intensity of this force field. The *electric field intensity vector* **E** for this charge distribution is defined as the force exerted on a positive charge introduced into the field per unit of that positive charge. To determine this field intensity, we introduce a small test charge q into the field and divide the resulting force on q by the magnitude of q:

$$\mathbf{E} = \lim_{q \to 0} \frac{\mathbf{F}}{q} \tag{12}$$

We use the limiting operation in (12) so that the test charge will not disturb the original charge distribution. Thus, using the results given in (6) and (9), the electric field intensity about some point charge Q becomes

$$\mathbf{E} = \frac{\mathbf{F}}{q} = \frac{Q}{4\pi\epsilon_0 R^2}\,\mathbf{a}_R = \frac{Q(\mathbf{r} - \mathbf{r}')}{4\pi\epsilon_0 |\mathbf{r} - \mathbf{r}'|^3} \tag{13}$$

where $\mathbf{a}_R$ is directed radially away from Q. If Q is negative, the electric field intensity is directed toward the charge. The electric field intensity vector about some distribution of charge can be similarly obtained by dividing (11) by q:

$$\mathbf{E} = \frac{\mathbf{F}}{q} = \int_v \frac{\rho_v}{4\pi\epsilon_0 R^2}\,\mathbf{a}_R\, dv = \int_v \frac{\rho_v\, |\mathbf{r} - \mathbf{r}'|}{4\pi\epsilon_0 |\mathbf{r} - \mathbf{r}'|^3}\, dv \tag{14}$$

which also applies in the usual fashion to line and surface charge distributions. Once the electric field intensity vector is determined for a particular distribution of charge, the force exerted on some other charge Q^* introduced into the field is obtained simply by multiplying the electric field vector by the charge introduced: $\mathbf{F} = Q^*\mathbf{E}$.

The units of the electric field intensity vector (or, briefly, the electric field) are the unit of force divided by the unit of charge, or newtons per coulomb (N/C). In a subsequent section the unit of voltage will be defined as the volt (V), which has the units of newton-meters per coulomb (N·m/C). Anticipating this definition, let us note that the units of electric field intensity will be volts per meter [V/m = (N·m/C)/m = N/C].

EXAMPLE 3.7 Consider the problem of a cylindrical distribution of charge shown in Fig. 3.6*a*. The cylindrical surface is infinite in length with radius r_c and has a uniform charge distribution ρ_s C/m^2 at its surface. Determine the electric field at a distance R away from the center of the cylinder.

Solution We first observe that because of the infinite length of the cylinder and the uniform nature of the charge distribution, the electric field will be directed solely in the radial direction (perpendicular to the cylindrical surface); therefore, a cross-sectional view of the cylinder (Fig. 3.6*b*) is sufficient. We also

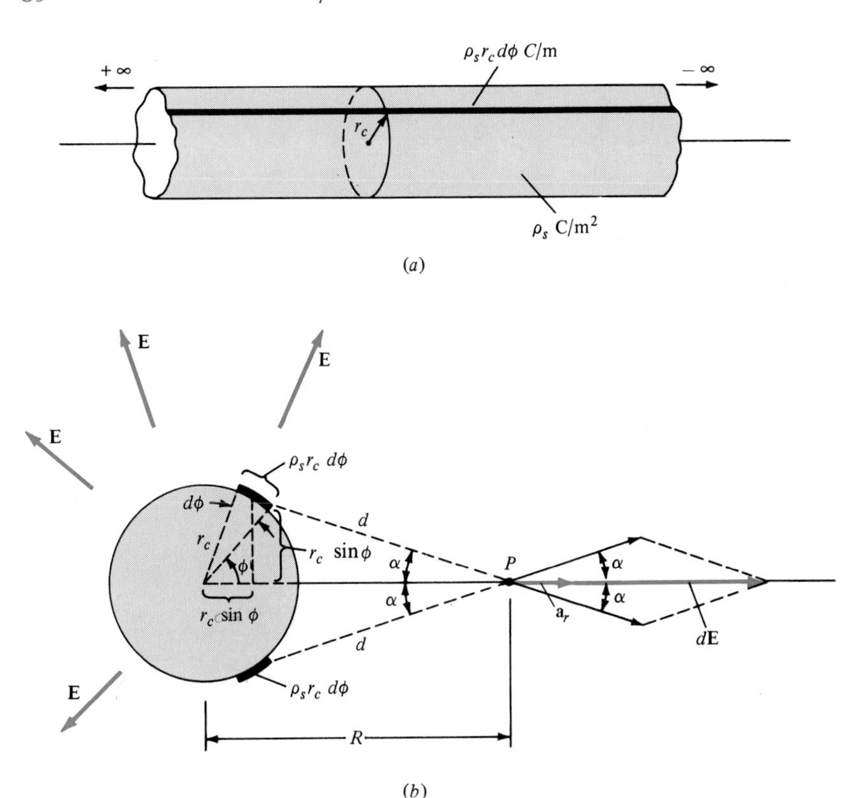

FIGURE 3.6
Example 3.7. A uniform cylindrical distribution of charge and the resulting electric field.

observe that the field is independent of the angular position of the point at which we are determining the resultant field (the field point). To simplify this problem, let us use a previously derived result. We found that for the infinite line charge with a uniform charge distribution ρ_l C/m, the force exerted on a point charge q a distance d away from the line charge is given by $F = q\rho_l/2\pi\epsilon_0\, d$. The electric field for this infinite line charge is therefore $E = F/q = \rho_l/2\pi\epsilon_0\, d$. We can use this result to solve the problem at hand if we consider the charge distribution on the cylindrical surface to be be composed of infinitesimal line charges occupying a circumferential angle $d\phi$. The charge distribution for these line charge distributions can be obtained by multiplying the surface charge distribution by the arc length around the cylinder, $r_c\, d\phi$, resulting in $\rho_l = \rho_s r_c\, d\phi$ C/m. Now consider two symmetrically disposed line charges on the cylinder shown in Fig. 3.6*b*. The net electric field is the vector combination of the fields from symmetrically disposed line charges; that is,

$$d\mathbf{E} = 2\,\frac{\rho_s}{2\pi\epsilon_0 d}\cos\alpha\; r_c\; d\phi\; \mathbf{a}_r$$

where $\mathbf{a}_r$ is the radial unit vector. The distance d between the field point and the differential line charge is found from the law of cosines to be

$$d = \sqrt{R^2 + r_c^2 - 2Rr_c \cos \phi}$$

and $\cos \alpha$ is

$$\cos \alpha = \frac{R - r_c \cos \phi}{d}$$

Therefore, the differential contribution to the field due to symmetrically disposed segments of charge is

$$d\mathbf{E} = 2 \frac{\rho_s}{2\pi\epsilon_0} \frac{R - r_c \cos \phi}{d^2} r_c \, d\phi \, \mathbf{a}_r$$

The total electric field can then be obtained by integrating this result from $\phi = 0°$ to $\phi = 180°$; that is,

$$\mathbf{E} = \int_{\phi=0}^{\pi} \frac{2\rho_s r_c}{2\pi\epsilon_0} \frac{R - r_c \cos \phi}{R^2 + r_c^2 - 2Rr_c \cos \phi} \, d\phi \, \mathbf{a}_r$$

This integral can be evaluated, using a table of integrals, to be†

$$\mathbf{E} = \frac{\rho_s r_c}{\epsilon_0 R} \mathbf{a}_r \qquad \text{V/m} \qquad R \geq r_c$$

For $R < r_c$ this integral evaluates to zero, so that

$$\mathbf{E} = 0 \qquad R < r_c$$

3.4 Gauss' Law and the Electric Flux Density Vector

Notice from Eq. (13) that the electric field depends on the permittivity ϵ_0. If we define a quantity $\mathbf{D}$ by

$$\mathbf{D} = \epsilon_0 \mathbf{E} \tag{15}$$

then we may rewrite (13) as

$$\mathbf{D} = \frac{Q}{4\pi R^2} \mathbf{a}_R \tag{16}$$

† An exceptionally thorough table of integrals can be found in H. B. Dwight, *Tables of Integrals and Other Mathematical Functions*, Macmillan, New York, 1961. See p. 228, integral 859.124.

We define **D** as *electric flux density*, which is independent of the permittivity. Referring to (16), we may visualize lines of force emanating from (or directed toward) the charge Q. A light source produces a similar effect, which is referred to as *light flux*. Therefore, it is common to classify these lines of electric force as *electric flux* denoted as ψ_e.

To compute the total electric flux produced by some distribution of charge, we integrate $\mathbf{D}\cdot d\mathbf{s}$ over a closed surface surrounding these charges. For an isolated point charge Q (shown in Fig. 3.7), the flux through some differential element of a general surface s is, from (16),

$$\mathbf{D}\cdot d\mathbf{s} = \frac{Q}{4\pi R^2}\,\mathbf{a}_R\cdot d\mathbf{s} \tag{17}$$

where $d\mathbf{s} = ds\ \mathbf{a}_n$ and $\mathbf{a}_n$ is the unit vector normal to the surface. In solid geometry, the element of solid angle $d\Omega$ is the ratio of the surface area subtended to the square of the radius:

$$d\Omega = \frac{ds \cos\theta}{R^2} \tag{18}$$

and (17) integrated over the closed surface is

$$\oint_s \mathbf{D}\cdot d\mathbf{s} = \frac{Q}{4\pi}\oint_s d\Omega = \frac{Q}{4\pi}4\pi = Q \tag{19}$$

since the total solid angle subtended by any closed surface is 4π steradians (sr).

Note that the total flux ψ_e passing through a closed surface containing a point charge is equal to the point charge enclosed. Suppose that the surface encloses several point charges or a distribution of charge. By considering the charge distribution to be composed of differential elements of charge or point charges and by using superposition, we see that (19) again results where Q in

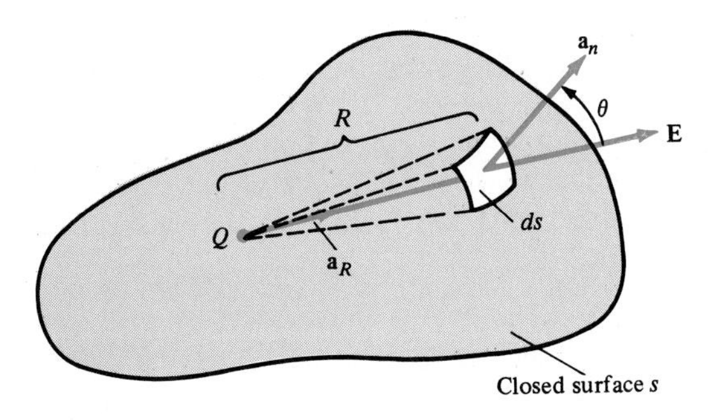

FIGURE 3.7

Determination of the net flux of a point charge through a closed surface.

(19) now becomes the total charge enclosed by s, $Q_{\text{enclosed by } s}$. Hence, we obtain Gauss' law for charges in free space:

$$\oint_s \mathbf{D} \cdot d\mathbf{s} = Q_{\text{enclosed by } s} \tag{20}$$

Stated in words, according to Gauss' law, the total outward flux from a closed surface equals the total (or net) charge enclosed by the surface.

This is a remarkable result since the surface s (referred to as the *gaussian surface*) can be of arbitrary shape and size. (It must, however, be large enough to enclose the charge.) Note, however, that the result depends on the electric flux vector from a point charge or charge distribution varying inversely with the square of the distance from the charge. This inevitably relates back to the $1/R^2$ dependence in Coulomb's law.

According to the definition of electric flux density $\mathbf{D}$, the electric flux through some closed surface is given by

$$\psi_e = \oint_s \mathbf{D} \cdot d\mathbf{s} \tag{21}$$

The unit of electric flux is the coulomb; one coulomb of charge produces one line of electric flux. Since flux lines emanate from positive charges and terminate on negative charges, Gauss' law allows us to determine only the *net* positive charge enclosed by s; a surface containing more negative than positive charge will have a net flux of vector $\mathbf{D}$ into the surface, and ψ_e will be negative.

Gauss' law given by (20) was obtained for charges in free space but will be shown in Sec. 3.6 to hold for charges in certain other material media. The relation between $\mathbf{D}$ and $\mathbf{E}$ in these material media will be given by $\mathbf{D} = \epsilon\mathbf{E}$, where ϵ is the permittivity of the particular medium.

Gauss' law is a very powerful tool for solving certain field problems. This is a result of the fact that the choice of the gaussian surface s is at our disposal. A judicious choice can greatly simplify the problem, as the following example shows.

EXAMPLE 3.8 As a simple example of the application of Gauss' law, consider the case of a cylindrical surface of infinite length that supports a uniform charge distribution (which was considered in Example 3.7). Determine the electric field intensity produced by the charge distribution.

Solution Since the cylinder is infinite in length, the total charge contained on it is infinite, but the charge contained per unit length is $\rho_l = \rho_s 2\pi r_c$ C/m and the flux density vector $\mathbf{D}$ is directed in the radial direction. Therefore, if we choose a surface over which to evaluate Gauss' law (the gaussian surface) which

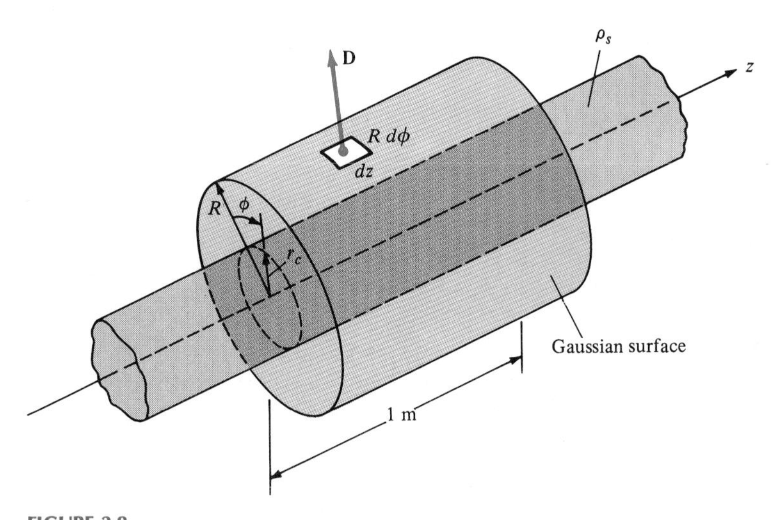

FIGURE 3.8
Example 3.8. Gauss' law for a cylindrical uniform-charge distribution.

is in the form of another cylinder of radius R and unity length, as shown in Fig. 3.8, then $\mathbf{D}\cdot d\mathbf{s} = DR\,d\phi\,dz$ and

$$\oint_s \mathbf{D}\cdot d\mathbf{s} = \int_{\phi=0}^{2\pi}\int_{z=0}^{1} DR\,d\phi\,dz = 2\pi RD$$

(The integral over the ends of this gaussian surface yields zero since, by symmetry, $\mathbf{D}$ is radially directed.) The total charge enclosed is $\rho_s 2\pi r_c$, so that Gauss' law provides $\rho_s 2\pi r_c = D2\pi R$ and

$$D = \frac{\rho_s r_c}{R}$$

The electric field is related to $\mathbf{D}$ by $\mathbf{D} = \epsilon_0\mathbf{E}$; thus, the magnitude of the electric field vector becomes

$$E = \frac{\rho_s r_c}{\epsilon_0 R}$$

which was obtained in Example 3.7 by evaluating a fairly complicated integral. We see that the electric field produced by the cylindrical distribution of charge ρ_s is the same as if the charge distributed over this surface, $\rho_s 2\pi r_c$, were concentrated on the axis of the cylinder as a line charge distribution of linear density $\rho_l = \rho_s 2\pi r_c$ since any gaussian surface about the cylinder of charge will enclose the same amount of charge regardless of whether this charge is concentrated on the axis of the cylinder or distributed uniformly over the surface of the cylinder.

Also, we see immediately that the electric field interior to the cylindrical surface is zero; no net charge is enclosed in any gaussian surface constructed within the interior of the cylinder!

If we apply the divergence theorem developed in Chap. 2 to Gauss' law, we obtain

$$\oint_s \mathbf{D} \cdot d\mathbf{s} = \int_v (\nabla \cdot \mathbf{D}) dv = Q_{\text{enclosed by } s} \tag{22}$$

where v is the volume enclosed by s. The total charge enclosed can be written as

$$Q_{\text{enclosed by } s} = \int_v \rho_v \, dv \tag{23}$$

Comparing (22) and (23), we find that

$$\nabla \cdot \mathbf{D} = \rho_v \tag{24}$$

since s is arbitrary. This is the *point form* of Gauss' law. It is a logical result since it shows that the divergence (or emanation) of electric flux lines from some charge is related to the strength and net polarity of the charge.

3.5 Electrostatic Potential

Consider the movement of a point charge q through an electric field $\mathbf{E}$, as shown in Fig. 3.9. The work required to move q from point a to point b along the

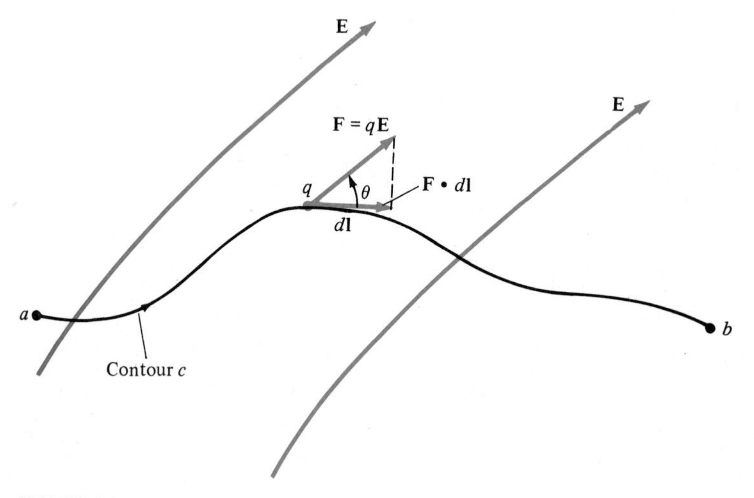

FIGURE 3.9

Movement of a point charge in an electric field.

contour c is given by

$$W_{ab} = -\int_a^b |\mathbf{F}| \cos\theta \, dl = -\int_a^b \mathbf{F} \cdot d\mathbf{l} = -q \int_a^b \mathbf{E} \cdot d\mathbf{l} \tag{25}$$

Note the presence of the minus sign. If $90° > \theta > 270°$, then $\cos\theta$ in (25) is positive, indicating that the component of $\mathbf{F}$ along the path is in the direction of $d\mathbf{l}$. Thus, the field that provides $\mathbf{F}$ also provides the energy to move the charge and W_{ab} is negative. This implies that work is done by the electric field and that the energy of the system is decreased by the amount of work done. This type of a situation occurs in cathode-ray tubes and microwave high-power oscillators where electrons (that is, electric charges) travel in regions of electric fields.

Now consider the closed path or contour c shown in Fig. 3.10a, of contour c_1 from a to b and of c_2 from b to a. Movement of a point charge q around the

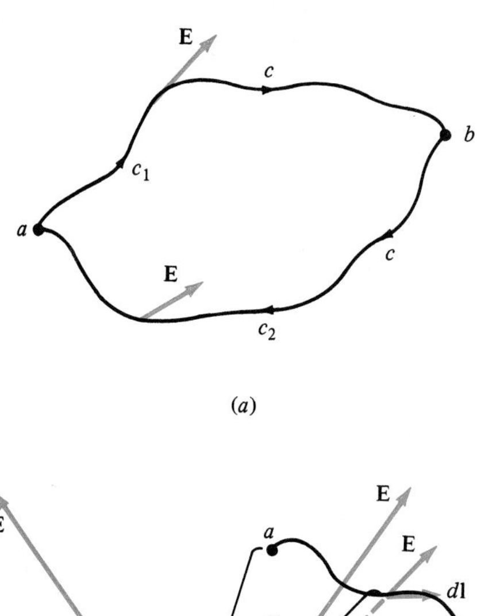

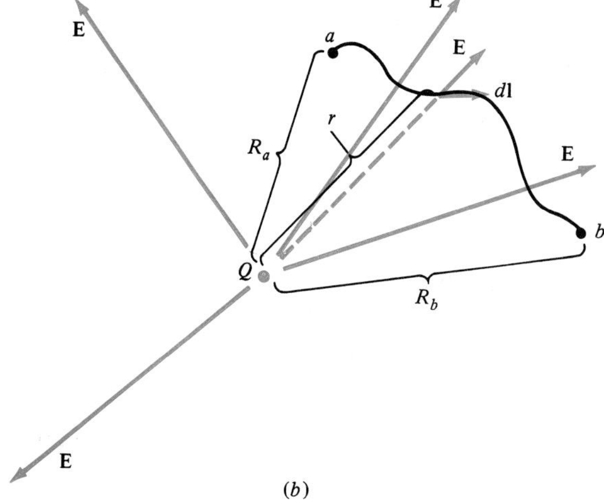

FIGURE 3.10
The conservative nature of the static electric field.

closed contour c should result in no work being done, as (25) shows that the work done is a function of initial and final positions. Thus, we should find that

$$W_{ab}\Big|_{c1} + W_{ba}\Big|_{c2} = 0 \tag{26}$$

or

$$-q \int_a^b \mathbf{E} \cdot d\mathbf{l} - q \int_b^a \mathbf{E} \cdot d\mathbf{l} = 0 \tag{27}$$

But (27) is equivalent to

$$\oint_c \mathbf{E} \cdot d\mathbf{l} = 0 \tag{28}$$

Equation (28) is referred to as the *conservative property of the electrostatic field.* Applying Stokes' theorem to (28) yields

$$\oint_c \mathbf{E} \cdot d\mathbf{l} = \int_s (\nabla \times \mathbf{E}) \cdot d\mathbf{s} = 0 \tag{29}$$

Since the surface s is arbitrary, we conclude that

$$\nabla \times \mathbf{E} = 0 \tag{30}$$

for the electrostatic field. This symbolizes that the electrostatic field has no circulation.

The conservative property of the electric field resulting from a stationary distribution of charge (the electrostatic field) given in (28) can be proved directly as follows. Consider the field of a point charge shown in Fig. 3.10*b*. Suppose we evaluate the line integral of **E** from point a to point b along some general path between these two points. We place the point charge at the origin of a spherical coordinate system, and the differential path length $d\mathbf{l}$ becomes

$$d\mathbf{l} = dr\,\mathbf{a}_r + r\,d\theta\,\mathbf{a}_\theta + r\sin\theta\,d\phi\,\mathbf{a}_\phi \tag{31}$$

Thus, with R becoming the spherical coordinate r,

$$\mathbf{E} \cdot d\mathbf{l} = \left(\frac{Q}{4\pi\epsilon_0 r^2}\,\mathbf{a}_r\right) \cdot d\mathbf{l} = \frac{Q}{4\pi\epsilon_0 r^2}\,dr \tag{32}$$

and we obtain

$$\int_c \mathbf{E} \cdot d\mathbf{l} = \frac{Q}{4\pi\epsilon_0}\left(\frac{1}{R_a} - \frac{1}{R_b}\right) \tag{33}$$

Thus, it is clear that the line integral of **E** along some general path in the presence of a point charge is independent of the path chosen between the endpoints of the path and depends only on the radial distances from the point charge to the endpoints of the path. By superposition, we may extend this result to the case of several point charges. Similarly, this result is seen to hold for line, surface, or volume charge distributions by considering these distributions to be made up of differential elements of charge. Since we have seen that the line integral of **E** for a general charge distribution is independent of the chosen path, it follows directly that the electrostatic field obeys the conservative property given in (28).

The reader should be cautioned that the electric field is conservative only for the static electric fields. In Chap. 5 we will find that for time-varying fields, the electric field is not necessarily conservative.

We found in Chap. 2 that if the line integral of a vector field around a closed contour is identically zero, then that vector field can be written as the gradient of some scalar field. Therefore, for the electrostatic field, (28) shows that we may write

$$\mathbf{E}(x, y, z) = -\nabla V(x, y, z) \tag{34}$$

where the minus sign is introduced to define positive work as that done on the positive test change used to define **E** via Coulomb's law. The scalar field $V(x, y, z)$ will be called the *electrostatic potential function.* If we evaluate the line integral of **E** along some contour c between points a and b, we obtain

$$\int_a^b \mathbf{E} \cdot d\mathbf{l} = -\int_a^b \nabla V \cdot d\mathbf{l} = -V(b) + V(a) \tag{35}$$

by the definition of the gradient function: $dV = \nabla V \cdot d\mathbf{l}$. In (35), $V(b)$ and $V(a)$ are the values of the potential function at b and a, respectively.

The value of the scalar potential function at a point will be referred to as the *voltage* of that point and denoted as

$$\begin{aligned} V_b &= V(b) \\ V_a &= V(a) \end{aligned} \tag{36}$$

The physical interpretation of voltage is now apparent. Note from (35) that the voltage difference between two points is

$$V_{ab} = V_b - V_a = -\int_a^b \mathbf{E} \cdot d\mathbf{l} = \frac{W_{ab}}{q} \tag{37}$$

which is the work per unit charge required to move a charge q from a to b. The potentials of a point such as in (36) are referred to as the *absolute potentials* of those points since

$$V_b = -\int_{\infty}^{b} \mathbf{E} \cdot d\mathbf{l} \tag{38}$$

which represents the work per unit charge required to move a unit charge from infinity to the point. Notice that the negative sign associated with $\mathbf{E} \cdot d\mathbf{l}$ gives a positive potential at b, and compare (38) with (25). In (38) we have implicitly taken the potential at infinity to be zero:

$$V_b = V(b) - \cancelto{0}{V}(\infty) \tag{39}$$

Thus, infinity is the arbitrarily chosen reference for the absolute potential function. Any other arbitrarily chosen reference would be suitable since the important quantity is the difference in potential between two points.

The absolute potential of a point P at a distance R_P from a point charge Q was found to be

$$V_P = -\int_{\infty}^{R_P} \frac{Q}{4\pi\epsilon_0 r^2}\, dr = \frac{Q}{4\pi\epsilon_0 R_P} \tag{40}$$

We may generalize this result to the case of some general charge distribution ρ_v by considering this distribution to be composed of point charges, $dQ = \rho_v\, dv$. The potential of some point P then becomes

$$V_P = \int_v \frac{\rho_v\, dv}{4\pi\epsilon_0 R} \tag{41}$$

where R is the distance between $dQ = \rho_v\, dv$ and the point P. The volume v must enclose the entire charge distribution ρ_v.

We have chosen (arbitrarily) the zero reference potential to be at infinity. Any other arbitrary choice of reference potential will do, so that (41) can be written as

$$V_P = \int_v \frac{\rho_v\, dv}{4\pi\epsilon_0 R} + C \tag{42}$$

where C is some constant that depends on the choice of reference.

EXAMPLE 3.9 Now consider the problem presented in Example 3.7 and shown in Fig. 3.6 of a cylindrical surface of radius r_c that is infinite in length and supports a uniform surface charge distribution ρ_s C/m^2. Determine the potential difference between two points at radial distances R_a and R_b from the cylinder, with $R_b > R_a$.

Solution For this problem, the electric field intensity is directed radially away from the cylinder and is given by $\mathbf{E} = (\rho_s r_c/\epsilon_0 R)\mathbf{a}_r$ for points outside the cylinder and $\mathbf{E} = 0$ for points within the cylinder. The potential difference between two points a and b at radial distances R_a and R_b from the axis of the cylinder is

$$V_{ab} = -\int_{R_a}^{R_b} \mathbf{E} \cdot d\mathbf{l} = -\frac{\rho_s r_c}{\epsilon_0} \ln \frac{R_b}{R_a} = \frac{\rho_s r_c}{\epsilon_0} \ln \frac{R_a}{R_b}$$

with $R_b > R_a$. Clearly this is a sensible result since positive work is required to move a positive charge toward the cylinder ($V_{ba} = -V_{ab}$) and negative work is required to move a charge away from the cylinder. Note that since the electric field is zero at points within the cylinder, the potential difference between a point interior to the cylinder and a point on the surface of the cylinder is zero. Therefore, all points interior to the cylinder are at the same potential as the cylinder surface. The above result can be written in terms of the charge distribution per unit of cylinder length, $\rho_l = \rho_s 2\pi r_c$, as

$$V_{ab} = \frac{\rho_l}{2\pi\epsilon_0} \ln \frac{R_a}{R_b}$$

It will be helpful to visualize surfaces on which the potential is constant. These are called *equipotential surfaces.* As an example, consider the case of the point charge shown in Fig. 3.11. Note that no work is required to move along the portion of contour c_2 from a to c. The reason is that at all points along this segment of the contour, the electric field intensity is perpendicular to the path. If we now visualize a spherical surface of radius R surrounding Q, we also see that $\mathbf{E}$ is perpendicular to this surface. Therefore, no work is required to move between any two points on this spherical surface; it is an equipotential surface. Thus, the equipotential surfaces for a point charge are spheres. Our choice of path c_2 for determining the potential difference between points a and b in Fig. 3.11 will obviously simplify the calculation. We could have chosen path c_1 and obtained the same result, but the mathematical details would have been considerably more involved.

In Example 3.9 the surfaces of equipotential are cylinders concentric with the cylinder bearing the uniform charge distribution. Movement along the surface of one of these equipotentials results in no expenditure of energy. Clearly, the maximum rate of expenditure of energy will be encountered if we move perpendicular to the equipotential surfaces (in the radial direction). This fact is easily demonstrated since $\mathbf{E} = -\nabla V$ and the direction of ∇V (which is opposite to the direction of $\mathbf{E}$) is the direction of the maximum rate of change of V (or work per unit charge).

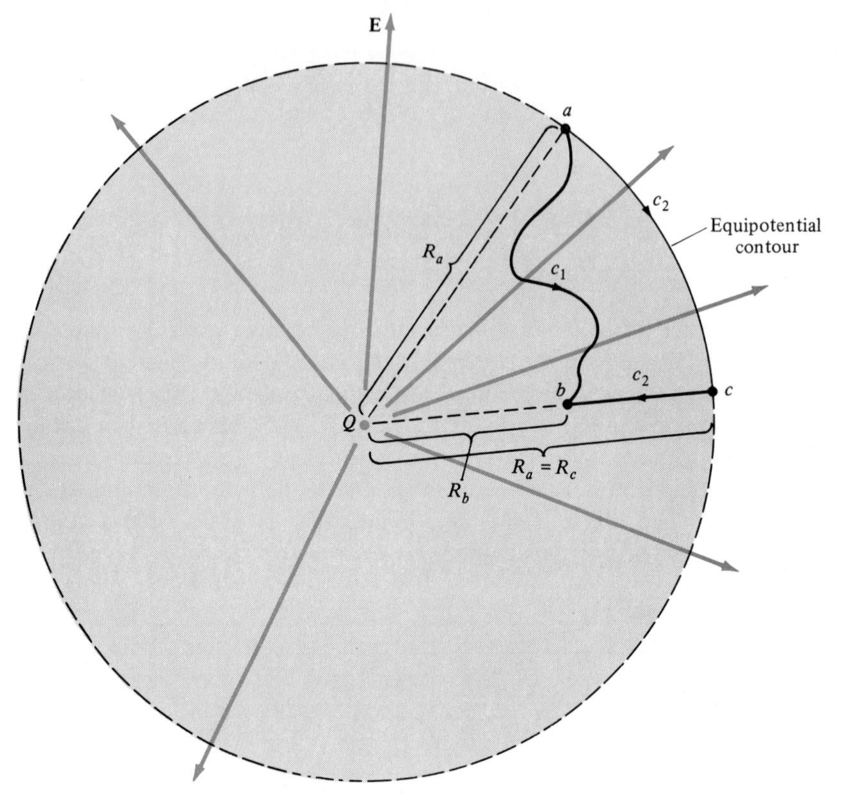

FIGURE 3.11

An equipotential contour about a point charge: computation of the difference in potential between two points.

3.6 Conductors and Dielectrics

In all our previous discussions, we have considered the surrounding medium to be free space; however, fields may exist in material media as well. The purpose of this section is briefly to examine how these material media affect our previous results and, more important, to determine how to characterize quantitatively the effect of these materials on the field quantities.

3.6.1 Conductors and Ohm's Law

Conductors are materials which, although electrically neutral, possess a relatively large number of free (mobile) charges that are free to move through the material. Our previous discussions have concentrated on free charges in free space. In conductive materials such as metals, these free charges are the conduction electrons associated with the unfilled outer atomic orbits (valence

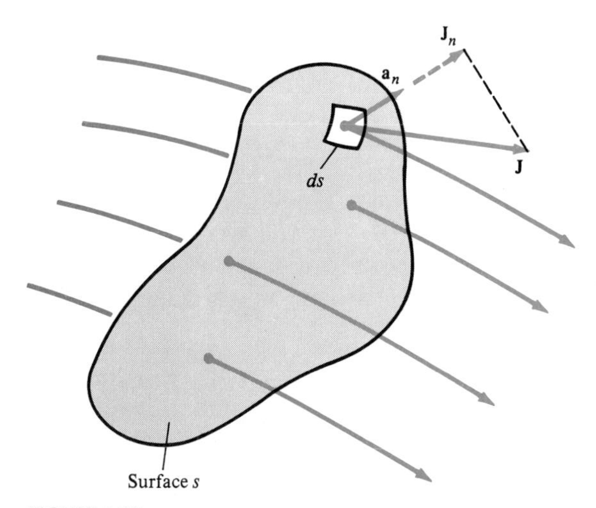

FIGURE 3.12
Current density and the calculation of the total current passing through the open surface s.

bands) of the material. A simple but useful model of a conductive material is a lattice of relatively massive positive ions which is fixed in position and which has an associated electron cloud that is free to move about. Ordinarily, these free electrons are in a state of random motion throughout the lattice due to thermal agitation. They collide with the ion lattice, resulting in new, random directions of movement. The average charge density in any region of the material at any time, however, is zero, so the material remains neutral.

In the presence of an applied electric field, it is found that a coordinated movement of these free electrons occurs but that the average space charge density remains zero at any point in the material. This movement of charge (free charge) is referred to as an *electric current*, and the electric current can be conveniently described as a current-density vector $\mathbf{J}$, whose units are amperes per square meter (A/m^2). In conductive materials, the relationship between $\mathbf{J}$ and the applied electric field $\mathbf{E}$ is found to be a simple one and is known as *Ohm's law*:

$$\mathbf{J} = \sigma \mathbf{E} \qquad \text{A/m}^2 \tag{43}$$

The constant of proportionality, σ, in Ohm's law is the conductivity of the material. The units of σ are (A/m^2)/(V/m) = A/(V · m), a conductance per meter. The SI unit of conductance is the siemen (S), and thus the units of σ are siemens per meter (S/m). The electric current I passing through some open surface s (as depicted in Fig. 3.12) is defined as the net positive charge passing through this surface per unit of time:

$$I = \lim_{\Delta t \to 0} \frac{\Delta Q}{\Delta t} = \frac{dQ}{dt} \tag{44}$$

This current passing through s is related to the current-density vector as

$$I = \int_s \mathbf{J} \cdot d\mathbf{s} \tag{45}$$

Some representative values of conductivity are: copper, $\sigma = 5.8 \times 10^7$ S/m; seawater, $\sigma = 4$ S/m; and dry earth, $\sigma = 10^{-5}$ S/m. The conductivity of a material is generally not constant but becomes a function of frequency. However, for typical good conductors such as metals, this variation of σ with frequency occurs above frequencies on the order of 10^{14} Hz.

EXAMPLE 3.10 Consider a circular, cylindrical, copper conductor of radius 1.025 mm. (This is equivalent to a 12-gauge wire used in household wiring.) If the wire carries a current of 0.2 A dc, determine the current density in the wire and the voltage drop across a 1-m length. Determine the resistance per unit of wire length.

Solution Because the current distribution over the wire's cross section is uniform, we compute the current density from

$$\int_s \mathbf{J} \cdot d\mathbf{s} = I$$

where $d\mathbf{s}$ is the vector cross-sectional area of the wire. Assuming the current density to be directed along the wire axis at all points in the wire, we obtain

$$J\pi r_w^2 = I$$

or

$$J = \frac{I}{\pi r_w^2}$$

$$= 60{,}594.4 \text{ A/m}^2$$

where r_w is the wire radius. If we now neglect any fringing of the electric field, that is, if we assume that the resulting electric field is confined to the wire and directed along the wire axis, we obtain

$$E = \frac{J}{\sigma}$$

$$= 1.045 \times 10^{-3} \text{ V/m}$$

Thus, the voltage drop across a 1-m length of the wire is approximately 1 mV. A formula for the resistance of a wire of length l, radius r_w, and conductivity σ can be obtained under the above assumptions as

$$R = \frac{V}{I} = \frac{El}{J\pi r_w^2} = \frac{l}{\sigma\pi r_w^2}$$

3.6.2 The Electric Dipole

Before we consider the properties of materials that are classified as dielectrics, we will study the field of a pair of point charges of unlike sign: a dipole of charge. Consider the pair of point charges $+q$ and $-q$ separated by a distance l, as shown in Fig. 3.13. From (40), the absolute potential of a point P at a radius r from the center of the coordinate system is

$$V_P = \frac{q}{4\pi\epsilon_0 r_+} - \frac{q}{4\pi\epsilon_0 r_-} \tag{46}$$

The distances r_+ and r_- are the appropriate distances from each point charge to point P. Now imagine that point P is at a much greater distance from the center

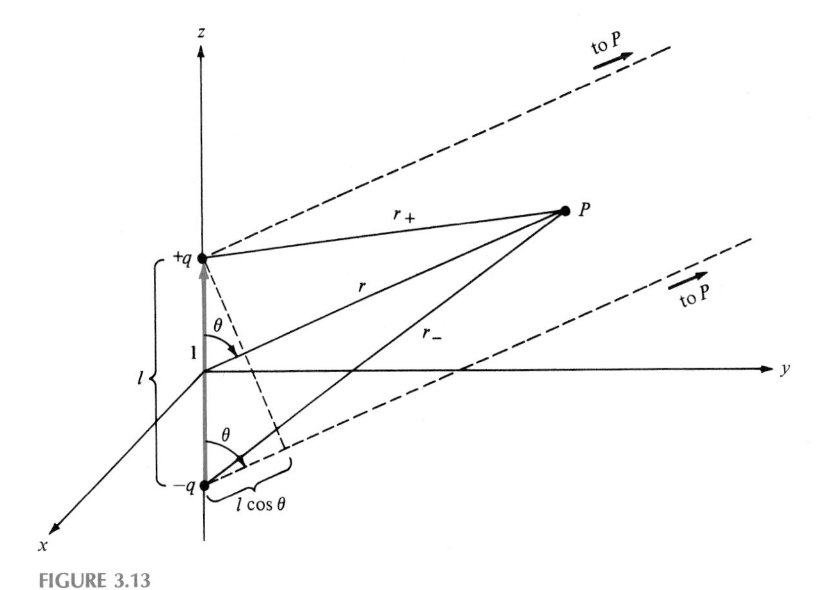

FIGURE 3.13

The electric dipole.

of the coordinate system than the distance between the two charges, i.e., $r \gg l$. In this case, the lines joining each charge and point P (shown as dotted in Fig. 3.13) are almost parallel, so that

$$r_+ = r - \frac{l}{2} \cos \theta \tag{47a}$$

and

$$r_- = r + \frac{l}{2} \cos \theta \tag{47b}$$

resulting in

$$V_P \simeq \frac{ql \cos \theta}{4\pi\epsilon_0 r^2} \tag{48}$$

Using the expression for the gradient in spherical coordinates (see Appendix A), we obtain

$$\mathbf{E} = -\nabla V_P = \frac{ql}{4\pi\epsilon_0 r^3} (2 \cos \theta \, \mathbf{a}_r + \sin \theta \, \mathbf{a}_\theta) \tag{49}$$

Thus, the electric field varies inversely as the cube of the distance from the dipole. We will find it helpful to define a *vector dipole moment* $\mathbf{p}$ as

$$\mathbf{p} = q\mathbf{l} \tag{50}$$

where $\mathbf{l}$ is the vector pointing from $-q$ to $+q$, i.e., $\mathbf{l} = l\mathbf{a}_z$ in Fig. 3.13. The reason for using the term *dipole moment* is that in the presence of some applied electric

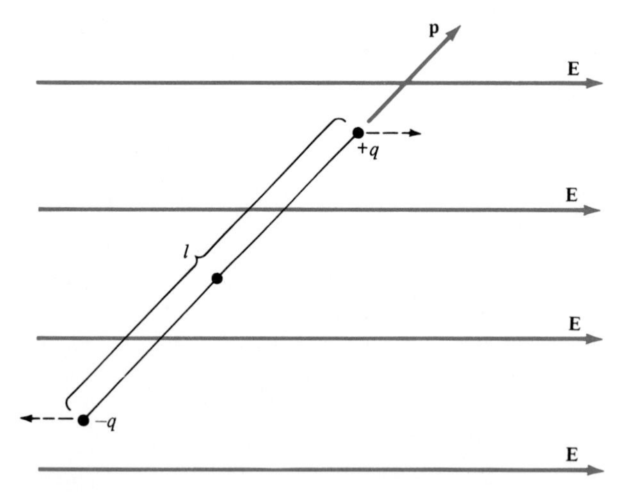

FIGURE 3.14
Alignment of a dipole with an electric field by rotation.

field, the dipole will tend to rotate so as to align the vector $\mathbf{p} = q\mathbf{l}$ with the direction of the applied field, as shown in Fig. 3.14. The larger the separation l, the larger the moment arm and consequently the force of rotation, and the term *dipole moment* arises quite naturally. These concepts of dipole moment and dipole rotation to align with an applied field will be of considerable assistance in helping us understand dielectric material properties (discussed in Sec. 3.6.3).

3.6.3 Bound Charge and Dielectrics

As opposed to conductors, dielectrics (perfect dielectrics) have no free (mobile) charge. A "good dielectric" has some free charge in the form of free electrons, as in a conductor; however, it is a very small amount relative to the amount of free charge in a "good conductor." The charges in a perfect dielectric (or, to be brief, a dielectric) are strongly bound to the parent atoms.

Even though a dielectric is electrically neutral, an externally applied electric field may cause microscopic separations of the centers of positive and negative charges which thus behave like dipoles of charges considered in the previous section. These charge separation distances are on the order of atomic dimensions, but the vast numbers of dipoles may provide a significant effect. This phenomenon is referred to as the *polarization* of the dielectric.

The polarization of a dielectric material due to some externally applied electric field may occur as a result of three effects: electronic polarization, ionic polarization, or orientational polarization. *Electronic polarization* occurs when the externally applied electric field causes a shift in the atom's positive and negative charges. Equilibrium is attained when the internal coulomb attractive force produced by the charge separation balances the applied force. When the charge separation occurs, we essentially have a microscopic electric dipole. *Ionic polarization* occurs in molecules composed of positively and negatively charged ions. An externally applied electric field again results in a microscopic separation of charge centers thus resembling a dipole of charge. *Orientational polarization* occurs in materials that possess permanent, microscopic separations of charge centers. A solid dielectric possessing persistent polarization is called an *electret*, which is an analog of a magnet. In the absence of an applied electric field, these permanent dipoles are randomly oriented. In the presence of an applied electric field, these permanent dipoles tend to rotate to align with the applied field, as discussed in Sec. 3.6.2. Materials (such as water) possessing these permanent dipoles in the absence of a field are said to be *polar* substances. In many substances the polarization occurs as a combination of electronic, ionic, and orientational polarization effects. However, each of these effects results in the production of a dipole of charge, and we may view the result as a large number of dipoles in free space, each composed of charge $-q_b$ and $+q_b$. The charge q_b is referred to as *bound charge* since it is not normally available for conduction except under the application of very large electric fields.

Each of these microscopic dipoles has a dipole moment,

$$\mathbf{p} = q_b\mathbf{l} \tag{51}$$

where each dipole is composed of bound charges $-q_b$ and $+q_b$. This is illustrated for the case of orientational polarization in Fig. 3.15*a*. A polarization vector **P** is defined as the dipole moment per unit volume; that is,

$$\mathbf{P} = \lim_{\Delta v \to 0} \frac{\sum_i \mathbf{p}_i}{\Delta v} = \frac{d\mathbf{p}}{dv} \tag{52}$$

where we sum (vectorially) all of the individual dipole moments in the volume Δv. In the absence of an externally applied electric field, the polarization vector would be zero: $\mathbf{P} = 0$. For nonpolar substances, there would be no shift of charge centers and no resultant dipoles. For polar substances, the permanent dipoles would be oriented randomly, resulting in no net polarization. Note that the individual dipole moments are, by definition, directed from $-q_b$ to $+q_b$ (in opposition to the direction of the electric field between these charges that is produced by the charge separation).

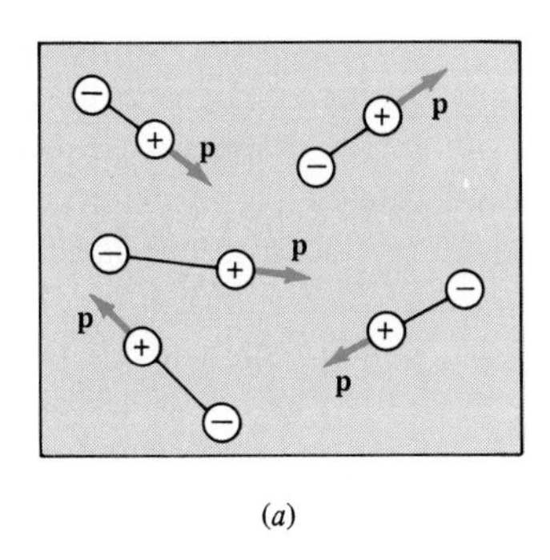

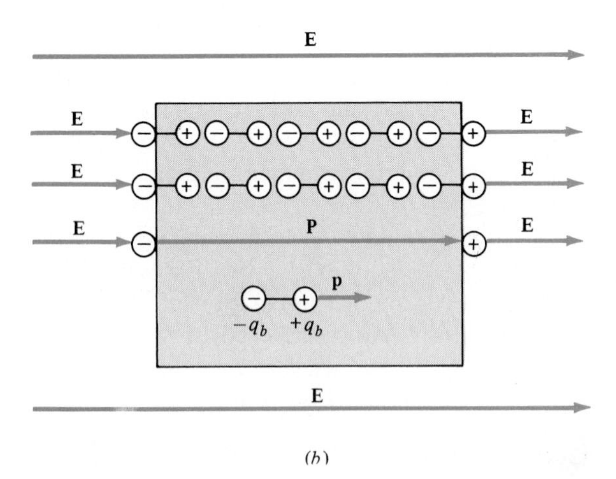

FIGURE 3.15
Orientational polarization of a dielectric material.

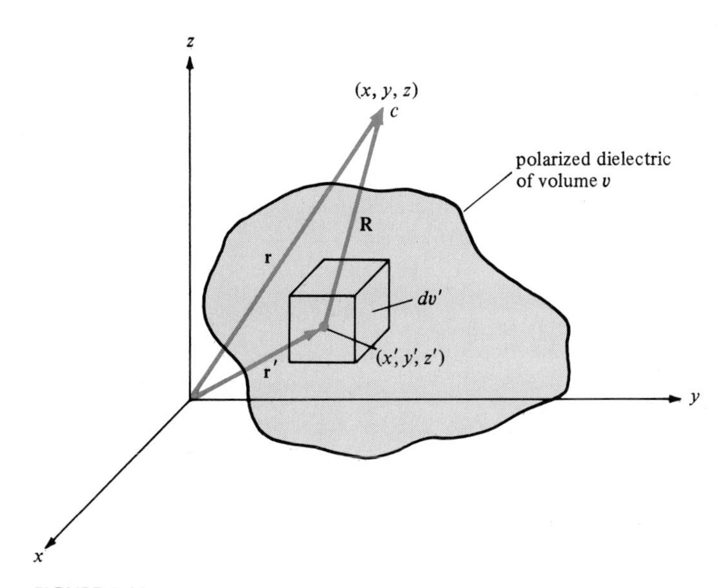

FIGURE 3.16
A dielectric with a nonuniform polarization.

For the block of material immersed in an electric field shown in Fig. 3.15*b*, the tendency of the dipoles to align with the field causes charges to appear at the surfaces of the material. Thus, it appears that an overall separation of charge has been achieved between the surfaces of the material, as shown in Fig. 3.15*b*. This is one of the important results in the polarization process.

In order to understand the effects of the presence of polarized dielectrics, let us see how the polarization vector **P** is related to potentials and charge distributions. Notice that the polarization vector **P** depends on the dipole moment **p**, which in turn is related to the potential V. So, we may combine (48) and (50) as

$$V_P = \frac{\mathbf{p} \cdot \mathbf{a}_r}{4\pi\epsilon_0 r^2} \tag{53}$$

(Fig. 3.13), where $\mathbf{a}_r$ is a unit vector directed from the origin toward point P. Now consider an elemental volume dv' within a polarized dielectric of volume v, as shown in Fig. 3.16. The vector $\mathbf{r}'$ is directed from the origin to dv' (the "source point") and is given by

$$\mathbf{r}' = x'\mathbf{a}_x + y'\mathbf{a}_y + z'\mathbf{a}_z \tag{54}$$

We wish to determine the potential at a point c located by the vector $\mathbf{r}$ where

$$\mathbf{r} = x\mathbf{a}_x + y\mathbf{a}_y + z\mathbf{a}_z \tag{55}$$

From (53) we have

$$dV_c = \frac{d\mathbf{p} \cdot \mathbf{a}_R}{4\pi\epsilon_0 R^2} \tag{56}$$

where $d\mathbf{p}$ is the dipole moment of the volume dv' and $\mathbf{a}_R$ is the unit vector in the direction of $\mathbf{R}$, which is shown in Fig. 3.16.

Now, from (52) we have $d\mathbf{p} = \mathbf{P}\,dv'$. Hence, (56) may be written as

$$dV_c = \frac{\mathbf{P} \cdot \mathbf{a}_R dv'}{4\pi\epsilon_0 R^2} \tag{57}$$

Since $\mathbf{R} = \mathbf{r} - \mathbf{r}'$, we can show that $\mathbf{\nabla}'(1/R) = R/R^3$, where ∇' is the gradient with respect to the primed coordinates. Therefore, (57) becomes

$$dV_c = \frac{1}{4\pi\epsilon_0}\left[\mathbf{P} \cdot \mathbf{\nabla}'\left(\frac{1}{R}\right)dv'\right] \tag{58}$$

If we use the vector identity $\nabla \cdot (\psi \mathbf{A}) = \psi \mathbf{\nabla} \cdot \mathbf{A} + \mathbf{A} \cdot \mathbf{\nabla}\psi$, (58) becomes

$$dV_p = \frac{1}{4\pi\epsilon_0}\left[\mathbf{\nabla}' \cdot \left(\frac{\mathbf{P}}{R}\right) - \frac{1}{R}\mathbf{\nabla}' \cdot \mathbf{P}\right]dv' \tag{59}$$

Hence, the potential at c is given by

$$V_c = \frac{1}{4\pi\epsilon_0}\int_v \left[\mathbf{\nabla}' \cdot \left(\frac{\mathbf{P}}{R}\right) - \frac{1}{R}\mathbf{\nabla}' \cdot \mathbf{P}\right]dv' \tag{60}$$

The first term on the right-hand side of (60) may be transformed into a surface integral by the divergence theorem so that

$$V_c = \oint_s \frac{\mathbf{P} \cdot d\mathbf{s}'}{4\pi\epsilon_0 R} - \int_v \frac{\mathbf{\nabla}' \cdot \mathbf{P}}{4\pi\epsilon_0 R}\,dv' \tag{61}$$

Comparing (61) with (41), we observe that the potential at c arises from equivalent bound surface and volume charge densities, ρ_{sb} and ρ_{vb}, respectively, such that

$$V_c = \oint_s \frac{\rho_{sb}\,ds'}{4\pi\epsilon_0 R} + \int_v \frac{\rho_{vb}\,dv'}{4\pi\epsilon_0 R} \tag{62}$$

For (61) and (62) to be equivalent, we must have

$$\rho_{sb} = P_n \tag{63}$$

where P_n is the component of **P** normal to the surface and

$$\rho_{vb} = -\nabla \cdot \mathbf{P} \tag{64}$$

Equation (62) shows that a polarized dielectric can be replaced by equivalent surface and volume charge distributions for the purpose of determining the resultant electric field and electric potentials.

Clearly, *the source of the polarization vector* **P** *is bound charge.* The units of **P** are coulombs per square meter, as is clear from (51) and (52), and it is logical to expect that this charge per unit surface area would be related to **P** at the surface.

Note the similarity of (64) to Gauss' law in (24). The negative sign in (64) is a result of the dipole moments **p** being directed from the negative charge of the dipole to the positive charge. This is opposite to the direction of an electric field produced by these two charges.

We could remove the dielectric (conceptually) and replace it with the dipoles in free space. Now consider a region of space that may contain free charge as well as bound charge. In our discussion of electric field intensity, the source of **E** was charge of any type, with no distinction being made between free and bound charge. Now that we have replaced the dielectric with the bound charges in free space, the total charge that influences the resultant electric field is $\rho_T = \rho_{vf} + \rho_{vb}$, where ρ_{vf} is the volume charge density of the free charge. Consequently, using Gauss' law in point form for this free-space region, we may write

$$\nabla \cdot (\epsilon_0 \mathbf{E}) = \rho_T = \rho_{vf} + \rho_{vb} \tag{65}$$

On the other hand, (64) shows that this may be written as

$$\nabla \cdot (\epsilon_0 \mathbf{E}) = \rho_{vf} - \nabla \cdot \mathbf{P} \tag{66}$$

and (66) becomes

$$\nabla \cdot (\epsilon_0 \mathbf{E} + \mathbf{P}) = \rho_{vf} \tag{67}$$

With the result in (67), we may now define a vector $\epsilon_0 \mathbf{E} + \mathbf{P}$ whose source is free charge only. This may be defined as the electric flux density vector **D**; that is,

$$\mathbf{D} = \epsilon_0 \mathbf{E} + \mathbf{P} \tag{68}$$

Note that for a region of free space containing no dielectric, $\mathbf{P} = 0$ and $\mathbf{D} = \epsilon_0 \mathbf{E}$, as we had defined for free space in Sec. 3.4. Therefore, no inconsistency results from defining **D** as in (68). The source of the electric flux density vector **D** is therefore free charge. Equation (67) becomes, by integrating over some closed surface s and applying the divergence theorem,

$$\oint_s \mathbf{D} \cdot d\mathbf{s} = \int_v \rho_{vf} \tag{69}$$

Equation (69) is the general form of Gauss' law, not restricted to free space. In the remainder of the text ρ_{vf} will be written simply as ρ, with the understanding that ρ symbolizes free charge only.

It is important that we now review these results. Consider the pair of parallel conducting plates shown in Fig. 3.17*a*, which are very large in extent and connected by a battery. The medium between the two plates is free space. When the battery is connected, free charge will be transferred to these plates from the battery. Now consider the insertion of a block of dielectric between the plates as shown in Fig. 3.17*b*. The electric field polarizes the dielectric, causing net positive and negative bound charges (shown by small circles) to appear on opposite surfaces on the dielectric. In order to balance this induced charge, an additional amount of free charge must be transferred from the battery to each plate, resulting in an increased amount of free charge on each plate due to the

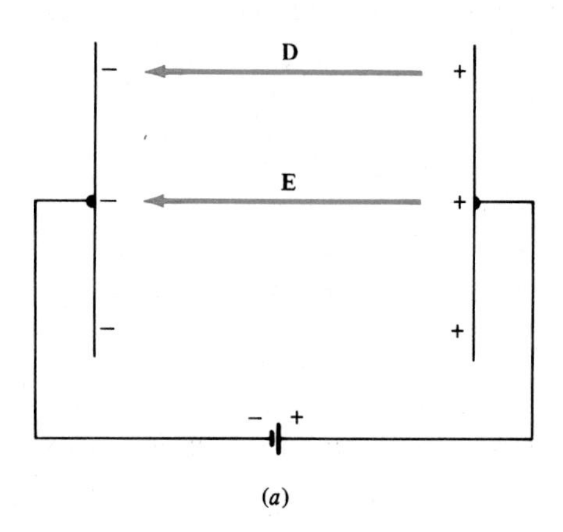

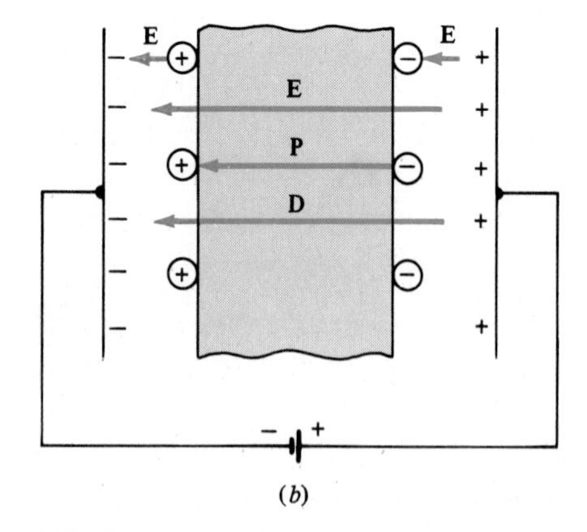

FIGURE 3.17
Increase of free charge by the polarization of a dielectric.

polarization of the dielectric. Note that the **P** lines begin and terminate on the bound charges: the source of **P**. The lines of **D** begin and terminate on free charges: the source of **D**. Note also that the lines of **E** (or more properly $\epsilon_0\mathbf{E}$) begin and terminate on charges of both types. This visual picture confirms the vector addition in (68).

Clearly, the strength of **P** and the strength of **E** are related. For most common dielectric materials, **P** and **E** are also parallel. Therefore, for these materials we may relate **P** and **E** as

$$\mathbf{P} = \chi_e \epsilon_0 \mathbf{E} \tag{70}$$

The constant χ_e is called the *electric susceptibility* of the material. Substituting this into (68) we obtain

$$\mathbf{D} = \epsilon_0(1 + \chi_e)\mathbf{E} \tag{71a}$$

The quantity $\epsilon_0(1 + \chi_e)$ is given a symbol of $\epsilon = \epsilon_0(1 + \chi_e)$ and we obtain

$$\mathbf{D} = \epsilon\mathbf{E} \tag{71b}$$

The constant ϵ is known as the *permittivity* of the dielectric. Note that for free space, χ_e must be zero since $\mathbf{P} = 0$ (no bound charge and hence no polarization vector); thus, $\epsilon = \epsilon_0$.

It is also common to classify materials according to a relative permittivity or dielectric contant $\epsilon_r = \epsilon/\epsilon_0$, which gives a measure of the polarizability of a material relative to free space. As an example, air has a relative permittivity of $\epsilon_r = 1.0006$. Distilled water has $\epsilon_r = 80$, and dry soil typically has $\epsilon_r = 3$. Typical metals, such as copper and aluminum, have dielectric properties similar to those of free space: $\epsilon_r \simeq 1$.

The permittivity of dielectrics may also depend on frequency, so the alternative term—*dielectric constant*—is not preferred for ϵ_r. A simple way of understanding this is to note that if the battery in Fig. 3.17 is replaced by a sinusoidal voltage source, the free charge on the plates tends to alternate sign with each half-cycle change in polarity of the source. The dipoles in the dielectric also tend to align with each resulting change in direction of the field produced by the alternating source; however, atomic and molecular restoring forces prevent an instantaneous alignment. As the frequency of the voltage source is increased, the tendency of the dipole alignment to lag behind the directional changes of the field becomes more pronounced. For most common dielectrics, such as Teflon, this frequency dependence begins to show up only above some very high frequency, for example, 10^{10} Hz.

We have tacitly assumed that the electric susceptibility χ_e in (70) is a scalar constant. If the electric susceptibility is independent of the applied electric field, the medium is said to be *linear*. Thus, the polarization vector **P** and the applied electric field vector **E** are linearly related. Nonlinear media can also be thought of as having permittivities that depend on the strength of the applied electric field, i.e., $\epsilon(\mathbf{E})$. In addition, we have assumed that **P** and **E** are parallel and point in the same direction. Material media having this property are said to be *isotropic*. For anisotropic materials, such as crystals, **P** and **E** are not parallel. In this case the scalar permittivity must be replaced by a matrix relating the three components of **P** to the three components of **E**, such as

$$\begin{bmatrix} P_x \\ P_y \\ P_z \end{bmatrix} = \begin{bmatrix} \epsilon_{xx} & \epsilon_{xy} & \epsilon_{xz} \\ \epsilon_{yx} & \epsilon_{yy} & \epsilon_{yz} \\ \epsilon_{zx} & \epsilon_{zy} & \epsilon_{zz} \end{bmatrix} \begin{bmatrix} E_x \\ E_y \\ E_z \end{bmatrix} \tag{72}$$

And finally we have assumed that the susceptibility (and permittivity) is the same at all points in the dielectric. These materials are said to be *homogeneous.* In some materials, the value of ϵ depends on location in the material, and these materials are said to be *inhomogeneous*; for these, the permittivity must be written as $\epsilon(x, y, z)$ to denote its dependence on position. Materials that are linear, homogeneous, and isotropic are said to be simple media. Fortunately, most of the dielectrics that we will deal with will be simple, although nonsimple media are quite useful in constructing devices that have useful properties not possible with simple media.

The typical metallic conductors that we will consider, in which $\mathbf{J} = \sigma\mathbf{E}$, are reasonably linear, homogeneous, and isotropic—in the sense, respectively, that σ is independent of the magnitude of **E**, that σ is the same at all points of the material, and that **J** and **E** are parallel (and point in the same direction) at all points in the material. Given abnormally large values of applied electric field, however, many materials that are linear in moderate fields break down, resulting in a nonlinear behavior.

3.6.4 Conservation of Charge and Charge Relaxation

We have been considering electric current as the flow of free charge in conducting materials. Strictly speaking, the movement of bound charge in dielectrics also constitutes a current or a charge flow, although *current* will be intended to mean the flow of free charge only. Movement of free charge may also occur in certain nonmetallic media. For example, in a vacuum tube, electrons are liberated by heating a metallic plate (the cathode) that is coated with an oxide. These free electrons are drawn to and collected by another plate (the anode) by an electric field established between the plates. This is an example of a *convection current.* In the case of convection currents, the current density is related to the volume charge density of positive (free) charge, ρ_v^+, and negative (free) charge, ρ_v^-, and their respective vector velocities, $\mathbf{v}^+$ and $\mathbf{v}^-$, as

$$\mathbf{J} = \rho_v^+\mathbf{v}^+ + \rho_v^-\mathbf{v}^- \tag{73}$$

For either conduction or convection currents, the rate of decrease of free charge within a closed surface s is equal to the current passing through that surface:

$$\oint_s \mathbf{J} \cdot d\mathbf{s} = -\frac{d}{dt} \int_v \rho_{\text{enclosed}}\, dv \tag{74}$$

From the divergence theorem, we observe that

$$\oint_s \mathbf{J} \cdot d\mathbf{s} = \int_v \nabla \cdot \mathbf{J}\, dv \tag{75}$$

Comparing (74) and (75) we find that

$$\nabla \cdot \mathbf{J} = -\frac{\partial \rho}{\partial t} \tag{76}$$

Equation (74) and its point form in (76) are referred to as the *equations of continuity* or the *law of conservation of charge*. Accordingly, the rate of movement of free charge out of a region (the divergence of $\mathbf{J}$) is equal to the rate of decrease of the free charge contained within the region. For steady (dc) currents, $\nabla \cdot \mathbf{J} = 0$.

Consider the case of a conductor in which an excess free charge is placed. The resulting coulomb forces due to this charge imbalance will force the excess charges to move. If the medium surrounding the conductor is a perfect dielectric (meaning that none of this free charge can leave the conductor), we find that this excess free charge accumulates on the surface of the conductor as a surface density of charge. The removal of charge from the interior of the conductor continues until the interior is once again devoid of excess charge and is electrically neutral. The time required for resumption of neutrality within the conductor is quite rapid for good conductors, and the majority of the excess charge is moved to the surface in a time known as the *relaxation time*, which is on the order of 10^{-19} s for typical metals. Once neutrality has been reestablished, no net charge exists within the conductor interior, and by Gauss' law we find that

$$\mathbf{E} = 0$$

in the conductor.

In order to determine the relaxation time, we substitute $\mathbf{J} = \sigma \mathbf{E}$ into the continuity equation:

$$\nabla \cdot (\sigma \mathbf{E}) = -\frac{\partial \rho}{\partial t} \tag{77}$$

and compare (77) to Gauss' law:

$$\nabla \cdot \mathbf{D} = \nabla \cdot (\epsilon \mathbf{E}) = \rho \tag{78}$$

From (77) and (78) we obtain

$$\frac{\partial \rho}{\partial t} + \frac{\sigma}{\epsilon} \rho = 0 \tag{79}$$

Since the conductor is assumed to be linear, homogeneous, and isotropic, σ and ϵ are constants and may be factored out of (77) and (78).

The solution to (79) is

$$\rho = \rho_0 e^{(-\sigma/\epsilon)t} \tag{80}$$

If at $t = 0$ we introduce a net excess of free charge into the interior of a conductor, then evaluating (80) at $t = 0$ shows that the unknown constant ρ_0 is this excess charge. We may write (80) as

$$\rho = \rho_0 e^{-t/\tau} \tag{81}$$

where

$$\tau = \frac{\epsilon}{\sigma} \tag{82}$$

is the relaxation time. From (81) it is clear that the excess charge in the interior of the conductor will eventually decay to zero.

For a time equal to the relaxation time, the excess charge will have been reduced to 37 percent ($1/e$) of its original value. For a perfect conductor, $\sigma = \infty$, the relaxation time is 0 s. For copper, $\epsilon = \epsilon_0 = 1/36\pi \times 10^{-9}$ F/m and $\sigma = 5.8 \times 10^7$ S/m, and the relaxation time is $\tau = 1.5 \times 10^{-19}$ s. On the other hand, for fused quartz, the relaxation time is approximately 10 days. For a metallic conductor, there can be essentially no *net* free charge within its interior; thus, the electric field in a metallic conductor is, for all practical purposes, zero.

3.7 Boundary Conditions

Often we will have to solve field problems consisting of two (or more) regions with different material properties. To solve electromagnetic field problems involving a boundary between two different materials, we need to determine the transitional properties of the field in the two regions at this boundary. These are called *boundary conditions*.

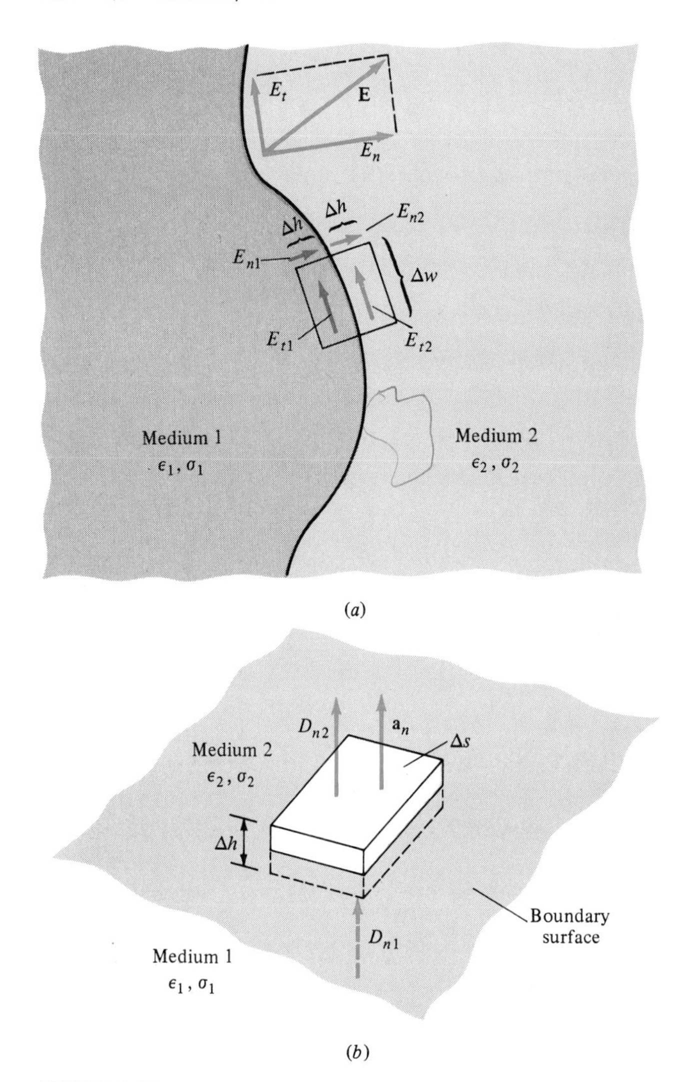

FIGURE 3.18
The boundary conditions.

First consider the electric field intensity vector **E**. At the boundary between two different media, we may decompose the total electric field **E** into a component tangent to the boundary surface, E_t, and a component normal to the boundary surface, E_n, as shown in Fig. 3.18*a*. Now consider a small, rectangular path parallel to a cross section of the surface, with sides of width Δw parallel to the surface and sides of length Δh perpendicular to the surface, as shown in Fig. 3.18*a*. According to Eq. (28), the electrostatic field is conservative:

$$\oint_c \mathbf{E} \cdot d\mathbf{l} = 0 \tag{28}$$

Evaluating this integral around the rectangular contour, we obtain

$$E_{t2}\,\Delta w - E_{n2}\,\Delta h - E_{n1}\,\Delta h - E_{t1}\,\Delta w + E_{n1}\,\Delta h + E_{n2}\,\Delta h = 0 \tag{83}$$

In the limit as the rectangular path approaches the surface, i.e., as $\Delta h \to 0$, this becomes

$$(E_{t2} - E_{t1})\Delta w = 0 \tag{84}$$

Since $\Delta w \neq 0$, from (84) we conclude that

$$E_{t2} = E_{t1} \tag{85}$$

Therefore, *the components of the electric field vector* **E** *that are tangent to the surface of a boundary between two materials must be continuous across that boundary.*

Similarly, consider the boundary conditions on the electric flux density vector **D**. This vector obeys Gauss' law:

$$\oint_s \mathbf{D} \cdot d\mathbf{s} = \int_v \rho\, dv \tag{86}$$

In order to evaluate this integral at the boundary, let us construct a small gaussian surface in the form of a box, as shown in Fig. 3.18*b*. As the height of this box approaches zero, i.e., as $\Delta h \to 0$, only the components of **D** normal to the boundary contribute to Gauss' law, which becomes

$$D_{n2}\,\Delta s - D_{n1}\,\Delta s = \lim_{\Delta h \to 0} \rho(\Delta h\,\Delta s) \tag{87}$$

Dividing both sides by Δs, we obtain

$$D_{n2} - D_{n1} = \lim_{\Delta h \to 0} \rho\,\Delta h \tag{88}$$

The quantity on the right-hand side of this result in the limit as $\Delta h \to 0$ is simply the free surface charge distribution on the boundary, ρ_s, i.e.,

$$\rho_s = \lim_{\Delta h \to 0} \rho\,\Delta h \tag{89}$$

The units of ρ_s are $(\mathrm{C/m^3})\mathrm{m} = \mathrm{C/m^2}$. Therefore, we find that

$$D_{n2} - D_{n1} = \rho_s \tag{90}$$

which can be stated as follows: *The difference in the components of the electric flux density vector* **D** *at the boundary between two regions that are normal to the* *boundary is equal to any free surface charge density at that boundary.* Note that D_{n2} is assumed to be pointing away from the surface and that D_{n1} is assumed to be pointing toward the surface. These assumed directions serve to fix the value and sign of the resulting surface charge on the interface. Equation (90) simply states that the *net* electric flux normal to and pointing *away from* the interface is equal to the net positive surface charge density at the interface.†

Now that we have derived relations governing tangential **E** and normal **D**, let us find the appropriate results for normal **E** and tangential **D**. Substituting the relation $\mathbf{D} = \epsilon\mathbf{E}$ into (85) we obtain

$$\frac{D_{t2}}{\epsilon_2} = \frac{D_{t1}}{\epsilon_1} \tag{91}$$

or

$$\frac{D_{t2}}{D_{t1}} = \frac{\epsilon_2}{\epsilon_1} \tag{92}$$

Similarly, from (90) we obtain

$$\epsilon_2 E_{n2} - \epsilon_1 E_{n1} = \rho_s \tag{93}$$

Note that the properties of the two materials enter into the conditions in (92) and (93) and did not enter into those in (85) and (90) since (85) and (90) were derived strictly from the conservative property of **E** and Gauss' law, which are independent of the properties of the medium.

Now let us specialize these boundary conditions to the cases in which one or both of the materials is a perfect conductor ($\sigma = \infty$) or a perfect dielectric ($\sigma = 0$). Suppose that medium 1 is a perfect conductor ($\sigma_1 = \infty$) and that medium 2 is a perfect dielectric. We know that the electric field **E** and consequently the electric flux density **D** must be zero in a perfect conductor; therefore, for this case, we obtain

$$E_{t2} = 0$$

$$D_{n2} = \rho_s$$

$$D_{t2} = 0 \qquad \text{Medium 1 is a perfect conductor} \tag{94}$$

$$E_{n2} = \frac{\rho_s}{\epsilon_2}$$

† In a dielectric material the divergence of the polarization vector **P** is related to the density of bound charge ρ_{vb} as $\nabla \cdot \mathbf{P} = -\rho_{vb}$. [See (64).] A similar derivation will show that the discontinuity in the normal components of the polarization vector at a dielectric boundary is equal to the negative of the bound surface charge; that is, $P_{n2} - P_{n1} = -\rho_{sb}$.

The first equation in (94) is a very important result and will be used in numerous instances. It shows that *the component of the electric field in a medium that is tangent to the surface of a perfect conductor* (*or, for practical purposes, a good conductor*) *is zero.* Note that any electric field at the surface of a perfect conductor must be normal to that surface and that this normal electric field is numerically equal to ρ_s/ϵ_2.

Similarly, let us consider the case in which medium 1 and medium 2 are perfect dielectrics. The first problem that arises is how we must interpret the surface charge density ρ_s in (90). It is clear that this is strictly a free charge since the source of **D** is a free charge. How can a free charge arise at the surface of two perfect dielectrics? It certainly did not "leak through" the dielectrics, since perfect dielectrics do not possess a free charge that would permit a conduction current. Clearly, any surface charge at the boundary must have been intentionally placed on the boundary, such as may occur by rubbing the boundary with cat's fur. Therefore, ρ_s in the boundary condition in (90) is taken to be zero unless it is placed on the boundary, and we have

$$\begin{aligned} E_{t2} &= E_{t1} \\ D_{n2} &= D_{n1} \\ \frac{D_{t2}}{D_{t1}} &= \frac{\epsilon_2}{\epsilon_1} \qquad \text{medium 1 and medium 2 are perfect dielectrics} \qquad (95) \\ \frac{E_{n2}}{E_{n1}} &= \frac{\epsilon_1}{\epsilon_2} \end{aligned}$$

EXAMPLE 3.11 Consider the boundary between two perfect dielectrics, as shown in Fig. 3.19. Determine a relation between the angles θ_1 and θ_2.

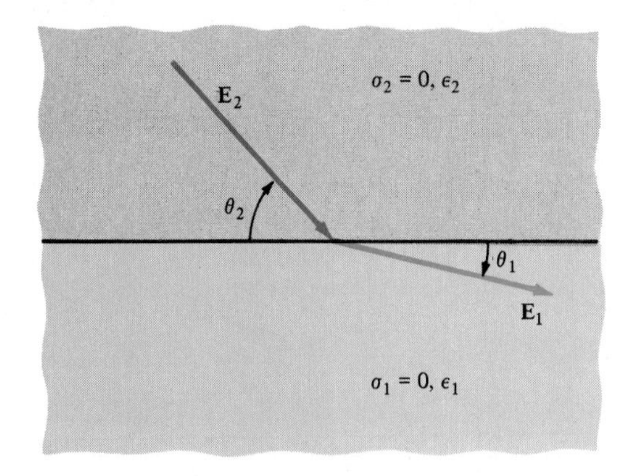

FIGURE 3.19

Example 3.11. The application of the boundary conditions at an interface.

Solution The tangential components $E_2 \cos\theta_2$ and $E_1 \cos\theta_1$ must be equal, from which we obtain

$$\frac{E_2}{E_1} = \frac{\cos\theta_1}{\cos\theta_2}$$

The normal components of $\mathbf{D}_2 = \epsilon_2 \mathbf{E}_2$ and $\mathbf{D}_1 = \epsilon_1 \mathbf{E}_1$ must also be continuous, from which we obtain

$$\epsilon_2 E_2 \sin\theta_2 = \epsilon_1 E_1 \sin\theta_1$$

or

$$\frac{\epsilon_2 E_2}{\epsilon_1 E_1} = \frac{\sin\theta_1}{\sin\theta_2}$$

Combining these two expressions, we obtain

$$\frac{\tan\theta_1}{\tan\theta_2} = \frac{\epsilon_2}{\epsilon_1}$$

Note that the angles θ_1 and θ_2 are measured with respect to the surface. If the angles are measured with respect to a normal to the surface, i.e., $\theta_1' = \pi/2 - \theta_1$, $\theta_2' = \pi/2 - \theta_2$, then

$$\frac{\tan\theta_1'}{\tan\theta_2'} = \frac{\epsilon_1}{\epsilon_2}$$

If a region has finite but nonzero conductivity, the current flow and the electric field in the region are related by Ohm's law, $\mathbf{J} = \sigma\mathbf{E}$. Consequently, we should obtain corresponding boundary conditions on the normal and tangential components of the current at the boundary. Consider medium 1 and medium 2 as conductive regions having finite, nonzero conductivities σ_1 and σ_2, respectively. For static currents, the continuity equation in (76) reduces to $\nabla \cdot \mathbf{J} = 0$. Comparing this to Gauss' law, we obtain

$$J_{n2} = J_{n1} \tag{96a}$$

Since the tangential components of the electric field must again be continuous across the boundary, we have $E_{t2} = E_{t1}$, or

$$\frac{J_{t2}}{\sigma_2} = \frac{J_{t1}}{\sigma_1} \tag{96b}$$

relating the tangential components of the current density at the boundary.

Since the conductivities of the regions are not both zero, we cannot argue that the surface charge density at the interface is zero; thus, (90) applies to the normal components of **D**. Substituting $\mathbf{D} = \epsilon\mathbf{E}$ into (90) and $\mathbf{J} = \sigma\mathbf{E}$ into (96*a*) we obtain

$$\epsilon_2 E_{n2} - \epsilon_1 E_{n1} = \rho_s \tag{97a}$$

$$\sigma_2 E_{n2} = \sigma_1 E_{n1} \tag{97b}$$

or

$$\rho_s = \left(\epsilon_2 - \epsilon_1 \frac{\sigma_2}{\sigma_1}\right) E_{n2} = \left(\epsilon_2 \frac{\sigma_1}{\sigma_2} - \epsilon_1\right) E_{n1} \tag{98}$$

and the surface charge at the boundary can be found from a knowledge of the normal component of the electric field in either region. Note the result in (98) when either $\sigma_1 = 0$ and $\sigma_2 \neq 0$ or $\sigma_2 = 0$ and $\sigma_1 \neq 0$. Also consider the result when either σ_1 or σ_2 is infinite.

It is important to point out that the relation for continuity of the normal components of **J** in (96*a*) was derived for static currents, $\nabla \cdot \mathbf{J} = 0$, as was the equation for ρ_s in (98). If the current density is a function of time, the general result given in (76), $\nabla \cdot \mathbf{J} = -\partial\rho/\partial t$, applies and (96*a*) becomes $J_{n2} - J_{n1} = -\partial\rho_s/\partial t$, so that the result in (98) no longer applies.

EXAMPLE 3.12 Consider the pair of parallel conducting plates of area A and very large extent shown in Fig. 3.20. Suppose a voltage of V_0 is applied between the plates, which are separated a distance d, with air as the intervening medium. The battery is then removed and a dielectric slab ($\epsilon = \epsilon_r \epsilon_0$) of thickness d is inserted between the plates. The airgap between each plate and the dielectric surface is assumed to be of negligible thickness but is shown in exaggerated proportions in the sketches. Determine E, D, and P before and after the insertion of the dielectric; also determine the free surface charge density ρ_{sf} on the plates and the bound surface charge density ρ_{sb} on the surface of the dielectric. By symmetry and the assumption that the plates are very large in extent, the vector fields will be in one direction at all points, as shown in Fig. 3.20. We will assume these vector fields to be directed from the positive plate to the negative plate and will therefore need only deal with their magnitudes. Superscripts b and a will be used on all calculated quantities to symbolize *before* and *after* the insertion of the dielectric, respectively.

Solution Before the insertion of the dielectric, the battery establishes an electric field between the plates (which is essentially normal to the plate surfaces, neglecting the fringing of the field at the plate edges) of

$$E^b = \frac{V_0}{d} \qquad \text{V/m}$$

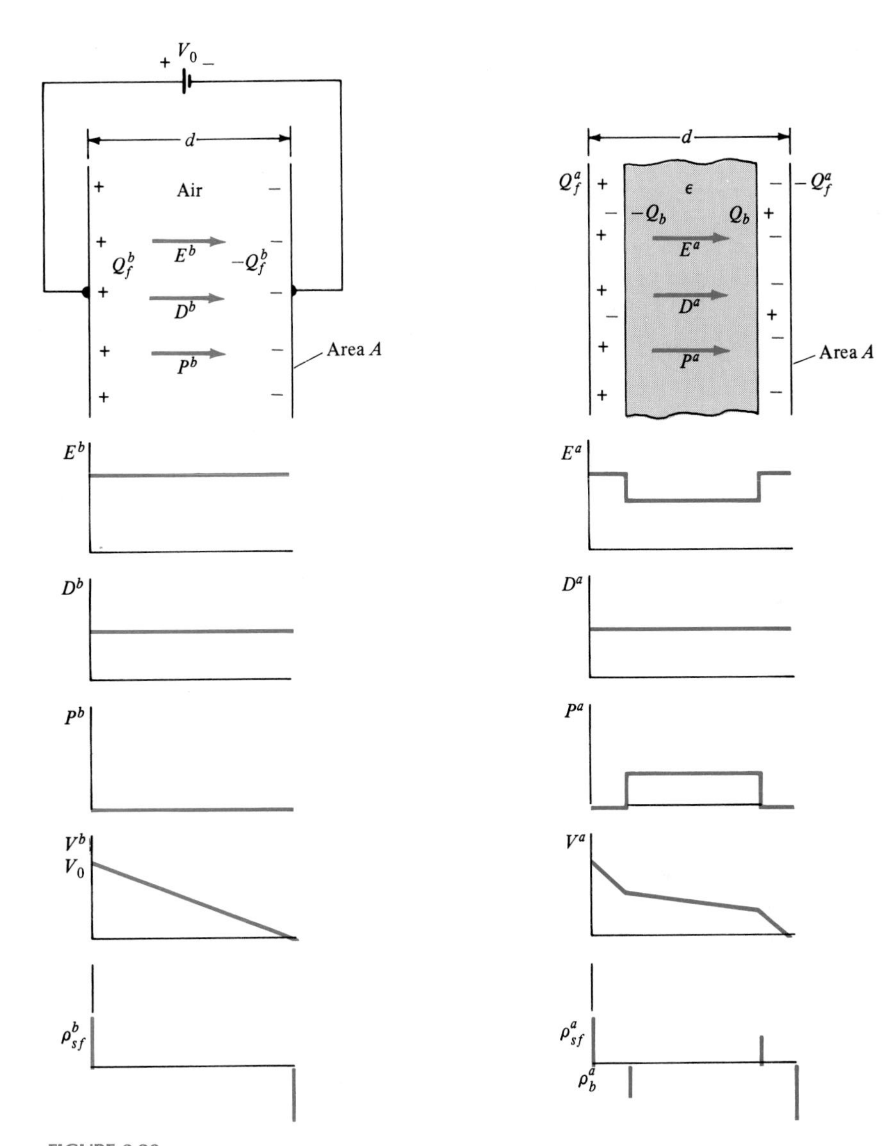

FIGURE 3.20

Example 3.12. The effect of a dielectric.

The electric flux density vector is

$$D^b = \epsilon_0 E^b$$

$$= \frac{\epsilon_0 V_0}{d} \qquad \text{C/m}^2$$

According to the boundary conditions, the free surface charge density on each plate is numerically equal to the component of D that is normal to these (perfectly conducting) plates. On the positive plate this becomes

$$\rho_{sf}^b = D^b$$

$$= \frac{\epsilon_0 V_0}{d} \qquad \text{C/m}^2$$

The total charge on the positive plate having area A is

$$Q_f^b = \rho_{sf}^b A$$

$$= \frac{\epsilon_0 A V_0}{d}$$

Since air has $\epsilon \simeq \epsilon_0$, $P^b = 0$. When the battery is removed and the dielectric inserted, the total free charge on each plate remains unchanged, and so $Q_f^a = Q_f^b$. Thus, the free surface charge density is unchanged. Therefore, the D field between the plates remains unchanged, $D^a = D^b$, since no free charge has been added by the dielectric insertion. In the airgaps the electric field is

$$E_{\text{air}}^a = \frac{D^b}{\epsilon_0}$$

$$= \frac{V_0}{d}$$

and in the dielectric it is

$$E_{\text{dielectric}}^a = \frac{D^b}{\epsilon}$$

$$= \frac{\epsilon_0}{\epsilon} \frac{V_0}{d}$$

$$= \frac{\epsilon_0}{\epsilon} E^b$$

Thus, the electric field in the dielectric is reduced from its value prior to the insertion of the dielectric ($\epsilon_r \geq 1$). In the airgaps we still have

$$P^a_{\text{air}} = 0$$

but in the dielectric we have

$$\begin{aligned} P^a_{\text{dielectric}} &= D^a_{\text{dielectric}} - \epsilon_0 E^a_{\text{dielectric}} \\ &= D^b - \epsilon_0 E^a_{\text{dielectric}} \\ &= \left(1 - \frac{\epsilon_0}{\epsilon}\right)\left(\frac{\epsilon_0 V_0}{d}\right) \\ &= \left(\frac{\epsilon_r - 1}{\epsilon_r}\right)\left(\frac{\epsilon_0 V_0}{d}\right) \end{aligned}$$

For $\epsilon_r \geq 1$, P in the dielectric is in the direction of E and D. The bound surface charge density can be found from the component of P normal to the dielectric surface. On the surface adjacent to the positive conductor, P is directed into the dielectric, whereby

$$\begin{aligned} \rho^a_{sb} &= -P^a_{\text{dielectric}} \\ &= -\left(\frac{\epsilon_r - 1}{\epsilon_r}\right)\left(\frac{\epsilon_0 V_0}{d}\right) \end{aligned}$$

Thus, a negative bound surface charge appears on the surface adjacent to the positive plate. In Sec. 3.9 we will characterize the ability of this structure to store free charge per unit of voltage between the plates as capacitance. Even though the free charge stored on the plates remains the same after the insertion of the dielectric (since the battery was removed), the electric field in the dielectric is reduced. Hence, the potential difference between the plates has been reduced, and the capacitance is increased by the addition of the dielectric (as it turns out, by a factor of ϵ_r). The results of the example are plotted in Fig. 3.20.

3.8 Energy Density in the Electrostatic Field

Consider the case of two point charges Q_1 and Q_2 held fixed in space in close proximity to each other. Clearly, some energy is being stored in the resulting field since the two charges tend to move with respect to each other because of the coulomb force. It is therefore of interest to examine the energy stored in the field (much in the same way as potential energy is stored when we lift an object). Clearly, we cannot "pin down" the exact location of the stored electric energy any more than we can in the case of gravitational potential energy; we simply observe that it exists and is a property of the force field. We will find it natural to speak of an energy density in the field, as the following development shows.

Consider three point charges Q_1, Q_2, and Q_3 that are initially stored at infinity. We wish to find the energy required to move these charges from infinity to some final positions. (For the purposes of this calculation, we assume that the region is initially devoid of charge.) Bringing the first charge Q_1 from infinity to a point P_1 requires no energy expenditure since we assume that the region initially has no charge and hence that no force is exerted on Q_1. Bringing the second charge Q_2 from infinity to a point P_2 in the vicinity of Q_1 (held in a fixed position) requires an energy expenditure of

$$W_{21} = Q_2 \frac{Q_1}{4\pi\epsilon_0 R_{21}} = Q_2 V_{21} \tag{99}$$

where R_{21} is the distance between the final positions of Q_1 and Q_2 and where V_{21} is the absolute potential of point P_2 due to Q_1. Now, with Q_1 and Q_2 fixed, the energy required to move Q_3 to some point P_3 is

$$W_{31} + W_{32} = Q_3 V_{31} + Q_3 V_{32} \tag{100}$$

where

$$W_{ij} = Q_i \frac{Q_j}{4\pi\epsilon_0 R_{ij}} = Q_i V_{ij} \tag{101}$$

The total energy required to assemble these charges is

$$W_e = W_{21} + W_{31} + W_{32} = Q_2 V_{21} + Q_3 V_{31} + Q_3 V_{32} \tag{102}$$

Note, however, that $W_{ij} = W_{ji}$ in (101); the energy required to move Q_i to P_i in the presence of Q_j is the same as the energy required to move Q_j to P_j in the presence of Q_i. Therefore, the expenditure of energy given in (102) could have been written as

$$W_e = W_{12} + W_{13} + W_{23} = Q_1 V_{12} + Q_1 V_{13} + Q_2 V_{23} \tag{103}$$

Adding (102) to (103) and collecting like terms, we obtain

$$2W_e = Q_1(V_{12} + V_{13}) + Q_2(V_{21} + V_{23}) + Q_3(V_{31} + V_{32}) \tag{104}$$

Note, for example, that $V_2 = V_{21} + V_{23}$ is the absolute potential of P_2 (in the absence of Q_2 at that point) due to the absolute potentials established by Q_1 and Q_3 individually. Therefore,

$$V_i = \sum_{\substack{j=1 \\ i \neq j}}^{3} V_{ij} \tag{105}$$

and (104) may be written as

$$W_e = \frac{1}{2} \sum_{i=1}^{3} Q_i V_i \tag{106}$$

Extending this result to the assembly of N point charges or, more generally, to the assembly of some volume distribution of charge, ρ_v, we obtain†

$$W_e = \frac{1}{2} \int_v \rho_v V \, dv \tag{107}$$

Equation (107) can be written in an alternative form. Gauss' law in point form, $\nabla \cdot \mathbf{D} = \rho_v$, can be substituted into (107) to yield

$$W_e = \frac{1}{2} \int_v (\nabla \cdot \mathbf{D}) V \, dv \tag{108}$$

To simplify this integral, we use the identity (see Appendix A)

$$(\nabla \cdot \mathbf{D}) V = \nabla \cdot (V\mathbf{D}) - \mathbf{D} \cdot (\nabla V) \tag{109}$$

Substituting the vector identity, we obtain

$$W_e = \frac{1}{2} \int_v \nabla \cdot (V\mathbf{D}) dv - \frac{1}{2} \int_v \mathbf{D} \cdot (\nabla V) dv \tag{110}$$

Applying the divergence theorem to the first of these volume integrals yields

$$W_e = \frac{1}{2} \oint_s V\mathbf{D} \cdot d\mathbf{s} - \frac{1}{2} \int_v \mathbf{D} \cdot (\nabla V) dv \tag{111}$$

If we allow the volume v in (107) to include all space, the surface s in (111) goes to infinity. The absolute potential V decays as $1/r$, and $\mathbf{D}$ decays as $1/r^2$. The surface area, however, increases as r^2, so that the first integral in (111) goes to zero as s tends to infinity. Substituting $\mathbf{E} = -\nabla V$ in the remaining integral yields‡

$$W_e = \frac{1}{2} \int_v \mathbf{D} \cdot \mathbf{E} \, dv \tag{112}$$

† Equation (106) does not include the self-energy of the charges, that is, the energy required to assemble each charge. Only the interaction energy between the charges is included in (106). For a more detailed discussion, see M. Zahn, *Electromagnetic Field Theory*, Wiley, New York, 1979, sec. 3-8.

‡ Note that volume v in (112) includes all space.

The result in (112) indicates that the quantity $w_e = \frac{1}{2}\mathbf{D}\cdot\mathbf{E}$ is a volume energy density, or density of the energy stored in the field produced by the charge, since the units of $\mathbf{D}\cdot\mathbf{E}$ are joules per cubic meter (J/m^3). It must be emphasized that the location of the energy stored in the field cannot be determined with any more confidence than we could locate the potential energy stored in a gravitational field. Equation (112) simply states that if we integrate the quantity $w_e = \frac{1}{2}\mathbf{D}\cdot\mathbf{E}$ (which has dimensions of energy density) over all space, we will obtain the total energy stored in the field. For a linear, homogenous, and isotropic medium, $\mathbf{D} = \epsilon\mathbf{E}$, and (112) becomes

$$W_e = \frac{1}{2}\int_v \epsilon|\mathbf{E}|^2\,dv \tag{113}$$

EXAMPLE 3.13 A coaxial cable is shown in Fig. 3.21. A conductor of radius a is centered on the axis of a cylindrical shell of interior radius b. The structure is assumed to be infinite in length. A charge per unit of line length of q is deposited on the inner conductor, and $-q$ is on the inner surface of the outer conductor. Determine the energy stored in the field per unit of line length.

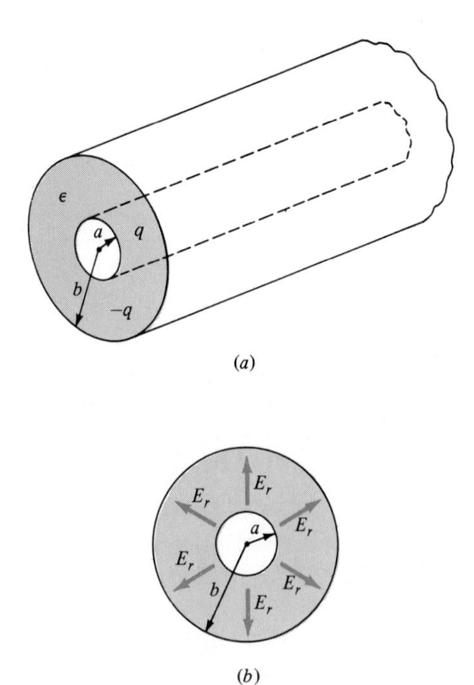

FIGURE 3.21

Example 3.13. A coaxial cable.

Solution By symmetry, the electric field will be directed in the radial direction, as shown in Fig. 3.21*b*. Constructing a gaussian surface in the shape of a cylinder of radius r and 1-m length about the inner cylinder, we obtain

$$\oint_s \mathbf{D} \cdot \mathbf{ds} = q$$

or

$$\epsilon E_r 2\pi r = q$$

(since the field is radial and therefore normal to the sides of the gaussian cylinder). From this result we obtain the radial electric field as

$$E_r = \frac{q}{2\pi\epsilon r}$$

Integrating (113) throughout a volume between $r = a$ and $r = b$ over a 1-m length of line, we obtain the energy stored per unit length:

$$\begin{aligned} W_e &= \frac{1}{2}\epsilon \int_{z=0}^{1} \int_{r=a}^{b} \int_{\phi=0}^{2\pi} |E_r|^2\, r\, dr\, d\phi\, dz \\ &= \frac{q^2}{8\pi^2\epsilon} \int_{r=a}^{b} \int_{\phi=0}^{2\pi} \frac{1}{r}\, dr\, d\phi \\ &= \frac{q^2}{4\pi\epsilon} \ln \frac{b}{a} \qquad \text{J/m} \end{aligned}$$

3.9 Capacitance

Consider the pair of parallel, perfectly conducting plates shown in Fig. 3.22. The plates have surface area A and are separated by a distance d in free space. A battery maintains a potential difference V between the two plates. Before the battery is connected, we presume that the plates are electrically neutral. When the battery is connected, charge is transferred to the plates of the capacitor, resulting in $+Q$ on one plate and $-Q$ on the other plate. We assume that the area of the plates, A, is much larger than the plate separation d, so that the fringing at the edges can be neglected. Thus, we assume that the lines of **E** are perpendicular to the plate surfaces. In addition, this assumption permits the total charge Q on each plate to be uniformly distributed over the surface of that plate (certainly an approximation whose validity is examined in Chap. 10).

If we construct a rectangular gaussian surface around one plate (and neglect fringing), we conclude from Gauss' law that the magnitude of the electric field is given by $E = Q/\epsilon_0 A$ and that the potential difference between the two plates is therefore $V = Qd/\epsilon_0 A$. Thus, we may define a capacity to store charge or

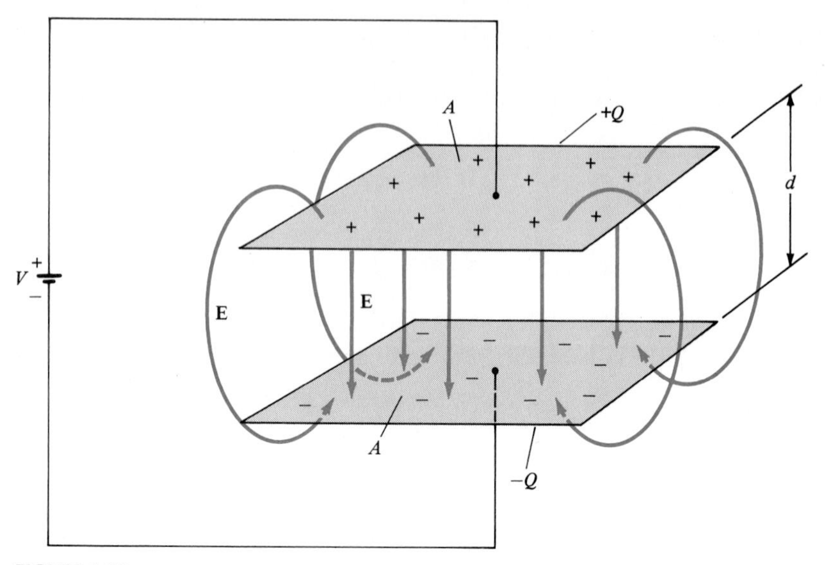

FIGURE 3.22
A parallel-plate capacitor.

capacitance of this structure to be the ratio of the charge stored to the potential difference between the two conductors; that is

$$C = \frac{Q}{V} = \frac{\epsilon_0 A}{d} \quad \text{F} \tag{114}$$

The units of capacitance are coulombs per volt. This ratio of units will be given the special name of the farad (F), where one farad is one coulomb per volt (F = C/V).

Suppose we now insert a block of dielectric having permittivity ϵ between the plates, so that it completely fills the space. The resulting polarization of the dielectric will result in an increase of free charge on the plate surface to counter the induced, bound charge on the surfaces of the dielectric (see Fig. 3.17). In this case, we must be careful to state which type of charge is being considered in (114). Since C is to be the capacity of the structure to store charge (extracted from the battery), then clearly the charge Q in (114) must be free charge:

$$C = \frac{Q_f}{V} \tag{115}$$

Again applying Gauss' law to this problem, we determine that between the plates $D = \epsilon E = Q_f/A$, so that $E = Q_f/\epsilon A$. The potential difference between the plates is again (neglecting fringing) $V = Ed$, so that (115) becomes

$$C = \frac{\epsilon A}{d} \tag{116}$$

Although stated in connection with a parallel-plate capacitance, (115) is the general definition of capacitance of any other structure consisting of two conducting bodies. For example, consider Fig. 3.23*a*, in which two conducting bodies (assumed to be perfect conductors) of general shape are immersed in a dielectric (linear, homogeneous, isotropic) having permittivity ϵ. Applying a voltage V between the two bodies results in a charge separation; one conductor

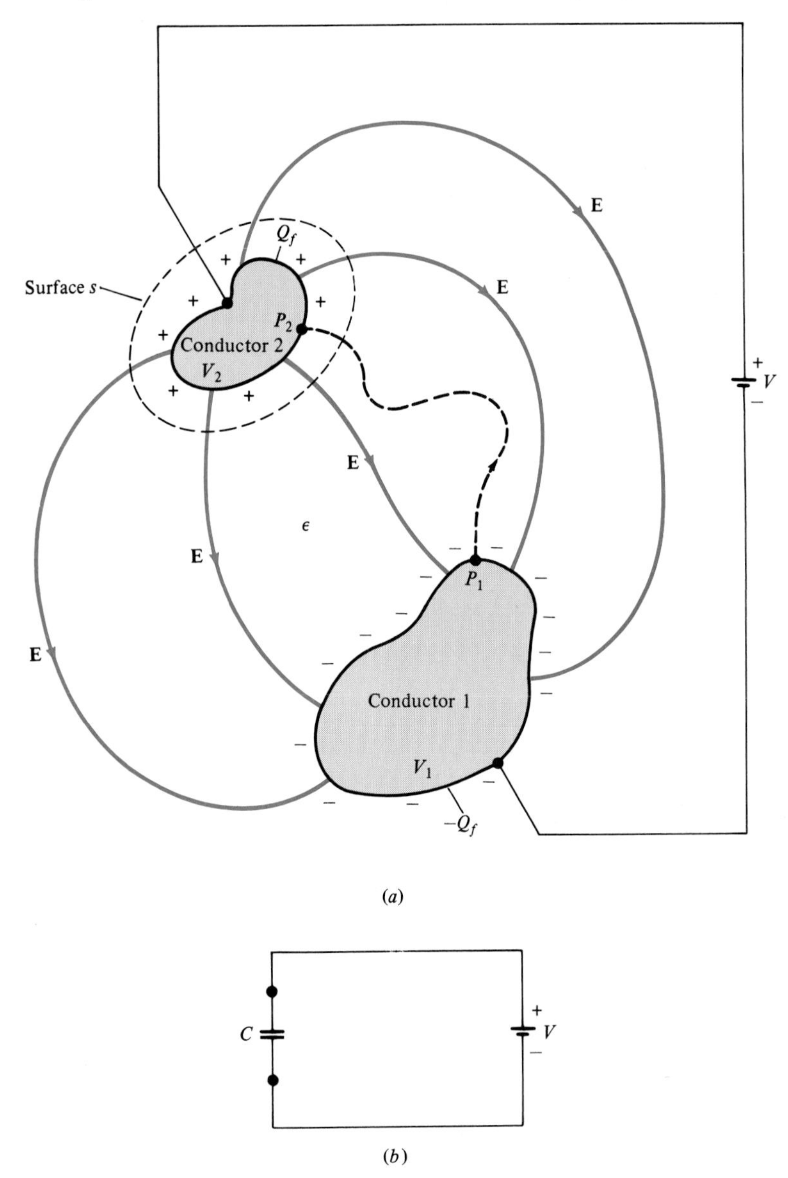

FIGURE 3.23

Capacitance of a general two-conductor structure. (*a*) Field model and (*b*) lumped-circuit model.

bears a total free charge of $+Q_f$ on its surface, and the other conductor bears a total free charge of $-Q_f$ on its surface. The electric field lines, and consequently the electric flux density lines, terminate normal to the surface of the bodies. The total charge on the surface of the positively charged body is found from Gauss' law and the boundary condition in (94) as

$$Q_f = \oint_s \mathbf{D} \cdot d\mathbf{s} \tag{117}$$

where s is the surface just off the surface of the positively charged conductor. The potential difference between the two conductors, V, established by the battery is obtained by integrating the electric field along a contour c between two points P_1 and P_2 on the surfaces (equipotential surfaces) of the conductors as

$$V = -\int_{P_1}^{P_2} \mathbf{E} \cdot d\mathbf{l} \tag{118}$$

Substituting (117) and (118) into (115), we obtain a general expression for the capacitance of this structure:

$$C = \frac{Q_f}{V} = \frac{\oint_s \mathbf{D} \cdot d\mathbf{s}}{-\int_{P_1}^{P_2} \mathbf{E} \cdot d\mathbf{l}} = -\epsilon \frac{\oint_s \mathbf{E} \cdot d\mathbf{s}}{\int_{P_1}^{P_2} \mathbf{E} \cdot d\mathbf{l}} \tag{119}$$

Since an electric field is established in the region surrounding the two bodies in Fig. 3.23*a*, it is clear from the previous section that energy is stored in this field, and the total energy is given by (113). In order to evaluate (113), we must integrate the resulting electric field throughout the entire volume surrounding the two bodies. A more convenient method of evaluating the energy stored in the field that avoids this integration is as follows: Let us assume that body 1 and body 2 are at absolute potentials V_1 and V_2, respectively, so that $V = V_2 - V_1$. The energy stored in the field can be obtained by evaluating (107) over the surface of each body. For example, the energy stored in the field due to the charge on body 1 at potential V_1 is

$$W_1 = \tfrac{1}{2} \int_{s_1} \rho_{s1} V_1 \, ds \tag{120}$$

where s_1 is the surface of body 1 and ρ_{s1} is the surface charge distribution over that body. Similarly, for body 2 we obtain

$$W_2 = \tfrac{1}{2} \int_{s_2} \rho_{s2} V_2 \, ds \tag{121}$$

Note that both bodies are equipotential surfaces; thus, V_1 and V_2 are independent of position on these surfaces and may consequently be removed from the integrals in (120) and (121). Thus, we obtain

$$W_1 = \tfrac{1}{2}V_1 \int_{S_1} \rho_{s1}\, ds = \tfrac{1}{2}V_1(-Q_f) \tag{122}$$

and

$$W_2 = \tfrac{1}{2}V_2(Q_f) \tag{123}$$

The total energy stored in the field is the sum of (122) and (123):

$$W_e = \tfrac{1}{2}Q_f(V_2 - V_1) = \tfrac{1}{2}Q_f V \tag{124}$$

Thus knowing the charge stored on one body and the potential difference between the two bodies, we may immediately evaluate the stored energy. This bypasses the involved integral given in (113). A further simplification can be obtained if we substitute the capacitance relation given in (115) into (124), resulting in

$$W_e = \tfrac{1}{2}(CV)V = \tfrac{1}{2}CV^2 \tag{125}$$

or

$$W_e = \frac{1}{2}\frac{Q_f^2}{C} \tag{126}$$

Therefore, we need only to know any two of the three quantities—capacitance (C), stored free charge (Q_f), and established potential difference (V)—in order to compute the energy stored in the electric field of a capacitor.

EXAMPLE 3.14 Consider two cylindrical conductors that are infinite in length, as shown in Fig. 3.24*a*. Determine the capacitance per unit of length of this structure.

Solution Each conductor is a perfect conductor and carries a charge distribution of ρ_l C/m: positive on one conductor and negative on the other. Note that ρ_l is the distribution of charge in the direction of the cylinder axes. Thus, the total charge contained in a 1-m length of one cylinder is ρ_l, and of the other cylinder is $-\rho_l$. Since the cylinders are infinite in length, the distribution of charge is uniform along the cylinder axes. However, the distribution is not uniform around the periphery of each cylinder, as shown in the cross-sectional view of Fig. 3.24*b*. Clearly, the closer the cylinders, the more nonuniform the charge distribution around the peripheries of the cylinders. This is called *proximity effect*. If, however, we assume that the separation between the cylinders, d, is much larger than the radius of each cylinder, r^+ and r^-, then it is reasonable to assume a uniform distribution of charge around the cylinder

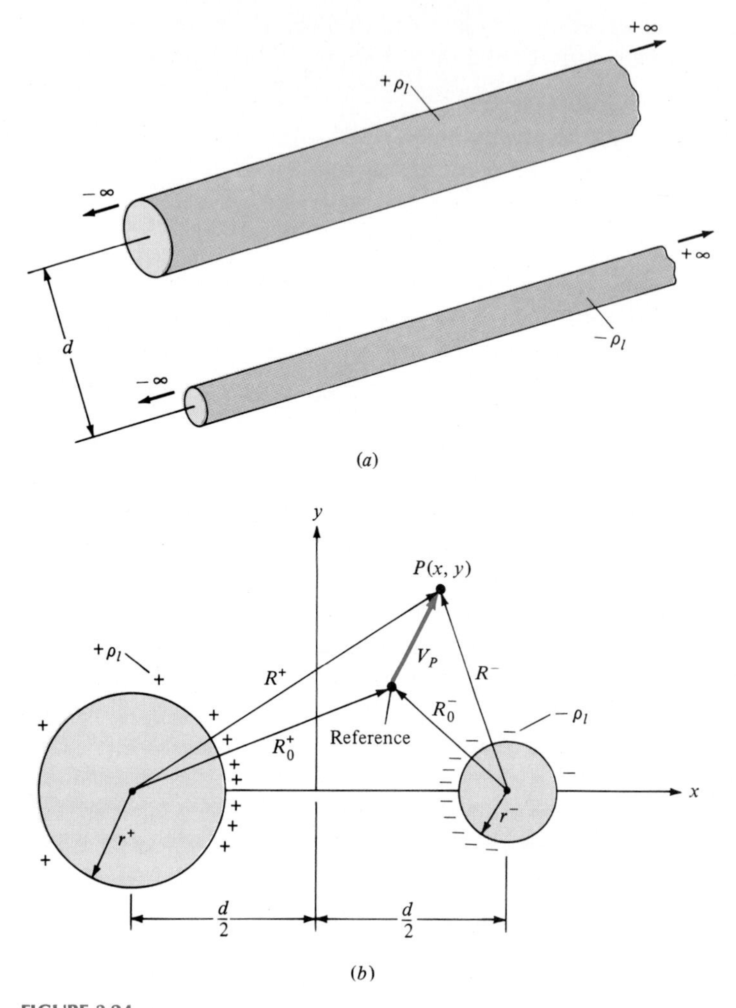

FIGURE 3.24

Example 3.14. A two-wire transmission line.

peripheries. This turns out to be a reasonable assumption if the smallest ratio of d/r^+ and d/r^- is greater than approximately 10. The exact result, independent of proximity effect, will be derived in Chap. 10. In this case, we may define uniform surface charge distributions on each cylinder of $\rho_l/2\pi r^+$ and $-\rho_l/2\pi r^-$. If we assume that the cylinders are sufficiently separated so that the charge distribution is uniform around the conductor peripheries, then we may compute the potential at a point P using previously obtained results. We may superimpose the potentials due to each charge distribution. The potential of point P with respect to the reference point shown in Fig. 3.24b is (see Example 3.9)

$$V_P = \frac{\rho_l}{2\pi\epsilon_0} \ln \frac{R_0^+}{R^+} - \frac{\rho_l}{2\pi\epsilon_0} \ln \frac{R_0^-}{R^-} = \frac{\rho_l}{2\pi\epsilon_0} \ln \frac{R_0^+ R^-}{R^+ R_0^-}$$

The potential difference between the surfaces of the cylinder may be found by moving P to the surface of the positively charged conductor and moving the reference point to the surface of the negatively charged conductor. Since separation between the cylinders was assumed to be much greater than the radii, we may use the approximations $R^+ = r^+$, $R_0^+ = d$, $R^- = d$, and $R_0^- = r^-$. Thus, we obtain

$$V \simeq \frac{\rho_l}{2\pi\epsilon_0} \ln \frac{d^2}{r^+ r^-}$$

In order to determine the capacitance of this structure, we must determine the free charge stored on the conductors. However, we note that the total free charge on each cylinder is infinite since the cylinders are infinite in length. However, the total free charge per unit of cylinder length may be readily determined. For the cylinder bearing the positive charge distribution, the total charge contained on a 1-m length is ρ_l; therefore, the net free charge stored on a 1-m line is $q_f = \rho_l$. Thus, the per-unit-length capacitance (per-unit-length quantities will be denoted by lowercase symbols) becomes

$$c = \frac{q_f}{V} = \frac{2\pi\epsilon_0}{\ln (d^2/r^+ r^-)} \qquad \text{F/m}$$

The particular type of structure in Fig. 3.24 is often referred to as a *transmission line*. We will find in Chap. 7 that the two conductors serve to guide the propagation of energy. A common example is the pair of wires (cylindrical conductors) connecting a home television set to its antenna. In most cases of two-wire transmission lines, the wires will be identical. Substituting $r^+ = r^- = r$ yields the per-unit-length capacitance of this structure:

$$c = \frac{\pi\epsilon_0}{\ln (d/r)} \qquad \text{F/m}$$

EXAMPLE 3.15 Consider the coaxial cable shown in Fig. 3.21. Using the relation between capacitance, voltage, and stored energy in (125) and the results of Example 3.13, determine the capacitance per unit length.

Solution The stored energy per unit of line length was found from Example 3.13 to be

$$W_e = \frac{q^2}{4\pi\epsilon} \ln \frac{b}{a} \qquad \text{J/m}$$

and the radial electric field was determined to be

$$E_r = \frac{q}{2\pi\epsilon r} \qquad \text{V/m}$$

The voltage between the conductors may then be computed as

$$V = -\int_b^a \mathbf{E} \cdot d\mathbf{l}$$

$$= -\int_b^a \frac{q}{2\pi\epsilon r}\, dr$$

$$= \frac{q}{2\pi\epsilon} \ln \frac{b}{a}$$

Thus, the capacitance per unit length is found from (125) to be

$$c = \frac{2W_e}{V^2}$$

$$= \frac{2\pi\epsilon}{\ln (b/a)} \qquad \text{F/m}$$

This can also be found in a more direct fashion as

$$c = \frac{q}{V}$$

$$= \frac{2\pi\epsilon}{\ln (b/a)} \qquad \text{F/m}$$

3.10 Resistance

In the previous section we discussed the definition and calculation of capacitance between two conductors. In this section we present a related topic—resistance. Consider two perfect conductors of arbitrary shape situated in a lossy medium, as shown in Fig. 3.25*a*. The medium in which these two conductors are immersed is characterized by permittivity ϵ and conductivity σ.

Applying a voltage V between the two conductors will cause a current I to flow in the surrounding medium. An electric field $\mathbf{E}$ will be established by the applied voltage in the region between the conductors. Because of the nonzero conductivity of this region, a current density $\mathbf{J} = \sigma\mathbf{E}$ will exist in the region. The current I can be obtained by integrating $\mathbf{J}$ over the surface of one of the conductors:

$$I = \oint_s \mathbf{J} \cdot d\mathbf{s} \tag{127}$$

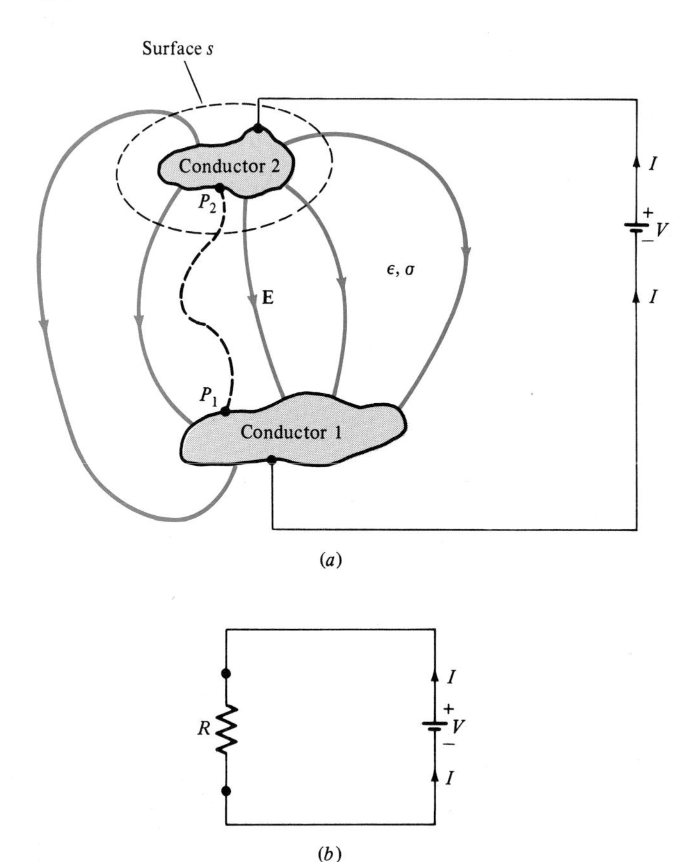

FIGURE 3.25

Resistance of a general two-conductor structure. (*a*) Field model and (*b*) lumped-circuit model.

where the normal for $d\mathbf{s}$ is directed outward for conductor 2 and inward if s surrounds conductor 1. From Eq. (37), the voltage V may be defined as

$$V = -\int_{P_1}^{P_2} \mathbf{E} \cdot d\mathbf{l} \tag{128}$$

where $d\mathbf{l}$ is a path between points on the two conductors. The resistance between the two conductors presented by this lossy medium is then defined as

$$R = \frac{V}{I} = -\frac{\int_{P_1}^{P_2} \mathbf{E} \cdot d\mathbf{l}}{\oint_s \mathbf{J} \cdot d\mathbf{s}} = \frac{1}{\sigma} \frac{\int_{P_2}^{P_2} \mathbf{E} \cdot d\mathbf{l}}{\oint \mathbf{E} \cdot d\mathbf{s}} \tag{129}$$

and represented as in Fig. 3.25*b*.

Substituting $\mathbf{J} = \sigma\mathbf{E}$ into (129) and comparing the result to (119), we obtain an interesting relationship:

$$RC = \frac{\epsilon}{\sigma} \tag{130}$$

In terms of conductance $G = 1/R$, we have

$$\frac{G}{C} = \frac{\sigma}{\epsilon} \tag{131}$$

Therefore, the resistance and capacitance associated with two conducting bodies are related, and one can be found from a knowledge of the other.

EXAMPLE 3.16 Consider the coaxial cable shown in Fig. 3.21. Assume the medium between the perfectly conducting cylinders to have conductivity σ and permittivity ϵ. Determine the conductance per unit of cable length.

Solution The voltage between the two cylinders was obtained in Example 3.15 in terms of the charge per unit length on each cylinder q as

$$V = \frac{q}{2\pi\epsilon} \ln \frac{b}{a} \quad \text{V}$$

The electric field is, by symmetry, radial and is given by

$$E_r = \frac{q}{2\pi\epsilon r}$$

where $a < r < b$. The current density is therefore radial and given by

$$\begin{aligned} J_r &= \sigma E_r \\ &= \frac{\sigma q}{2\pi\epsilon r} \quad \text{A/m}^2 \end{aligned}$$

Integrating this over a surface of a cylinder of radius $a < R < b$ and unit length gives

$$\begin{aligned} I &= \int_{\phi=0}^{2\pi} J_r R \, d\phi \\ &= \frac{\sigma q}{\epsilon} \quad \text{A/m} \end{aligned}$$

Hence, the conductance per unit of cable length is

$$g = \frac{I}{V}$$

$$= \frac{2\pi\sigma}{\ln(b/a)} \qquad \text{S/m}$$

The capacitance per unit length was determined in Example 3.15 to be

$$c = \frac{2\pi\epsilon}{\ln(b/a)}$$

Note that

$$\frac{g}{c} = \frac{\sigma}{\epsilon}$$

as expected.

3.11 Power Dissipation

The presence of an electric field in a medium having nonzero conductivity causes the movement of free charge in that region, giving rise to a current density of

$$\mathbf{J} = \sum_i n_i q_i \mathbf{u}_i \tag{132}$$

where n_i is the number of charges of one type (electrons, ions, etc.), q_i, per unit volume and $\mathbf{u}_i$ is the vector velocity of these charges. The movement of a charge q over a distance $\Delta\mathbf{l}$ requires an expenditure of energy by the electric field of $\Delta w = q\mathbf{E}\cdot\Delta\mathbf{l}$. The power required is

$$p = \lim_{\Delta t \to 0} \frac{\Delta w}{\Delta t}$$

$$= \lim_{\Delta t \to 0} q\mathbf{E}\cdot\frac{\Delta\mathbf{l}}{\Delta t}$$

$$= q\mathbf{E}\cdot\mathbf{u} \tag{133}$$

The total power required in moving all charged carriers in an infinitesimal volume dv is

$$dP = \sum p_i$$

$$= \mathbf{E}\cdot\left(\sum n_i q_i \mathbf{u}_i\right) dv$$

$$= \mathbf{E}\cdot\mathbf{J}\, dv \tag{134}$$

Thus, a power density per unit volume is defined as

$$\frac{dP}{dv} = \mathbf{E} \cdot \mathbf{J} \qquad \text{W/m}^3 \tag{135}$$

The total power delivered to the charges is therefore

$$P = \int_v \mathbf{E} \cdot \mathbf{J} \, dv \qquad \text{W} \tag{136}$$

This relation is known as *Joule's law.*

Consider a conductor of uniform cross-sectional area s and length l. Assuming that the electric field and current density are, at all points, directed along the conductor length, Eq. (136) may be written as

$$P = \int_L E \, dl \int_S J \, ds$$

$$= VI \qquad \text{W} \tag{137}$$

If the resistance of that conductor is computed according to the method of the previous section, we obtain

$$P = \frac{V^2}{R}$$

$$= I^2 R \qquad \text{W} \tag{138}$$

This, of course, conforms to the usual lumped-circuit expression for power dissipated in a resistor.

EXAMPLE 3.17 Determine the power dissipated in a coaxial cable per unit length of the cable.

Solution The electric field and current density for the coaxial cable shown in Fig. 3.21 are radial and are given by

$$E_r = \frac{q}{2\pi\epsilon r}$$

$$J_r = \frac{\sigma q}{2\pi\epsilon r}$$

Integrating these expressions according to (136) yields

$$p = \int_{r=a}^{b} \int_{\phi=0}^{2\pi} \int_{z=0}^{1} E_r J_r r \, dz \, d\phi \, dr$$

$$= \frac{\sigma q^2}{2\pi\epsilon^2} \ln \frac{b}{a}$$

$$= \frac{2\pi\sigma}{\ln(b/a)} V^2$$

$$= gV^2 \qquad \text{W/m}$$

3.12 Poisson's and Laplace's Equations

In previous sections we have examined methods for determining the electric field and electrostatic potential function, given the charge distribution producing these quantities. We now briefly examine alternative techniques for determining these quantities when the charge distribution is not known. Such problems belong to the class of boundary-value problems since the known quantity is the specification of the potential (or some form of the potential) over the boundaries of some region.

We have seen that the electric field and the electrostatic potential function $V(x, y, z)$ in a region are related by

$$\mathbf{E}(x, y, z) = -\nabla V(x, y, z) \tag{139}$$

If we take the divergence of both sides of (139) we obtain

$$\nabla \cdot \mathbf{E} = -\nabla \cdot \nabla V \tag{140}$$

In terms of the electric flux density vector for an isotropic medium, $\mathbf{D} = \epsilon \mathbf{E}$, (140) becomes

$$\nabla \cdot \left(\frac{1}{\epsilon} \mathbf{D} \right) = -\nabla \cdot \nabla V \tag{141}$$

If the region is homogeneous, the permittivity may be removed from the left-hand side of (141), resulting in

$$\frac{1}{\epsilon} \nabla \cdot D = -\nabla \cdot \nabla V \tag{142}$$

Substituting Gauss' law into (142) we obtain Poisson's equation:

$$\nabla \cdot \nabla V = -\frac{\rho_v}{\epsilon} \tag{143}$$

The operator $\nabla \cdot \nabla$ will be given the special notation of ∇^2, which is the laplacian (see Sec. 2.14), so that (143) becomes

$$\nabla^2 V = -\frac{\rho_v}{\epsilon} \tag{144}$$

and

$$\nabla^2 V = \nabla \cdot \nabla V = \frac{\partial^2 V}{\partial x^2} + \frac{\partial^2 V}{\partial y^2} + \frac{\partial^2 V}{\partial z^2} \tag{145}$$

If the region contains no free charge, i.e., $\rho_v = 0$, then Poisson's equation becomes Laplace's equation:

$$\nabla^2 V = 0 \qquad \rho_v = 0 \tag{146}$$

The solution to Poisson's and Laplace's equations for the electrostatic potential function $V(x, y, z)$ in some region, given V over the boundary of that region, will be considered in detail in Chap. 10.

EXAMPLE 3.18 As an example of the application of Laplace's equation, consider the coaxial cable in Fig. 3.21. If the inner conductor is at a potential of V_0 and the outer conductor at zero potential (or, equivalently, the potential difference between the conductors is V_0), use Laplace's equation to determine the electric field between the conductors. Also, determine the capacitance per unit length of the cable.

Solution Clearly, use of cylindrical coordinates is preferred, with the z axis aligned with the axis of the cable. From Appendix A, Laplace's equation in cylindrical coordinates is

$$\begin{aligned} \nabla^2 V &= \frac{1}{r}\frac{\partial}{\partial r}\left(r\frac{\partial V}{\partial r}\right) + \frac{1}{r^2}\frac{\partial^2 V}{\partial \phi^2} + \frac{\partial^2 V}{\partial z^2} \\ &= 0 \end{aligned}$$

For this problem, symmetry shows that V is independent of z and ϕ and can depend only on r. Thus

$$\begin{aligned} \nabla^2 V &= \frac{1}{r}\frac{\partial}{\partial r}\left(r\frac{\partial V}{\partial r}\right) = \frac{1}{r}\frac{\partial V}{\partial r} + \frac{\partial^2 V}{\partial r^2} \\ &= 0 \end{aligned}$$

A solution to this differential equation can be determined as

$$V = K_1 \ln r + K_2$$

where K_1 and K_2 are constants to be determined. At $r = b$, $V = 0$, so that

$$0 = K_1 \ln b + K_2$$

At $r = a$, $V = V_0$, so that

$$V_0 = K_1 \ln a + K_2$$

Solving these two equations for K_1 and K_2 yields

$$K_2 = -K_1 \ln b$$

$$K_1 = \frac{V_0}{\ln (a/b)}$$

so that

$$V = \frac{V_0 \ln r}{\ln (a/b)} - \frac{V_0 \ln b}{\ln (a/b)}$$

$$= \frac{V_0}{\ln (a/b)} \ln \frac{r}{b}$$

The per-unit-length capacitance can be found by determining the charge per unit length in terms of the voltage. The electric field is

$$\mathbf{E} = -\nabla V$$

and from Appendix A we obtain

$$\mathbf{E} = -\frac{\partial V}{\partial r}\mathbf{a}_r - \frac{1}{r}\frac{\partial V}{\partial \phi}\mathbf{a}_\phi - \frac{\partial V}{\partial z}\mathbf{a}_z$$

$$= \frac{V_0}{r \ln (b/a)}\mathbf{a}_r$$

since $b > a$. The free surface charge density on the surface of the interior conductor is numerically equal to the normal component of **D** at the surface ($r = a$):

$$\rho_f = \epsilon E\bigg|_{r=a}$$

$$= \frac{\epsilon V_0}{a \ln (b/a)} \qquad \text{C/m}^2$$

The total charge per unit length on the inner conductor is

$$q = \rho_f 2\pi a \qquad \text{C/m}$$

since, by symmetry, the charge is uniformly distributed over the surface of the conductor. Therefore, the capacitance per unit length is

$$c = \frac{q}{V}$$

$$= \frac{\epsilon V_0/[a \ln (b/a)]2\pi a}{V_0}$$

$$= \frac{2\pi\epsilon}{\ln (b/a)} \qquad \text{F/m}$$

which is a result derived in Example 3.15.

3.13 Mechanical Force of Electrostatic Origin

The fact that a mechanical force is developed in an electrostatic system (such as a parallel-plate capacitor) when the electrostatic field is disturbed is utilized in electrostatic transducers (such as an electrostatic microphone). The force of electrostatic origin can be evaluated from the principle of energy conservation, which is schematically represented in Fig. 3.26, and is stated as

Input electric energy	+	input mechanical energy	=	increase in stored energy	+	energy dissipated as heat

or

Input electric energy	=	mechanical work done	+	increase in stored energy	+	energy dissipated as heat

For a conservative (or lossless) system, the above may also be written as

Input electric energy	=	mechanical work done	+	increase in stored energy	(147)

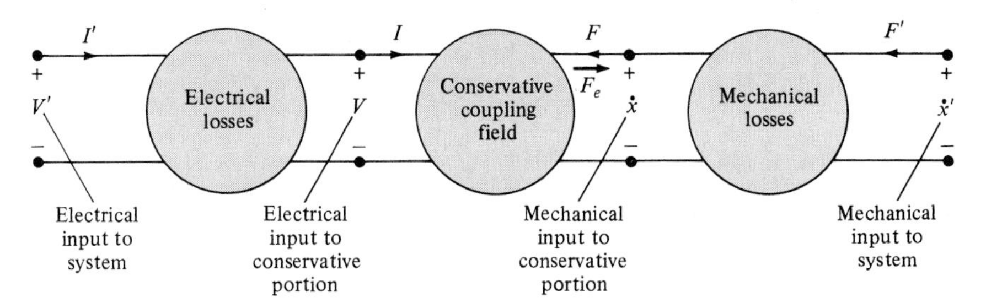

FIGURE 3.26

A representation of an electromechanical system in terms of energy.

Thus, in a conservative electrostatic system, the input electric energy divides into a portion utilized doing mechanical work and a portion resulting in increasing the energy stored in the electric field. If we refer to Fig. 3.26, (147) implies that

$$VI\,dt = -F\,dx + dW_e \tag{148a}$$

or

$$F\,dx + VI\,dt = dW_e \tag{148b}$$

where $F\ dx$ = mechanical energy input, $VI\ dt$ = electric energy input, and dW_e = increase in stored energy in the electric field. Now if F_e is the force of electrostatic origin and acts against F (Fig. 3.26), (148) may be rewritten as

$$F_e\,dx = -dW_e + VI\,dt \tag{149}$$

Substituting $I = dQ/dt$ in (149) gives

$$F_e\,dx = -dW_e + V\,dQ \tag{150}$$

In an electrostatic system we have the following functional relationships:

$$W_e = W_e(V, x) \tag{151}$$

$$Q = Q(V, x) \tag{152}$$

which simply means that the energy stored W_e and the electric charge Q depend on the voltage V and mechanical position x. Now (151) and (152) yield

$$-dW_e = -\frac{\partial W_e}{\partial x}\,dx - \frac{\partial W_e}{\partial V}\,dV \tag{153}$$

and

$$dQ = \frac{\partial Q}{\partial x}\,dx + \frac{\partial Q}{\partial V}\,dV \tag{154}$$

which when substituted in (150) give

$$F_e\,dx = \left(-\frac{\partial W_e}{\partial x} + V\frac{\partial Q}{\partial x}\right)dx + \left(-\frac{\partial W_e}{\partial V} + V\frac{\partial Q}{\partial V}\right)dV \tag{155}$$

or by dividing both sides by dx,

$$F_e = \left(-\frac{\partial W_e}{\partial x} + V\frac{\partial Q}{\partial x}\right) + \left(-\frac{\partial W_e}{\partial V} + V\frac{\partial Q}{\partial V}\right)\frac{dV}{dx} \tag{156}$$

Because the incremental changes dx and dV are arbitrary, F_e must be independent of these changes. Thus, in (156) the coefficient of dV/dx must be zero. Hence, (156) finally becomes

$$F_e = -\frac{\partial W_e(V, x)}{\partial x} + V\frac{\partial Q(V, x)}{\partial x} \tag{157}$$

EXAMPLE 3.19 Apply (157) to determine the force between the plates of a parallel-plate capacitor, the capacitance of which is $C(x) = \epsilon A/x$.

Solution Recall that

$$Q = CV$$

Assume that the voltage between the plates is fixed, i.e., V is independent of x. Then

$$V\frac{\partial Q}{\partial x} = V^2\frac{\partial C}{\partial x}$$

and

$$W_e = \tfrac{1}{2}CV^2$$

so that

$$\frac{\partial W_e}{\partial x} = \tfrac{1}{2}V^2\frac{\partial C}{\partial x}$$

Hence, (157) for this specific case becomes

$$F_e = -\tfrac{1}{2}V^2\frac{\partial C}{\partial x} + V^2\frac{\partial C}{\partial x} = \tfrac{1}{2}V^2\frac{\partial C}{\partial x} = -\frac{V^2\epsilon A}{2x^2}$$

3.14 Summary

We began this discussion of the electrostatic field by considering the well-known result that charges exert forces on each other; the basic law governing these forces is Coulomb's law. On the basis of this result, we postulated the existence of a force field, the electric field, existing in the region surrounding these charges. The electric field at a point was defined as the force exerted on an infinitesimal test charge located at the point per unit of that charge. This led to the concept of lines of electric force and the electric flux density vector.

Gauss' law gives the net positive charge enclosed by a surface as the net electric flux penetrating and directed out of that surface. Gauss' law was found to be a very powerful tool in determining the electric field produced by a specified distribution of charge. However, this application of Gauss' law relies on the existence of certain symmetries in the problem. When these symmetries do not exist, other techniques must be used (as discussed in Chap. 10).

The concept of electrostatic potential, or voltage, as the work per unit charge required to move a charge through an electrostatic field was considered. It was found that the potential difference between two points in an electrostatic field is independent of the path taken between those two points. This was referred to as the conservative property of the electrostatic field. We will find in Chap. 5 that a time-varying field does not necessarily have this conservative property. For the time-varying fields, a definition of voltage between two points as the line integral of the electric field will depend on the chosen path; thus, the voltage cannot be uniquely defined in the time-varying case.

We next briefly considered the effect of materials on the field properties. We found that it is possible to condense the microscopic effects of some field in certain types of dielectrics into a single parameter—relative permittivity. The distinction between bound and free charge was made. It was found that the source of the electric flux density vector **D** is free charge, the source of the polarization vector **P** is bound charge, and the source of ϵ_0 **E** is charge of both types. The concept of polarization of a dielectric as the formation or orientation of bound charge dipoles was discussed.

Boundary conditions on the field vectors were derived. It was shown that the tangential components of **E** are continuous across a boundary. It was also shown that the normal components of **D** are discontinuous by the amount of free surface charge at the boundary.

The concept of the energy density of an electrostatic field was discussed, and the corresponding circuit quantity, capacitance, was shown to be related to this stored energy. The capacitance of a system of two conducting bodies was related to either the charge separation (storage) capability of the system or, equivalently, to the energy storage of the system. Similarly, the resistance of a system of two conducting bodies was related to the ability to conduct current between them or, equivalently, to the power dissipation of the system. Mechanical forces resulting from electrostatic fields were also discussed in terms of the rate of change of energy stored in the system.

Although these results seem to be quite general, the reader is, once again, cautioned that many of these electrostatic properties and concepts will have to be modified when the fields vary with time. These topics must therefore be viewed in terms of their applicability, namely, fixed distributions of charge. Chapter 4 details the concepts for magnetic fields, which are, in a certain sense, dual to the electrostatic field concepts of this chapter. The magnetostatic field properties discussed in the next chapter are the result of steady charge movement. The study of the general case in which the charge movement varies with time will begin in Chap. 5.

Problems

3-1 A volume charge distribution $\rho_v = 2z$ C/m^3 is contained in a region defined in cylindrical coordinates as $0 \le z \le 2$ m, $0 \le r \le 1$ m, and $45° \le \phi \le 90°$. Determine the total charge contained in the region.

3-2 A surface charge distribution is contained in a flat, wedge-shaped surface whose corners are defined in a rectangular coordinate system by [2, 1, 2] m, [1, 1, 2] m, and [1, 3, 2] m. The charge distribution is given by $\rho_s = 3xyz$ C/m^2. Determine the total charge contained on the surface.

3-3 A 100-μC point charge is located in a rectangular coordinate system at [1, 1, 1] m, and another point charge of 50 μC is at [−1, 0, −2] m. Find the vector force on the first charge.

3-4 Four 100-μC point charges are located on the corners of a square that are defined in a rectangular coordinate system by [1, 0, 0] m, [0, 1, 0] m, [−1, 0, 0] m, and [0, −1, 0] m. Determine the vector force exerted on another 100-μC charge that is located at:

(a) [0, 0, 0] m
(b) [0, 0, 1] m
(c) [1, 1, 0] m

3-5 Three point charges A, B, and C, having 18 μC, -8μC, and 72 μC charges, respectively, are located in free space in a straight line, as shown in Fig. P3.5.

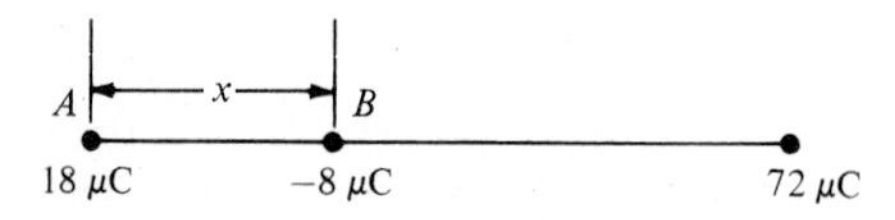

FIGURE P3.5

Determine the distance between the charges A and B such that the mutual force between A and B is the same as the force between B and C and equal to the force between C and A. The charges A and C are separated from each other by 3 cm. Observe that for the calculated distance and for the given polarities, the system of charges will be in equilibrium.

3-6 Two equal and opposite sign 35-μC charges are separated in air by a distance of 20 cm, and an electron is located midway between these charges. Calculate the force experienced by the electron.

3-7 A surface charge of 100 μC is uniformly distributed over a circular disk having a radius of 2 m. The disk lies in the xy plane and is centered at the origin. Determine the vector force exerted on a 50-μC point charge located on the z axis at $z = 4$ m.

3-8 Two point charges Q are suspended with identical, massless strings of length l from a common point. If each point charge has mass m, determine the charge Q, given the lengths of the strings and given the angle between each string and a vertical line.

3-9 A circular loop of radius R has a uniform line charge density $+\rho_l$ distributed over one half of the loop and $-\rho_l$ over the other half. Determine the electric field at the center of the loop.

3-10 An infinite sheet carries a uniform surface charge density ρ_s C/m^2. Determine the electric field at every point above the sheet. Determine this result with and without Gauss' law.

3-11 Two infinite sheets of charge, each with a uniform charge density ρ_s C/m^2, are separated by a distance d. Determine the electric field everywhere.

3-12 An infinitely long, uniform line charge ρ_l C/m produces an electric field. Show that the divergence of this field is zero everywhere (except at the line). Justify this result from a sketch of the field.

3-13 Two point charges $A = 0.02$ μC and $B = 0.01$ μC are separated from each other by a distance of 25 cm in free space. Calculate the electric field at a point P that is 15 cm away from A and 20 cm from B.

3-14 Two infinite line charges, one having charge distribution ρ_l C/m and the other $-\rho_l$ C/m, are separated by a distance a. Determine the electric field intensity at a point P between the charges which is located on a line joining the charges and which is a distance b from the positive line charge.

3-15 Charge is uniformly distributed throughout a spherical volume of radius R with distribution ρ_v C/m^3. Determine a vector expression for the electric field intensity at all points that are a distance r from the origin. Obtain your result by direct integration and by using Gauss' law.

3-16 A charge Q is uniformly distributed throughout a spherical volume of radius R. A point charge q is located in a rectangular coordinate system at $[z, 0, 0]$ m, and the origin of the sphere is at $[0, 0, 0]$ m. Determine the vector force exerted on the point charge. Does your result agree with the results of Prob. 3-15?

3-17 Charge is uniformly distributed over a strip which is infinite in length and which has a width W with a surface distribution ρ_s C/m^2. Determine the electric field at a point P which is on a line perpendicular to the strip and which is a distance D away from the center of the strip.

3-18 The volume charge density within a cube with 1-m sides is $\rho_v = 16xyz$ μC/m^3. What is the total outward electric flux from the cube?

3-19 The electric flux density due to a volume charge distribution (in spherical coordinates) is given by

$$\mathbf{D} = k\frac{\sin ar}{r^2}\mathbf{a}_r$$

Determine the volume charge density.

3-20 A volume charge density $\rho_v = k/r(r \neq 0$, and $k =$ a constant) exists within a sphere of radius a. This charge distribution produces a certain electric field at $r > a$. Determine that value of a point charge placed at the origin which will produce the same electric field at $r > a$.

3-21 A circular strip of radius $1 \text{ m} \le r \le 2 \text{ m}$ supports a surface charge density that decays inversely as the distance away from the center: $\rho_s = 100/r\ \mu\text{C}$. Determine the resulting electric field produced by this charge distribution at a point which is on a line perpendicular to the strip and which is 10 m away from the center of the circular strip.

3-22 A point charge $Q = 100\ \mu\text{C}$ is located at the origin of a rectangular coordinate system. Determine the total electric flux passing through the following surfaces:

(a) A flat, square surface in the xy plane centered at $z = 2$ having area $= 4\text{ m}^2$. (*Hint*: Consider the flux emanating from a cube having 2-m sides.)

(b) A hemisphere of radius 2 m defined by $0 \le \theta \le \pi/2$, $0 \le \phi \le 2\pi$, and $r = R$.

(c) A spherical shell defined by $\theta_1 \le \theta \le \theta_2$ and $r = R$.

3-23 In a certain region in free space, the electric potential has the distribution shown in Fig. P3.23. Sketch the destruction of the corresponding E_x.

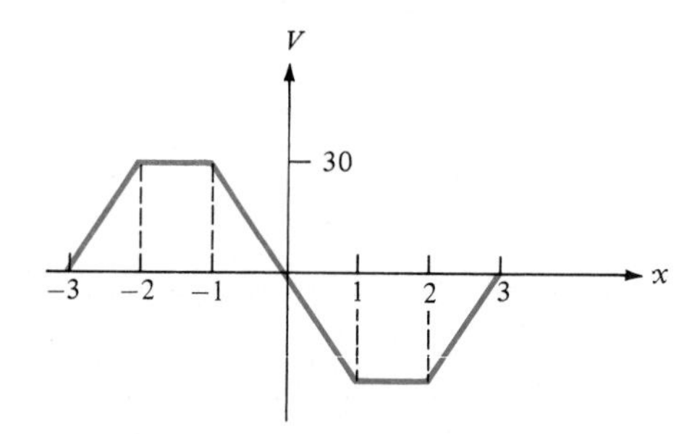

FIGURE P3.23

3-24 A sphere of radius R supports a uniform charge distribution of $\rho_s = 10\ \mu\text{C/m}^2$ on its surface. Determine the electric field produced by utilizing

(a) direct integration, and

(b) Gauss' law.

Determine the potential distribution everywhere. Sketch the electric field and potential as functions of radial distance from the center of the sphere.

3-25 Consider a spherical conducting shell with inner radius a and outer radius b. If the shell is initially uncharged and a point charge Q is placed at the center of the shell, show by Gauss' law that a charge of $-Q$ must be induced on the interior of the shell. What charge is induced on the outer surface? Why?

3-26 Determine the electric field produced by an infinite sheet of charge ρ_s C/m^2 by using Gauss' law. Compare your result to that obtained by direct integration in Prob. 3-10.

3-27 **(a)** Find the electric field $E_x(x)$ inside a layer of charge of uniform density ρ. At $x = x_1$ (Fig. P3.27), $E = E_1$ and $V = V_1$. There is no variation in y and z directions.

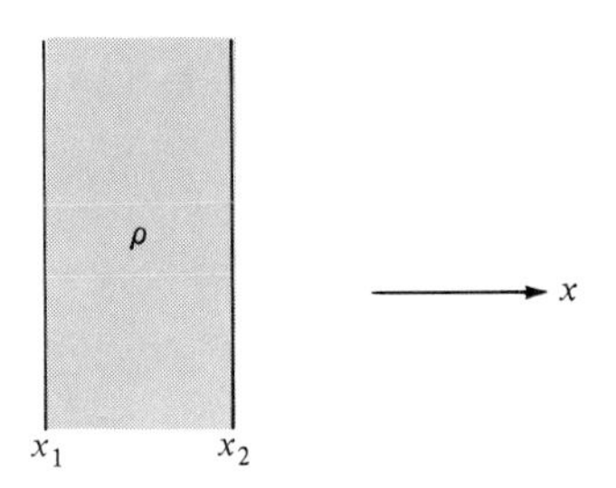

FIGURE P3.27

(b) Find the potential $V(x)$ in the layer of charge.

(c) If the layer is infinitesimally thin, show that the potential is continuous through the layer of charge.

3-28 An infinitely long cylinder of radius a is filled with a charge of uniform density ρ. If the potential on the axis of the cylinder is V_0, determine the potential variation within the cylinder.

3-29 A circular plate of radius a has a uniform charge density ρ_s C/m^2 distributed over its surface. Evaluate the potential at a point on the axis of the plate h m away using

$$V = \frac{1}{4\pi\epsilon}\int_v \frac{\rho_s}{r}\,dv$$

3-30 Given a line density $\rho_l = 5\ \mu$C/m, determine the potential difference between the points (1 m, π, 0) and (3 m, π, 4 m).

3-31 A point charge of 1 μC is located at the origin of a coordinate system. Determine the potential difference between two points 1 m and 2 m radially away from the origin.

3-32 A very thin wire carrying a uniform charge density of 1 μC/m is formed in the shape of a circular loop of 6-m diameter. What is the potential at a point on its axis 4 m away from the center of the loop?

3-33 The potential V at the surface of an infinitely long conical conductor is given by $V = \ln \cot(\theta/2)$, with $\theta = \alpha$ (see Fig. P3.33). Determine the charge on the conductor surface per unit length from the origin.

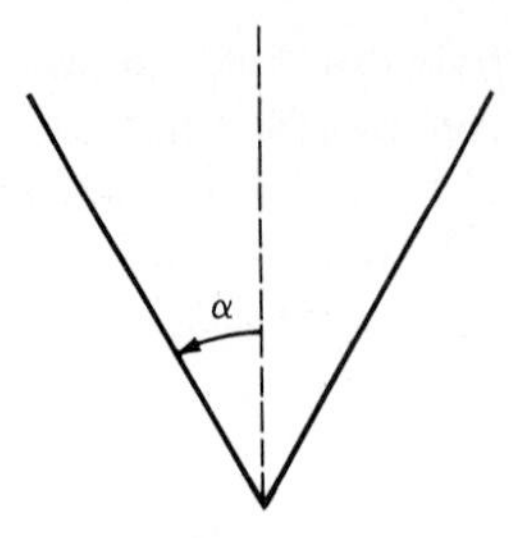

FIGURE P3.33

3-34 A pair of parallel line charges of uniform charge density ρ_l are separated by a distance d, as shown in Fig. P3.34. Derive an expression for equipotential lines, and sketch the equipotentials on a plane perpendicular to the charges.

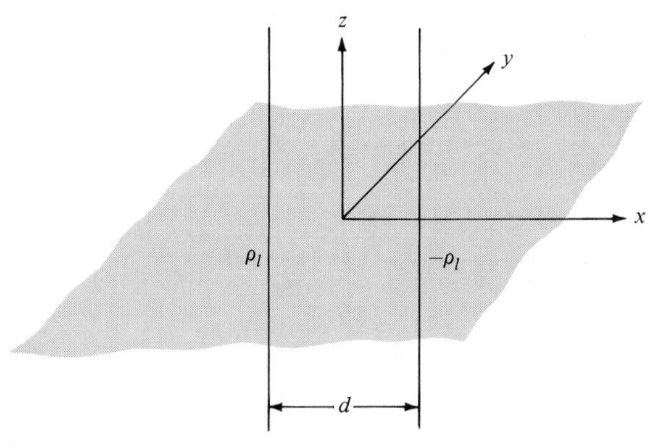

FIGURE P3.34

3-35 Consider two concentric spheres. The inner sphere has radius a, and a charge Q is distributed uniformly over its outer surface. The outer sphere has radius b, and charge $-Q$ is distributed uniformly over its inner surface. Determine the potential difference between the two spheres and the capacitance of the structure.

3-36 A dielectric sphere of permittivity ϵ and of radius R is centered at the origin of a spherical coordinate system and polarized radially such that the polarization vector is given by $\mathbf{P} = kr\mathbf{a}_r$, where k is a constant. Evaluate the electric potential at the center of the sphere.

3-37 Two cubes of dielectric materials are in contact so that the z axis of rectangular coordinates is perpendicular to the planar interface of the cubes. An electric field $\mathbf{E} = 3\mathbf{a}_x + 4\mathbf{a}_y - 12\mathbf{a}_z$ exists in the top cube, the material of which has a relative permittivity of 3.

(a) Obtain an expression for **D** in the bottom cube, where the relative permittivity is 1.5.

(b) What is the potential difference between the origin and the point $(0, 0, -2)$ in the bottom of cube?

(c) What is the energy density in the top cube?

3-38 A capacitor is constructed in the form shown in Fig. P3.38. The capacitor consists of two large parallel plates 3 m apart, with a dielectric slab 1 m thick located midway between the plates. The electric field in region I is $\mathbf{E} = (\rho_s/\epsilon_0)\mathbf{a}_x$, where ρ_s is the surface charge density on one plate. The surface area of each plate is A. Write the boundary conditions at the two interfaces and determine the capacitance in terms of A, ϵ_0, and ϵ.

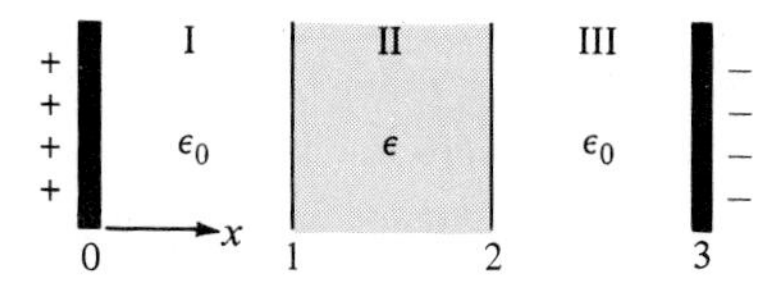

FIGURE P3.38

3-39 Calculate the capacitance of an 8-km-long coaxial cable (Fig. P3.39) having a core diameter of 1 cm and an insulation thickness of 0.5 cm, for which $\epsilon_r = 3.5$. Also, find the potential gradient at the surface of the core that is at a voltage of 1000 V with respect to the outer shield.

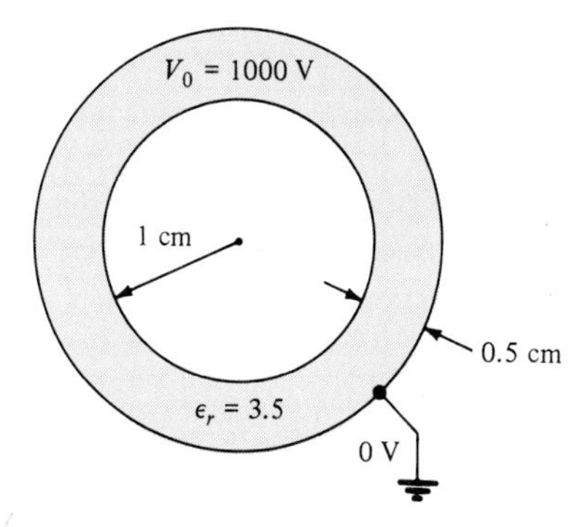

FIGURE P3.39

3-40 A coaxial power cable, having a core (conductor) radius of r_1, is filled with two concentric layers of dielectrics ϵ_1 and ϵ_2, as shown in Fig. P3.40.

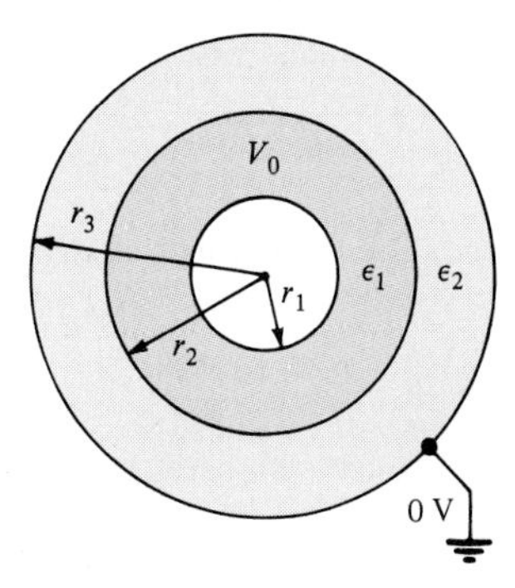

FIGURE P3.40

(a) Determine the capacitance of the cable per unit length.

(b) If the conductor is at a potential V_0 and the outer shield is grounded, determine the maximum potential gradient in each dielectric from the following data: $V_0 = 1200$ V, $\epsilon_{r1} = 1.5$, $\epsilon_{r2} = 4.5$, and $r_3 = 2r_2 = 4r_1 = 4$ cm.

3-41 An electric field is produced by a line charge of density ρ_l. In this field a point charge Q is moved from a distance r to a distance $4r$ from the line charge. Evaluate the work done, and hence show that it is independent of r.

3-42 Calculate the work done in carrying a charge of -2 C from the point [2, 1, -1] to the point [8, 2, 1] in an electric field $\mathbf{E} = y\mathbf{a}_x + x\mathbf{a}_y$ along the (parabolic) path $x = 2y^2$. Hence, determine the potential difference between the two given points.

3-43 A certain subatomic (or elementary) particle occurs in the neutral or charged state. Because the particle is so small, its mass is expressed as energy in electron volts (eV $= mc^2$, where m is the mass in kg and c is the velocity of light). The charged particle has electrostatic stored energy and is consequently heavier in the charged than in the neutral state. From experiments it has been found that the difference between the charged and the neutral particle is 4.6 MeV (million electron volts) due to a 1.6×10^{-19} C charge on the charged particle. Assuming that the particle is a perfect sphere and that the charge is uniformly distributed within the sphere, calculate the radius of the particle. Given: 1 eV = joule/charge on an electron.

3-44 In the classical Millikan oil-drop experiment, a charged oil drop is placed in an electric field in the vertical direction. Two forces acting in opposite directions are experienced by the electron; one force is due to the electric field, and the other is the gravitational force. In the experiment, the electric field is so adjusted that the drop attains equilibrium. In a certain experiment it was found that a 1.6×10^{-14}-kg oil drop could be balanced with a 2×10^5-V/m electric field. Calculate the charge on the charged oil drop.

3-45 A pair of 20-cm-long concentric cylindrical conductors of radii 5 and 10 cm is filled with a dielectric $\epsilon = 10\epsilon_0$. A voltage is applied between the conductors to establish an electric field $\mathbf{E} = (10^6/r)\mathbf{a}_r$ between the cylinders. Calculate

(a) the energy stored,
(b) the capacitance, and
(c) the applied voltage between the cylinders.

3-46 Determine the capacitance of a parallel-plate capacitor filled with two dielectrics, as shown in Fig. P3.46.

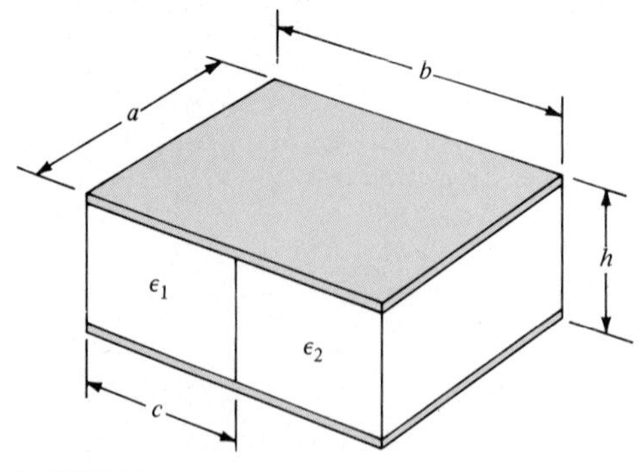

FIGURE P3.46

3-47 A parallel-plate capacitor has two layers of dielectrics, as shown in Fig. P3.47. For a 900-V potential difference applied between the plates, calculate how this voltage is divided across the dielectrics; that is, find V_1 and V_2. Given: $\epsilon_{r2} = \epsilon_{r1} = 6\epsilon_0$; $A = 100$ cm^2; $b = 2a = 2$ mm.

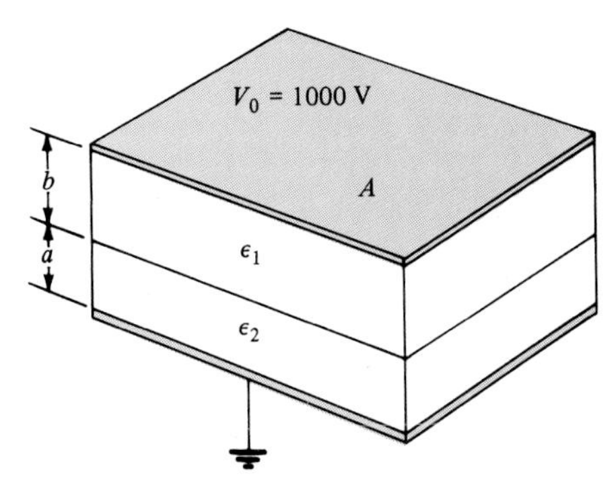

FIGURE P3.47

3-48 A capacitance is formed from a segment of two coaxial cylinders, as shown in Fig. P3.48. The vertical planes form the plates of the capacitor. For the given dimensions, find the capacitance. Neglect fringing.

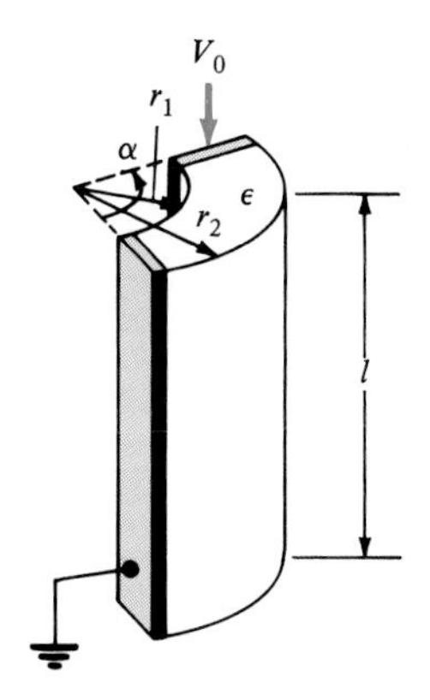

FIGURE P3.48

3-49 In a cylindrical electron beam of radius R and unit length, the volume charge density is given by $\rho = ar$, where a is a constant. Determine the energy stored in the beam.

3-50 The potential distribution within a hydrogen atom is given by

$$V(r) = \frac{Ae^{-\alpha r}}{r}\left(1 + \frac{\alpha r}{2}\right)$$

where α and A are constants. Determine the charge distribution required to produce this potential. Exclude the origin from the calculations and assume that the atom is a perfect sphere.

3-51 A 10-m-long conductor of circular cross section has a diameter of 3 cm. It carries a current of 3000 A dc with 10 V across the ends. Calculate the conductivity of the conductor material.

3-52 Determine the conductance G of a spherical capacitor of inner sphere radius a and outer sphere radius b that is filled with a lossy dielectric having parameters σ and $\epsilon_r \epsilon_0$. Relate your result to the capacitance of this structure.

3-53 Determine the capacitance of an isolated sphere of radius a.

3-54 A parallel-plate capacitor having a plate area A is filled with a dielectric of permittivity ϵ. The capacitor is charged to a value Q, which remains constant regardless of the separation between the plates. Obtain an expression for the force (of attraction) between the plates, and hence show that this force is independent of the separation between the plates.

3-55 A coaxial power cable is partially filled with a dielectric ϵ. With only air as a dielectric between the two conductors (Fig. P3.55), the cable can withstand a potential difference of V_0 V before breakdown. Determine the breakdown voltage

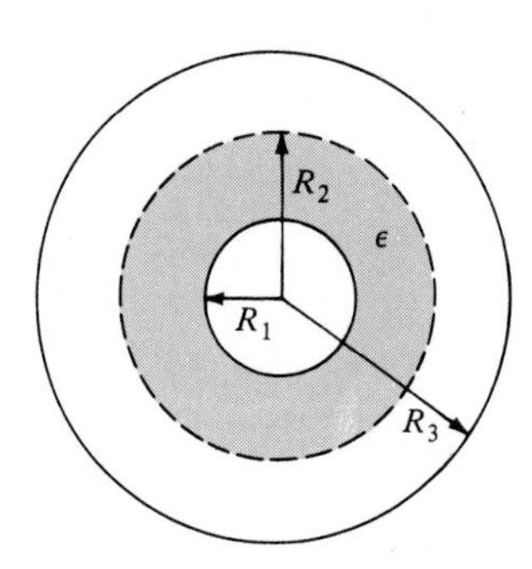

FIGURE P3.55

in the presence of the dielectric ϵ ($\epsilon > \epsilon_0$) for the dimensions shown in Fig. P3-55. Given: $V_0 = 600$ kV, $\epsilon = 10\epsilon_0$, $R_3 = 3$ cm, $R_2 = 2$ cm, and $R_1 = 1$ cm.

3-56 For the cable of Prob. 3-40, determine all the free and bound charge densities.

3-57 A dielectric slab is attracted into the space between the plates of a parallel-plate capacitor. The voltage-charge (V-Q), relationship is nonlinear and is given by $Q = V^3/(k_1 + k_2 x)$, where k_1 and k_2 are constants and the dielectric slab is x m away from the plates. Determine the force on the slab.

3-58 The expression for stored energy in an electrostatic field in Eq. (112) indicates that stored energy exists in a region in which an electric field exists even though no charge is contained in that region. On the other hand, (107) indicates that stored energy exists in a region only if the region contains charge. Discuss these seemingly contradictory expressions. Can we isolate stored energy in a particular region?

3-59 For the parallel-plate capacitor shown in Fig. P3-47, determine **E**, **D**, **P**, and the bound and free surface charge densities in terms of some general potential difference between the plates of V_0. Neglect fringing.

3-60 For the spherical capacitor of Prob. 3-35, determine **E**, **D**, **P**, and the bound and free surface charge densities. The space between the spheres is filled with a dielectric having permittivity ϵ.

CHAPTER 4

Magnetostatic Fields

The preceding chapter introduced the concept of an electric field on the basis of the observation that an electric charge brought into the vicinity of other charges experiences a force. It can similarly (experimentally) be demonstrated that a current element will experience a force if it is brought into a region containing another current. Such a region of forces is termed a region of *magnetic field.* Another source of a magnetic field is the permanent magnet. Because the magnetic field is measured by the forces between current-carrying conductors, just as the electric field is measured by forces between charges (Chap. 3), we now proceed to formulate the laws that will enable us to determine these forces and hence the corresponding magnetic fields.

4.1 Ampère's Force Law

In Chap. 3 we saw how Coulomb's law can be used to determine forces between stationary distributions of charges. This led us to the definition of the electric field.

Ampère's law for forces between current-carrying conductors located in free space is similar to Coulomb's law for forces between stationary electric charges. Consider two current-carrying elements $I_1\, dl_1$ and $I_2\, dl_2$ separated by a distance R, as shown in Fig. 4.1. It was found experimentally by Ampère that there is a mutual force of attraction or repulsion between the two current-carrying elements. In terms of the variables shown in Fig. 4.1, the force $d\mathbf{F}_1$ on the

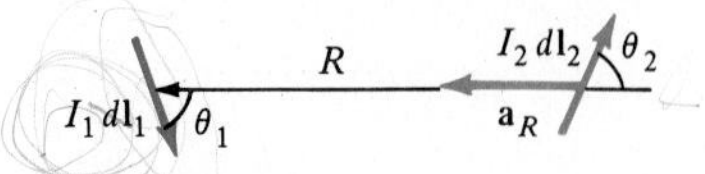

FIGURE 4.1

Current elements $I_1\,dl_1$ and $I_2\,dl_2$, and their respective orientation.

element $I_1\,d\mathbf{l}_1$, interacting with the element $I_2\,d\mathbf{l}_2$, is proportional to: $I_1\,dl_1$, $I_2\,dl_2$, $\sin\theta_1$, $\sin\theta_2$, and $1/R^2$. Hence, in vector notation, we have

$$d\mathbf{F}_1 = \alpha \frac{I_1\,d\mathbf{l}_1 \times (I_2\,d\mathbf{l}_2 \times \mathbf{a}_R)}{R^2} = \alpha \frac{I_1\,d\mathbf{l}_1 \times (I_2\,d\mathbf{l}_2 \times (\mathbf{r}-\mathbf{r}'))}{|\mathbf{r}-\mathbf{r}'|^3} \tag{1}$$

The second form of this equation gives a more general form in terms of the position vectors $\mathbf{r}$ and $\mathbf{r}'$. Vector $\mathbf{r}$ is directed from the origin to $I_1\,d\mathbf{l}_1$ (the field point) whereas $\mathbf{r}'$ is directed from the origin to $I_2\,d\mathbf{l}_2$ (the source point). Both forms are similar to Coulomb's law for electric forces. In (1), the constant of proportionality, α, in SI units is $\mu_0/4\pi$, where μ_0 is defined as the *permeability of free space* and has value of $4\pi \times 10^{-7}$ N/A^2 (or henry/meter, as we shall see later). Hence, (1) becomes

$$d\mathbf{F}_1 = \frac{\mu_0 I_1\,d\mathbf{l}_1 \times (I_2\,d\mathbf{l}_2 \times \mathbf{a}_R)}{4\pi R^2} \tag{2}$$

If the current elements $I_1\,d\mathbf{l}_1$ and $I_2\,d\mathbf{l}_2$ belong to two current loops (Fig. 4.2), then, from (2), the force $\mathbf{F}_1$ experienced by loop 1 as a result of the current in loop 2 is given by

$$\mathbf{F}_1 = \frac{\mu_0 I_1 I_2}{4\pi} \oint_{l_1} \oint_{l_2} \frac{d\mathbf{l}_1 \times (d\mathbf{l}_2 \times \mathbf{a}_R)}{R^2} \tag{3}$$

In the following example we now apply Ampère's law, as given by (2), to find the force between two straight conductors.

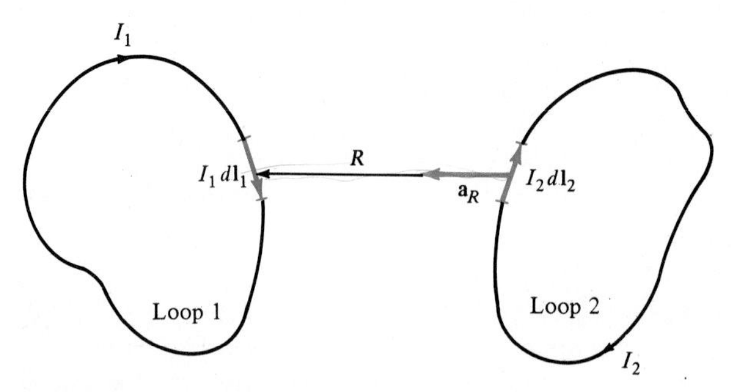

FIGURE 4.2

Two current-carrying loops.

EXAMPLE 4.1 Two infinitely long straight conductors, which are parallel to each other, are separated by a distance b. The conductors carry currents I_1 and I_2 in the directions shown in Fig. 4.3. Determine the mutual force per unit length of the conductors.

Solution Consider (2) term by term. Thus, from Fig. 4.3,

$$|I_2\, d\mathbf{l}_2 \times \mathbf{a}_R| = I_2\, dl_2 \sin\phi = I_2\, dl_2 \cos\theta$$

where

$$l_2 = b \tan\theta$$

Hence,

$$dl_2 = b \sec^2\theta\, d\theta$$

and

$$I_2\, dl_2 \cos\theta = bI_2 \sec\theta\, d\theta$$

and since $R^2 = b^2 \sec^2\theta$,

$$\frac{I_2\, dl_2 \cos\theta}{R^2} = \frac{I_2}{b} \cos\theta\, d\theta$$

The direction of $I_2\, d\mathbf{l}_2 \times \mathbf{a}_R$ is perpendicular to the plane containing the two wires. Hence, the direction of $I_1\, d\mathbf{l}_1 \times (I_2\, d\mathbf{l}_2 \times \mathbf{a}_R)$, that is, the direction of the

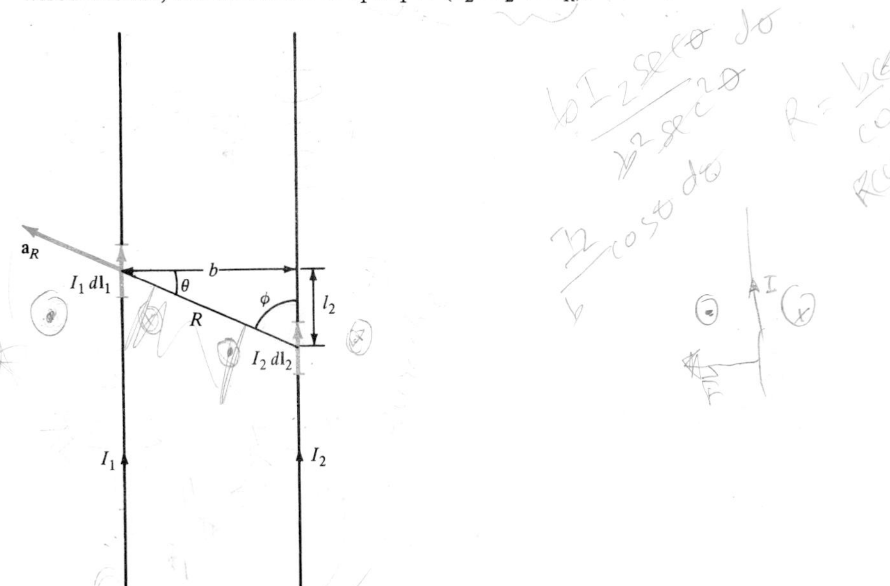

FIGURE 4.3
Example 4.1. Force exerted on one current by another.

force $\mathbf{F}_1$, is toward the other wire. In other words, the force is attractive rather than repulsive, and the magnitude of this force is given by

$$dF_1 = \frac{\mu_0 I_1 I_2 \, dl_1 \cos\theta \, d\theta}{4\pi b}$$

Integrating over the wire length gives

$$F_1 = \int_{\theta=-\pi/2}^{\pi/2} \frac{\mu_0 I_1 I_2 \, dl_1 \cos\theta \, d\theta}{4\pi b}$$

$$= \frac{\mu_0 I_1 I_2 \, dl_1}{2\pi b}$$

Thus, the force per unit length is

$$F_1 = \frac{\mu_0 I_1 I_2}{2\pi b}$$

Let us now return to (3) and express it in an alternative form. For this purpose, we use the vector identity $\mathbf{A} \times (\mathbf{B} \times \mathbf{C}) = \mathbf{B}(\mathbf{A} \cdot \mathbf{C}) - \mathbf{C}(\mathbf{A} \cdot \mathbf{B})$ (Appendix A) to obtain

$$d\mathbf{l}_1 \times (d\mathbf{l}_2 \times \mathbf{a}_R) = d\mathbf{l}_2(d\mathbf{l}_1 \cdot \mathbf{a}_R) - \mathbf{a}_R(d\mathbf{l}_1 \cdot d\mathbf{l}_2) \tag{4}$$

Substituting (4) in (3) yields

$$\mathbf{F}_1 = \frac{\mu_0 I_1 I_2}{4\pi}\left[\oint_{l_1}\oint_{l_2} \frac{d\mathbf{l}_2(d\mathbf{l}_1 \cdot \mathbf{a}_R)}{R^2} - \oint_{l_1}\oint_{l_2} \frac{\mathbf{a}_R(d\mathbf{l}_1 \cdot d\mathbf{l}_2)}{R^2}\right] \tag{5}$$

Recalling from Chap. 3 that $\nabla(1/R) = (1/R^2)\mathbf{a}_R$, we may write the first integral in (5) as

$$\oint_{l_1}\oint_{l_2} \frac{d\mathbf{l}_2(d\mathbf{l}_1 \cdot \mathbf{a}_R)}{R^2} = -\oint_{l_1}\oint_{l_2} d\mathbf{l}_2\left[d\mathbf{l}_1 \cdot \nabla\left(\frac{1}{R}\right)\right]$$

$$= -\oint_{l_2} d\mathbf{l}_2 \int_{s_1} \nabla \times \nabla\left(\frac{1}{R}\right) \cdot d\mathbf{s} = 0 \tag{6}$$

This last result is obtained by applying Stokes' theorem and using the result (used repeatedly in Chap. 3) that the curl of the gradient of any scalar field is zero: $\nabla \times \nabla\phi = 0$. Finally, from (5) and (6) we obtain

$$\mathbf{F}_1 = -\frac{\mu_0 I_1 I_2}{4\pi}\oint_{l_1}\oint_{l_2} \frac{\mathbf{a}_R}{R^2} \, d\mathbf{l}_1 \cdot d\mathbf{l}_2 \tag{7}$$

Proceeding in a similar fashion, we may show that

$$\mathbf{F}_2 = -\mathbf{F}_1 \tag{8}$$

how?

4.2 Magnetic Flux Density and the Biot-Savart Law

Let us recall Ampère's force law as given by (2) and define a quantity $d\mathbf{B}$ such that

$$d\mathbf{B}_2 = \frac{\mu_0 I_2 \, d\mathbf{l}_2 \times \mathbf{a}_R}{4\pi R^2} \tag{9}$$

and (2) then becomes

$$d\mathbf{F}_1 = I_1 \, d\mathbf{l}_1 \times d\mathbf{B}_2 \tag{10}$$

or

$$\mathbf{F}_1 = I_1 \oint d\mathbf{l}_1 \times \mathbf{B}_2 \tag{11}$$

where

$$\mathbf{B}_2 = \frac{\mu_0 I_2}{4\pi} \oint \frac{d\mathbf{l}_2 \times \mathbf{a}_R}{R^2} \tag{12}$$

The quantity $\mathbf{B}_2$ is defined as the magnetic flux density at a certain point due to the current I_2. The units are webers per square meter (Wb/m^2) or tesla (T). Dropping the subscript 2, we may generalize (9) [or (12)] as

$$d\mathbf{B} = \frac{\mu_0 I \, d\mathbf{l} \times \mathbf{a}_R}{4\pi R^2} \tag{13}$$

Equation (13) is a mathematical form of the *Biot–Savart* law. In words, the Biot–Savart law states that due to a differential current element $I \, d\mathbf{l}$, the magnitude of the magnetic flux density at a point P (Fig. 4.4) is directly proportional to the product of the current I, the length of the current element, dl, and the sine of the angle between the current element and the line PQ joining the point P to the current element. But the flux density is inversely proportional to the square of the distance R between the point and the current element. The law further states that the direction of the magnetic flux density is normal to the plane containing $d\mathbf{l}$ and P, that is, into the paper. (Notice that the field is not in the direction of $\mathbf{a}_R$, Fig. 4.4.) This normal is in the same direction as that of the cross product $I \, d\mathbf{l} \times \mathbf{a}_R$, and the vectors $d\mathbf{l}$, $\mathbf{a}_R$, and $d\mathbf{B}$ form a right-handed

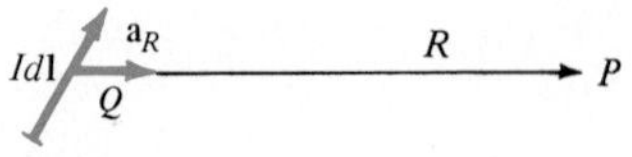

FIGURE 4.4
Geometry for the Biot-Savart law.

coordinate system. The total magnetic flux density is obtained by integrating (13) over the path of the current flow and is given by

$$\mathbf{B} = \int \frac{\mu_0 I \, d\mathbf{l} \times \mathbf{a}_R}{4\pi R^2} \tag{14}$$

In (14) we chose the origin at Q (Fig. 4.4). In terms of an arbitrary choice of origin (Fig. 4.5), (14) can also be written as

$$\mathbf{B} = \int \frac{\mu_0 I \, d\mathbf{l} \times (\mathbf{r} - \mathbf{r}')}{4\pi |\mathbf{r} - \mathbf{r}'|^3} \tag{15}$$

Application of the Biot–Savart law is illustrated by the following examples.

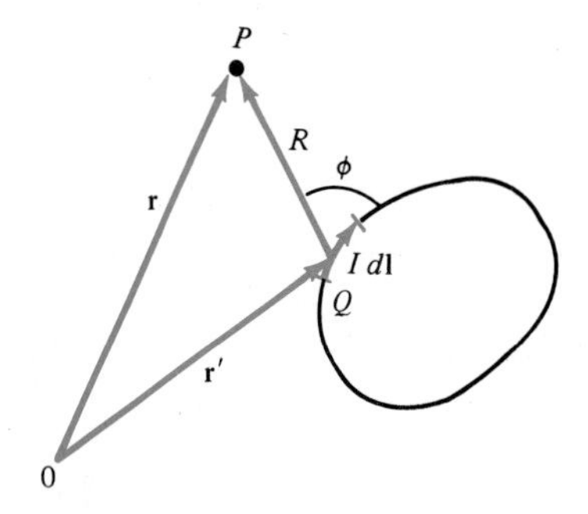

FIGURE 4.5
Magnetic flux density at P with reference to a general coordinate system, having the origin at O.

EXAMPLE 4.2 Determine the magnetic flux density, at point P, due to a very long wire carrying a current I. The point P is at a distance r away from the conductor.

Solution The geometry is shown in Fig. 4.6. According to (13) we have

$$d\mathbf{B} = \frac{\mu_0 I \, d\mathbf{l} \times \mathbf{a}_R}{4\pi R^2}$$

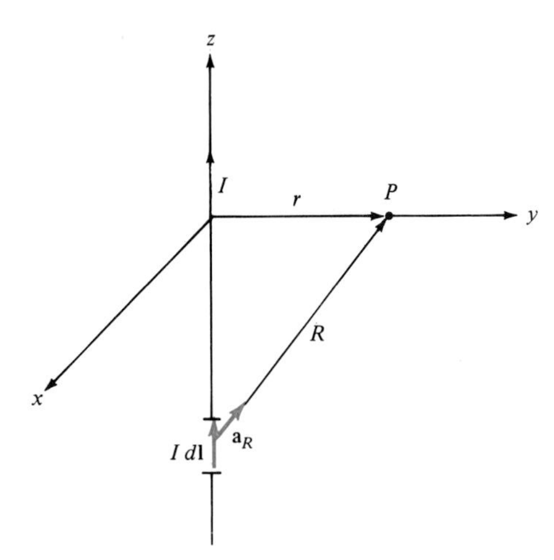

FIGURE 4.6

Example 4.2. Magnetic flux density due to a straight conductor of infinite length.

We choose the origin so that the point P is in the xy ($z = 0$) plane. In this case

$$\mathbf{R} = z\mathbf{a}_z + r\mathbf{a}_r \qquad |\mathbf{R}| = \sqrt{z^2 + r^2}$$

and

$$\mathbf{a}_R = \frac{z\mathbf{a}_z + r\mathbf{a}_r}{\sqrt{z^2 + r^2}}$$

Thus

$$d\mathbf{B} = \frac{\mu_0 I\, dz\, \mathbf{a}_z \times (z\mathbf{a}_z + r\mathbf{a}_r)}{4\pi(z^2 + r^2)^{3/2}}$$

$$= \frac{\mu_0 I\, dz\, r\mathbf{a}_\phi}{4\pi(r^2 + z^2)^{3/2}} \qquad \text{since } \mathbf{a}_z \times \mathbf{a}_z = 0 \qquad \text{and} \qquad \mathbf{a}_z \times \mathbf{a}_r = \mathbf{a}_\phi$$

Consequently,

$$\mathbf{B} = \left[\frac{\mu_0 I r}{4\pi} \int_{-\infty}^{\infty} \frac{dz}{(r^2 + z^2)^{3/2}}\right] \mathbf{a}_\phi = \frac{\mu_0 I r}{4\pi} \left(\frac{z}{r^2\sqrt{r^2 + z^2}}\right)_{-\infty}^{\infty} \mathbf{a}_\phi$$

$$= \frac{2\mu_0 I r}{4\pi r^2}\,\mathbf{a}_\phi = \frac{\mu_0 I}{2\pi r}\,\mathbf{a}_\phi \tag{16}$$

which is the required flux density.

EXAMPLE 4.3 Next we consider a straight wire of length l carrying a current I, shown in Fig. 4.7. The problem is to determine the magnetic flux density $\mathbf{B}$ at the point P.

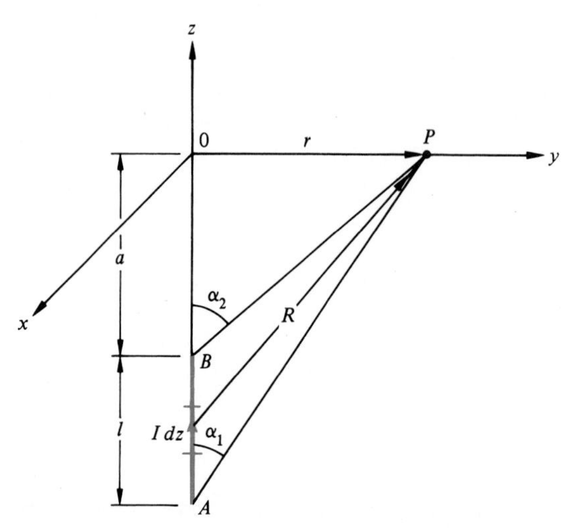

FIGURE 4.7

Example 4.3. Magnetic flux density due to a straight conductor of finite length.

Solution Comparing with the last example, we may again use (16) to express the total flux density as

$$\mathbf{B} = \left[\frac{\mu_0 Ir}{4\pi}\int_{-(a+l)}^{-a}\frac{dz}{(r^2+z^2)^{3/2}}\right]\mathbf{a}_\phi = \frac{\mu_0 Ir}{4\pi}\left(\frac{z}{r^2\sqrt{r^2+z^2}}\right)_{-(a+l)}^{-a}\mathbf{a}_\phi$$

$$= \frac{\mu_0 I}{4\pi r}\left[\frac{-a}{\sqrt{a^2+r^2}} + \frac{a+l}{\sqrt{(a+l)^2+r^2}}\right]\mathbf{a}_\phi$$

$$= \frac{\mu_0 I}{4\pi r}\left(\frac{OA}{AP} - \frac{OB}{BP}\right)\mathbf{a}_\phi = \frac{\mu_0 I}{4\pi r}(\cos\alpha_1 - \cos\alpha_2)\mathbf{a}_\phi \tag{17}$$

Notice that for an infinitely long wire, $\alpha_1 = 0$ and $\alpha_2 = \pi$, in which case (16) becomes a special case of (17).

EXAMPLE 4.4 Determine the magnetic flux density at the center of a square loop of side l carrying a current I as shown in Fig. 4.8.

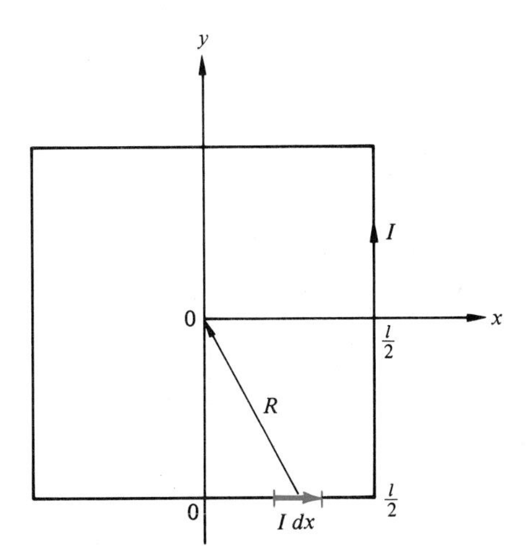

FIGURE 4.8
Example 4.4. A square loop of current.

Solution In this problem we make use of the geometrical symmetry. Thus, we have to determine the flux density due to each half-side. Using the results of the preceding example, we obtain

$$d\mathbf{B} = \frac{\mu_0(I\,dx\,\mathbf{a}_x) \times [-x\mathbf{a}_x + (l/2)\mathbf{a}_y]}{4\pi[x^2 + (l/2)^2]^{3/2}}$$

$$= \frac{\mu_0 Il\,dx}{8\pi[x^2 + (l/2)^2]^{3/2}}\mathbf{a}_z \qquad \text{since } \mathbf{a}_x \times \mathbf{a}_x = 0 \qquad \text{and} \qquad \mathbf{a}_x \times \mathbf{a}_y = \mathbf{a}_z$$

The total flux density at the center then becomes

$$\mathbf{B} = 8\int_0^{l/2} \frac{\mu_0 Il\,dx}{8\pi[x^2 + (l/2)^2]^{3/2}}\mathbf{a}_z$$

$$= \frac{\mu_0 Il}{\pi}\left[\frac{x}{(l/2)^2\sqrt{x^2 + (l/2)^2}}\right]_0^{l/2}\mathbf{a}_z = \frac{\mu_0 Il}{\pi}\left[\frac{l/2}{(l/2)^2\sqrt{2(l/2)^2}}\right]\mathbf{a}_z$$

$$= \frac{\mu_0 2\sqrt{2}I}{\pi l}\mathbf{a}_z \quad \text{A/m} \tag{18}$$

Notice that up to this point we have considered the Biot–Savart law for linear current elements. The form (14) is also valid for surface current-density distributions. In terms of surface current density $\mathbf{J}_s$, (14) is written as

$$\mathbf{B} = \int_s \frac{\mu_0 \mathbf{J}_s \times \mathbf{a}_R\,ds}{4\pi R^2} \tag{19}$$

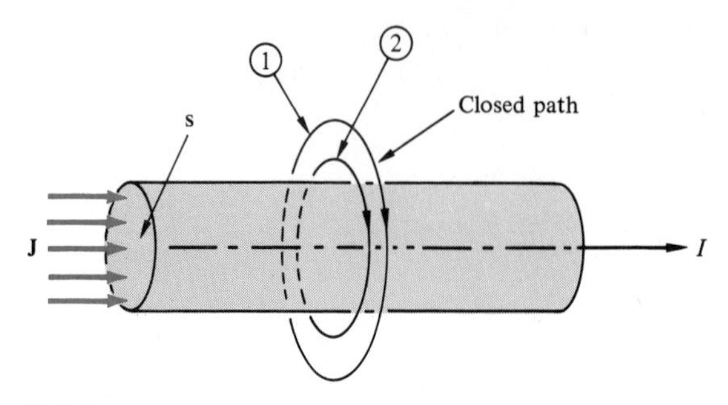

FIGURE 4.9
Illustration of Ampère's law: current enclosed by a closed path.

4.3 Ampère's Circuital Law

Consider a very long wire carrying a current I which was considered in Example 4.2. Evaluate the line integral $\oint \mathbf{B} \cdot d\mathbf{l}$ around a circular path of radius r. From (16) we have

$$\oint \mathbf{B} \cdot d\mathbf{l} = \frac{\mu_0 I}{2\pi r} \oint dl \qquad \text{since } \mathbf{B} = B_\phi \mathbf{a}_\phi \qquad \text{and} \qquad d\mathbf{l} = dl\, \mathbf{a}_\phi$$

$$= \frac{\mu_0 I}{2\pi r}(2\pi r) = \mu_0 I \tag{20}$$

Notice from (20) that the line integral of a static magnetic flux density around any given closed path must equal the product of μ_0 and the total current enclosed by that path. The preceding is a statement of *Ampère's circuital law.* Notice that this law is somewhat analogous to Gauss' law of electrostatics. Equation (20) may be more formally written as

$$\oint_c \mathbf{B} \cdot d\mathbf{l} = \int_s \mu_0 \mathbf{J} \cdot d\mathbf{s} = \mu_0 I \tag{21}$$

where $\mathbf{J}$ is the current density penetrating the surface s (Fig. 4.9). Regarding the line integral, (21) is valid regardless of the path of integration, 1 or 2 in Fig. 4.9, and the sign convention for the current is that the current is positive if it advances like a right-hand screw rotated in the direction of the path of integration. The directions shown in Fig. 4.9 correspond to the positive sense.

Before considering the derivation of Ampère's circuital law, let us consider an example of its application and discover some new concepts.

EXAMPLE 4.5 A straight conductor of cylindrical cross section and of radius a carries a current I, which is uniformly distributed over the conductor cross section. Using Ampère's circuital law, find the magnetic flux density within and outside of the conductor.

Solution Within the conductor, let us locate a point at a radius b ($b < a$) from the axis of the conductor. To apply (21), for the given point we have $d\mathbf{l} = b\,d\phi\,\mathbf{a}_\phi$ and $\mathbf{B} = B_\phi \mathbf{a}_\phi$. Hence,

$$\oint_c \mathbf{B}\cdot d\mathbf{l} = \int_{\phi=0}^{\phi=2\pi} B_\phi b\,d\phi = 2\pi b B_\phi$$

The right-hand side of (21) becomes

$$\mu_0 I = \int \mu_0 \mathbf{J}\cdot d\mathbf{s} = \frac{\mu_0 I}{\pi a^2}(\pi b^2) = \frac{b^2 \mu_0 I}{a^2}$$

Hence,

$$2\pi b B_\phi = \frac{b^2 \mu_0 I}{a^2}$$

or

$$B_\phi = \frac{b\mu_0 I}{2\pi a^2} \qquad b < a$$

Outside the conductor, proceeding as above, we locate a point at a radius c ($a < c$) from the axis of the conductor. Then from Ampère's circuital law we have

$$B_\phi 2\pi c = \mu_0 I$$

or

$$B_\phi = \frac{\mu_0 I}{2\pi c}$$

Obviously the result obtained in this example agrees with (16), but notice the reduction in the amount of mathematical manipulations to obtain this result as compared with the derivation of (16). Such a simplification of procedure is possible mainly because of the geometrical symmetry in the problem.

In the preceding discussion the magnetic flux density $\mathbf{B}$ has been defined. Recalling (11), we may write it as

$$\mathbf{F}_1 = \mu_0 I_1 \oint d\mathbf{l}_1 \times \mathbf{H}_2 \tag{22}$$

where

$$\mathbf{H}_2 = \frac{I_2}{4\pi}\oint \frac{d\mathbf{l}_2 \times \mathbf{a}_R}{R^2} \tag{23}$$

In (23), $\mathbf{H}_2$ is defined as the magnetic field intensity whose units are Amperes per meter (A/m). Comparing (12) and (23), we find that

$$\mathbf{B}_2 = \mu_0 \mathbf{H}_2 \tag{24}$$

Or, dropping the subscripts

$$\mathbf{B} = \mu_0 \mathbf{H} \tag{25}$$

In free space, the Biot-Savart law can be written either in terms of **B** or in terms of **H**. In terms of **H**, Ampère's circuital law can also be written for free space by dividing (21) by μ_0 to yield

$$\oint_c \mathbf{H} \cdot d\mathbf{l} = \int_s \mathbf{J} \cdot d\mathbf{s} = I \tag{26}$$

The results above have been obtained for free space in which we may use either **B** or **H**, converting from one form to the other by using (25). In Sec. 4.6 we will investigate magnetic fields in magnetic materials. Ferrous materials such as iron are common examples of such materials. We will find that magnetic materials support bound currents, $\mathbf{J}_b$, in addition to free currents, **J**, which we have previously considered. These bound currents give rise to magnetic dipoles just as bound charges give rise to electric dipoles in dielectrics. We will also find that the source of **H** is free current, whereas the source of **B** (more properly $\mathbf{B}/\mu_0$) is current of both types, $\mathbf{J} + \mathbf{J}_b$. In this context, an analogy may be drawn between electric and magnetic fields, wherein **E** and **B** are duals and **D** and **H** are duals with regard to their sources. However, in terms of similarity in units, we might infer a duality between **E** and **H** and between **D** and **B**.

In material media other than free space, we will find that **B** and **H** are also related, and this relationship will be written as

$$\mathbf{B} = \mu \mathbf{H} \tag{27}$$

where μ is said to be the permeability of the material. However, for magnetic materials, this relationship is typically more complicated than the relationship between **D** and **E** in dielectric materials. It is typically a nonlinear relation (and, in some cases, anisotropic), so that $\mathbf{B} = \mu \mathbf{H}$ is only a symbolic relation, and μ is not a simple, scalar constant. For magnetic materials we will be able to modify these previous results, derived for free space, in a rather straightforward fashion.

We will learn more about the magnetic properties of materials in Sec. 4.6. The magnetic flux density may also be defined as the force on a conductor which is carrying unit current and which is oriented at right angles to the flux lines. Although from such a definition we do not gain any additional insight beyond that contained in the Biot-Savart law or in Ampère's force law, this alternative definition enables us to determine the force on a current-carrying conductor located in a magnetic field, as we shall see in Sec. 4.9.

4.4 Ampère's Circuital Law in Point Form

We recall that in the last chapter Gauss' law was first stated in the integral form and then in differential (or point) form. In order to go from the integral to the differential form, we used the concept of divergence. To accomplish a similar transformation of Ampère's circuital law from the integral to the differential form, we use the vector operation—*curl*, introduced in Chap. 2. Recall that the curl is defined in terms of a line integral around an infinitesimal path and the area enclosed by the path:

$$\text{curl } \mathbf{F} = \lim_{\Delta s \to 0} \frac{\oint_c \mathbf{F} \cdot d\mathbf{l}}{\Delta s} \mathbf{a}_n \tag{28}$$

We might also refer to (28) as the circulation per unit area as the area tends to zero (or shrinks to a point), the orientation of the area being such that the circulation is a maximum. As in Ampère's circuital law, the line integral in (28) is taken in the right-hand sense for positive unit normal $\mathbf{a}_n$. In Chap. 2, it was also seen that (28) is equivalent to

$$\text{curl } \mathbf{F} = \nabla \times \mathbf{F} \tag{29}$$

where ∇ is the del operator defined in rectangular coordinates as

$$\nabla = \mathbf{a}_x \frac{\partial}{\partial y} + \mathbf{a}_y \frac{\partial}{\partial y} + \mathbf{a}_z \frac{\partial}{\partial z}$$

We may now apply the definition of curl and its mathematical form (29) to express Ampère's circuital law given in (26) in terms of $\mathbf{H}$ in the differential form. For the sake of illustration, we let the current be only z-directed. Thus, from (26) and (28) we get

$$\text{curl } \mathbf{H} = \lim_{\Delta s \to 0} \frac{\oint_c \mathbf{H} \cdot d\mathbf{l}}{\Delta s} \mathbf{a}_z = \lim_{\Delta s \to 0} \frac{\int_{\Delta s} J_z \, ds}{\Delta s} \mathbf{a}_z \tag{30}$$

But $\Delta s = dx\,dy$ in the limiting case ($\Delta s \to 0$), so that (30) becomes

$$(\text{curl } \mathbf{H})_z = \frac{J_z \, dx\,dy}{dx\,dy} = J_z \tag{31}$$

Combining (29) and (31) yields

$$(\nabla \times \mathbf{H})_z = J_z \tag{32}$$

and similarly for all other components of **J**. Therefore, in general we have

$$\nabla \times \mathbf{H} = \mathbf{J} \tag{33}$$

This expression is Ampère's circuital law in the differential (or point) form.

EXAMPLE 4.6 Suppose we wish to produce a magnetic field of the form $\mathbf{H} = K \sin x \, \mathbf{a}_y$, where K is a constant. Find the current density that will produce this field.

Solution From (29) and (33) we find that

$$\nabla \times \mathbf{H} = \frac{\partial}{\partial x}(H_y)\mathbf{a}_z = J\,\mathbf{a}_z$$

Substituting the given H_y above yields

$$\mathbf{J} = K \cos x \, \mathbf{a}_z$$

We will refer to Ampère's circuital law as simply Ampère's law in all of our future discussions.

4.5 Potentials in Magnetostatics

In our discussions so far, we have defined the static magnetic field, identified the sources that produce magnetic fields, and formulated mathematical expressions that relate fields to sources. Thus, using the Biot–Savart law or Ampère's law, we were able to find the magnetic field at a point for a given current. We can also use the concept of "potentials" to evaluate the fields in a region. In fact, the use of potential functions offers a very powerful method of solving both static and dynamic electromagnetic field problems (as demonstrated in later chapters).

We were introduced earlier to the scalar potential to determine the static electric field. As we will see in the next subsections, in determining the static magnetic fields we may use both scalar and vector potentials. By scalar and vector potentials, we mean that the potential could be respectively a scalar or a vector quantity. Why both are feasible and where these are applicable are now discussed.

4.5.1 Magnetic Scalar Potential

Let us consider a source-free region, for which (33) becomes

$$\nabla \times \mathbf{H} = 0 \tag{34}$$

This is a relationship similar to the one we obtained for the electric field. Therefore, we can express the vector **H** as a gradient of a scalar quantity $\mathcal{F}$, because

$$\nabla \times \nabla \mathcal{F} = 0 \tag{35}$$

which is a general identity for any scalar field (see Appendix A). Or from (34) and (35) we may write

$$\mathbf{H} = -\nabla \mathcal{F} \tag{36}$$

The negative sign has been chosen in order to retain the similarity with the electrostatic potential V. The quantity $\mathcal{F}$ is called the *magnetic scalar potential.* From (36) we also have†

$$\mathcal{F}_2 - \mathcal{F}_1 = -\int_1^2 \mathbf{H} \cdot d\mathbf{l} \tag{37}$$

In (36) and (37) $\mathcal{F}$ has the dimension of amperes. For most practical cases, the ampere-turn (abbreviated At) is frequently used as a unit for $\mathcal{F}$. As a potential rise or source of magnetic field, the term *magnetomotive force*, or mmf, is often used for (37), and potential drop is expressed as a reluctance drop, as we shall see in the study of magnetic circuits in Sec. 4.10.

To establish another useful property of the magnetic scalar potential, we observe that over a closed surface s

$$\psi_m = \oint_s \mathbf{B} \cdot d\mathbf{s} = 0 \tag{38}$$

where ψ_m is the magnetic flux. This relationship is true because in a region of magnetic flux lines, if we consider a closed surface, the flux entering the surface must all come out, magnetic monopoles being nonexistent. Recall that (38) is analogous to Gauss' law of electrostatics for a source-free region. In a region containing a charge, the electric flux lines will terminate on the charge. But no physical magnetic quantity is analogous to an isolated electric charge. Magnets have pole pairs called *north* and *south poles*, and we do not have a magnet that has either a north or a south pole.

Equation (38) can be expressed in the differential form by using the divergence theorem. Thus, we have

$$\oint_s \mathbf{B} \cdot d\mathbf{s} = \int_v (\nabla \cdot \mathbf{B}) dv = 0 \tag{39}$$

---

† This should be clear by direct analogy to the electrostatic field in which

$$\mathbf{E} = -\nabla V \qquad V_2 - V_1 = -\int_1^2 \mathbf{E} \cdot d\mathbf{l}$$

In (39) the integrand must be zero, and dv not being zero, we must have

$$\nabla \cdot \mathbf{B} = 0 \tag{40}$$

But, substituting (27) in (40) yields

$$\nabla \cdot \mathbf{H} = 0 \tag{41}$$

and combining (36) and (41), we obtain

$$\nabla \cdot \mathbf{H} = \nabla \cdot (-\nabla \mathcal{F}) = -\nabla^2 \mathcal{F} = 0$$

Or, finally,

$$\nabla^2 \mathcal{F} = 0 \tag{42}$$

This is Laplace's equation for the magnetic scalar potential in a source-free region. As we shall see in Chap. 10, in many problems of practical interest we can first solve Laplace's equation for the potential function, from which we can then obtain the desired field quantities by the use of the defining equation, such as (36).

4.5.2 Magnetic Vector Potential

In the preceding discussion we considered a source-free region for which we derived the magnetic scalar potential in order to study the magnetic field in the region. We were able to derive (36) by virtue of (34), which is a curl equation valid only if $\mathbf{J} = 0$. If we now consider the divergence equation (40), which is valid everywhere in a region (with or without sources), and use the vector identity $\nabla \cdot \nabla \times \mathbf{A} \equiv 0$, we may write

$$\mathbf{B} = \nabla \times \mathbf{A} \tag{43}$$

where $\mathbf{A}$ is known as the magnetic vector potential. Here, we wish to point out that although $\mathbf{A}$ has a physical significance, we will be better off by considering $\mathbf{A}$ as a concept in the intermediate step in obtaining the magnetic field than by wondering about the physical meaning of $\mathbf{A}$. For the time being, we might as well treat (43) as a mathematical operation.

Writing Ampère's law in terms of $\mathbf{B}$, we have

$$\nabla \times \mathbf{B} = \mu_0 \mathbf{J} \tag{44}$$

in free space having a source of current density $\mathbf{J}$. Now taking the curl of both sides of (43), we get (see Appendix A)

$$\nabla \times \nabla \times \mathbf{A} = \nabla(\nabla \cdot \mathbf{A}) - \nabla^2 \mathbf{A} = \nabla \times \mathbf{B} \tag{45}$$

In (43) no restriction was placed on **A**, except that it is a vector. We may put a constraint on **A** such that $\nabla \cdot \mathbf{A} = 0$. Later we will see that it is not an arbitrary restriction for convenience, rather, it can be proved that $\nabla \cdot \mathbf{A} = 0$ is consistent with (43). Substituting this condition in (45) and combining with (44) yields

$$\nabla^2 \mathbf{A} = -\mu_0 \mathbf{J} \tag{46}$$

which is Poisson's equation for the magnetic vector potential. Clearly, if the region is source-free, $\mathbf{J} = 0$, and (46) reduces to Laplace's equation

$$\nabla^2 \mathbf{A} = 0 \tag{47}$$

Having established the fact that **A** satisfies (46) in free space containing a current density **J**, and (47) in a current-free region, we may solve these for **A** and then use (43) to find the magnetic flux density. The process is illustrated by the following example.

EXAMPLE 4.7 In a certain region the magnetic vector potential is given by $\mathbf{A} = ke^{-\alpha y} \sin \alpha x \, \mathbf{a}_z$, where k and α are constants. Find **B**. This form of a magnetic vector potential may be produced by a current sheet having a linear density (A/m) of sinusoidal variation.

Solution From (43) we have

$$\nabla \times \mathbf{A} = \frac{\partial A_z}{\partial y} \mathbf{a}_x - \frac{\partial A_z}{\partial x} \mathbf{a}_y = B_x \mathbf{a}_x + B_y \mathbf{a}_y = \mathbf{B}$$

Therefore, knowing that $A_z = ke^{-\alpha y} \sin \alpha x$, we obtain

$$\mathbf{B} = -k\alpha e^{-\alpha y}(\sin \alpha x \, \mathbf{a}_x + \cos \alpha x \, \mathbf{a}_y)$$

The magnetic vector potential is related to the current by (46). We can also obtain an integral relationship between **A** and **J**. We shall do this rather intuitively by considering the similarities between equations governing the static electric field and the static magnetic field. Thus, we observe that the electric potential V and the magnetic vector potential **A** both satisfy Poisson's equation. Also, V can be related to a volume charge density by

$$V = \int_v \frac{\rho_v \, dv}{4\pi\epsilon R} \tag{48}$$

Similar to (48) we let **A** be related to a current element by

$$\mathbf{A} = \int_v \frac{\mu_0 \mathbf{J} \, dv}{4\pi R} \tag{49}$$

If this form of **A** is consistent with the definition of **H** obtained from the Biot–Savart law, our conjecture that **A** is given by (49) is correct. We rewrite (23) below:

$$\mathbf{H} = \int_c \frac{I\, d\mathbf{l} \times \mathbf{a}_R}{4\pi R^2} \tag{50}$$

Also, for convenience, we repeat (25) and (43) as

$$\mathbf{B} = \mu_0 \mathbf{H} = \nabla \times \mathbf{A} \tag{51}$$

From (49) and (51) we have

$$\mathbf{H} = \frac{1}{\mu_0} \nabla \times \mathbf{A} = \frac{1}{4\pi} \int_v \nabla \times \frac{\mathbf{J}'\, dv'}{R} \tag{52}$$

Notice that in (52) we used $\mathbf{J}'\, dv'$ to indicate that $\mathbf{J}'\, dv'$ is located at some point (x', y', z'), whereas the field **H** is to be found at (x, y, z) and ∇ is a derivative with respect to (x, y, z). Using the vector identity (Appendix A)

$$\nabla \times (\phi \mathbf{A}) = \nabla \phi \times \mathbf{A} + \phi \nabla \times \mathbf{A}$$

in (52) yields

$$\mathbf{H} = \frac{1}{4\pi} \int_v \left[\left(\nabla \frac{1}{R} \right) \times \mathbf{J}' + \frac{1}{R} \nabla \times \mathbf{J}' \right] dv' \tag{53}$$

Clearly $\nabla \times \mathbf{J}' = 0$, because the partial derivative of a function of (x', y', z') with respect to (x, y, z) is zero, which implies that (53) becomes

$$\mathbf{H} = \frac{1}{4\pi} \int_v \left[\left(\nabla \frac{1}{R} \right) \times \mathbf{J}' \right] dv' \tag{54}$$

Because $\nabla(1/R) = -\mathbf{a}_R/R^2$, (54) can also be written as

$$\mathbf{H} = -\frac{1}{4\pi} \int_v \frac{\mathbf{a}_R \times \mathbf{J}'}{R^2}\, dv' = \int_v \frac{\mathbf{J}' \times \mathbf{a}_R}{4\pi R^2}\, dv' \tag{55}$$

If $\mathbf{J}'$ is a line current, we can write $\mathbf{J}'\, dv' = I\, d\mathbf{l}$ and (55) then becomes

$$\mathbf{H} = \int_v \frac{I\, d\mathbf{l} \times \mathbf{a}_R}{4\pi R^2} \tag{56}$$

Notice that (50) and (56) are identical, indicating that **A** is correctly defined by (49).

We may now use the definition of the magnetic vector potential to derive Ampère's law, and also show that $\nabla \cdot \mathbf{A} = 0$, as presumed in (45) and (46).

Let us consider the divergence of $\mathbf{A}$ first. We have, from (49),

$$\nabla \cdot \mathbf{A} = \frac{\mu_0}{4\pi} \int_v \nabla \cdot \frac{\mathbf{J}'}{R}\, dv' \tag{57}$$

where $\mathbf{J}' = \mathbf{J}(x', y', z')$, $\mathbf{A} = \mathbf{A}(x, y, z)$, and $\nabla = \nabla(x, y, z)$. We now use the vector identity (see Appendix A)

$$\nabla \cdot (\psi \mathbf{A}) = \mathbf{A} \cdot \nabla \psi + \psi \nabla \cdot \mathbf{A}$$

in (57) to obtain

$$\nabla \cdot \mathbf{A} = \frac{\mu_0}{4\pi} \int_v \left[\mathbf{J}' \cdot \left(\nabla \frac{1}{R} \right) + \frac{1}{R} \nabla \cdot \mathbf{J}' \right] dv' \tag{58}$$

Because ∇ is a partial derivative with respect to (x, y, z) and $\mathbf{J}'$ is a function of (x', y', z'), $\nabla \cdot \mathbf{J}' = 0$. Thus, (58) becomes

$$\nabla \cdot \mathbf{A} = \frac{\mu_0}{4\pi} \int_v \mathbf{J}' \cdot \left(\nabla \frac{1}{R} \right) dv' \tag{59}$$

Now

$$\nabla \frac{1}{R} = -\nabla' \frac{1}{R} \tag{60}$$

Substituting (60) in (59) and expanding, we get

$$\begin{aligned} \nabla \cdot \mathbf{A} &= -\frac{\mu_0}{4\pi} \int_v \mathbf{J}' \cdot \left(\nabla' \frac{1}{R} \right) dv' \\ &= \frac{\mu_0}{4\pi} \int_v \left[\frac{1}{R} \nabla' \cdot \mathbf{J}' - \nabla' \cdot \frac{\mathbf{J}'}{R} \right] dv' \end{aligned} \tag{61}$$

From the continuity equation, $\nabla' \cdot \mathbf{J}' = 0$, and (61) reduces to

$$\nabla \cdot \mathbf{A} = -\frac{\mu_0}{4\pi} \int_v \left(\nabla' \cdot \frac{\mathbf{J}'}{R} \right) dv'$$

which can be expressed in the following form with the aid of the divergence theorem:

$$\nabla \cdot \mathbf{A} = -\frac{\mu_0}{4\pi} \oint_{s'} \frac{\mathbf{J}'}{R} \cdot d\mathbf{s}' \tag{62}$$

Because the surface s' must include all currents, there can be no current flow through the surface. Consequently,

$$\mathbf{J}' \cdot d\mathbf{s}' = 0$$

and finally,

$$\nabla \cdot \mathbf{A} = 0 \tag{63}$$

To derive Ampère's law reconsider (45) as

$$\mu_0 \nabla \times \mathbf{H} = \nabla(\nabla \cdot \mathbf{A}) - \nabla^2 \mathbf{A} \tag{64}$$

and (63) and (64) give

$$\nabla^2 \mathbf{A} = -\mu_0 \nabla \times \mathbf{H} \tag{65}$$

Returning now to the similarity between static electric and magnetic fields, (48) and (49), we recall that (48) satisfies Poisson's equation. Thus, by analogy, (49) must also satisfy Poisson's equation (in terms of different variables); that is,

$$\nabla^2 \mathbf{A} = -\mu_0 \mathbf{J} \tag{66}$$

Comparing (65) and (66) yields

$$\nabla \times \mathbf{H} = \mathbf{J}$$

which is Ampère's law.

Let us summarize the essence of this section. We first introduced the concepts of magnetic scalar and vector potentials. We saw that in source-free regions both potentials satisfy Laplace's equation. Both aid in the determination of magnetic fields. The choice between the scalar and vector potentials depends on the nature of the problem to be solved. In a source-free region both potentials are equally applicable, but in a region containing currents the use of vector potential is necessary. For instance, the correct definition of the magnetic vector potential led us to the derivation of Ampère's circuital law.

EXAMPLE 4.8 Find the magnetic field, at a point P, produced by a current element of length l and carrying a current I, as shown in Fig. 4.10.

Solution This problem can be solved in a number of ways, but we shall use the magnetic vector potential $\mathbf{A}$ to find the fields. From (49) we have

$$\mathbf{A} = A_z \mathbf{a}_z = \frac{\mu_0 I}{4\pi} \int_{-l/2}^{l/2} \frac{dl}{R} \mathbf{a}_z \tag{67}$$

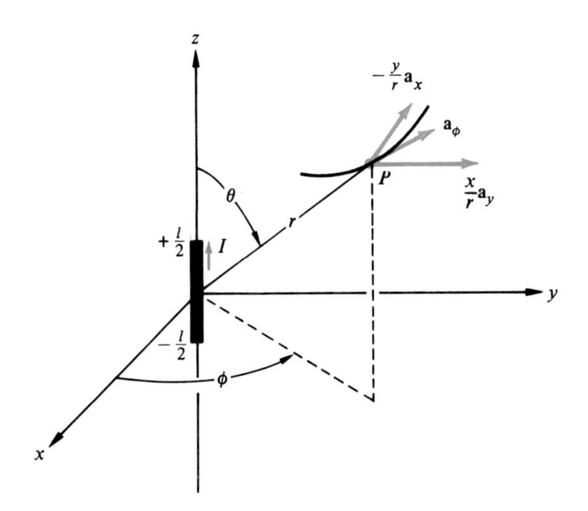

FIGURE 4.10
Example 4.8. Magnetic field due to a current element of finite length.

If we assume that $l \ll r$, then (67) becomes

$$A_z = \frac{\mu_0 Il}{4\pi r} \tag{68}$$

From (43) we have

$$B_x = \frac{\partial A_z}{\partial y} \tag{69}$$

and

$$B_y = -\frac{\partial A_z}{\partial x} \tag{70}$$

Substituting $r = \sqrt{x^2 + y^2 + z^2}$ in (68) and then using (69) and (70) yields (after some simplification)

$$\mathbf{B} = \frac{\mu_0 Il}{4\pi r^2}\left(-\frac{y}{r}\,\mathbf{a}_x + \frac{x}{r}\,\mathbf{a}_y\right) \tag{71}$$

Or, using the geometry of Fig. 4.10, (71) can be expressed as

$$\mathbf{B} = \frac{\mu_0 Il \sin\theta}{4\pi r^2}(-\sin\phi\,\mathbf{a}_x + \cos\phi\,\mathbf{a}_y) = \frac{\mu_0 Il \sin\theta}{4\pi r^2}\,\mathbf{a}_\phi$$

EXAMPLE 4.9 A circular loop of radius a is located in free space and carries a current I. Determine the magnetic vector potential at a point P located at a distance r from the center of the loop. Given: $r \gg a$. Also find the magnetic flux density at P.

Solution Referring to Fig. 4.11, we pair off current elements at equal $\pm\phi$. Thus, from symmetry in spherical coordinates, **A** is only ϕ-directed at the point P. From (49), therefore,

$$\mathbf{A} = A_\phi \mathbf{a}_\phi = \frac{\mu_0 I}{4\pi} \oint \frac{dl_\phi}{R} \mathbf{a}_\phi \tag{72}$$

For simplicity we choose P in the xz plane, for which $\phi = 0$. Then

$$dl_\phi = dl \cos \phi = a \, d\phi \cos \phi$$

and

$$R^2 = a^2 + r_1^2 + z^2 - 2ar_1 \cos \phi$$

Hence, (72) becomes

$$\mathbf{A} = \frac{2\mu_0 I}{4\pi} \int_0^\pi \frac{a \cos \phi \, d\phi}{\sqrt{a^2 + r_1^2 + z^2 - 2ar_1 \cos \phi}} \mathbf{a}_\phi \tag{73}$$

Since $r \gg a$ and $r^2 \gg 2ar_1$, we may write

$$\frac{1}{\sqrt{a^2 + r_1^2 + z^2 - 2ar_1 \cos \phi}} \simeq \frac{1}{\sqrt{r^2 - 2ar \cos \phi}} = r\left(1 - \frac{2ar_1}{r^2} \cos \phi\right)^{-1/2}$$

$$\simeq \frac{1}{r}\left(1 + \frac{ar_1}{r^2} \cos \phi\right)$$

Consequently, (73) becomes

$$\mathbf{A} = \frac{\mu_0 I}{2\pi} \int_0^\pi \frac{a \cos \phi}{r} \left(1 + \frac{ar_1}{r^2} \cos \phi\right) d\phi \, \mathbf{a}_\phi = \frac{\mu_0 I}{2\pi} \frac{a^2 r_1}{r^3} \frac{\pi}{2} \mathbf{a}_\phi = \frac{\mu_0 I a^2 r_1}{4r^3} \mathbf{a}_\phi$$

$$= \frac{\mu_0 I a^2 \sin \theta}{4r^2} \mathbf{a}_\phi \tag{74}$$

We define the quantity

$$m = \pi a^2 I \tag{75}$$

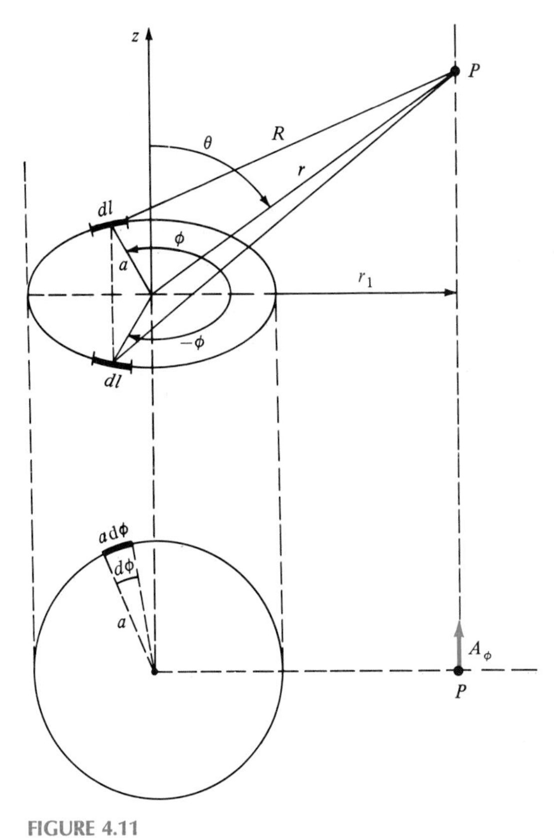

FIGURE 4.11
Example 4.9. Magnetic field of a current loop.

as the *magnetic-dipole moment*. Then (74) and (75) yield

$$\mathbf{A} = \frac{\mu_0 m \sin\theta}{4\pi r^2}\,\mathbf{a}_\phi \tag{76}$$

Hence, from (43) and (76) it follows that

$$B_r = \frac{\mu_0 m}{2\pi r^3}\cos\theta$$

$$B_\theta = \frac{\mu_0 m}{4\pi r^3}\sin\theta$$

and

$$B_\phi = 0$$

4.6 Macroscopic Properties of Magnetic Materials

An exact approach to the study of the magnetic properties of materials involves the use of quantum mechanics, but here we shall take a much more simplified approach.

It was shown in Example 4.9 that the magnetic field at a point (r, θ, ϕ), in spherical coordinates, due to a circular loop of radius a carrying a current I lying in the xy plane and centered at the origin is given by

$$B_r = \mu_0 \frac{I\pi a^2}{2\pi r^3} \cos\theta$$

$$B_\theta = \mu_0 \frac{I\pi a^2}{4\pi r^3} \sin\theta$$

$$B_\phi = 0$$

The quantity $I\pi a^2$ (defined in Example 4.9 as the magnetic-dipole moment) acts along $\theta = 0$. Thus, a current-carrying circular loop can be considered to be a magnetic dipole. Now, returning to our simplified approach, we consider an atom to consist of a positive nucleus surrounded by electrons that spin about their axes in addition to orbiting around the nucleus. An orbiting electron is similar to a current loop and can thus be considered to have magnetic-dipole moments. The magnetic properties of a material depend on the contributions of the various components of the magnetic moments.

We have just mentioned that orbiting electrons are analogous to circular current loops which give rise to magnetic dipoles. To visualize the current loops in the material, we replace the magnetic dipoles by circulating currents, as

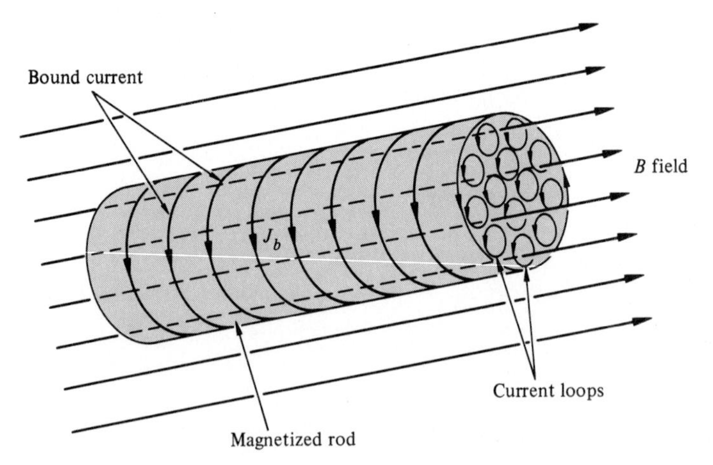

FIGURE 4.12

A magnetized rod in a magnetic field. Magnetic dipoles have been replaced by current loops, resulting in a bound current sheet.

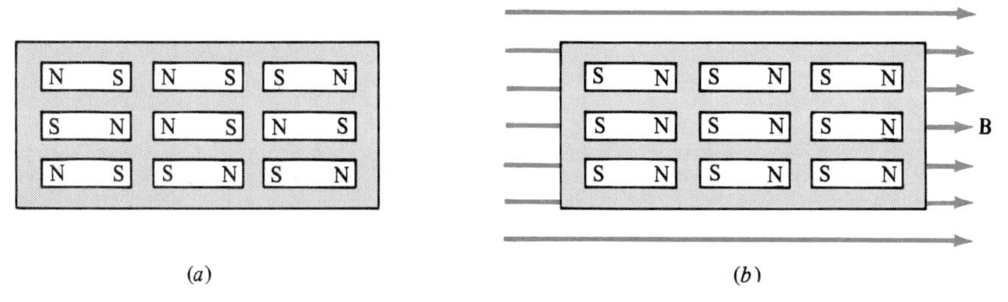

FIGURE 4.13
Magnetic dipoles in a magnetic material: (*a*) randomly oriented dipoles; (*b*) magnetic dipoles in a magnetic field magnetizing the material.

shown in Fig. 4.12. Notice that these small current loops result in a uniform circulating current around the periphery of the material, as shown in Fig. 4.12. The resulting surface current is known as the *bound current* and denoted as J_b. It is clear from Fig. 4.12 that the bound current is in the same direction as the current producing the magnetizing field.

Thus, a magnetic material may be considered to consist of randomly distributed magnetic dipoles (Fig. 4.13*a*) (similar to electric dipoles in a dielectric material, as discussed in Chap. 3). If a magnetic material is brought into a region of a magnetic field, the effect of the field on magnetic dipoles is similar to the effect of an electric field on electric dipoles in that we obtain a net magnetic-dipole moment, as shown in Fig. 4.13*b*. The net magnetic-dipole moment per unit volume is defined as the *magnetization*, or *magnetic polarization*, M of the material. By magnetization, we imply that because of magnetic dipole moments, a magnetic field is established. In other words, just as a magnetic intensity $\mathbf{H}$ produces a magnetic flux density, a magnetic-dipole moment per unit volume $\mathbf{M}$ also gives rise to a flux density. In this regard $\mathbf{M}$ is analogous to $\mathbf{H}$. We now recall Ampère's law expressed as (33), $\nabla \times \mathbf{H} = \mathbf{J}$, and conclude from the similarity between $\mathbf{M}$ and $\mathbf{H}$ that magnetization is produced by certain currents known as bound, or amperean, currents. Therefore, on the basis of Ampère's law, we write in point form

$$\nabla \times \mathbf{M} = \mathbf{J}_b \tag{77}$$

where $\mathbf{J}_b$ is the bound current density. Notice that we distinguish the bound current density $\mathbf{J}_b$ from the free current density $\mathbf{J}_f$. The total current density $\mathbf{J}$ is defined as

$$\mathbf{J} = \mathbf{J}_b + \mathbf{J}_f \tag{78}$$

If $\mathbf{B}$ is the field produced in free space by $\mathbf{J}$, then from Ampère's law and (78) we obtain

$$\nabla \times \frac{\mathbf{B}}{\mu_0} = \mathbf{J} = \mathbf{J}_b + \mathbf{J}_f \tag{79}$$

Equations (77) and (79) yield

$$\nabla \times \left(\frac{\mathbf{B}}{\mu_0} - \mathbf{M} \right) = \mathbf{J}_f \tag{80}$$

Thus, the new vector $(\mathbf{B}/\mu_0) - \mathbf{M}$ is produced by the free current, and we may define $\mathbf{H}$ such that

$$\frac{\mathbf{B}}{\mu_0} - \mathbf{M} = \mathbf{H} \tag{81}$$

which may also be written as

$$\mathbf{B} = \mu_0(\mathbf{M} + \mathbf{H}) \tag{82}$$

where $\mathbf{M}$ is the magnetic polarization, or magnetization, and is due to the magnetic dipoles of the material. For a linear magnetic material, $\mathbf{M}$ is directly proportional to $\mathbf{H}$ such that

$$\mathbf{M} = \chi_m \mathbf{H} \tag{83}$$

We may combine (82) and (83) to obtain

$$\mathbf{B} = \mu \mathbf{H} = \mu_0(1 + \chi_m)\mathbf{H} \tag{84}$$

where

$$\mu = \mu_0(1 + \chi_m) = \mu_0 \mu_r \tag{85}$$

where χ_m is known as *magnetic susceptibility*. The quantity μ is called the permeability of the material and μ_r is defined as the *relative permeability* of the material.

A material is classified according to the nature of its relative permeability μ_r, which is actually related to the internal atomic structure of the material. Most "nonmagnetic" materials are classified as either *paramagnetic*, for which μ_r is slightly greater than 1.0, or *diamagnetic*, in which μ_r is slightly less than 1.0. For all practical purposes, μ_r can be considered to be equal to 1.0 for all of these materials. However, there is one interesting case of diamagnetism that is becoming of interest in certain types of electromagnetic devices. This is "perfect diamagnetism" (the Meissner effect) and occurs in certain types of materials known as *superconductors* at temperatures near absolute zero. In such materials $\mathbf{B} = 0$ and μ is essentially zero; that is, no magnetic field can be established in the superconducting material.

There are several further classifications of materials that exhibit greater degrees of magnetism, but only two classes are discussed in detail here: ferromagnetic and ferrimagnetic materials. (1) Ferromagnetic materials are subgrouped into hard and soft materials, this classification roughly corresponding to the physical hardness of the materials. Soft ferromagnetic materials include the elements iron, nickel, cobalt, and some rare earth elements; most soft steels; and many alloys of the four elements. Hard ferromagnetic materials include the permanent magnet materials (such as the alnicos), several alloys of cobalt with the rare earth elements, chromium steels, certain copper-nickel alloys, and many other metal alloys. (2) Ferrimagnetic materials are the ferrites and are composed of iron oxides having the formula $MeOFe_2O_3$, where Me represents a metallic iron. Ferrites are likewise subgrouped into hard and soft ferrites, the former being the permanent magnetic ferrites, usually barium or strontium ferrite. Soft ferrites include the nickel-zinc and manganese-zinc ferrites and are used in microwave devices, delay lines, transformers, and other generally high-frequency applications. (3) A third class of magnetic materials of growing importance is made from powdered iron (or other magnetic material) particles suspended in a nonferrous matrix, such as epoxy or plastic. Sometimes termed *superparamagnetic*, powdered-iron parts are formed by compression or injection molding techniques.

Having classified the various types of magnetic materials, we turn to a very interesting characteristic of ferromagnetic (and certain other types of) materials. These materials exhibit the phenomenon of *hysteresis*, whereby the magnetization (or the **B** field) in the material saturates beyond a certain magnetic field strength (**H**) and the material retains a net remanent magnetization after **H** has been reduced to zero. A negative **H** is needed to reduce the magnetization to zero, and this negative **H** is known as the *coercive force*. A complete cyclic change of **B** with **H** for a ferromagnetic material is shown in Fig. 4.14. Notice that the *B-H* curve is a closed loop, known as the *hysteresis loop*.

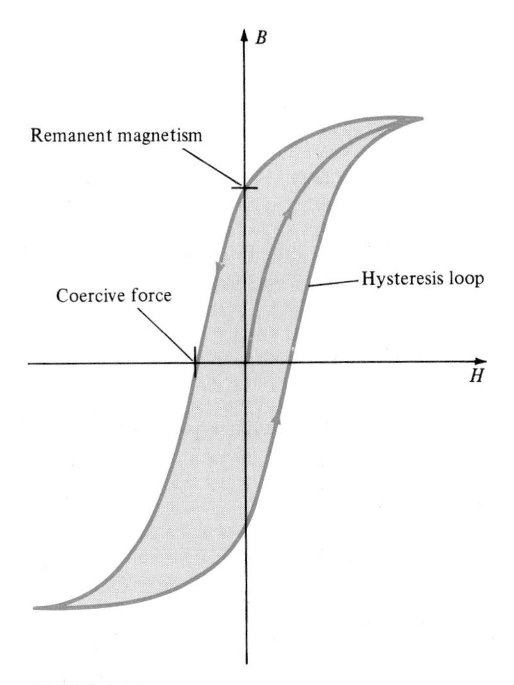

FIGURE 4.14

A typical *B-H* curve.

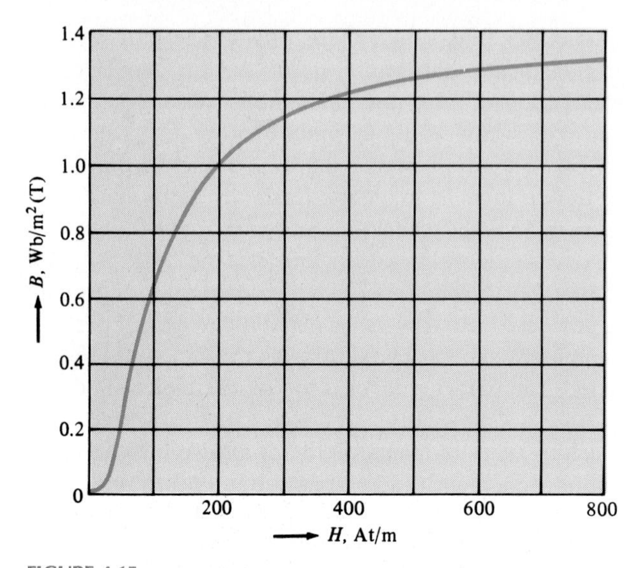

FIGURE 4.15
Saturation curve for a silicon-sheet steel example.

For a unidirectional magnetization of a ferromagnetic material, the phenomenon of magnetic saturation is illustrated in Fig. 4.15, which shows that **B** is linearly proportional to **H** up to a certain point. Beyond this point, **B** does not increase substantially, regardless of an increase in **H**. It is evident from Fig. 4.15 that the permeability of a ferromagnetic material, in general, is not a constant. Rather, it is a function of **H** (that is, the operating point), and the material is nonlinear.

EXAMPLE 4.10 A cylindrical conductor of radius a and permeability μ carries a z-directed current I uniformly distributed over the conductor cross section. Evaluate the magnetization **M** within the conductor. Also, determine the bound current density.

Solution From Ampère's law, within the conductor we have

$$\oint_c \mathbf{H} \cdot d\mathbf{l} = I$$

Thus, for $0 < r < a$,

$$H_\phi 2\pi r = \frac{I\pi r^2}{\pi a^2}$$

or

$$B_\phi = \mu H_\phi = \frac{\mu I r}{2\pi a^2}$$

and

$$M_\phi = \frac{B_\phi}{\mu_0} - H_\phi = \frac{I}{2\pi a^2}\left(\frac{\mu}{\mu_0} - 1\right) r \qquad \text{A/m}$$

From $\mathbf{J}_b = \nabla \times \mathbf{M}$ we obtain, for the z-directed bound current,

$$J_b = \frac{1}{r}\frac{\partial}{\partial r}(rM_\phi) = \frac{I}{\pi a^2}\left(\frac{\mu}{\mu_0} - 1\right) \qquad \text{A/m}^2$$

4.7 Boundary Conditions for Magnetic Fields

The preceding discussisons lead us to formulate the boundary conditions for static magnetic fields. Consider the interface shown in Fig. 4.16. The material properties of region 1 are different from those of region 2, and a sheet of current flows at the interface, as shown. This current is a surface current and has a linear current density of K A/m width. Let the field components in region 1 be B_{n1} and H_{t1}, and those in region 2 be B_{n2} and H_{t2}, where the subscripts denote normal and tangential components.

To obtain the relationships between H_{t1} and H_{t2}, we evaluate the line integral around the path shown in Fig. 4.16 and apply Ampère's law to obtain

$$\oint_c \mathbf{H}\cdot d\mathbf{l} = H_{t1}\,\Delta l - H_{t2}\,\Delta l = K\,\Delta l \tag{86}$$

In (86) Δl is small and thus H_t is uniform over Δl. At the interface, the portions 2 to 3 and 4 to 1 in the integration path are of infinitesimal length. Therefore (86) becomes

$$H_{t1} - H_{t2} = K \tag{87}$$

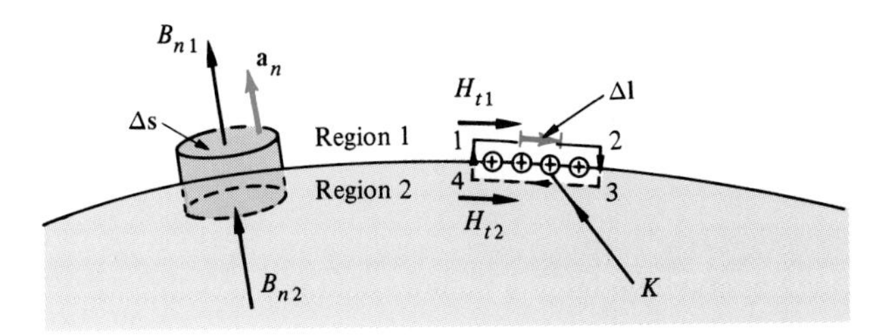

FIGURE 4.16

Magnetic field components at a boundary: determining the boundary conditions.

The relationship between the normal components of the **B** field can be found by considering the infinitesimal volume at the interface, as shown in Fig. 4.16. From (38) we obtain, over the surface Δs,

$$\oint_s \mathbf{B} \cdot d\mathbf{s} = B_{n1}\,\Delta s - B_{n2}\,\Delta s = 0 \tag{88}$$

Here, the assumption is that Δs is small enough that B_n does not change over it. Therefore, (88) yields

$$B_{n1} = B_{n2} \tag{89}$$

Equations (87) and (89) constitute the boundary conditions for the field components. We see from (89) that the normal component of the **B** field is continuous at the interface, and (87) shows that the tangential component of the **H** field is continuous if there are no currents at the boundary; otherwise, there is a discontinuity in H_t at the interface, as given in (87).

4.8 Energy Stored in the Magnetic Field

In this chapter we have touched on the analogy between the static electric and magnetic fields. In particular, the electric field **E** can be viewed as being analogous to the magnetic flux density **B**, whereas the electric flux density **D** is similar to the magnetic field intensity **H**. We have seen in Chap. 3 that the energy stored in an electric field is given by

$$W_e = \tfrac{1}{2} \int_v \mathbf{D} \cdot \mathbf{E}\, dv$$

Correspondingly, we write the energy stored in the magnetic field W_m as†

$$W_m = \tfrac{1}{2} \int_v \mathbf{B} \cdot \mathbf{H}\, dv \tag{90}$$

Using the relationship $\mathbf{B} = \mu \mathbf{H}$, (90) can also be expressed as

$$W_m = \tfrac{1}{2}\mu \int_v |\mathbf{H}|^2\, dv \tag{91}$$

or

$$W_m = \frac{1}{2\mu} \int_v |\mathbf{B}|^2\, dv \tag{92}$$

† A rigorous derivation of (90) to (92) is possible with the aid of Faraday's law and the Lorentz force equation. See R. S. Elliott, *Electromagnetics*, McGraw-Hill, New York, 1966, pp. 283–285.

4.9 Forces in Electric and Magnetic Fields

In connection with the definition of the electric field, we saw that the force $\mathbf{F}_e$ on a test charge Q in an electric field $\mathbf{E}$ is given by

$$\mathbf{F}_e = Q\mathbf{E} \tag{93}$$

and thus we defined the electric field. Although in this chapter we defined the magnetic field in terms of Ampère's force law, we could use an alternative formulation to define the magnetic field. Accordingly, if a test charge moves with a velocity $\mathbf{u}$ in a region and experiences a force $\mathbf{F}_m$, then the region is characterized by a magnetic field $\mathbf{B}$, and the force is given by

$$\mathbf{F}_m = Q\mathbf{u} \times \mathbf{B} \tag{94}$$

Notice from (93) and (94) that $\mathbf{F}_e$ is in the direction of $\mathbf{E}$, whereas $\mathbf{F}_m$ is in a direction at right angles to both $\mathbf{u}$ and $\mathbf{B}$. In a region having an electric field $\mathbf{E}$ and a magnetic field $\mathbf{B}$, the total force $\mathbf{F}$ is the sum of $\mathbf{F}_e$ and $\mathbf{F}_m$. Therefore, from (93) and (94) we obtain

$$\mathbf{F} = \mathbf{F}_e + \mathbf{F}_m = Q(\mathbf{E} + \mathbf{u} \times \mathbf{B}) \qquad \text{N} \tag{95}$$

This equation is known as the *Lorentz force equation* and is often expressed as a force-density equation. Consider a volume dv small enough that the force density $\mathbf{f}$ is uniform within dv, and let ρ be the uniform charge density; then

$$\mathbf{F} = \mathbf{f}\, dv \qquad Q = \rho\, dv$$

Substituting these in (95) yields

$$\mathbf{f}\, dv = (\rho\mathbf{E} + \rho\mathbf{u} \times \mathbf{B})\, dv$$

or

$$\mathbf{f} = \rho(\mathbf{E} + \mathbf{u} \times \mathbf{B}) \qquad \text{N/m}^3 \tag{96}$$

But $\rho\mathbf{u} = \mathbf{J}$, the current density in amperes per square meter, so that (96) finally becomes

$$\mathbf{f} = \rho\mathbf{E} + \mathbf{J} \times \mathbf{B} \tag{97}$$

An alternative nature of force production in a magnetic field is discussed in Sec. 4.12.

We will now consider a few examples to illustrate some of the theory developed in this section.

EXAMPLE 4.11 A conductor of cross section A and carrying a current I is oriented along the y axis (as shown in Fig. 4.17) in a magnetic field $\mathbf{B} = B_0\mathbf{a}_x + B_0\mathbf{a}_y$. What is the force density exerted on the conductor?

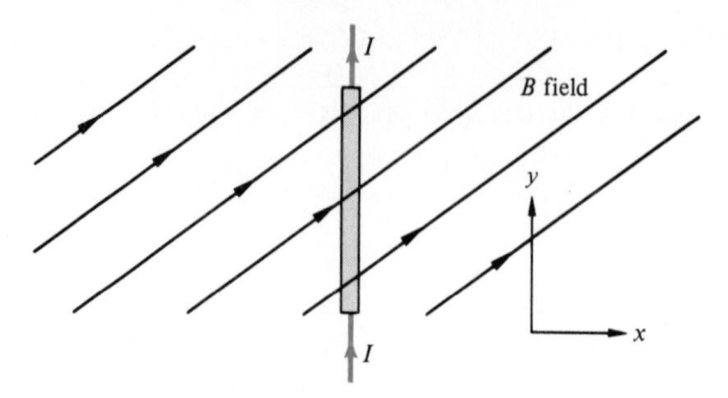

FIGURE 4.17
Example 4.11. Force on a current-carrying conductor in a magnetic field.

Solution The current density **J** is given by

$$\mathbf{J} = \frac{I}{A}\,\mathbf{a}_y$$

From the Lorentz force equation, (97) we have

$$\mathbf{f} = \mathbf{J} \times \mathbf{B} = \frac{I}{A}\,\mathbf{a}_y \times (B_0\mathbf{a}_x + B_0\mathbf{a}_y)$$

or

$$\mathbf{f} = -\frac{B_0 I}{A}\,\mathbf{a}_z \qquad \text{N/m}^3$$

EXAMPLE 4.12 If the B field of Example 4.11 exists in free space, find the magnetic energy density in the region.

Solution The energy density is given by

$$w_m = \frac{B^2}{2\mu_0} \qquad \text{J/m}^3$$

In the given problem, $|\mathbf{B}| = \sqrt{B_0^2 + B_0^2} = B_0\sqrt{2}$. Hence, the energy stored is

$$w_m = \frac{B_0^2}{\mu_0} \qquad \text{J/m}^3$$

Typically, let us say, $B_0 = 0.5$ Wb/m^2 (0.5 T); when

$$w_m = \frac{10^7}{16\pi} \simeq 2 \times 10^5 \qquad \text{J/m}^3$$

It should be emphasized that magnetic energy density cannot be ascribed to a particular region with any more confidence than can electric energy density. However, if the energy density is integrated throughout all space, we will obtain the total magnetic energy.

4.10 Magnetic Circuits

In Sec. 4.6 we briefly considered some of the properties of ferromagnetic materials. We are now ready to consider magnetic circuits consisting of ferromagnetic cores and airgaps. We shall also consider magnetic circuits containing permanent magnets. It is important to emphasize that a magnetic field is a *distributed-parameter* phenomenon; i.e., it is distributed over a region of space. As such, rigorous analysis requires the use of the space variables as contained in the divergence and curl operations. However, under the proper conditions it is possible to apply *lumped-parameter* analysis to certain classes of magnetic field problems just as it is applied in electric circuit analysis—although the accuracy and precision of such analysis in the magnetic circuit problem is much less than in electric circuit problems because of the relatively small difference in permeability between magnetic conductors and insulators (as compared to the difference in conductivity between electric conductors and insulators). This section briefly describes lumped-circuit analysis as applied to magnetic systems, often called *magnetic circuit analysis.* The laws for magnetic circuits are similar to those for dc electric circuits, except for the phenomena of hysteresis and saturation that occur in the former circuits. A typical saturation curve is shown in Fig. 4.15. Example 4.13 will illustrate how to take into account the effects of saturation in magnetic circuit calculations. But first of all, let us review briefly the laws of magnetic circuits.

Magnetomotive force (mmf) $\mathfrak{F}$ between two points 1 and 2 was defined in (37) as

$$\mathfrak{F} = -\int_1^2 \mathbf{H} \cdot d\mathbf{l} \tag{98}$$

The magnetic flux ψ_m can be defined by

$$\psi_m = \int_s \mathbf{B} \cdot d\mathbf{s} \tag{99}$$

Notice that ψ_m is similar to the ψ_e in electrostatics. The ratio of mmf to the total flux is called *reluctance* $\mathfrak{R}$. The "Ohm's law" for the magnetic circuit can therefore be written as

$$\mathfrak{F} = \mathfrak{R}\psi_m \tag{100}$$

The reciprocal of reluctance is called *permeance* $\mathcal{P}$. Extending the analogy with the electric circuit a little further, we see that the reluctance of a magnetic circuit of length l and area of cross section A is given by

$$\mathcal{R} = \frac{l}{\mu A} \tag{101}$$

and the resistance R is given by

$$R = \frac{l}{\sigma A} \tag{102}$$

where μ and σ are permeability and conductivity, respectively.

Because we know from (26) that

$$\oint_c \mathbf{H} \cdot d\mathbf{l} = I \tag{103}$$

we can express the mmf for an N-turn coil carrying a current I from (103) as

$$\mathcal{F} = \oint_c \mathbf{H} \cdot d\mathbf{l} = NI \tag{104}$$

The following example illustrates the application of the above concepts.

EXAMPLE 4.13 A simple magnetic circuit with an airgap is shown in Fig. 4.18. The mean length of both the flux path and the cross section s are as indicated.

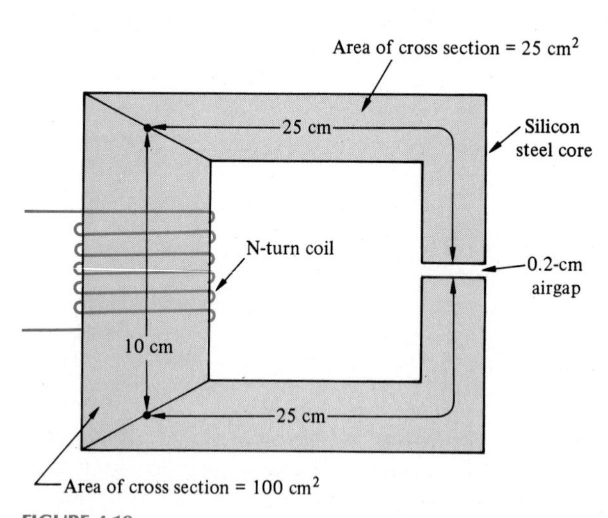

FIGURE 4.18
Example 4.13. A magnetic circuit.

How many turns should the exciting coil have in order to establish a 1.0-Wb/m^2 flux density in the airgap? The maximum allowable current through the coils is 10 A. The core is of silicon steel. Neglect any fringing of the magnetic field at the airgap.

Solution We see from Fig. 4.15 that at $B = 1.0$ Wb/m^2 the B-H relationship is nonlinear. Therefore, it is unwise to begin our reluctance calculations with the iron. The total mmf can be expressed as

$$\mathcal{F} = \mathcal{F}_{\text{steel}} + \mathcal{F}_{\text{air}}$$

The reluctance of the airgap is, from (101),

$$\mathcal{R}_{\text{air}} = \frac{2 \times 10^{-3}}{4\pi \times 10^{-7} \times 25 \times 10^{-4}} = 6.36 \times 10^5 \text{ At/Wb}$$

The total flux is

$$\psi_m = 1.0 \times 25 \times 10^{-4} = 2.5 \times 10^{-3} \text{ Wb}$$

which should be the same for the entire magnetic circuit because no flux leaks out. We have from (100)

$$\mathcal{F}_{\text{air}} = 6.36 \times 10^5 \times 2.5 \times 10^{-3} = 1590 \text{ At}$$

From Fig. 4.15: for $B = 1.0$ Wb/m^2, $H = 200$ At/m. Thus, for the length (25 + 25) cm, $\mathcal{F} = 100$ At. The total flux in the limb 10 cm long is still 2.5×10^{-3} Wb; or, flux density $= 2.5 \times 10^{-3}/10^{-2} = 0.25$ Wb/m^2. From Fig. 4.15 again: for $B = 0.25$ Wb/m^2, $H = 70$ At/m; or, for 10 cm, $\mathcal{F} = 7.0$ At. Thus

$$\mathcal{F}_{\text{steel}} = 100 + 7.0 = 107 \text{ At}$$

or

$$\mathcal{F} = NI = 107 + 1590 = 1697 \text{ At}$$

The maximum allowable current is $I = 10$ A, and thus the required number of turns is

$$N = 169.7 \simeq 170 \text{ turns}$$

EXAMPLE 4.14 From the last example it is evident that $\mathcal{F}_{\text{steel}} \ll \mathcal{F}_{\text{air}}$. Assuming an ideal core ($\mu_i \to \infty$), calculate the flux density and the magnetic energy stored in the airgap of the magnetic circuit shown in Fig. 4.19a, which also shows the various dimensions. Also, draw an electric equivalent circuit for the given magnetic circuit.

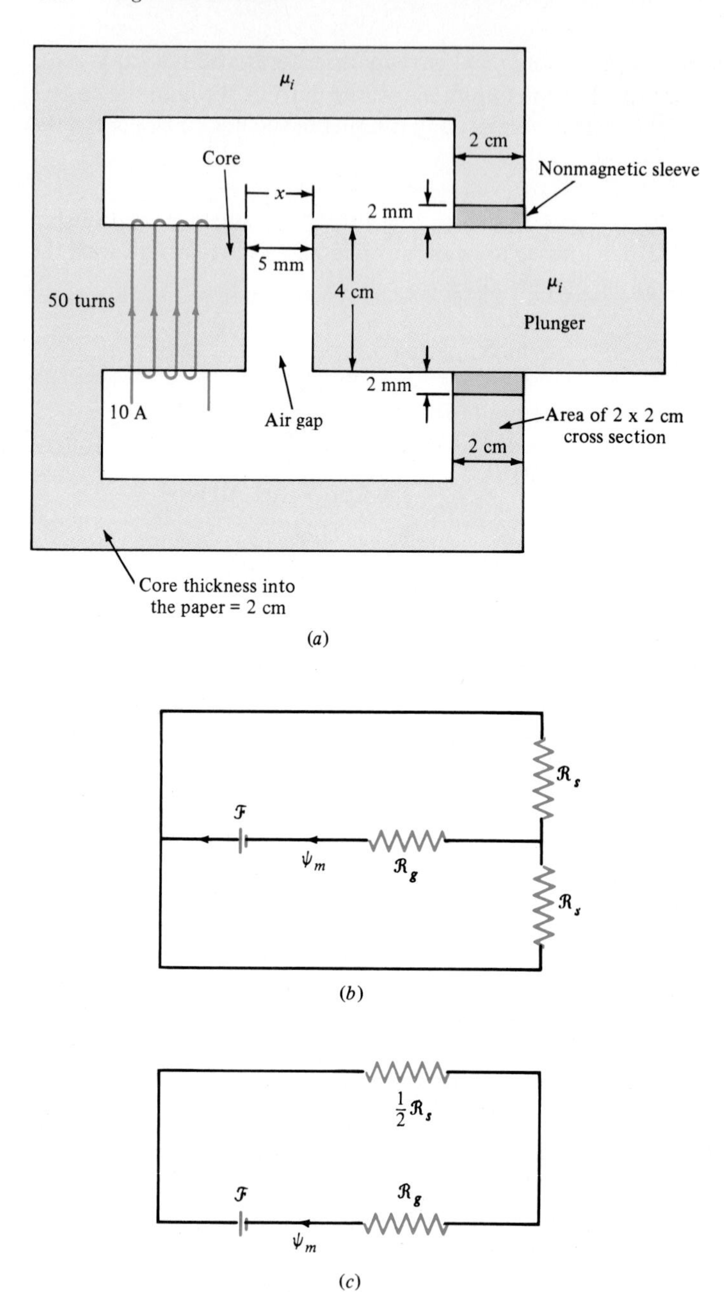

FIGURE 4.19

Example 4.14. (*a*) A magnetic circuit and its electrical analogs (*b*) and (*c*).

Solution The electrical analogs are shown in Figs. 4.19*b* and *c*. From these, the given data, and (101) we have

Airgap reluctance $$\mathcal{R}_g = \frac{5 \times 10^{-3}}{\mu_0 \times 2 \times 4 \times 10^{-4}} = \frac{50}{8\mu_0}$$

Sleeve reluctance $$\mathcal{R}_s = \frac{2 \times 10^{-3}}{\mu_0 \times 2 \times 2 \times 10^{-4}} = \frac{20}{4\mu_0}$$

Total reluctance $$\mathcal{R}_t = \mathcal{R}_g + \frac{1}{2}\mathcal{R}_s = \frac{70}{8\mu_0}$$

Airgap flux $$\psi_m = \frac{\mathcal{F}}{\mathcal{R}_t} = \frac{50 \times 10}{70/8\mu_0} = \frac{400\mu_0}{7}$$

Airgap flux density $$B_g = \frac{\psi_m}{A_g} = \frac{400\mu_0}{7 \times 2 \times 4 \times 10^{-4}}$$

Substituting $\mu_0 = 4\pi \times 10^{-7}$ H/m, we obtain

$$B_g = \frac{400 \times 4\pi \times 10^{-7}}{7 \times 2 \times 4 \times 10^{-4}} = 0.0898 \text{ T}$$

The energy stored in the airgap is thus

$$W_m = \frac{1}{2\mu_0} B_g^2 \times \text{vol}_{\text{gap}} = 0.0128 \text{ J}$$

So far we have considered only electric currents as sources of magnetic fields. As mentioned in the beginning of this chapter, another source of a magnetic field is the permanent magnet. Permanent magnet excitation is generally chosen for a given magnetic circuit with the aid of the second quadrant B-H curve, as shown in Fig. 4.20. The *retentivity*, or remanent magnetism, is expressed by B_r in Fig. 4.20, which also shows the coercive force $-H_c$. The area OB_rH_cO under the curve measures the effectiveness of the magnet.

Magnetic circuits containing permanent magnets usually also contain ferromagnetic parts and airgaps. In such a circuit, the operating condition of the permanent magnet depends a great deal on the external magnetic circuit. Furthermore, the operating point and subsequent performance of the magnet are determined by the physical installation of the magnet and whether it is magnetized before or after installation. In designing a magnetic circuit containing a permanent magnet, we make use of the demagnetization curve (Fig. 4.20) and of the energy-product curve, which is a plot of the product of B and H versus

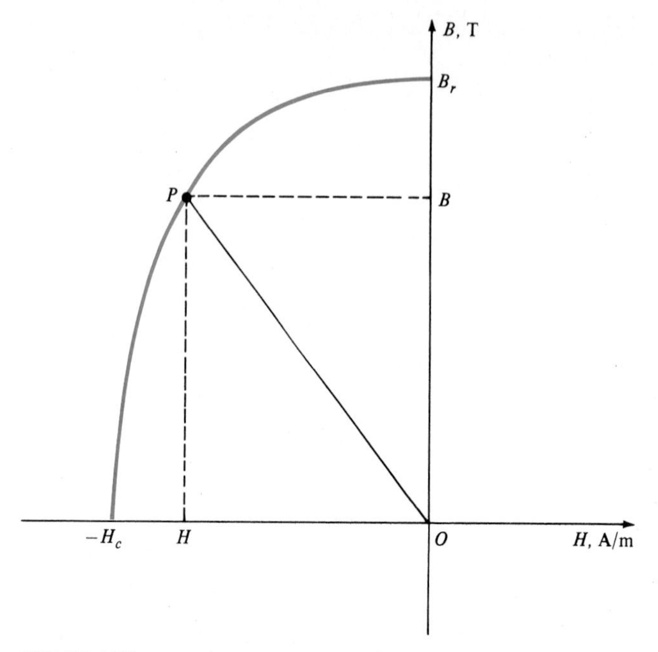

FIGURE 4.20
Second-quadrant *B-H* characteristic of a permanent magnet.

H. A permanent magnet is generally most efficiently used when operated at conditions of B and H that result in the maximum energy product. The concept of *leakage factor*, which is the ratio of magnetic flux leaving the magnet to the flux in the airgap, is very useful in designing magnetic circuits with permanent magnets. But this factor is calculated by relatively cumbersome empirical formulas that are not included here. In other words, we shall neglect leakage fields in our discussions.

If we neglect leakage, the magnetic circuit calculations are fairly straightforward. Consider a simple magnetic circuit containing a permanent magnet (shown in Fig. 4.21) for which we have, from Ampère's law,

$$\oint_c \mathbf{H} \cdot d\mathbf{l} = Hl + H_g l_g + \mathcal{F}_i = 0 \tag{105}$$

where H is the magnetic field intensity of the magnet, l is the length of the magnet, H_g is the field intensity across the airgap, l_g is the length of the airgap, and $\mathcal{F}_i$ is the magnetic potential drop in the iron portion of the circuit. The magnet dimensions are found from (105) and from

$$\psi_m = BA = B_g A_g \tag{106}$$

where B is the flux density of the magnet and where A is the cross-sectional area of the magnet. Similarly, B_g and A_g correspond to airgap quantities. For a first

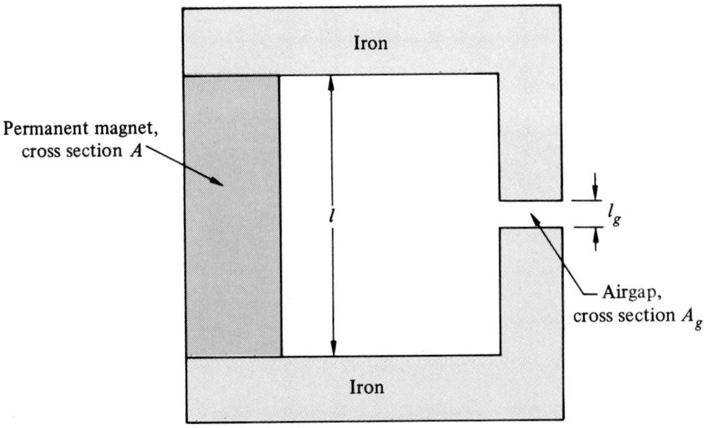

Iron

Permanent magnet, cross section A

l

l_g

Airgap, cross section A_g

Iron

FIGURE 4.21

Example 4.15. A magnetic circuit with a permanent magnet.

approximation, $\mathcal{F}_i$ in (105) may be generally neglected. Thus, solving for H in terms of B, (105) and (106) yield

$$H = -\frac{Al_g}{\mu_0 A_g l} B \tag{107}$$

which is plotted as the straight line OP in Fig. 4.20.

Neglecting the effect of leakage, the volume V of the magnet can be expressed in terms of the airgap dimensions as

$$V = \frac{B_g H_g}{BH} A_g l_g \tag{108}$$

which is obtained by combining (105) to (107), and with $\mathcal{F}_i = 0$. And, as mentioned earlier, the volume of the magnet is a minimum for a given airgap when the energy product BH is a maximum. We now illustrate the procedure for magnet-size calculations by the following example.

EXAMPLE 4.15 The magnetic circuit of Fig. 4.21 uses an alnico V magnet having the B-H characteristic shown in Fig. 4.22. The various dimensions are: $l = 20$ cm, $A = 3$ cm^2, $l_g = 0.5$ cm, and $A_g = 2.5$ cm^2. Determine the airgap flux density and the energy product of the magnet. Neglect leakage and the reluctance of iron.

Solution From (107)

$$H = -\frac{(3 \times 10^{-4} \times 0.5 \times 10^{-2})B}{4\pi \times 10^{-7} \times 2.5 \times 10^{-4} \times 20 \times 10^{-2}} = -23{,}873B$$

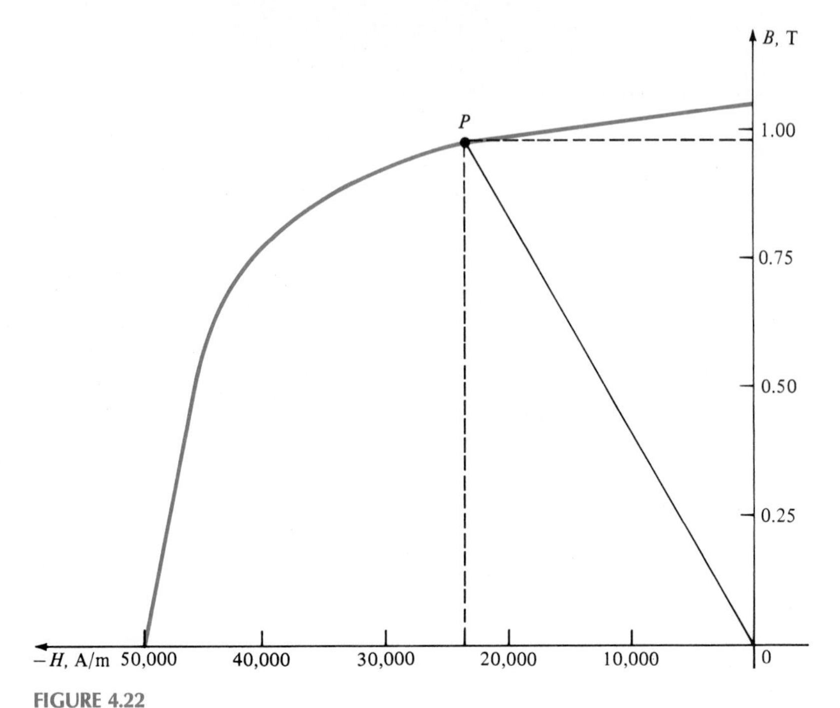

FIGURE 4.22

Example 4.15. Demagnetization curve for an alnico V magnet.

This is plotted as the line OP in Fig. 4.22, from which $B = 0.975$ T and $H = 23{,}300$ At/m.

Energy product $\qquad 0.975 \times 23{,}300 = 22{,}717$

Airgap flux $\qquad \psi_m = BA = 0.975 \times 3 \times 10^{-4} = 29.25$ mWb

Airgap flux density $\qquad \dfrac{29.25}{2.5} \times 10^{-1} = 1.17$ T

Two comments of practical importance relating to the preceding calculations are in order. First, the B-H characteristics of permanent magnets available from commercial manufacturers are invariably in centimeter-gram-second (cgs) units; that is, B is expressed in gauss (G) and H in oersted (Oe). The conversion to SI units is given by

$$B: \qquad 1\ \text{T} = 10^4\ \text{G}$$

$$H: \qquad 1\ \text{A/m} = 0.004\pi\ \text{Oe}$$

Second, in practical calculations it is important to account for leakage and fringing factors. Formulas for determining these factors are available in specifi-

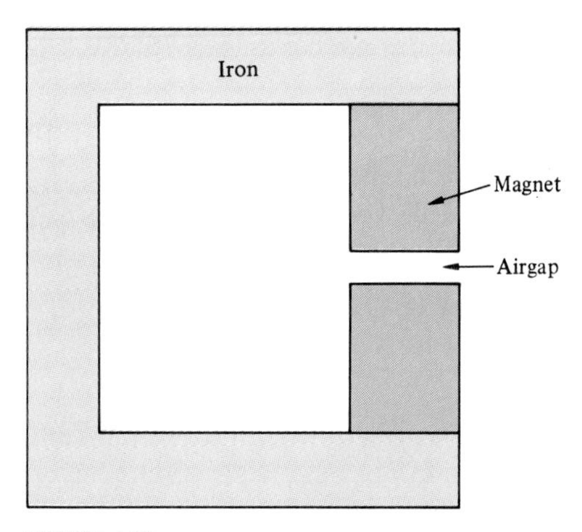

FIGURE 4.23

Example 4.15. A better utilization of the magnet in the circuit of Fig. 4.21.

cation and design brochures supplied by magnet manufacturers. In this connection it may be shown that the magnet of Fig. 4.21 is better utilized if located as in Fig. 4.23 in that the leakage is reduced.

4.11 Inductance

The concept of inductance can be understood by referring to Fig. 4.24, which shows a magnetic circuit of mean length l, area of cross section A, and permeability μ. Let an N-turn coil be wound on this core and carry a current I. According to the laws of magnetic circuits, a flux ψ_m would be established in the core, as shown in Fig. 4.24. Notice that the flux lines "link" the N-turn coil. Thus, we define the quantity *flux linkage* λ by

$$\lambda = N\psi_m \tag{109}$$

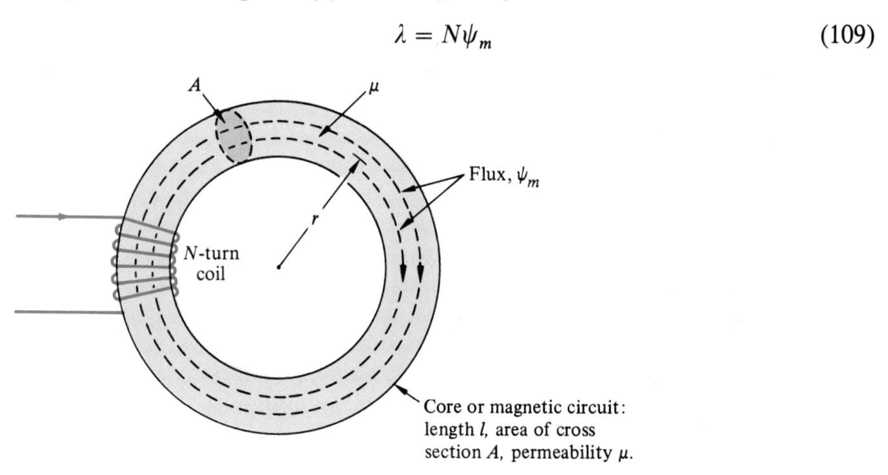

FIGURE 4.24

A magnetic circuit with an N-turn coil.

Inductance L of the coil is then defined as flux linkage per ampere; that is,

$$L = \frac{\lambda}{I} = \frac{N\psi_m}{I} \tag{110}$$

Using (110) in conjunction with (100), (101), and (104), we can readily verify that the inductance of the N-turn coil is

$$L = \mu \frac{N^2 A}{l}$$

Now consider the magnetic toroid around which are wound n distinct coils electrically isolated from each other, as shown in Fig. 4.25. The coils are linked magnetically by the flux ψ_m, some portion of which links each of the coils. A number of inductances can be defined for this system:

$$L_{ij} = \frac{\text{flux linking the } i\text{th coil due to the current in the } j\text{th coil}}{\text{current in the } j\text{th coil}}$$

Mathematically, this can be stated as

$$L_{ij} = \frac{N_i k \psi_{mij}}{I_j} \tag{111}$$

where k is the portion of the flux due to coil j that links coil i and is known as the *coupling coefficient*. By definition, its maximum value is 1.0. A value of k less

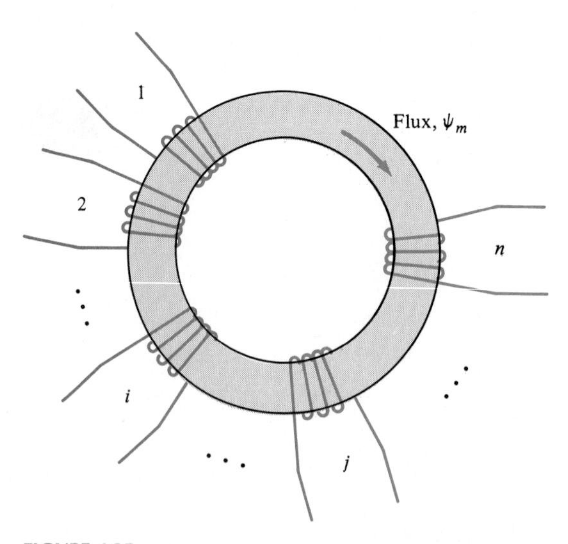

FIGURE 4.25
Coils (numbered 1, 2, ..., *i*, ..., *j*, ..., *n*) wound on a toroid.

than 1.0 is due to leakage flux in the regions between the location of coil i and coil j. We can now see that when the two subscripts in (111) are identical, the inductance is termed *self-inductance*; when different, the inductance is termed *mutual inductance* between coils i and j. Mutual inductances will later be shown to be symmetrical; that is,

$$L_{ij} = L_{ji} \tag{112}$$

From the preceding remarks, it is clear that inductance is related to magnetic field quantities. Thus, an alternative method of determining the inductance of a circuit is obtained by equating the energy stored in the inductance to the corresponding energy stored in the magnetic field. This point can be illustrated by considering the configuration of Fig. 4.24.

First, we know from electric circuit theory that W_m, the energy stored in an inductance L, carrying a current I, is given by

$$W_m = \tfrac{1}{2}LI^2 \tag{113}$$

Now, from (91) the same energy is expressed as

$$W_m = \tfrac{1}{2}\mu \int_v |\mathbf{H}|^2 \, dv \tag{114}$$

But, for the given problem,

$$\oint_c \mathbf{H} \cdot d\mathbf{l} = |\mathbf{H}|l = NI$$

or

$$|\mathbf{H}| = \frac{NI}{l} \tag{115}$$

and the volume of the core is

$$\int_v dv = lA \tag{116}$$

Therefore, (113) to (116) yield

$$LI^2 = \frac{\mu N^2 I^2}{l^2} lA$$

or

$$L = \frac{\mu N^2 A}{l}$$

EXAMPLE 4.16 Determine the inductance of a toroid shown in Fig. 4.24. The mean radius of the toroid is r and its cross-sectional area is A. Assume a uniform flux density within the toroid.

Solution The flux linkage is

$$\lambda = N\psi_m = NBA$$

and

$$BA = \frac{NI}{2\pi r/\mu A} = \frac{\mu ANI}{2\pi r}$$

Thus

$$L = \frac{\lambda}{I} = \frac{\mu AN^2}{2\pi r}$$

EXAMPLE 4.17 We recall from Chap. 3 that capacitance can be determined by evaluating the energy stored in the electric field. In this example we illustrate the energy-storage method of finding the inductance of a coaxial cable, per unit length, shown in Fig. 4.26.

Solution Consider a cable of infinite length. It follows from Ampère's law that

$$H_\phi = \frac{I}{2\pi r} \qquad a < r < b$$

and the magnetic energy stored per unit of length is

$$W_m = \tfrac{1}{2}LI^2 = \tfrac{1}{2}\mu_0 \int_a^b H_\phi^2 2\pi r = \frac{\mu_0 I^2}{4\pi} \ln\frac{b}{a}$$

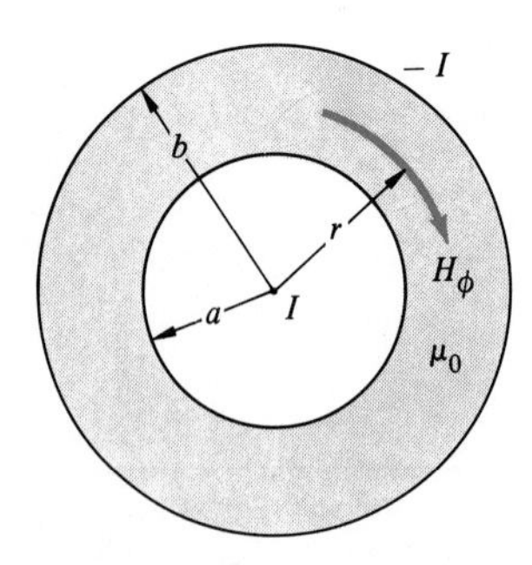

FIGURE 4.26
Example 4.17. A coaxial cable carrying a current I.

Consequently, the per-unit-length inductance is

$$L = \frac{\mu_0}{2\pi} \ln \frac{b}{a} \qquad \text{H/m}$$

It can be easily verified that the same result is obtained by the flux-linkage method used in Example 4.16.

An alternative method of computing inductances that is particularly useful for filamentary conductors is the Neumann formula. Consider two loops composed of filamentary conductors, as shown in Fig. 4.2. The mutual inductance between these loops is defined by (111). The flux linking loop 2 as a result of the current in loop 1 is

$$\psi_{m21} = \int_{s2} \mathbf{B}_{21} \cdot d\mathbf{s}_2 \tag{117}$$

so that the mutual inductance between the two loops becomes

$$L_{21} = \frac{\psi_{m21}}{I_1} \tag{118}$$

The magnetic flux density can be written in terms of the magnetic vector potential via (43):

$$\mathbf{B}_{21} = \nabla \times \mathbf{A}_{21} \tag{119}$$

Substituting (119) into (117) gives

$$\psi_{m21} = \int_{s2} (\nabla \times \mathbf{A}_{21}) \cdot d\mathbf{s}_2 \tag{120}$$

Using Stokes' theorem gives

$$\psi_{m21} = \oint_{l_2} \mathbf{A}_{21} \cdot d\mathbf{l}_2 \tag{121}$$

The vector magnetic potential is

$$\mathbf{A}_{21} = \frac{\mu}{4\pi} \oint_{l_1} \frac{I_1\, d\mathbf{l}_1}{R_{21}} \tag{122}$$

where R_{21} is the distance between $d\mathbf{l}_1$ and $d\mathbf{l}_2$. Substituting (122) into (121) and using (118) gives the Neumann formula:

$$L_{21} = \frac{\mu}{4\pi} \oint_{l_1} \oint_{l_2} \frac{d\mathbf{l}_1 \cdot d\mathbf{l}_2}{R_{21}} \tag{123}$$

This clearly shows that $L_{21} = L_{12}$. Implicitly it has been assumed (1) that I_1 is the same at all points around loop 1 and (2) that μ is a scalar (isotropic medium) and independent of position (homogeneous medium).

EXAMPLE 4.18 Consider two coaxial loops composed of filamentary conductors, as shown in Fig. 4.27. Determine the mutual inductance between the loops. Assume that the spacing between the loops, D, is much larger than the loop radii.

Solution The magnetic vector potential at a point P on the second loop due to the current I_1 in the first loop is (see Example 4.9)

$$\mathbf{A}_{21} = \frac{\mu_0 r_1^2 I_1 \sin\theta}{4R^2} \mathbf{a}_\phi$$

$$= \frac{\mu_0 r_1^2 I_1 r_2}{4R^3} \mathbf{a}_\phi$$

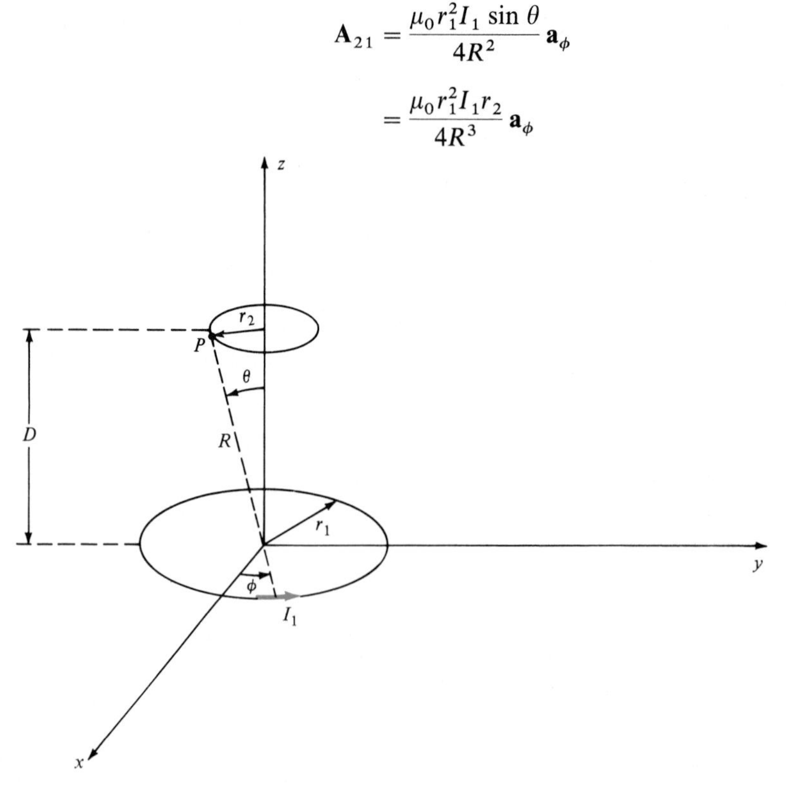

FIGURE 4.27

Example 4.18. Determining inductance using the magnetic vector potential.

Assuming that $D \gg r_2, r_1$, we have $R \simeq D$, or

$$\mathbf{A}_{21} = \frac{\mu_0 r_1^2 r_2 I_1}{4D^3}\mathbf{a}_\phi$$

Thus

$$L_{21} = \frac{\psi_{m21}}{I_1}$$

$$= \frac{1}{I_1}\oint_{l_2} \mathbf{A}_{21} \cdot d\mathbf{l}_2$$

$$= \frac{\mu_0 r_1^2 r_2}{4D^3}\int_{l_2} dl_2$$

$$= \frac{\pi \mu_0 r_1^2 r_2^2}{2D^3}$$

4.12 Mechanical Forces of Electromagnetic Origin: A Further Discussion

In Sec. 4.9 we saw that interactions between magnetic fields and current-carrying conductors result in the production of mechanical forces. Thus, we formulated the Lorentz force equation. Another magnetic field effect resulting in a mechanical force is alignment of flux lines. Examples of "alignment" are shown in Figs. 4.28*a* and *b*. In Fig. 4.28*a* the force on the ferromagnetic pieces causes them to align with the flux lines, thus shortening the magnetic flux path and reducing the reluctance. Figure 4.28*b* shows an elementary form of a reluctance motor, in which the force tends to align the rotor axis with that of the stator. Quantitative evaluation of such a force can be made by following a procedure similar to that of Sec. 3.13. Thus, for a lossless system, conservation of energy requires that

Input electric energy	=	mechanical work done	+	increase in stored energy

Expressed mathematically, this is

$$I\,d\lambda = F_e\,dx + dW_m \tag{124}$$

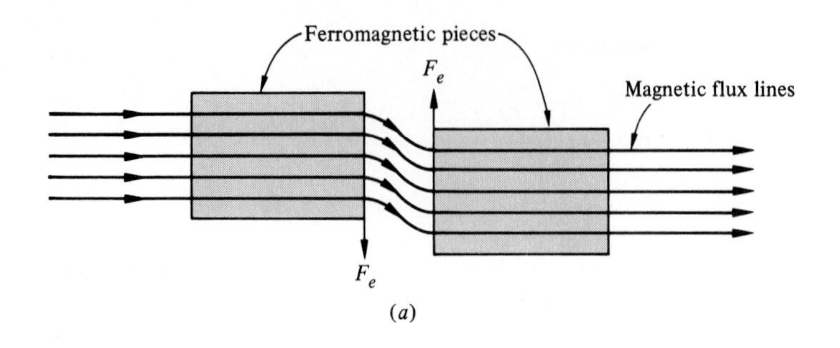

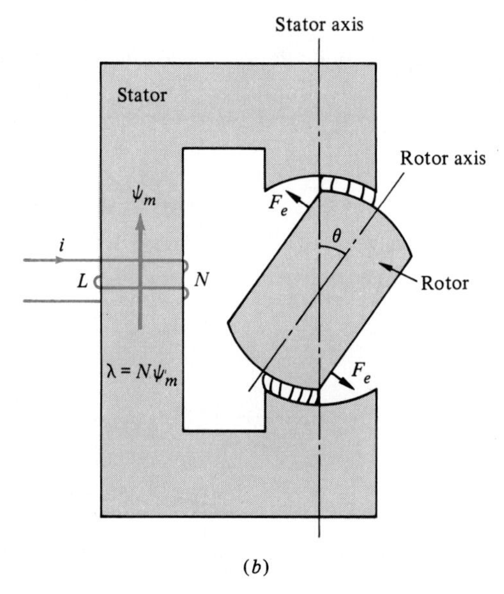

FIGURE 4.28

Alignment of magnetic flux lines to produce a force (a) between two ferromagnetic pieces and (b) in a reluctance motor.

where dW_m is the increase in magnetic stored energy, $I\,d\lambda$ is the input electric energy, and F_e is the force of electromagnetic origin. If we choose the current I to be the independent variable, then

$$\lambda = \lambda(I, x)$$

or

$$d\lambda = \frac{\partial \lambda}{\partial I} dI + \frac{\partial \lambda}{\partial x} dx \tag{125}$$

And

$$W_m = W_m(I, x)$$

or

$$dW_m = \frac{\partial W_m}{\partial I} dI + \frac{\partial W_m}{\partial x} dx \tag{126}$$

Substituting (125) and (126) in (124) yields

$$F_e\,dx = \left(-\frac{\partial W_m}{\partial x} + I\frac{\partial \lambda}{\partial x}\right)dx + \left(-\frac{\partial W_m}{\partial I} + I\frac{\partial \lambda}{\partial I}\right)dI \tag{127}$$

Because dI and dx are arbitrary, F_e must be independent of these changes. Thus, the coefficient of dI in (127) must be zero, and we finally have

$$F_e = -\frac{\partial W_m(I, x)}{\partial x} + I\frac{\partial \lambda(I, x)}{\partial x} \tag{128}$$

which is the force equation. Its application is illustrated by the following example.

EXAMPLE 4.19 Calculate the force on the solenoid plunger shown in Fig. 4.19*a*.

Solution First, let the plunger be at a distance x mm away from the core. Then, as in Example 4.14,

$$\mathcal{R}_g = \frac{x \times 10^{-3}}{\mu_0 \times 2 \times 4 \times 10^{-4}} = \frac{10x}{8\mu_0}$$

$$\mathcal{R}_s = \frac{2 \times 10^{-3}}{\mu_0 \times 2 \times 2 \times 10^{-4}} = \frac{20}{4\mu_0}$$

$$\mathcal{R}_t = \mathcal{R}_g + \tfrac{1}{2}\mathcal{R}_s = \frac{10x + 20}{8\mu_0} = \frac{5x + 10}{4\mu_0}$$

$$\text{Inductance } L(x) = \frac{N^2}{\mathcal{R}_t} = \frac{4\mu_0 \times 50}{5x + 10} = \frac{4\pi \times 10^{-3}}{5x + 10}$$

Now,

$$W_m = \tfrac{1}{2}LI^2$$

$$\lambda = LI$$

and

$$I\,d\lambda = I\,dL$$

Substituting these in (128) yields

$$F_e = \tfrac{1}{2}I^2\frac{dL}{dx}$$

For the given data, therefore,

$$F_e = \tfrac{1}{2} \times 10^2 \frac{d}{dx}\left(\frac{4\pi \times 10^{-3}}{5x + 10}\right) = 2\pi \times 10^{-1}\left[\frac{-5}{(5x + 10)^2}\right]$$

For $x = 5$ mm, we finally obtain

$$F_e = -\frac{\pi}{35^2} \quad \text{N}$$

where the negative sign indicates that the force tends to decrease x.

4.13 Summary

In this chapter we considered the production of static magnetic fields by electric currents and permanent magnets. Similarities with electrostatic fields were shown as far as possible. We introduced the concept of magnetic fields through Ampère's force law. On this basis, we formulated the Biot–Savart law. We studied Ampère's circuital law and saw that in certain cases the field calculations are much simpler by Ampère's law than by the Biot–Savart law. Next, we considered the use of potential functions, such as the magnetic scalar and vector potentials, in evaluating magnetic fields. We saw that in source-free regions both potentials satisfy Laplace's equation. With the aid of the definition of the magnetic vector potential, we derived Ampère's circuital law. We discussed the boundary conditions for magnetic fields, and we briefly discussed certain macroscopic properties of magnetic materials. In this connection, we introduced properties such as magnetic polarization, susceptibility, and relative permeability. Finally, we discussed the topics of force and energy in magnetic fields. We introduced the idea of a magnetic circuit, and magnetic circuit calculations were presented for magnetic circuits having ferromagnetic materials, permanent magnets, and air. Inductance was defined, various methods of its evaluation were presented, the relationship between the field concept H and the circuit concept L was shown, and an alternative form of force production in a magnetic field was considered.

Problems

4-1 A square loop measuring 1.5 by 1.5 m carries a 7.5-A steady current. Choose the coordinates to locate the loop in the xz plane, the origin coinciding with a corner of the square. Calculate the **B** field at a point on the y axis 0.35 m from the origin.

4-2 Using the Biot-Savart law, find the magnetic field intensity **H** at a point on the axis of a circular loop of radius a carrying a current I. The point is at a distance h (on the axis) from the center of the loop.

4-3 A semicircular loop of radius R carries a current I. The loop is located in air. Find the magnetic field intensity and the magnetic flux density at the center of the loop.

4-4 A plexiglass disk of radius R is charged with a uniform surface charge density ρ_s. The disk is rotated at a constant speed of N r/min. Thus, we have circular loops of electric charges in motion, which may be considered to be loops of currents. Determine the magnetic field intensity at the center of the loop.

4-5 From Prob. 4-2 we find that the **B** field at the axis of a current-carrying circular loop is given by

$$\mathbf{B} = \frac{\mu_0 I a^2}{2(z^2 + a^2)^{3/2}} \mathbf{a}_z$$

where I = current in the loop and where a = loop radius and is the distance between the center of the loop and a point located at the axis of the loop. Determine B_y.

4-6 Within a cylindrical conductor of radius a, the current density exponentially decreases with the radius such that $\mathbf{J} = Ae^{-kr}\mathbf{a}_z$, where A and k are constants. Determine the resulting magnetic field intensity everywhere.

4-7 Given

$$\mathbf{H} = \frac{Ir}{2\pi a^2} \mathbf{a}_\phi$$

within a solid conductor of radius a, determine the current density within the conductor.

4-8 A rectangular loop is placed in the field of a very long straight conductor carrying a current I, as shown in Fig. P4.8, which also shows the various dimensions. What is the total magnetic flux passing through the loop?

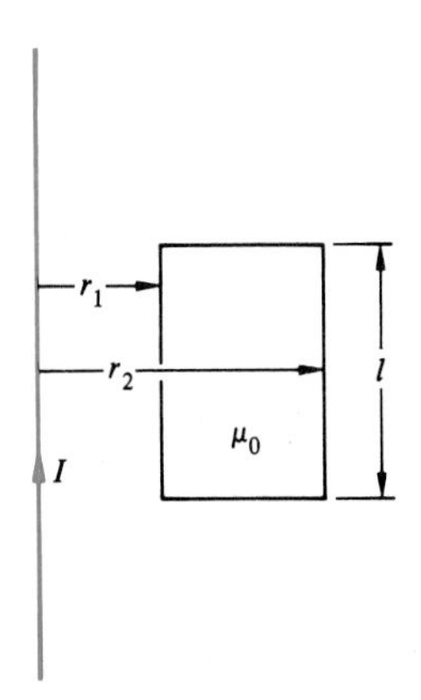

FIGURE P4.8

4-9 For the air-core toroid shown in Fig. P4.9,

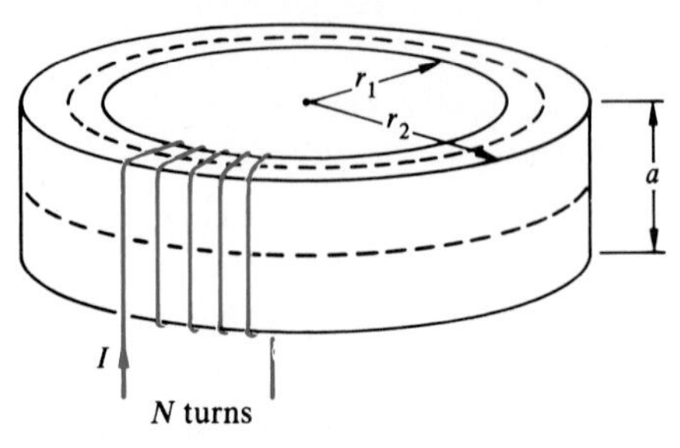

FIGURE P4.9

(a) determine the core flux.

(b) If the core flux density is assumed to be uniform and equal to its value at the arithmetic mean radius, what percent error would be made in the computation of the core flux by this approximation as compared with the answer to part **(a)**?

(c) If the geometric instead of arithmetic mean radius were used, calculate the percent error. Illustrate the answers to parts **(b)** and **(c)** for $r_2/r_1 = 2$.

4-10 Alternating current flows through a coaxial cable, the inner conductor of which is solid and is of radius a. Because of the skin effect (to be discussed in Chaps. 6 and 7), the current-density distribution within the conductor is nonuniform and may be approximated by $J(r) = J_0 e^{-(a-r)}$, where J_0 is the current density at the surface of the conductor. The return current is through the outer conductor, which consists of an infinitely thin cylindrical shell of radius b. Determine the magnetic field intensity everywhere.

4-11 The operation of a magnetic compass depends on the presence of the earth's magnetic field. At a certain location the tangential component of the earth's magnetic field is 0.02 mT, and this location is under a 345-kV transmission line carrying a 1000-A current. If the height of the line is 8 m, calculate the approximate net field affecting the reading of the compass. Consider the extreme case in which the field of the transmission line is in the same direction as the direction of the tangential component of the earth's magnetic field.

4-12 An annular cylindrical space has an axial length of l and inner and outer radii of r_1 and r_2. The magnetic vector potential in this region is

$$\mathbf{A} = -k \ln r \, \mathbf{a}_z$$

where k is a constant. Determine the total magnetic flux in the annular space.

4-13 As shown in Fig. P4.13, the magnetic field intensity under the pole (in the airgap) of an electric motor is given by

$$\mathbf{H} = \frac{10^5}{r} \sin 2\phi \, \mathbf{a}_r \qquad \text{A/m}$$

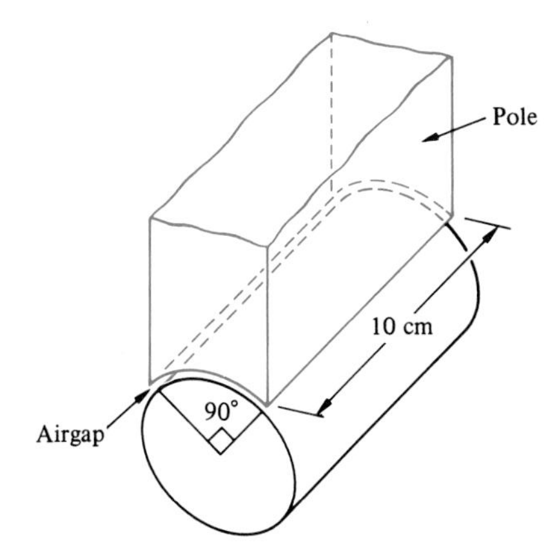

FIGURE P4.13

The axial length of the pole is 10 cm, and the arc of the pole subtends an angle of 90° at the axis of rotation of the motor. Calculate the flux per pole.

4-14 Within a cylindrical conductor the magnetic vector potential, due to a current of density J_0 A/m^2, is given by $\mathbf{A} = -(\mu_0 J_0/4)(x^2 + y^2)\mathbf{a}_z$. Obtain the magnetic field intensity within the conductor, and verify that the result is consistent with that obtained from Ampère's law.

4-15 In Prob. 4.14 we observe that the result obtained from a given magnetic vector potential is valid. Alternatively, show that given the vector potential $[\mathbf{A} = -\frac{1}{4}\mu_0 J_0(x^2 + y^2)\mathbf{a}_z]$ is a valid function in that it satisfies Poisson's equation.

4-16 Find the magnetic vector potential in the hole in the infinitely long cylindrical wire shown in Fig. P4.16.

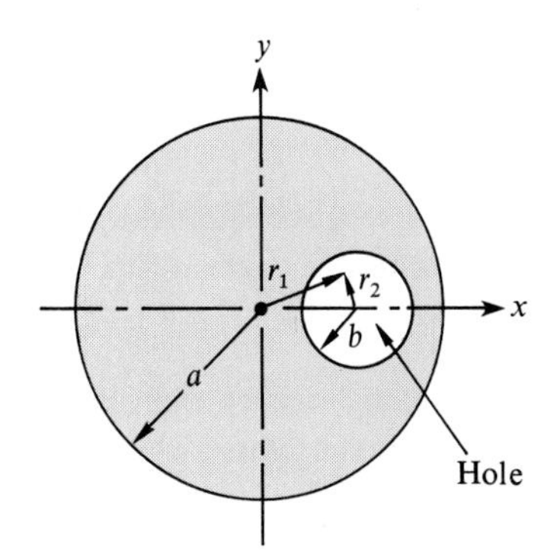

FIGURE P4.16

(a) Express **A** in terms of r_1 and r_2 (at an arbitrary point P). The wire carries a current I in the z direction.

(b) Express **A** in the hole in rectangular coordinates.

(c) Find the flux density in the hole, and show that **B** is uniform and is entirely in the y direction.

4-17 The magnetic vector potential in a region in space is given by

$$\mathbf{A} = e^{-y} \cos x \, \mathbf{a}_x + (1 + \sin x)\mathbf{a}_z$$

Calculate the magnetic flux density at the origin.

4-18 Two infinitely permeable semi-infinite blocks are separated from each other by a distance g, the planes constituting the surfaces of the blocks being parallel to each other (Fig. P4.18). With reference to the coordinates shown, the magnetic vector potential within the airgap is given by

$$\mathbf{A} = \mathbf{a}_z(k_1 \cosh y + k_2 \sinh y) \sin x$$

Evaluate the constants k_1 and k_2, and determine the **B** field at the center of the airgap if $B_x = 0.8$ Wb at $y = 0$.

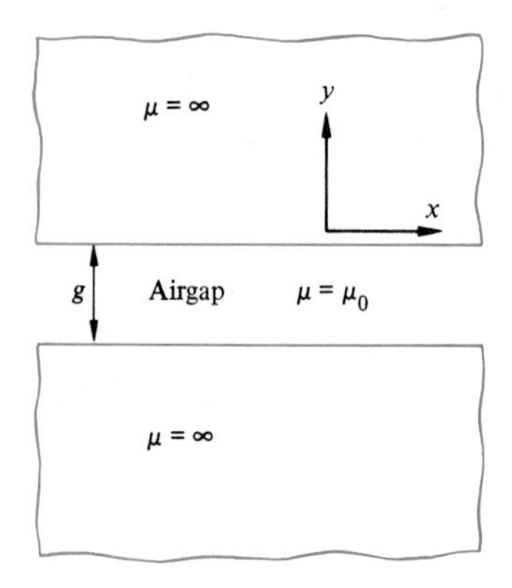

FIGURE P4.18

4-19 Magnetic flux lines enter at an angle of 45° from free space into a ferromagnetic region having a relative permeability $20\mu_0$. Determine the angle at which the flux lines would emerge from the interface.

4-20 Sometimes the B-H curve of a ferromagnetic material can be expressed by the Froelich equation $B = aH/(b + H)$, where a and b are constants of the material. In a particular case, $a = 1.5$ T and $b = 100$ A/m. A magnetic circuit consists of two parts: lengths l_1 and l_2, and respective cross-sectional areas A_1 and A_2. If $A_1 = 2A_2 = 25$ cm^2 and $2l_1 = l_2 = 25$ cm, and if the magnetic circuit carries an mmf of 1000 At, calculate the flux through the circuit.

4-21 A composite magnetic circuit of varying cross section, shown in Fig. P4.21*a*, made of air and iron has the BH characteristic of Fig. P4.21*b*. Given: $N = 100$ turns; $l_1 = 4l_2 = 40$ cm; $A_1 = 2A_2 = 10$ cm^2; $l_g = 2$ mm. Calculate I to establish an airgap flux density of 0.6 T. Neglect leakage and fringing.

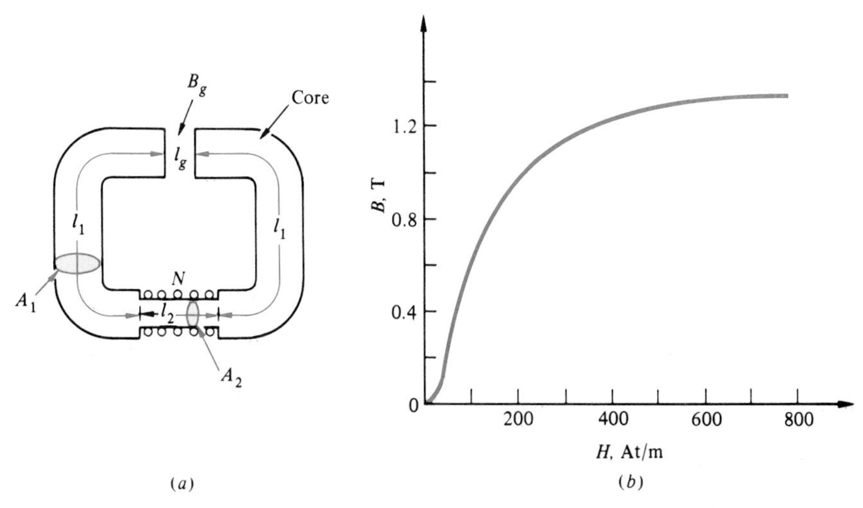

FIGURE P4.21

4-22 Calculate the magnetic energy stored in

(a) the iron, and
(b) the airgap of the magnetic circuit of Prob. 4.21.

4-23 Calculate the inductance of the coil shown in Fig. P4.21.

4-24 The **B** field in a certain region in space is given by

$$\mathbf{B} = 0.03\mathbf{a}_x + 0.2\mathbf{a}_y \qquad \text{T}$$

In this field a conductor, carrying a 20-A current, is located in the xz plane. The coordinates of the ends of the conductor are at (0, 0, 0) and (1, 0, 1) m. What is the force on the conductor?

4-25 Two infinitely long parallel wires constitute a dc transmission line and are separated from each other by a distance of 30 cm. What is the current in the wires if the resulting force per unit length between the wires is 0.6 N/m? The two wires carry currents in opposite directions.

4-26 An iron-core toroid has a uniform rectangular cross section (as shown in Fig. P4.26) and is wound with an N-turn coil.

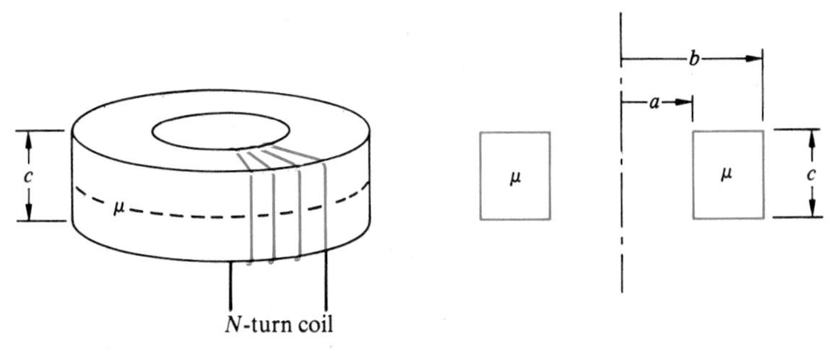

FIGURE P4.26

(a) Find the inductance of the coil.

(b) Determine the approximate inductance of a thin ring (for which $b - a \ll a$).

(c) Calculate the inductance of the coil from the following data: $N = 100$ turns, $\mu = 500\mu_0$, $b = 10$ cm, $a = 9.2$ cm, and $c = 2$ cm.

4-27 If the coil of Prob. 4.26 carries a 10-A current, calculate the core flux for the data of Prob. 4-26**(c)**.

4-28 What is the core flux density and the energy stored in the field of the toroid of Prob. 4.26? Assume a uniform flux-density distribution over the core cross section.

4-29 A toroid has a core of square cross section of 25-cm^2 area and a mean diameter of 25 cm. The core material has a relative permeability of 1000. Calculate the number of turns to be wound on the core to obtain a 1-H inductance. What is the effect of doubling the number of turns on the inductance?

4-30 What are the values of B and H at the mean radius of the core if the coil of the toroid of Prob. 4.29 has a magnetomotive force of 500 At?

4-31 Determine the inductance per unit length of the coaxial conductor system shown in Fig. P4.31.

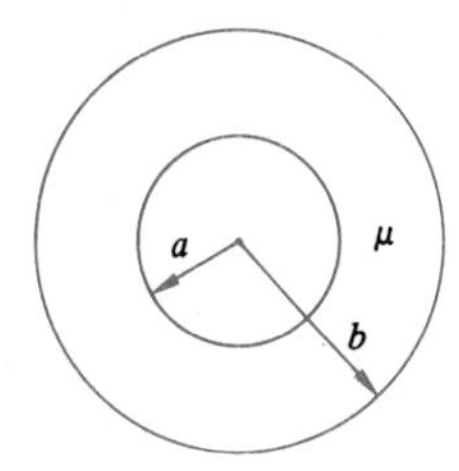

FIGURE P4.31

4-32 The (i, ψ_m) relationship for the electromagnet shown in Fig. P4.32 is given by $i = a\psi_m^2 + \psi_m(x - b)^2$, where a and b are constants. Evaluate the force on the iron mass at $x = g$.

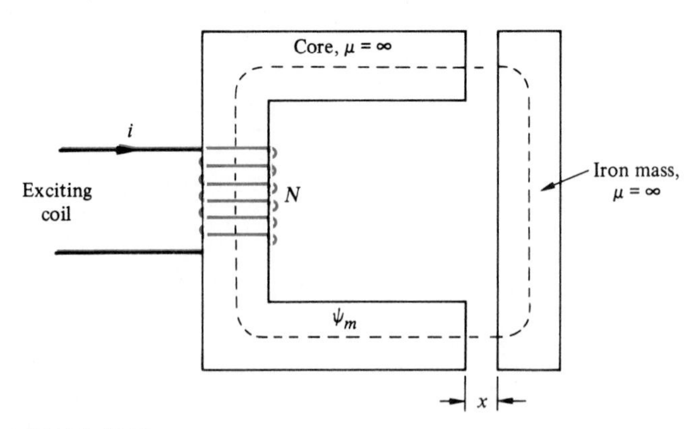

FIGURE P4.32

4-33 An elementary reluctance motor is shown in Fig. P4.33. The exciting coil inductance is given by $L(\theta) = k_0 + k_1 \cos 2\theta$, where k_0 and k_1 are constants. The coil current is $i = I_m \sin \omega t$. What is the instantaneous torque produced by the motor? If the motor runs at a particular speed $\omega_m = \omega$, what is the average torque?

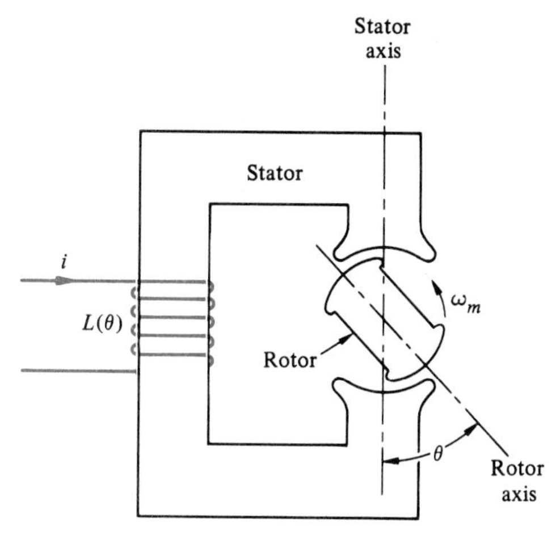

FIGURE P4.33

4-34 A solenoid of cylindrical geometry is shown in Fig. P4.34.

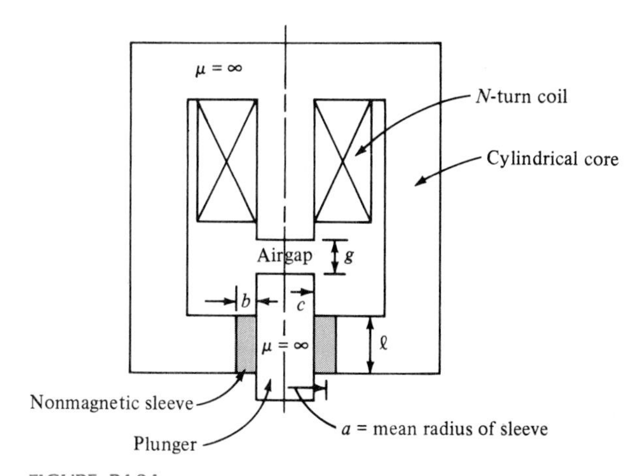

FIGURE P4.34

(a) If the exciting coil carries a dc steady current I, derive an expression for the force on the plunger.

(b) For the numerical values $I = 10$ A, $N = 500$ turns, $g = 5$ mm, $a = 20$ mm, $b = 2$ mm, and $l = 40$ mm, what is the magnitude of the force? Assume $\mu_{\text{core}} = \infty$ and neglect leakage.

CHAPTER 5

Maxwell's Equations

The study of static electric fields in Chap. 3 pertained to the fields due to stationary charge distributions, although the concept of a current as charge movement was briefly considered. In Chap. 4 we studied static magnetic fields due to a steady movement of charge (a steady current). We will now consider a more general case of fields resulting from charge movement in which the charge movement (and the resulting fields) may vary with time. This will lead to the propagation of energy in the form of electromagnetic waves (discussed in subsequent chapters).

We found in Chap. 3 that stationary distributions of charge, ρ, produce only two of the four basic field vectors, **E** and **D**. In Chap. 4 we observed that the steady movement of charge, current density **J**, resulted in the other two basic vectors, **H** and **B**. The equations governing these static field vectors were found to be:†

Integral form	*Point form*	
$\oint_c \mathbf{E} \cdot d\mathbf{l} = 0$	$\nabla \times \mathbf{E} = 0$	(1*a*)
$\oint_c \mathbf{H} \cdot d\mathbf{l} = \int_s \mathbf{J} \cdot d\mathbf{s}$	$\nabla \times \mathbf{H} = \mathbf{J}$	(1*b*)
$\oint_s \mathbf{D} \cdot d\mathbf{s} = \int_v \rho \, dv$	$\nabla \cdot \mathbf{D} = \rho$	(1*c*)
$\oint_s \mathbf{B} \cdot d\mathbf{s} = 0$	$\nabla \cdot \mathbf{B} = 0$	(1*d*)

† We will use the symbol ρ to denote volume charge density unless specified otherwise.

Note that for the static case, each of the field vectors appears in only one equation; thus, the field vectors **E** and **D** and the field vectors **H** and **B** are uncoupled sets. If $\sigma \neq 0$, then $\mathbf{J} = \sigma\mathbf{E}$ and Eq. (1*b*) becomes $\nabla \times \mathbf{H} = \mathbf{J} = \sigma\mathbf{E}$. In this case, **H** and **E** appear to be coupled. However, a static electric field in a region with $\sigma \neq 0$ will cause a current to flow, and this current will then produce a magnetic field. Nevertheless, the static electric field can be completely determined from the static charge distribution, and the magnetic field is a consequence. For time-varying fields, we will find that the field vectors are always coupled and that each one affects the others.

In this chapter we will find that, in the general case of time-varying charge movement, the equations of the static field must be modified. These field equations for the general time-varying case are collectively known as *Maxwell's equations* after James Clerk Maxwell, a Scottish physicist and mathematician of the 1800s, who is credited with their compilation. Actually, Maxwell did not "discover" these equations. He compiled the known results obtained by Ampère, Faraday, Gauss, Coulomb, and others and made an important addition to one of these results (Ampère's law).

We will find that certain properties of the static field are no longer true when the field is time-varying. For example, it was found in Chap. 3 that the electrostatic field is conservative; that is, the line integral of **E** around any closed path is zero, as shown in (1*a*). This permitted the unique definition of voltage as the line integral of **E** between two points independent of the path taken. We will find that the time-varying electric field is no longer necessarily conservative and that voltage cannot, in general, be uniquely defined in a fashion similar to the static case. Similar properties of the magnetostatic field—in particular, Ampère's law given in (1*b*)—must also be modified. However, these modifications are not without benefit since they lead to many interesting and useful phenomena and devices that are not possible with static fields. These will be discussed in subsequent chapters. We will find in Chap. 6 that Maxwell's equations predict the transmission of energy in the form of waves. Means of guiding or focusing electromagnetic waves will be considered in Chap. 7 ("Transmission Lines"), Chap. 8 ("Waveguides"), and Chap. 9 ("Antennas").

5.1 Maxwell's Equations

We will use a boldface script notation for the time-varying field vectors: $\boldsymbol{\mathcal{E}}(x, y, z, t)$, $\boldsymbol{\mathcal{D}}(x, y, z, t)$, $\boldsymbol{\mathcal{H}}(x, y, z, t)$, and $\boldsymbol{\mathcal{B}}(x, y, z, t)$. The argument lists of these vectors include the dependence on time t as well as the spatial dependence in terms of rectangular coordinate system variables x, y, and z. Similar notations may be used for other coordinate systems. The static field vectors are denoted in the previous fashion as nonscript quantities. Table 5.1 summarizes the notation.

TABLE 5.1

Quantity	*Time-Varying Field*	*Static Field*	*SI Units*
Electric field intensity	$\mathcal{E}(x, y, z, t)$	$\mathbf{E}(x, y, z)$	V/m
Electric flux density	$\mathcal{D}(x, y, z, t)$	$\mathbf{D}(x, y, z)$	C/m^2
Magnetic field intensity	$\mathcal{H}(x, y, z, t)$	$\mathbf{H}(x, y, z)$	A/m
Magnetic flux density	$\mathcal{B}(x, y, z, t)$	$\mathbf{B}(x, y, z)$	Wb/m^2 or T
Current density	$\mathcal{J}(x, y, z, t)$	$\mathbf{J}(x, y, z)$	A/m^2
Volume charge density†	$\rho(x, y, z, t)$	$\rho(x, y, z)$	C/m^3

† The symbol ρ will be used to denote both static and time-varying charge densities.

5.1.1 Faraday's Law

In Chap. 3 we found that the electrostatic field lines originate from positive charge and terminate on negative charge and thus do not close on themselves. In this section we will find that a time-varying magnetic field will also produce an electric field and that these electric field lines close on themselves. Thus, for time-varying fields it is no longer correct to say that all electric field lines begin and end on charges.

We found in Chap. 4 that a steady electric current produces a magnetic field. In 1831 Michael Faraday observed that a *time-varying* magnetic field also produces a current in a closed loop of wire.† Consider Fig. 5.1*a*, which shows a circular loop of wire and a time-varying magnetic field $\mathcal{B}$ that penetrates the flat surface bounded by the loop. The changing magnetic field induces a current i in the wire loop, and this current has a direction such that it induces another magnetic field $\mathcal{B}_{\text{ind}}$ that tends to oppose the change in the original magnetic field. If the reverse were true, the induced current would induce a magnetic field $\mathcal{B}_{\text{ind}}$ that aids $\mathcal{B}$, which would induce an additional current that would further aid $\mathcal{B}$, and so on. Thus, conservation of energy would not result. The time-varying magnetic flux penetrating the loop is said to induce an electromotive force (emf) in the loop, and this emf drives the current. If we open the loop as shown in Fig. 5.1*b*, this emf will appear in the form of a separation of charge at the terminals. This relationship between the direction of the induced emf and the direction of the induced current is referred to as *Lenz' law*; the emf induces a current in the closed loop, whose magnetic field then opposes the *change* in the original magnetic field.

Faraday's law relates this emf to the rate of change of the magnetic flux ψ_m penetrating the loop as

$$\text{emf} = -\frac{d\psi_m}{dt} \tag{2}$$

† Faraday actually observed that a sudden interruption of the magnetic field linking the coils of a toroid produces a deflection of a galvanometer (voltmeter) attached to the ends of the toroid wires. From this we infer that current would be passed through a load attached to these ends of the wires.

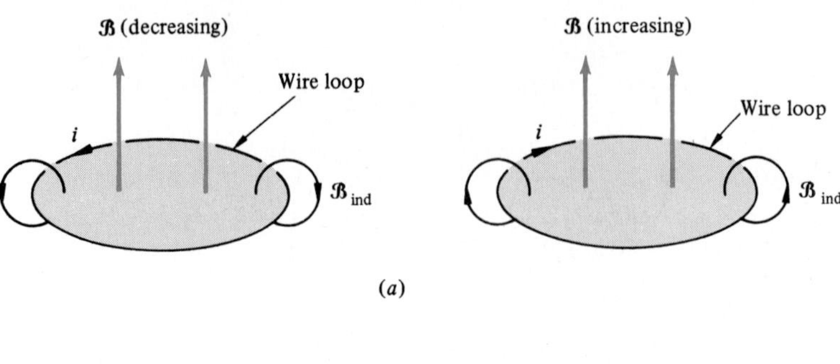

(a)

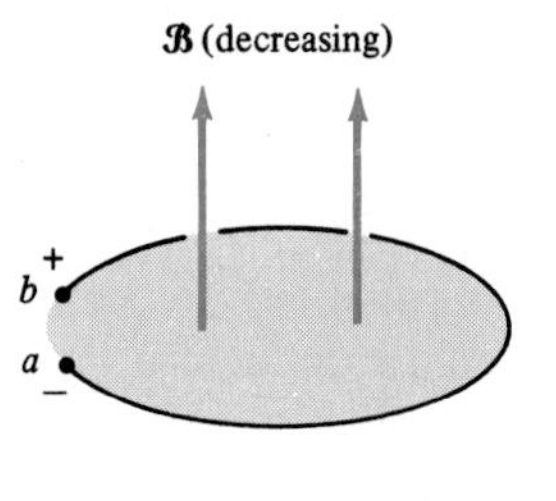

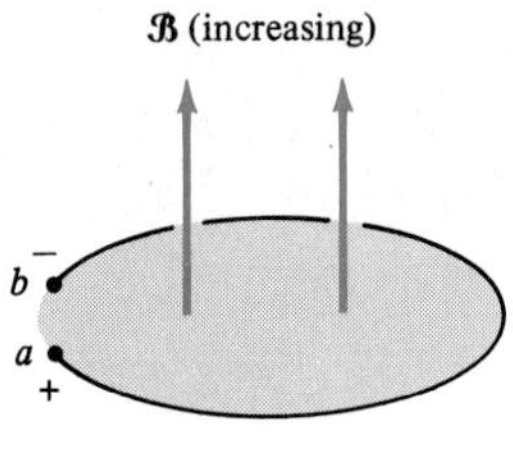

(b)

FIGURE 5.1
Illustration of Faraday's law.

The terms in Faraday's law are

$$\text{emf} = \oint_c \boldsymbol{\mathcal{E}} \cdot d\mathbf{l} \tag{3}$$

and

$$\psi_m = \int_s \boldsymbol{\mathcal{B}} \cdot d\mathbf{s} \tag{4}$$

where contour c is around the loop and the open surface s is bounded by c. Thus, Faraday's law becomes

$$\oint_c \boldsymbol{\mathcal{E}} \cdot d\mathbf{l} = -\frac{d}{dt}\int_s \boldsymbol{\mathcal{B}} \cdot d\mathbf{s} \tag{5}$$

Although stated for the specific case of a wire loop, Faraday's law applies to more general contours. The contour c and surface s are related as shown in Fig. 5.2. An illustrative example is to think of c as the mouth of a balloon and s as the surface area of that balloon. As we inflate the balloon, the surface area changes but the contour c remains the same. The total magnetic flux penetrating the surface s remains the same regardless of its shape so long as the contour c remains fixed. In addition, we need to fix the direction of this open surface s. This is done with the right-hand rule. If we place the fingers of our right hand in the direction of contour c, the direction of our thumb will define the direction of $d\mathbf{s}$

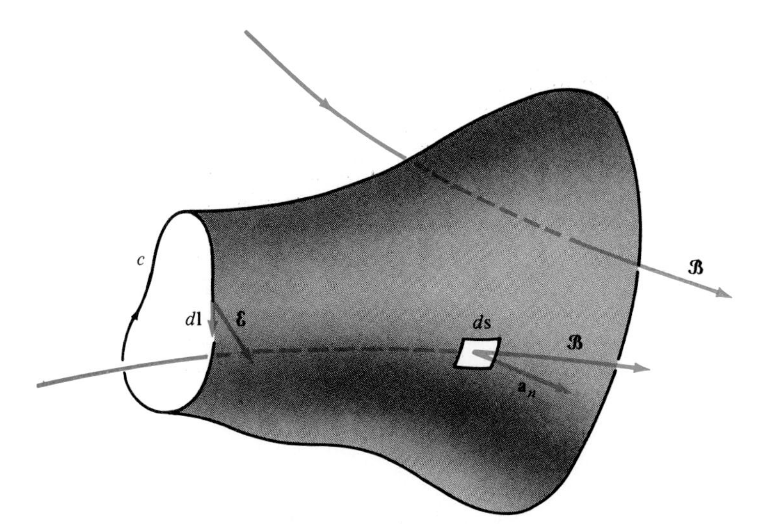

FIGURE 5.2
Illustration of the relationship between the contour c and the open surface s in Faraday's law.

or, equivalently, the unit normal to the surface, $\mathbf{a}_n$, where $d\mathbf{s} = ds\,\mathbf{a}_n$. There is no restriction on the size or shape of the contour or surface. However, c must be the closed contour that bounds the open surface s; thus, even though either one is arbitrary, c and s are intimately related.

Faraday's law indicates that a time-varying magnetic field $\mathcal{B}$ penetrating a surface s produces (induces) an electric field $\mathcal{E}$ along a closed contour c that will exert a force on any charges q that *may be present* along the contour. Charges do not have to be present along c; an electric field $\mathcal{E}$ is nevertheless the result of a time-varying magnetic field $\mathcal{B}$. Therefore, c need not be a physical contour (such as the wire loop) but may be an imaginary contour. In the case of the wire loop with an infinitesimal gap, as shown in Fig. 5.1*b*, this induced electric field $\mathcal{E}$ exerts a force on the free electrons of the wire, resulting in a charge separation at the terminals of the gap. Integrating (3) around the wire loop and across the gap, we find that the only contribution is across the gap since the total electric field along the conducting wires is essentially zero. Thus, the induced emf appears across the terminals of the gap as a separation of charge.

The direction of the induced emf or induced electric field $\mathcal{E}$ can be obtained in the following manner. For a contour c (physical or imaginary), if the flux linking the contour, ψ_m, is increasing with time, $-d\psi_m/dt$ is negative; thus, the emf is negative—whereby $\mathcal{E}$ is opposite the direction of the contour c [as shown by (5)] and of the right-hand rule relating contour c and surface s. On the other hand, if ψ_m is decreasing with time, $-d\psi_m/dt$ is positive and the induced electric field $\mathcal{E}$ is in the direction of the contour c. Alternatively, this direction of $\mathcal{E}$ can be found with Lenz' law by imagining a closed wire loop along contour c. The induced electric field produces a force $\mathcal{F} = q\,\mathcal{E}$ on positive charges along the contour

($-\mathcal{F}$ on the free electrons of the wire); thus, $\boldsymbol{\mathcal{E}}$ is in the direction of the resulting induced current (direction of net positive charge movement).

At this point we may be tempted to conclude that a *unique* voltage for time-varying fields may be related to the line integral of $\boldsymbol{\mathcal{E}}$ between two points in the same fashion that potential (voltage) was defined for static fields as the line integral of **E**. We will now show that such a definition depends, in general, on the path of integration between the two points: an important distinction between time-varying fields and static fields. Consider Fig. 5.3, in which a closed contour c consists of two sections c_1 and c_2 such that $c = c_1 - c_2$. Faraday's law shows that

$$\oint_c \boldsymbol{\mathcal{E}} \cdot d\mathbf{l} = \int_{c1} \boldsymbol{\mathcal{E}} \cdot d\mathbf{l} - \int_{c2} \boldsymbol{\mathcal{E}} \cdot d\mathbf{l} = -\frac{d\psi_m}{dt} \tag{6}$$

Therefore,

$$\int_{c1} \boldsymbol{\mathcal{E}} \cdot d\mathbf{l} \neq \int_{c2} \boldsymbol{\mathcal{E}} \cdot d\mathbf{l} \tag{7}$$

unless the magnetic flux penetrating the surface bounded by c does not vary with time. Hence, a definition of voltage as the line integral of $\boldsymbol{\mathcal{E}}$ between two points depends on the path chosen.

Consider the wire loop of Fig. 5.1*b*. The emf between the two terminals of the gap can (under certain conditions) be interpreted as a voltage in the same fashion as for static fields. However, we saw that for time-varying fields, the definition of voltage as the line integral of $\boldsymbol{\mathcal{E}}$ between two points is not unique, but depends on the chosen path. We, of course, use voltmeters to provide reliable readings in ac circuits, but if we accept the fact that voltage for time-varying circuits is not unique, then this use of voltmeters in ac circuits must rely on some approximations and therefore imposes some limits on its validity.

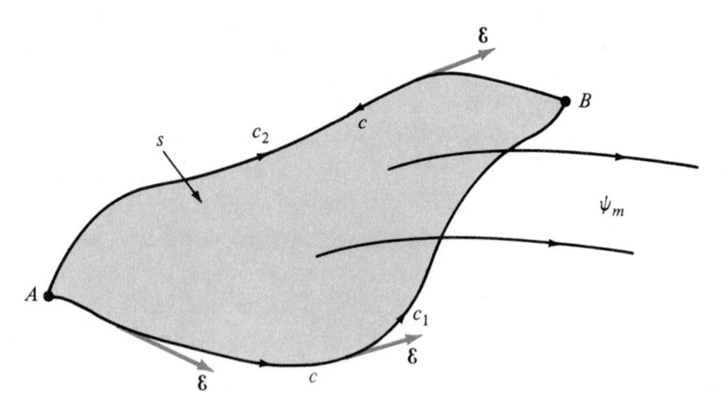

FIGURE 5.3

Illustration of the dependence of voltage definitions on the chosen path for time-varying fields.

Consider the enlarged view of the wire loop shown in Fig. 5.4. Suppose we place two voltmeters across the terminals of the gap. The leads (wires) of the voltmeter are connected to the gap terminals, and we will assume that the actual voltmeters are infinitesimal in size. The voltmeters are considered ideal in that they measure the integral $\int \mathcal{E} \cdot d\mathbf{l}$ between their respective terminals and in that they are effectively open circuits, thus drawing negligible current. We would like to determine the required conditions on the positions of the voltmeter leads so that the voltmeter readings will be nearly identical. Applying Faraday's law to the closed contour consisting of the first voltmeter, its leads (c_1), and the wire loop (c), we obtain

$$\mathcal{V}_1 + \int_{a,c}^{b} \mathcal{E} \cdot d\mathbf{l} + \int_{b,c_1}^{a} \mathcal{E} \cdot d\mathbf{l} = -\frac{d}{dt}\left(\int_{\Delta s} \mathcal{B} \cdot d\mathbf{s} + \int_{s} \mathcal{B} \cdot d\mathbf{s}\right) \tag{8}$$

where Δs is the surface between the two positions of the voltmeter leads and s is the surface bounded by the wire loop and the leads of the second voltmeter. Since the voltmeter leads and the wire loop are considered to be perfectly conducting wires, $\mathcal{E}$ along the wire surfaces is zero and this result reduces to

$$\mathcal{V}_1 = -\frac{d}{dt}\left(\int_{\Delta s} \mathcal{B} \cdot d\mathbf{s} + \int_{s} \mathcal{B} \cdot d\mathbf{s}\right) \tag{9}$$

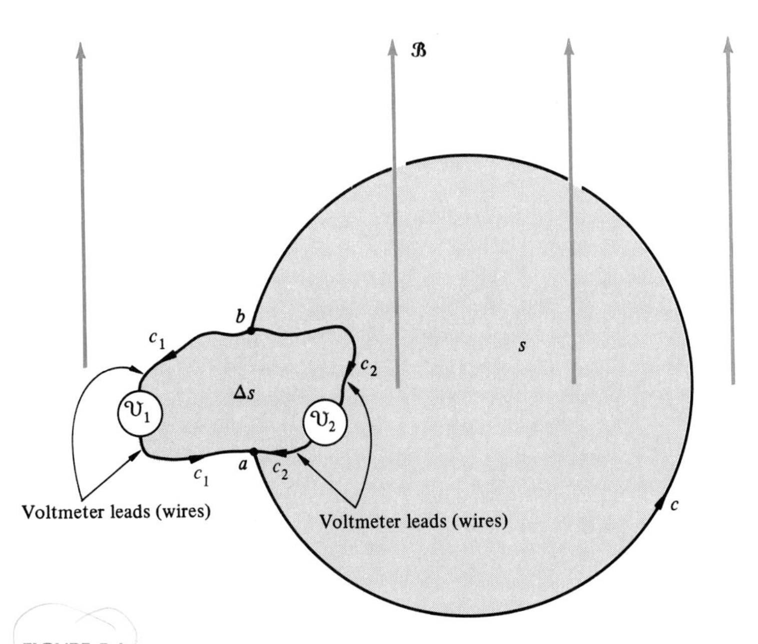

FIGURE 5.4
Illustration of the dependence of measured voltage on the positions of the voltmeter leads for time-varying fields.

Similarly, for the second voltmeter, we obtain

$$\mathcal{V}_2 = -\frac{d}{dt}\int_s \mathcal{B}\cdot ds \tag{10}$$

Combining these results, we obtain

$$\mathcal{V}_1 - \mathcal{V}_2 = -\frac{d}{dt}\int_{\Delta s} \mathcal{B}\cdot ds \tag{11}$$

Thus, $\mathcal{V}_1 \simeq \mathcal{V}_2$ if Δs is "small enough" or the time variation of $\mathcal{B}$ is "small enough." This explains why the positions and lengths of the voltmeter leads become more important in ac circuits with increasing frequencies.

Perhaps a more illustrative example is shown in Fig. 5.5. In Fig. 5.5*a* a wire loop with voltmeter leads is arranged so that the closed loop encircles some time-varying magnetic field $\mathcal{B}$. The voltage that the voltmeter reads is given by

$$\mathcal{V}(t) + \int_4^1 \overset{0}{\cancel{\mathcal{E}}}\cdot d\mathbf{l} + \int_1^2 \overset{0}{\cancel{\mathcal{E}}}\cdot d\mathbf{l} + \int_2^3 \overset{0}{\cancel{\mathcal{E}}}\cdot d\mathbf{l} = -\frac{d\psi_m}{dt} \tag{12}$$

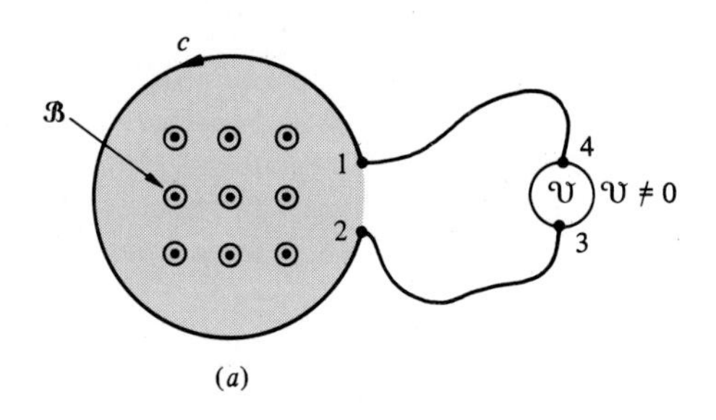

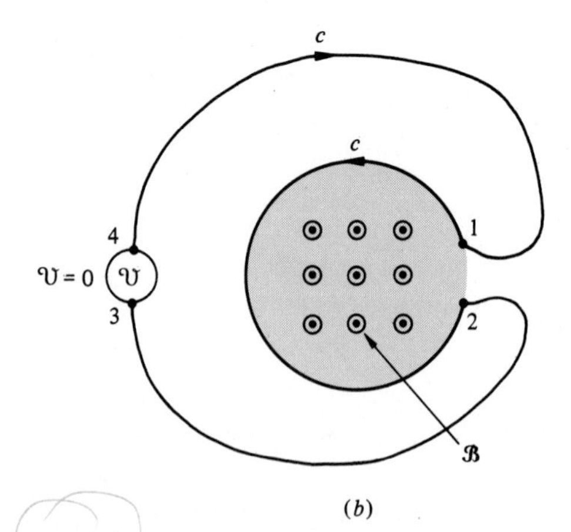

FIGURE 5.5
Illustration of the dependence of voltage on the path for time-varying fields.

If the leads are arranged as shown in Fig. 5.5*b*, the net magnetic flux penetrating this closed loop is zero and the voltmeter reads zero. Clearly, the position of the voltmeter leads is important for the time-varying case.

The integral form of Faraday's law in (5) applies to a region of space. We may also obtain a point form of Faraday's law in the following manner. Take the time derivative under the integral to obtain†

$$\oint_c \mathcal{E}\cdot d\mathbf{l} = -\int_s \frac{\partial}{\partial t}(\mathcal{B}\cdot d\mathbf{s}) = -\int_s \left(\frac{\partial \mathcal{B}}{\partial t}\cdot d\mathbf{s} + \mathcal{B}\cdot\frac{\partial\, d\mathbf{s}}{\partial t}\right) \tag{13}$$

If the surface s is stationary, $\partial\, d\mathbf{s}/\partial t = 0$, and (13) reduces to

$$\oint_c \mathcal{E}\cdot d\mathbf{l} = -\int_s \frac{\partial \mathcal{B}}{\partial t}\cdot d\mathbf{s} \tag{14}$$

and thus we may interchange the order of integration and differentiation of $\mathcal{B}$. The case of moving contours is considered in Appendix B and in the problems at the end of this chapter. We will restrict our discussions to stationary contours throughout the remaining chapters. Stokes' theorem allows us to write

$$\oint_c \mathcal{E}\cdot d\mathbf{l} = \int_s (\nabla\times\mathcal{E})\cdot d\mathbf{s} \tag{15}$$

and (14) becomes

$$\int_s (\nabla\times\mathcal{E})\cdot d\mathbf{s} = -\int_s \frac{\partial \mathcal{B}}{\partial t}\cdot d\mathbf{s} \tag{16}$$

As the surface s becomes infinitesimally small, we obtain

$$\nabla\times\mathcal{E} = -\frac{\partial \mathcal{B}}{\partial t} \tag{17}$$

which is the point form of Faraday's law. For static fields, both the integral form of Faraday's law in (5) and the point form in (17) reduce to the static case given in (1*a*).

EXAMPLE 5.1 Suppose that a time-varying magnetic field is defined in space in a cylindrical coordinate system as

$$\mathcal{B} = \begin{cases} B_0 \sin\omega t\,\mathbf{a}_z & r \le r_0 \\ 0 & r > r_0 \end{cases}$$

as shown in Fig. 5.6. Determine the induced electric field via Faraday's law.

---

† Note that a partial derivative with respect to time, $\partial/\partial t$, is required in (13) since $\mathcal{B}$ is a function of spatial coordinates as well as of time. In (5), the integral of $\mathcal{B}$ over s removes this spatial dependence and the ordinary derivative d/dt is sufficient.

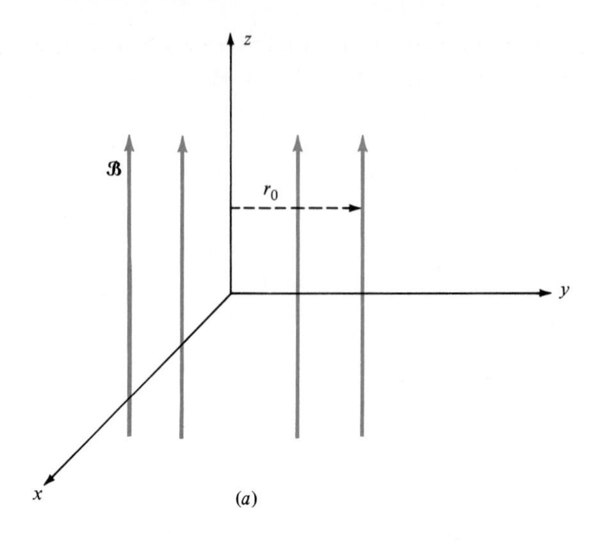

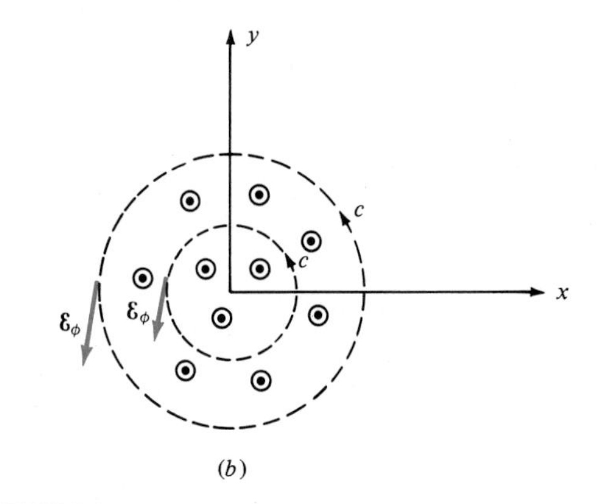

FIGURE 5.6
Example 5.1. Determination of induced electric field due to a time-varying magnetic field.

Solution Note that the magnetic field is uniform for $r \le r_0$ and is zero elsewhere. Since the magnetic field has circular symmetry about the z axis, it follows that the induced electric field will also have circular symmetry about the z axis and must therefore be independent of ϕ. Also, since $\mathcal{B}$ is independent of z, the induced electric field is independent of z. Thus, the induced electric field must be of the form

$$\boldsymbol{\mathcal{E}} = \mathcal{E}_\phi(r)\mathbf{a}_\phi$$

Since the induced electric field will possess circular symmetry, we will choose the contour c in (5) to be a circular contour about the z axis, and the surface s will be chosen, for purposes of simplifying the integration, to be the flat surface (in the xy plane) bounded by c. The magnetic flux penetrating s is given by

$$\psi_m = \int_s \mathcal{B} \cdot d\mathbf{s}$$

$$= \int_{\phi=0}^{2\pi} \int_{r=0}^{r} B_0 \sin \omega t \, r \, d\phi \, dr$$

$$= \begin{cases} \pi r^2 B_0 \sin \omega t & r \leq r_0 \\ \pi r_0^2 B_0 \sin \omega t & r \geq r_0 \end{cases}$$

The right-hand side of (5) becomes

$$-\frac{d}{dt} \int_s \mathcal{B} \cdot d\mathbf{s} = \begin{cases} -\pi r^2 \omega B_0 \cos \omega t & r \leq r_0 \\ -\pi r_0^2 \omega B_0 \cos \omega t & r \geq r_0 \end{cases}$$

$$= -\int_s \frac{\partial \mathcal{B}}{\partial t} \cdot d\mathbf{s}$$

since the contour c is stationary. The left-hand side of (5) evaluated along contour c yields

$$\oint_c \mathcal{E} \cdot d\mathbf{l} = \int_{\phi=0}^{2\pi} \mathcal{E}_\phi r \, d\phi$$

$$= 2\pi r \, \mathcal{E}_\phi$$

Thus, the induced electric field becomes

$$\mathcal{E}_\phi = \begin{cases} -\dfrac{\omega r B_0}{2} \cos \omega t & r \leq r_0 \\ -\dfrac{\omega r_0^2 B_0}{2r} \cos \omega t & r \geq r_0 \end{cases}$$

Using the form of $\nabla \times \mathcal{E}$ in cylindrical coordinates (Appendix A), the reader can (and should) show that this field satisfies the point form of Faraday's law in (17).

The direct application of Faraday's law is also possible in the case of moving contours: moving with respect to $\mathcal{B}$. The following example illustrates this.

EXAMPLE 5.2 A pair of perfectly conducting parallel wires in the xy plane forms a set of rails along which a shorting bar moves with velocity u, as shown in

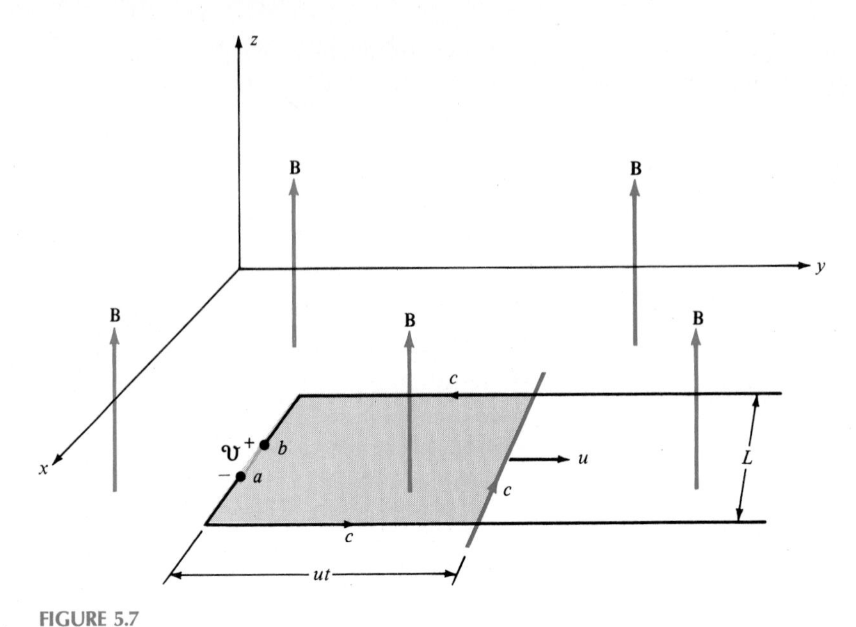

FIGURE 5.7
Example 5.2. Illustration of induced voltage caused by moving contours.

Fig. 5.7. If the structure is immersed in a dc magnetic field that is perpendicular to the *xy* plane and given by

$$\mathbf{B} = B_0\mathbf{a}_z \qquad \text{Wb/m}^2$$

determine the voltage $\mathcal{V}$, with polarity as indicated, that is induced across a small gap in the wires.

Solution In applying (5), we choose the contour *c* to be along the rails and the shorting bar, as shown in Fig. 5.7. The surface *s* is chosen to be the flat surface bounded by *c*, as shown. The total magnetic flux penetrating *s* is

$$\psi_m = \int_s \mathbf{B} \cdot d\mathbf{s}$$

$$= B_0 L u t \qquad \text{Wb}$$

where

$$d\mathbf{s} = dx\, dy\mathbf{a}_z$$

Thus

$$\text{emf} = \oint_c \boldsymbol{\mathcal{E}} \cdot d\mathbf{l}$$

$$= -\frac{d\psi_m}{dt}$$

$$= -B_0 L u$$

Clearly, if the loop were closed, a current would flow in a direction *opposite* to the direction of c. Thus, positive charge would accumulate at terminal a and the induced voltage would be

$$\mathcal{V} = -B_0 L u$$

5.1.2 Gauss' Laws for Electric and Magnetic Fields†

Gauss' laws for electric and magnetic fields for the static case in (1c) and (1d) remain the same for the time-varying case; that is,

$$\nabla \cdot \mathcal{D} = \rho \qquad \oint_s \mathcal{D} \cdot d\mathbf{s} = \int_v \rho \, dv \tag{18}$$

$$\nabla \cdot \mathcal{B} = 0 \qquad \oint_s \mathcal{B} \cdot d\mathbf{s} = 0 \tag{19}$$

Electric flux lines that begin on positive charges terminate on negative charges. Equation (18) is a simple statement of this fact. Similarly, the lines of magnetic flux form closed paths with no known sources or sinks. Equation (19) is a simple statement of this observation.

It is important to note that lines of electric field intensity (or electric flux density) may be generated by a time-varying magnetic field, as is clear from Faraday's law. Thus, it would be *incorrect* to state that *all* electric field lines begin and terminate on electric charges. Those electric field lines which are produced by a time-varying magnetic field form closed paths; those produced by electric charges do not.

5.1.3 Ampère's Law and Displacement Current

First consider Ampère's law for static fields given in (1b). Suppose we try to use this for time-varying fields:

$$\nabla \times \mathcal{H} \stackrel{?}{=} \mathcal{J} \tag{20}$$

If we take the divergence of both sides of (20), we obtain

$$\nabla \cdot (\nabla \times \mathcal{H}) = 0$$
$$= \nabla \cdot \mathcal{J} \tag{21}$$

---

† It is not customary to refer to (19) as one of Gauss' laws. Strictly speaking, only (18) is Gauss' law. We will, however, refer to both as Gauss' laws for easy reference.

since we have the vector identity that the divergence of the curl of any vector field yields zero (Chap. 2). This result, however, requires that the divergence of the current density vector be zero, which is obviously incorrect for time-varying fields since the continuity equation shows that

$$\nabla \cdot \mathfrak{J} = -\frac{\partial \rho}{\partial t} \tag{22a}$$

$$\oint_s \mathfrak{J} \cdot ds = -\frac{d}{dt}\int_v \rho \, dv \tag{22b}$$

That is, the divergence or net outflow of $\mathfrak{J}$ from some region is the rate of decrease of charge contained in the region, which is necessarily zero only for static currents, i.e., $\nabla \cdot \mathbf{J} = 0$.

Thus, a term is missing from (20). Obviously, this term should be the time rate of change of some vector field so that, for static fields, the new equation would reduce to (1*b*). Perhaps Maxwell's most important contribution was the determination of this missing term. Maxwell modified (20) by adding $\partial \mathfrak{D}/\partial t$ to the right-hand side so that

$$\nabla \times \mathfrak{H} = \mathfrak{J} + \frac{\partial \mathfrak{D}}{\partial t} \tag{23}$$

where $\mathfrak{D}$ is the electric flux density vector, so no inconsistency results. We can see this by again taking the divergence of both sides of (23), resulting in

$$0 = \nabla \cdot \mathfrak{J} + \nabla \cdot \frac{\partial \mathfrak{D}}{\partial t}$$

$$= \nabla \cdot \mathfrak{J} + \frac{\partial}{\partial t}(\nabla \cdot \mathfrak{D}) \tag{24}$$

since space derivatives ($\nabla \cdot$) and time derivatives ($\partial/\partial t$) are interchangeable. Substituting (18) into (24), we obtain the continuity equation in (22*a*) and no inconsistency results.

Equation (23) is Ampère's law in point form. The integral form of this law is obtained by integrating both sides of (23) over some open surface s:

$$\int_s (\nabla \times \mathfrak{H}) \cdot ds = \int_s \mathfrak{J} \cdot ds + \int_s \frac{\partial \mathfrak{D}}{\partial t} \cdot ds \tag{25}$$

and applying Stokes' theorem to yield Ampère's law in integral form:

$$\oint_c \mathfrak{H} \cdot d\mathbf{l} = \int_s \mathfrak{J} \cdot ds + \int_s \frac{\partial \mathfrak{D}}{\partial t} \cdot ds \tag{26}$$

If we assume a stationary contour c, then $\partial/\partial t$ may be removed from the integrand, resulting in

$$\oint_c \mathcal{H} \cdot d\mathbf{l} = \int_s \mathcal{J} \cdot d\mathbf{s} + \frac{d}{dt}\int_s \mathcal{D} \cdot d\mathbf{s} \tag{27}$$

Clearly, the first portion of the right-hand side of (27) is free current (either conduction or convection current), which we denote by I_c;

$$I_c = \int_s \mathcal{J} \cdot d\mathbf{s} \tag{28}$$

The other term also has the unit of current, and this term will be given the name of *displacement current* I_d:

$$I_d = \frac{d}{dt}\int_s \mathcal{D} \cdot d\mathbf{s} \tag{29}$$

Strictly speaking, the displacement current is not a current—in the sense that it does not represent the flow of free charge through the surface s. How, then, do we interpret this quantity?

First let us interpret Ampère's law in the integral form given in (27). As with Faraday's law, the closed contour c bounds the open surface s, and $d\mathbf{l}$ and $d\mathbf{s}$ are related by the right-hand rule, as shown in Fig. 5.8*a*. The contour c and surface s in Ampère's law in (27) are again intimately related.

To interpret the meaning of displacement current, consider the electric circuit shown in Fig. 5.8*b* consisting of a sinusoidal voltage source connected to a parallel-plate capacitor. Construct two "balloonlike" surfaces, as shown in Fig. 5.8*b*. The first surface s_1 has the wire and consequently free current (conduction current in this case) penetrating it, and the related contour c encircles the wire. The second surface s_2 is constructed so that it encloses one of the capacitor plates. No free current penetrates this surface, but the contour c again encircles the wire. As the polarity of the sinusoidal generator changes, free charge will be transferred to and from each capacitor plate via the wires of the circuit. For surface s_1 and related contour c, we observe that

$$\oint_c \mathcal{H} \cdot d\mathbf{l} = \int_{s_1} \mathcal{J} \cdot d\mathbf{s} = I_c \tag{30}$$

where $\mathcal{J}$ is the density of this free current in the wire and I_c refers to conduction current in the wire. Now consider surface s_2 and related contour c. Again

$$\oint_c \mathcal{H} \cdot d\mathbf{l} = \int_{s_2} \mathcal{J} \cdot d\mathbf{s} + \frac{d}{dt}\int_{s_2} \mathcal{D} \cdot d\mathbf{s} \tag{31}$$

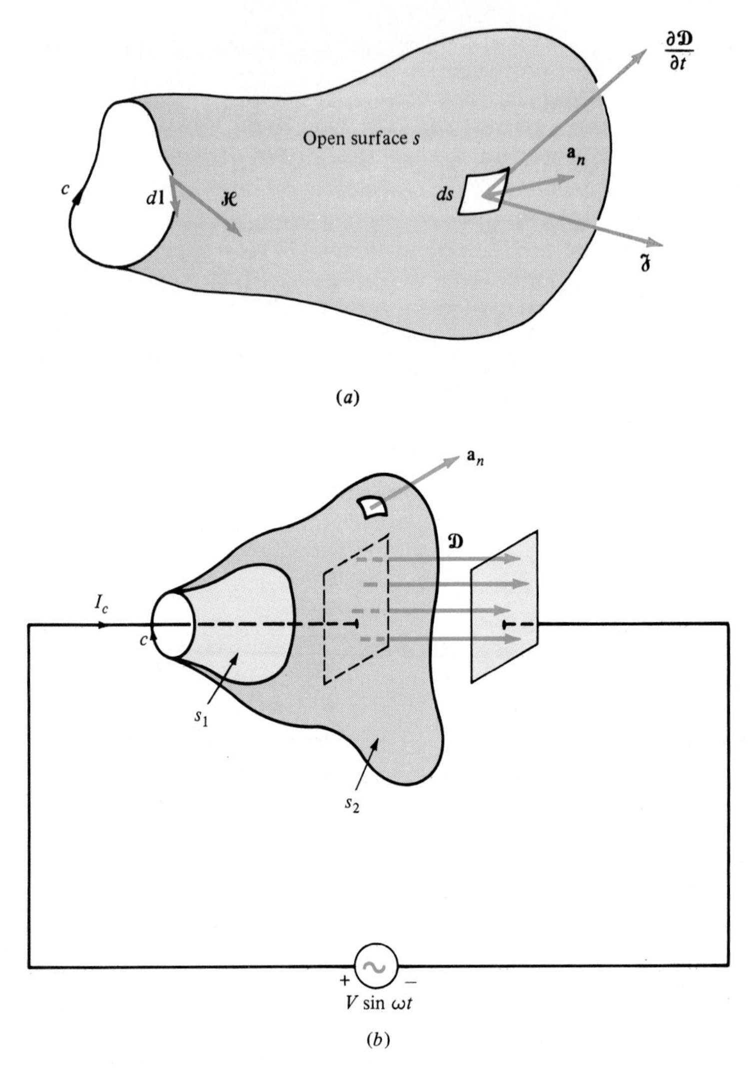

FIGURE 5.8
Illustration of Ampère's law and displacement current: (*a*) relation of contour *c* and open surface *s*; (*b*) continuity of conduction and displacement currents.

But over the surface s_2

$$\int_{s_2} \mathfrak{J} \cdot d\mathbf{s} = 0 \tag{32}$$

since no free current penetrates this surface. However, as free charge is being stored and removed from each capacitor plate, a time-varying field $\mathfrak{D} = \epsilon_0 \mathfrak{E}$ will be developed between the plates. These lines of electric flux penetrate the surface s_2, thus

$$\oint_c \mathcal{H}\cdot d\mathbf{l} = \frac{d}{dt}\int_{s2} \mathfrak{D}\cdot d\mathbf{s}$$

$$= I_d \tag{33}$$

Thus, it would appear that the rate of change of $\mathfrak{D}$ in the region between the capacitor plates is intimately related to the "displacement" of electric charge. Hence, $\mathfrak{D}$ is often referred to as the *displacement vector*, and the rate of change of electric flux is referred to as a *displacement current density*, i.e., $\mathfrak{J}_d = \partial\mathfrak{D}/\partial t$.

Note that the contours bounding the two surfaces in Fig. 5.8*b* are chosen to be the same. If the displacement current term were not present on the right-hand side of the equation for Ampère's law, we would have an immediate inconsistency; choosing the two surfaces having the *same* contour would yield different results. Therefore, we can consider the displacement current as "completing the circuit;" where conduction current ends, displacement current takes over to complete the circuit.

EXAMPLE 5.3 Compare the conduction and displacement current densities in copper ($\epsilon \simeq \epsilon_0$, $\mu \simeq \mu_0$, and $\sigma = 5.8 \times 10^7$ S/m) at a frequency of 1 MHz. Repeat for Teflon, which has $\epsilon \simeq 2.1\epsilon_0$, $\mu \simeq \mu_0$, and $\sigma \simeq 3 \times 10^{-8}$ S/m at 1 MHz.

Solution Assuming sinusoidal variation of the electric field in the material,

$$\mathcal{E} = E_0 \sin \omega t \qquad \text{V/m}$$

where $\omega = 2\pi f$ and $f = 10^6$, the conduction current density is

$$\mathfrak{J}_c = \sigma\mathcal{E}$$

$$= \sigma E_0 \sin \omega t \qquad \text{A/m}^2$$

The displacement current density is

$$\mathfrak{J}_d = \frac{\partial\mathfrak{D}}{\partial t}$$

$$= \epsilon\frac{\partial\mathcal{E}}{\partial t}$$

$$= \omega\epsilon E_0 \cos \omega t \qquad \text{A/m}^2$$

The ratio of the magnitudes of these currents is

$$\frac{|\mathcal{J}_c|}{|\mathcal{J}_d|} = \frac{\sigma}{\omega\epsilon}$$

For copper at 1 MHz, we find

$$\frac{\sigma}{\omega\epsilon} = \frac{5.8 \times 10^7}{2\pi \times 10^6 \times 1/36\pi \times 10^{-9}}$$
$$= 10^{12}$$

Even if the frequency is raised to an extremely high value (for example, 100 GHz or 10^{11} Hz), the conduction current dominates the displacement current by an enormous amount. Thus, for copper (and most other "conductors') it is reasonable to neglect displacement current. For Teflon at 1 MHz, we find that

$$\frac{\sigma}{\omega\epsilon} = \frac{3 \times 10^{-8}}{2\pi \times 10^6 \times 2.1 \times 1/36\pi \times 10^{-9}}$$
$$= 2.57 \times 10^{-4}$$

Consequently, in Teflon, a reasonably good insulator, at 1 MHz the conduction current may be neglected. However, at much lower frequencies the ratio $\sigma/\omega\epsilon$ may approach unity, and the conduction and displacement currents may become comparable.

5.1.4 Summary of Maxwell's Equations

The results of combining Faraday's law, Ampère's law, and Gauss' laws are referred to as Maxwell's equations:

$$\oint_c \mathcal{E} \cdot d\mathbf{l} = -\frac{d}{dt}\int_s \mathcal{B} \cdot ds \qquad \nabla \times \mathcal{E} = -\frac{\partial \mathcal{B}}{\partial t} \tag{34a}$$

$$\oint_c \mathcal{H} \cdot d\mathbf{l} = \int_s \mathcal{J} \cdot ds + \frac{d}{dt}\int_s \mathcal{D} \cdot ds \qquad \nabla \times \mathcal{H} = \mathcal{J} + \frac{\partial \mathcal{D}}{\partial t} \tag{34b}$$

$$\oint_s \mathcal{D} \cdot ds = \int_v \rho \, dv \qquad \nabla \cdot \mathcal{D} = \rho \tag{34c}$$

$$\oint_s \mathcal{B} \cdot ds = 0 \qquad \nabla \cdot \mathcal{B} = 0 \tag{34d}$$

These deceptively simple equations are the result of many years of research and study of the phenomena associated with electricity and magnetism. Although these equations have not been derived analytically, they are reasonable and no experiments have shown them to be invalid. In the absence of any such data, we may accept them as a valid characterization of electromagnetic phenomena. However, these equations are applicable only where the dimensions are large compared to atomic dimensions.

Associated with Maxwell's equations, we have the Lorentz force equation relating mechanical and electromagnetic phenomena:

$$\mathcal{F} = q(\mathcal{E} + \mathbf{u} \times \mathcal{B}) \qquad \text{N} \tag{35}$$

The quantity $\mathcal{F}$ is the force exerted on a point charge q by an electric field $\mathcal{E}$ and a magnetic field $\mathcal{B}$. The charge q is moving with velocity u, and $\mathbf{u}$ is the velocity vector in the direction of movement of the charge. For a continuous distribution of charge, ρ C/m^3, the Lorentz equation becomes

$$f = \rho(\mathcal{E} + \mathbf{u} \times \mathcal{B}) \qquad \text{N/m}^3 \tag{36}$$

where f is the force vector per unit volume. Noting that the movement of this charge distribution constitutes a current, $\mathcal{J} = \rho\mathbf{u}$, we also obtain

$$f = \rho\mathcal{E} + \mathcal{J} \times \mathcal{B} \qquad \text{N/m}^3 \tag{37}$$

The equation of continuity or conservation of charge,

$$\nabla \cdot \mathcal{J} = -\frac{\partial \rho}{\partial t} \tag{38}$$

is also implicit in Maxwell's equations, as was shown in Sec. 5.1.3.

It should be noted that the four equations of Maxwell in (34) are not all independent. For example, if we take the divergence of the point form of Faraday's law in (34*a*), we find that

$$\frac{\partial}{\partial t}(\nabla \cdot \mathcal{B}) = 0$$

since $\nabla \cdot (\nabla \times \mathcal{E}) = 0$. This implies that $\nabla \cdot \mathcal{B}$ is independent of time, or

$$\nabla \cdot \mathcal{B} = \text{a constant}$$

In the absence of proof of the existence of any isolated magnetic sources, we take this constant to be zero and we obtain (34*d*). Similarly, taking the divergence of the point form of Ampère's law in (34*b*), we obtain, since $\nabla \cdot (\nabla \times \mathcal{H}) = 0$,

$$\nabla \cdot \mathcal{J} = -\frac{\partial}{\partial t}(\nabla \cdot \mathcal{D})$$

But according to the continuity equation in (38), this implies (34*c*).

5.2 Constitutive Properties of the Medium

We have considered the properties of linearity, isotropy, and homogeneity of a material medium in previous chapters in discussions of static fields. For time-varying fields, there are essentially no changes in these definitions or ideas. Although the free current density $\mathfrak{J}$ in Maxwell's equations could be either conduction current (in a conducting material) or convection current (as in a vacuum), we will henceforth consider $\mathfrak{J}$ to denote only conduction current.

Maxwell's equations in (34) contain 12 unknowns: $\mathcal{E}_x$, $\mathcal{E}_y$, $\mathcal{E}_z$, $\mathcal{H}_x$, $\mathcal{H}_y$, $\mathcal{H}_z$, $\mathfrak{D}_x$, $\mathfrak{D}_y$, $\mathfrak{D}_z$, $\mathfrak{B}_x$, $\mathfrak{B}_y$, and $\mathfrak{B}_z$. Since only the curl equations (34*a*) and (34*b*) are independent, we have six equations (each vector equation contains three scalar equations in terms of components). Thus, we need six additional equations relating these components. These are provided by the *constitutive relations* of the medium:

$$\mathfrak{D} = f_{\mathfrak{D}}(\mathcal{E}) \tag{39a}$$

$$\mathfrak{B} = f_{\mathfrak{B}}(\mathcal{H}) \tag{39b}$$

Each equation relates three field vector components, giving a total of six equations. The type of medium will determine these specific functional relationships. In addition, if we treat $\mathfrak{J}$ in (34*b*) as unknown, we introduce three additional unknowns, $\mathfrak{J}_x$, $\mathfrak{J}_y$, and $\mathfrak{J}_z$. Thus, we need three additional equations given by

$$\mathfrak{J} = f_{\mathfrak{J}}(\mathcal{E}) \tag{39c}$$

The material medium determines these specific functional relationships between the various field vectors. Throughout the remainder of this text we will consider only media that are *linear*, *homogeneous*, and *isotropic* such that

$$\mathfrak{D} = \epsilon\mathcal{E} \tag{40a}$$

$$\mathfrak{B} = \mu\mathcal{H} \tag{40b}$$

$$\mathfrak{J} = \sigma\mathcal{E} \tag{40c}$$

Such media are said to be *simple*. The scalars ϵ (permittivity), μ (permeability), and σ (conductivity) in (40) are assumed to be constants.

For sinusoidal variation of the fields in linear, homogeneous, and isotropic media, we will find that these scalars often depend on frequency f, so that we would write $\epsilon(f)$, $\mu(f)$, and $\sigma(f)$. In this case it would be improper to write (40) as shown, since the field vectors there are time-domain quantities: i.e., $\mathfrak{D}(x, y, z, t)$, etc. In later sections we will concentrate on sinusoidal variation of the fields and will investigate *phasor* field vectors. In this case, (40) will relate those phasor field vectors and it would be proper to use frequency-dependent parameters. In future we will continue to write (40) as shown, with this understanding.

A medium is said to be *linear* if the above relationships are independent on the magnitudes, or levels, of the fields. An example of a *nonlinear* medium is a ferromagnetic material in which the relationship between $\mathcal{B}$ and $\mathcal{H}$ is specified by a nonlinear, hysteresis curve instead of by a scalar constant μ.

An isotropic medium is one in which $\mathcal{D}$ is parallel to $\mathcal{E}$, $\mathcal{B}$ is parallel to $\mathcal{H}$, and $\mathcal{J}$ is parallel to $\mathcal{E}$. Isotropic media exhibit the same properties in all directions. An *anisotropic* dielectric would have each of the components of $\mathcal{D}$ related to those of $\mathcal{E}$ by a 3 × 3 matrix:

$$\begin{bmatrix} \mathcal{D}_x \\ \mathcal{D}_y \\ \mathcal{D}_z \end{bmatrix} = \begin{bmatrix} \epsilon_{xx} & \epsilon_{xy} & \epsilon_{xz} \\ \epsilon_{yx} & \epsilon_{yy} & \epsilon_{yz} \\ \epsilon_{zx} & \epsilon_{zy} & \epsilon_{zz} \end{bmatrix} \begin{bmatrix} \mathcal{E}_x \\ \mathcal{E}_y \\ \mathcal{E}_z \end{bmatrix} \tag{41}$$

A *homogeneous* medium is one in which the medium properties are the same at all points in the medium. An *inhomogeneous* medium would have, for example, ϵ, μ, and σ functions of the spatial parameters: e.g., $\epsilon(x, y, z)$, $\mu(x, y, z)$, and $\sigma(x, y, z)$.

Specification of ϵ and μ as in (40*a*) and (40*b*) is not the only possible way of characterizing the medium. In Chaps. 3 and 4 we wrote

$$\mathcal{D} = \epsilon_0 \mathcal{E} + \mathcal{P} \tag{42a}$$

$$\mathcal{H} = \frac{\mathcal{B}}{\mu_0} - \mathcal{M} \tag{42b}$$

where $\mathcal{P}$ is the *electric polarization vector* and $\mathcal{M}$ is the *magnetic polarization vector*. Furthermore, we may write

$$\mathcal{P} = \epsilon_0 \chi_e \mathcal{E} \tag{43a}$$

$$\mathcal{M} = \chi_m \mathcal{H} \tag{43b}$$

where χ_e and χ_m are the electric and magnetic susceptibilities, respectively. Thus, $\mathcal{D}$ and $\mathcal{H}$ consist of the superposition of free space contributions, $\epsilon_0 \mathcal{E}$ and $\mathcal{B}/\mu_0$, and the contributions due to the *dipoles* (electric and magnetic) of the material in terms of the electric dipole moment per unit volume, $\mathcal{P}$, and the magnetic dipole moment per unit volume, $\mathcal{M}$. The properties of linearity, isotropy, and homogeneity of the medium could then have been related to $\mathcal{P}$ and $\mathcal{M}$.

For example, a material is linear if the susceptibilities χ_e and χ_m are constants independent of the applied fields. A material is isotropic if $\mathcal{P}$ and $\mathcal{E}$ are in the same direction and $\mathcal{M}$ and $\mathcal{B}$ are in the same direction. Finally, a material is homogeneous if χ_e and χ_m are independent of position in the medium.

However, we will use the simpler relations in (40) throughout the remainder of this text with no loss in generality. Our future interest will be in examining the behavior of fields in material media, not in the characterization of those media.

For a linear, isotropic, homogeneous medium, Maxwell's equations reduce, by substituting (40) into (34), to

Integral form | *Point form*

$$\oint_c \mathcal{E}\cdot d\mathbf{l} = -\mu \frac{d}{dt}\int_s \mathcal{H}\cdot d\mathbf{s} \qquad \nabla\times\mathcal{E} = -\mu\frac{\partial \mathcal{H}}{\partial t} \tag{44a}$$

$$\oint_c \mathcal{H}\cdot d\mathbf{l} = \sigma\int_s \mathcal{E}\cdot d\mathbf{s} + \epsilon\frac{d}{dt}\int_s \mathcal{E}\cdot d\mathbf{s} \qquad \nabla\times\mathcal{H} = \sigma\mathcal{E} + \epsilon\frac{\partial \mathcal{E}}{\partial t} \tag{44b}$$

$$\oint_s \mathcal{E}\cdot d\mathbf{s} = \frac{1}{\epsilon}\int_v \rho\, dv \qquad \nabla\cdot\mathcal{E} = \rho/\epsilon \tag{44c}$$

$$\oint_s \mathcal{H}\cdot d\mathbf{s} = 0 \qquad \nabla\cdot\mathcal{H} = 0 \tag{44d}$$

and only two of the four general field vectors need be determined.

EXAMPLE 5.4 In later chapters we will find that an important class of waves whose propagation is predicted by Maxwell's equations have the following form. Thus, show that the following field vectors in free space ($\mu = \mu_0$, $\epsilon = \epsilon_0$, $\sigma = 0$, $\rho = 0$, and $\mathcal{J} = 0$) satisfy all of Maxwell's equations:

$$\mathcal{E} = E_0 \cos(\omega t - \beta z)\mathbf{a}_x$$

$$\mathcal{H} = \frac{E_0}{\eta}\cos(\omega t - \beta z)\mathbf{a}_y$$

Solution From Faraday's law we must have

$$\nabla\times\mathcal{E} = -\mu_0\frac{\partial \mathcal{H}}{\partial t}$$

Expanding the curl (using the simple mnemonic device described in Sec. 2.12), we obtain

$$\begin{aligned}\nabla\times\mathcal{E} &= \left(\frac{\partial \mathcal{E}_z}{\partial y} - \frac{\partial \mathcal{E}_y}{\partial z}\right)\mathbf{a}_x + \left(\frac{\partial \mathcal{E}_x}{\partial z} - \frac{\partial \mathcal{E}_z}{\partial x}\right)\mathbf{a}_y + \left(\frac{\partial \mathcal{E}_y}{\partial x} - \frac{\partial \mathcal{E}_x}{\partial y}\right)\mathbf{a}_z \\ &= \frac{\partial \mathcal{E}_x}{\partial z}\mathbf{a}_y \\ &= \beta E_0 \sin(\omega t - \beta z)\mathbf{a}_y \end{aligned} \tag{45}$$

(In the first line every term except $\partial \mathcal{E}_x/\partial z$ is marked 0.)

since $\boldsymbol{\mathcal{E}}$ has only an x component and is independent of the x and y coordinates. Also,

$$-\mu_0 \frac{\partial \boldsymbol{\mathcal{H}}}{\partial t} = \frac{\omega \mu_0 E_0}{\eta} \sin(\omega t - \beta z)\mathbf{a}_y \tag{46}$$

In order for Faraday's law to be satisfied, (45) and (46) must be equal, which requires that

$$\beta = \frac{\omega \mu_0}{\eta} \tag{47}$$

Thus, β and η are related by (47). Ampère's law is

$$\nabla \times \boldsymbol{\mathcal{H}} = \cancelto{0}{\boldsymbol{\mathcal{J}}} + \epsilon_0 \frac{\partial \boldsymbol{\mathcal{E}}}{\partial t}$$

Expanding the curl, we obtain

$$\begin{aligned} \nabla \times \boldsymbol{\mathcal{H}} &= -\frac{\partial \mathcal{H}_y}{\partial z}\mathbf{a}_x \\ &= -\frac{\beta E_0}{\eta} \sin(\omega t - \beta z)\mathbf{a}_x \end{aligned} \tag{48}$$

since $\boldsymbol{\mathcal{H}}$ is in the y direction and is independent of the x and y coordinates. Forming the right-hand side of Ampère's law, we obtain

$$\epsilon_0 \frac{\partial \boldsymbol{\mathcal{E}}}{\partial t} = -\omega \epsilon_0 E_0 \sin(\omega t - \beta z)\mathbf{a}_x \tag{49}$$

In order for Ampère's law to be satisfied, (48) and (49) must be equal, which requires that

$$\beta = \omega \epsilon_0 \eta \tag{50}$$

Thus, β and η must also satisfy (50). Combining (47) and (50), we obtain

$$\frac{\omega \mu_0}{\eta} = \omega \epsilon_0 \eta$$

or

$$\eta^2 = \frac{\mu_0}{\epsilon_0}$$

so that η must be

$$\eta = \pm \sqrt{\frac{\mu_0}{\epsilon_0}} \tag{51}$$

Substituting η into (47) or (50), we find that β must be

$$\beta = \pm\omega\sqrt{\mu_0\epsilon_0} \tag{52}$$

We finally check Gauss' laws. First

$$\nabla \cdot \mathcal{D} = \epsilon_0 \nabla \cdot \mathcal{E} = 0$$

or

$$\frac{\partial \mathcal{E}_x}{\partial x} + \frac{\partial \mathcal{E}_y}{\partial y} + \frac{\partial \mathcal{E}_z}{\partial z} = 0$$

which we observe to be true since the only component of $\mathcal{E}$, $\mathcal{E}_x$, is independent of x. Similarly, let us check the other law of Gauss:

$$\nabla \cdot \mathcal{B} = \mu_0 \nabla \cdot \mathcal{H} = 0$$

or

$$\frac{\partial \mathcal{H}_x}{\partial x} + \frac{\partial \mathcal{H}_y}{\partial y} + \frac{\partial \mathcal{H}_z}{\partial z} = 0$$

We find this to be true since the only component of $\mathcal{H}$, $\mathcal{H}_y$, is independent of y. Therefore, the fields are valid ones—but *only if* η and β satisfy the constraints in (51) and (52).

5.3 Boundary Conditions on the Field Vectors

The relations between the corresponding field vectors in two different media at the boundary between the two media are essentially unchanged from the static case. For example, Fig. 5.9*a* shows a cross section of the boundary between two media. Applying Faraday's law in integral form to a small contour c_b which consists of sides Δl and Δh and which bounds the flat surface s_b at the contour, we obtain

$$\oint_{c_b} \mathcal{E} \cdot d\mathbf{l} = -\frac{d}{dt}\int_{s_b} \mathcal{B} \cdot d\mathbf{s} \tag{53}$$

In the limit as $\Delta h \to 0$, the area of s_b will become vanishingly small, so that

$$\oint_{c_b} \mathcal{E} \cdot d\mathbf{l} = 0$$

which is identical to the static case. Consequently, the boundary condition on $\mathcal{E}$ is that the tangential components must be continuous across the boundary:

$$\mathcal{E}_{t2} = \mathcal{E}_{t1} \tag{54}$$

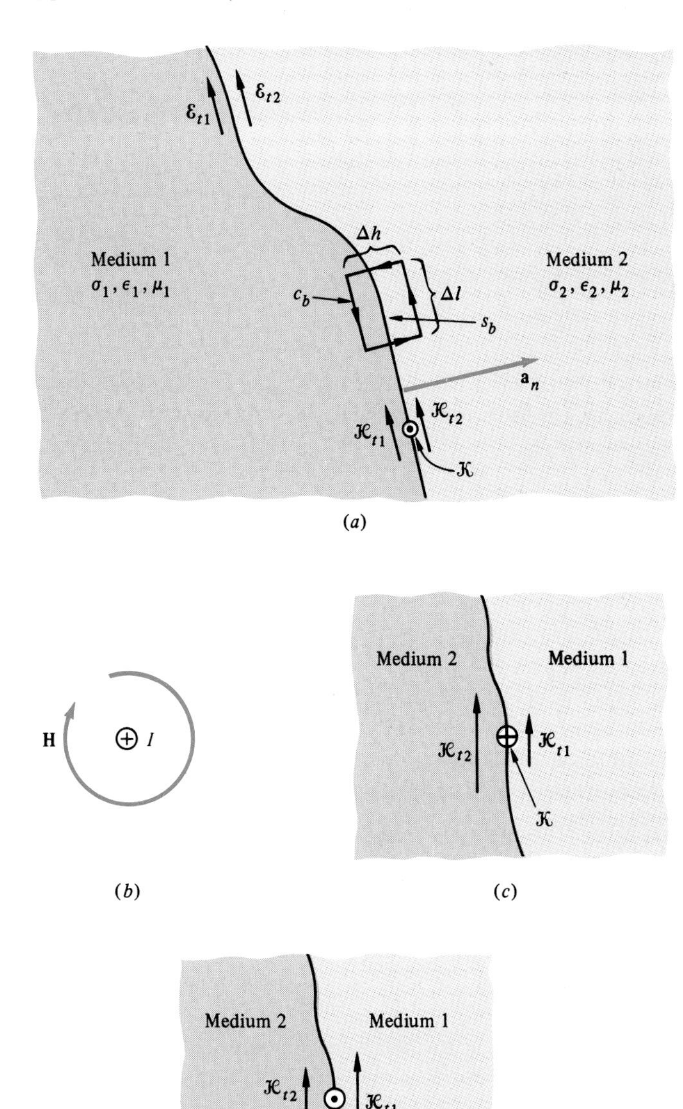

FIGURE 5.9

Illustration of the boundary conditions relating the tangential components of the field vectors at an interface.

Similarly, apply Ampère's law in integral form to this contour:

$$\oint_{c_b} \mathcal{H} \cdot d\mathbf{l} = \int_{s_b} \mathcal{J} \cdot d\mathbf{s} + \frac{d}{dt}\int_{s_b} \mathcal{D} \cdot d\mathbf{s} \tag{55}$$

Note that in the limit as $\Delta h \to 0$,

$$\frac{d}{dt}\int_{s_b} \mathcal{D} \cdot d\mathbf{s} \to 0$$

with the resulting equation being identical to Ampère's law for the static case. Consequently, the boundary condition on the $\mathcal{H}$ field is identical to the static case:

$$\mathcal{H}_{t2} - \mathcal{H}_{t1} = \mathcal{K} \qquad \text{A/m} \tag{56}$$

where $\mathcal{K}$ is any linear current density existing at the interface, as discussed in Chap. 4.

Since the direction of the contour c_b is taken to be counterclockwise in Fig. 5.9*a*, the direction of $\mathcal{K}$ is out of the paper. A simple mnemonic device for remembering the direction of $\mathcal{K}$ is obtained from the following: The direction of the static magnetic field **H** about a wire carrying a static electric current is related to the current direction in the wire by the familiar right-hand rule, as shown in Fig. 5.9*b*. Now consider Fig. 5.9*c*, in which $\mathcal{H}_{t2}$ is shown to be larger in magnitude than $\mathcal{H}_{t1}$. For this case, there will be a net *circulation* of $\mathcal{H}$ at the boundary in the clockwise direction. The linear current density $\mathcal{K}$ is then into the paper and is given by $\mathcal{K} = \mathcal{H}_{t2} - \mathcal{H}_{t1}$. This corresponds analogously to the case of a current-carrying wire in Fig. 5.9*b*. Now consider the case shown in Fig. 5.9*d*, in which there is a net circulation of $\mathcal{H}$ in the counter-clockwise direction at the boundary. In this case, the linear current density is directed out of the page and is given by $\mathcal{K} = \mathcal{H}_{t1} - \mathcal{H}_{t2}$.

The derivation leading to (56) was performed for one possible cross section of the boundary. To investigate other cross sections, define the unit normal vector $\mathbf{a}_n$ perpendicular to the interface and pointing into medium 2, as shown in Fig. 5.9*a*. Consider the contour c_b and surface s_b rotated 90° about $\mathbf{a}_n$. If we again apply Ampère's law to this new contour, we obtain a relationship that is of the form of (56) and which characterizes the discontinuity of the tangential components of $\mathcal{H}$ in this plane, which is perpendicular to the plane used to derive (56). Combining these two results, which are orthogonal, we may write a general expression for the boundary condition on $\mathcal{K}$ at the interface:

$$\mathbf{a}_n \times (\mathcal{H}_2 - \mathcal{H}_1) = \mathcal{K} \tag{57}$$

where $\mathcal{H}_2$ and $\mathcal{H}_1$ are the magnetic field intensity vectors at (but not necessarily tangent to) the interface. The quantity $\mathbf{a}_n \times (\mathcal{H}_2 - \mathcal{H}_1)$ gives the net resultant magnetic field tangent to the boundary (magnitude and direction), which is numerically equal to the vector linear current density $\mathcal{K}$ at the boundary. This linear current density is perpendicular to the plane containing $\mathcal{H}_2 - \mathcal{H}_1$ and $\mathbf{a}_n$ according to the right-hand rule. In a similar fashion, the general vector relationship on continuity of the tangential components of the electric field can be obtained by rotating the contour in Fig. 5.9*a* by 90° about $\mathbf{a}_n$, resulting in

$$\mathbf{a}_n \times (\mathcal{E}_2 - \mathcal{E}_1) = 0 \tag{58}$$

at the surface.

Note that time variations of the field vectors do not change the forms of either of Gauss' laws:

$$\nabla \cdot \mathcal{D} = \rho \tag{59a}$$

$$\nabla \cdot \mathcal{B} = 0 \tag{59b}$$

Thus the boundary conditions on the normal components of $\mathcal{D}$ and $\mathcal{B}$ that were derived from these two laws for the static case are unchanged for the time-varying case. Applying Gauss' laws to an infinitesimal, rectangular volume at the boundary (as shown in Fig. 5.10*a*), we obtain, as $\Delta h \to 0$,

$$\mathcal{D}_{n2} - \mathcal{D}_{n1} = \rho_s \quad \text{C/m}^2 \tag{60a}$$

$$\mathcal{B}_{n2} - \mathcal{B}_{n1} = 0 \tag{60b}$$

where ρ_s is the free surface charge density at the boundary. Note that $\mathcal{D}_{n2}$ is the component of the electric flux density vector in region 2 that is normal to and directed away from the boundary and that $\mathcal{D}_{n1}$ is the component of the electric flux density vector in region 1 that is normal to and directed toward the boundary, as shown in Fig. 5.10*a*. These directions serve to fix the sign and magnitude of the surface charge. Note that $-\mathcal{D}_{n1}$ is the component of the electric flux density in region 1 that is normal to and directed *away from* the boundary. Thus, Eq. (60*a*) simply states that the *net* electric flux that is normal to and directed *away from* the boundary is equal to the net *positive* surface charge density at the boundary. An example is shown in Fig. 5.10*b*. Similarly, (60*b*) states that the components of the magnetic flux density vector that are normal to the boundary are continuous across the boundary.

Equations (60*a* and *b*) may be written in equivalent forms:

$$\mathbf{a}_n \cdot (\mathcal{D}_2 - \mathcal{D}_1) = \rho_s \tag{61a}$$

$$\mathbf{a}_n \cdot (\mathcal{B}_2 - \mathcal{B}_1) = 0 \tag{61b}$$

where $\mathbf{a}_n$ is the unit normal perpendicular to the interface and pointing into region 2, as shown in Fig. 5.10*a*. Note that $\mathbf{a}_n \cdot \mathcal{D}_2$ is the component of $\mathcal{D}_2$

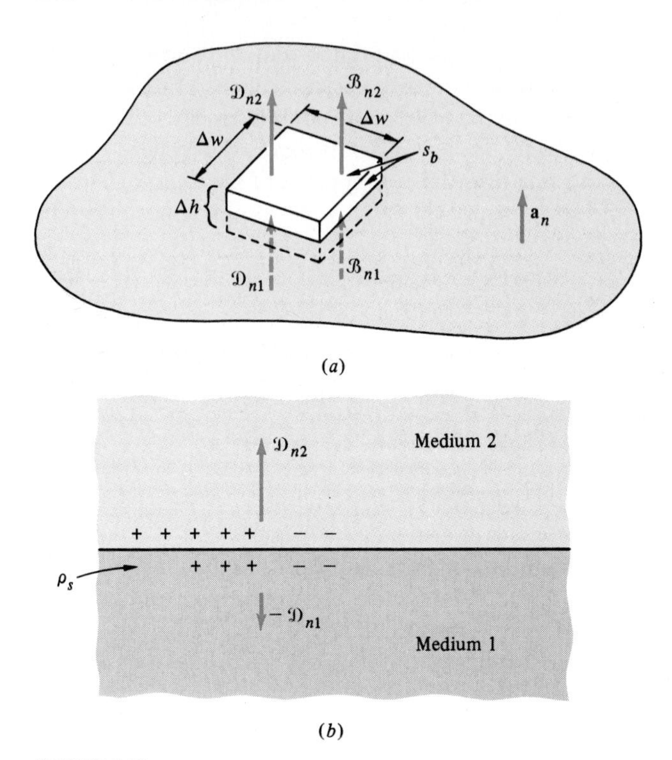

FIGURE 5.10

Illustration of the boundary conditions relating the normal components of the field vectors at an interface.

normal to the boundary and pointing away from the boundary. Similarly, $\mathbf{a}_n \cdot \mathfrak{D}_1$ is the component of $\mathfrak{D}_1$ normal to and pointing toward the boundary.

The boundary conditions on the normal components of $\mathfrak{D}$ and $\mathfrak{B}$ in (61*a*) and (61*b*) were obtained from the divergence equations in (34*c*) and (34*d*). Similarly, the boundary conditions on the tangential components of $\mathcal{E}$ and $\mathcal{H}$ in (58) and (57) were obtained from the curl equations (34*a*) and (34*b*). However the divergence and curl equations are not independent, as was shown previously; thus, we expect that the above four boundary conditions are not all independently specifiable—and thus we cannot independently specify all four boundary conditions, or else contradictions will result. For the time-varying case, the condition on tangential $\mathcal{E}$ is equivalent to the condition on normal $\mathfrak{B}$. Similarly, the condition on tangential $\mathcal{H}$ is equivalent to the condition on normal $\mathfrak{D}$; therefore, specifying tangential $\mathcal{H}$ and normal $\mathfrak{D}$ could result in contradictions. Similarly, specifying tangential $\mathcal{E}$ and normal $\mathfrak{B}$ could result in contradictions.

EXAMPLE 5.5 The magnetic flux density in region 1 of an interface between two materials is given by

$$\mathfrak{B}_1 = 0.6\mathbf{a}_x + 1.1\mathbf{a}_y \qquad \text{Wb/m}^2$$

The permeability of region 1 is μ_1 and that of region 2 is μ_2. Determine the $\mathcal{B}$ field in region 2 just across the interface. The interface is in the xz plane.

Solution From $\mathcal{B} = \mu\mathcal{H}$ and the given $\mathcal{B}$ field in region 1 we have

$$\mathcal{H}_1 = \frac{0.6}{\mu_1}\mathbf{a}_x + \frac{1.1}{\mu_1}\mathbf{a}_y \qquad \text{A/m}$$

Next, we use the continuity conditions. The normal components of the $\mathcal{B}$ field must be continuous: $\mathcal{B}_{n1} = \mathcal{B}_{n2}$. Assuming no linear current at the boundary, the tangential components of the $\mathcal{H}$ field must be continuous: $\mathcal{H}_{t2} = \mathcal{H}_{t1}$. (More will be said in Sec. 5.3.1 about omitting the surface current at the boundary.) Because the interface is in the xz plane, $\mathcal{B}_n$ corresponds to $\mathcal{B}_y$. Hence,

$$\mathcal{B}_{n1} = \mathcal{B}_{n2} = \mathcal{B}_{y1} = \mathcal{B}_{y2} = 1.1 \text{ Wb/m}^2$$

and

$$\mathcal{H}_{t1} = \mathcal{H}_{t2} = \mathcal{H}_{x1} = \mathcal{H}_{x2} = \frac{0.6}{\mu_1} \text{ A/m}$$

But

$$\mathcal{H}_{x2} = \frac{\mathcal{B}_{x2}}{\mu_2}$$

so that

$$\mathcal{B}_{x2} = \frac{\mu_2}{\mu_1} \times 0.6$$

Consequently,

$$\mathcal{B}_2 = 0.6\frac{\mu_2}{\mu_1}\mathbf{a}_x + 1.1\mathbf{a}_y \qquad \text{Wb/m}^2$$

The above boundary conditions are quite general in the sense that no restrictions on the properties of the two media were imposed in the development. These boundary conditions were simply a consequence of Maxwell's equations. There are, however, important and frequently encountered special cases that deserve repeated emphasis. The majority of our applications of the boundary conditions will be for these special cases.

5.3.1 Boundary Conditions for Perfect Conductors

The special case of primary interest is that of a perfect conductor for which the conductivity is assumed to be infinite, $\sigma = \infty$. The permittivity ϵ and permeability μ, however, are taken to be finite. A perfect conductor is not so ideal as it may seem. For example, there exist materials which, when cooled to temperatures approaching absolute zero (0 K, or $-273°C$), exhibit an abrupt drop in their resistivity (the inverse of conductivity) to a value of zero. These materials are called *superconductors*. For our discussions, however, we will use the term *perfect conductor* simply to mean an idealized material having an infinite conductivity. Alternatively, we will see that a perfect conductor is an idealized material that may support a linear current density $\mathcal{K}$ on its surface.

In the case of static electric fields, we determined that there can be essentially no net free (mobile) charge within a conductor that has a finite but nonzero conductivity. Free charge will certainly exist within these conductors, but any excess charge will move by mutual repulsion to the surface of the conductor. The time required to establish equilibrium within the conductor is related to the relaxation time, $\tau = \epsilon/\sigma$, as discussed in Chap. 3. For the case of a good conductor, such as copper, this relaxation time is on the order of 10^{-19} s. The relaxation time is the time required for the excess charge density to decay to $1/e$, or 37 percent, of its original value; therefore, strictly speaking, the excess charge will require infinite time to decay to zero. However, for a good conductor, equilibrium would essentially be established in a relatively short time. For a perfect conductor with $\sigma = \infty$, equilibrium would theoretically be established immediately. Once equilibrium is established, there will be no *net* free charge in the interior of the conductor to establish an electric field, and thus $\mathbf{E} = 0$ within the conductor.

For time-varying fields, we may also show that the electric field $\mathcal{E}$ within a perfect conductor is zero. To do this, let us presume that the conductivity again relates the current density $\mathcal{J}$ to the electric field $\mathcal{E}$ within the perfect conductor as $\sigma = \mathcal{J}/\mathcal{E}$. If $\sigma = \infty$, then either $\mathcal{J}$ is infinite or $\mathcal{E}$ is zero. An infinite current density is not reasonable, since this would imply either that an infinite amount of charge is being transported in a finite time or that a finite amount of charge is being transported in zero time. Thus, we conclude that $\mathcal{E} = 0$ within a perfect conductor. If we presume that $\mathcal{D} = \epsilon\mathcal{E}$ within the perfect conductor with ϵ finite, then we observe that $\mathcal{D} = 0$ within a perfect conductor also. From a practical standpoint, we may approximate conductors as perfect conductors so long as the rate of change of the field is slow compared to the inverse of the relaxation time.

Now consider Faraday's law within this perfect conductor:

$$\nabla \times \mathcal{E} = -\frac{\partial \mathcal{B}}{\partial t}$$

Since $\mathcal{E} = 0$ within the perfect conductor, we see that

$$\frac{\partial \mathcal{B}}{\partial t} = 0$$

Therefore, $\mathcal{B}$ in the perfect conductor must be independent of time—and consequently there can be no *time-varying* magnetic field within a perfect conductor. There remains, however, the possibility of a static magnetic field within a perfect conductor.† Assuming that $\mathcal{B} = \mu\mathcal{H}$ in this perfect conductor, with μ finite, we also obtain $\mathcal{H} = 0$.

Therefore, *in a perfect conductor, all time-varying fields are zero, and the static electric field is zero.* This may also be thought of as the definition of a perfect conductor. Good conductors, however, approximate perfect conductors quite well, as we shall see in Chap. 6, Sec. 6.4, where we consider the concept of skin depth. Consider the boundary between two different media, one of which is a perfect conductor; suppose that medium 1 is the perfect conductor, i.e., $\sigma_1 = \infty$. In this case, the boundary conditions become

$$\begin{aligned} \mathcal{E}_{t2} &= 0 \\ \mathcal{H}_{t2} &= \mathcal{K} \\ \mathcal{D}_{n2} &= \rho_s \\ \mathcal{B}_{n2} &= 0 \end{aligned} \qquad \sigma_1 = \infty \tag{62}$$

EXAMPLE 5.6 The electromagnetic fields in a rectangular waveguide (Chap. 8) shown in Fig. 5.11 are given by

$$\mathcal{E}_x = 0$$

$$\mathcal{E}_y = C\frac{\omega\mu_0 a}{\pi}\sin\left(\frac{\pi x}{a}\right)\sin(\omega t - \beta z)$$

$$\mathcal{E}_z = 0$$

$$\mathcal{H}_x = -C\frac{\beta a}{\pi}\sin\left(\frac{\pi x}{a}\right)\sin(\omega t - \beta z)$$

$$\mathcal{H}_y = 0$$

$$\mathcal{H}_z = C\cos\left(\frac{\pi x}{a}\right)\cos(\omega t - \beta z)$$

where C is a constant and $\omega = 2\pi f$, with f the frequency of excitation. The walls of the waveguide are assumed to be perfect conductors. Show that these fields satisfy all of Maxwell's equations within the guide, the boundary conditions on the walls, and determine the surface charge densities and currents on those walls.

---

† Experiments have shown that a static magnetic field cannot exist inside a superconductor.

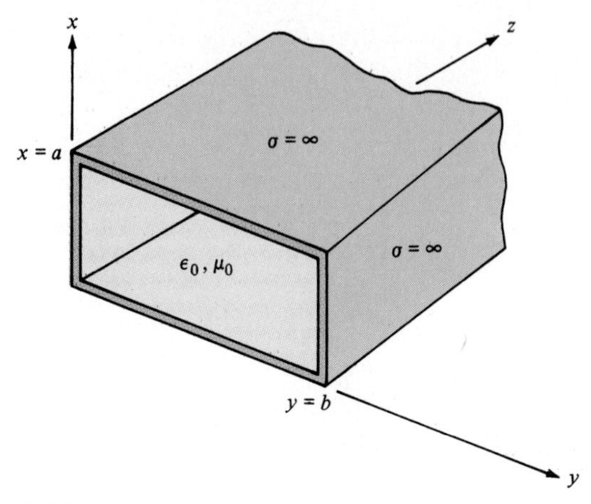

FIGURE 5.11
Example 5.6. Application of the boundary conditions to fields within a waveguide.

Solution First we check Faraday's law:

$$\nabla \times \mathcal{E} = -\mu_0 \frac{\partial \mathcal{H}}{\partial t}$$

or

$$-\frac{\partial \mathcal{E}_y}{\partial z}\mathbf{a}_x + \frac{\partial \mathcal{E}_y}{\partial x}\mathbf{a}_z = -\mu_0 \frac{\partial \mathcal{H}_x}{\partial t}\mathbf{a}_x - \mu_0 \frac{\partial \mathcal{H}_z}{\partial t}\mathbf{a}_z$$

We obtain

$$-\frac{\partial \mathcal{E}_y}{\partial z} = \beta C \frac{\omega\mu_0 a}{\pi} \sin\left(\frac{\pi x}{a}\right) \cos(\omega t - \beta z)$$

$$\frac{\partial \mathcal{E}_y}{\partial x} = \frac{\pi}{a} C \frac{\omega\mu_0 a}{\pi} \cos\left(\frac{\pi x}{a}\right) \sin(\omega t - \beta z)$$

$$-\mu_0 \frac{\partial \mathcal{H}_x}{\partial t} = \omega\mu_0 C \frac{\beta a}{\pi} \sin\left(\frac{\pi x}{a}\right) \cos(\omega t - \beta z)$$

$$-\mu_0 \frac{\partial \mathcal{H}_z}{\partial t} = \omega\mu_0 C \cos\left(\frac{\pi x}{a}\right) \sin(\omega t - \beta z)$$

Matching components, we see that Faraday's law is satisfied. Checking Ampère's law ($\mathcal{J} = 0$),

$$\nabla \times \mathcal{H} = \epsilon_0 \frac{\partial \mathcal{E}}{\partial t}$$

or

$$\frac{\partial \mathcal{H}_z}{\partial y}\mathbf{a}_x + \left(\frac{\partial \mathcal{H}_x}{\partial z} - \frac{\partial \mathcal{H}_z}{\partial x}\right)\mathbf{a}_y - \frac{\partial \mathcal{H}_x}{\partial y}\mathbf{a}_z = \epsilon_0 \frac{\partial \mathcal{E}_y}{\partial t}\mathbf{a}_y$$

We obtain

$$\frac{\partial \mathcal{H}_z}{\partial y} = 0$$

$$\frac{\partial \mathcal{H}_x}{\partial z} - \frac{\partial \mathcal{H}_z}{\partial x} = \beta C \frac{\beta a}{\pi} \sin\left(\frac{\pi x}{a}\right) \cos(\omega t - \beta z) + \frac{\pi}{a} C \sin\left(\frac{\pi x}{a}\right) \cos(\omega t - \beta z)$$

$$-\frac{\partial \mathcal{H}_x}{\partial y} = 0$$

$$\epsilon_0 \frac{\partial \mathcal{E}_y}{\partial t} = \epsilon_0 \omega C \frac{\omega \mu_0 a}{\pi} \sin\left(\frac{\pi x}{a}\right) \cos(\omega t - \beta z)$$

Matching y components, we see that we must have

$$\beta^2 \frac{a}{\pi} + \frac{\pi}{a} = \omega^2 \epsilon_0 \mu_0 \frac{a}{\pi}$$

for Ampère's law to be satisfied. This requires that β be

$$\beta = \pm \sqrt{\omega^2 \mu_0 \epsilon_0 - \left(\frac{\pi}{a}\right)^2}$$

Checking Gauss' law for the electric field ($\rho = 0$),

$$\nabla \cdot (\epsilon_0 \mathcal{E}) = 0$$

we have

$$\frac{\partial \mathcal{E}_x}{\partial x} + \frac{\partial \mathcal{E}_y}{\partial y} + \frac{\partial \mathcal{E}_z}{\partial z} = 0$$

This we see to be true, since the only nonzero component of $\mathcal{E}$, $\mathcal{E}_y$, is independent of y. Similarly, we may show that Gauss' law for the magnetic field

$$\nabla \cdot (\mu_0 \mathcal{H}) = 0$$

or

$$\frac{\partial \mathcal{H}_x}{\partial x} + \frac{\partial \mathcal{H}_y}{\partial y} + \frac{\partial \mathcal{H}_z}{\partial z} = 0$$

is also satisfied since

$$\frac{\partial \mathcal{H}_x}{\partial x} = -C\beta \cos\left(\frac{\pi x}{a}\right) \sin(\omega t - \beta z)$$

$$\frac{\partial \mathcal{H}_z}{\partial z} = C\beta \cos\left(\frac{\pi x}{a}\right) \sin(\omega t - \beta z)$$

The boundary conditions are that tangential $\mathcal{E}$ equal zero and that normal $\mathcal{B}$ equal zero along the (perfectly conducting) walls of the waveguide. The tangential $\mathcal{H}$ condition gives a linear current that we do not know but will later determine. Similarly, the normal $\mathcal{D}$ condition gives a surface charge density that we do not know but will also later determine. Thus, we can only check tangential $\mathcal{E}$ and normal $\mathcal{B}$. Since $\mathcal{E}$ only has a y component, we check that it is zero along the top ($x = a$) and bottom ($x = 0$) walls:

$$\mathcal{E}_y \bigg|_{x=0} = 0$$

$$\mathcal{E}_y \bigg|_{x=a} = 0$$

Substitution of $x = 0$ and $x = a$ into the field equation for $\mathcal{E}_y$ does indeed yield zero, so this boundary condition is satisfied. Similarly, the magnetic flux density $\mathcal{B} = \mu_0 \mathcal{H}$ has only x and z components. The only component normal to a wall is the x component. Thus, we check

$$\mu_0 \mathcal{H}_x \bigg|_{x=0} = 0$$

$$\mu_0 \mathcal{H}_x \bigg|_{x=a} = 0$$

Substitution of $x = 0$ and $x = a$ into the field equation for $\mathcal{H}_x$ shows that those boundary conditions are satisfied. Thus, the boundary conditions are satisfied.

The linear current density on the walls is numerically equal to the component(s) of $\mathcal{H}$ tangent to the walls. Along the left ($y = 0$) and right ($y = b$) walls, the total tangential field is the sum of the x and z components evaluated on those walls. On the left wall ($y = 0$)

$$\begin{aligned}
\mathcal{K} &= -\mathcal{H}_x \bigg|_{y=0} \mathbf{a}_z + \mathcal{H}_z \bigg|_{y=0} \mathbf{a}_x \\
&= C \frac{\beta a}{\pi} \sin\left(\frac{\pi x}{a}\right) \sin(\omega t - \beta z)\mathbf{a}_z \\
&\quad + C \cos\left(\frac{\pi x}{a}\right) \cos(\omega t - \beta z)\mathbf{a}_x
\end{aligned}$$

Note the use of the right-hand rule discussed previously for determining the direction of this current. On the right wall ($y = b$) we have a similar result since $\mathcal{H}_x$ and $\mathcal{H}_z$ are independent of y:

$$\mathbf{\mathcal{K}} = \mathcal{H}_x\Big|_{y=b} \mathbf{a}_z - \mathcal{H}_z\Big|_{y=b} \mathbf{a}_x$$

$$= -C\frac{\beta a}{\pi}\sin\left(\frac{\pi x}{a}\right)\sin(\omega t - \beta z)\mathbf{a}_z$$

$$-C\cos\left(\frac{\pi x}{a}\right)\cos(\omega t - \beta z)\mathbf{a}_x$$

Note that this current is opposite in direction to that on the left wall. This is a result of the fact that the fields are on the interior walls and that we require the net circulation caused by the tangential fields on each side of the boundary. Along the top wall ($x = a$) only the z component of $\mathbf{\mathcal{H}}$ is tangent to the wall, so that

$$\mathbf{\mathcal{K}} = \mathcal{H}_z\Big|_{x=a} \mathbf{a}_y$$

$$= -C\cos(\omega t - \beta z)\mathbf{a}_y$$

Along the bottom wall ($x = 0$)

$$\mathbf{\mathcal{K}} = -\mathcal{H}_z\Big|_{x=0} \mathbf{a}_y$$

$$= -C\cos(\omega t - \beta z)\mathbf{a}_y$$

Once again, note the use of the right-hand rule in determining the net circulation of the tangential $\mathbf{\mathcal{H}}$ field and the resulting direction of the surface current.

The surface charge density on the walls is numerically equal to the component of $\mathbf{\mathcal{D}} = \epsilon_0\mathbf{\mathcal{E}}$ normal to the wall. Along the left wall ($y = 0$)

$$\rho_s = \epsilon_0\,\mathcal{E}_y\Big|_{y=0}$$

$$= C\epsilon_0\mu_0\frac{\omega a}{\pi}\sin\left(\frac{\pi x}{a}\right)\sin(\omega t - \beta z)$$

Along the right wall ($y = b$)

$$\rho_s = -\epsilon_0 \mathcal{E}_y \Big|_{y=b}$$
$$= -C\epsilon_0 \mu_0 \frac{\omega a}{\pi} \sin\left(\frac{\pi x}{a}\right) \sin(\omega t - \beta z)$$

Along the top wall ($x = a$)

$$\rho_s = -\epsilon_0 \mathcal{E}_x \Big|_{x=a}$$
$$= 0$$

and along the bottom wall ($x = 0$)

$$\rho_s = \epsilon_0 \mathcal{E}_x \Big|_{x=0}$$
$$= 0$$

5.3.2 Boundary Conditions for Material Media

On the other hand, suppose that both media have finite conductivity. In this case, current cannot exist solely on the boundary but will penetrate into the media. Thus, we presume that there can be no isolated, linear current density on the boundary between two media both of which have finite conductivity, and we obtain

$$\begin{aligned} \mathcal{E}_{t2} &= \mathcal{E}_{t1} \\ \mathcal{H}_{t2} &= \mathcal{H}_{t1} \\ \mathcal{D}_{n2} - \mathcal{D}_{n1} &= \rho_s \\ \mathcal{B}_{n2} - \mathcal{B}_{n1} &= 0 \end{aligned} \qquad \sigma_1, \sigma_2 \text{ finite} \tag{63}$$

One additional point should also be noted. The discontinuity in the normal components of $\mathcal{D}$ is the free surface charge density at the boundary. If both media are perfect dielectrics, this surface charge could not have arisen without having been intentionally placed on the boundary. So we conclude that in the absence of any intentionally placed charge,

$$\mathcal{D}_{n2} = \mathcal{D}_{n1} \qquad \begin{aligned} \sigma_1 &= 0 \\ \sigma_2 &= 0 \end{aligned} \tag{64}$$

EXAMPLE 5.7 A perfectly conducting sphere of radius R in free space has a charge Q uniformly distributed over its surface. Utilizing the boundary conditions, determine the electric field at the surface of the sphere. Show by using Gauss' law that the result is correct.

Solution Since the charge Q is uniformly distributed over the surface, a surface charge density of

$$\rho_s = \frac{Q}{4\pi R^2} \qquad \text{C/m}^2$$

resides on the surface. At the perfectly conducting surface, the electric field must, by the boundary conditions given in (62), be normal to the sphere and equal to

$$\begin{aligned}\mathbf{E} &= \frac{\rho_s}{\epsilon_0}\,\mathbf{a}_r \\ &= \frac{Q}{4\pi\epsilon_0 R^2}\,\mathbf{a}_r\end{aligned}$$

Gauss' law may be utilized to obtain the electric field at points away from the sphere. Enclosing the sphere with a spherical gaussian surface of radius r, we obtain

$$\oint_s \epsilon_0 \mathbf{E} \cdot d\mathbf{s} = Q$$

or, by symmetry,

$$\mathbf{E} = \frac{Q}{4\pi\epsilon_0 r^2}\,\mathbf{a}_r$$

Evaluating this just off the surface, we obtain our previous result.

5.4 Power Flow and the Poynting Vector

The units of $\mathcal{E}$ are volts per meter and the units of $\mathcal{H}$ are amperes per meter. Therefore, the product of their magnitudes, $|\mathcal{E}|\,|\mathcal{H}|$, has the units of $\text{V/m} \cdot \text{A/m} = \text{VA/m}^2$, or watts per square meter. Thus, this product implies a distribution of power in the field over some surface area: an item that will be of interest in our future investigations. How shall we define the product of the two field vectors? Should we use the dot product $\mathcal{E} \cdot \mathcal{H}$ or the cross product $\mathcal{E} \times \mathcal{H}$? Although it may seem that the dot product of the two vectors would be a possibility, we will actually be interested in the flow of power, and $\boldsymbol{\mathcal{E}} \cdot \boldsymbol{\mathcal{H}}$ has no direction. Thus, let us define the power density vector as

$$\boldsymbol{\mathcal{S}} = \boldsymbol{\mathcal{E}} \times \boldsymbol{\mathcal{H}} \qquad \text{W/m}^2 \tag{65}$$

and show that this vector relates to power. This vector is given the name of the *Poynting vector* after an English physicist, John H. Poynting, who is credited with the following development. Note that the units of the Poynting vector are watts per square meter and that the direction of $\boldsymbol{\mathcal{S}}$ is perpendicular to the plane containing $\boldsymbol{\mathcal{E}}$ and $\boldsymbol{\mathcal{H}}$ (and according to the right-hand rule for the cross product).

Since the Poynting vector seems to indicate a power flow, we are naturally led to investigate the divergence of this vector, $\nabla \cdot \boldsymbol{\mathcal{S}}$. To compute this quantity, we make use of the vector identity (Appendix A)

$$\nabla \cdot (\mathbf{A} \times \mathbf{B}) = \mathbf{B} \cdot (\nabla \times \mathbf{A}) - \mathbf{A} \cdot (\nabla \times \mathbf{B}) \tag{66}$$

We therefore obtain

$$\begin{aligned} \nabla \cdot \boldsymbol{\mathcal{S}} &= \nabla \cdot (\boldsymbol{\mathcal{E}} \times \boldsymbol{\mathcal{H}}) \\ &= \boldsymbol{\mathcal{H}} \cdot (\nabla \times \boldsymbol{\mathcal{E}}) - \boldsymbol{\mathcal{E}} \cdot (\nabla \times \boldsymbol{\mathcal{H}}) \end{aligned} \tag{67}$$

Substituting (34*a*) and (34*b*) into (67) yields

$$\nabla \cdot \boldsymbol{\mathcal{S}} = -\boldsymbol{\mathcal{H}} \cdot \frac{\partial \boldsymbol{\mathcal{B}}}{\partial t} - \boldsymbol{\mathcal{E}} \cdot \boldsymbol{\mathcal{J}} - \boldsymbol{\mathcal{E}} \cdot \frac{\partial \boldsymbol{\mathcal{D}}}{\partial t} \tag{68}$$

or

$$-\nabla \cdot \boldsymbol{\mathcal{S}} = \boldsymbol{\mathcal{E}} \cdot \boldsymbol{\mathcal{J}} + \left(\boldsymbol{\mathcal{E}} \cdot \frac{\partial \boldsymbol{\mathcal{D}}}{\partial t} + \boldsymbol{\mathcal{H}} \cdot \frac{\partial \boldsymbol{\mathcal{B}}}{\partial t} \right) \qquad \text{W/m}^3 \tag{69}$$

which is known as the *point form of Poynting's theorem*. Integrating both sides of (69) over some volume v and applying the divergence theorem, we obtain the integral form of Poynting's theorem; that is,

$$-\oint_s \boldsymbol{\mathcal{S}} \cdot d\mathbf{s} = \int_v \boldsymbol{\mathcal{E}} \cdot \boldsymbol{\mathcal{J}}\, dv + \int_v \left(\boldsymbol{\mathcal{E}} \cdot \frac{\partial \boldsymbol{\mathcal{D}}}{\partial t} + \boldsymbol{\mathcal{H}} \cdot \frac{\partial \boldsymbol{\mathcal{B}}}{\partial t} \right) dv \qquad \text{W} \tag{70}$$

The term on the left of (70) is the net *inward* flux of $\boldsymbol{\mathcal{S}}$ into the volume v. Therefore, (70) indicates that the net flux of $\boldsymbol{\mathcal{S}}$ into some volume is the sum of two contributions. The first term on the right-hand side of (70)

$$\mathcal{P}_{\text{diss}} = \int_v \boldsymbol{\mathcal{E}} \cdot \boldsymbol{\mathcal{J}}\, dv \tag{71}$$

is a *power dissipation term* in that it represents the rate of expenditure of energy by the electric field in moving the charges of the current density. In a region containing convection currents,

$$\mathcal{J} = \rho^+\mathbf{u}^+ + \rho^-\mathbf{u}^- \tag{72}$$

where $\rho^\pm$ and $\mathbf{u}^\pm$ are the density and velocity of $\pm$ charges. For a medium containing only conduction currents, $\mathcal{J} = \sigma\mathcal{E}$, and (71) becomes

$$\mathcal{P}_{\text{diss}} = \sigma \int_v |\mathcal{E}|^2 \, dv \tag{73}$$

which clearly indicates an ohmic power dissipation.

The second integral on the right-hand side of (70) involves a time rate of change. For an isotropic medium in which $\mathcal{D} = \epsilon\mathcal{E}$ and $\mathcal{B} = \mu\mathcal{H}$, we obtain

$$\begin{aligned} \mathcal{E}\cdot\frac{\partial\mathcal{D}}{\partial t} &= \epsilon\mathcal{E}\cdot\frac{\partial\mathcal{E}}{\partial t} \\ &= \tfrac{1}{2}\epsilon\frac{\partial|\mathcal{E}|^2}{\partial t} \end{aligned} \tag{74}$$

and similarly

$$\mathcal{H}\cdot\frac{\partial\mathcal{B}}{\partial t} = \tfrac{1}{2}\mu\frac{\partial|\mathcal{H}|^2}{\partial t} \tag{75}$$

since for any vector $\mathbf{A}$,

$$\begin{aligned} \mathbf{A}\cdot\frac{\partial\mathbf{A}}{\partial t} &= A_x\frac{\partial A_x}{\partial t} + A_y\frac{\partial A_y}{\partial t} + A_z\frac{\partial A_z}{\partial t} \\ &= \frac{1}{2}\frac{\partial A_x^2}{\partial t} + \frac{1}{2}\frac{\partial A_y^2}{\partial t} + \frac{1}{2}\frac{\partial A_z^2}{\partial t} \\ &= \frac{1}{2}\frac{\partial}{\partial t}|\mathbf{A}|^2 \end{aligned} \tag{76}$$

We saw in previous chapters that

$$w_e = \tfrac{1}{2}\epsilon|\mathcal{E}|^2 \qquad \text{J/m}^3 \tag{77a}$$

and

$$w_m = \tfrac{1}{2}\mu|\mathcal{H}|^2 \qquad \text{J/m}^3 \tag{77b}$$

represent electric and magnetic energy densities in the field, respectively. Therefore (74) and (75) are the time rates of change of energy stored in the field:

$$\frac{\partial w_e}{\partial t} \quad \text{and} \quad \frac{\partial w_m}{\partial t}$$

Therefore, the integral form of Poynting's theorem given in (70) states that the net inward flux of the Poynting vector through some closed surface is the sum of the power dissipated in the volume enclosed by the surface and the rate of change of energy stored in the volume enclosed by the surface. The Poynting vector seems to indicate power flow. It should be emphasized, however, that the Poynting vector only implies a distribution of power in the field. We cannot pinpoint the precise location of this power with any more confidence than we can pinpoint gravitational energy. Equation (70) shows only that if we integrate $\mathbf{S}$ over some closed surface, we will obtain the power dissipated and the rate of change of stored energy in the region enclosed by that surface.

EXAMPLE 5.8 An antenna in free space (Chap. 9) is centered at the origin of a spherical coordinate system, as shown in Fig. 5.12. The fields produced by the antenna at a radial distance r are given by

$$\boldsymbol{\mathcal{E}} = \frac{E_0}{r} \sin\theta \sin\omega\left(t - \frac{r}{u_0}\right)\mathbf{a}_\theta$$

$$\boldsymbol{\mathcal{H}} = \frac{E_0}{r\sqrt{\mu_0/\epsilon_0}} \sin\theta \sin\omega\left(t - \frac{r}{u_0}\right)\mathbf{a}_\phi$$

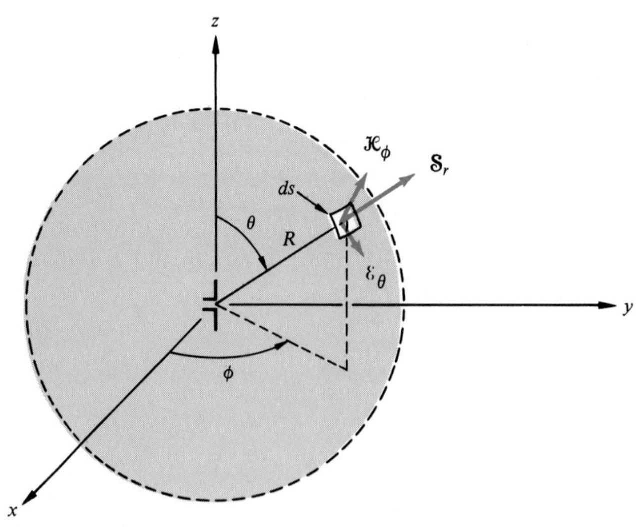

FIGURE 5.12

Example 5.8. Illustration of the Poynting vector and power flow for an antenna.

Determine the Poynting vector and the total average power leaving (or radiated by) the antenna.

Solution The electric field is in the θ direction and the magnetic field is in the ϕ direction. Thus, the Poynting vector is in the radial direction:

$$\begin{aligned}\boldsymbol{\mathcal{S}} &= \boldsymbol{\mathcal{E}} \times \boldsymbol{\mathcal{H}}\\ &= \frac{E_0^2}{r^2\sqrt{\mu_0/\epsilon_0}} \sin^2\theta \sin^2\omega\left(t - \frac{r}{u_0}\right)\mathbf{a}_r \qquad \text{W/m}^2\end{aligned}$$

The power leaving the antenna can be obtained by integrating $\oint_s \boldsymbol{\mathcal{S}} \cdot d\mathbf{s}$ over some suitably chosen closed surface s. Since $\boldsymbol{\mathcal{S}}$ is in the radial direction, the appropriate choice of this surface is a sphere of radius R so that $\boldsymbol{\mathcal{S}} \cdot d\mathbf{s}$ becomes simply

$$\begin{aligned}\boldsymbol{\mathcal{S}} \cdot d\mathbf{s} &= \frac{E_0^2}{R^2\sqrt{\mu_0/\epsilon_0}} \sin^2\theta \sin^2\omega\left(t - \frac{R}{u_0}\right)\underbrace{R^2 \sin\theta\, d\phi\, d\theta}_{ds}\\ &= \frac{E_0^2}{\sqrt{\mu_0/\epsilon_0}} \sin^2\omega\left(t - \frac{R}{u_0}\right)\sin^3\theta\, d\phi\, d\theta\end{aligned}$$

The power radiated by the antenna then becomes

$$\begin{aligned}P_{\text{rad}} &= \int_{\phi=0}^{2\pi}\int_{\theta=0}^{\pi} \boldsymbol{\mathcal{S}} \cdot d\mathbf{s}\\ &= \frac{E_0^2}{\sqrt{\mu_0/\epsilon_0}} \sin^2\omega\left(t - \frac{R}{u_0}\right)\int_{\phi=0}^{2\pi}\int_{\theta=0}^{\pi} \sin^3\theta\, d\phi\, d\theta\\ &= \frac{8\pi}{3}\frac{E_0^2}{\sqrt{\mu_0/\epsilon_0}} \sin^2\omega\left(t - \frac{R}{u_0}\right) \qquad \text{W}\end{aligned}$$

The average power radiated is obtained by time-averaging this result over one period of the sinusoid:

$$\begin{aligned}P_{\substack{\text{rad}\\ \text{av}}} &= \frac{1}{T}\int_0^T P_{\text{rad}}\, dt\\ &= \frac{4\pi}{3}\frac{E_0^2}{\sqrt{\mu_0/\epsilon_0}} \qquad \text{W}\end{aligned}$$

since $\sin^2 A = \frac{1}{2}(1 - \cos 2A)$ and the integral of $\cos 2A$ over 2π rad is zero.

EXAMPLE 5.9 Consider a wire of conductivity σ and radius r_w carrying a dc current I, as shown in Fig. 5.13. Determine the net power entering a wire of length L by integrating the Poynting vector over the wire surface. Show directly that this is the power dissipated by the wire resistance. Also compute the volume integral of $\mathbf{J} \cdot \mathbf{E}$ to verify the previous result.

Solution We center the wire on the z axis of a cylindrical coordinate system. Since $J_z = \sigma E_z$ and $J_z = I/A$, where $A = \pi r_w^2$ is the area of the wire end, we obtain

$$E_z = \frac{I}{A\sigma}$$

Also from Ampère's law, the magnetic field intensity on the wire surface is

$$\mathbf{H} = H_\phi \mathbf{a}_\phi$$

where

$$H_\phi = \frac{I}{2\pi r_w}$$

The Poynting vector at points on the wire surface is

$$\begin{aligned}\mathbf{S} &= \mathbf{E} \times \mathbf{H} \\ &= \frac{I}{A\sigma}\mathbf{a}_z \times \frac{I}{2\pi r_w}\mathbf{a}_\phi \\ &= -\frac{I^2}{2\pi r_w A\sigma}\mathbf{a}_r\end{aligned}$$

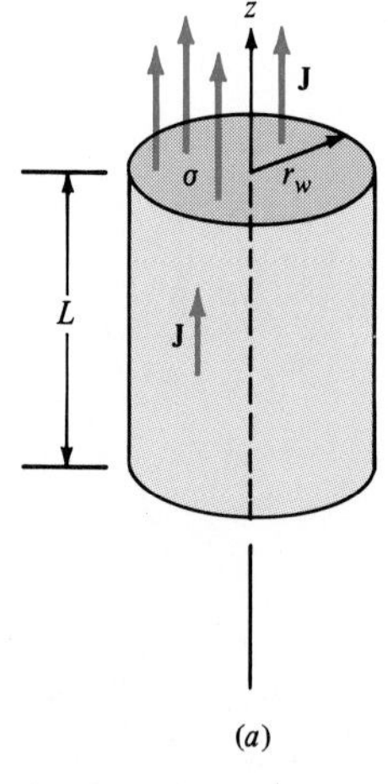

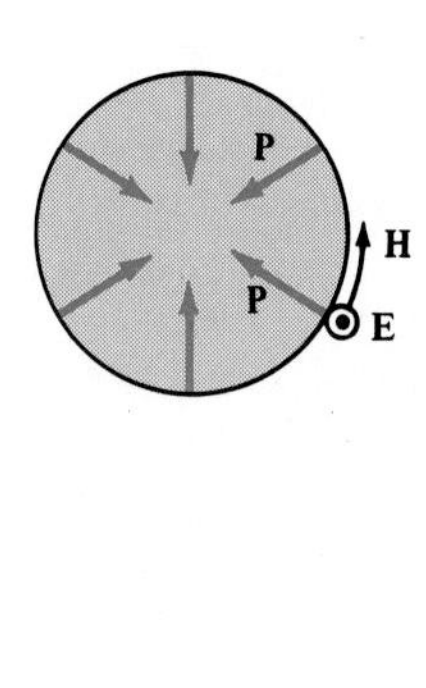

FIGURE 5.13
Example 5.9. Illustration of power flow in a resistor.

The Poynting vector is tangent to the end caps. Thus, the net power entering the volume enclosed by the sides of the wire and the end caps is

$$
\begin{aligned}
P &= -\oint_s \mathbf{S} \cdot d\mathbf{s} \\
&= -\int_{\substack{s \\ \text{side}}} \mathbf{S} \cdot d\mathbf{s} - \int_{\substack{s \\ \text{ends}}} \cancelto{0}{\mathbf{S}} \cdot d\mathbf{s} \\
&= \frac{I^2}{2\pi r_w A\sigma} \int_{z=0}^{L} \int_{\phi=0}^{2\pi} \mathbf{a}_r \cdot r\, d\phi\, dz\, \mathbf{a}_r \bigg|_{r=r_w} \\
&= \frac{I^2 2\pi r_w L}{2\pi r_w A\sigma} \\
&= I^2 \frac{L}{A\sigma} \\
&= I^2 R \qquad \text{W}
\end{aligned}
$$

where the resistance of the wire is $R = L/A\sigma$.

This result can be obtained in an alternative fashion by computing the power dissipated from (71):

$$
\begin{aligned}
P_{\text{diss}} &= \int_v \mathbf{J} \cdot \mathbf{E}\, dv \\
&= \int_v \sigma |\mathbf{E}|^2\, dv \\
&= \sigma \frac{I^2}{A^2\sigma^2} \int_{z=0}^{L} \int_{\phi=0}^{2\pi} \int_{r=0}^{r_w} r\, dr\, d\phi\, dz \\
&= \frac{I^2}{A^2\sigma} \pi r_w^2 L \\
&= I^2 R \qquad \text{W}
\end{aligned}
$$

as before.

5.5 The Sinusoidal Steady State

The major emphasis in this study of Maxwell's equations for time-varying fields will be concerned with the sinusoidal, steady-state behavior of the field vectors. For this case, the field vectors will be written in the form of phasors, which will provide a considerable simplification of many of the mathematical details.

To illustrate this important concept, let us consider the electric field intensity vector written in terms of components as

$$\mathcal{E}(x, y, z, t) = \mathcal{E}_x(x, y, z, t)\mathbf{a}_x + \mathcal{E}_y(x, y, z, t)\mathbf{a}_y + \mathcal{E}_z(x, y, z, t)\mathbf{a}_z \tag{78}$$

Suppose that each of these components has a sinusoidal time variation (which we will arbitrarily take to be cosinusoidal) of the form

$$\mathcal{E}_x = E_x \cos(\omega t + \theta_x) \tag{79a}$$

$$\mathcal{E}_y = E_y \cos(\omega t + \theta_y) \tag{79b}$$

$$\mathcal{E}_z = E_z \cos(\omega t + \theta_z) \tag{79c}$$

where the magnitudes E_x, E_y, E_z and phase angles θ_x, θ_y, θ_z of the components are independent of time t but may depend on the spatial coordinates, e.g., $E_x(x, y, z,)$, $\theta_x(x, y, z)$. Each of these time forms will be written in phasor form by defining complex phasor quantities (denoted with a "hat," or caret). For example, the phasor form of $\mathcal{E}_x$ becomes

$$\begin{aligned} \hat{E}_x &= E_x e^{j\theta_x} \\ &= E_x \angle \theta_x \end{aligned} \tag{80}$$

and similarly for the phasors $\hat{E}_y$ and $\hat{E}_z$. The time-domain forms of the components can then be found from the phasor forms by multiplying by $e^{j\omega t}$ and taking the real part of the result:

$$\begin{aligned} \mathcal{E}_x &= \text{Re}\,(\hat{E}_x e^{j\omega t}) \\ &= \text{Re}\,(E_x e^{j(\omega t + \theta_x)}) \\ &= E_x \cos(\omega t + \theta_x) \end{aligned} \tag{81}$$

where Re $(\cdot)$ denotes the real part of the enclosed complex quantity.

The complete field vector may be written in a similar manner as

$$\begin{aligned} \mathcal{E}(x, y, z, t) &= E_x \cos(\omega t + \theta_x)\mathbf{a}_x \\ &\quad + E_y \cos(\omega t + \theta_y)\mathbf{a}_y + E_z \cos(\omega t + \theta_z)\mathbf{a}_z \\ &= \text{Re}\,(\hat{E}_x e^{j\omega t}\mathbf{a}_x + \hat{E}_y e^{j\omega t}\mathbf{a}_y + \hat{E}_z e^{j\omega t}\mathbf{a}_z) \\ &= \text{Re}\,[(\hat{E}_x\mathbf{a}_x + \hat{E}_y\mathbf{a}_y + \hat{E}_z\mathbf{a}_z)e^{j\omega t}] \end{aligned} \tag{82}$$

From this result we may define the *phasor form* of the complete field vector $\mathcal{E}$ as

$$\hat{\mathbf{E}} = \hat{E}_x\mathbf{a}_x + \hat{E}_y\mathbf{a}_y + \hat{E}_z\mathbf{a}_z \tag{83}$$

and (82) may be written as

$$\mathcal{E}(x, y, z, t) = \text{Re}\,(\hat{\mathbf{E}}e^{j\omega t}) \tag{84}$$

In order to solve problems involving sinusoidal variation of the field vectors, we replace the field vectors with their phasor forms multiplied by $e^{j\omega t}$:

$$\mathcal{E}(x, y, z, t) \Rightarrow \hat{\mathbf{E}}(x, y, z)e^{j\omega t} \tag{85a}$$

$$\mathcal{H}(x, y, z, t) \Rightarrow \hat{\mathbf{H}}(x, y, z)e^{j\omega t} \tag{85b}$$

$$\mathcal{B}(x, y, z, t) \Rightarrow \hat{\mathbf{B}}(x, y, z)e^{j\omega t} \tag{85c}$$

$$\mathcal{D}(x, y, z, t) \Rightarrow \hat{\mathbf{D}}(x, y, z)^{j\omega t} \tag{85d}$$

Note that differentiation of the forms in (85) with respect to time is equivalent to multiplying the form by $j\omega$; that is,

$$\frac{\partial}{\partial t}\hat{\mathbf{E}}(x, y, z)e^{j\omega t} = j\omega\hat{\mathbf{E}}e^{j\omega t} \tag{86}$$

This very important property allows a considerable simplification in the solution of these problems. Substituting the forms of the field vectors in (85) into Maxwell's equations and canceling the $e^{j\omega t}$ term common to both sides of the equations results in

Integral form	*Point form*	
$\oint_c \hat{\mathbf{E}}\cdot d\mathbf{l} = -j\omega \int_s \hat{\mathbf{B}}\cdot d\mathbf{s}$	$\nabla \times \hat{\mathbf{E}} = -j\omega\hat{\mathbf{B}}$	(87*a*)
$\oint_c \hat{\mathbf{H}}\cdot d\mathbf{l} = \int_s \hat{\mathbf{J}}\cdot d\mathbf{s} + j\omega \int_s \hat{\mathbf{D}}\cdot d\mathbf{s}$	$\nabla \times \hat{\mathbf{H}} = \hat{\mathbf{J}} + j\omega\hat{\mathbf{D}}$	(87*b*)
$\oint_s \hat{\mathbf{B}}\cdot d\mathbf{s} = 0$	$\nabla \cdot \hat{\mathbf{B}} = 0$	(87*c*)
$\oint_s \hat{\mathbf{D}}\cdot d\mathbf{s} = \int_v \hat{\rho}\, dv$	$\nabla \cdot \hat{\mathbf{D}} = \hat{\rho}$	(87*d*)

The product of each phasor field vector and $e^{j\omega t}$ can be viewed as being composed of two parts. For example, the x component of $\hat{\mathbf{E}}e^{j\omega t}$ can be written as

$$\begin{aligned}\hat{E}_x e^{j\omega t} &= E_x \cos(\omega t + \theta_x) + jE_x \sin(\omega t + \theta_x)\\ &= \text{Re}\,(\hat{E}_x e^{j\omega t}) + j\,\text{Im}\,(\hat{E}_x e^{j\omega t})\end{aligned} \tag{88}$$

where Im $(\cdot)$ denotes the imaginary part of the enclosed complex quantity. Thus, $\hat{\mathbf{E}}e^{j\omega t}$ is the sum of two terms:

$$\hat{\mathbf{E}}e^{j\omega t} = \text{Re}\,(\hat{\mathbf{E}}e^{j\omega t}) + j\,\text{Im}\,(\hat{\mathbf{E}}e^{j\omega t}) \tag{89}$$

Since each of Maxwell's equations is linear, each equation in (87), when multiplied by $e^{j\omega t}$, may be factored into the sum of two such equations: one for the Re $(\cdot)$ parts (the parts we desire) and the other for the j Im $(\cdot)$ parts. Thus, we solve (87) and use the desired portion of the solution. Therefore, *solving* (87) *for the complex phasor quantities* $\hat{\mathbf{E}}$, $\hat{\mathbf{H}}$, $\hat{\mathbf{B}}$, $\hat{\mathbf{D}}$, *we obtain the time-domain forms of the solutions simply by multiplying each phasor by* $e^{j\omega t}$ *and taking the real part of the result.*

If the medium is linear, homogeneous, and isotropic (which we assume throughout the remainder of this text), (87) becomes:

Integral form	*Point form*	
$\oint_c \hat{\mathbf{E}}\cdot d\mathbf{l} = -j\omega\mu \int_s \hat{\mathbf{H}}\cdot d\mathbf{s}$	$\nabla \times \hat{\mathbf{E}} = -j\omega\mu\hat{\mathbf{H}}$	(90*a*)
$\oint_c \hat{\mathbf{H}}\cdot d\mathbf{l} = (\sigma + j\omega\epsilon) \int_s \hat{\mathbf{E}}\cdot d\mathbf{s}$	$\nabla \times \hat{\mathbf{H}} = (\sigma + j\omega\epsilon)\hat{\mathbf{E}}$	(90*b*)
$\oint_s \hat{\mathbf{H}}\cdot d\mathbf{s} = 0$	$\nabla\cdot\hat{\mathbf{H}} = 0$	(90*c*)
$\oint_s \hat{\mathbf{E}}\cdot d\mathbf{s} = \frac{1}{\epsilon}\int_v \hat{\rho}\, dv$	$\nabla\cdot\hat{\mathbf{E}} = \hat{\rho}/\epsilon$	(90*d*)

Here the permittivity, permeability, and conductivity may be functions of frequency [i.e., $\epsilon(f)$, $\mu(f)$ and $\sigma(f)$], as they usually are for material media.

We will use the hat (ˆ) notation to designate not only the complex phasor quantities but also other quantities that are complex. This will serve to distinguish between complex quantities and real quantities.

EXAMPLE 5.10 Rework Example 5.4 by using phasors; that is, show that the following field vectors in free space ($\mu = \mu_0$, $\epsilon = \epsilon_0$, $\sigma = 0$, $\rho = 0$, $\mathfrak{J} = 0$) satisfy all of Maxwell's equations:

$$\mathcal{E} = E_0 \cos(\omega t - \beta z)\mathbf{a}_x$$

$$\mathcal{H} = \frac{E_0}{\eta}\cos(\omega t - \beta z)\mathbf{a}_y$$

Solution Converting to phasors, we obtain

$$\hat{\mathbf{E}} = E_0 e^{-j\beta z}\mathbf{a}_x$$

$$\hat{\mathbf{H}} = \frac{E_0}{\eta} e^{-j\beta z}\mathbf{a}_y$$

Checking Faraday's law:

$$\begin{aligned}\nabla \times \hat{\mathbf{E}} &= -j\omega\hat{\mathbf{B}} \\ &= -j\omega\mu_0\hat{\mathbf{H}}\end{aligned}$$

we form

$$\begin{aligned}\nabla \times \hat{\mathbf{E}} &= \left(\frac{\partial \cancelto{0}{\hat{E}_z}}{\partial y} - \frac{\partial \cancelto{0}{\hat{E}_y}}{\partial z}\right)\mathbf{a}_x + \left(\frac{\partial \hat{E}_x}{\partial z} - \frac{\partial \cancelto{0}{\hat{E}_z}}{\partial x}\right)\mathbf{a}_y + \left(\frac{\partial \cancelto{0}{\hat{E}_y}}{\partial x} - \cancelto{0}{\frac{\partial \hat{E}_x}{\partial y}}\right)\mathbf{a}_z \\ &= \frac{\partial \hat{E}_x}{\partial z}\mathbf{a}_y \\ &= -j\beta E_0 e^{-j\beta z}\mathbf{a}_y\end{aligned} \tag{91}$$

The right-hand side of Faraday's law becomes

$$-j\omega\mu_0\hat{\mathbf{H}} = -j\omega\frac{\mu_0 E_0}{\eta} e^{-j\beta z}\mathbf{a}_y \tag{92}$$

In order for Faraday's law to be satisfied, (91) and (92) must be equal, thus

$$\beta = \frac{\omega\mu_0}{\eta} \tag{93}$$

which we obtained in Example 5.4. Similarly, checking Ampère's law:

$$\nabla \times \hat{\mathbf{H}} = \cancelto{0}{\hat{\mathbf{J}}} + j\omega\epsilon_0\hat{\mathbf{E}}$$

we obtain

$$\begin{aligned}\nabla \times \hat{\mathbf{H}} &= -\frac{\partial \hat{H}_y}{\partial z}\mathbf{a}_x \\ &= j\frac{\beta E_0}{\eta} e^{-j\beta z}\mathbf{a}_x\end{aligned} \tag{94}$$

and

$$j\omega\epsilon_0\hat{\mathbf{E}} = j\omega\epsilon_0 E_0 e^{-j\beta z}\mathbf{a}_x \tag{95}$$

Thus, (94) must equal (95) for Ampère's law to be satisfied, and we obtain

$$\beta = \omega \epsilon_0 \eta \tag{96}$$

which was also obtained in Example 5.4. Similarly, we find that Gauss' laws are satisfied:

$$\begin{aligned}\nabla \cdot \hat{\mathbf{D}} &= \epsilon_0 \nabla \cdot \hat{\mathbf{E}} \\ &= \epsilon_0 \left(\frac{\partial \hat{E}_x}{\partial x} + \frac{\partial \hat{E}_y}{\partial y} + \frac{\partial \hat{E}_z}{\partial z} \right) \\ &= 0\end{aligned}$$

and

$$\begin{aligned}\nabla \cdot \hat{\mathbf{B}} &= \mu_0 \nabla \cdot \hat{\mathbf{H}} \\ &= \mu_0 \left(\frac{\partial \hat{H}_x}{\partial x} + \frac{\partial \hat{H}_y}{\partial y} + \frac{\partial \hat{H}_z}{\partial z} \right) \\ &= 0\end{aligned}$$

The constraints on β and η,

$$\beta = \pm \omega \sqrt{\mu_0 \epsilon_0}$$

$$\eta = \pm \sqrt{\frac{\mu_0}{\epsilon_0}}$$

are obtained once again. Although the computational details involved in working in the time domain in Example 5.4 or with phasors as in this example are about the same, we will encounter numerous other cases in which the use of phasors provides a considerable simplification over working directly in the time domain.

For the sinusoidal steady state, we can also define a phasor form of the Poynting vector. The Poynting vector $\mathbf{S}$, derived in the previous section, represents *instantaneous* power. In sinusoidal, steady-state problems, we will be interested in average power rather than instantaneous power. To determine the average power flow, we define the phasor Poynting vector as

$$\hat{\mathbf{S}} = \hat{\mathbf{E}} \times \hat{\mathbf{H}}^* \tag{97}$$

where the complex conjugate of a phasor $\hat{A}$ is denoted by $\hat{A}^*$. We will now show that the density of *average power* is given by

$$\begin{aligned}\mathbf{S}_{\text{av}} &= \tfrac{1}{2} \operatorname{Re} \hat{\mathbf{S}} \qquad \text{W/m}^2 \\ &= \tfrac{1}{2} \operatorname{Re} (\hat{\mathbf{E}} \times \hat{\mathbf{H}}^*) \\ &= \tfrac{1}{2} \operatorname{Re} (\hat{\mathbf{E}}^* \times \hat{\mathbf{H}})\end{aligned} \tag{98}$$

The reader will observe that (98) is the vector counterpart to the average power calculation in sinusoidal, steady-state circuit-analysis problems; that is, $P_{av} = \frac{1}{2}$ Re $(\hat{V}\hat{I}^*)$, where $\hat{V}$ and $\hat{I}$ are the phasor voltage and current associated with a two-terminal circuit element (or portion of a circuit) and P_{av} is the average power delivered to the element (or portion of the circuit).

To show that (98) is true, we write

$$\mathcal{E} = \text{Re}\,(\hat{\mathbf{E}}e^{j\omega t}) = \tfrac{1}{2}(\hat{\mathbf{E}}e^{j\omega t} + \hat{\mathbf{E}}^*e^{-j\omega t}) \tag{99a}$$

$$\mathcal{H} = \text{Re}\,(\hat{\mathbf{H}}e^{j\omega t}) = \tfrac{1}{2}(\hat{\mathbf{H}}e^{j\omega t} + \hat{\mathbf{H}}^*e^{-j\omega t}) \tag{99b}$$

where we have arbitrarily assumed cosinusoidal field variation. To show that (99) is true, let us consider the complex number $\hat{C} = A + jB$. The real part is A, which is equivalent to $\frac{1}{2}(\hat{C} + \hat{C}^*) = A$. The instantaneous power flow is obtained by substituting (99) into (65):

$$\mathcal{S} = \mathcal{E} \times \mathcal{H} = \tfrac{1}{4}(\hat{\mathbf{E}} \times \hat{\mathbf{H}}^* + \hat{\mathbf{E}}^* \times \hat{\mathbf{H}}) + \tfrac{1}{4}(\hat{\mathbf{E}} \times \hat{\mathbf{H}}e^{j2\omega t} + \hat{\mathbf{E}}^* \times \hat{\mathbf{H}}^*e^{-j2\omega t}) \tag{100}$$

Since $(\hat{\mathbf{E}} \times \hat{\mathbf{H}}^*)^* = \hat{\mathbf{E}}^* \times \hat{\mathbf{H}}$ and $\hat{\mathbf{E}}^* \times \hat{\mathbf{H}}^* = (\hat{\mathbf{E}} \times \hat{\mathbf{H}})^*$, (100) may be written as

$$\mathcal{S} = \tfrac{1}{4}(\hat{\mathbf{M}} + \hat{\mathbf{M}}^*) + \tfrac{1}{4}(\hat{\mathbf{N}}e^{j2\omega t} + \hat{\mathbf{N}}^*e^{-j2\omega t}) \tag{101}$$

where $\hat{\mathbf{M}} = \hat{\mathbf{E}} \times \hat{\mathbf{H}}^*$ and $\hat{\mathbf{N}} = \hat{\mathbf{E}} \times \hat{\mathbf{H}}$, and we obtain

$$\mathcal{S} = \tfrac{1}{2}\,\text{Re}\,\hat{\mathbf{M}} + \tfrac{1}{2}\,\text{Re}\,(\hat{\mathbf{N}}e^{j2\omega t}) = \tfrac{1}{2}\,\text{Re}\,(\hat{\mathbf{E}} \times \hat{\mathbf{H}}^*) + \tfrac{1}{2}\,\text{Re}\,(\hat{\mathbf{E}} \times \hat{\mathbf{H}}e^{j2\omega t}) \tag{102}$$

The first term on the right-hand side of (102) is independent of time, and the second term can be written as

$$\tfrac{1}{2}\,\text{Re}\,(\hat{\mathbf{N}}e^{j2\omega t}) = \tfrac{1}{2}[N_x \cos(2\omega t + \theta_x)\mathbf{a}_x + N_y \cos(2\omega t + \theta_y)\mathbf{a}_y + N_z \cos(2\omega t + \theta_z)\mathbf{a}_z] \tag{103}$$

The density of average power is

$$\mathbf{S}_{av} = \frac{1}{T}\int_0^T \mathcal{S}\,dt \tag{104}$$

where T is the period of the sinusoid, i.e., $T = 1/f$. Applying (104) to (102) yields the result in (98) since the time average of the second term in (102) is zero.

EXAMPLE 5.11 Suppose that the field vectors of a wave in free space are given by

$$\mathcal{E} = 100\cos\left(\omega t + \frac{4\pi}{3}x\right)\mathbf{a}_z \qquad \text{V/m}$$

$$\mathcal{H} = \frac{100}{120\pi}\cos\left(\omega t + \frac{4\pi}{3}x\right)\mathbf{a}_y \qquad \text{A/m}$$

where $\omega = 2\pi f$ and $f = 200$ MHz. Determine the direction of power flow and the average power crossing the surface area bounded by $y = 2$ m, $y = 0$, $z = 2$ m, and $z = 0$.

Solution In phasor form, the field vectors become

$$\hat{\mathbf{E}} = 100e^{(j4\pi/3)x}\mathbf{a}_z$$

$$\hat{\mathbf{H}} = \frac{100}{120\pi}e^{(j4\pi/3)x}\mathbf{a}_y$$

The phasor Poynting vector is

$$\begin{aligned}\hat{\mathbf{S}} &= \hat{\mathbf{E}} \times \hat{\mathbf{H}}^* \\ &= -\hat{E}_z\hat{H}_y^*\mathbf{a}_x \\ &= -\frac{(100)^2}{120\pi}\mathbf{a}_x\end{aligned}$$

so that power flow is in the negative x direction. The average power density vector is

$$\begin{aligned}\mathbf{S}_{\text{av}} &= \tfrac{1}{2}\,\text{Re}\,\hat{\mathbf{S}} \\ &= \frac{1}{2}\left[-\frac{(100)^2}{120\pi}\right]\mathbf{a}_x \\ &= -13.26\mathbf{a}_x \qquad \text{W/m}^2\end{aligned}$$

and the average power crossing the designated surface is

$$\begin{aligned}P_{\text{av}} &= \int_{z=0}^{2}\int_{y=0}^{2}\mathbf{S}_{\text{av}}\cdot(-\mathbf{a}_x)dy\,dz \\ &= 13.26 \times 4 \\ &= 53.04 \text{ W}\end{aligned}$$

where we have taken the unit normal to the surface to be in the negative x direction. Thus 53.04 W is the average power crossing the surface in the negative x direction.

EXAMPLE 5.12 Rework Example 5.8 using phasor notation; that is, for the following fields, determine the Poynting vector and the total average power radiated by the antenna:

$$\mathcal{E} = \frac{E_0}{r} \sin\theta \sin\omega\left(t - \frac{r}{u_0}\right)\mathbf{a}_\theta \qquad \text{V/m}$$

$$\mathcal{H} = \frac{E_0}{r\sqrt{\mu_0/\epsilon_0}} \sin\theta \sin\omega\left(t - \frac{r}{u_0}\right)\mathbf{a}_\phi \qquad \text{A/m}$$

Solution In phasor form, the field vectors become

$$\hat{\mathbf{E}} = \frac{E_0}{r} \sin\theta e^{-j\omega r/u_0}\mathbf{a}_\theta$$

$$\hat{\mathbf{H}} = \frac{E_0}{r\sqrt{\mu_0/\epsilon_0}} \sin\theta e^{-j\omega r/u_0}\mathbf{a}_\phi$$

In this problem,

$$\mathcal{E} = \text{Im}\,(\hat{\mathbf{E}}e^{j\omega t})$$

$$\mathcal{H} = \text{Im}\,(\hat{\mathbf{H}}e^{j\omega t})$$

The phasor Poynting vector becomes

$$\hat{\mathbf{S}} = \hat{\mathbf{E}} \times \hat{\mathbf{H}}^*$$

$$= \frac{E_0^2}{r^2\sqrt{\mu_0/\epsilon_0}} \sin^2\theta\, \mathbf{a}_r$$

and the average power density vector is

$$\mathbf{S}_{\text{av}} = \tfrac{1}{2}\,\text{Re}\,\hat{\mathbf{S}}$$

$$= \frac{E_0^2}{2r^2\sqrt{\mu_0/\epsilon_0}} \sin^2\theta\, \mathbf{a}_r$$

Integrating $\mathbf{S}_{av} \cdot d\mathbf{s}$ over a sphere of radius R, we obtain the total average power radiated by the antenna:

$$P_{rad} = \oint_s \mathbf{S}_{av} \cdot d\mathbf{s}$$

$$= \frac{E_0^2}{2\sqrt{\mu_0/\epsilon_0}} \int_{\phi=0}^{2\pi} \int_{\theta=0}^{\pi} \sin^3\theta \, d\theta \, d\phi$$

$$= \frac{4\pi}{3} \frac{E_0^2}{\sqrt{\mu_0/\epsilon_0}} \quad \text{W}$$

as before.

EXAMPLE 5.13 For the waveguide problem in Example 5.6, determine the average power density Poynting vector and the total average power in the z direction (the guide axis) over the guide cross section.

Solution The phasor forms of the field vectors are

$$\hat{E}_x = 0$$

$$\hat{E}_y = -jC \frac{\omega\mu_0 a}{\pi} \sin\left(\frac{\pi x}{a}\right) e^{-j\beta z}$$

$$\hat{E}_z = 0$$

$$\hat{H}_x = jC \frac{\beta a}{\pi} \sin\left(\frac{\pi x}{a}\right) e^{-j\beta z}$$

$$\hat{H}_y = 0$$

$$\hat{H}_z = C \cos\left(\frac{\pi x}{a}\right) e^{-j\beta z}$$

The average power density Poynting vector is

$$\mathbf{S}_{av} = \tfrac{1}{2}\,\mathrm{Re}(\hat{\mathbf{E}} \times \hat{\mathbf{H}}^*)$$

$$= \tfrac{1}{2}\,\mathrm{Re}(\hat{E}_y \hat{H}_z^* \mathbf{a}_x - \hat{E}_y \hat{H}_x^* \mathbf{a}_z)$$

$$= \tfrac{1}{2}\,\mathrm{Re}\left[-jC^2 \frac{\omega\mu_0 a}{\pi} \sin\left(\frac{\pi x}{a}\right) \cos\left(\frac{\pi x}{a}\right) \mathbf{a}_x + C^2 \frac{\omega\mu_0 \beta a^2}{\pi^2} \sin^2\left(\frac{\pi x}{a}\right) \mathbf{a}_z\right]$$

$$= \frac{C^2}{2} \frac{\omega\mu_0 \beta a^2}{\pi^2} \sin^2\left(\frac{\pi x}{a}\right) \mathbf{a}_z \quad \text{W/m}^2$$

Therefore, average power flow is only along the axis of the guide. The total average power in the z direction is obtained by integrating $\int_s \mathbf{S}_{av} \cdot d\mathbf{s}$ over the guide cross section:

$$P_{av} = \int_{S_{xy}} \mathbf{S}_{av} \cdot d\mathbf{s}$$

$$= \frac{C^2}{2} \frac{\omega\mu_0 \beta a^2}{\pi^2} \int_{x=0}^{a} \int_{y=0}^{b} \sin^2\left(\frac{\pi x}{a}\right) dy\, dx$$

$$= \frac{C^2 \omega\mu_0 \beta a^3 b}{4\pi^2} \quad \text{W}$$

As an example, consider a typical guide with dimensions $a = 0.9$ in (2.29 cm) and $b = 0.4$ in (1.02 cm) operated at a frequency of 7 GHz. If the amplitude of the electric field is 1000 V/m, then

$$E_y = -j1000 \sin\left(\frac{\pi x}{a}\right) e^{-j\beta z}$$

so that

$$\frac{C\omega\mu_0 a}{\pi} = 1000$$

Using the result obtained for β in Example 5.6,

$$\beta = \sqrt{\omega^2 \mu_0 \epsilon_0 - \left(\frac{\pi}{a}\right)^2}$$

we find that

$$P_{av} = \left(\frac{C\omega\mu_0 a}{\pi}\right)^2 \frac{\beta ab}{4\omega\mu_0}$$

$$= (1000)^2 \frac{ab}{4} \sqrt{\frac{\epsilon_0}{\mu_0} - \left(\frac{\pi}{\omega\mu_0 a}\right)^2}$$

Substituting numerical values, we obtain

$$P_{av} = 54.6 \text{ mW}$$

5.6 Summary

The topics considered in this chapter form the basis, or cornerstone, for all our subsequent studies. Any valid electromagnetic field must satisfy Maxwell's equations (all of them simultaneously) and the associated boundary conditions. Faraday's law, Ampère's law, and the two laws of Gauss are therefore essential ingredients in our further studies.

The reader should show that the fields in Example 5.1 satisfy only three of the four equations of Maxwell: Faraday's law and the two laws of Gauss. They do not satisfy Ampère's law and are therefore not valid fields. However, if the time variation of the field is small enough, they represent first-order approximations. Fields of this type are referred to as quasi-static fields, such as are typically assumed in the analysis of lumped electric circuits at frequencies other than dc. On the other hand, the field vectors in Examples 5.4, 5.6, 5.10, and 5.11 satisfy all of Maxwell's equations, as we have seen, and therefore represent a valid electromagnetic field.

The concept of power flow and the Poynting vector illustrate the distributive nature of the fields problem. In lumped electric circuit, this distributed nature of the field is neglected on the assumption that the electrical dimensions (a topic considered in subsequent chapters) are small. The lumped-circuit concepts are therefore approximate representations of electromagnetic phenomena. Thus, we are embarking on a more general study of electrical phenomena than is provided by lumped-circuit theory

The remaining chapters will address some of the important phenomena predicted by Maxwell's equations. In Chap. 6 we will study the propagation of waves that is predicted by Maxwell's equations. Wave propagation is one of the more important consequences (but certainly not the only consequence) of these equations. Obviously, our society would have a completely different standard of living if propagation of electromagnetic energy in the form of waves were not possible. The important techniques for guiding or focusing these waves will be considered in Chaps. 7, 8, and 9.

Problems

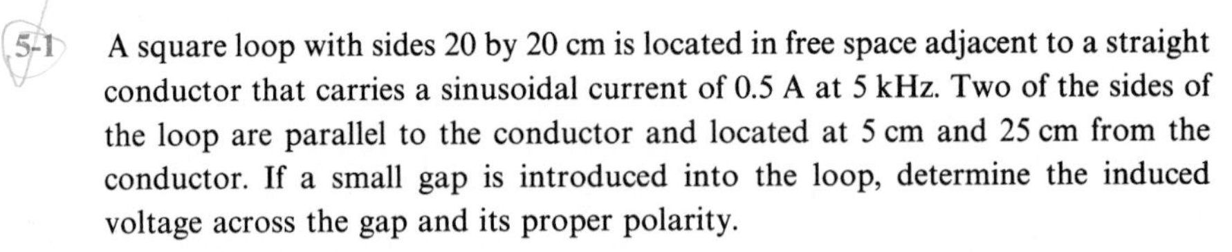

5-1 A square loop with sides 20 by 20 cm is located in free space adjacent to a straight conductor that carries a sinusoidal current of 0.5 A at 5 kHz. Two of the sides of the loop are parallel to the conductor and located at 5 cm and 25 cm from the conductor. If a small gap is introduced into the loop, determine the induced voltage across the gap and its proper polarity.

5-2 A circular loop of 10-cm radius replaces the square loop in Prob. 5-1. Recompute the voltage induced across a gap in this loop if the center of the loop is 15 cm away from the conductor.

5-3 An inductor is formed by winding 10 turns of a thin wire around a wooden dowel rod that has a radius of 2 cm. If a uniform, sinusoidal magnetic field with a magnitude of 0.01 Wb/m^2 and a frequency of 10,000 Hz is directed along the axis of the dowel rod, determine the voltage induced between the two ends of the wire, assuming that the two ends are close together.

5-4 A uniform and constant magnetic field of 0.01 Wb/m^2 is directed along the z axis of a rectangular coordinate system. A circular contour in the xy plane centered at the origin has a radius that is decreasing at a rate of 100 m/s. If the initial radius is 10 cm, determine the induced emf in the path as a function of time.

5-5 Repeat Example 5.2 if the magnetic field varies with time as

$$\mathcal{B} = 0.01 \cos 3t \, \mathbf{a}_z \qquad \text{Wb/m}^2$$

5-6 In Example 5.2, suppose that the magnetic field is constant and given by

$$\mathbf{B} = 0.01\mathbf{a}_z \qquad \text{Wb/m}^2$$

but that the velocity of the bar varies with time as

$$\mathbf{u} = 100 \cos 10t \, \mathbf{a}_y \qquad \text{m/s}$$

Determine the induced voltage.

5-7 In the circuit of Fig. P5.7 a rectangular wire loop with a resistance of 0.02 Ω rotates (as shown) in a constant magnetic field given by

$$\mathbf{B} = 0.01\mathbf{a}_y \qquad \text{Wb/m}^2$$

One side of the loop lies along the z axis while another side rotates at an angular speed of $\omega = 2$ rad/s. Determine the induced current with the direction shown. The loop lies in the xz plane at $t = 0$.

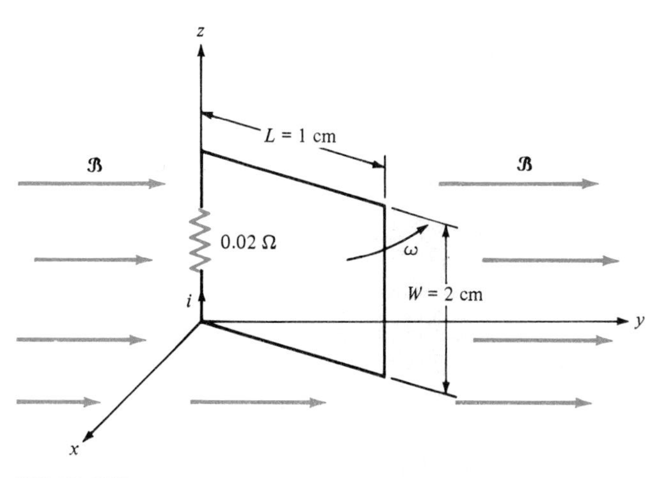

FIGURE P5.7

5-8 A loop of wire is inserted into a time-varying magnetic field, as shown in Fig. P5.8. Determine the current i circulating around the loop. From this, determine the voltages V_2 and V_1 across the resistors. Are they equal? If not, why not?

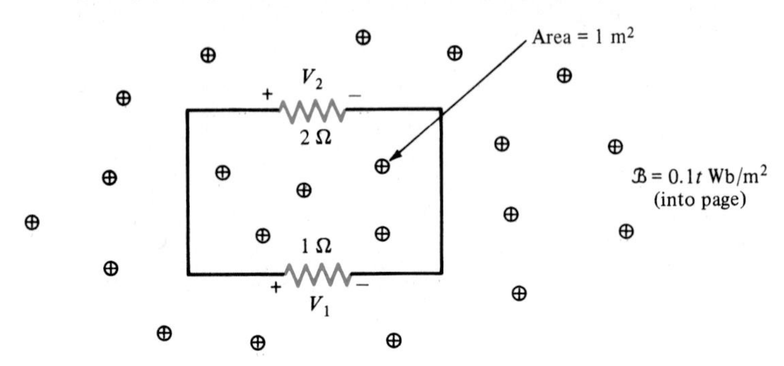

FIGURE P5.8

5-9 A 1-m² loop of wire completely encloses a time-varying magnetic field that is uniformly distributed over the loop, as shown in Fig. P5.9. A voltmeter that draws negligible current is placed in the three positons shown. Determine (and give reasons for) the voltmeter readings in each of the three positions.

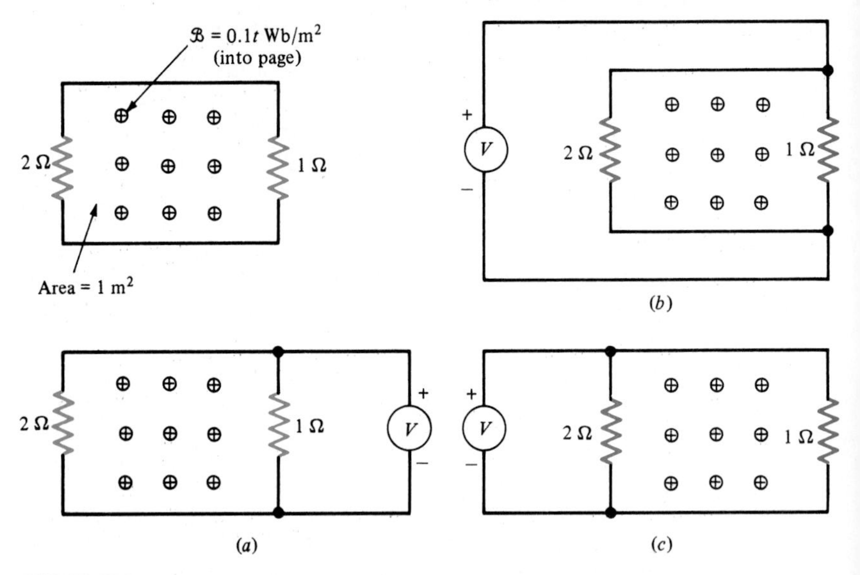

FIGURE P5.9

5-10 A material having a conductivity σ and permittivity ϵ is placed in a sinusoidal, time-varying electric field having a frequency ω. At what frequency will the conduction current equal the displacement current? If $\sigma = 10^{-2}$ S/m and $\epsilon = 3\epsilon_0$, determine this frequency.

5-11 An air-filled parallel-plate capacitor has plates of 10 cm² that are separated by a distance of 2 mm. If the capacitance is connected to a 50-V 1-MHz sinusoidal voltage source, calculate the magnitude of the displacement current, neglecting fringing.

5-12 Show that the fields

$$\mathcal{E} = E_m \sin x \sin t \, \mathbf{a}_y$$

$$\mathcal{H} = \frac{E_m}{\mu_0} \cos x \cos t \, \mathbf{a}_z$$

in free space satisfy Faraday's law and the two laws of Gauss but do not satisfy Ampère's law, and thus are not valid solutions of Maxwell's equations.

5-13 An electric field in free space is given by

$$\mathcal{E} = E_m \sin \alpha x \cos (\omega t - \beta z)\mathbf{a}_y$$

Find the corresponding magnetic field from Faraday's law. Under what conditions (α, β, ω) do these fields satisfy all of Maxwell's equations?

5-14 An electric field of the form

$$\mathcal{E}(z_1 t) = E_o \sin \beta z \cos \omega t \, \mathbf{a}_x$$

exists in free space. Determine

(a) the corresponding charge density and
(b) the associated magnetic field consistent with all of Maxwell's equations.

5-15 Obtain the instantaneous and the average power density in the y direction for the electric and magnetic fields of Prob. 5-14.

5-16 Show that the electric field of Example 5.1, which was derived from the magnetic field via Faraday's law, does not satisfy all of Maxwell's equations. Under what conditions would you expect this to be an adequate approximation to the fields?

5-17 A thin wire carrying a current i is immersed in a magnetic field $\mathcal{B}$. Show that the vector force exerted on a differential segment of the wire of length dl is given by

$$d\mathcal{F} = i \, d\mathbf{l} \times \mathcal{B}$$

where $d\mathbf{l}$ is directed in the direction of i. (*Hint*: Suppose that the segment supports a linear charge density of ρ_l that is moving with velocity u. The current is $i = \rho_l u$. Use the Lorentz force equation.) Evaluate this result for the force per unit length if the wire carries a current of 10 A in a constant magnetic field of 0.1 Wb/m^2 that is directed perpendicular to the wire.

5-18 A section of straight conducting wire of length l is moving with velocity u through a uniform and constant magnetic field **B**, as shown in Fig. P5.18. Determine the voltage induced between the endpoints of the wire. Show the proper polarity of this voltage. (*Hint*: The force on the free electrons of the wire is given by the

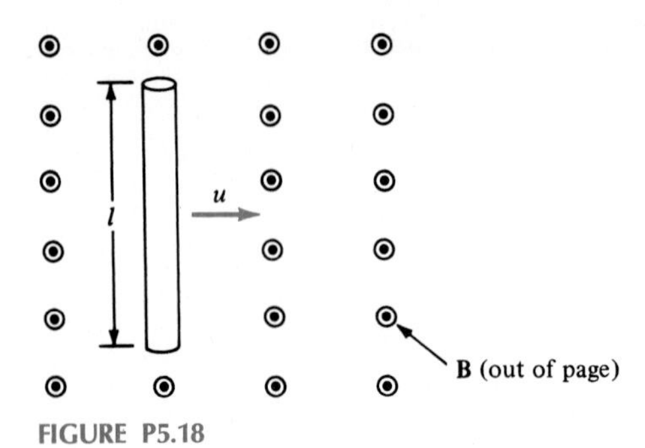

FIGURE P5.18

Lorentz force equation. This results in a separation of charge at the ends of the wire.)

5-19 An interface between two media lies in the yz plane at $x = 0$, as shown in Fig. P5.19. Medium 1 has parameters ϵ_1, μ_1, and σ_1, and medium 2 has parameters ϵ_2,

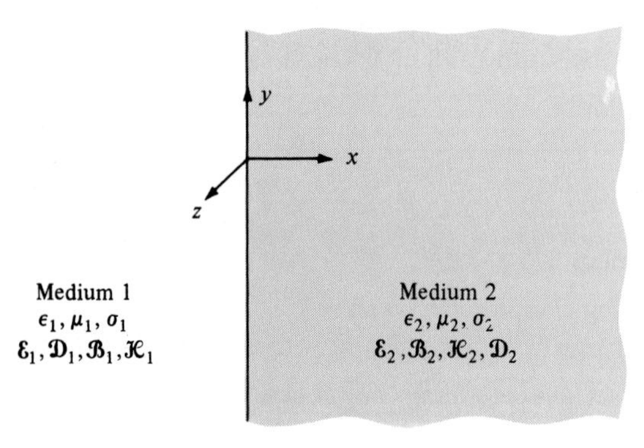

FIGURE P5.19

μ_2, and σ_2. If the electric field intensity vector in region 1 at the interface ($x = 0$) is given by

$$\boldsymbol{\mathcal{E}}_1 = \alpha \mathbf{a}_x + \beta \mathbf{a}_y + \delta \mathbf{a}_z \qquad \text{V/m}$$

determine $\boldsymbol{\mathcal{E}}_2$ at $x = 0$ if the regions are perfect dielectrics, that is, $\sigma_1 = \sigma_2 = 0$.

5-20 For the same interface between two regions in Fig. P5.19, find the magnetic flux density vector $\boldsymbol{\mathcal{B}}_2$ at $x = 0$ if

$$\boldsymbol{\mathcal{B}}_1 = \alpha \mathbf{a}_x + \beta \mathbf{a}_y + \delta \mathbf{a}_z \qquad \text{Wb/m}^2$$

at $x = 0$. The region conductivities are nonzero but finite.

5-21 At the interface between two regions in Fig. P5.19, find the electric flux density vector $\boldsymbol{\mathcal{D}}_2$ at $x = 0$ if

$$\boldsymbol{\mathcal{D}}_1 = \alpha \mathbf{a}_x + \beta \mathbf{a}_y + \delta \mathbf{a}_z \qquad \text{C/m}^2$$

at $x = 0$. Both regions are perfect dielectrics.

5-22 At the interface between two regions in Fig. P5.19 find the magnetic field intensity vector $\mathcal{H}_2$ at $x = 0$ if

$$\mathcal{H}_1 = \alpha\mathbf{a}_x + \beta\mathbf{a}_y + \delta\mathbf{a}_z \qquad \text{A/m}$$

at $x = 0$. The conductivities of the regions are nonzero but finite.

5-23 In Fig. P5.19, region 2 is a perfect conductor ($\sigma_2 = \infty$) and region 1 is free space. Suppose that the electric and magnetic fields in region 1 at the interface are given by

$$\mathcal{E}_1 = \alpha_e\mathbf{a}_x + \beta_e\mathbf{a}_y + \gamma_e\mathbf{a}_z \qquad \text{V/m}$$

$$\mathcal{B}_1 = \alpha_h\mathbf{a}_x + \beta_h\mathbf{a}_y + \gamma_h\mathbf{a}_z \qquad \text{Wb/m}^2$$

Determine α_e, β_e, γ_e, α_h, β_h, and γ_h if the surface charge density on the perfect conductor is ρ_s C/m^2 and the linear current density is $\mathcal{K}\mathbf{a}_y$ A/m.

5-24 Two regions are separated by a plane defined by $x + y = 1$, as shown in Fig. P5.24. If the electric and magnetic fields at the interface in region 1 are given by

$$\mathcal{E}_1 = 2\mathbf{a}_y + 3\mathbf{a}_z \qquad \text{V/m}$$

$$\mathcal{H}_1 = 0.1\mathbf{a}_x + 0.2\mathbf{a}_z \qquad \text{A/m}$$

determine $\mathcal{D}_2$, $\mathcal{B}_2$, $\mathcal{E}_2$, and $\mathcal{H}_2$ at the interface in region 2.

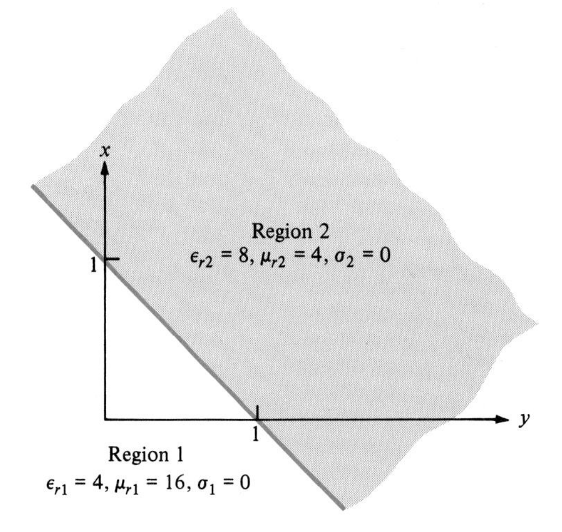

FIGURE P5.24

5-25 In Fig. P5.24, suppose that region 2 is a perfect conductor and that region 1 is unchanged. If the fields in region 1 at the interface are given by

$$\mathcal{E}_1 = 2\mathbf{a}_x + 2\mathbf{a}_y \qquad \text{V/m}$$

$$\mathcal{B}_1 = -2\mathbf{a}_x + 2\mathbf{a}_y + \mathbf{a}_z \qquad \text{Wb/m}^2$$

determine the surface charge density and vector linear current density on the surface of the perfect conductor.

5-26 The electric and magnetic fields in free space in a spherical coordinate system are given by

$$\mathcal{E} = \frac{10}{r} \sin\theta \cos\left(\omega t - \frac{4\pi}{3} r\right)\mathbf{a}_\theta \qquad \text{V/m}$$

$$\mathcal{H} = \frac{10}{120\pi r} \sin\theta \cos\left(\omega t - \frac{4\pi}{3} r\right)\mathbf{a}_\phi \qquad \text{A/m}$$

Determine the Poynting vector. What is the direction of power flow? Determine the total average power leaving spherical, closed regions of radius 100 m and 10,000 m.

5-27 Write the fields in Prob. 5-26 in phasor form. Using this result recompute the average power using Eqs. (97) and (98) and verify the result in Prob. 5.26.

5-28 The electric and magnetic fields in a region are given in a rectangular coordinate system by

$$\mathcal{E} = 10e^{-200x} \cos(\omega t - 200x)\mathbf{a}_z \qquad \text{V/m}$$

$$\mathcal{H} = -\tfrac{1}{2}e^{-200x} \cos\left(\omega t - 200x - \frac{\pi}{4}\right)\mathbf{a}_y \qquad \text{A/m}$$

Write the phasor expressions for these fields. Determine the phasor Poynting vector. Compute the average power dissipated in a cube with sides 1 cm in length. The corners of the cube are at [0, 0, 0], [0, 1 cm, 0], [1 cm, 0, 0], [1 cm, 1cm, 0], [0, 0, 1 cm], [0, 1 cm, 1 cm], [1 cm, 0, 1 cm], and [1 cm, 1 cm, 1 cm].

5-29 Write the following phasors in time-domain form. The frequency in each case is 10 MHz.

(a) $\hat{\mathbf{E}} = -j30\mathbf{a}_x - \dfrac{10}{j}\mathbf{a}_y$

(b) $\hat{\mathbf{H}} = 10e^{j4\pi/5}\mathbf{a}_z$

(c) $\hat{\rho} = j4e^{-j\pi/3}$

(d) $\hat{\mathbf{B}} = j2e^{(-j4\pi/3)z}e^{-20z}\mathbf{a}_x$

(e) $\hat{\mathbf{A}} \times \hat{\mathbf{B}}^*$

where $\hat{\mathbf{A}} = j3e^{-j2x}\mathbf{a}_y + 2e^{-3x}\mathbf{a}_z$

$$\hat{\mathbf{B}} = -je^{-jx}\mathbf{a}_x - (1 + j)e^{-jx}\mathbf{a}_z$$

5-30 Repeat Prob. 5-12 using the phasor forms of the given field vectors.

5-31 Repeat Prob. 5-13 using the phasor forms of the field vectors.

5-32 Repeat Prob. 5-16 using the phasor forms of the field vectors.

5-33 A lossy dielectric has $\mu = 4\pi \times 10^{-7}$ H/m, $\epsilon = 10^{-8}/36\pi$ F/m, and $\sigma = 2 \times 10^{-8}$ S/m. An electric field $\mathbf{\mathcal{E}} = 200 \sin \omega t \, \mathbf{a}_z$ V/m exists at a certain point in the dielectric.

- **(a)** At what frequency will the conduction current density and displacement current density have equal magnitudes?
- **(b)** At this frequency, calculate the instantaneous displacement current density.
- **(c)** What is the phase angle between the conduction current and the displacement current?

5-34 A certain type of electric motor, known as the sleeve-rotor induction motor, may be modeled by a conducting plate of thickness h and conductivity σ, backed by an iron core of infinite permeability, as shown in Fig. P5.34. The $\mathbf{\mathcal{B}}$ field at the surface of the plate is $\mathbf{\mathcal{B}} = \mathcal{B}_m \cos (\omega t - \beta x)\mathbf{a}_y$. With the axes of reference as shown, and

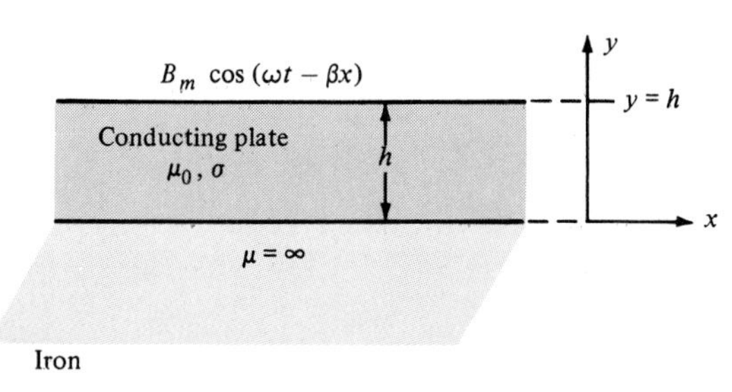

FIGURE P5.34

assuming that the $\mathcal{E}$ field is only z-directed, determine the $\mathcal{E}$-field distribution within the plate. Further assume that there is no relative motion between the $\mathbf{\mathcal{B}}$ field and the conducting plate.

5-35 In a certain region of free space, the $\mathbf{\mathcal{B}}$ field is given by $\mathbf{\mathcal{B}} = B_0 \cos (\omega t - \beta_0 x)\mathbf{a}_y$, where $\beta_0 = \omega\sqrt{\mu_0 \epsilon_0}$ and the corresponding $\mathcal{E}$ field is constrained to be z-directed.

- **(a)** Determine the ratios of the stored energy densities in the electric and magnetic fields. Show that the energy stored in the electric field is negligible at power frequencies (60 Hz).
- **(b)** If w_m is the energy density in the magnetic field and $\mathbf{S}_x$ is the power density in the x direction, show that

$$2\frac{\partial w_m}{\partial t} + \frac{\partial \mathbf{S}_x}{\partial x} = 0$$

5-36 A two-winding transformer is shown in Fig. P5.36. The primary has N_1 turns and carries a current I_1 at a radian frequency ω. Assume that $\mu_{\text{core}} \gg \mu_0$, and for the given dimensions evaluate the total reactive power (imaginary part of $\frac{1}{2}\hat{\mathbf{E}} \times \hat{\mathbf{H}}^*$) in

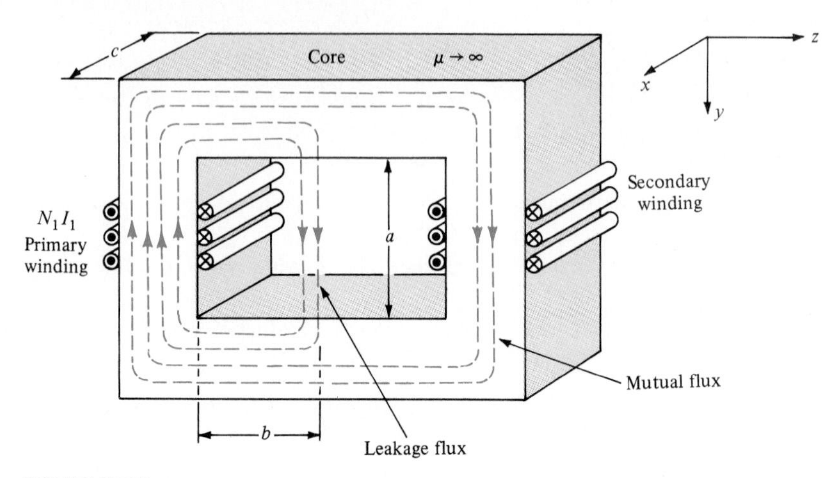

FIGURE P5.36

the z direction in the leakage flux field. Hence, determine the value of the leakage inductance if $N_1 = 200$ turns, $a = 5$ cm, $b = 2$ cm, and $c = 5$ cm. Assume uniform flux density in the leakage field.

CHAPTER 6

Propagation of Uniform Plane Waves

In this chapter we begin a study of the applications and implications of Maxwell's equations for time-varying electromagnetic fields, and our interest will center on various methods of energy propagation. One of the more controversial aspects of Maxwell's equations at the time they were published in 1864 was that they predicted the propagation of energy in the form of waves. Heinrich Hertz verified, experimentally, the existence of these waves in 1887.

This study of electromagnetic wave propagation will begin with an investigation of uniform plane waves, which perhaps represent the simplest form of wave propagation. We initially study uniform plane waves not only because they are simple but also for the practical reason that waves propagated from radio antennas resemble uniform plane waves at sufficiently distant points from the antenna, as we will find in Chap. 9. We will also find that many of the properties and characteristics of uniform plane waves have close parallels in the waves propagated on transmission lines (Chap. 7) and within hollow waveguides (Chap. 8). Thus, this study of uniform plane waves will not be an end in itself but will provide a simple introduction to the study of many other types of wave propagation.

6.1 The Wave Equation

First we derive the wave equation that governs the propagation of all electromagnetic waves. Consider a linear, isotropic, homogeneous medium. We will assume that the net free charge in the region is zero ($\rho = 0$) and that any currents in the region are conduction currents ($\mathcal{J} = \sigma\mathcal{E}$). These types of regions are quite general ones and include the practical cases of free space ($\sigma = 0$) as well as most conductors and dielectrics. Maxwell's equations in point form for this region become

$$\nabla \times \mathcal{E} = -\mu \frac{\partial \mathcal{H}}{\partial t} \tag{1a}$$

$$\nabla \times \mathcal{H} = \sigma\mathcal{E} + \epsilon \frac{\partial \mathcal{E}}{\partial t} \tag{1b}$$

$$\nabla \cdot \mathcal{H} = 0 \tag{1c}$$

$$\nabla \cdot \mathcal{E} = 0 \tag{1d}$$

Taking the curl of (1*a*) and substituting (1*b*), we obtain

$$\nabla \times \nabla \times \mathcal{E} = -\mu\sigma \frac{\partial \mathcal{E}}{\partial t} - \mu\epsilon \frac{\partial^2 \mathcal{E}}{\partial t^2} \tag{2}$$

We similarly obtain, by taking the curl of (1*b*) and substituting (1*a*),

$$\nabla \times \nabla \times \mathcal{H} = -\mu\sigma \frac{\partial \mathcal{H}}{\partial t} - \mu\epsilon \frac{\partial^2 \mathcal{H}}{\partial t^2} \tag{3}$$

In order to interpret these results, we use the vector identity

$$\nabla \times \nabla \times \mathbf{A} = \nabla(\nabla \cdot \mathbf{A}) - \nabla^2\mathbf{A} \tag{4}$$

where $\nabla^2\mathbf{A}$ is the vector laplacian. In rectangular coordinates, the vector laplacian is calculated from (4) to be

$$\nabla^2\mathbf{A} = \nabla^2 A_x \mathbf{a}_x + \nabla^2 A_y \mathbf{a}_y + \nabla^2 A_z \mathbf{a}_z \tag{5}$$

where each scalar component is the scalar laplacian of that component; for example,

$$\nabla^2 A_x = \frac{\partial^2 A_x}{\partial x^2} + \frac{\partial^2 A_x}{\partial y^2} + \frac{\partial^2 A_x}{\partial z^2} \tag{6}$$

The forms of the vector laplacian in cylindrical and spherical coordinate systems are derived strictly from (4) and are not as simple as for a rectangular coordinate system. Substituting (4) into (2) and (3), we obtain

$$\nabla^2 \boldsymbol{\mathcal{E}} = \mu\sigma \frac{\partial \boldsymbol{\mathcal{E}}}{\partial t} + \mu\epsilon \frac{\partial^2 \boldsymbol{\mathcal{E}}}{\partial t^2} \tag{7a}$$

$$\nabla^2 \boldsymbol{\mathcal{H}} = \mu\sigma \frac{\partial \boldsymbol{\mathcal{H}}}{\partial t} + \mu\epsilon \frac{\partial^2 \boldsymbol{\mathcal{H}}}{\partial t^2} \tag{7b}$$

since $\nabla \cdot \boldsymbol{\mathcal{E}} = \nabla \cdot \boldsymbol{\mathcal{H}} = 0$ for this medium. The vector differential equations in (7) are called the *wave equations*, or *Helmholtz equations*. Each equation is composed of three scalar differential equations in terms of the components of the vectors. For example, by matching components, we obtain

$$\nabla^2 \mathcal{E}_x = \mu\sigma \frac{\partial \mathcal{E}_x}{\partial t} + \mu\epsilon \frac{\partial^2 \mathcal{E}_x}{\partial t^2} \tag{8}$$

and similarly for the other two components of $\boldsymbol{\mathcal{E}}$ and the three components of $\boldsymbol{\mathcal{H}}$.

Our interest in the wave equations will be from the standpoint of sinusoidal, steady-state variation of the field vectors. For this case, the wave equations in (7) in terms of the phasor field vectors become

$$\begin{aligned}\nabla^2 \hat{\mathbf{E}} &= \mu\sigma(j\omega\hat{\mathbf{E}}) + \mu\epsilon(j\omega)^2\hat{\mathbf{E}} \\ &= j\omega\mu(\sigma + j\omega\epsilon)\hat{\mathbf{E}}\end{aligned} \tag{9a}$$

$$\nabla^2 \hat{\mathbf{H}} = j\omega\mu(\sigma + j\omega\epsilon)\hat{\mathbf{H}} \tag{9b}$$

We will use a special symbol γ^2 for the quantity $j\omega\mu(\sigma + j\omega\epsilon)$ such that

$$\gamma^2 = j\omega\mu(\sigma + j\omega\epsilon) \tag{10}$$

The positive square root of γ^2, γ, will be referred to as the *propagation constant* of the medium for reasons that will become clear in the following section. Since γ^2 is a complex number, the square root of γ^2 will also be a complex number, which we write as†

$$\begin{aligned}\gamma &= \alpha + j\beta \\ &= \sqrt{j\omega\mu(\sigma + j\omega\epsilon)}\end{aligned} \tag{11}$$

† Calculation of the square root of a complex number is quite simple, especially with the aid of an electronic calculator having the provision for converting complex numbers between rectangular and polar forms. For example, if $\hat{c} = a + jb = M\underline{/\theta}$, then $\sqrt{\hat{c}} = \sqrt{M}\underline{/\theta/2}$; that is, $\sqrt{\hat{c}}$ has a magnitude that is the square root of the magnitude of $\hat{c}$ and an angle that is one-half the angle of $\hat{c}$. Since $\sqrt{\hat{c}}$ is defined to be the number which, multiplied times itself, yields $\hat{c}$, there are several other alternative forms:

$$\sqrt{\hat{c}} = \pm\sqrt{M}\underline{/\frac{\theta}{2} \pm n\pi} \qquad \text{where } n = 1, 2, 3, \ldots$$

Expanding (9) in terms of components, the wave equations for the phasor components of the field vectors become

$$\frac{\partial^2 \hat{E}_x}{\partial x^2} + \frac{\partial^2 \hat{E}_x}{\partial y^2} + \frac{\partial^2 \hat{E}_x}{\partial z^2} = \gamma^2 \hat{E}_x \tag{12a}$$

$$\frac{\partial^2 \hat{E}_y}{\partial x^2} + \frac{\partial^2 \hat{E}_y}{\partial y^2} + \frac{\partial^2 \hat{E}_y}{\partial z^2} = \gamma^2 \hat{E}_y \tag{12b}$$

$$\frac{\partial^2 \hat{E}_z}{\partial x^2} + \frac{\partial^2 \hat{E}_z}{\partial y^2} + \frac{\partial^2 \hat{E}_z}{\partial z^2} = \gamma^2 \hat{E}_z \tag{12c}$$

$$\frac{\partial^2 \hat{H}_x}{\partial x^2} + \frac{\partial^2 \hat{H}_x}{\partial y^2} + \frac{\partial^2 \hat{H}_x}{\partial z^2} = \gamma^2 \hat{H}_x \tag{12d}$$

$$\frac{\partial^2 \hat{H}_y}{\partial x^2} + \frac{\partial^2 \hat{H}_y}{\partial y^2} + \frac{\partial^2 \hat{H}_y}{\partial z^2} = \gamma^2 \hat{H}_y \tag{12e}$$

$$\frac{\partial^2 \hat{H}_z}{\partial x^2} + \frac{\partial^2 \hat{H}_z}{\partial y^2} + \frac{\partial^2 \hat{H}_z}{\partial z^2} = \gamma^2 \hat{H}_z \tag{12f}$$

6.2 Uniform Plane Waves

We first need to define what is meant by the term *uniform plane waves*. The term *plane* is meant to indicate that the field vectors $\boldsymbol{\mathcal{E}}$ and $\boldsymbol{\mathcal{H}}$ at each point in space lie in a plane, with the planes at any two different points being parallel to each other. The term *uniform* is included to indicate that the phasor field vectors (magnitude and phase) are independent of position in each of these planes.

Without any restriction on the validity of our results, we may orient our coordinate system so that the field vectors lie in the xy plane, as shown in Fig. 6.1*a*. The phasor Poynting vector is then in the z direction. Again, without any loss of generality, we may arbitrarily assume the direction of $\hat{\mathbf{E}}$ to be in the positive x direction; that is,

$$\hat{\mathbf{E}} = \hat{E}_x(z)\mathbf{a}_x \tag{13}$$

This x component of $\hat{\mathbf{E}}$ is a function of only z since the field is to be uniform over the xy plane and is thus independent of the x and y coordinates. Therefore, we have

$$\frac{\partial \hat{E}_x}{\partial x} = \frac{\partial \hat{E}_x}{\partial y} = 0 \tag{14}$$

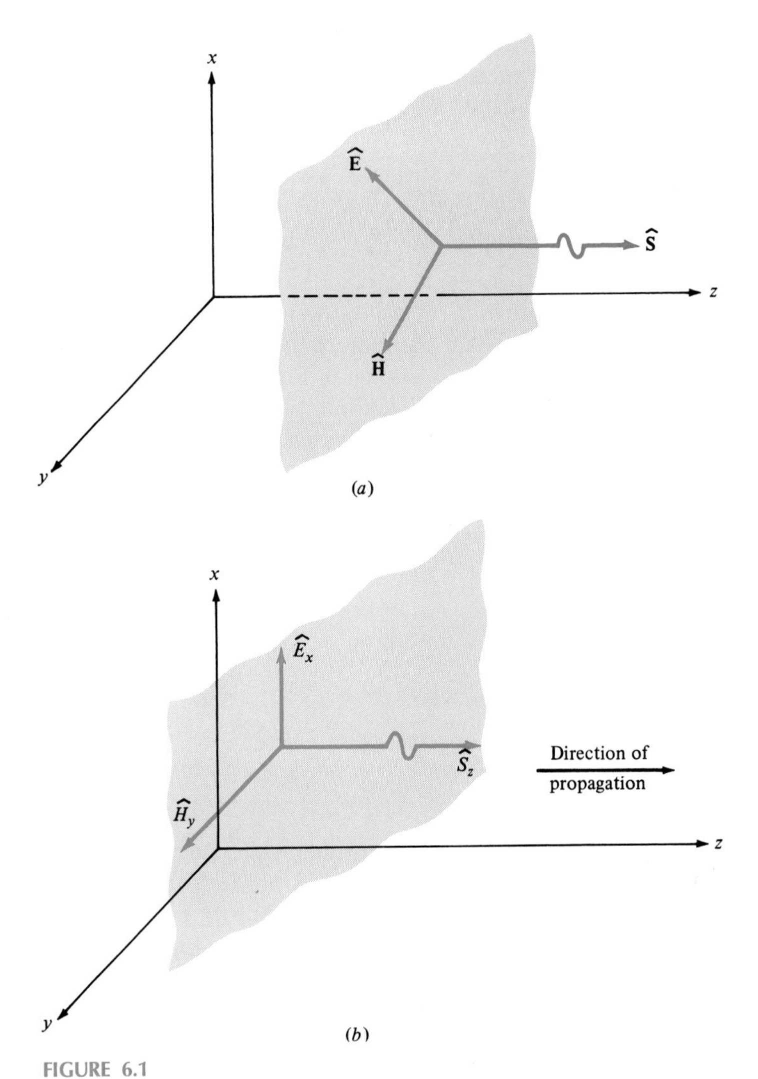

FIGURE 6.1
Propagation of uniform plane waves.

From Faraday's law

$$\nabla \times \hat{\mathbf{E}} = -j\omega\mu\hat{\mathbf{H}} \tag{15}$$

Equations (13) and (14) show that

$$\begin{aligned}\nabla \times \hat{\mathbf{E}} &= \left(\frac{\partial \hat{E}_z}{\partial y} - \frac{\partial \hat{E}_y}{\partial z}\right)\mathbf{a}_x + \left(\frac{\partial \hat{E}_x}{\partial z} - \frac{\partial \hat{E}_z}{\partial x}\right)\mathbf{a}_y + \left(\frac{\partial \hat{E}_y}{\partial x} - \frac{\partial \hat{E}_x}{\partial y}\right)\mathbf{a}_z \\ &= \frac{\partial \hat{E}_x}{\partial z}\,\mathbf{a}_y\end{aligned} \tag{16}$$

Therefore, by matching components of (15), we see that the only nonzero component of $\hat{\mathbf{H}}$ is the y component, as shown by (16); that is,

$$\begin{aligned}\hat{\mathbf{H}} &= \hat{H}_y(z)\mathbf{a}_y \\ &= -\frac{1}{j\omega\mu}\frac{\partial \hat{E}_x}{\partial z}\mathbf{a}_y\end{aligned} \tag{17}$$

This y component of $\hat{\mathbf{H}}$ is also shown as being independent of x and y. This follows directly from (17) since $\hat{E}_x$ is independent of x and y. Therefore, $\hat{\mathbf{E}}$ and $\hat{\mathbf{H}}$ for a uniform plane wave are *orthogonal*, as shown in Fig. 6.1*b*. Note also that the Poynting vector is in the z direction: the direction of $\mathcal{E} \times \mathcal{H}$. Thus, the power flow is in the z direction.†

The resulting components of the wave equations given in (12) become particularly simple for this case. Since $\hat{E}_y = \hat{E}_z = 0$ and $\hat{H}_x = \hat{H}_z = 0$, (12*b*), (12*c*), (12*d*), and (12*f*) are zero. The remaining two equations, (12*a*) and (12*e*), become

$$\frac{d^2\hat{E}_x}{dz^2} = \gamma^2\hat{E}_x \tag{18a}$$

$$\frac{d^2\hat{H}_y}{dz^2} = \gamma^2\hat{H}_y \tag{18b}$$

where partial derivatives have been replaced by ordinary derivatives since $\hat{H}_y$ and $\hat{E}_x$ are functions of only one variable, z.

Before we solve the wave equations in (18), it is worthwhile to recognize that these could have been derived in a more straightforward manner for this special case of uniform plane waves. Substituting the forms of the field vectors given in (13) and (17) directly into Faraday's law and Ampère's law results in

$$\frac{d\hat{E}_x}{dz} = -j\omega\mu\hat{H}_y \tag{19a}$$

$$\frac{d\hat{H}_y}{dz} = -(\sigma + j\omega\epsilon)\hat{E}_x \tag{19b}$$

Differentiating (19*a*) with respect to z and substituting (19*b*) yields (18*a*). Similarly, differentiating (19*b*) with respect to z and substituting (19*a*) yields (18*b*).

† The power flow will be in either the $+z$ or $-z$ direction, depending on the signs of $\hat{E}_x$ and $\hat{H}_y$.

The solutions of the equations in (18) are of the form

$$\hat{E}_x = \hat{E}_m^+ e^{-\gamma z} + \hat{E}_m^- e^{\gamma z}$$
$$= \hat{E}_m^+ e^{-\alpha z} e^{-j\beta z} + \hat{E}_m^- e^{\alpha z} e^{j\beta z} \tag{20a}$$

$$\hat{H}_y = \hat{H}_m^+ e^{-\gamma z} + \hat{H}_m^- e^{\gamma z}$$
$$= \hat{H}_m^+ e^{-\alpha z} e^{-j\beta z} + \hat{H}_m^- e^{\alpha z} e^{j\beta z} \tag{20b}$$

where we have substituted

$$\gamma = \sqrt{j\omega\mu(\sigma + j\omega\epsilon)}$$
$$= \alpha + j\beta \tag{21}$$

and $\hat{H}_m^+$, $\hat{H}_m^-$, $\hat{E}_m^+$, and $\hat{E}_m^-$ are (as yet) undetermined (and possibly complex) constants. Before interpreting these solutions, we will see that these undetermined constants are related. For example, substituting (20) into (19*a*), we obtain

$$-\gamma \hat{E}_m^+ e^{-\gamma z} + \gamma \hat{E}_m^- e^{\gamma z} = -j\omega\mu(\hat{H}_m^+ e^{-\gamma z} + \hat{H}_m^- e^{\gamma z}) \tag{22}$$

Matching appropriate terms in this equation, we obtain

$$\frac{\hat{E}_m^+}{\hat{H}_m^+} = \frac{j\omega\mu}{\gamma} = \hat{\eta} \tag{23a}$$

$$\frac{\hat{E}_m^-}{\hat{H}_m^-} = -\frac{j\omega\mu}{\gamma} = -\hat{\eta} \tag{23b}$$

The quantity $j\omega\mu/\gamma$ has the units of ohms since it is a ratio of electric field intensity (volts per meter) to magnetic field intensity (amperes per meter). It will be called the *intrinsic impedance* of the medium and denoted by the symbol $\hat{\eta}$:

$$\hat{\eta} = \frac{j\omega\mu}{\gamma}$$
$$= \sqrt{\frac{j\omega\mu}{\sigma + j\omega\epsilon}}$$
$$= \eta \underline{/\theta_\eta} \tag{24}$$

where the magnitude of $\hat{\eta}$ is denoted by η and the angle by θ_η. Substituting (24) and (23) into (20) yields

$$\hat{E}_x = \hat{E}_m^+ e^{-\alpha z} e^{-j\beta z} + \hat{E}_m^- e^{\alpha z} e^{j\beta z} \tag{25a}$$

$$\begin{aligned}\hat{H}_y &= \frac{\hat{E}_m^+}{\hat{\eta}} e^{-\alpha z} e^{-j\beta z} - \frac{\hat{E}_m^-}{\hat{\eta}} e^{\alpha z} e^{j\beta z} \\ &= \frac{\hat{E}_m^+}{\eta} e^{-\alpha z} e^{-j\beta z} e^{-j\theta_\eta} - \frac{\hat{E}_m^-}{\eta} e^{\alpha z} e^{j\beta z} e^{-j\theta_\eta}\end{aligned} \tag{25b}$$

If $\hat{E}_m^+$ and $\hat{E}_m^-$ are complex constants written as

$$\hat{E}_m^+ = E_m^+ \underline{/\theta^+} = E_m^+ e^{j\theta^+} \tag{26a}$$

$$\hat{E}_m^- = E_m^- \underline{/\theta^-} = E_m^- e^{j\theta^-} \tag{26b}$$

where E_m^+ and E_m^- are real numbers, then (25) becomes

$$\hat{E}_x = E_m^+ e^{-\alpha z} e^{-j\beta z} e^{j\theta^+} + E_m^- e^{\alpha z} e^{j\beta z} e^{j\theta^-} \tag{27a}$$

$$\hat{H}_y = \frac{E_m^+}{\eta} e^{-\alpha z} e^{-j\beta z} e^{j\theta^+} e^{-j\theta_\eta} - \frac{E_m^-}{\eta} e^{\alpha z} e^{j\beta z} e^{j\theta^-} e^{-j\theta_\eta} \tag{27b}$$

The time-domain forms become

$$\begin{aligned}\mathcal{E}_x &= \operatorname{Re}(\hat{E}_x e^{j\omega t}) \\ &= E_m^+ e^{-\alpha z} \cos(\omega t - \beta z + \theta^+) + E_m^- e^{\alpha z} \cos(\omega t + \beta z + \theta^-)\end{aligned} \tag{28a}$$

$$\begin{aligned}\mathcal{H}_y &= \operatorname{Re}(\hat{H}_y e^{j\omega t}) \\ &= \frac{E_m^+}{\eta} e^{-\alpha z} \cos(\omega t - \beta z + \theta^+ - \theta_\eta) \\ &\quad - \frac{E_m^-}{\eta} e^{\alpha z} \cos(\omega t + \beta z + \theta^- - \theta_\eta)\end{aligned} \tag{28b}$$

6.2.1 Lossless Media ($\sigma = 0$)

The physical interpretation of the time-domain results in (28) is particularly important. First consider the case of a lossless medium, $\sigma = 0$. For a lossless medium,

$$\begin{aligned}\gamma &= \sqrt{j\omega\mu(j\omega\epsilon)} \\ &= j\omega\sqrt{\mu\epsilon} \qquad \sigma = 0\end{aligned}$$

so that

$$\alpha = 0 \tag{29a}$$

$$\beta = \omega\sqrt{\mu\epsilon} \tag{29b}$$

Also, the intrinsic impedance becomes a real number, so that

$$\hat{\eta} = \sqrt{\frac{j\omega\mu}{j\omega\epsilon}}$$

$$= \sqrt{\frac{\mu}{\epsilon}} \qquad \sigma = 0$$

Thus

$$\eta = \sqrt{\frac{\mu}{\epsilon}} \tag{30a}$$

$$\theta_\eta = 0 \tag{30b}$$

For a lossless medium, the field components in (28) become

$$\mathcal{E}_x = E_m^+ \cos(\omega t - \beta z + \theta^+) + E_m^- \cos(\omega t + \beta z + \theta^-) \tag{31a}$$

$$\mathcal{H}_y = \frac{E_m^+}{\eta} \cos(\omega t - \beta z + \theta^+) - \frac{E_m^-}{\eta} \cos(\omega t + \beta z + \theta^-) \tag{31b}$$

since $\alpha = 0$ and $\theta_\eta = 0$.

Consider the first term of $\mathcal{E}_x$ in (31*a*), $E_m^+ \cos(\omega t - \beta z + \theta^+)$. This portion of $\mathcal{E}_x$ represents a wave traveling in the positive z direction; this can be seen from Fig. 6.2*a*, in which $E_m^+ \cos(\omega t - \beta z + \theta^+)$ has been plotted as a function of z for two different instants t_0 and $t_1 > t_0$. Note that *corresponding points on the waveforms occur at positions and times such that the argument of the cosine has the same value*; that is,

$$\omega t_0 - \beta z_0 + \theta^+ = \omega t_1 - \beta z_1 + \theta^+ \tag{32}$$

Thus, we observe that a point on the waveform must move in the positive z direction for increasing time, so that

$$\omega t - \beta z + \theta^+ = \text{a constant} \tag{33}$$

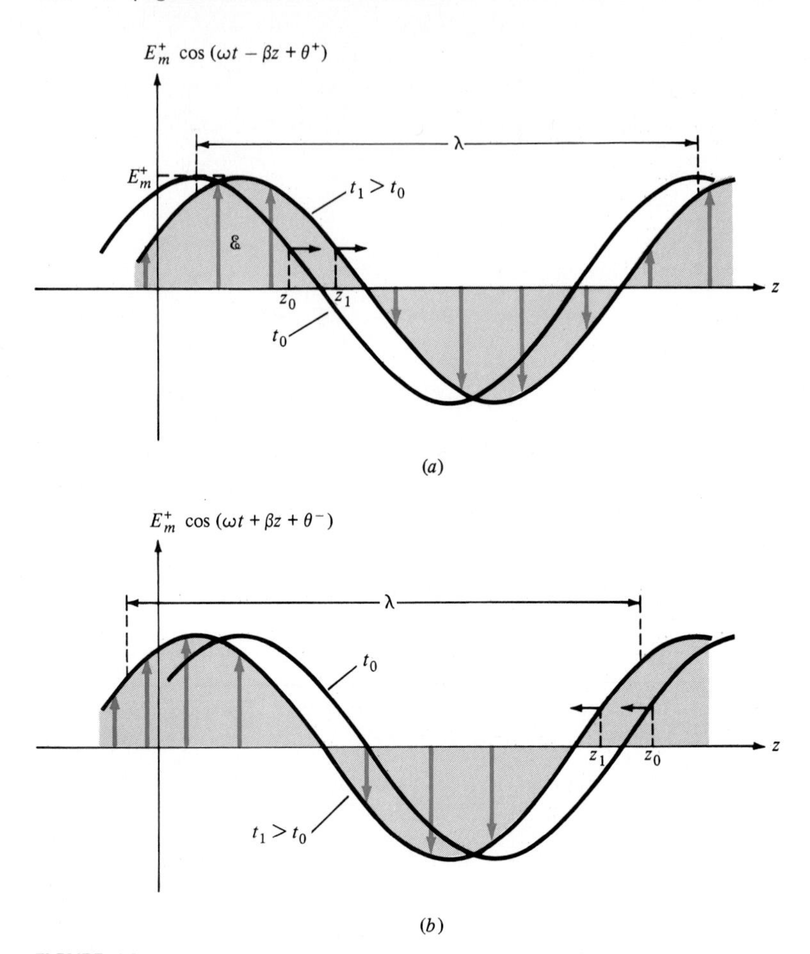

FIGURE 6.2

Components of uniform plane waves. *(a)* Forward-traveling waves; *(b)* backward-traveling waves.

Taking the derivative of this expression with respect to time yields the velocity at which the points of constant phase travel. This is referred to as the *phase velocity* of propagation of the wave, which becomes

$$u = \frac{dz}{dt} = \frac{\omega}{\beta} = \frac{1}{\sqrt{\mu\epsilon}} \qquad \text{m/s} \tag{34}$$

Note that the phase angle θ^+—introduced on the assumption that the undetermined constant $\hat{E}_m^+$ in (26*a*) may be, in general, complex—has nothing whatever to do with the function $E_m^+ \cos(\omega t - \beta z + \theta^+)$ representing a traveling wave, nor does it affect the velocity of propagation of the wave. Similarly, we observe that the term $E_m^- \cos(\omega t + \beta z + \theta^-)$ in (31*a*) represents a wave traveling in the negative z direction since a point on the waveform must move in the negative z direction for increasing time in order for the argument of the cosine function to remain constant; that is, to track the movement of corresponding point on the waveform:

$$\omega t_0 + \beta z_0 + \theta^- = \omega t_1 + \beta z_1 + \theta^- \tag{35}$$

This is illustrated in Fig. 6.2*b*. Note that similar observations hold for the magnetic field given in (31*b*). However, the magnetic field in the backward-traveling wave is in the negative y direction.

The quantity $\beta = \omega\sqrt{\mu\epsilon}$ is referred to as the *phase constant*, or *wave number*. The units of β are radians per meter, as is clear from (31), so that β is the change in phase of the wave with distance. The distance between corresponding adjacent points on the wave is known as the *wavelength* and is denoted by λ. From Fig. 6.2, we see that $\beta\lambda = 2\pi$ or

$$\beta = \frac{2\pi}{\lambda} \tag{36}$$

Since $\beta = \omega\sqrt{\mu\epsilon}$ and $u = 1/\sqrt{\mu\epsilon}$ for this lossless medium,

$$\beta = \frac{\omega}{u} \tag{37}$$

and

$$\lambda = \frac{2\pi}{\beta} = \frac{u}{f} \tag{38}$$

Increased frequencies result in shorter wavelengths. Note that the wavelength is also a function of the properties of the medium since $u = 1/\sqrt{\mu\epsilon}$. For typical materials, $\mu \geq \mu_0$ and $\epsilon \geq \epsilon_0$; thus, the phase velocity of propagation is slower than in free space and the wavelength is shorter.

The complete solution in (31) can be written as the sum of forward and backward traveling waves as

$$\mathcal{E}_x = \mathcal{E}_x^+ + \mathcal{E}_x^- \tag{39a}$$

$$\mathcal{H}_y = \mathcal{H}_y^+ + \mathcal{H}_y^- \tag{39b}$$

where

$$\mathcal{E}_x^+ = E_m^+ \cos(\omega t - \beta z + \theta^+) \tag{40a}$$

$$\mathcal{H}_y^+ = \frac{E_m^+}{\eta} \cos(\omega t - \beta z + \theta^+) \tag{40b}$$

and
$$\mathcal{E}_x^- = E_m^- \cos(\omega t + \beta z + \theta^-) \tag{41a}$$

$$\mathcal{H}_y^- = -\frac{E_m^-}{\eta} \cos(\omega t + \beta z + \theta^-) \tag{41b}$$

These individual waves are plotted for $t = 0$ in Fig. 6.3. The complete wave is the superposition of these forward- and backward-traveling waves. Observe that the Poynting vector for the forward-traveling components $\mathcal{E}_x^+$ and $\mathcal{H}_y^+$ is in the $+z$ direction, indicating power flow in that direction. However, $\mathcal{H}_y^-$ is in the $-y$ direction, and thus the Poynting vector for the backward-traveling components is in the $-z$ direction, indicating power flow in that direction. Thus, power flows in the direction of wave travel: a sensible result.

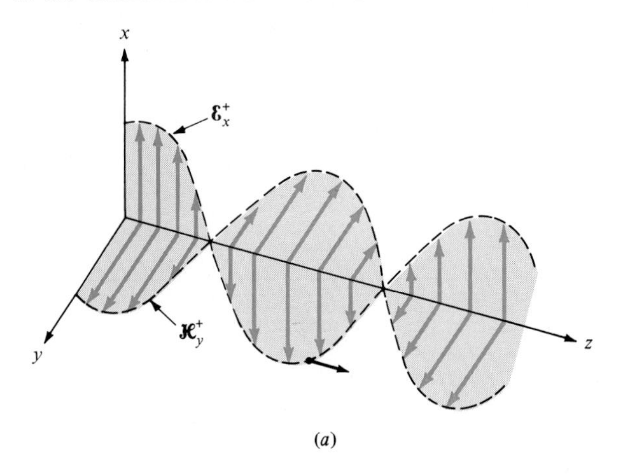

(a)

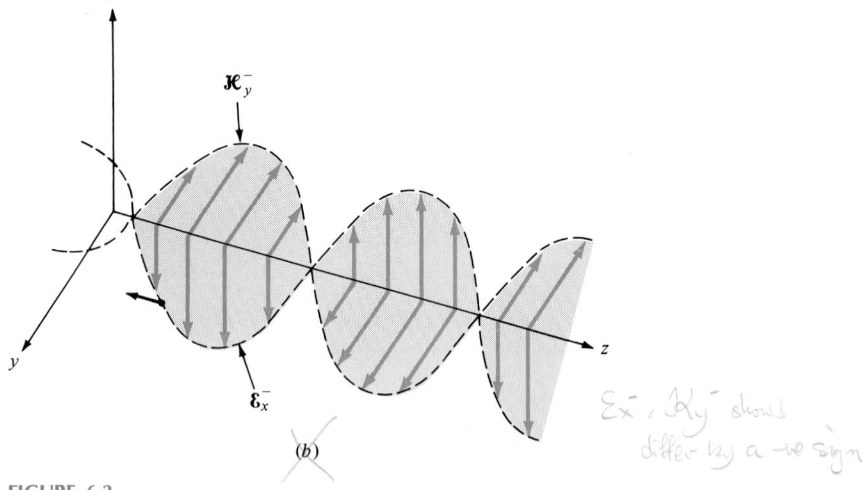

(b)

FIGURE 6.3
The individual forward- and backward-traveling waves as a function of position at $t = 0$. *(a)* Forward-traveling waves; *(b)* backward-traveling waves.

In free space $\mu_0 = 4\pi \times 10^{-7}$ and $\epsilon_0 \simeq 1/36\pi \times 10^{-9}$, so that

$$u_0 \simeq 3 \times 10^8 \text{ m/s}$$

At a frequency of 300 MHz,

$$\lambda_0 = \frac{u_0}{f}$$

$$= 1 \text{ m} \qquad (f = 300 \text{ MHz})$$

Similarly,

$$\lambda_0 = 1 \text{ cm} \qquad (f = 30 \text{ GHz})$$

and

$$\lambda_0 = 3107 \text{ miles} \qquad (f = 60 \text{ Hz})$$

The intrinsic impedance becomes

$$\eta_0 = \sqrt{\frac{\mu_0}{\epsilon_0}}$$

$$\simeq 120\,\pi$$

$$= 377\ \Omega$$

For any other lossless material medium with $\epsilon = \epsilon_r \epsilon_0$ and $\mu = \mu_r \mu_0$,

$$u = \frac{u_0}{\sqrt{\mu_r \epsilon_r}} \quad \text{m/s} \tag{42a}$$

$$\eta = \eta_0 \sqrt{\frac{\mu_r}{\epsilon_r}} \quad \Omega \tag{42b}$$

$$\beta = \beta_0 \sqrt{\mu_r \epsilon_r} \quad \text{rad/m} \tag{42c}$$

$$\lambda = \frac{\lambda_0}{\sqrt{\mu_r \epsilon_r}} \quad \text{m} \tag{42d}$$

In calculating these quantities in material media, one should translate the corresponding results in free space to those in the medium by using these relationships. One should never again calculate u_0 and η_0! Similarly, one can calculate λ_0 by scaling the result $\lambda_0 = 1$ m at $f = 300$ MHz. Similarly, one can obtain β_0 from this result and the relation $\beta_0 = 2\pi/\lambda_0$.

EXAMPLE 6.1 A uniform plane wave at a frequency of 1 GHz is traveling in a large block of Teflon ($\epsilon_r \simeq 2.1$, $\mu_r \simeq 1$, and $\sigma \simeq 0$). Determine u, η, β, and λ.

Solution

$$u = \frac{u_0}{\sqrt{\mu_r \epsilon_r}}$$

$$= \frac{3 \times 10^8}{\sqrt{2.1}}$$

$$= 2.07 \times 10^8 \text{ m/s}$$

Thus, in Teflon the velocity of propagation is 69 percent of that in free space. Also,

$$\eta = \eta_0 \sqrt{\frac{\mu_r}{\epsilon_r}}$$

$$= \frac{377}{\sqrt{\epsilon_r}}$$

$$= 260\ \Omega$$

Similarly, the intrinsic impedance is 69 percent of that in free space, as is the wavelength:

$$\lambda = \frac{\lambda_0}{\sqrt{\mu_r \epsilon_r}}$$

$$= \frac{\lambda_0}{\sqrt{\epsilon_r}}$$

$$= \frac{30 \text{ cm}}{\sqrt{2.1}}$$

$$= 20.7 \text{ cm}$$

Also,

$$\beta = \beta_0 \sqrt{\mu_r \epsilon_r}$$

$$= \frac{2\pi}{\lambda}$$

$$= 30.4 \text{ rad/m}$$

$$= 1740 \text{ deg/m}$$

Thus, at 1 GHz the field vectors reverse direction approximately every 10 centimeters along the direction of propagation.

6.2.2 Lossy Media ($\sigma \neq 0$)

There are two important differences between uniform plane waves in lossless media and in lossy media. The first difference is that the propagation constant γ has a nonzero real part:

$$\begin{aligned}\gamma &= \sqrt{j\omega\mu(\sigma + j\omega\epsilon)} \\ &= \alpha + j\beta\end{aligned} \tag{43}$$

This results in the waves for the lossless case being multiplied by the exponentials $e^{-\alpha z}$ and $e^{\alpha z}$:

$$\mathcal{E}_x^+ = E_m^+ e^{-\alpha z} \cos(\omega t - \beta z + \theta^+) \tag{44a}$$

$$\mathcal{E}_x^- = E_m^- e^{\alpha z} \cos(\omega t + \beta z + \theta^-) \tag{44b}$$

These are obviously still forward- and backward-traveling waves, but the *amplitude* of the forward-traveling wave is $E_m^+ e^{-\alpha z}$, which is reduced for increasing z (in the direction of propagation). This is shown for a fixed time as a function of z in Fig. 6.4*a*. Similarly, the amplitude of the backward-traveling wave, $E_m^- e^{\alpha z}$, is also reduced since it is propagating in the $-z$ direction (decreasing z).

The real part of γ, α, is referred to as the *attenuation constant* for the above reasons. The SI unit of α is nepers per meter (Np/m). Note that the phase constant β is the imaginary part of γ. It is not the same value as for a lossless medium:

$$\begin{aligned}\beta &= \text{Im}(\gamma) \\ &\neq \omega\sqrt{\mu\epsilon}\end{aligned} \tag{45}$$

The second difference between lossless and lossy media is in the intrinsic impedance of the medium:

$$\begin{aligned}\hat{\eta} &= \sqrt{\frac{j\omega\mu}{\sigma + j\omega\epsilon}} \\ &= \eta\underline{/\theta_\eta}\end{aligned} \tag{46}$$

For a lossy medium, $\hat{\eta}$ is complex and does not have the same value as for a lossless medium:

$$\eta \neq \sqrt{\frac{\mu}{\epsilon}} \tag{47}$$

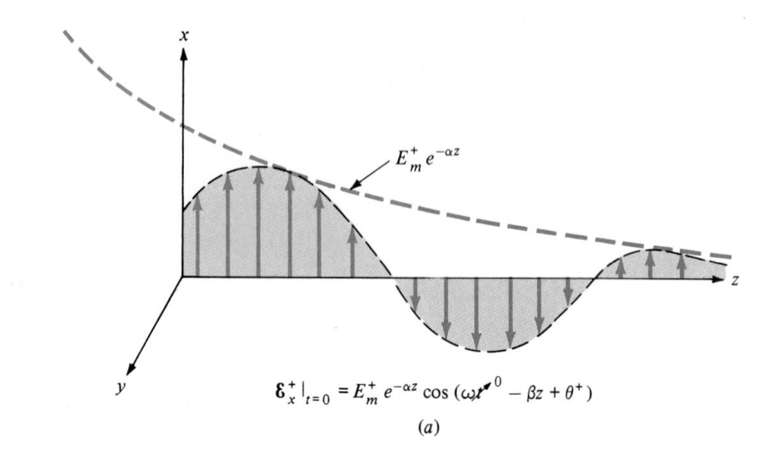

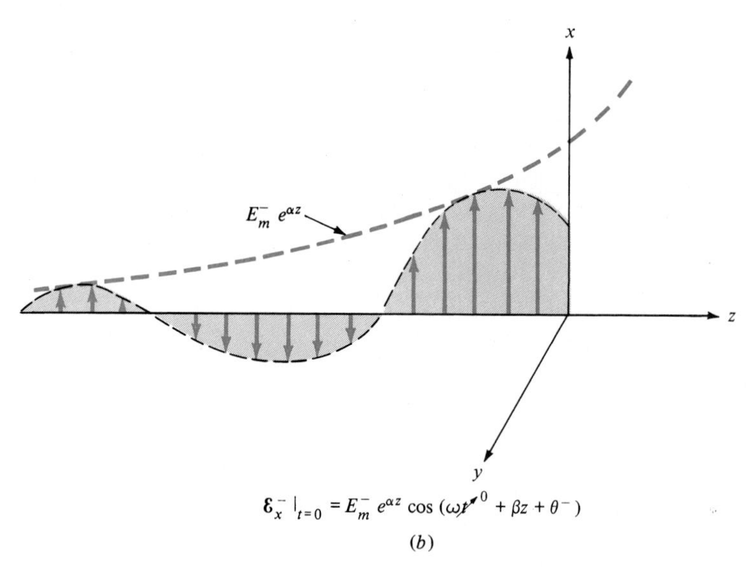

FIGURE 6.4

Uniform plane waves in lossy media as a function of position at $t = 0$. *(a)* Forward-traveling waves; *(b)* backward-traveling waves.

For the lossless case, $\theta_\eta = 0$ and we observe that the electric and magnetic fields of the forward-traveling wave are in time phase, as are those in the backward-traveling waves. However, for the lossy case, the phase angle of the intrinsic impedance, θ_η, results in the electric and magnetic fields of each traveling wave being out of time phase by the phase angle θ_η. Note from (46) that θ_η will be such that $0 \leq \theta_\eta \leq 45°$. Therefore, the magnetic field of the forward-traveling wave lags the forward-traveling electric field by θ_η. Similar results hold for the backward-traveling components of the wave.

The defining relationships for the phase velocity of propagation and wavelength are obviously the same as for a lossless medium:

$$u = \frac{\omega}{\beta} \tag{48}$$

$$\lambda = \frac{2\pi}{\beta}$$

$$= \frac{u}{f} \tag{49}$$

However, β is the imaginary part of the propagation constant γ and is no longer equal to $\omega\sqrt{\mu\epsilon}$ for the lossy case.

EXAMPLE 6.2 Determine the phase velocity of propagation, attenuation constant, phase constant, and intrinsic impedance for a forward-traveling wave in a large block of copper at 1 MHz ($\sigma = 5.8 \times 10^7$, $\epsilon_r \simeq 1$, and $\mu_r \simeq 1$). Determine the phase shift between the electric and magnetic fields and the distance that the wave must travel to be attenuated (reduced in amplitude) by a factor of 100 (40 dB).

Solution The propagation constant is

$$\gamma = \sqrt{j\omega\mu(\sigma + j\omega\epsilon)}$$

$$= \sqrt{j2\pi \times 10^6 \times 4\pi \times 10^{-7} \times (5.8 \times 10^7 + j2\pi \times 10^6 \times 1/36\pi \times 10^{-9})}$$

$$= 2.14 \times 10^4 \underline{/45^\circ}$$

$$= 1.513 \times 10^4 + j1.513 \times 10^4$$

From this result we obtain

$$\alpha = 1.513 \times 10^4 \text{ Np/m}$$

$$\beta = 1.513 \times 10^4 \text{ rad/m}$$

The wavelength in copper at this frequency becomes

$$\lambda = \frac{2\pi}{\beta}$$

$$= 4.153 \times 10^{-4} \text{ m}$$

The velocity of propagation is

$$u = \frac{\omega}{\beta}$$

$$= \frac{2\pi \times 10^6}{1.513 \times 10^4}$$

$$= 415.3 \text{ m/s}$$

and the wave propagates considerably slower in copper than in free space ($u_0 = 3 \times 10^8$ m/s). The intrinsic impedance becomes

$$\hat{\eta} = \sqrt{\frac{j\omega\mu}{\sigma + j\omega\epsilon}}$$

$$= \sqrt{\frac{j2\pi \times 10^6 \times 4\pi \times 10^{-7}}{5.8 \times 10^7 + j2\pi \times 10^6 \times 1/36\pi \times 10^{-9}}}$$

$$= 3.689 \times 10^{-4} \underline{/45^\circ} \ \Omega$$

Thus, we identify

$$\eta = 3.689 \times 10^{-4} \quad \Omega$$

$$\theta_\eta = 45^\circ$$

From (28) the magnetic field of the forward-propagating wave lags the electric field by θ_η or 45°. Also, from (28) the electric field and magnetic field are attenuated by a factor of $e^{-\alpha z}$. Thus, the fields are attenuated by a factor of 100 in a distance d such that

$$e^{-\alpha d} = \tfrac{1}{100}$$

or

$$\alpha d = \ln 100$$

giving

$$d = 3.04 \times 10^{-4}$$

or 0.3 mm.

6.2.3 Power Flow

Let us again write the form of the fields given in (28) in the form of (39):

$$\mathcal{E}_x = \mathcal{E}_x^+ + \mathcal{E}_x^- \tag{50a}$$

$$\mathcal{H}_y = \mathcal{H}_y^+ + \mathcal{H}_y^- \tag{50b}$$

where

$$\mathcal{E}_x^+ = E_m^+ e^{-\alpha z} \cos(\omega t - \beta z + \theta^+) \tag{51a}$$

$$\mathcal{E}_x^- = E_m^- e^{\alpha z} \cos(\omega t + \beta z + \theta^-) \tag{51b}$$

$$\mathcal{H}_y^+ = \frac{E_m^+}{\eta} e^{-\alpha z} \cos(\omega t - \beta z + \theta^+ - \theta_\eta) \tag{51c}$$

$$\mathcal{H}_y^- = -\frac{E_m^-}{\eta} e^{\alpha z} \cos(\omega t + \beta z + \theta^- - \theta_\eta) \tag{51d}$$

Note in (50) that we are writing the electric and magnetic fields as sums of forward- and backward-traveling waves. The vectors of the forward and backward electric field waves are in the same direction, whereas those of the magnetic field are in opposite directions. This, of course, actually makes sense, because $\mathcal{E}_x^+ \mathbf{a}_x \times \mathcal{H}_y^+ \mathbf{a}_y = \mathcal{E}_x^+ \mathcal{H}_y^+ \mathbf{a}_z$ and the direction of power flow for the forward-traveling wave is in the $+z$ direction, as shown in Fig. 6.3*a*. Similarly, from Fig. 6.3*b*, the direction of power flow for the negative-traveling waves is in the $-z$ direction since $\mathcal{E}_x^- \mathbf{a}_x \times \mathcal{H}_y^- \mathbf{a}_y = \mathcal{E}_x^- \mathcal{H}_y^- \mathbf{a}_z$ but $\mathcal{H}_y^-$ contains a negative sign.

The average power density for the complete wave is

$$\begin{aligned}\mathbf{S}_{\text{av}} &= \tfrac{1}{2} \operatorname{Re}(\hat{\mathbf{E}} \times \hat{\mathbf{H}}^*)\\ &= \tfrac{1}{2} \operatorname{Re}(\hat{E}_x \mathbf{a}_x \times \hat{H}_y^* \mathbf{a}_y)\\ &= \tfrac{1}{2} \operatorname{Re}(\hat{E}_x \hat{H}_y^*)\mathbf{a}_z \end{aligned} \tag{52}$$

Substituting (27) into (52), we obtain

$$\begin{aligned}\mathbf{S}_{\text{av}} = \Bigg[&\frac{1}{2}\frac{(E_m^+)^2}{\eta} e^{-2\alpha z} \cos\theta_\eta\\ &- \frac{1}{2}\frac{(E_m^-)^2}{\eta} e^{2\alpha z} \cos\theta_\eta\\ &- \frac{E_m^+ E_m^-}{\eta} \sin\theta_\eta \sin(2\beta z + \theta^- - \theta^+)\Bigg]\mathbf{a}_z \end{aligned} \tag{53}$$

The first term of (53), denoted as $\mathbf{S}_{av}^{+}$,

$$\mathbf{S}_{av}^{+} = \frac{1}{2}\frac{(E_m^{+})^2}{\eta} e^{-2\alpha z} \cos\theta_\eta \, \mathbf{a}_z \tag{54}$$

is the average power density in the forward-traveling wave (in the absence of the backward-traveling wave), as can be checked by applying (52) to only that portion of the wave. The second term, denoted by $\mathbf{S}_{av}^{-}$,

$$\mathbf{S}_{av}^{-} = -\frac{1}{2}\frac{(E_m^{-})^2}{\eta} e^{2\alpha z} \cos\theta_\eta \, \mathbf{a}_z \tag{55}$$

is the average power density in the backward-traveling wave. The third and final term in (53) is a cross-coupling term that disappears for lossless media where $\theta_\eta = 0$.

For the lossless case, we may then write

$$\mathbf{S}_{av} = \mathbf{S}_{av}^{+} + \mathbf{S}_{av}^{-} \qquad \sigma = 0 \tag{56}$$

where

$$\mathbf{S}_{av}^{+} = \frac{1}{2}\frac{(E_m^{+})^2}{\eta} \mathbf{a}_z \qquad \sigma = 0 \tag{57a}$$

$$\mathbf{S}_{av}^{-} = -\frac{1}{2}\frac{(E_m^{-})^2}{\eta} \mathbf{a}_z \qquad \sigma = 0 \tag{57b}$$

Note that (56) implies that the net average power traveling in the $+z$ direction is conserved for lossless media, as it should be. This is not the case for a lossy medium because the cross-coupling term in (53) represents power dissipated in the medium.

EXAMPLE 6.3 Consider a 100-V/m uniform plane wave of frequency 300 MHz traveling in an infinite, lossless medium having $\epsilon_r = 9$, $\mu_r = 1$, and $\sigma = 0$. Write complete time-domain expressions for the field vectors of the forward-traveling wave and determine the average power density vector.

Solution The intrinsic impedance and propagation constants are

$$\begin{aligned}\eta &= \sqrt{\frac{\mu}{\epsilon}} = \sqrt{\frac{\mu_r}{\epsilon_r}}\sqrt{\frac{\mu_0}{\epsilon_0}} \\ &= \tfrac{1}{3}\eta_0 = 126\ \Omega\end{aligned}$$

$$\begin{aligned}\beta &= \omega\sqrt{\mu\epsilon} = \omega\sqrt{\mu_r\epsilon_r}\sqrt{\mu_0\epsilon_0} \\ &= 3\beta_0 \\ &= 6\pi \text{ rad/m}\end{aligned}$$

The electric field has an amplitude of 100 V/m and is oriented in the x direction, thus the forms of the field vectors become

$$\mathcal{E} = E_m^+ \cos(\omega t - \beta z)\mathbf{a}_x$$
$$= 100 \cos(600\pi \times 10^6 t - 6\pi z)\mathbf{a}_x \qquad \text{V/m}$$
$$\mathcal{H} = \frac{E_m^+}{\eta} \cos(\omega t - \beta z)\mathbf{a}_y$$
$$= 0.8 \cos(600\pi \times 10^6 t - 6\pi z)\mathbf{a}_y \qquad \text{A/m}$$

where we may arbitrarily take θ^+ to be zero. The average power density vector is

$$\mathbf{S}_{\text{av}} = \frac{1}{2}\frac{(E_m^+)^2}{\eta}\mathbf{a}_z$$
$$= 39.79\mathbf{a}_z \qquad \text{W/m}^2$$

Next we will consider an example of a lossy medium.

EXAMPLE 6.4 Consider the problem in Example 6.3, where the medium is now lossy, with $\sigma = 10$ S/m.

Solution The intrinsic impedance becomes

$$\hat{\eta} = \sqrt{\frac{j\omega\mu}{\sigma + j\omega\epsilon}}$$
$$= \sqrt{236.87\underline{/89.14^\circ}}$$
$$= 15.4\underline{/44.57^\circ}\ \Omega$$

The propagation constant becomes

$$\gamma = \sqrt{j\omega\mu(\sigma + j\omega\epsilon)}$$
$$= \sqrt{23{,}689.72\underline{/90.86^\circ}}$$
$$= 153.91\underline{/45.43^\circ}$$
$$= 108.01 + j109.65$$

so that $\alpha = 108.01$ and $\beta = 109.65$. The forms of the field vectors become

$$\mathcal{E} = E_m^+ e^{-\alpha z} \cos(\omega t - \beta z)\mathbf{a}_x$$
$$= 100e^{-108z} \cos(600\pi \times 10^6 t - 109.65z)\mathbf{a}_x \qquad \text{V/m}$$
$$\mathcal{H} = \frac{E_m^+}{\eta} e^{-\alpha z} \cos(\omega t - \beta z - \theta_\eta)\mathbf{a}_y$$
$$= 6.49e^{-108z} \cos(600\pi \times 10^6 t - 109.65z - 44.57^\circ)\mathbf{a}_y \qquad \text{A/m}$$

The average power density vector is

$$\mathbf{S}_{av} = \frac{1}{2}\frac{(E_m^+)^2}{\eta} e^{-2\alpha z} \cos\theta_\eta \, \mathbf{a}_z$$

$$= 231.3e^{-216z}\mathbf{a}_z \qquad \text{W/m}^2$$

6.3 Conductors and Dielectrics Revisited

We will find it convenient at this point to step back and reconsider some fundamental notions concerning the distinction between conductors and dielectrics. Note that the only change in Maxwell's equations introduced by losses in the medium occurs in Ampère's law.† For $\sigma \neq 0$, Ampère's law may be written as

$$\nabla \times \hat{\mathbf{H}} = (\sigma + j\omega\epsilon)\hat{\mathbf{E}}$$

$$= j\omega\epsilon\left(1 - j\frac{\sigma}{\omega\epsilon}\right)\hat{\mathbf{E}} \tag{58}$$

Comparing this expression to the lossless case

$$\nabla \times \hat{\mathbf{H}} = j\omega\epsilon\hat{\mathbf{E}} \qquad \sigma = 0 \tag{59}$$

we see that we could derive results for the lossless case and "fix up" those results to include losses by replacing ϵ in the lossless results with a complex permittivity

$$\hat{\epsilon} = \epsilon\left(1 - j\frac{\sigma}{\omega\epsilon}\right) \tag{60}$$

as a comparison of (58) and (59) shows.

The term $\sigma/\omega\epsilon$ in (58) and (60) is referred to as the *loss tangent* of the material and is a function of frequency. Values of the loss tangent are tabulated for various materials at several frequencies in handbooks. The origin of the name "loss tangent" is related to the following observation. Note that there are two components of current involved in Ampère's law in (58): a conduction current

$$\hat{\mathbf{J}}_c = \sigma\hat{\mathbf{E}} \tag{61}$$

---

† For the purposes of the present discussion, we are considering losses to be conductive-type losses associated with a nonzero conductivity of the medium. We will also find that there is another type of loss mechanism, polarization loss, that is predominant in dielectric materials. However, this polarization loss will be shown to be easily accounted for from the standpoint of an effective conductivity.

and a displacement current

$$\hat{\mathbf{J}}_d = j\omega\epsilon\hat{\mathbf{E}} \tag{62}$$

The conduction current represents an energy loss mechanism and the displacement current represents energy storage, as we have observed previously. The ratio of these two currents is a measure of the lossy nature of the material. Ampère's law can then be written as

$$\nabla \times \hat{\mathbf{H}} = \hat{\mathbf{J}}_{\text{total}} \tag{63a}$$

where

$$\hat{\mathbf{J}}_{\text{total}} = \hat{\mathbf{J}}_c + \hat{\mathbf{J}}_d \tag{63b}$$

The conduction and displacement current phasors are 90° out of time phase. If we plot the conduction and displacement current phasors in the complex plane as shown in Fig. 6.5, we observe that

$$\begin{aligned} \tan \phi &= \frac{|\hat{\mathbf{J}}_c|}{|\hat{\mathbf{J}}_d|} \\ &= \frac{\sigma}{\omega\epsilon} \end{aligned} \tag{64}$$

where ϕ is the angle shown in Fig. 6.5. Thus, the term *loss tangent* arises naturally.

The notion of the loss tangent as the ratio of conduction current to displacement current provides a meaningful way of distinguishing between conductors and dielectrics. We will classify materials according to whether the

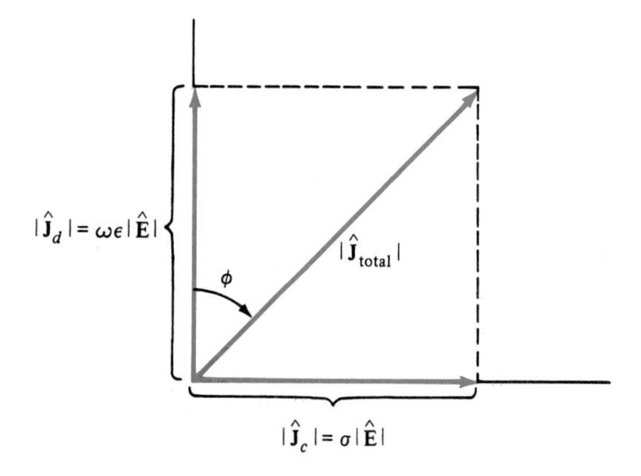

FIGURE 6.5

Illustration of the loss tangent relating conduction current densities, $\hat{\mathbf{J}}_c$, and displacement current densities, $\hat{\mathbf{J}}_d$.

conduction current is larger than the displacement current (good conductor) or whether the displacement current is larger than the conduction current (good dielectric); that is,

$$\frac{\sigma}{\omega\epsilon} \ll 1 \qquad \text{good dielectric} \tag{65a}$$

$$\frac{\sigma}{\omega\epsilon} \gg 1 \qquad \text{good conductor} \tag{65b}$$

Calculations of the intrinsic impedance and propagation constants for lossless media, $\eta = \sqrt{\mu/\epsilon}$ and $\gamma = j\omega\sqrt{\mu\epsilon}$, are particularly simple. Determination of these quantities for lossy media is slightly more involved due to the requirement that we find the square root of complex numbers, as in (21) and (24).† For the case of good conductors or good dielectrics, however, we can obtain simple approximate expressions for γ and $\hat{\eta}$.

6.3.1 Good Dielectrics ($\sigma \ll \omega\epsilon$)

First consider good dielectrics, where $\sigma/\omega\epsilon \ll 1$. The propagation constant can be written as

$$\begin{aligned}\gamma &= \alpha + j\beta \\ &= \sqrt{j\omega\mu(\sigma + j\omega\epsilon)} \\ &= j\omega\sqrt{\mu\epsilon}\sqrt{1 - j\frac{\sigma}{\omega\epsilon}}\end{aligned} \tag{66}$$

Using the binomial expansion:

$$(1 + x)^n = 1 + \frac{nx}{1!} + \frac{n(n-1)x^2}{2!} + \frac{n(n-1)(n-2)x^3}{3!} + \cdots \tag{67}$$

which is valid for $|x| < 1$, we may write

$$\left(1 - j\frac{\sigma}{\omega\epsilon}\right)^{1/2} = 1 - j\frac{\sigma}{2\omega\epsilon} + \frac{\sigma^2}{8\omega^2\epsilon^2} - \cdots \tag{68}$$

† See the footnote on the calculation of the square root of a complex number in Sec. 6.1.

and

$$\gamma = j\omega\sqrt{\mu\epsilon}\sqrt{1 - j\frac{\sigma}{\omega\epsilon}}$$

$$\simeq j\omega\sqrt{\mu\epsilon}\left(1 - j\frac{\sigma}{2\omega\epsilon} + \frac{\sigma^2}{8\omega^2\epsilon^2}\right) \tag{69}$$

From this result we identify

$$\beta \simeq \omega\sqrt{\mu\epsilon}\left(1 + \frac{\sigma^2}{8\omega^2\epsilon^2}\right) \tag{70a}$$

$$\alpha \simeq \frac{\sigma}{2}\sqrt{\frac{\mu}{\epsilon}} \tag{70b}$$

The velocity of propagation becomes

$$u = \frac{\omega}{\beta}$$

$$\simeq \frac{1}{\sqrt{\mu\epsilon}(1 + \sigma^2/8\omega^2\epsilon^2)} \tag{71}$$

Thus, the velocity of propagation of the wave is reduced slightly by the losses in the medium. The intrinsic impedance becomes†

$$\hat{\eta} = \sqrt{\frac{j\omega\mu}{\sigma + j\omega\epsilon}}$$

$$= \sqrt{\frac{\mu}{\epsilon}\frac{1}{1 - j(\sigma/\omega\epsilon)}}$$

$$= \sqrt{\frac{\mu}{\epsilon}}\left(1 - j\frac{\sigma}{\omega\epsilon}\right)^{-1/2}$$

$$\simeq \sqrt{\frac{\mu}{\epsilon}}\left(1 + j\frac{\sigma}{2\omega\epsilon} - \frac{3}{8}\frac{\sigma^2}{\omega^2\epsilon^2}\right) \tag{72}$$

† Note that (72) can also be written as

$$\hat{\eta} = \sqrt{\frac{\mu}{\epsilon}}\left(1 - j\frac{\sigma}{\omega\epsilon}\right)^{-1/2}$$

Using the binomial expansion once again, we may write

$$\left(1 - j\frac{\sigma}{\omega\epsilon}\right)^{-1/2} = 1 + j\frac{\sigma}{2\omega\epsilon} - \frac{3\sigma^2}{8\omega^2\epsilon^2}\cdots$$

These expressions may be realistically approximated for good dielectrics by

$$\left.\begin{aligned} \alpha &\simeq \frac{\sigma}{2}\sqrt{\frac{\mu}{\epsilon}} \\ \beta &\simeq \omega\sqrt{\mu\epsilon} \\ u &\simeq \frac{1}{\sqrt{\mu\epsilon}} \\ \eta &\simeq \sqrt{\frac{\mu}{\epsilon}} \end{aligned}\right\} \tag{73}$$

From these results we see that the phase constant, velocity of propagation, and intrinsic impedance are essentially unchanged from the lossless case.

6.3.2 Good Conductors ($\sigma \gg \omega\epsilon$)

For a good conductor ($\sigma/\omega\epsilon \gg 1$), we obtain

$$\begin{aligned} \gamma &= \sqrt{j\omega\mu(\sigma + j\omega\epsilon)} \\ &= \sqrt{j\omega\mu\sigma\left(1 + j\frac{\omega\epsilon}{\sigma}\right)} \\ &\simeq \sqrt{j\omega\mu\sigma} = \sqrt{\omega\mu\sigma}\ \underline{/45^\circ} \end{aligned} \tag{74}$$

so that

$$\alpha = \beta \simeq \sqrt{\frac{\omega\mu\sigma}{2}} \tag{75}$$

The velocity of propagation is

$$\begin{aligned} u &= \frac{\omega}{\beta} \\ &\simeq \sqrt{\frac{2\omega}{\mu\sigma}} \end{aligned} \tag{76}$$

and the intrinsic impedance becomes

$$\begin{aligned}\hat{\eta} &= \sqrt{\frac{j\omega\mu}{\sigma + j\omega\epsilon}} \\ &= \sqrt{\frac{j\omega\mu/\sigma}{1 + j\omega\epsilon/\sigma}} \\ &\simeq \sqrt{\frac{j\omega\mu}{\sigma}} = \sqrt{\frac{\omega\mu}{\sigma}}\,\underline{/45^\circ} \\ &= \sqrt{\frac{\omega\mu}{2\sigma}}(1 + j1) \end{aligned} \tag{77}$$

The reader should compare the results of these approximations to the results of Example 6.4.

EXAMPLE 6.5 A large copper conductor ($\sigma = 5.8 \times 10^7$ S/m, $\epsilon_r \simeq 1$, and $\mu_r \simeq 1$) supports a uniform plane wave at 60 Hz. Determine the ratio of conduction current to displacement current. Compute the attenuation constant, propagation constant, intrinsic impedance, wavelength, and phase velocity of propagation. Repeat for 10 GHz.

Solution At 60 Hz

$$\begin{aligned}\frac{\sigma}{\omega\epsilon} &= \frac{5.8 \times 10^7}{2\pi \times 60 \times 1/36\pi \times 10^{-9}} \\ &= 1.74 \times 10^{16}\end{aligned}$$

so at this frequency, copper is an unqualified good conductor. The other quantities to be calculated become

$$\begin{aligned}\alpha = \beta &\simeq \sqrt{\frac{\omega\mu\sigma}{2}} \\ &= 117.21\end{aligned}$$

$$\begin{aligned}\hat{\eta} &\simeq \sqrt{\frac{\omega\mu}{\sigma}}\,\underline{/45^\circ} \\ &= 2.86 \times 10^{-6}\underline{/45^\circ}\ \Omega\end{aligned}$$

$$\begin{aligned}\lambda &= \frac{2\pi}{\beta} \\ &\simeq 5.36 \times 10^{-2}\ \text{m}\end{aligned}$$

$$\begin{aligned}u &\simeq \sqrt{\frac{2\omega}{\mu\sigma}} \\ &= 3.216\ \text{m/s}\end{aligned}$$

At 10 GHz

$$\frac{\sigma}{\omega\epsilon} = \frac{5.8 \times 10^7}{2\pi \times 10^{10} \times 1/36\pi \times 10^{-9}}$$
$$= 1.044 \times 10^8$$

so that copper remains a very good conductor. At 10 GHz we compute

$$\alpha = \beta \simeq 1.513 \times 10^6$$

$$\hat{\eta} \simeq 3.69 \times 10^{-2} \underline{/45^\circ}\ \Omega$$

$$\lambda \simeq 4.153 \times 10^{-6} \text{ m}$$

$$u \simeq 4.153 \times 10^4 \text{ m/s}$$

It should also be pointed out that materials generally classified as dielectrics in terms of the absence (or small amount) of free charge also exhibit an additional type of loss mechanism. This loss mechanism, although different from that occurring in materials classified as conductors because they have relatively large amounts of free charge, may be handled in a fashion similar to conductive materials.

As was pointed out in Chap. 3, in the presence of a sinusoidally varying electric field, the permanent dipoles of a dielectric material tend to rotate to align with the changing direction of the field. In materials that do not possess permanent dipoles, there may also occur "induced dipoles" created by a net shift of the electron cloud with respect to the positive nucleus in the atoms or by a shift of positive and negative ions of the material when an electric field is applied. As the frequency of the applied field is increased, there is a tendency for the formation or alignment of these dipoles to lag behind the changes in direction of the sinusoidal field. For either permanent dipoles or induced dipoles, the polarization of the material is therefore not in time phase with the applied electric field. As the frequency is increased this effect becomes more pronounced. This loss phenomenon is also accounted for by ascribing a complex permittivity to the substance:†

$$\hat{\epsilon} = \epsilon' - j\epsilon'' \tag{78}$$

If this material has zero conductivity, Ampère's law in the material becomes

$$\begin{aligned} \nabla \times \hat{\mathbf{H}} &= j\omega\hat{\epsilon}\hat{\mathbf{E}} \\ &= \omega\epsilon''\hat{\mathbf{E}} + j\omega\epsilon'\hat{\mathbf{E}} \end{aligned} \tag{79}$$

† A corresponding characterization of the frequency-dependent behavior of magnetic materials in which $\mu \neq \mu_0$ can be obtained via a complex permeability as $\hat{\mu} = \mu' - j\mu''$.

Thus, these loss properties of the material may be treated in exactly the same fashion as conductive loss by replacing σ in (58) with $\omega\epsilon''$ and ϵ in (58) with ϵ'. The loss tangent becomes

$$\tan\phi = \frac{\epsilon''}{\epsilon'} \tag{80}$$

Although this phenomenon generally occurs at very high frequencies (in the gigahertz range), we can include the effect in the same fashion as for conductive materials. In fact, even if a dielectric material has a nonzero conductivity (realistically quite small, however), we may write Ampère's law as

$$\begin{aligned}\nabla \times \hat{\mathbf{H}} &= (\sigma + j\omega\hat{\epsilon})\hat{\mathbf{E}} \\ &= (\sigma + \omega\epsilon'')\hat{\mathbf{E}} + j\omega\epsilon'\hat{\mathbf{E}}\end{aligned} \tag{81}$$

and use an effective conductivity of $\sigma + \omega\epsilon''$ and effective permittivity of ϵ' in our previous results to handle this situation.

In future discussions we will therefore consider σ to be the effective conductivity of the material, $\sigma + \omega\epsilon''$, and ϵ to be the effective permittivity of the material, ϵ'. It is important, however, to note at this point that the effective conductivity $\sigma + \omega\epsilon''$ and the effective permittivity ϵ' will, in general, vary with frequency. For example, polyvinyl chloride has $\epsilon_r' = 4.3$ at 10 kHz but decreases to $\epsilon_r' = 3$ at 100 MHz where $\epsilon' = \epsilon_r'\epsilon_0$. On the other hand, polystyrene has $\epsilon_r' \simeq 2.7$ from dc up to around 10^{10} Hz (10 GHz). Polystyrene, however, has a loss tangent

$$\tan\phi = \frac{\sigma + \omega\epsilon''}{\omega\epsilon'} \tag{82}$$

of 0.7×10^{-4} at 1 MHz and 4.3×10^{-4} at 10 GHz. We will therefore assume that the values of (effective) conductivity σ and (effective) permittivity ϵ are those for the frequency being considered and not simply the dc values.

6.4 Skin Depth

Consider a forward-traveling wave in a lossy material having material constants, ϵ, μ, and σ. The electric field may be written as

$$\mathcal{E} = E_m e^{-\alpha z}\cos(\omega t - \beta z)\mathbf{a}_x \tag{83}$$

where E_m is the magnitude of the field. As the wave travels through the material, its amplitude will be attenuated by the factor $e^{-\alpha z}$. (See Fig. 6.4*a*.) Over a distance of

$$\delta = \frac{1}{\alpha} \tag{84}$$

the magnitude of the wave will have been reduced by $1/e$, or 37 percent. The quantity δ is termed the *skin depth*, or *depth of penetration*, of the material. Substituting the relation for α in a good conductor given in (75), we obtain

$$\delta = \sqrt{\frac{2}{\omega\mu\sigma}} \qquad \text{good conductor}$$

$$= \frac{1}{\sqrt{\pi f \mu\sigma}} \tag{85}$$

For example, the values of skin depth for copper are

Frequency	*Skin depth* δ
60 Hz	8.53 mm
100 MHz	0.0066 mm
10 GHz	6.61×10^{-7} m

For good conductors, the skin depth becomes extremely small as the frequency is increased. Effective eletromagnetic shielding of electronic devices, as well as rooms, from external fields that may cause interference can be obtained with conductive enclosures having wall thicknesses greater than several skin depths. An incident wave that enters the wall will be rapidly attenuated before exiting it, thus reducing the wave's interfering effects.

For good conductors, the wave equations in (9) simplify to

$$\nabla^2\hat{\mathbf{E}} = j\omega\mu\sigma\hat{\mathbf{E}} \qquad \sigma \gg \omega\epsilon$$

$$= j\frac{2}{\delta^2}\hat{\mathbf{E}} \tag{86a}$$

$$\nabla^2\hat{\mathbf{H}} = j\omega\mu\sigma\hat{\mathbf{H}} \qquad \sigma \gg \omega\epsilon$$

$$= j\frac{2}{\delta^2}\hat{\mathbf{H}} \tag{86b}$$

The current density $\hat{\mathbf{J}}$ and the electric field are related by conductivity, so that

$$\nabla^2\hat{\mathbf{J}} = j\omega\mu\sigma\hat{\mathbf{J}} \qquad \sigma \gg \omega\epsilon$$

$$= j\frac{2}{\delta^2}\hat{\mathbf{J}} \tag{86c}$$

These equations are referred to as the *diffusion equations*. In good conductors, such as metals, the fields undergo significant attenuation. We may thus think of these fields as "diffusing" into the material.

6.5 Polarization of Uniform Plane Waves

Consider the electric field vector of a uniform plane wave that is traveling in a lossless medium:

$$\mathcal{E} = E_m \cos(\omega t - \beta z)\mathbf{a}_x \tag{87a}$$

The magnetic field vector is given by

$$\mathcal{H} = \frac{E_m}{\eta} \cos(\omega t - \beta z)\mathbf{a}_y \tag{87b}$$

An instantaneous plot of these two vectors as a function of z in the medium is shown in Fig. 6.6*a*. Note that as time increases, the envelope (shown as a dashed line) moves to the right with velocity $u = 1/\sqrt{\mu\epsilon}$.

On the other hand, let us observe the variation of the wave at some fixed position, $z = 0$, as time varies. At $z = 0$ we obtain from (87)

$$\mathcal{E}\Big|_{z=0} = E_m \cos \omega t \; \mathbf{a}_x \tag{88a}$$

$$\mathcal{H}\Big|_{z=0} = \frac{E_m}{\eta} \cos \omega t \; \mathbf{a}_y \tag{88b}$$

If we take a cross-sectional view of the wave in the xy plane at $z = 0$, we observe the magnitude of the field vectors varying sinusoidally according to (88), as shown in Fig. 6.6*b*. Note that the field vectors reverse direction with each half-cycle but that this variation is always along a line (the $\pm x$ direction for $\mathcal{E}$ and the $\pm y$ direction for $\mathcal{H}$). This wave is said to be linearly polarized in the sense that the direction of each of the field vectors at some point in space and every point in time lies along a line in a plane perpendicular to the direction of propagation. Thus, the uniform plane waves that we have considered are said to be *linearly polarized*.

Now consider another uniform plane wave formed by the rotation of the field vectors in (87) 90° in space and the addition of a phase angle θ to the arguments of the cosine terms:

$$\mathcal{E} = E_m \cos(\omega t - \beta z + \theta)\mathbf{a}_y \tag{89a}$$

$$\mathcal{H} = -\frac{E_m}{\eta} \cos(\omega t - \beta z + \theta)\mathbf{a}_x \tag{89b}$$

Note the minus sign in (89*b*), which is necessary to ensure that the power flow ($\mathcal{E} \times \mathcal{H}$) is in the proper ($+z$) direction. We can readily verify that the field vectors in (89) also satisfy Maxwell's equations and are also linearly polarized.

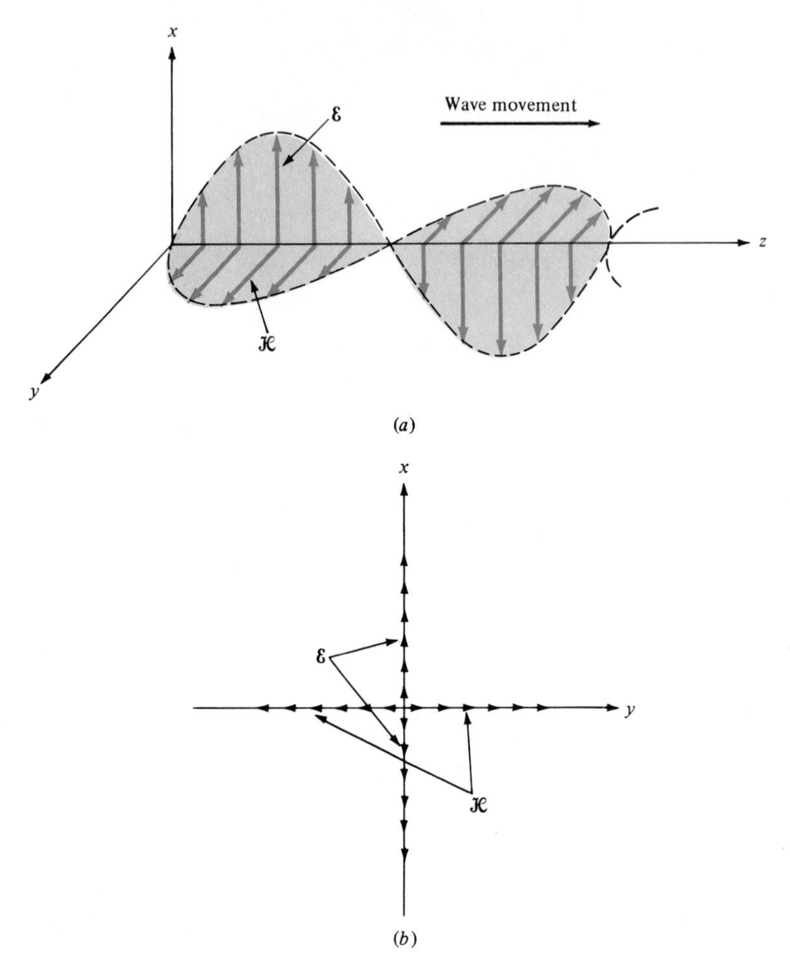

FIGURE 6.6
The linearly polarized uniform plane wave *(a)* as a function of distance at $t = 0$, and *(b)* as a function of time at $z = 0$.

In addition, since Maxwell's equations are linear (for linear media), the sum of the field vectors in (87) and (89) are also valid solutions. Adding these solutions, we obtain

$$\mathcal{E} = E_{m1} \cos(\omega t - \beta z)\mathbf{a}_x + E_{m2} \cos(\omega t - \beta z + \theta)\mathbf{a}_y \tag{90a}$$

$$\mathcal{H} = \frac{E_{m1}}{\eta} \cos(\omega t - \beta z)\mathbf{a}_y - \frac{E_{m2}}{\eta} \cos(\omega t - \beta z + \theta)\mathbf{a}_x \tag{90b}$$

where we allow for the possibility of having different magnitudes of the components of the composite electric field through E_{m1} and E_{m2}.

Let us now consider, in more detail, the combined electric field in (90a) at a fixed point in space, $z = 0$, as time progresses. Substituting $z = 0$ into (90a), we obtain

$$\mathcal{E}\Big|_{z=0} = E_{m1} \cos \omega t \, \mathbf{a}_x + E_{m2} \cos (\omega t + \theta)\mathbf{a}_y \tag{91}$$

Several cases in terms of the values of the magnitudes E_{m1} and E_{m2} and phase angle θ will now be considered.

Case 1 $E_{m1} = E_{m2}$, $\theta = 0$. For this case, (91) reduces to

$$\mathcal{E}\Big|_{z=0} = E_{m1} \cos \omega t \, (\mathbf{a}_x + \mathbf{a}_y) \tag{92}$$

which is plotted in Fig. 6.7a. The composite wave is linearly polarized.

Case 2 $E_{m1} \neq E_{m2}$, $\theta = 0$. For this case, (91) reduces to

$$\mathcal{E}\Big|_{z=0} = E_{m1} \cos \omega t \left(\mathbf{a}_x + \frac{E_{m2}}{E_{m1}} \mathbf{a}_y \right) \tag{93}$$

which is also a linearly polarized wave, as shown in Fig. 6.7b.

Case 3 $E_{m1} = E_{m2}$, $\theta = -90°$. Equation (91) reduces to

$$\begin{aligned} \mathcal{E}\Big|_{z=0} &= E_m[\cos \omega t \, \mathbf{a}_x + \cos (\omega t - 90°)\mathbf{a}_y] \\ &= E_m(\cos \omega t \, \mathbf{a}_x + \sin \omega t \, \mathbf{a}_y) \end{aligned} \tag{94}$$

The length of this vector is E_m at all times, and the angle between the vector and the x axis is

$$\begin{aligned} \phi &= \tan^{-1} \left[\frac{\sin \omega t}{\cos \omega t} \right] \\ &= \omega t \end{aligned} \tag{95}$$

Thus, the tip of the resultant vector traces out a circle. This is referred to as *circular polarization.* The vector rotates in the clockwise direction and is referred to as a *right-hand circularly polarized* wave, since if one places the fingers of the right hand in the direction of rotation, the thumb will point in the direction of propagation. If we had chosen $\theta = +90°$, the tip of the resultant vector would again trace a circle but the vector would rotate in the counterclockwise direction; this would be referred to as *left-hand circular polarization.*

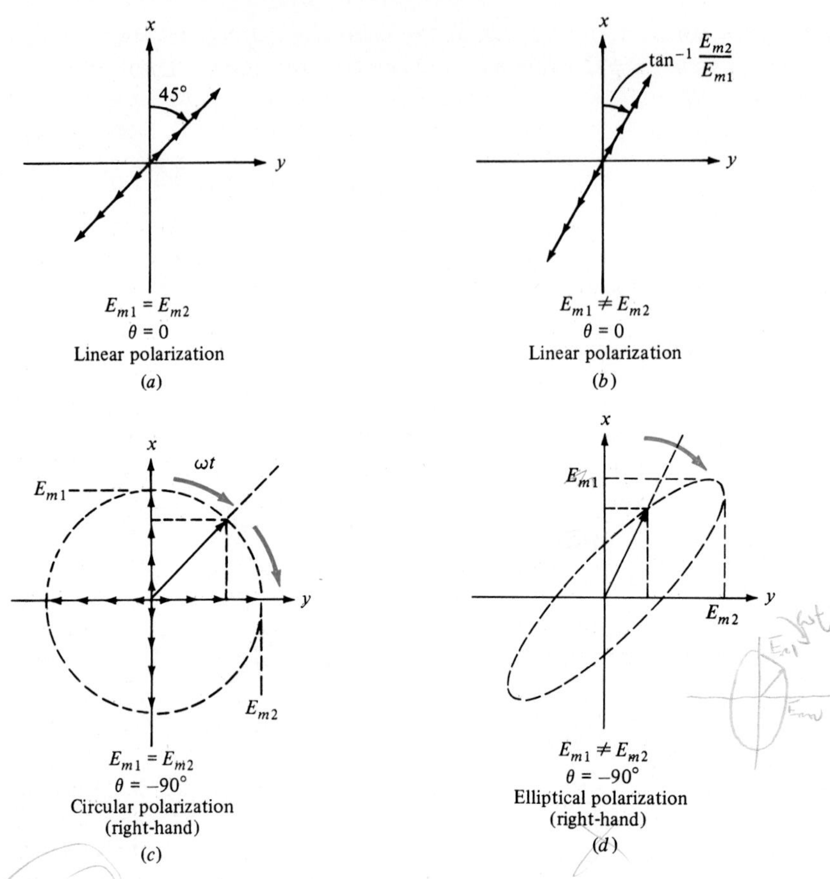

FIGURE 6.7
Illustration of polarization of uniform plane waves viewed at $z = 0$ in the direction of propagation as a function of time.

Case 4 $E_{m1} \neq E_{m2}$, $\theta = -90°$. For this case,

$$\boldsymbol{\mathcal{E}}\bigg|_{z=0} = E_{m1}\left[\cos \omega t\, \mathbf{a}_x + \frac{E_{m2}}{E_{m1}} \sin \omega t\, \mathbf{a}_y\right] \tag{96}$$

At any particular time we find that (96) yields the equation of an ellipse:

$$\left(\frac{\mathcal{E}_x}{E_{m1}}\right)^2 + \left(\frac{\mathcal{E}_y}{E_{m2}}\right)^2 = 1 \tag{97}$$

The tip of the resultant vector thus traces out an ellipse and rotates in the clockwise direction. The wave is said to have *right-hand elliptical polarization.* If $E_{m1} = E_{m2}$, (97) is the equation of a circle.

It is important to note from the above development that a uniform plane wave having some general (elliptical) polarization can be viewed (or analyzed) as the superposition of two orthogonal, linearly polarized waves, as shown by (90). Therefore, any analysis of a uniform plane wave having some general polarization can be accomplished by the superposition of the analyses of each linearly polarized (and orthogonal) component since Maxwell's equations are linear (for linear media). We will use this concept on numerous occasions in order to simplify our analysis of complex wave propagation. Thus, the analysis of linearly polarized waves is important and fundamental.

6.6 Group Velocity

Thus far we have been considering waves in which the field vectors vary sinusoidally at only one frequency. Electromagnetic waves are often used to transmit information (speech, digital data, etc.) in a signal referred to as a baseband signal. We will find in later chapters that in order to propagate waves efficiently, such as with waveguides (Chap. 8) or antennas (Chap. 9), the frequencies of the wave must be quite large since the physical dimensions of these propagation structures must be on the order of a wavelength. Baseband signals consisting of voice or music extend from dc to around 15 kHz, whereas television or digital data range from dc to several megahertz. Thus, large propagation structures would be required in order to transmit these baseband signals directly.

To circumvent this problem, the low-frequency baseband signal is translated about some higher carrier frequency, thus permitting more reasonable sizes of waveguides and antennas. A common method of doing this is with amplitude modulation (AM), as shown in Fig. 6.8. A baseband signal $m(t)$ varies the amplitude of a carrier signal whose frequency is f_c as

$$s(t) = \mathrm{A}\,\mathrm{m}(t) \cos(2\pi f_c t) \tag{98}$$

In the frequency domain, the baseband signal consists of a band of frequencies $M(f)$ which is translated via amplitude modulation about the higher carrier frequency as $S(f) = M(f - f_c)$.

We may consider this composite signal as being composed of a multitude of discrete-frequency components. Transmission of this signal through a *linear* medium can be viewed, by superposition, as the transmission of the individual frequency components. If the phase velocity for the transmission medium is independent of frequency, then all component waves travel at the same velocity. Such is the case for uniform plane waves in lossless media. Thus, the component waves arrive at the receiver with the same relative phase as in the transmitted wave. Recombining these received components then yields a time-domain received signal of the same form as the transmitted signal in (98). In this case, we say that no distortion has occurred. The medium is said to have no dispersion since all frequency components travel at the same velocity.

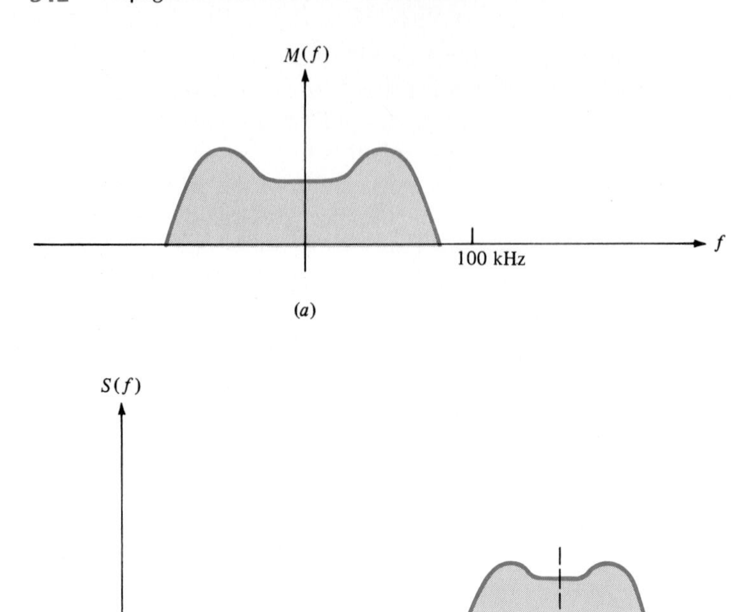

FIGURE 6.8
Frequency translation of information for efficient transmission.

On the other hand, suppose that the phase velocity of the trasmission medium is a function of frequency: $u_p(\omega)$. In this case, the individual frequency components travel at different phase velocities, so the individual frequency components in the received signal do not maintain the same relative phase as in the transmitted signal. Consequently, the time-domain received signal will be a distorted version of the transmitted signal. The wave is said to have suffered *dispersion*, and the transmission medium is said to be *dispersive*.

The phase velocity is given by the fundamental relation

$$u_p = \frac{\omega}{\beta} \tag{99}$$

where β is the phase constant. In free space, the phase constant of uniform plane waves is $\beta = \omega\sqrt{\mu_0\epsilon_0}$; thus, $u_p = 1/\sqrt{\mu_0\epsilon_0}$ and is independent of frequency. On the other hand, in a lossy medium, the phase constant for uniform plane waves is the imaginary part of the propagation constant:

$$\begin{aligned} \beta &= \text{Im}\,(\gamma) \\ &= \text{Im}\,(\sqrt{j\omega\mu(\sigma + j\omega\epsilon)}) \end{aligned} \tag{100}$$

Consequently, in a lossy medium, uniform plane waves at different frequencies travel at different phase velocities and the medium is said to be dispersive. There are transmission structures, such as waveguides (Chap. 8), that exhibit dispersion even though the transmission medium is lossless.

Because of the fundamental equation for phase velocity given in (99), it is important to investigate the relationship between ω and β. This is usually displayed as the so-called ω-β diagram, where β is plotted as a function of ω. Examples are shown in Fig. 6.9 for free space, a lossy medium, and a waveguide.

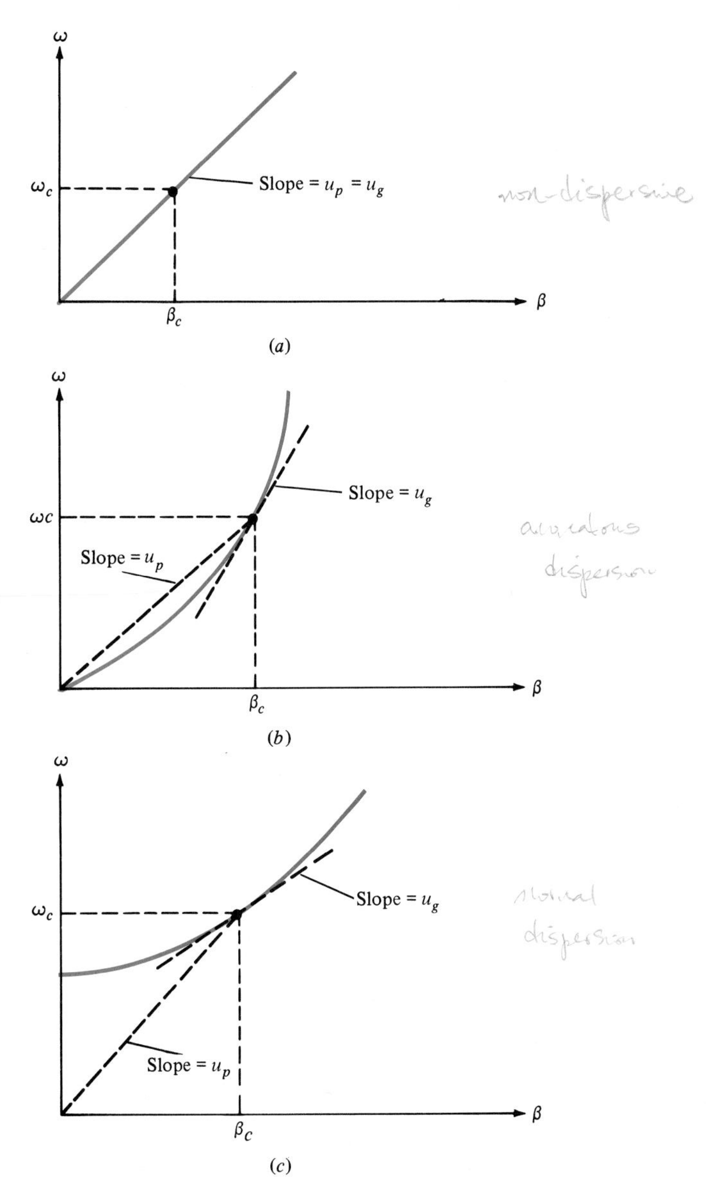

FIGURE 6.9
Illustration of ω-β diagrams showing phase and group velocities.

According to (99), the phase velocity of a certain frequency component is the slope of a line drawn from the origin to the appropriate point on the diagram. For signals containing a band of frequency components, such as in the AM signal above, an additional important concept is that of *group velocity*, which is the slope of the curve at the carrier frequency:

$$u_g = \left.\frac{d\omega}{d\beta}\right|_{\omega=\omega_c} \tag{101}$$

To see how this concept of group velocity applies to information-bearing signals, consider an amplitude-modulated signal in which the modulation is a single-frequency sinusoid at frequency f_m. The AM signal becomes

$$\begin{aligned} s(t) &= A \cos(\omega_m t) \cos(\omega_c t) \\ &= \frac{A}{2} \cos(\omega_c - \omega_m)t + \frac{A}{2} \cos(\omega_c + \omega_m)t \end{aligned} \tag{102}$$

Thus, the composite signal consists of two frequency components at $\omega_c - \omega_m$ and $\omega_c + \omega_m$. If this signal is transmitted as a linearly polarized, uniform plane wave, we would write (based on our previous results)

$$\begin{aligned} \mathcal{E}(z, t) &= E_m \cos[(\omega_c - \omega_m)t - \beta_- z] \\ &\quad + E_m \cos[(\omega_c + \omega_m)t - \beta_+ z] \end{aligned} \tag{103}$$

where β_- and β_+ are the phase constants at frequencies $\omega_c - \omega_m$ and $\omega_c + \omega_m$, respectively, as shown in Fig. 6.10*a*. Now suppose that the modulation frequency is much smaller than the carrier frequency: $\omega_m \ll \omega_c$. This is typical of the components of baseband signals. In this case, $\beta_+ - \beta_c$ and $\beta_c - \beta_-$ are approximately equal, and thus we may rewrite (103) approximately as

$$\begin{aligned} \mathcal{E}(z, t) &= E_m \cos[(\omega_c - \omega_m)t - (\beta_c - \Delta\beta)z] \\ &\quad + E_m \cos[(\omega_c + \omega_m)t - (\beta_c + \Delta\beta)z] \\ &= 2E_m \cos(\Delta\omega t - \Delta\beta z) \cos(\omega_c t - \beta_c z) \end{aligned} \tag{104}$$

where

$$\Delta\beta = \beta_+ - \beta_c \simeq \beta_c - \beta_- \quad \text{and} \quad \Delta\omega = (\omega_c + \omega_m) - \omega_c = \omega_c - (\omega_c - \omega_m) = \omega_m.$$

This may be rewritten as

$$\mathcal{E}(z, t) = 2E_m \cos \omega_m \left(t - \frac{z}{u_g}\right) \cos \omega_c \left(t - \frac{z}{u_c}\right) \tag{105a}$$

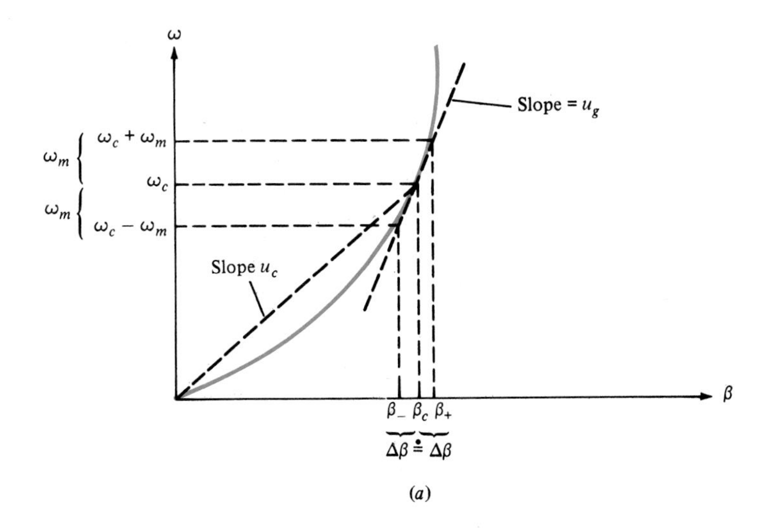

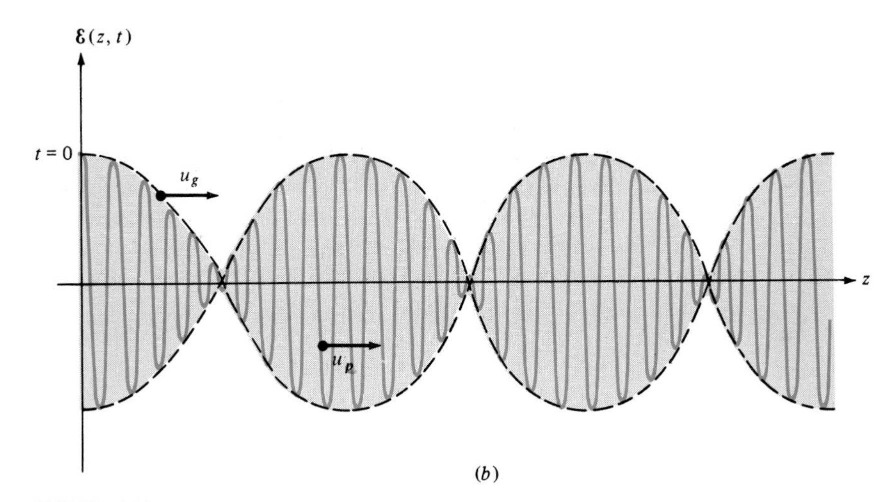

FIGURE 6.10

Group velocity of a wave packet composed of the sum of two waves at slightly different frequencies.

where
$$u_g = \frac{\Delta\omega}{\Delta\beta} \tag{105b}$$

and
$$u_c = \frac{\omega_c}{\beta_c} \tag{105c}$$

Equation (105*a*) is plotted for $t = 0$ in Fig. 6.10*b*. This result shows that the composite wave consists of the carrier traveling at a velocity u_c and a modulation envelope traveling at the "group velocity" u_g.

This result can be seen more readily by viewing (104) as a linear combination of two waves: one traveling at a phase velocity of $(\omega_c - \omega_m)/(\beta_c - \Delta\beta)$ and the other traveling at a phase velocity of $(\omega_c + \omega_m)/(\beta_c + \Delta\beta)$. The two waves, traveling at slightly different velocities, slide past each other, producing constructive interference at certain points (the peaks in the modulation envelope) and destructive interference at other points (the nulls in the modulation envelope).

For other narrowband modulation signals, group velocity is similarly defined as

$$\begin{aligned} u_g &= \lim_{\Delta\omega \to 0} \frac{\Delta\omega}{\Delta\beta} \\ &= \frac{d\omega}{d\beta} \\ &= \frac{1}{d\beta/d\omega} \end{aligned} \tag{106}$$

The latter form in (106) is usually preferred since the phase constant is usually written as an explicit function of ω.

As noted previously, transmission media in which the phase constant is independent of frequency are said to be nondispersive. An example is shown in Fig. 6.9*a*. In this case, the ω-β diagram is a straight line and the phase and group velocities are equal. The medium is said to exhibit anomalous dispersion if $u_g > u_p$, as shown in Fig. 6.9*b*. Normal dispersion is said to exist if $u_g < u_p$, as shown in Fig. 6.9*c*. It is important to note that this concept of group velocity is only valid for narrowband signals, so that the slope of the ω-β diagram for all frequency components of the signal can be regarded as being approximately the same.

EXAMPLE 6.6 An air-filled rectangular waveguide (Chap. 8) has the field vectors given in Example 5.13 of Chap. 5. Each of these phasor field vectors is dependent on the z coordinate (the guide axis) through the multiplicative factor $e^{-j\beta z}$. This represents wave propagation in the z direction, as we can see by multiplying by $e^{j\omega t}$ and taking the real part of the result, giving the time-domain field vectors in Example 5.6. It was found that in order to satisfy Maxwell's equations, β must be given by

$$\beta = \sqrt{\omega^2 \mu_0 \epsilon_0 - \left(\frac{\pi}{a}\right)^2}$$

where a is one of the side dimensions of the guide. (See Fig. 5.11.) If this guide is carrying a narrowband information signal centered around a carrier frequency of 7 GHz and the guide dimensions are $a = 0.9$ in (2.29 cm) and $b = 0.4$ in (1.02 cm), determine the phase and group velocities of this signal. Compare these to the phase velocity of a uniform plane wave in this (free-space) medium.

Solution At 7 GHz we obtain

$$\beta = 51.7 \text{ rad/m}$$

The phase velocity is

$$u_p = \frac{\omega}{\beta}$$

$$= 8.51 \times 10^8 \text{ m/s}$$

which is considerably greater than $u_0 = 3 \times 10^8$ m/s. To compute the group velocity, we write

$$\beta = \left[\omega^2 \mu_0 \epsilon_0 - \left(\frac{\pi}{a}\right)^2\right]^{1/2}$$

so that

$$\frac{d\beta}{d\omega} = \frac{1}{2}\left[\omega^2 \mu_0 \epsilon_0 - \left(\frac{\pi}{a}\right)^2\right]^{-1/2} (2\omega\mu_0\epsilon_0)$$

Thus the group velocity is

$$u_g = \frac{1}{d\beta/d\omega}$$

$$= \frac{u_0^2}{u_p}$$

$$= 1.058 \times 10^8 \text{ m/s}$$

6.7 Normal Incidence of Uniform Plane Waves on Plane Boundaries

We now consider uniform plane waves that are incident on a boundary between two media. Consider Fig. 6.11, which shows a plane boundary between two different media having material properties ϵ_1, μ_1, and σ_1 for medium 1 and ϵ_2, μ_2, and σ_2 for medium 2. A (forward-traveling) uniform plane wave traveling to

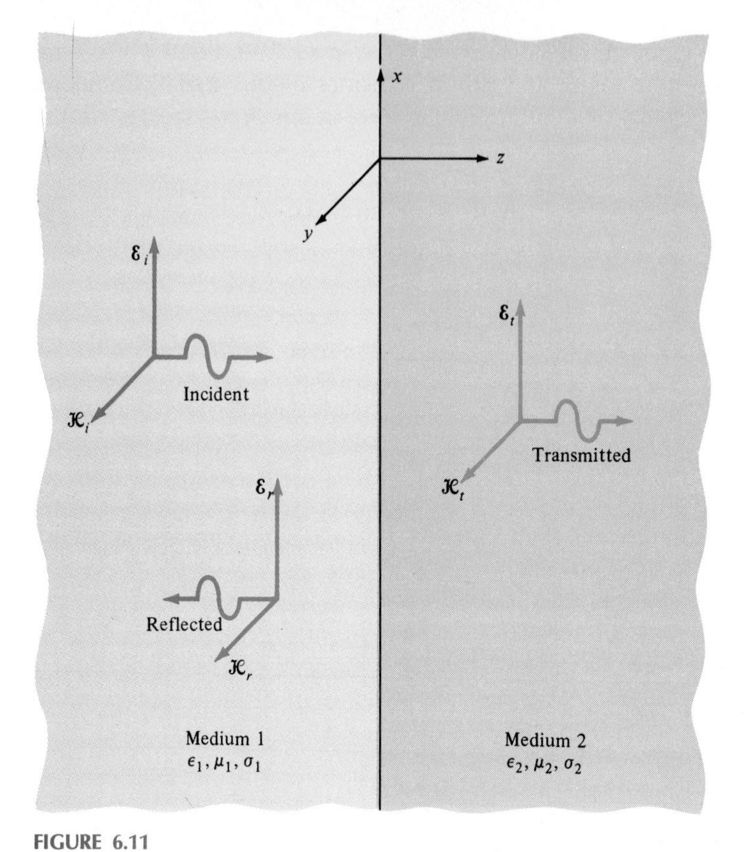

FIGURE 6.11

Normal incidence of a uniform plane wave on a plane boundary showing incident, reflected, and transmitted waves.

the right in medium 1 is incident on the interface normal to the boundary. We denote this incident field in phasor form as

$$\begin{aligned}\hat{\mathbf{E}}_i &= \hat{E}_i e^{-\gamma_1 z}\mathbf{a}_x \\ &= \hat{E}_i e^{-\alpha_1 z} e^{-j\beta_1 z}\mathbf{a}_x \end{aligned} \tag{107a}$$

$$\begin{aligned}\hat{\mathbf{H}}_i &= \frac{\hat{E}_i}{\hat{\eta}_1} e^{-\gamma_1 z}\mathbf{a}_y \\ &= \frac{\hat{E}_i}{\eta_1} e^{-\alpha_1 z} e^{-j\beta_1 z} e^{-j\theta_{\eta 1}}\mathbf{a}_y \end{aligned} \tag{107b}$$

where

$$\begin{aligned}\gamma_1 &= \alpha_1 + j\beta_1 \\ &= \sqrt{j\omega\mu_1(\sigma_1 + j\omega\epsilon_1)} \end{aligned} \tag{108}$$

and
$$\hat{\eta}_1 = \eta_1 \underline{/\theta_{\eta 1}}$$
$$= \sqrt{\frac{j\omega\mu_1}{\sigma_1 + j\omega\epsilon_1}} \tag{109}$$

The boundary gives rise to a reflected wave, which we represent as

$$\hat{\mathbf{E}}_r = \hat{E}_r e^{\gamma_1 z} \mathbf{a}_x$$
$$= \hat{E}_r e^{\alpha_1 z} e^{j\beta_1 z} \mathbf{a}_x \tag{110a}$$

$$\hat{\mathbf{H}}_r = -\frac{\hat{E}_r}{\hat{\eta}_1} e^{\gamma_1 z} \mathbf{a}_y$$
$$= -\frac{\hat{E}_r}{\eta_1} e^{\alpha_1 z} e^{j\beta_1 z} e^{-j\theta_{\eta 1}} \mathbf{a}_y \tag{110b}$$

in accordance with the general solution of the wave equation in this material. A portion of the incident wave will be transmitted into the second medium, which we represent as

$$\hat{\mathbf{E}}_t = \hat{E}_t e^{-\gamma_2 z} \mathbf{a}_x$$
$$= \hat{E}_t e^{-\alpha_2 z} e^{-j\beta_2 z} \mathbf{a}_x \tag{111a}$$

$$\hat{\mathbf{H}}_t = \frac{\hat{E}_t}{\hat{\eta}_2} e^{-\gamma_2 z} \mathbf{a}_y$$
$$= \frac{\hat{E}_t}{\eta_2} e^{-\alpha_2 z} e^{-j\beta_2 z} e^{-j\theta_{\eta 2}} \mathbf{a}_y \tag{111b}$$

where
$$\gamma_2 = \alpha_2 + j\beta_2$$
$$= \sqrt{j\omega\mu_2(\sigma_2 + j\omega\epsilon_2)} \tag{112}$$

and
$$\hat{\eta}_2 = \eta_2 \underline{/\theta_{\eta 2}}$$
$$= \sqrt{\frac{j\omega\mu_2}{\sigma_2 + j\omega\epsilon_2}} \tag{113}$$

We reason that there will be no backward-traveling component of the transmitted wave, since we assume the second medium to be infinite in extent, and thus that there will be no reflections at its rightmost surface. Nevertheless, the general solutions in (107), (110), and (111) satisfy Maxwell's equations in the respective regions. If the unknowns in these equations, $\hat{E}_i$, $\hat{E}_r$, and $\hat{E}_t$, can be found such that the boundary conditions at the boundary $z = 0$ are satisfied, we will have determined a valid solution.

At the boundary $z = 0$, the boundary conditions require that the tangential components of the total electric and magnetic fields be continuous. Since $\hat{\mathbf{E}}_i$, $\hat{\mathbf{E}}_r$, and $\hat{\mathbf{E}}_t$ are defined to be in the x direction,

$$\hat{\mathbf{E}}_i + \hat{\mathbf{E}}_r = \hat{\mathbf{E}}_t \qquad \text{at } z = 0 \tag{114a}$$

and similarly for $\hat{\mathbf{H}}_i$, $\hat{\mathbf{H}}_r$, and $\hat{\mathbf{H}}_t$, which are defined to be in the y direction:

$$\hat{\mathbf{H}}_i + \hat{\mathbf{H}}_r = \hat{\mathbf{H}}_t \qquad \text{at } z = 0 \tag{114b}$$

Substituting the forms of the field vectors for $z = 0$ into (114), we obtain

$$\begin{aligned} \frac{\hat{E}_r}{\hat{E}_i} &= \frac{\hat{\eta}_2 - \hat{\eta}_1}{\hat{\eta}_2 + \hat{\eta}_1} \\ &= \hat{\Gamma} \end{aligned} \tag{115}$$

and

$$\begin{aligned} \frac{\hat{E}_t}{\hat{E}_i} &= \frac{2\hat{\eta}_2}{\hat{\eta}_2 + \hat{\eta}_1} \\ &= \hat{T} \end{aligned} \tag{116}$$

We can also prove an important result:

$$1 + \hat{\Gamma} = \hat{T} \tag{117}$$

by direct substitution of (115) and (116). The quantities $\hat{\Gamma}$ and $\hat{T}$ are the *reflection* and *transmission coefficients*, respectively, of the boundary. It is a simple matter to show that $|\hat{\Gamma}| \leq 1$. The magnitude of $\hat{T}$ may exceed unity. Note that $\hat{\Gamma}$ and $\hat{T}$ will be real only if both regions are lossless, i.e., $\sigma_1 = \sigma_2 = 0$; otherwise, $\hat{\Gamma}$ and $\hat{T}$ will in general be complex, which we write as

$$\hat{\Gamma} = \Gamma \angle \theta_\Gamma \tag{118a}$$

$$\hat{T} = T \angle \theta_T \tag{118b}$$

We now assume that the form of the incident wave is $\hat{\mathbf{E}}_i = E_m e^{-\gamma_1 z}\mathbf{a}_x$, where the magnitude of this incident wave is denoted by E_m. Thus, the phasor forms of the field vectors become, in terms of the incident field magnitude, using (115) and (116),

$$\hat{\mathbf{E}}_i = E_m e^{-\gamma_1 z}\mathbf{a}_x \tag{119a}$$

$$\hat{\mathbf{H}}_i = \frac{E_m}{\hat{\eta}_1} e^{-\gamma_1 z}\mathbf{a}_y \tag{119b}$$

$$\hat{\mathbf{E}}_r = \hat{\Gamma} E_m e^{\gamma_1 z}\mathbf{a}_x \tag{119c}$$

$$\hat{\mathbf{H}}_r = -\frac{\hat{\Gamma} E_m}{\hat{\eta}_1} e^{\gamma_1 z}\mathbf{a}_y \tag{119d}$$

$$\hat{\mathbf{E}}_t = \hat{T} E_m e^{-\gamma_2 z}\mathbf{a}_x \tag{119e}$$

$$\hat{\mathbf{H}}_t = \frac{\hat{T} E_m}{\hat{\eta}_2} e^{-\gamma_2 z}\mathbf{a}_y \tag{119f}$$

Multiplying (119) by $e^{j\omega t}$ and taking the real part of the result, we obtain the time-domain forms of the field vectors:

$$\mathcal{E}_i = E_m e^{-\alpha_1 z} \cos(\omega t - \beta_1 z)\mathbf{a}_x \tag{120a}$$

$$\mathcal{H}_i = \frac{E_m}{\eta_1} e^{-\alpha_1 z} \cos(\omega t - \beta_1 z - \theta_{\eta 1})\mathbf{a}_y \tag{120b}$$

$$\mathcal{E}_r = \Gamma E_m e^{\alpha_1 z} \cos(\omega t + \beta_1 z + \theta_\Gamma)\mathbf{a}_x \tag{120c}$$

$$\mathcal{H}_r = -\frac{\Gamma E_m}{\eta_1} e^{\alpha_1 z} \cos(\omega t + \beta_1 z + \theta_\Gamma - \theta_{\eta 1})\mathbf{a}_y \tag{120d}$$

$$\mathcal{E}_t = T E_m e^{-\alpha_2 z} \cos(\omega t - \beta_2 z + \theta_T)\mathbf{a}_x \tag{120e}$$

$$\mathcal{H}_t = \frac{T E_m}{\eta_2} e^{-\alpha_2 z} \cos(\omega t - \beta_2 z + \theta_T - \theta_{\eta 2})\mathbf{a}_y \tag{120f}$$

The power density vector of the transmitted wave is

$$\begin{aligned}\mathbf{S}_{\mathrm{av},t} &= \tfrac{1}{2}\,\mathrm{Re}\,(\hat{\mathbf{E}}_t \times \hat{\mathbf{H}}_t^*) \\ &= \tfrac{1}{2}\,\mathrm{Re}\left(\hat{T} E_m e^{-\gamma_2 z}\,\frac{\hat{T}^* E_m e^{-\gamma_2^* z}}{\hat{\eta}_2^*}\right)\mathbf{a}_z \\ &= \frac{1}{2}\frac{E_m^2 T^2}{\eta_2} e^{-2\alpha_2 z} \cos\theta_{\eta 2}\,\mathbf{a}_z \qquad \mathrm{W/m^2}\end{aligned} \tag{121}$$

where we denote $|\hat{T}| = T, |\hat{\eta}_2| = \eta_2$. Note that this is a simple calculation in the second medium because there is only one wave in this medium. In medium 1 there are two waves—a forward-traveling wave and a backward-traveling wave. In this medium the Poynting vector contains a cross-coupling term, as shown by (53).

EXAMPLE 6.7 Because of the large losses encountered in sea water by higher-frequency waves, low-frequency radio waves are used to communicate with submarines. To illustrate this, consider an airplane flying over the surface of the ocean, for which $\sigma \simeq 4$ S/m, $\epsilon_r \simeq 81$, and $\mu_r \simeq 1$. The airplane transmits a signal at 1 MHz using a trailing, long-wire antenna. Assume for simplification that this transmitted wave is in the form of a uniform plane wave with an electric field intensity of 1000 V/m and is incident normal to the ocean surface. If the submarine requires a minimum signal level of 10 μV/m for adequate reception, determine the maximum communication depth of the submarine. If the coordinate system is arranged as in Fig. 6.12 such that the ocean surface is the yz plane, write complete expressions for all field vectors.

Solution We first determine the propagation constant and intrinsic impedance of the ocean. The loss tangent at 1 MHz is

$$\frac{\sigma_2}{\omega\epsilon_2} = 889$$

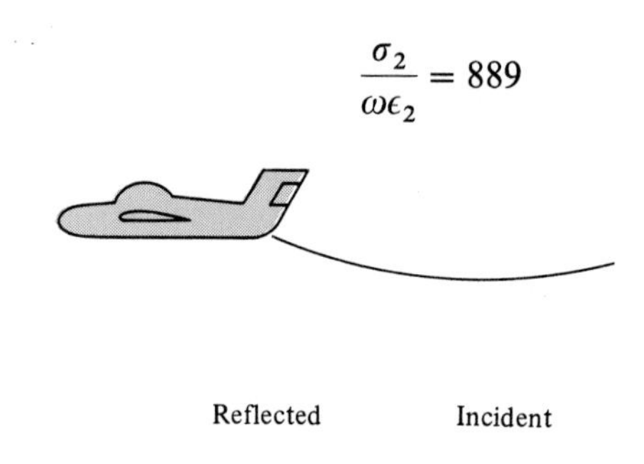

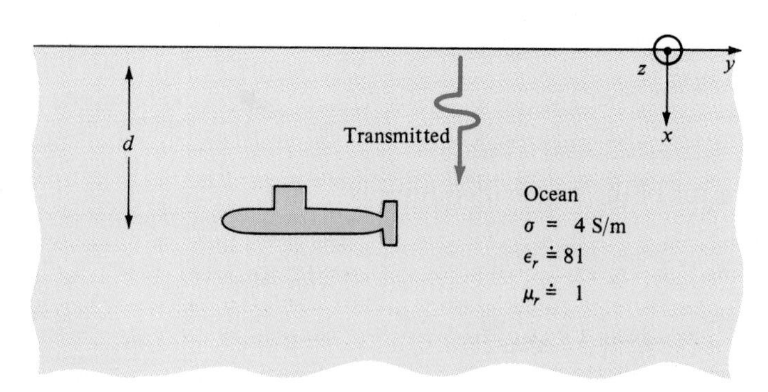

FIGURE 6.12

Example 6.7. Communication with a submarine.

so that the ocean may be classified as a good conductor at this frequency. Thus, we obtain from the results of Sec. 6.3.2

$$\hat{\eta}_2 \simeq \sqrt{\frac{\omega\mu}{\sigma}}\,\underline{/45^\circ}$$
$$= 1.4\underline{/45^\circ}$$

$$\alpha_2 = \beta_2 \simeq \sqrt{\frac{\omega\mu\sigma}{2}}$$
$$= 3.97$$

The corresponding terms for free space are

$$\beta_1 = \omega\sqrt{\mu_0\epsilon_0} \qquad \eta_1 = \sqrt{\frac{\mu_0}{\epsilon_0}}$$
$$= 2.09 \times 10^{-2} \text{ rad/m} \qquad = 120\pi\Omega$$

Consequently, $\alpha_1 = 0$ and $\theta_{\eta 1} = 0$. The reflection coefficient becomes

$$\hat{\Gamma} = \frac{\hat{\eta}_2 - \hat{\eta}_1}{\hat{\eta}_2 + \hat{\eta}_1}$$
$$= \frac{1.4\underline{/45^\circ} - 120\pi}{1.4\underline{/45^\circ} + 120\pi}$$
$$= 0.995\underline{/180^\circ}$$

The transmission coefficient becomes

$$\hat{T} = \frac{2\hat{\eta}_2}{\hat{\eta}_2 + \hat{\eta}_1}$$
$$= \frac{2.8\underline{/45^\circ}}{1.4\underline{/45^\circ} + 120\pi}$$
$$= 7.43 \times 10^{-3}\underline{/44.8^\circ}$$

Calculation will again show that $1 + \hat{\Gamma} = \hat{T}$. Consequently, the electric field transmitted into the ocean will be $1000 \times |\hat{T}| = 7.43$ V/m. This will be reduced in magnitude over a distance d as it propagates into the ocean by the factor $e^{-\alpha_2 d}$. Thus, we require d such that

$$10 \times 10^{-6} = 7.43e^{-3.97d}$$

or
$$d = \frac{1}{3.97} \ln (7.43 \times 10^5)$$
$$= 3.41 \text{ m}$$

Assuming the electric field to be polarized in the z direction (out of the page), we obtain from Fig. 6.12

$$\mathcal{E}_i = E_m \cos (\omega t - \beta_1 x)\mathbf{a}_z$$
$$= 1000 \cos (2\pi \times 10^6 t - 2.09 \times 10^{-2} x)\mathbf{a}_z \qquad \text{V/m}$$

$$\mathcal{E}_r = \Gamma E_m \cos (\omega t + \beta_1 x + \theta_\Gamma)\mathbf{a}_z$$
$$= 995 \cos (2\pi \times 10^6 t + 2.09 \times 10^{-2} x + 180°)\mathbf{a}_z \qquad \text{V/m}$$

$$\mathcal{E}_t = T E_m e^{-\alpha_2 x} \cos (\omega t - \beta_2 x + \theta_T)\mathbf{a}_z$$
$$= 7.43 e^{-3.97x} \cos (2\pi \times 10^6 t - 3.97x + 44.8°)\mathbf{a}_z \qquad \text{V/m}$$

The magnetic fields are now determined to be

$$\mathcal{H}_i = -\frac{E_m}{\eta_1} \cos (\omega t - \beta_1 x)\mathbf{a}_y$$
$$= -2.65 \cos (2\pi \times 10^6 t - 2.09 \times 10^{-2} x)\mathbf{a}_y \qquad \text{A/m}$$

$$\mathcal{H}_r = \frac{\Gamma E_m}{\eta_1} \cos (\omega t + \beta_1 x + \theta_\Gamma)\mathbf{a}_y$$
$$= 2.64 \cos (2\pi \times 10^6 t + 2.09 \times 10^{-2} x + 180°)\mathbf{a}_y \qquad \text{A/m}$$

$$\mathcal{H}_t = -\frac{T E_m}{\eta_2} e^{-\alpha_2 x} \cos (\omega t - \beta_2 x + \theta_T - \theta_{\eta 2})\mathbf{a}_y$$
$$= -5.31 e^{-3.97x} \cos (2\pi \times 10^6 t - 3.97x + 44.8° - 45°)\mathbf{a}_y \qquad \text{A/m}$$

In determining the proper orientations of the field vectors, we assumed (as was the case in our previous derivations) that the electric field vectors are all in the same direction (the x direction). The magnetic field vectors are then orthogonal to these, with a sign included such that the resulting Poynting vector for each wave is in the correct direction for that wave.

The average power transmitted through a 1-m^2 area of the surface is computed from (121) for $x = 0$:

$$P_{\text{av},t}\big|_{x=0} = \frac{1}{2} \frac{E_m^2 T^2}{\eta_2} \cos \theta_{\eta 2}$$
$$= \frac{1}{2} \frac{(1000)^2 (7.43 \times 10^{-3})^2}{1.4} \cos 45°$$
$$= 13.9 \text{ W}$$

This could also have been computed as the difference in average powers in the incident and reflected waves:

$$\frac{E_m^2}{2\eta_1}(1 - |\hat{\Gamma}|^2) = 13.9 \text{ W}$$

The average power dissipated in the surface of the ocean which is one skin depth in thickness and which has an area of 1 m^2 is ($\delta = 1/\alpha_2 = 25.2$ cm)

$$\begin{aligned} P_{\text{diss}} &= P_{\text{av},t}|_{x=0} - P_{\text{av},t}|_{x=\delta} \\ &= P_{\text{av},t}|_{x=0}[1 - e^{-2\alpha_2\delta}] \\ &= 12 \text{ W} \end{aligned}$$

since the transmitted average power density varies as $e^{-2\alpha_2 x}$. Because of the lossy nature of the ocean, the power in the transmitted wave is dissipated rather rapidly.

6.7.1 Lossless Media ($\sigma_1 = \sigma_2 = 0$)

If both media are lossless, i.e., $\sigma_1 = \sigma_2 = 0$, the above results become quite simple. The reflection coefficient, transmission coefficient, and intrinsic impedances become real numbers:

$$\eta_1 = \sqrt{\frac{\mu_1}{\epsilon_1}} \tag{122a}$$

$$\eta_2 = \sqrt{\frac{\mu_2}{\epsilon_2}} \tag{122b}$$

$$\hat{T} = T\angle 0^\circ = \frac{\eta_2 - \eta_1}{\eta_2 + \eta_1} \tag{123}$$

$$\hat{\Gamma} = \Gamma\angle 0^\circ = \frac{2\eta_2}{\eta_2 + \eta_1} \tag{124}$$

In addition, the attenuation constants α_1 and α_2 are zero, so that

$$\gamma_1 = j\beta_1 \tag{125a}$$

$$\gamma_2 = j\beta_2 \tag{125b}$$

where

$$\beta_1 = \omega\sqrt{\mu_1\epsilon_1} \tag{126a}$$

$$\beta_2 = \omega\sqrt{\mu_2\epsilon_2} \tag{126b}$$

The phasor forms of the field vectors in (119) become

$$\hat{\mathbf{E}}_i = E_m e^{-j\beta_1 z}\mathbf{a}_x \tag{127a}$$

$$\hat{\mathbf{H}}_i = \frac{E_m}{\eta_1} e^{-j\beta_1 z}\mathbf{a}_y \tag{127b}$$

$$\hat{\mathbf{E}}_r = \Gamma E_m e^{j\beta_1 z}\mathbf{a}_x \tag{127c}$$

$$\hat{\mathbf{H}}_r = -\frac{\Gamma E_m}{\eta_1} e^{j\beta_1 z}\mathbf{a}_y \tag{127d}$$

$$\hat{\mathbf{E}}_t = T E_m e^{-j\beta_2 z}\mathbf{a}_x \tag{127e}$$

$$\hat{\mathbf{H}}_t = \frac{T E_m}{\eta_2} e^{-j\beta_2 z}\mathbf{a}_y \tag{127f}$$

For this case, the time-domain forms in (120) simplify to

$$\mathcal{E}_i = E_m \cos(\omega t - \beta_1 z)\mathbf{a}_x \tag{128a}$$

$$\mathcal{H}_i = \frac{E_m}{\eta_1} \cos(\omega t - \beta_1 z)\mathbf{a}_y \tag{128b}$$

$$\mathcal{E}_r = \Gamma E_m \cos(\omega t + \beta_1 z)\mathbf{a}_x \tag{128c}$$

$$\mathcal{H}_r = -\frac{\Gamma E_m}{\eta_1} \cos(\omega t + \beta_1 z)\mathbf{a}_y \tag{128d}$$

$$\mathcal{E}_t = T E_m \cos(\omega t - \beta_2 z)\mathbf{a}_x \tag{128e}$$

$$\mathcal{H}_t = \frac{T E_m}{\eta_2} \cos(\omega t - \beta_2 z)\mathbf{a}_y \tag{128f}$$

The total phasor fields in region 1 are, from (127),

$$\begin{aligned}\hat{\mathbf{E}}_1 &= \hat{\mathbf{E}}_i + \hat{\mathbf{E}}_r \\ &= [E_m e^{-j\beta_1 z} + E_m \Gamma e^{j\beta_1 z}]\mathbf{a}_x \\ &= E_m e^{-j\beta_1 z}[1 + \Gamma e^{j2\beta_1 z}]\mathbf{a}_x\end{aligned} \tag{129a}$$

$$\begin{aligned}\hat{\mathbf{H}}_1 &= \hat{\mathbf{H}}_i + \hat{\mathbf{H}}_r \\ &= \frac{E_m}{\eta_1} e^{-j\beta_1 z}[1 - \Gamma e^{j2\beta_1 z}]\mathbf{a}_y\end{aligned} \tag{129b}$$

The magnitude of the total electric field in region 1 varies with z according to the variation of the bracketed term in (129a):

$$|\hat{E}_1| = E_m|1 + \Gamma e^{j2\beta_1 z}| \tag{130}$$

Note that $|1 + \Gamma e^{j2\beta_1 z}|$ may be interpreted as the sum of two complex numbers: $1\angle 0$ and $\Gamma\angle 2\beta_1 z$. The magnitude of this term may be observed as the vector addition of these as z varies, as shown in the so-called crank diagram in Fig. 6.13. For distances away from the boundary, z is increasingly negative, so that the vector $\Gamma e^{j2\beta_1 z}$ rotates in the clockwise direction about the tip of the $1\angle 0^\circ$ vector.

For Γ positive, the maximum will occur at $z = 0$ as well as at $2\beta_1 z = -2n\pi$ for $n = 1, 2, \ldots$. Since $\beta_1 = 2\pi/\lambda_1$, the maxima occur at $z = -n\lambda_1/2$, or at multiples of one-half wavelength away from the boundary:

$$|\hat{E}_1|_{\max} = E_m(1 + \Gamma) \qquad z = \frac{-\eta\lambda_1}{2} \quad n = 0, 1, 2, \ldots \quad \Gamma > 0 \tag{131}$$

The minima will occur when $2\beta_1 z = -m\pi$ for $m = 1, 3, 5, \ldots$ or $z = -m\lambda_1/4$; that is, at odd multiples of a quarter wavelength away from the boundary:

$$|\hat{E}_1|_{\min} = E_m(1 - \Gamma) \qquad z = \frac{-m\lambda_1}{4} \quad m = 1, 3, 5 \ldots \quad \Gamma > 0 \tag{132}$$

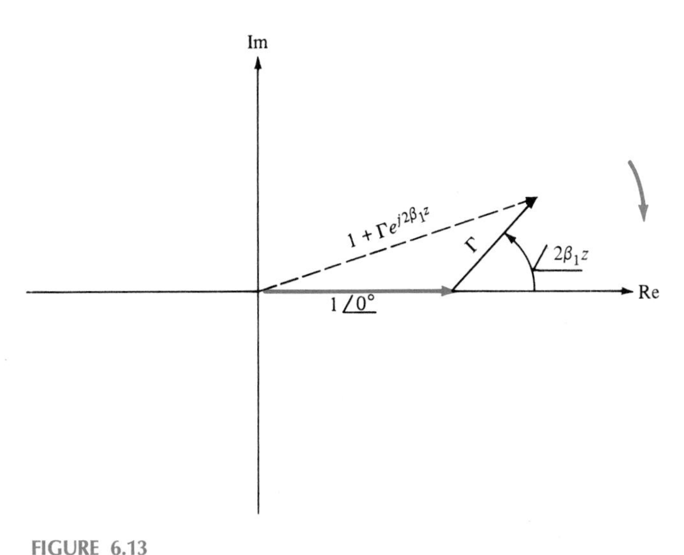

FIGURE 6.13
The crank diagram.

This is plotted (for a fixed time) in Fig. 6.14. Note from (129*b*) that the total magnetic field has the similar variation but is 180° out of phase with the electric field. Note also that the adjacent maxima on each waveform are separated by $\lambda_1/2$ and that a maximum and its adjacent minimum are separated by $\lambda_1/4$.

In the time domain at each z, these magnitudes vary sinusoidally. The result is shown in Fig. 6.14 by the dashed lines. The combination of a forward-traveling wave and a backward-traveling wave results in a waveform that simply pulsates, yielding a "standing wave."

For a reflection coefficient that is negative, $\Gamma < 0$, the above results reverse roles; that is, $|\hat{H}_1|_{max}$ and $|\hat{E}_1|_{min}$ occur at $z = 0$. Thus, $|\hat{E}_1|_{max} = 1 - \Gamma$ and $|\hat{E}_1|_{min} = 1 + \Gamma$. For either case, $\Gamma < 0$ or $\Gamma > 0$, we see that

$$|\hat{E}_1|_{max} = 1 + |\Gamma| \tag{133a}$$

$$|\hat{E}_1|_{min} = 1 - |\Gamma| \tag{133b}$$

The relation between adjacent maxima and minima, however, is the same for both cases; that is, adjacent maxima as well as minima are separated by $\lambda_1/2$ and an adjacent maximum and minimum are separated by $\lambda_1/4$. In fact, adjacent corresponding points (not necessarily maxima or minima) are separated by $\lambda_1/2$. We will find these results to be quite general. In fact, the above results—wherein a forward-traveling wave and backward-traveling wave combine to yield a standing wave—will occur again with transmission lines (Chap. 7) and waveguides (Chap. 8). The reader should be alert for these similarities.

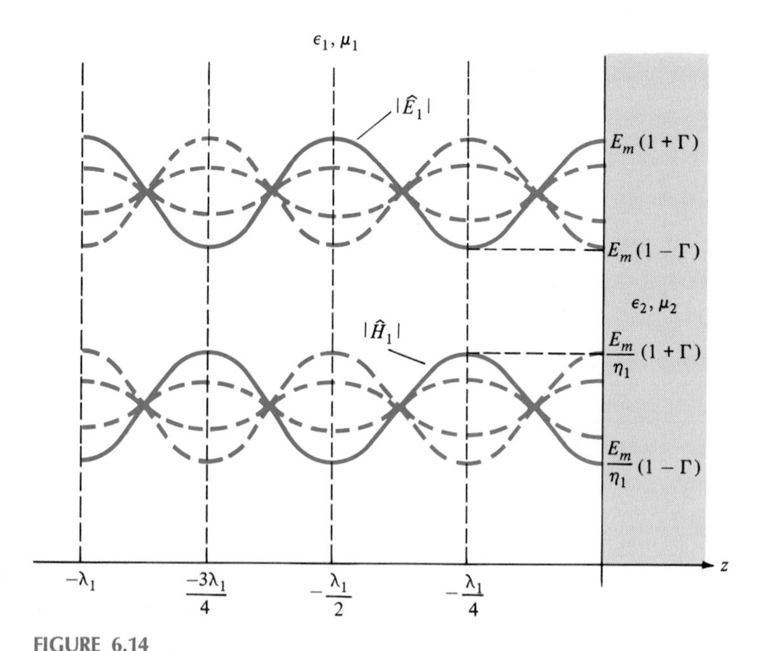

FIGURE 6.14

Variation of the total fields in region 1, $\Gamma > 0$.

The ratio of the maximum to minimum values of the field is referred to as the standing wave ratio S. From (133)

$$S = \frac{|\hat{E}_1|_{\max}}{|\hat{E}_1|_{\min}} = \frac{1 + |\Gamma|}{1 - |\Gamma|} \tag{134}$$

This can be inversely solved for $|\Gamma|$ in terms of S as

$$|\Gamma| = \frac{S - 1}{S + 1} \tag{135}$$

For the case of lossless media ($\sigma_1 = \sigma_2 = 0$), we may compute the average power density in the waves as

$$\begin{aligned} \mathbf{S}_{\text{av},i} &= \tfrac{1}{2} \operatorname{Re} (\hat{\mathbf{E}}_i \times \hat{\mathbf{H}}_i^*) \\ &= \frac{1}{2} \frac{E_m^2}{\eta_1} \mathbf{a}_z \qquad \sigma_1 = \sigma_2 = 0 \end{aligned} \tag{136a}$$

and similarly

$$\mathbf{S}_{\text{av},r} = -\frac{1}{2} \frac{E_m^2 \Gamma^2}{\eta_1} \mathbf{a}_z \qquad \sigma_1 = \sigma_2 = 0 \tag{136b}$$

$$\mathbf{S}_{\text{av},t} = \frac{1}{2} \frac{E_m^2 T^2}{\eta_2} \mathbf{a}_z \qquad \sigma_1 = \sigma_2 = 0 \tag{136c}$$

From these results, we can prove the relation

$$|\mathbf{S}_{\text{av},i}| = |\mathbf{S}_{\text{av},r}| + |\mathbf{S}_{\text{av},t}| \qquad \sigma_1 = \sigma_2 = 0 \tag{137}$$

Equation (137) can be proved by direct substitution of (136), resulting in

$$1 = \Gamma^2 + \frac{\eta_1}{\eta_2} T^2 \qquad \sigma_1 = \sigma_2 = 0 \tag{138}$$

and (138) can be verified by substituting (123) and (124) (which are real quantities for this situation). The vector counterpart to (137) is

$$\mathbf{S}_{\text{av},i} + \mathbf{S}_{\text{av},r} = \mathbf{S}_{\text{av},t} \qquad \sigma_1 = \sigma_2 = 0 \tag{139}$$

which indicates conservation of net power traveling in the $+z$ direction—a logical result since both media are lossless. Note that the percentage of total power in region 1 which resides in the reflected wave is proportional to Γ^2.

EXAMPLE 6.8 As an example of the application of these results, consider the boundary between free space and glass having $\sigma \simeq 0$, $\epsilon_r \simeq 4$, and $\mu_r = 1$. If a uniform plane wave with $E_m = 1$ V/m and a frequency of 200 MHz is incident from free space normal to the glass, determine (*a*) the time-domain forms of the incident, reflected, and transmitted fields, (*b*) the average power transmitted through a 5-m^2 surface of the glass, and (*c*) the standing wave ratio in free space.

Solution The phase constants and intrinsic impedances of the two media are

$$\beta_1 = \omega\sqrt{\mu_0\epsilon_0} \qquad \eta_1 = \sqrt{\frac{\mu_0}{\epsilon_0}}$$

$$= \frac{4\pi}{3} \text{ rad/m} \qquad = 120\pi\ \Omega$$

$$\beta_2 = \omega\sqrt{\mu_2\epsilon_2} \qquad \eta_2 = \sqrt{\frac{\mu_2}{\epsilon_2}}$$

$$= \omega\sqrt{\epsilon_{r2}\epsilon_0\mu_{r2}\mu_0} \qquad = \sqrt{\frac{\mu_{r2}\mu_0}{\epsilon_{r2}\epsilon_0}}$$

$$= 2\beta_1 \quad \text{rad/m} \qquad = \frac{\eta_1}{2}\ \Omega$$

$$= 60\pi\ \Omega$$

The transmission and reflection coefficients are

$$\Gamma = \frac{\eta_2 - \eta_1}{\eta_2 + \eta_1} \qquad T = \frac{2\eta_2}{\eta_2 + \eta_1}$$

$$= \frac{\eta_1/2 - \eta_1}{\eta_1/2 + \eta_1} \qquad = \frac{2\eta_1/2}{\eta_1/2 + \eta_1}$$

$$= -\frac{1}{3} \qquad = \frac{2}{3}$$

Note that $T = 1 + \Gamma$. The time-domain forms of the incident fields are

$$\mathcal{E}_i = \mathbf{E}_m \cos(\omega t - \beta_1 z)\mathbf{a}_x$$

$$= 1 \cos\left(\omega t - \frac{4\pi}{3} z\right)\mathbf{a}_x \qquad \text{V/m}$$

$$\mathcal{H}_i = \frac{E_m}{\eta_1} \cos(\omega t - \beta_1 z)\mathbf{a}_y$$

$$= \frac{1}{120\pi} \cos\left(\omega t - \frac{4\pi}{3} z\right)\mathbf{a}_y \qquad \text{A/m}$$

The reflected fields are

$$\boldsymbol{\mathcal{E}}_r = \Gamma E_m \cos(\omega t + \beta_1 z)\mathbf{a}_x$$
$$= -\frac{1}{3}\cos\left(\omega t + \frac{4\pi}{3} z\right)\mathbf{a}_x \qquad \text{V/m}$$

$$\boldsymbol{\mathcal{H}}_r = -\frac{\Gamma E_m}{\eta_1}\cos(\omega t + \beta_1 z)\mathbf{a}_y$$
$$= \frac{1}{360\pi}\cos\left(\omega t + \frac{4\pi}{3} z\right)\mathbf{a}_y \qquad \text{A/m}$$

The transmitted fields are

$$\boldsymbol{\mathcal{E}}_t = T E_m \cos(\omega t - \beta_2 z)\mathbf{a}_x$$
$$= \frac{2}{3}\cos\left(\omega t - \frac{8}{3}\pi z\right)\mathbf{a}_x \qquad \text{V/m}$$

$$\boldsymbol{\mathcal{H}}_t = \frac{T E_m}{\eta_2}\cos(\omega t - \beta_2 z)\mathbf{a}_y$$
$$= \frac{2}{180\pi}\cos\left(\omega t - \frac{8}{3}\pi z\right)\mathbf{a}_y \qquad \text{A/m}$$

The average power transmitted through a 5-m^2 area of the dielectric surface is

$$P_{\text{trans}} = \frac{1}{2}\frac{E_t^2}{\eta_2} \times 5$$
$$= \frac{1}{2}\frac{T^2 E_m^2}{\eta_2} \times 5$$
$$= 5.9 \times 10^{-3}\ \text{W}$$

Note also that, according to (137),

$$P_{\text{trans}} = P_{\text{inc}} - P_{\text{refl}}$$
$$= \frac{1}{2}\frac{E_m^2}{\eta_1} \times 5 - \frac{1}{2}\frac{\Gamma^2 E_m^2}{\eta_1} \times 5$$
$$= \frac{1}{2}\frac{E_m^2}{\eta_1} \times 5(1 - \Gamma^2)$$
$$= 6.63 \times 10^{-3}(1 - 0.011)$$
$$= 5.9 \times 10^{-3}\ \text{W}$$

The standing wave ratio in medium 1 is

$$S = \frac{1 + |\Gamma|}{1 - |\Gamma|}$$
$$= 2$$

6.7.2 Incidence on Perfect Conductors ($\sigma_2 = \infty$)

Suppose that medium 2 is a perfect conductor ($\sigma_2 = \infty$). Since the fields in a perfect conductor are zero, there is no transmitted wave and the boundary conditions at $z = 0$ become

$$\hat{\mathbf{E}}_i + \hat{\mathbf{E}}_r = 0 \qquad \text{at } z = 0 \tag{140}$$

thus,

$$\hat{\mathbf{E}}_r = -\hat{\mathbf{E}}_i \qquad \text{at } z = 0 \tag{141}$$

Since

$$\hat{\mathbf{E}}_i = E_m e^{-\gamma_1 z}\mathbf{a}_x \tag{142}$$

then

$$\hat{\mathbf{E}}_r = -E_m e^{\gamma_1 z}\mathbf{a}_x \tag{143}$$

Note that this could have been obtained more directly by realizing that $\hat{\eta}_2 = 0$, and thus the reflection coefficient in (115) becomes

$$\hat{\Gamma} = -1 \tag{144}$$

and the transmission coefficient in (116) becomes

$$\hat{T} = 0 \tag{145}$$

Thus, the total fields in region 1 become

$$\hat{\mathbf{E}}_1 = \hat{\mathbf{E}}_i + \hat{\mathbf{E}}_r$$
$$= E_m(e^{-\gamma_1 z} - e^{\gamma_1 z})\mathbf{a}_x \tag{146a}$$

$$\hat{\mathbf{H}}_1 = \hat{\mathbf{H}}_i + \hat{\mathbf{H}}_r$$
$$= \frac{E_m}{\eta_1}(e^{-\gamma_1 z} + e^{\gamma_1 z})\mathbf{a}_y \tag{146b}$$

Now suppose that medium 1 is lossless, $\sigma_1 = 0$. The total fields in medium 1 become

$$\begin{aligned}\hat{\mathbf{E}}_1 &= \hat{\mathbf{E}}_i + \hat{\mathbf{E}}_r \\ &= E_m(e^{-j\beta_1 z} - e^{j\beta_1 z})\mathbf{a}_x \\ &= -2jE_m \sin \beta_1 z \, \mathbf{a}_x \end{aligned} \tag{147a}$$

$$\begin{aligned}\hat{\mathbf{H}}_1 &= \hat{\mathbf{H}}_i + \hat{\mathbf{H}}_r \\ &= \left(\frac{E_m}{\eta_1} e^{-\beta_1 z} + \frac{E_m}{\eta_1} e^{j\beta_1 z}\right)\mathbf{a}_y \\ &= 2\frac{E_m}{\eta_1} \cos \beta_1 z \, \mathbf{a}_y \end{aligned} \tag{147b}$$

The time-domain expressions become

$$\begin{aligned}\mathcal{E}_1 &= \text{Re}\,(\hat{\mathbf{E}}_1 \, e^{j\omega t}) \\ &= 2E_m \sin \beta_1 z \sin \omega t \, \mathbf{a}_x \end{aligned} \tag{148a}$$

$$\mathcal{H}_1 = \frac{2E_m}{\eta_1} \cos \beta_1 z \cos \omega t \, \mathbf{a}_y \tag{148b}$$

The total fields represent, once again, standing waves.

The magnitudes of the phasor fields are

$$\begin{aligned}|\hat{E}_1| &= 2E_m|\sin \beta_1 z| \\ &= 2E_m\left|\sin \frac{2\pi z}{\lambda_1}\right| \end{aligned} \tag{149a}$$

$$|\hat{H}_1| = 2\frac{E_m}{\eta_1}\left|\cos \frac{2\pi z}{\lambda_1}\right| \tag{149b}$$

These are plotted in Fig. 6.15. Note that the properties of separation of adjacent maxima and minima are the same as for the general case. The only difference here is that the minima are *zero*; that is, the total electric field is zero at multiples of $\lambda_1/2$ away from the surface and the magnetic field is zero at odd multiples of $\lambda_1/4$ away from the surface. Since the minima are zero, the standing wave ratio S is infinite.

At the surface of the (perfect) conductor ($z = 0$), the tangential magnetic field becomes

$$\mathcal{H}_1\Big|_{z=0} = \frac{2E_m}{\eta_1} \cos \omega t \, \mathbf{a}_y \tag{150}$$

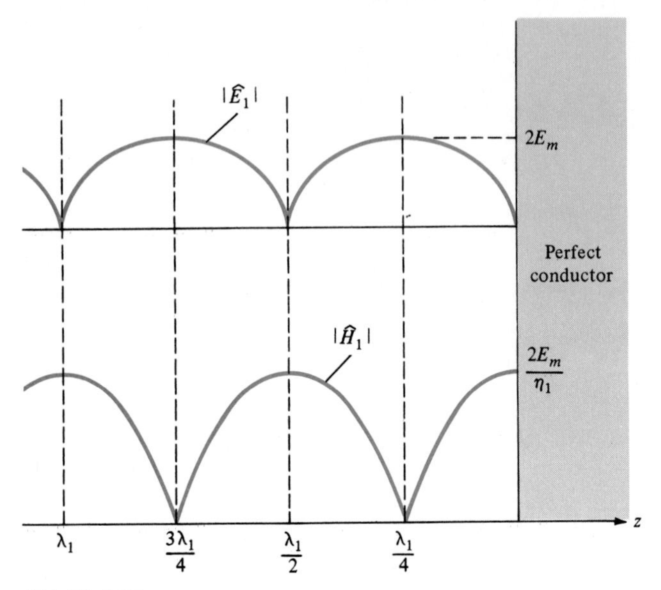

FIGURE 6.15
Variation of the total fields incident normal to the surface of a perfect conductor.

The resulting linear current density at the surface is determined from the boundary condition on the tangential component of the magnetic field [Eq. (62) of Chap. 5]:

$$\mathcal{K} = \frac{2E_m}{\eta_1} \cos \omega t \, \mathbf{a}_x \qquad \text{A/m} \tag{151}$$

with the direction being determined as described in Sec. 5.3.

6.8 Oblique Incidence of Uniform Plane Waves on Plane Boundaries

In the previous sections we considered uniform plane waves that are incident on plane boundaries normal to those boundaries. In the following sections of this chapter we will consider uniform plane waves that are incident on plane boundaries at *arbitrary* angles of incidence. We will again assume that the boundary between the two media lies in the xy plane. The Poynting vectors of the incident, reflected, and transmitted waves will be assumed to lie in the xz plane; this is referred to as the *plane of incidence*. The angle of incidence of the incident wave, θ_i, will be measured with respect to a normal to the surface, as shown in Fig. 6.16. Similarly, the angle of reflection of the reflected wave, θ_r, will

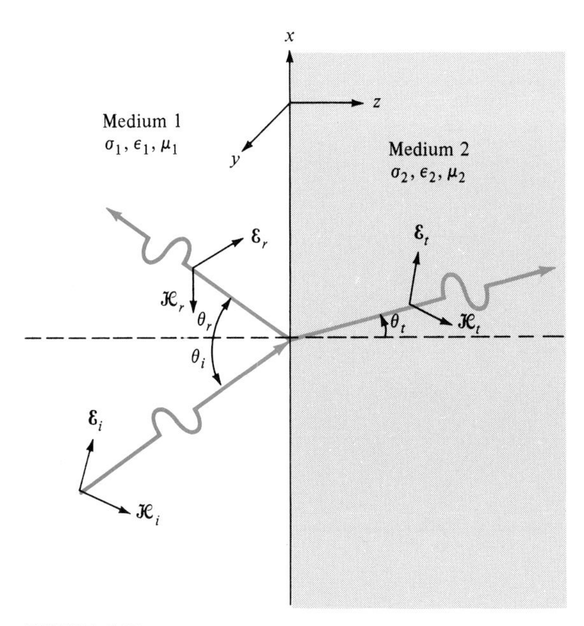

FIGURE 6.16
Oblique incidence of uniform plane waves on plane boundaries.

also be measured with respect to this normal. A portion of the incident wave will be transmitted into medium 2, and the angle of transmission of this transmitted wave, θ_t, will also be measured with respect to a normal to the boundary, as shown in Fig. 6.16.

In the case of normal incidence considered previously, we could assume, without loss of generality, that the electric field of the incident wave is polarized in the x direction. For oblique incidence, we have an infinite number of distinct possibilities for polarization of the incident-wave electric field. In order to consider all these cases, we will decompose the incident wave into two linearly polarized waves with the electric fields of these waves orthogonal to each other. It was shown in Sec. 6.5 that a wave with arbitrary polarization could be handled in this fashion. The result for a wave having some arbitrary polarization will then be the superposition of the results for these two individual waves (assuming linear media).

Before considering these cases, we will prove some general properties for this case of oblique incidence. In order to show these, as well as to analyze the cases for specific polarizations, we must consider the description of uniform plane waves propagating in arbitrary space directions.

6.8.1 Propagation in Arbitrary Space Directions

Consider a uniform plane wave propagating in the z' direction in an orthogonal coordinate system composed of x', z', and z' axes, as shown in Fig.

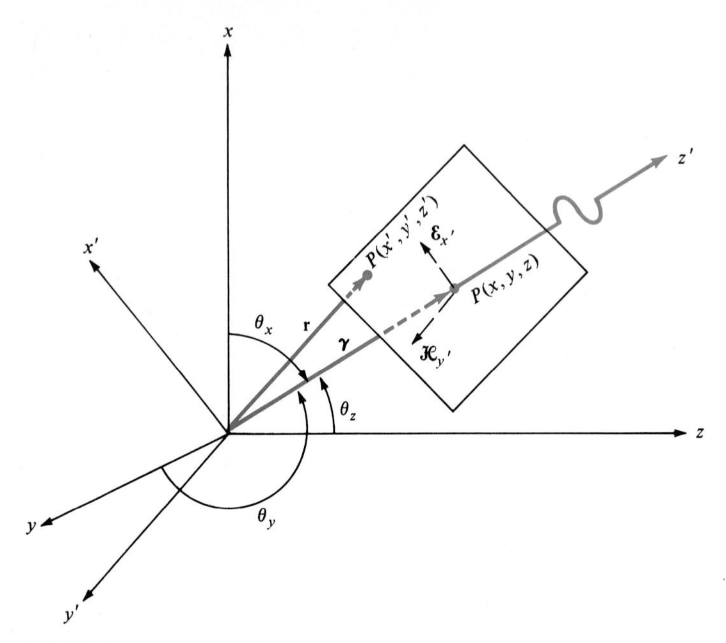

FIGURE 6.17

Propagation in arbitrary space directions.

6.17. The phasor fields in this coordinate system are, from our previous results, immediately written as

$$\hat{\mathbf{E}} = E_m e^{-\gamma z'} \mathbf{a}_{x'} \tag{152a}$$

$$\hat{\mathbf{H}} = \frac{E_m}{\hat{\eta}} e^{-\gamma z'} \mathbf{a}_{y'} \tag{152b}$$

We would like to describe these in the unprimed system consisting of x, y, and z coordinate axes.

Essentially, we are interested in a rotation of the two coordinate systems. This can be accomplished by using direction cosines. Note that

$$z' = \cos\theta_x x + \cos\theta_y y + \cos\theta_z z \tag{153}$$

Thus, we may write

$$e^{-\gamma z'} = e^{-\gamma(\cos\theta_x x + \cos\theta_y y + \cos\theta_z z)} \tag{154}$$

This result may be written in a more compact form by defining the propagation vector

$$\boldsymbol{\gamma} = \gamma \mathbf{a}_{z'} \tag{155}$$

which has a magnitude γ and a direction in the direction of propagation of the wave. Also, we may define the radius vector directed from the common origin of the coordinate systems to a point in the plane containing the electric and magnetic field vectors of the wave, as shown in Fig. 6.17:

$$\begin{aligned}\mathbf{r} &= x'\mathbf{a}_{x'} + y'\mathbf{a}_{y'} + z'\mathbf{a}_{z'} \\ &= x\mathbf{a}_x + y\mathbf{a}_y + z\mathbf{a}_z\end{aligned} \tag{156}$$

The dot product of $\boldsymbol{\gamma}$ and $\mathbf{r}$ yields

$$\begin{aligned}\boldsymbol{\gamma}\cdot\mathbf{r} &= \gamma z' \\ &= \gamma x\mathbf{a}_{z'}\cdot\mathbf{a}_x + \gamma y\mathbf{a}_{z'}\cdot\mathbf{a}_y + \gamma z\mathbf{a}_{z'}\cdot\mathbf{a}_z\end{aligned} \tag{157}$$

This reduces to

$$\boldsymbol{\gamma}\cdot\mathbf{r} = \gamma(\cos\theta_x x + \cos\theta_y y + \cos\theta_z z) \tag{158}$$

since

$$\begin{aligned}\mathbf{a}'_z\cdot\mathbf{a}_x &= \cos\theta_x \\ \mathbf{a}_{z'}\cdot\mathbf{a}_y &= \cos\theta_y \\ \mathbf{a}_{z'}\cdot\mathbf{a}_z &= \cos\theta_z\end{aligned} \tag{159}$$

Therefore, the exponential terms in (152) may be written in the unprimed coordinate system in a compact manner as

$$\begin{aligned}e^{-\gamma z'} &= e^{-\boldsymbol{\gamma}\cdot\mathbf{r}} \\ &= e^{-\gamma_x x - \gamma_y y - \gamma_z z} \\ &= e^{-\gamma_x x}e^{-\gamma_y y}e^{-\gamma_z z}\end{aligned} \tag{160}$$

where

$$\begin{aligned}\gamma_x &= \gamma\cos\theta_x \\ \gamma_y &= \gamma\cos\theta_y \\ \gamma_z &= \gamma\cos\theta_z\end{aligned} \tag{161}$$

are the projections of the propagation vector on the unprimed coordinate axes. This is an important result. It appears that the attenuation and phase constants of the wave in the z' direction have components in the x, y, and z directions.

Note that, by construction, the projection of $\mathbf{r}$ on $\boldsymbol{\gamma}$ is constant regardless of the selection of point P. (Point P lies in a plane that is orthogonal to $\boldsymbol{\gamma}$.) This confirms that the plane containing the field vectors is a plane of constant phase and uniform amplitude for the wave.

The remaining task is to write the electric and magnetic field vectors in (152) in terms of the unprimed coordinate system axes. With the above result, the complete expressions in (152) should be in the form

$$\hat{\mathbf{E}} = (\hat{E}_x\mathbf{a}_x + \hat{E}_y\mathbf{a}_y + \hat{E}_z\mathbf{a}_z)e^{-\gamma_x x - \gamma_y y - \gamma_z z} \tag{162a}$$

$$\hat{\mathbf{H}} = (\hat{H}_x\mathbf{a}_x + \hat{H}_y\mathbf{a}_y + \hat{H}_z\mathbf{a}_z)e^{-\gamma_x x - \gamma_y y - \gamma_z z} \tag{162b}$$

where $\hat{E}_x$, $\hat{E}_y$, $\hat{E}_z$, and $\hat{H}_x$, $\hat{H}_y$, $\hat{H}_z$ are the projections of the corresponding field vectors in the primed coordinate system on the x, y, and z axes.

Now that we understand the final result, it will be a simple matter in any specific problem to transform the field vectors from the primed coordinate system to the unprimed coordinate system. The process consists of (1) determining the projections of the propagation vector on the unprimed axes to give γ_x, γ_y, and γ_z; (2) determining the projections of the electric and magnetic field vectors on the unprimed axes to yield $\hat{E}_x$, $\hat{E}_y$, $\hat{E}_z$, $\hat{H}_x$, $\hat{H}_y$, and $\hat{H}_z$; and (3) writing the form of these field vectors as in (162). This is the essence of the result. For the problems we will consider, finding the above projections will be a simple matter.

There is one additional, general result. If we define a unit vector in the direction of propagation as

$$\begin{aligned}\mathbf{a}_n &= \mathbf{a}_{z'} \\ &= \frac{\boldsymbol{\gamma}}{|\gamma|} \\ &= \cos\theta_x\mathbf{a}_x + \cos\theta_y\mathbf{a}_y + \cos\theta_y\mathbf{a}_z\end{aligned} \tag{163}$$

we can obtain a general expression for the magnetic field as

$$\hat{\mathbf{H}} = \frac{\mathbf{a}_n \times \hat{\mathbf{E}}}{\hat{\eta}} \tag{164}$$

Thus, once the electric field vector is written in the unprimed coordinate system as in (162a), it is a simple matter to compute the corresponding magnetic field vector given in (162b) as

$$(\hat{H}_x\mathbf{a}_x + \hat{H}_y\mathbf{a}_y + \hat{H}_z\mathbf{a}_z) = \frac{\mathbf{a}_n \times (\hat{E}_x\mathbf{a}_x + \hat{E}_y\mathbf{a}_y + \hat{E}_z\mathbf{a}_z)}{\hat{\eta}} \tag{165}$$

Rarely will we need to use this result, since a straightforward geometric observation will usually yield the desired components.

6.8.2 Snell's Laws

Returning to the general problem of oblique incidence depicted in Fig. 6.16, we will now show some general properties of the angles of incidence, reflection, and transmission, θ_i, θ_r, and θ_t, respectively. These relations between the various angles are governed by Snell's laws. To show these relations, we write the phasor incident, reflected, and transmitted field vectors as in (162*a*) and (162*b*). We assume, without loss of generality, that all three propagation vectors lie in the xz plane, which we refer to as the plane of incidence. Thus, the propagation vectors have components only in the x and z directions. The electric and magnetic field vectors may have components along all three coordinate axes. From the geometry in Fig. 6.16 we may write the phasor fields as

$$\hat{\mathbf{E}}_i = [\hat{E}^i_x \mathbf{a}_x + \hat{E}^i_y \mathbf{a}_y + \hat{E}^i_z \mathbf{a}_z] e^{-\gamma_1(\sin\theta_i x + \cos\theta_i z)} \tag{166a}$$

$$\hat{\mathbf{H}}_i = [\hat{H}^i_x \mathbf{a}_x + \hat{H}^i_y \mathbf{a}_y + \hat{H}^i_z \mathbf{a}_z] e^{-\gamma_1(\sin\theta_i x + \cos\theta_i z)} \tag{166b}$$

$$\hat{\mathbf{E}}_r = [\hat{E}^r_x \mathbf{a}_x + \hat{E}^r_y \mathbf{a}_y + \hat{E}^r_z \mathbf{a}_z] e^{\gamma_1(-\sin\theta_r x + \cos\theta_r z)} \tag{166c}$$

$$\hat{\mathbf{H}}_r = [\hat{H}^r_x \mathbf{a}_x + \hat{H}^r_y \mathbf{a}_y + \hat{H}^r_z \mathbf{a}_z] e^{\gamma_1(-\sin\theta_r x + \cos\theta_r z)} \tag{166d}$$

$$\hat{\mathbf{E}}_t = [\hat{E}^t_x \mathbf{a}_x + \hat{E}^t_y \mathbf{a}_y + \hat{E}^t_z \mathbf{a}_z] e^{-\gamma_2(\sin\theta_t x + \cos\theta_t z)} \tag{166e}$$

$$\hat{\mathbf{H}}_t = [\hat{H}^t_x \mathbf{a}_x + \hat{H}^t_y \mathbf{a}_y + \hat{H}^t_z \mathbf{a}_z] e^{-\gamma_2(\sin\theta_t x + \cos\theta_t z)} \tag{166f}$$

We next impose the boundary conditions at the interface. At $z = 0$, the tangential electric fields must be continuous. From (166) we obtain

$$\begin{aligned}[\hat{E}^i_x \mathbf{a}_x + \hat{E}^i_y \mathbf{a}_y] e^{-\gamma_1 \sin\theta_i x} &+ [\hat{E}^r_x \mathbf{a}_x + \hat{E}^r_y \mathbf{a}_y] e^{-\gamma_1 \sin\theta_r x} \\ &= [\hat{E}^t_x \mathbf{a}_x + \hat{E}^t_y \mathbf{a}_y] e^{-\gamma_2 \sin\theta_t x} \qquad z = 0\end{aligned} \tag{167}$$

Similarly at $z = 0$, the tangential magnetic field must be continuous:

$$\begin{aligned}[\hat{H}^i_x \mathbf{a}_x + \hat{H}^i_y \mathbf{a}_y] e^{-\gamma_1 \sin\theta_i x} &+ [\hat{H}^r_x \mathbf{a}_x + \hat{H}^r_y \mathbf{a}_y] e^{-\gamma_1 \sin\theta_r x} \\ &= [\hat{H}^t_x \mathbf{a}_x + \hat{H}^t_y \mathbf{a}_y] e^{-\gamma_2 \sin\theta_t x}\end{aligned} \tag{168}$$

Both (167) and (168) must hold *for all x*. Thus, the exponential terms must be equal:

$$e^{-\gamma_1 \sin\theta_i x} = e^{-\gamma_1 \sin\theta_r x} = e^{-\gamma_2 \sin\theta_t x} \tag{169}$$

This results in two conditions:

$$\theta_i = \theta_r \tag{170a}$$

$$\gamma_1 \sin\theta_i = \gamma_2 \sin\theta_t \tag{170b}$$

Equation (170a) is referred to as *Snell's law of reflection*: The angle of incidence must equal the angle of reflection. The second relation given by (170b) is referred to as *Snell's law of refraction.* Although interpretations can be obtained for lossy media, we will henceforth assume that both media are lossless ($\sigma_1 = \sigma_2 = 0$), so that (170a) becomes

$$j\beta_1 \sin\theta_i = j\beta_2 \sin\theta_t \tag{171}$$

or

$$\frac{\sin\theta_i}{\sin\theta_t} = \frac{\beta_2}{\beta_1} \tag{172}$$

Since $\beta = \omega\sqrt{\mu\epsilon}$, this becomes

$$\frac{\sin\theta_i}{\sin\theta_t} = \sqrt{\frac{\mu_2\epsilon_2}{\mu_1\epsilon_1}} \tag{173}$$

That is, the angle of the transmitted wave is related to the angle of incidence by the properties of the two media. For typical dielectric media that are not ferromagnetic, $\mu_r \simeq 1$, so that

$$\frac{\sin\theta_i}{\sin\theta_t} = \sqrt{\frac{\epsilon_2}{\epsilon_1}} \tag{174}$$

The *index of refraction* of a medium that has $\mu_r = 1$ is defined as the ratio of the velocity of light in free space to that in the medium:

$$n = \frac{u_0}{u_p} = \sqrt{\epsilon_r} \tag{175}$$

In terms of the index of refraction, Snell's law of refraction may be stated as

$$\sin\theta_t = \frac{n_1}{n_2}\sin\theta_i \tag{176}$$

For transmission from a rarer into a denser medium, $\epsilon_{r2} > \epsilon_{r1}$, we see that $\theta_t < \theta_i$ and that the transmitted wave is bent toward the normal. This is the case for transmission of light from free space into water, glass, etc.

EXAMPLE 6.9 Consider the problem of using a submerged light source to illuminate a surface of a body of water. If the light is placed at a depth $d = 1$ m below the surface, determine the surface area of light seen on the surface. The relative permittivity of water at optical frequencies is $\epsilon_r = 1.77$.

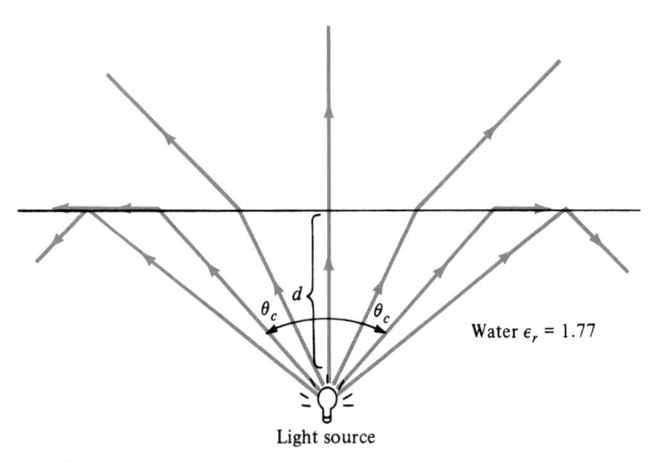

FIGURE 6.18

Example 6.9. Illustration of the critical angle.

Solution The problem is illustrated in Fig. 6.18. Only light within a cone of angle θ_c will be transmitted. At this angle, the angle of transmission is 90°. Thus, (176) yields

$$\sin\left(\frac{\pi}{2}\right) = \sqrt{1.77}\,\sin\theta_c$$

or

$$\theta_c = 48.7^\circ$$

Light will be seen within a circle of radius $d\tan\theta_c$. Thus, the surface illuminated will have an area of

$$\pi(d\tan\theta_c)^2 = 4.08\ \text{m}^2$$

6.8.3 Critical Angle of Total Reflection

In the previous example we determined an angle of incidence such that the angle of transmission is 90°, i.e., no wave is transmitted. This is referred to as the *critical angle* of total reflection. From (176) with $\theta_t = 90^\circ$, we obtain

$$\theta_c = \sin^{-1}\left(\frac{n_2}{n_1}\right)$$

$$= \sin^{-1}\sqrt{\frac{\epsilon_2}{\epsilon_1}} \tag{177}$$

Now let us suppose that the angle of incidence exceeds this critical angle, $\theta_i > \theta_c$. Re-examining Snell's law of refraction in (176), we see that

$$\begin{aligned} \sin\theta_t &= \frac{n_1}{n_2}\sin\theta_i \\ &> 1 \end{aligned} \tag{178}$$

implying that there is no real angle for θ_t. To understand this, we need to re-examine our previous assumptions. Note that

$$\begin{aligned} \cos\theta_t &= \sqrt{1-\sin^2\theta_t} \\ &= \pm j\sqrt{\frac{\epsilon_1}{\epsilon_2}\sin^2\theta_i - 1} \end{aligned} \tag{179}$$

Recall in our expressions for the transmitted wave in (166*e*) that we assumed medium 2 to be lossless, so that

$$\hat{\mathbf{E}}_t = \hat{\mathbf{E}}_m^t e^{-j\beta_2(\sin\theta_t x + \cos\theta_t z)} \tag{180}$$

Since for $\theta_i > \theta_c$ Eq. (179) shows that $\cos\theta_t$ is imaginary, we write

$$\beta_2 \cos\theta_t = -j\alpha_2 \tag{181}$$

where

$$\alpha_2 = \beta_2\sqrt{\frac{\epsilon_1}{\epsilon_2}\sin^2\theta_i - 1} \tag{182}$$

and we have chosen the negative result in (179) for reasons that will soon become clear. Substituting (181) into (180) yields

$$\hat{\mathbf{E}}_t = \hat{\mathbf{E}}_m^t e^{-\alpha_2 z} e^{-j\beta_2 \sin\theta_t x} \tag{183}$$

This result shows that the transmitted wave is propagating in the x direction (along the boundary) with propagation constant $\beta_2 \sin\theta_t$. However, the field is also multiplied by a term $e^{-\alpha_2 z}$ that is due to the angle of incidence exceeding the critical angle. Thus, we have a *nonuniform* plane wave propagating along the boundary in that the field is not constant over a plane perpendicular to the direction of propagation but is attenuated exponentially away from the boundary in medium 2. This is illustrated in Fig. 6.19. Waves of this types are also referred to as *surface waves* since they are tightly bound to the surface of the interface.

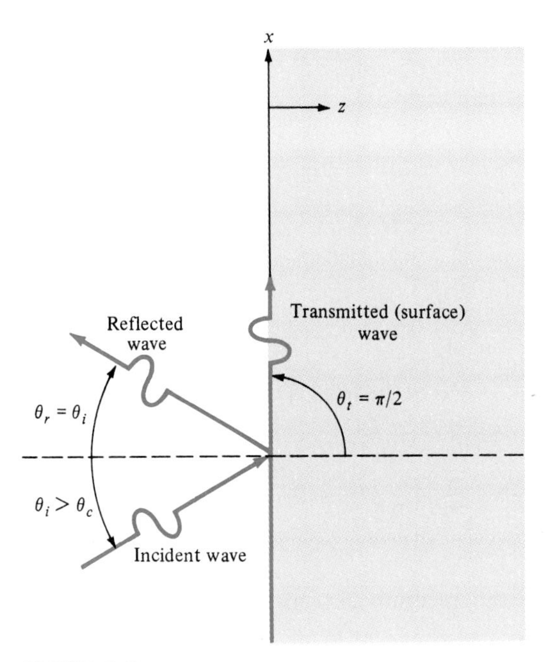

FIGURE 6.19
Illustration of surface waves and the critical angle of reflection.

EXAMPLE 6.10 A promising method of guiding electromagnetic waves from point to point makes use of fiber-optics. This technology utilizes a thin fiber of dielectric material, typically glass, to transmit information. The information-carrying capability of these fibers far exceeds some of the more conventional "transmission lines" constructed of parallel wires or coaxial cables (Chap. 7). Consider the fiber-optic "transmission line" shown in Fig. 6.20. Determine the minimum value of the dielectric constant ϵ_r of the fiber such that for any angle of incidence of light entering the one end of the fiber, the light will be totally contained within the fiber until it exits the other end.

Solution We essentially require that the angle θ shown in Fig. 6.20 be greater than or equal to the critical angle for any incident angle θ_i. Thus

$$\theta \geq \theta_c$$

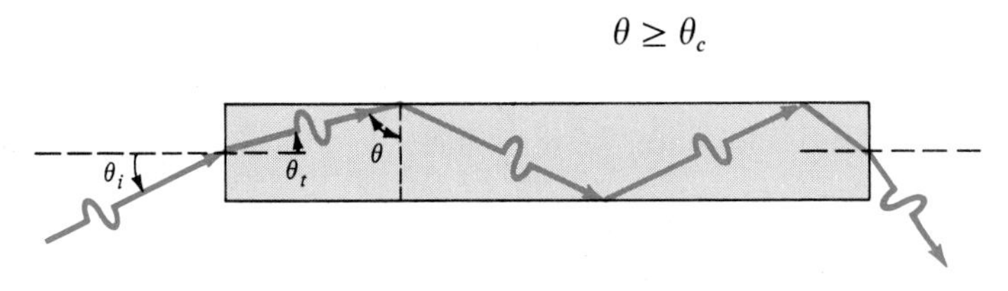

FIGURE 6.20
Example 6.10. An optical fiber transmission line.

or

$$\sin \theta \geq \sin \theta_c$$

Thus, we require

$$\sin \theta \geq \frac{1}{\sqrt{\epsilon_r}}$$

or

$$\sin^2 \theta \geq \frac{1}{\epsilon_r}$$

The light incident at an angle θ_i on the end of the fiber is transmitted at an angle of θ_t. Snell's law in (176) gives

$$\sin \theta_t = \frac{1}{\sqrt{\epsilon_r}} \sin \theta_i$$

where we assume free space for the medium surrounding the fiber. Since $\theta_t = 90^\circ - \theta$, we have

$$\cos \theta = \frac{1}{\sqrt{\epsilon_r}} \sin \theta_i$$

or

$$\cos^2 \theta = \frac{1}{\epsilon_r} \sin^2 \theta_i$$

Since $\cos^2 \theta = 1 - \sin^2 \theta$, we have

$$\sin^2 \theta = 1 - \frac{\sin^2 \theta_i}{\epsilon_r}$$

Combining these results gives

$$1 - \frac{\sin^2 \theta_i}{\epsilon_r} \geq \frac{1}{\epsilon_r}$$

or

$$\epsilon_r \geq \sin^2 \theta_i + 1$$

Thus, for all possible angles of incidence, we must have

$$\epsilon_r \geq 2$$

which is satisfied by typically used materials, such as quartz and glass.

6.9 Oblique Incidence, Lossless Media

As was indicated previously, we will consider any arbitrary polarization as the linear superposition of two waves, each being composed of linearly polarized waves whose electric fields are orthogonal. The general result for arbitrary polarization will be the superposition of these two results. The statement of the general problem is illustrated in Fig. 6.16. The plane of incidence is the plane containing the propagation vector of the incident wave and the normal to the boundary, which for this problem is the xz plane. We will consider two polarizations of the incident electric field. For *perpendicular polarization*, the incident electric field vector is perpendicular to the plane of incidence, as shown in 6.21*a*. For *parallel polarization*, the incident electric field vector is parallel to or in the plane of incidence, as shown in Fig. 6.21*b*.

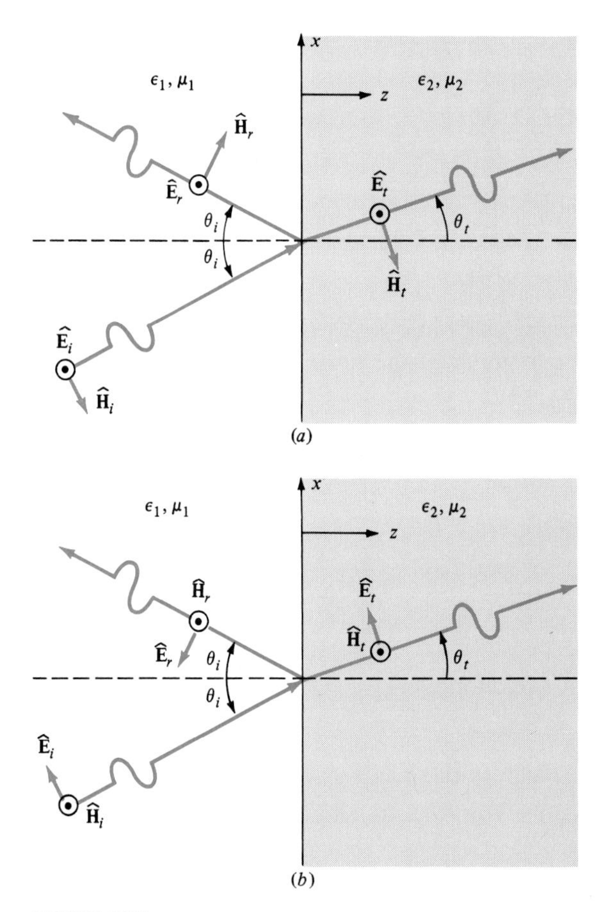

FIGURE 6.21
Oblique incidence. *(a)* Perpendicular polarization; *(b)* parallel polarization.

6.9.1 Perpendicular Polarization

For the case of perpendicular polarization shown in Fig. 6.21a, we may write the phasor field vectors in the form

$$\hat{\mathbf{E}}_i = \hat{E}_i e^{-j\beta_1(\sin\theta_i x + \cos\theta_i z)}\mathbf{a}_y \tag{184a}$$

$$\hat{\mathbf{H}}_i = \frac{\hat{E}_i}{\eta_1}(-\cos\theta_i\mathbf{a}_x + \sin\theta_i\mathbf{a}_z)e^{-j\beta_1(\sin\theta_i x + \cos\theta_i z)} \tag{184b}$$

$$\hat{\mathbf{E}}_r = \hat{E}_r e^{j\beta_1(-\sin\theta_i x + \cos\theta_i z)}\mathbf{a}_y \tag{184c}$$

$$\hat{\mathbf{H}}_r = \frac{\hat{E}_r}{\eta_1}(\cos\theta_i\mathbf{a}_x + \sin\theta_i\mathbf{a}_z)e^{j\beta_1(-\sin\theta_i x + \cos\theta_i z)} \tag{184d}$$

$$\hat{\mathbf{E}}_t = \hat{E}_t e^{-j\beta_2(\sin\theta_t x + \cos\theta_t z)}\mathbf{a}_y \tag{184e}$$

$$\hat{\mathbf{H}}_t = \frac{\hat{E}_t}{\eta_2}(-\cos\theta_t\mathbf{a}_x + \sin\theta_t\mathbf{a}_z)e^{-j\beta_2(\sin\theta_t x + \cos\theta_t z)} \tag{184f}$$

where we have substituted $\theta_i = \theta_r$ according to Snell's law of reflection. The boundary conditions require continuity of the tangential components of the electric and magnetic fields at $z = 0$:

$$\hat{E}_i e^{-j\beta_1 \sin\theta_i x} + \hat{E}_r e^{-j\beta_1 \sin\theta_i x} = \hat{E}_t e^{-j\beta_2 \sin\theta_t x} \tag{185a}$$

$$\frac{\hat{E}_i}{\eta_1}(-\cos\theta_i)e^{-j\beta_1 \sin\theta_i x} + \frac{\hat{E}_r}{\eta_1}(\cos\theta_i)e^{-j\beta_1 \sin\theta_i x} = \frac{\hat{E}_t}{\eta_2}(-\cos\theta_t)e^{-j\beta_2 \sin\theta_t x} \tag{185b}$$

Again we obtain Snell's law of refraction:

$$\frac{\sin\theta_i}{\sin\theta_t} = \frac{\beta_2}{\beta_1}$$

$$= \sqrt{\frac{\mu_2\epsilon_2}{\mu_1\epsilon_1}} \tag{186}$$

leaving

$$\hat{E}_i + \hat{E}_r = \hat{E}_t \tag{187a}$$

$$-\frac{\hat{E}_i}{\eta_1}\cos\theta_i + \frac{\hat{E}_r}{\eta_1}\cos\theta_i = -\frac{\hat{E}_t}{\eta_2}\cos\theta_t \tag{187b}$$

Solving (187) yields the reflection and transmission coefficients for perpendicular polarization:

$$\begin{aligned}\Gamma_\perp &= \frac{\hat{E}_r}{\hat{E}_i}\\ &= \frac{\eta_2\cos\theta_i - \eta_1\cos\theta_t}{\eta_2\cos\theta_i + \eta_1\cos\theta_t}\\ &= \frac{\eta_2\cos\theta_i - \eta_1\sqrt{1-(u_2/u_1)^2\sin^2\theta_i}}{\eta_2\cos\theta_i + \eta_1\sqrt{1-(u_2/u_1)^2\sin^2\theta_i}}\end{aligned} \tag{188a}$$

$$\begin{aligned}T_\perp &= \frac{\hat{E}_t}{\hat{E}_i}\\ &= \frac{2\eta_2\cos\theta_i}{\eta_2\cos\theta_i + \eta_1\cos\theta_t}\\ &= \frac{2\eta_2\cos\theta_i}{\eta_2\cos\theta_i + \eta_1\sqrt{1-(u_2/u_1)^2\sin^2\theta_i}}\end{aligned} \tag{188b}$$

The alternative forms of these equations in terms of θ_i only are obtained with Snell's law of refraction in (186) and by realizing that $\cos\theta = \sqrt{1-\sin^2\theta}$. The quantity $u = 1/\sqrt{\mu\epsilon}$ is the phase velocity in the appropriate medium. For normal incidence ($\theta_i = \theta_t = 0°$), these reduce, as they should, to our previously derived results for that case. Also, we find that

$$1 + \Gamma_\perp = T_\perp \tag{189}$$

as was the case for normal incidence. Equations (188*a*) and (188*b*) are often referred to as the *Fresnel equations.* We should note that θ_t is related to θ_i via Snell's law of refraction in (186), so that (188) may be calculated solely from the properties of the media and knowledge of the angle of incidence, θ_i.

6.9.2 Parallel Polarization

For the case of parallel polarization shown in Fig. 6.21*b*, we may write the phasor field vectors in the form

$$\hat{\mathbf{E}}_i = \hat{E}_i(\cos\theta_i \mathbf{a}_x - \sin\theta_i \mathbf{a}_z)e^{-j\beta_1(\sin\theta_i x + \cos\theta_i z)} \tag{190a}$$

$$\hat{\mathbf{H}}_i = \frac{\hat{E}_i}{\eta_1} e^{-j\beta_1(\sin\theta_i x + \cos\theta_i z)}\mathbf{a}_y \tag{190b}$$

$$\hat{\mathbf{E}}_r = -\hat{E}_r(\cos\theta_i \mathbf{a}_x + \sin\theta_i \mathbf{a}_z)e^{j\beta_1(-\sin\theta_i x + \cos\theta_i z)} \tag{190c}$$

$$\hat{\mathbf{H}}_r = \frac{\hat{E}_r}{\eta_1} e^{j\beta_1(-\sin\theta_i x + \cos\theta_i z)}\mathbf{a}_y \tag{190d}$$

$$\hat{\mathbf{E}}_t = \hat{E}_t(\cos\theta_t \mathbf{a}_x - \sin\theta_t \mathbf{a}_z)e^{-j\beta_2(\sin\theta_t x + \cos\theta_t z)} \tag{190e}$$

$$\hat{\mathbf{H}}_t = \frac{\hat{E}_t}{\eta_2} e^{-j\beta_2(\sin\theta_t x + \cos\theta_t z)}\mathbf{a}_y \tag{190f}$$

Again applying the boundary conditions of continuity of the tangential components of the electric and magnetic fields at $z = 0$ yields

$$\hat{E}_i \cos\theta_i e^{-j\beta_1 \sin\theta_i x} - \hat{E}_r \cos\theta_i e^{-j\beta_1 \sin\theta_i x} = \hat{E}_t \cos\theta_t e^{-j\beta_2 \sin\theta_t x} \tag{191a}$$

$$\frac{\hat{E}_i}{\eta_1} e^{-j\beta_1 \sin\theta_i x} + \frac{\hat{E}_r}{\eta_1} e^{-j\beta_1 \sin\theta_i x} = \frac{\hat{E}_t}{\eta_2} e^{-j\beta_2 \sin\theta_t x} \tag{191b}$$

Once again, Snell's law of refraction relates θ_i and θ_t as

$$\frac{\sin\theta_i}{\sin\theta_t} = \frac{\beta_2}{\beta_1} = \sqrt{\frac{\mu_2\epsilon_2}{\mu_1\epsilon_1}} \tag{192}$$

Thus, we obtain

$$\hat{E}_i \cos\theta_i - \hat{E}_r \cos\theta_i = \hat{E}_t \cos\theta_t \tag{193a}$$

$$\frac{\hat{E}_i}{\eta_1} + \frac{\hat{E}_r}{\eta_1} = \frac{\hat{E}_t}{\eta_2} \tag{193b}$$

Solving (193) yields the reflection and transmission coefficients for parallel polarization as

$$\Gamma_{\parallel} = \frac{\hat{E}_r}{\hat{E}_i} = \frac{\eta_1 \cos\theta_i - \eta_2 \cos\theta_t}{\eta_1 \cos\theta_i + \eta_2 \cos\theta_t}$$

$$= \frac{\eta_1 \cos\theta_i - \eta_2\sqrt{1 - (u_2/u_1)^2 \sin^2\theta_i}}{\eta_1 \cos\theta_i + \eta_2\sqrt{1 - (u_2/u_1)^2 \sin^2\theta_i}} \quad (194a)$$

$$T_{\parallel} = \frac{\hat{E}_t}{\hat{E}_i}$$

$$= \frac{2\eta_2 \cos\theta_i}{\eta_1 \cos\theta_i + \eta_2 \cos\theta_t}$$

$$= \frac{2\eta_2 \cos\theta_i}{\eta_1 \cos\theta_i + \eta_2\sqrt{1 - (u_2/u_1)^2 \sin^2\theta_i}} \quad (194b)$$

The alternative forms of these equations in terms of θ_i only are again obtained with Snell's law of refraction in (192) and by realizing that $\cos\theta = \sqrt{1 - \sin^2\theta}$. As in the previous development, the quantity $u = 1/\sqrt{\mu\epsilon}$ is the phase velocity in the appropriate medium. For this case it may be shown that

$$1 - \Gamma_{\parallel} = T_{\parallel}\left(\frac{\cos\theta_t}{\cos\theta_i}\right) \quad (195)$$

6.9.3 Brewster Angle of Total Transmission

For both parallel and perpendicular polarizations, the numerators of the reflection coefficients are the differences of two terms. This leads us to inquire whether a particular choice of the angle of incidence will lead to a reflection coefficient of zero, i.e., no reflected wave and therefore total transmission.

For perpendicular polarization, $\Gamma_{\perp} = 0$ if

$$\eta_2 \cos\theta_i = \eta_1 \cos\theta_t \quad (196)$$

Squaring both sides gives

$$\cos^2\theta_i = 1 - \sin^2\theta_i$$

$$= \left(\frac{\eta_1}{\eta_2}\right)^2 (1 - \sin^2\theta_t) \quad (197)$$

Substituting Snell's law of refraction yields

$$\sin^2 \theta_i = \frac{1 - (\eta_1/\eta_2)^2}{1 - (\eta_1 \beta_1/\eta_2 \beta_2)^2}$$

$$= \frac{1 - \mu_1 \epsilon_2/\mu_2 \epsilon_1}{1 - (\mu_1/\mu_2)^2} \tag{198}$$

The angle of incidence satisfying this equation is referred to as the *Brewster angle* and is denoted as $\theta_{B\perp}$. Note, however, that for materials which are not ferromagnetic, $\mu_1 = \mu_2 = \mu_0$, so that a Brewster angle for this polarization does not exist.

For parallel polarization, $\Gamma_{\parallel} = 0$ if

$$\eta_1 \cos \theta_i = \eta_2 \cos \theta_t \tag{199}$$

squaring both sides and substituting Snell's law of refraction yields

$$\sin \theta_i = \sqrt{\frac{1 - \mu_2 \epsilon_1/\mu_1 \epsilon_2}{1 - (\epsilon_1/\epsilon_2)^2}} \tag{200}$$

For typical dielectric materials, an angle of incidence satisfying this relation may exist. Such an angle is referred to as the Brewster angle for parallel polarization and is denoted as $\theta_{B\parallel}$.

For example, consider medium 1 as free space and medium 2 as glass having $\epsilon_{r2} = 4$ and $\mu_{r2} = 1$. The Brewster angle for parallel polarization is

$$\theta_{B\parallel} = \sin^{-1} \sqrt{\frac{1 - \frac{1}{4}}{1 - (\frac{1}{4})^2}}$$

$$= 63.4^\circ$$

FIGURE 6.22
Water-cooled krypton ion laser. Note the Brewster window at the end of the laser tube. *(Spectra-Physics, Laser Products Division.)*

Because of this ability to discriminate between parallel and perpendicular polarizations, quartz windows are placed at the ends of laser tubes, as shown in Fig. 6.22. These are placed at the Brewster angle $\theta_{B\parallel}$ to the axis of the tube such that the component of the unpolarized light within the tube that has parallel polarization is totally transmitted through the window. The part of the wave that has perpendicular polarization is partially reflected and partially transmitted; because of the reflection loss at the Brewster windows, this component will not resonate in the tube, resulting in an output that is linearly polarized.

6.10 Oblique Incidence on Perfect Conductors

Consider the oblique incidence of a wave traveling in a lossless medium on the surface of a perfect conductor ($\sigma_2 = \infty$). Once again we decompose a wave having some general polarization into parallel and perpendicular polarizations.

6.10.1 Perpendicular Polarization

From the results of Sec. 6.9.1, for $\sigma_2 = \infty$ we have, since $\eta_2 = 0$,

$$\Gamma_\perp = -1 \tag{201a}$$

$$T_\perp = 0 \tag{201b}$$

The incident and reflected fields can be written with these results from (184). Of particular interest are the total fields in medium 1. From (184*a*) to (184*d*) the total fields in medium 1 are

$$\begin{aligned}\hat{\mathbf{E}}_1 &= \hat{\mathbf{E}}_i + \hat{\mathbf{E}}_r \\ &= E_m e^{-j\beta_1 \sin\theta_i x}[e^{-j\beta_1 \cos\theta_i z} - e^{j\beta_1 \cos\theta_i z}]\mathbf{a}_y \\ &= -2jE_m e^{-j\beta_1 \sin\theta_i x} \sin(\beta_1 \cos\theta_i z)\mathbf{a}_y\end{aligned} \tag{202a}$$

$$\begin{aligned}\hat{\mathbf{H}}_1 &= -\frac{E_m}{\eta_1}\cos\theta_i e^{-j\beta_1 \sin\theta_i x}[e^{-j\beta_1 \cos\theta_i z} + e^{j\beta_1 \cos\theta_i z}]\mathbf{a}_x \\ &\quad + \frac{E_m}{\eta_1}\sin\theta_i e^{-j\beta_1 \sin\theta_i x}[e^{-j\beta_1 \cos\theta_i z} - e^{j\beta_1 \cos\theta_i z}]\mathbf{a}_z \\ &= -\frac{2E_m}{\eta_1}\cos\theta_i e^{-j\beta_1 \sin\theta_i x}\cos(\beta_1 \cos\theta_i z)\mathbf{a}_x \\ &\quad -2j\frac{E_m}{\eta_1}\sin\theta_i e^{-j\beta_1 \sin\theta_i x}\sin(\beta_1 \cos\theta_i z)\mathbf{a}_z\end{aligned} \tag{202b}$$

The time-domain results are obtained by multiplying these results by $e^{j\omega t}$ and taking the real part of the result:

$$\mathcal{E}_1 = 2E_m \sin(\beta_1 \cos\theta_i z)\sin(\omega t - \beta_1 \sin\theta_i x)\mathbf{a}_y \tag{203a}$$

$$\begin{aligned}\mathcal{H}_1 &= -2\frac{E_m}{\eta_1}\cos\theta_i \cos(\beta_1 \cos\theta_i z)\cos(\omega t - \beta_1 \sin\theta_i x)\mathbf{a}_x \\ &\quad + 2\frac{E_m}{\eta_1}\sin\theta_i \sin(\beta_1 \cos\theta_i z)\sin(\omega t - \beta_1 \sin\theta_i x)\mathbf{a}_z\end{aligned} \tag{203b}$$

The average power density vector becomes

$$\begin{aligned}\mathbf{S}_{av} &= \tfrac{1}{2}\,\mathrm{Re}\,[\hat{\mathbf{E}}_1 \times \hat{\mathbf{H}}_1^*] \\ &= \tfrac{1}{2}\,\mathrm{Re}\,[\hat{E}_y\mathbf{a}_y \times (\hat{H}_x^*\mathbf{a}_x + \hat{H}_z^*\mathbf{a}_z)] \\ &= -\tfrac{1}{2}\,\mathrm{Re}\,[\hat{E}_y\hat{H}_x^*]\mathbf{a}_z + \tfrac{1}{2}\,\mathrm{Re}\,[\hat{E}_y\hat{H}_z^*]\mathbf{a}_x\end{aligned} \tag{204}$$

Substituting the phasor field vectors from (202) yields

$$\mathbf{S}_{av} = 2\frac{E_m^2}{\eta_1}\sin\theta_i \sin^2(\beta_1 \cos\theta_i z)\mathbf{a}_x \tag{205}$$

Thus, the average power flow is in the $+x$ direction.

From the time-domain expressions in (203) we observe other interesting phenomena. The y component of $\mathcal{E}_1$ and the x component of $\mathcal{H}_1$ yield standing wave patterns in the z direction, varying with z as $\sin(\beta_1 \cos\theta_i z)$ and $\cos(\beta_1 \cos\theta_i z)$. Since these field components are 90° out of time phase, no average power is transmitted in the $\mathbf{a}_y \times \mathbf{a}_x = \mathbf{a}_z$ direction, as is shown by (205). The y component of $\mathcal{E}_1$ and the z-component of $\mathcal{H}_1$, however, are in phase and combine to yield a wave traveling in the x direction with phase velocity

$$\begin{aligned}u_x &= \frac{\omega}{\beta_1 \sin\theta_i} \\ &= \frac{u_1}{\sin\theta_i}\end{aligned} \tag{206}$$

This is a *nonuniform plane wave* since its magnitude varies with z according to $\sin(\beta_1 \cos\theta_i z)$.

Note that the y component of the electric field is zero at points away from the boundary where $\sin(\beta_1 \cos\theta_i z) = 0$; that, is,

$$\mathcal{E}_1 = 0 \qquad \text{at } \beta_1 \cos\theta_i z = -n\pi$$

for $n = 1, 2, 3, \ldots$. In terms of a wavelength in this medium, $\mathcal{E}_1 = 0$ at

$$z = -\frac{n\lambda_1}{2 \cos \theta_i} \tag{207}$$

Perfectly conducting planes could be inserted at these points without affecting the solution. We will find this interpretation to be useful in providing insight into the behavior of waveguides (as considered in Chap. 8).

The perfectly conducting surface supports a linear current density given by

$$\begin{aligned} \boldsymbol{\mathcal{K}} &= -\mathcal{H}_{1x}|_{z=0}\mathbf{a}_y \\ &= \frac{2E_m}{\eta_1} \cos \theta_i \cos (\omega t - \beta_1 \sin \theta_i x)\mathbf{a}_y \end{aligned} \tag{208}$$

Reflecting antennas utilize this result to reduce wind loading. These antennas are constructed of gratings consisting of parallel wires. The wires are oriented along the direction of $\boldsymbol{\mathcal{K}}$ so as to provide minimal disturbance to this surface current.

6.10.2 Parallel Polarization

From the results of Sec. 6.9.2, for $\sigma_2 = \infty$ we have, since $\eta_2 = 0$,

$$\Gamma_{\|} = +1 \tag{209a}$$

$$T_{\|} = 0 \tag{209b}$$

From Eqs. (190*a*) to (190*d*) the total fields in region 1 are

$$\begin{aligned} \hat{\mathbf{E}}_1 &= \hat{\mathbf{E}}_i + \hat{\mathbf{E}}_r \\ &= E_m \cos \theta_i e^{-j\beta_1 \sin \theta_i x}[e^{-j\beta_1 \cos \theta_i z} - e^{j\beta_1 \cos \theta_i z}]\mathbf{a}_x \\ &\quad - E_m \sin \theta_i e^{-j\beta_1 \sin \theta_i x}[e^{-j\beta_1 \cos \theta_i z} + e^{j\beta_1 \cos \theta_i z}]\mathbf{a}_z \\ &= -2jE_m \cos \theta_i e^{-j\beta_1 \sin \theta_i x} \sin (\beta_1 \cos \theta_i z)\mathbf{a}_x \\ &\quad - 2E_m \sin \theta_i e^{-j\beta_1 \sin \theta_i x} \cos (\beta_1 \cos \theta_i z)\mathbf{a}_z \end{aligned} \tag{210a}$$

$$\begin{aligned} \hat{\mathbf{H}}_1 &= \hat{\mathbf{H}}_i + \hat{\mathbf{H}}_r \\ &= \frac{E_m}{\eta_1} e^{-j\beta_1 \sin \theta_i x}[e^{-j\beta_1 \cos \theta_i z} + e^{j\beta_1 \cos \theta_i z}]\mathbf{a}_y \\ &= 2\frac{E_m}{\eta_1} \cos (\beta_1 \cos \theta_i z)e^{-j\beta_1 \sin \theta_i x}\mathbf{a}_y \end{aligned} \tag{210b}$$

The time-domain expressions are again obtained by multiplying these results by $e^{j\omega t}$ and taking the real part of the result:

$$\mathcal{E}_1 = 2E_m \cos\theta_i \sin(\beta_1 \cos\theta_i z)\sin(\omega t - \beta_1 \sin\theta_i x)\mathbf{a}_x$$
$$-2E_m \sin\theta_i \cos(\beta_1 \cos\theta_i z)\cos(\omega t - \beta_1 \sin\theta_i x)\mathbf{a}_z \tag{211a}$$

$$\mathcal{H}_1 = \frac{2E_m}{\eta_1}\cos(\beta_1 \cos\theta_i z)\cos(\omega t - \beta_1 \sin\theta_i x)\mathbf{a}_y \tag{211b}$$

The average power density vector becomes

$$\begin{aligned}\mathbf{S}_{av} &= \tfrac{1}{2}\,\text{Re}\,[\hat{\mathbf{E}}_1 \times \hat{\mathbf{H}}_1^*] \\ &= \tfrac{1}{2}\,\text{Re}\,[(\hat{E}_x\mathbf{a}_x + \hat{E}_z\mathbf{a}_z) \times \hat{H}_y^*\mathbf{a}_y] \\ &= \tfrac{1}{2}\,\text{Re}\,[\hat{E}_x\hat{H}_y^*]\mathbf{a}_z - \tfrac{1}{2}\,\text{Re}\,[\hat{E}_z\hat{H}_y^*]\mathbf{a}_x\end{aligned} \tag{212}$$

Substituting the phasor field vectors from (210) yields

$$\mathbf{S}_{av} = 2\frac{E_m^2}{\eta_1}\sin\theta_i \cos^2(\beta_1 \cos\theta_i z)\mathbf{a}_x \tag{213}$$

Once again we see that average power flow is in the x direction.

From the time-domain forms in (211) we see that the x component of $\mathcal{E}_1$ and the y component of $\mathcal{H}_1$ yield standing wave patterns in the z direction according to $\sin(\beta_1 \cos\theta_i z)$ and $\cos(\beta_1 \cos\theta_i z)$. Similarly, the z component of $\mathcal{E}_1$ and the y component of $\mathcal{H}_1$ yield a traveling wave propagating in the x direction with phase velocity

$$\begin{aligned}u_x &= \frac{\omega}{\beta_1 \sin\theta_i} \\ &= \frac{u_1}{\sin\theta_i}\end{aligned} \tag{214}$$

as for perpendicular polarization. This wave is again a nonuniform plane wave since its amplitude varies with z as $\cos(\beta_1 \cos\theta_i z)$. Again the x component of the electric field is zero in xy planes for z such that

$$\beta_1 \cos\theta_i z = -n\pi$$

for $n = 1, 2, 3 \ldots$. Thus, conducting planes may be inserted at

$$z = -\frac{n\lambda_1}{2\cos\theta_i} \tag{215}$$

away from the boundary without affecting the solutions, which is identical to the case for perpendicular polarization. Again, we will find this observation to be useful in explaining the operation of waveguides in Chap. 8.

Once again, the perfect conductor supports a linear current density given by

$$\mathcal{K} = \mathcal{H}_{1y}|_{z=0}\mathbf{a}_x$$
$$= \frac{2E_m}{\eta_1} \cos(\omega t - \beta_1 \sin\theta_i x)\mathbf{a}_x \tag{216}$$

6.11 Summary

In this chapter we have begun our study of electromagnetic wave propagation, which is one of the more important consequences of Maxwell's equations. Our study concentrated on a particularly important form of wave, the uniform plane wave. We will find in subsequent chapters that these uniform plane waves exhibit many of the characteristics of waves on transmission lines (Chap. 7) and in waveguides (Chap. 8), and waves radiated from antennas (Chap. 9).

An important consequence of this chapter is that the phasor electric and magnetic field vectors of a uniform plane wave propagating in the $+z$ direction (the forward-traveling wave) have a z dependence of the form

$$e^{-\gamma z}$$

where the propagation constant γ is

$$\gamma = \alpha + j\beta$$
$$= \sqrt{j\omega\mu(\sigma + j\omega\epsilon)}$$

and the medium is characterized by σ, μ, and ϵ. Similarly, waves propagating in the $-z$ direction (the backward-traveling wave) have a z dependence of

$$e^{\gamma z}$$

The electric field and magnetic field of each wave are related by the intrinsic impedance of the medium:

$$\hat{\eta} = \sqrt{\frac{j\omega\mu}{\sigma + j\omega\epsilon}}$$

The electric and magnetic fields of each wave are orthogonal in space, and the Poynting vector of each wave gives the direction of power flow for that wave.

Materials were characterized as good conductors or good dielectrics according to whether the conduction current was greater than or less than the displacement current, respectively.

The concept of skin depth as the effective depth of penetration of the fields into the conductor was introduced. This illustrated why effective electromagnetic shielding can be accomplished by using a metal to enclose a region. As the frequency is increased, the skin depth decreases, and the fields are effectively confined to a thin strip of the surface.

Polarization of uniform plane waves was considered. It was shown that circular and elliptical polarizations can be viewed as a combination of two linearly polarized waves. Our exclusive study of linearly polarized uniform plane waves is therefore quite general.

Normal incidence of uniform plane waves on plane boundaries between two media was also considered. The incident and reflected electric fields are related by the reflection coefficient $\hat{\Gamma}$, and the incident and transmitted electric fields are related by the transmission coefficient $\hat{T}$. The resulting magnetic fields are related to the appropriate electric fields by the intrinsic impedance of the appropriate medium. The polarity of these vectors results in power flow in the proper direction. For a wave incident on a perfect conductor, a linear current density is generated on the surface, which may be thought of as producing a reflected wave.

Oblique incidence of uniform plane waves on plane boundaries was considered. The general case of arbitrary polarization of the electric field of the incident wave was indirectly investigated by decomposing it into perpendicular and parallel polarizations and studying these cases separately. Snell's laws of reflection and refraction, which relate the angles of reflection and transmission to the angle of incidence, were obtained. It was also found that particular angles of incidence result in no transmitted wave (the critical angle) or no reflected wave (Brewster angle).

In the case of oblique incidence, we assumed each medium to be either a perfect dielectric ($\sigma = 0$) or a perfect conductor ($\sigma = \infty$). It is worth pointing out that lossy material media having finite but nonzero conductivities can easily be considered if we substitute a complex permittivity $\hat{\epsilon} = \epsilon(1 - j(\sigma/\omega\epsilon))$ for the real permittivity in all of these previous developments. In particular, one can show that Snell's law of reflection, $\theta_i = \theta_r$, holds if either or both of the media are lossy ($\sigma \neq 0$). Similarly, Snell's law of refraction holds if we substitute this complex permittivity, although one must be careful in interpreting the results.

In subsequent chapters we will investigate means of guiding or focusing these waves. Transmission lines (Chap. 7) and waveguides (Chap. 8) serve to guide waves or energy propagation from one point to another. Antennas (Chap. 9) serve to launch and focus these waves, which also facilitates energy propagation.

Problems

6-1 Prove the important vector identity

$$\nabla \times \nabla \times \mathbf{A} = \nabla(\nabla \cdot \mathbf{A}) - \nabla^2 \mathbf{A}$$

for any vector $\mathbf{A}$, where $\nabla^2 \mathbf{A}$ is vector laplacian expressed in rectangular coordinates as $\nabla^2 \mathbf{A} = \nabla^2 A_x \mathbf{a}_x + \nabla^2 A_y \mathbf{a}_y + \nabla^2 A_z \mathbf{a}_z$.

6-2 Show by direct substitution that the pair of field vectors

$$\boldsymbol{\mathcal{E}}_1 = E_{10}\cos(\omega t - \beta z)\mathbf{a}_x \qquad \beta = \omega\sqrt{\mu\epsilon}$$

$$\boldsymbol{\mathcal{H}}_1 = \frac{E_{10}}{\eta}\cos(\omega t - \beta z)\mathbf{a}_y \qquad \eta = \sqrt{\frac{\mu}{\epsilon}}$$

satisfies the time-domain Maxwell's equations for a lossless, linear, homogeneous, isotropic, charge-free medium characterized by μ, ϵ. Repeat for the pair of field vectors

$$\boldsymbol{\mathcal{E}}_2 = E_{20}\cos(\omega t + \beta z)\mathbf{a}_x$$

$$\boldsymbol{\mathcal{H}}_2 = -\frac{E_{20}}{\eta}\cos(\omega t + \beta z)\mathbf{a}_y$$

6-3 The amplitude and phase of a uniform plane wave is constant over any plane that is normal to the direction of propagation. These are not the only possible forms of waves that satisfy Maxwell's equations. For example, nonuniform waves have surfaces of constant amplitude and constant phase that are not the same. Show that the following nonuniform wave satisfies the phasor form of Maxwell's equations:

$$\hat{\mathbf{E}} = E_0 e^{-\alpha y} e^{-j\beta z}\mathbf{a}_x$$

$$\hat{\mathbf{H}} = \frac{\beta}{\omega\mu} E_0 e^{-\alpha y} e^{-j\beta z}\mathbf{a}_y + j\frac{\alpha}{\omega\mu} E_0 e^{-\alpha y} e^{-j\beta z}\mathbf{a}_z$$

where $$\alpha^2 - \beta^2 = -\omega^2\mu\epsilon$$

and the medium has $\sigma \neq 0$. Note that the plane of constant amplitude is normal to the y direction, whereas the plane of constant phase is normal to the z direction.

6-4 Suppose that a uniform plane wave is traveling in the x direction in a lossless medium, with the 100-V/m electric field in the z direction. If the wavelength is 25 cm and the velocity of propagation is 2×10^8 m/s, determine the frequency of the wave and the relative permittivity of the medium if the medium is characterized by free-space permeability. Write complete time-domain expressions for the electric and magnetic field vectors.

6-5 Write a time-domain expression for the electric field of a uniform plane wave if the magnetic field is given by

$$\boldsymbol{\mathcal{H}} = 0.1e^{-200y}\cos(2\pi \times 10^{10}t - 300y)\mathbf{a}_x \qquad \text{A/m}$$

and the medium is characterized by free-space permeability.

6-6 If a material has $\sigma = 2$ S/m, $\epsilon_r = 9$, and $\mu_r = 16$ at a frequency of 1 GHz, calculate the attenuation constant, phase constant, and intrinsic impedance for the material at this frequency.

6-7 A 5-GHz uniform plane wave is propagating in a material characterized by $\epsilon_r = 2.53$, $\mu_r = 1$, and $\sigma = 0$. If the electric field is given by $\mathcal{E} = 10 \cos(10\pi \times 10^9 t - \beta z)\mathbf{a}_x$, determine

(a) u,
(b) λ,
(c) β, and
(d) the magnitude of the magnetic field. Write the time-domain expression for the magnetic field.

6-8 Suppose that a 2-GHz uniform plane wave is propagating in a lossy medium that has a loss tangent of 10^{-2} at 2 GHz. At 2 GHz, the relative permittivity of the medium is 2.25 and the relative permeability is 1. Determine the conductivity, the attenuation constant, the phase constant, and the intrinsic impedance at 2 GHz. Compute these using approximations and then using exact equations, and compare the results.

6-9 Repeat Prob. 6-8 if the loss tangent is 10^2.

6-10 A uniform plane wave is propagating in a medium having the properties $\epsilon_r = 36$, $\mu_r = 4$, and $\sigma = 1$ S/m. The electric field is given by

$$\mathcal{E} = 100e^{-\alpha x} \cos(10\pi \times 10^8 t - \beta x)\mathbf{a}_z \qquad \text{V/m}$$

Determine α and β and write a time-domain expression for the associated magnetic field vector.

6-11 Determine the frequency range for which the conduction current exceeds the displacement current by a factor of at least 100 in seawater ($\sigma = 4$ S/m, $\mu_r \simeq 1$, and $\epsilon_r \simeq 81$). This may be considered to be the frequency range for which sea water is a good conductor.

6-12 Determine the phase velocity, attenuation constant, phase constant, skin depth, and intrinsic impedance of a uniform plane wave traveling in wet, marshy soil ($\sigma \simeq 10^{-2}$ S/m, and $\epsilon_r \simeq 15$, and $\mu_r \simeq 1$) at 60 Hz (power frequency), 1 MHz (AM radio broadcast frequency), 100 MHz (FM radio broadcast frequency), and 10 GHz (microwave relay frequency). Also compute the distance for the wave to travel at each frequency such that its amplitude is reduced by a factor of 20 dB (1/10).

6-13 A uniform plane wave has a wavelength of 2 cm in free space and 1 cm in a perfect dielectric ($\sigma = 0$, and $\mu_r \simeq 1$). Determine the relative permittivity of the dielectric.

6-14 Compare the distances required for a uniform plane wave to travel in sea water ($\sigma = 4$ S/m, $\mu_r \simeq 1$, and $\epsilon_r \simeq 81$) in order that the amplitude is reduced by 80 dB (a factor of 10,000) at the following frequencies:

(a) 1 kHz
(b) 10 kHz
(c) 100 kHz
(d) 1 MHz
(e) 10 MHz
(f) 100 MHz

This illustrates why low frequencies are used to communicate with submarines.

6-15 Consider a 100-V/m 7-GHz uniform plane wave traveling in sea water ($\sigma = 4$ S/m, and $\epsilon_r \simeq 81$) and compute the power dissipated in a surface block of sea water having a surface area of 10 cm^2 as the wave travels a distance of five skin depths. Repeat this calculation for a frequency of 10 kHz.

6-16 Show that the exact expressions for the attenuation and phase constants can be written explicitly as

$$\alpha = \frac{\beta'}{\sqrt{2}}[\sqrt{1 + \tan^2 \delta} - 1]^{1/2} \qquad \text{Np/m}$$

$$\beta = \frac{\beta'}{\sqrt{2}}[\sqrt{1 + \tan^2 \delta} + 1]^{1/2} \qquad \text{rad/m}$$

where $\beta' = \omega\sqrt{\mu\epsilon}$ and $\tan \delta = \sigma/\omega\epsilon$.

6-17 A 100-MHz uniform plane wave traveling in a lossy dielectric ($\mu_r \simeq 1$) has the following phasor expression for the magnetic field intensity vector:

$$\hat{\mathbf{H}} = (1\mathbf{a}_y + j2\mathbf{a}_z)e^{-0.2x}e^{-j2x}$$

Write complete time-domain expressions for the electric and magnetic field vectors.

6-18 Show that a linearly polarized wave can be written as the sum of two circularly polarized waves rotating in opposite directions but at the same angular rate.

6-19 Determine the polarization of the following uniform plane waves:

(a) $\boldsymbol{\mathcal{E}} = 1 \cos(\omega t + \beta z)\mathbf{a}_x + 1 \sin(\omega t + \beta z)\mathbf{a}_y$
(b) $\boldsymbol{\mathcal{E}} = 1 \cos(\omega t + \beta z)\mathbf{a}_x - 1 \sin(\omega t + \beta z)\mathbf{a}_y$
(c) $\boldsymbol{\mathcal{E}} = 1 \cos(\omega t + \beta z)\mathbf{a}_x - 2 \sin(\omega t + \beta z - 45°)\mathbf{a}_y$

6-20 A 100-kHz carrier is amplitude-modulated with a narrowband voice signal. This composite signal propagates as a wave in sea water ($\sigma = 4$ S/m, and $\epsilon_r \simeq 81$). Determine the phase and group velocities of this signal and comment on the type of dispersion exhibited.

6-21 A uniform plane wave traveling in free space is incident-normal to the surface of a perfect conductor. If the total electric field is zero at a distance of 1 m away from the surface of the perfect conductor, determine the lowest possible frequency of the incident wave.

6-22 A uniform plane wave whose electric field is given by

$$\boldsymbol{\mathcal{E}}_i = 100 \cos(\omega t - 6\pi x)\mathbf{a}_z \qquad \text{V/m}$$

is incident from a region having $\epsilon_r = 4$, $\mu_r = 1$, and $\sigma = 0$ normal to the plane surface of a material having $\epsilon_r = 9$, $\mu_r = 4$, and $\sigma = 0$. Write complete time-domain expressions for the incident, reflected, and transmitted electric and magnetic fields, and determine the average power transmitted through a 2-m^2 area of the surface.

6-23 A 10-V/m uniform plane wave of frequency 3 MHz is incident from free space normal to the surface of a material having $\epsilon_r = 4$, $\mu_r = 1$, and $\sigma = 10^3$ S/m. Determine the average power dissipated in a volume of the material that consists of a 2-m^2 area of the surface 1 mm deep.

6-24 A uniform plane wave of 200 MHz traveling in free space strikes a large block of a material having $\epsilon_r = 4$, $\mu_r = 9$, and $\sigma = 0$ normal to the surface. If the incident magnetic field intensity vector is given by

$$\mathcal{H}_i = 1 \cos(\omega t - \beta y)\mathbf{a}_z \qquad \text{A/m}$$

write complete time-domain expressions for the incident, reflected, and transmitted field vectors. Determine the average power crossing a 5-m^2 area of the surface.

6-25 A 300-MHz uniform plane wave traveling in free space strikes a large block of copper ($\mu_r \simeq 1$, $\epsilon_r \simeq 1$, and $\sigma = 5.8 \times 10^7$ S/m) normal to the surface. If the surface of the copper lies in the yz plane and the wave is propagating in the x direction, write complete time-domain expressions for the incident, reflected, and transmitted field vectors. Assume that the magnitude of the incident electric field is 1 V/m and that the field is in the z direction. Calculate the average power dissipated in a rectangular block of the copper that lies between the surface and three skin depths of the interior and has an area on the surface of the copper of 2m^2.

6-26 A uniform plane wave is incident-normal to the surface of a large conducting medium. The current density induced in the conducting medium by the transmitted electric field, $\mathcal{J}_t = \sigma \mathcal{E}_t$, decays exponentially with skin depth, of course, and is therefore not uniformly distributed in the medium. Show, however, that the total current per unit width in the plane of the surface (for an infinitely deep surface) is the same as would be obtained if $\mathcal{J}_t$ were constant from its value at the surface to a depth of one skin depth and zero elsewhere. (This result provides added importance to the concept of skin depth.) Also evaluate the surface impedance $\hat{Z}_S$, defined by $\hat{Z}_S = \hat{E}_0/\hat{J}_T$, where $\hat{E}_0$ is the tangential component of the electric field at the conductor surface and $\hat{J}_T$ is the total current density per unit of conductor width carried by the conductor.

6-27 A wave from free space strikes a coated dielectric region, as shown in Fig. P6.27. The coating material has $\mu_1 = \mu_0$ and $\epsilon_1 = 4\epsilon_0$ and is a quarter-wavelength in

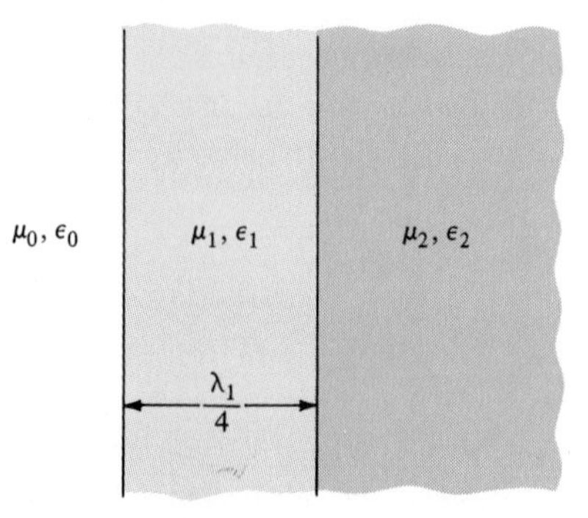

FIGURE P6.27

thickness referred to the velocity of propagation in that material. The coated material has constants $\mu_2 = \mu_0$ and $\epsilon_2 = \epsilon_2' \epsilon_0$ and may be considered infinite in extent. It is observed that in the space in front of the system, 1 percent of the incident plane-wave energy is reflected, and that an electric field maximum occurs at the surface of reflection. Determine ϵ_2'.

6-28 Airplanes use radar altimeters to determine their low-level altitude accurately. If an airplane is flying over the ocean ($\epsilon_r \simeq 81$, $\mu_r \simeq 1$, and $\sigma \simeq 4$ S/m) and the radar frequency is 7 GHz, determine the percent reduction in received power due to the ocean.

6-29 A light ray traveling in air is incident at an angle θ on a sheet of transparent material of thickness t having an index of refraction n, as shown in Fig. P6.29. Determine the distance d that the ray will be displaced from its path upon exiting the material.

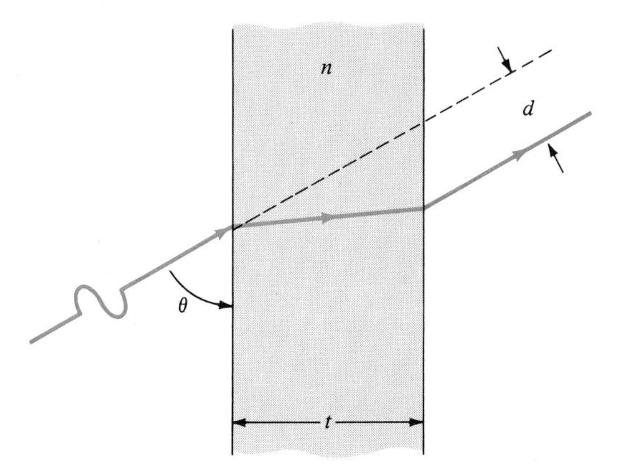

FIGURE P6.29

6-30 A glass isosceles prism is used to change the path of a light ray, as shown in Fig. P6.30. If the index of refraction of glass is 1.5, determine the ratio of transmitted and incident power densities.

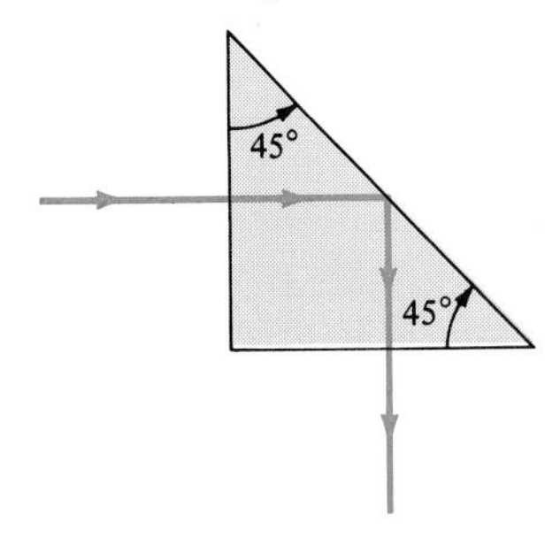

FIGURE P6.30

6-31 A person fishing from a boat observes a fish feeding on the bottom of a shallow lake. The person's height is 6 ft, and the lake depth at this location is 10 ft. The fish appears to be a distance of 20 ft from the boat. Determine the true distance of the fish from the boat.

6-32 A coating of lossless dielectric of thickness t is applied to the surface of a perfect conductor. Determine the effective reflection coefficient at the air interface of this dielectric for normal incidence. (*Hint*: Write the general expressions for uniform plane waves in the two regions, the air and the dielectric, and match the boundary conditions.) Show that if the thickness of the dielectric is chosen to be $\lambda/4$, where λ is the wavelength in the dielectric material, the dielectric appears to be invisible. Evaluate this thickness if the coating is glass ($\epsilon_r = 2.25$) and the incident wave is a ray of blue light ($\lambda = 450$ nm).

6-33 Effective shielding of rooms and electronic equipment from the effects of external electromagnetic fields can be obtained with conductive barriers. The shielding effectiveness (SE) is often expressed as the ratio of the magnitudes of the transmitted and incident electric fields, as shown in Fig. P6.33:

$$\text{SE} = \frac{|E_t|}{|E_i|}$$

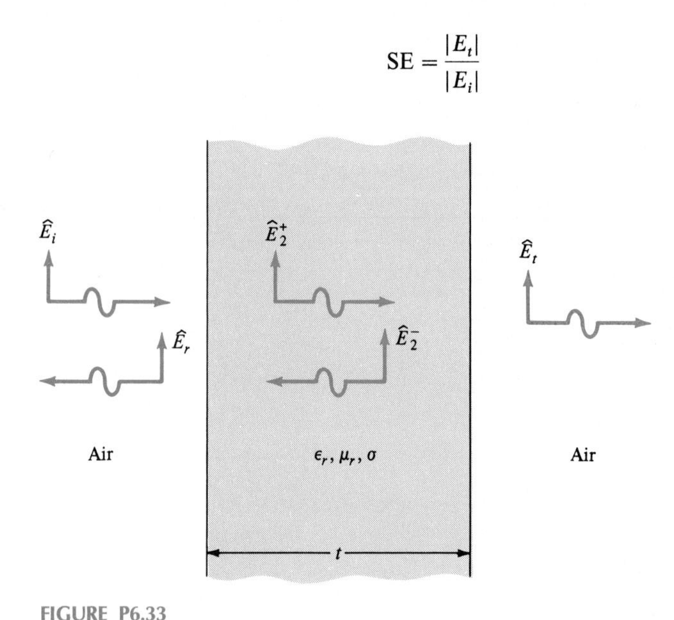

FIGURE P6.33

Assume that the barrier is a "good" conductor at the frequency of the incident wave and is of thickness t. Determine an expression for the shielding effectiveness. Use reasonable approximations. Compute the shielding effectiveness in decibels (dB):

$$\text{SE}_{\text{dB}} = 20 \log \left(\frac{1}{\text{SE}}\right)$$

for a copper plate 1 mil thick at 100 MHz.

6-34 A 100-MHz uniform plane wave traveling in free space is incident on the surface of a perfect conductor at an angle of 30° with the surface. The electric field is polarized parallel to the conductor surface. Determine the minimum nonzero distance away from the conductor at which the total electric field is zero.

6-35 A uniform plane wave is incident on a thick sheet of Teflon ($\epsilon_r = 2.1$, and $\sigma = 0$). The Teflon surface lies in the yz plane at $x = 0$. The incident electric field is given by

$$\mathcal{E}_i = 1.73 \cos(\omega t - 0.5\pi y - 0.866\pi x)\mathbf{a}_y - \cos(\omega t - 0.5\pi y - 0.866\pi x)\mathbf{a}_x \qquad \text{V/m}$$

Write complete time-domain expressions for the reflected, transmitted, and incident fields. Determine the frequency of the wave.

6-36 A uniform plane wave traveling in free space strikes the surface of a perfect conductor. The surface of the perfect conductor lies in the xz plane at $y = 0$ and extends for $y > 0$. The incident electric field is given by

$$\mathcal{E}_i = (2\mathbf{a}_x - 4\mathbf{a}_y + 3\mathbf{a}_z) \cos(\omega t - 1.5\pi z - 1.12\pi y) \qquad \text{V/m}$$

Write complete time-domain expressions for all the field vectors. Determine the frequency of the wave. (*Hint*: You will learn more and the solution will be simpler if you write the equations for the waves without trying to relate them to previously derived results.)

CHAPTER 7

Transmission Lines

Uniform plane waves, discussed in the previous chapter, are examples of unguided wave propagation in the sense that once they are propagated in an infinite block of material they continue to propagate in the same direction. On the other hand, transmission lines (along with waveguides, discussed in Chap. 8), serve to guide the propagation of energy from one point to another. Some common examples of transmission lines are shown in Fig. 7.1. Figure 7.1*a* shows a pair of parallel wires (solid, circular, cylindrical conductors), which serve to guide the propagation of energy from the source [characterized by $V_S(t)$ and R_S] to the load (characterized by R_L). In Fig. 7.1*b* the wire and the ground plane serve to guide the wave propagation, and in Fig. 7.1*c* the wave propagation takes place within the space enclosed by the overall circular, cylindrical shield. The transmission line in Fig. 7.1*c* is commonly referred to as a *coaxial cable.* There exist many other examples of transmission lines. Transmission lines consisting of more than two conductors are called *multiconductor transmission lines*; these are commonly found in numerous electronic systems in the form of bundles of wires or in power transmission systems in the form of high-voltage three-phase transmission lines situated above earth.

Transmission lines occur in many forms. It is becoming increasingly common to utilize conductors having rectangular cross sections. One of the more predominant ways of interconnecting electronic circuit elements is with printed-circuit boards (PCBs); a typical PCB is shown in Fig. 7.2*a*. Thin conductors having rectangular cross sections (*lands*) are etched on one side (single-sided) or both sides (double-sided) of a thin glass-epoxy board ($\epsilon_r \simeq 4.7$). Multilayer boards have these conductors imbedded in the board at various levels. Typical board thicknesses range from 47 to 64 mils (1 mil = 1/1000 in). Land widths vary from 10 to 100 mils, with typical thicknesses of 1.4 mils. PCBs have

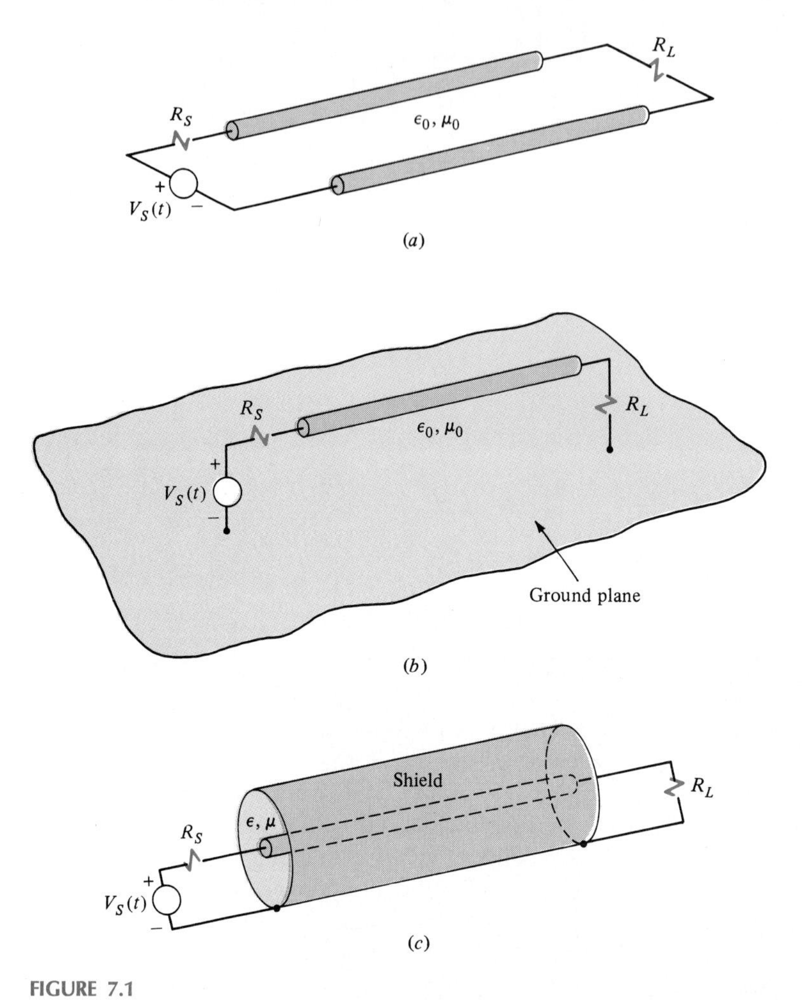

FIGURE 7.1

Typical transmission-line configurations. *(a)* A parallel-wire line; *(b)* a wire above a ground plane, *(c)* a coaxial line (cable).

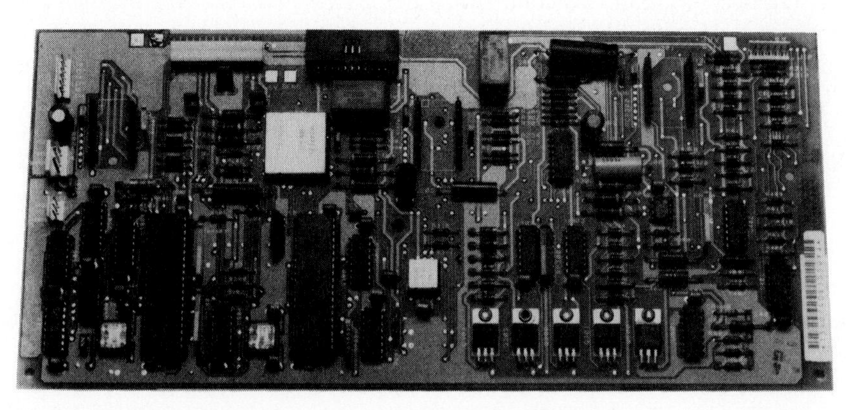

(a)

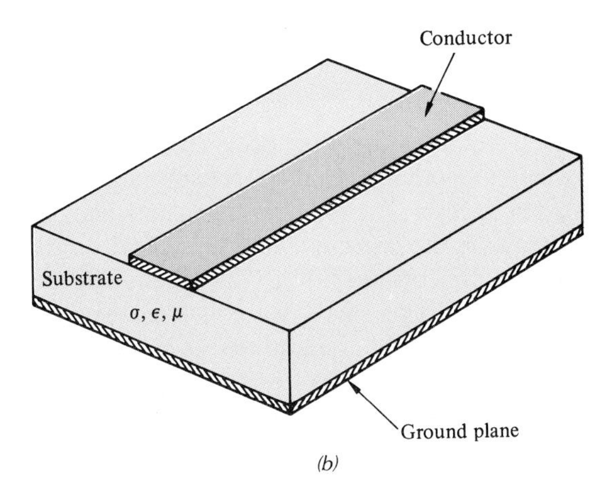

(b)

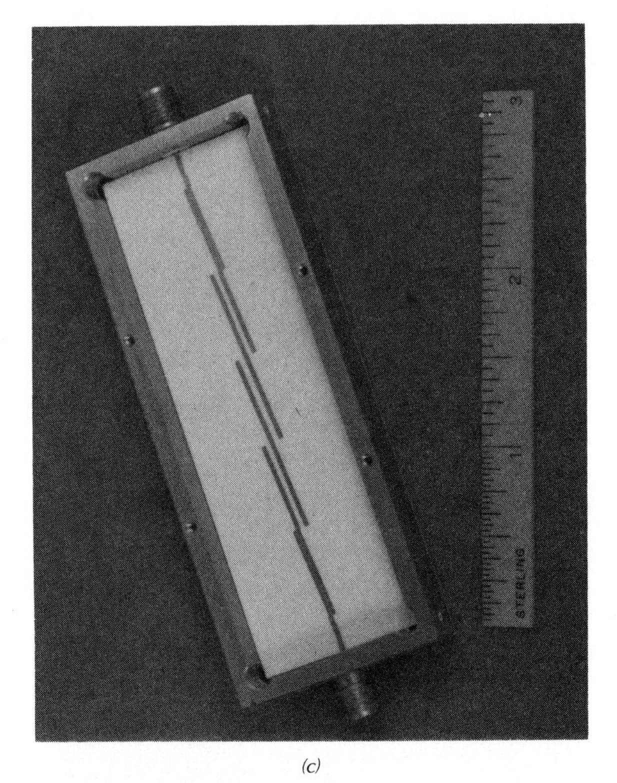

(c)

FIGURE 7.2

(a) A printed circuit board. Note the lands connecting the components. *(Information Products Division, International Business Machines Corporation.)* *(b)* Microstrip transmission line. *(c)* A 2.2- to 2.3-GHz telemetry bandpass filter using parallel, coupled, transmission lines. Insertion loss 1 dB with 40-dB rejection at 2.05 and 2.55 GHz. *(Trak Microwave Corporation.)*

eliminated many of the discrete wiring practices of the past and also serve to compact and interconnect very large numbers of electronic components in predominantly digital systems.

Another important example of a transmission line occurs in integrated circuits. Relatively small-scale electronic circuits can be constructed on a dielectric substrate. The conductors connecting the electronic elements may be deposited in the form of metallic strips on the substrate, and these devices are generally referred to as *microstrip* transmission lines (Fig. 7.2*b*). Although the physical dimensions of these circuits are quite small, we will find that the electrical dimensions (dimensions in wavelengths) determine the electromagnetic properties of the circuit. As the frequencies of operation continually increase, these "small" physical dimensions become increasingly important.

Transmission lines having large electrical dimensions (significant portions of a wavelength) can be designed to perform useful functions. For example, a microwave filter is shown in Fig. 7.2*c*. This filter can effectively isolate or filter out certain frequency components of a signal that are in the gigahertz frequency range.

All the transmission lines in Figs. 7.1 and 7.2*b* are said to be *uniform* lines in the sense that cross-sectional views of the line at any two points along the line yield the same picture. The lines in Fig. 7.1 are also said to be lines in a *homogeneous medium* since the medium surrounding the conductors is homogeneous. For the coaxial line in Fig. 7.1*c*, the propagation takes place within the space characterized by ϵ and μ, which may be considered to be a general dielectric, whereas the space surrounding the wires in Figs. 7.1*a* and *b* is logically considered to be free space characterized by ϵ_0 and μ_0. The microstrip line in Fig. 7.2*b*, on the other hand, is immersed in an inhomogeneous medium consisting of free space and the substrate. The conductors (lands) of a PCB are similarly immersed in an inhomogeneous medium consisting of free space and the board ($\epsilon_r \simeq 4.7$).

In the following sections we will investigate the propagation of energy on uniform two-conductor transmission lines in homogeneous media. The structures in Fig. 7.1 are common examples of these types of lines. Although we will only consider lines in a homogeneous medium, the general techniques that we will study also apply (with certain approximations) to lines in an inhomogeneous medium, such as microstrip lines and printed-circuit boards.

7.1 TEM Waves on Lossless Transmission Lines

We will begin our study by considering lossless transmission lines. A *lossless* transmission line is one in which the line conductors are perfect conductors and the medium surrounding the conductors is a perfect (lossless) dielectric.

The fundamental field structure on a transmission line is the *transverse electromagnetic* (*TEM*) field structure. A TEM field structure is one in which the electric and magnetic field vectors at each point in space have no components in the direction of propagation. Stated another way, a TEM field structure is one in which the electric and magnetic field vectors at each point in space lie in a plane transverse or orthogonal to the direction of propagation. The uniform plane wave considered in the previous chapter is a special case of a TEM wave. Uniform plane waves require not only that both field vectors lie in a plane perpendicular to the direction of propagation but also that these field vectors are independent of position in these planes (uniform). In a general TEM wave, the field vectors are not necessarily independent of position over these planes. If we assume the direction of propagation for a TEM wave to be the z direction, we require that $\mathcal{E}_z = \mathcal{H}_z = 0$ for a general TEM field structure.

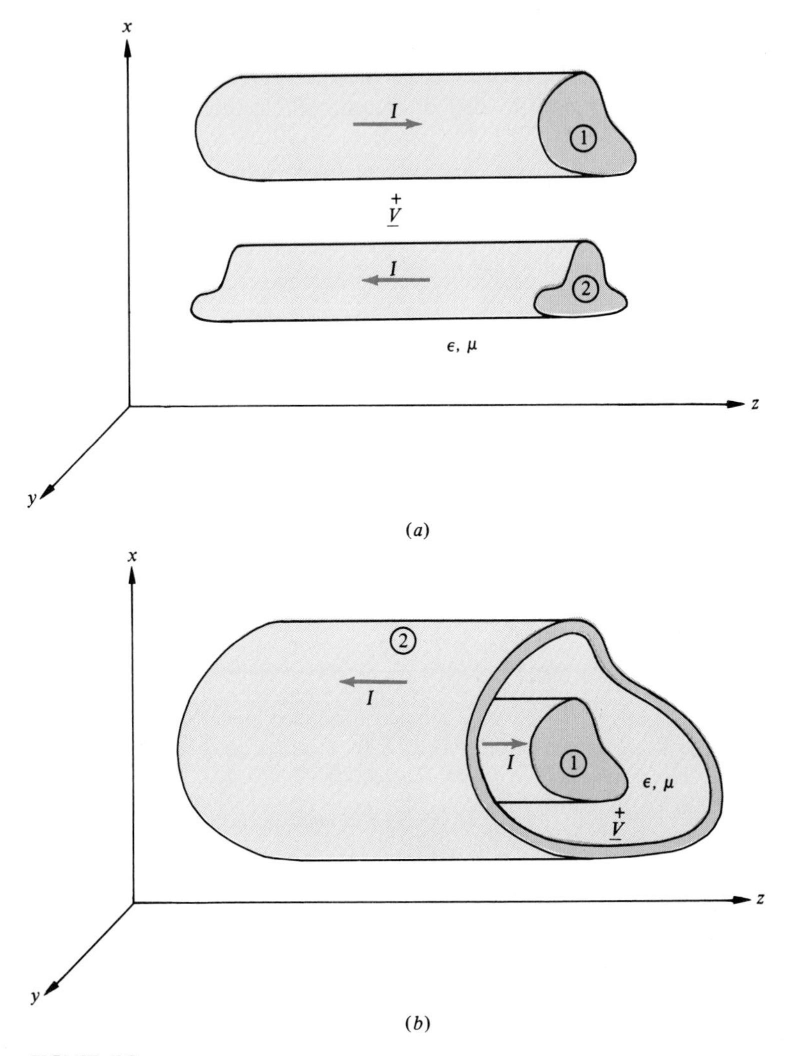

FIGURE 7.3

General, two-conductor transmission lines: *(a)* open-conductor line; *(b)* coaxial line.

To illustrate the generality of our results, we will consider general two-conductor uniform transmission lines having arbitrary cross sections, as shown in Fig. 7.3. The axis of the line is the z axis of a rectangular coordinate system, and the line cross section lies in the xy plane. If the two conductors of the line are infinite in length (so that we may neglect fringing of the fields at the conductor endpoints) and we apply a dc (0-Hz) voltage between the two conductors, the electric field lines will lie in the xy plane, transverse to the line (z) axis, as shown in Fig. 7.4; no component of the electric field will be directed in the z direction. A dc current along these conductors directed in the z direction will also produce magnetic field lines transverse to the line axis. As the frequency of excitation is increased, we would expect the field lines to maintain (to a reasonable approximation) this transverse structure up to some higher frequency. Energy obviously propagates along the line. (The Poynting vector is directed in the z direction, indicating power flow in this direction.) Thus, this is a TEM field structure.

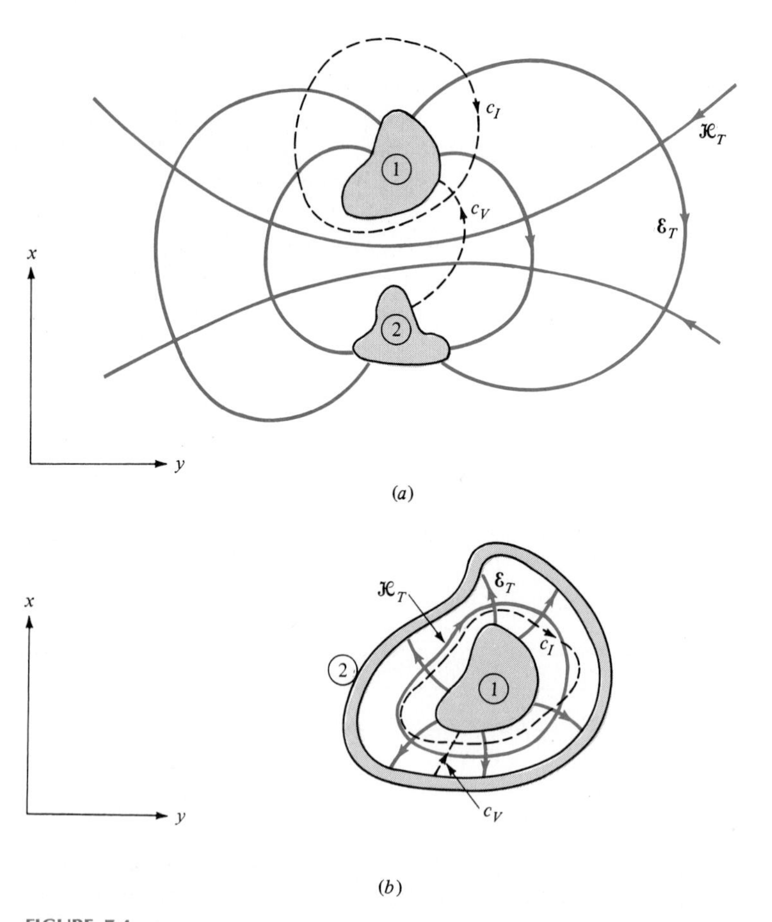

FIGURE 7.4
Transmission-line cross-sectional field distributions: *(a)* open-conductor line; *(b)* coaxial line.

There exist other possible field structures for which the electric and magnetic field vectors are not *both* transverse to the line axis, as we shall see in Chap. 8 when we consider waveguides. However, these field structures (or modes) have cutoff frequencies associated with them, above which these field structures propagate along the line and below which the fields are greatly attenuated (reduced in magnitude) in relatively short distances along the line. Thus, unless the frequency of excitation of the line is above the cutoff frequencies of these "higher-order" modes, energy propagation takes place via the TEM mode only (which clearly has a cutoff frequency of 0 Hz). The cutoff frequencies of these higher-order modes depend on the cross-sectional dimensions of the line (e.g., the conductor separation distances).† So long as the cross-sectional dimensions of the line are electrically small (i.e., much less than a wavelength), the higher-order modes are cut off and the TEM mode is the only mode of propagation on the line. For practical line dimensions, this allows us to assume that the TEM mode is the only mode of propagation up to frequencies on the order of 100 MHz to 1 GHz.

As a practical matter, if the conductors of the transmission line are not perfect conductors, we can see that the TEM mode is not a possible field structure even at dc. For example, a longitudinal (along the line axis) component of an electric field $\mathcal{E}_z$ will result from the conductor currents passing through the conductors. Therefore, considering imperfect line conductors and the TEM mode simultaneously is an approximation (usually a reasonable one for lines composed of good conductors). This will be discussed in more detail in Sec. 7.3. Until then, we will assume that the line conductors are perfect conductors. We will also assume that the medium surrounding the conductors is a perfect dielectric.

It was pointed out in Chap. 5 that when fields are time-varying, it is not generally possible to define a unique voltage between two points. If, however, we assume the TEM field structure (mode of propagation), it is possible to define a unique voltage between the two conductors of the transmission lines in Fig. 7.3 for nonzero frequencies, as the following shows; similarly, we will find that we may uniquely define currents on the conductors in a fashion similar to that for the case of static excitation. To illustrate this, we recall Faraday's law and Ampère's law in integral form in the lossless, homogeneous medium surrounding the line conductors:

$$\oint_c \boldsymbol{\mathcal{E}} \cdot d\mathbf{l} = -\mu \frac{d}{dt} \int_s \boldsymbol{\mathcal{H}} \cdot d\mathbf{s} \tag{1a}$$

$$\oint_c \boldsymbol{\mathcal{H}} \cdot d\mathbf{l} = \int_s \boldsymbol{\mathcal{J}} \cdot d\mathbf{s} + \epsilon \frac{d}{dt} \int_s \boldsymbol{\mathcal{E}} \cdot d\mathbf{s} \tag{1b}$$

† A discussion of higher-order modes on the coaxial line of Fig. 7.1*c* is given in S. Ramo, J. R. Whinnery, and T. VanDuzer, *Fields and Waves in Communication Electronics*, Wiley, New York, 1965, pp. 446–448. See also S. A. Schelkunoff, "The Electromagnetic Theory of Coaxial Transmission Lines and Cylindrical Shields," *Bell System Technical Journal*, vol. 13, October 1934, pp. 532–579.

If we take the contour c and surface s of both expressions to be in an xy plane transverse to the line axis denoted as c_{xy} and s_{xy}, respectively, we obtain

$$\oint_{c_{xy}} (\mathcal{E}_x\, dx + \mathcal{E}_y\, dy) = -\mu \frac{d}{dt} \int_{s_{xy}} \mathcal{H}_z\, dx\, dy \tag{2a}$$

$$\oint_{c_{xy}} (\mathcal{H}_x\, dx + \mathcal{H}_y\, dy) = \int_{s_{xy}} \mathcal{J}_z\, dx\, dy + \epsilon \frac{d}{dt} \int_{s_{xy}} \mathcal{E}_z\, dx\, dy \tag{2b}$$

But for the TEM mode, $\mathcal{E}_z = \mathcal{H}_z = 0$! Thus, we obtain

$$\oint_{c_{xy}} (\mathcal{E}_x\, dx + \mathcal{E}_y\, dy) = 0 \tag{3a}$$

$$\oint_{c_{xy}} (\mathcal{H}_x\, dx + \mathcal{H}_y\, dy) = \int_{s_{xy}} \mathcal{J}_z\, dx\, dy \tag{3b}$$

Equations (3*a*) and (3*b*) *are precisely the equations we obtain for the static case.* Thus, we arrive at the following important observations about the TEM mode:

1 The electric field of the TEM mode *in any transverse* (xy) *plane* is conservative and satisfies an electrostatic field distribution.
2 The magnetic field of the TEM mode satisfies a magnetostatic field distribution in *any transverse* (xy) *plane.*

Applying this result to the two-conductor lines in Fig. 7.3, we conclude that it is possible to uniquely define a voltage between the conductors at each point along the line as the line integral of $\boldsymbol{\mathcal{E}}$ along any path in the transverse xy plane; that is,

$$V(z, t) = -\int_{c_V} \boldsymbol{\mathcal{E}}_T \cdot d\mathbf{l} \tag{4}$$

where contour c_V is shown in Fig. 7.4 and $\boldsymbol{\mathcal{E}}_T$ is the transverse electric field:

$$\boldsymbol{\mathcal{E}}_T = \mathcal{E}_x \mathbf{a}_x + \mathcal{E}_y \mathbf{a}_y \tag{5}$$

Similarly, from (3*b*) we conclude that we may uniquely define the current on each conductor surface as the line integral of $\boldsymbol{\mathcal{H}}$ around any closed contour in the transverse xy plane which encircles that conductor; that is,

$$I(z, t) = \oint_{c_I} \boldsymbol{\mathcal{H}}_T \cdot d\mathbf{l} \tag{6}$$

where contour c_I is shown in Fig. 7.4 and $\boldsymbol{\mathcal{H}}_T$ is the transverse magnetic field:

$$\boldsymbol{\mathcal{H}}_T = \mathcal{H}_x \mathbf{a}_x + \mathcal{H}_y \mathbf{a}_y \tag{7}$$

Note that the voltage and current in (4) and (6) are shown as being functions of position along the line, z, as well as time t.

Consider some Δz section of the two-conductor line of Fig. 7.3*a*, as shown in Fig. 7.5. (The following results may also be easily seen to apply to the coaxial line shown in Fig. 7.3*b* in a similar fashion.) The current of one conductor and the returning current on the other conductor produce a transverse magnetic field $\mathcal{H}_T$, as shown in Fig. 7.5*a*. If this Δz section of the line is considered to be a

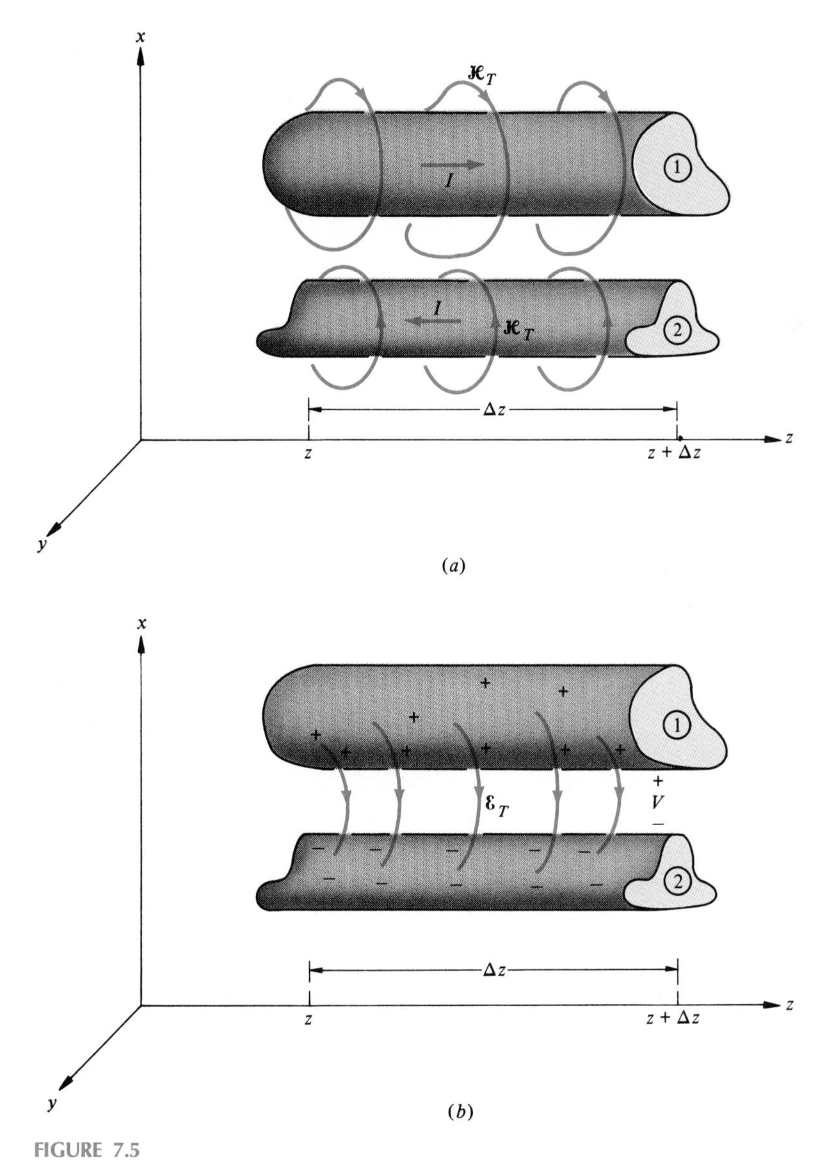

FIGURE 7.5

The TEM field structure on open-conductor lines: *(a)* magnetic fields; *(b)* electric fields.

loop, the magnetic flux passing between the conductors links the current of the loop which may be thought of as an inductance L. A per-unit-length inductance l of this section of line may be obtained by dividing the inductance of this section by the length of the section: $l = L/\Delta z$. This per-unit-length inductance will characterize all other sections of the line since the line is uniform. Similarly, the transverse electric field $\mathcal{E}_T$ results from the separation of charge on the conductor surfaces, as shown in Fig. 7.5*b*. This effect may be viewed as a capacitance. We may therefore ascribe a per-unit-length capacitance c to the line by dividing the capacitance C of some Δz section by the length of that section: $c = C/\Delta z$. All other sections of the line will be characterized by this per-unit-length capacitance since the line is uniform.

These observations follow from the fact that the transverse electric and magnetic fields satisfy static (0 Hz) *distributions in any transverse* (xy) *plane as shown by* (3). It is important to note that this shows that the per-unit-length parameters, c and l, are *static field parameters* that may be computed via the methods of Chaps. 3 and 4, respectively. The above results, however, show that these parameters represent the TEM field structure for *nonstatic* excitation—a remarkable result. The derivation of these per-unit-length parameters will be discussed in Sec. 7.4.

We may therefore characterize a Δz section of the line with a lumped capacitance and lumped inductance as shown in Fig. 7.6*a*. Note that the total-

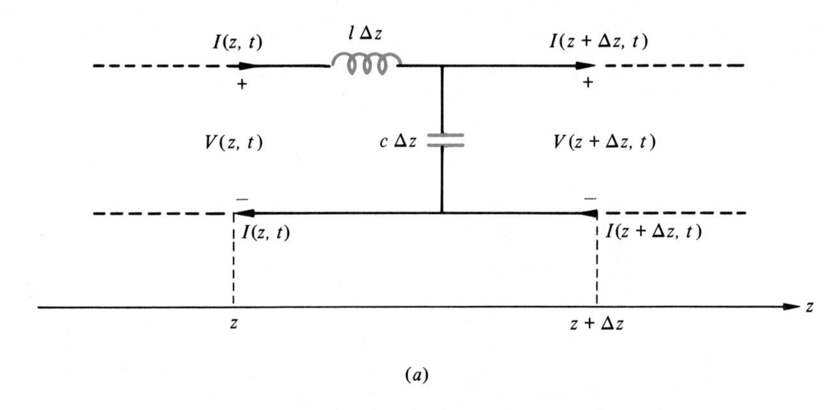

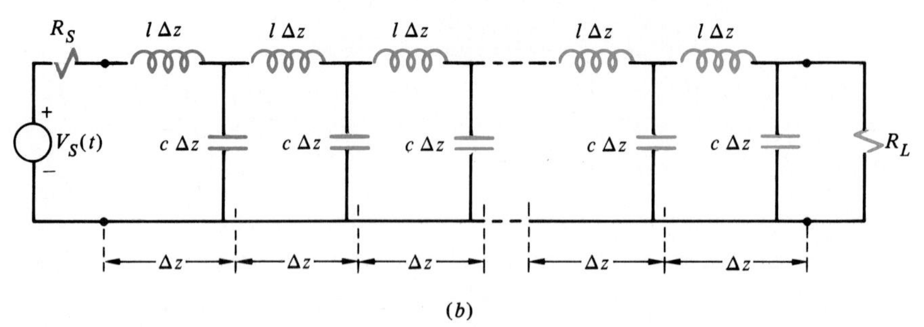

FIGURE 7.6
The per-unit-length equivalent circuit of a transmission line.

element value for a Δz section is the per-unit-length value multiplied by the section length. A representation of the total line is given in Fig. 7.6*b*. It is reasonable to expect that the voltages and currents of the line will *not* be discrete functions of position, z, along the line (as indicated by the lumped circuit of Fig. 7.6*b*) but will vary with distance along the line in a smooth fashion. Anticipating this behavior, we will allow the segment lengths Δz in the equivalent circuit to become infinitesimally small, $\Delta z \to 0$, and will at the same time add sections to keep the length of the total network equal to the total line length. In order to do this we write, from Fig. 7.6*a*,

$$V(z + \Delta z, t) - V(z, t) = -l\,\Delta z \frac{\partial}{\partial t} I(z, t) \tag{8a}$$

$$\begin{aligned} I(z + \Delta z, t) - I(z, t) &= -c\,\Delta z \frac{\partial}{\partial t} V(z + \Delta z, t) \\ &= -c\,\Delta z \frac{\partial}{\partial t}\left[V(z, t) - l\,\Delta z \frac{\partial}{\partial t} I(z, t) \right] \\ &= -c\,\Delta z \frac{\partial V(z, t)}{\partial t} + lc\,\Delta z^2 \frac{\partial^2 I(z, t)}{\partial t^2} \end{aligned} \tag{8b}$$

where we have substituted $V(z + \Delta z, t)$ from (8*a*) into (8*b*). Dividing both sides of (8) by Δz and taking the limit as $\Delta z \to 0$, we obtain the resulting *transmission-line equations*:

$$\frac{\partial V(z, t)}{\partial z} = -l \frac{\partial I(z, t)}{\partial t} \tag{9a}$$

$$\frac{\partial I(z, t)}{\partial z} = -c \frac{\partial V(z, t)}{\partial t} \tag{9b}$$

Problem 7-3 at the end of this chapter shows that the particular choice of the structure of the per-unit-length equivalent circuit in Fig. 7.6*a* is not unique: many others yield (9) in the limit as $\Delta z \to 0$.

The two equations in (9) may be combined in the following manner to yield two additional equations. Each of these new equations involves only the line voltage or the line current rather than both as in (9). Differentiate (9*a*) with respect to z and (9*b*) with respect to t to obtain

$$\frac{\partial^2 V(z, t)}{\partial z^2} = -l \frac{\partial^2 I(z, t)}{\partial z\,\partial t} \tag{10a}$$

$$\frac{\partial^2 I(z, t)}{\partial z\,\partial t} = -c \frac{\partial^2 V(z, t)}{\partial t^2} \tag{10b}$$

Substituting (10*b*) into (10*a*) yields

$$\frac{\partial^2 V(z,t)}{\partial z^2} = lc \frac{\partial^2 V(z,t)}{\partial t^2} \tag{11a}$$

A similar process results in

$$\frac{\partial^2 I(z,t)}{\partial z^2} = lc \frac{\partial^2 I(z,t)}{\partial t^2} \tag{11b}$$

A dimensional check will show that the product lc in (11) has the dimensions of the reciprocal of velocity squared. In fact, we note that the units of l and c are henrys per meter and farads per meter, respectively. The units of permeability μ and permittivity ϵ are also henrys per meter and farads per meter, respectively. We observed previously that uniform plane waves propagating in a homogeneous medium traveled at the (phase) velocity

$$u = \frac{1}{\sqrt{\mu\epsilon}} \quad \text{m/s} \tag{12}$$

Therefore, we conclude that $1/\sqrt{lc}$ has the dimensions of velocity also. Thus, we prescribe the symbol u to

$$u = \frac{1}{\sqrt{lc}} \quad \text{m/s} \tag{13}$$

We will see in Sec. 7.4 that, for uniform transmission lines immersed in a homogeneous medium characterized by μ and ϵ,

$$lc = \mu\epsilon \tag{14}$$

which is a reasonable result based on the above observations.

The solutions to (11) are

$$V(z,t) = V^+\left(t - \frac{z}{u}\right) + V^-\left(t + \frac{z}{u}\right) \tag{15a}$$

$$I(z,t) = I^+\left(t - \frac{z}{u}\right) + I^-\left(t + \frac{z}{u}\right) \tag{15b}$$

where the quantities V^+, V^-, I^+, and I^- are functions whose forms, e.g., $V^+(t - z/u) = 2\cos\omega(t - z/u)$, are not as yet determined. Nevertheless, whatever their forms, V^+ and I^+ are functions of t, z, and u only as $t - z/u$. Similarly, V^- and I^- depend on t, z, and u only as $t + z/u$.

This can be verified in the following manner. First observe that the equations in (11) are linear; thus, if the individual portions of the solution, V^+ and V^-, each satisfy the appropriate equation, (11*a*), then their sum will also satisfy the equation. Denote $s = t - z/u$ and evaluate

$$\frac{\partial V^+(s)}{\partial z} = \frac{\partial V^+(s)}{\partial s}\frac{\partial s}{\partial z} = -\frac{1}{u}\frac{\partial V^+(s)}{\partial s} \tag{16a}$$

$$\frac{\partial^2 V^+(s)}{\partial z^2} = \frac{1}{u^2}\frac{\partial^2 V^+(s)}{\partial s^2} \tag{16b}$$

Similarly, we obtain

$$\frac{\partial^2 V^+(s)}{\partial t^2} = \frac{\partial^2 V^+(s)}{\partial s^2} \tag{16c}$$

Substituting (16*b*) and (16*c*) into (11*a*), we obtain an identity. Similarly, the functions V^-, I^+, and I^- can be shown to satisfy (11), and thus Eqs. (15) represent valid solutions of (11).

From our knowledge of uniform plane waves, it is clear that the functions $V^+(t - z/u)$ and $I^+(t - z/u)$ (whose functional forms have not yet been determined) represent waves traveling with velocity u in the $+z$ direction: forward-traveling waves. This should be clear since corresponding points on each waveform must move in the $+z$ direction for increasing time in order to keep the arguments of the functions constant. Similarly, the functions $V^-(t + z/u)$ and $I^-(t + z/u)$ represent backward-traveling waves moving in the $-z$ direction.

As in the case of uniform plane waves, the forward-traveling current and voltage waves are related, as are those in the backward-traveling waves. The result is

$$I^+\left(t - \frac{z}{u}\right) = \frac{V^+(t - z/u)}{R_C} \tag{17a}$$

$$I^-\left(t + \frac{z}{u}\right) = -\frac{V^-(t + z/u)}{R_C} \tag{17b}$$

where

$$R_C = \sqrt{\frac{l}{c}} \tag{18}$$

has the units of ohms and is referred to as the *characteristic resistance* of the line. Note the similarity of these results to the uniform plane wave case. The characteristic resistance of the line, R_C, is analogous to the intrinsic impedance η of the medium for uniform plane waves, $\eta = \sqrt{\mu/\epsilon}$. Equations (17) may be verified by direct substitution into (9). The final result becomes

$$V(z, t) = V^+\left(t - \frac{z}{u}\right) + V^-\left(t + \frac{z}{u}\right) \tag{19a}$$

$$I(z, t) = \frac{1}{R_C} V^+\left(t - \frac{z}{u}\right) - \frac{1}{R_C} V^-\left(t + \frac{z}{u}\right) \tag{19b}$$

Consider Fig. 7.7, which shows a transmission line (consisting of two parallel wires for illustration) of total length $\mathscr{L}$ that is terminated in a resistance R_L and driven by a pulse voltage source having an open-circuit voltage waveform of $V_S(t)$ and an internal resistance R_S. We wish to determine the terminal voltages $V(0, t)$ and $V(\mathscr{L}, t)$ and terminal currents $I(0, t)$ and $I(\mathscr{L}, t)$ as functions of time.

First consider the portion of the line at the load shown in Fig. 7.8. At $z = \mathscr{L}$ we must have

$$V(\mathscr{L}, t) = R_L I(\mathscr{L}, t) \tag{20}$$

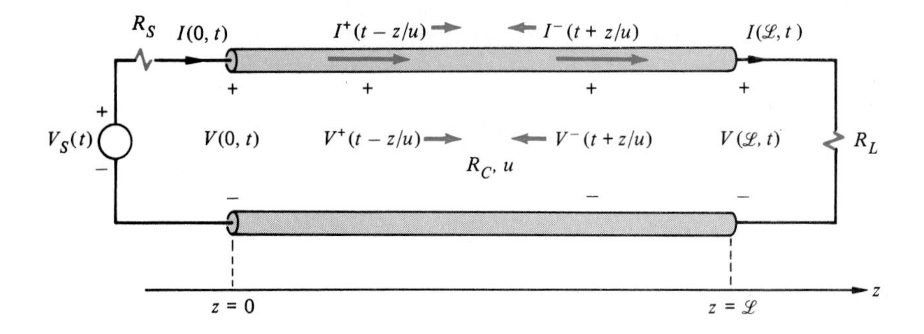

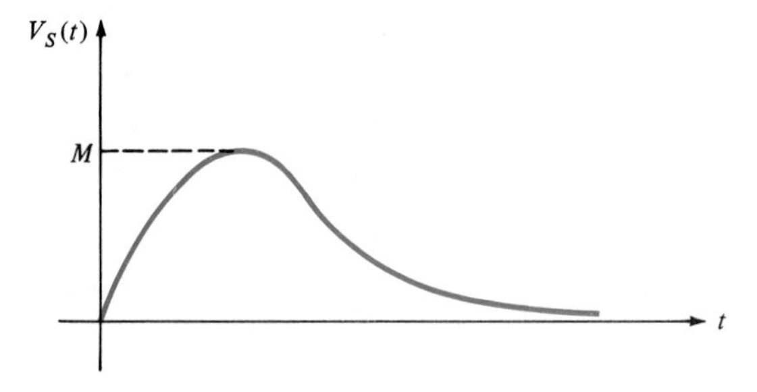

FIGURE 7.7
The terminated transmission line.

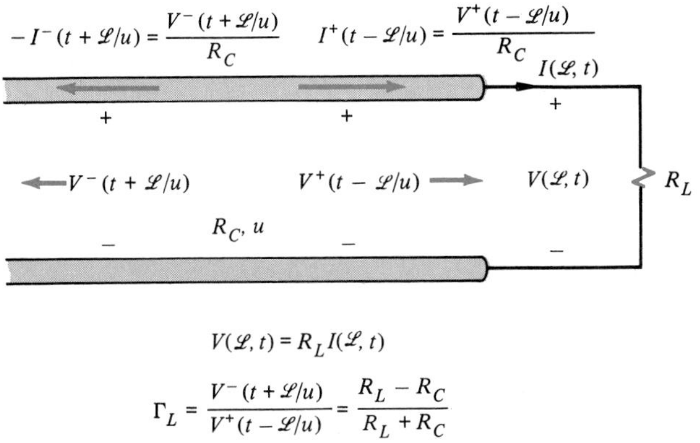

FIGURE 7.8
Voltage reflection coefficient at a termination.

It should be clear that this line discontinuity requires the existence of forward- and backward-traveling waves at $z = \mathscr{L}$ since if there are only forward-traveling waves existing at the load, then from (19)

$$V^+\left(t - \frac{\mathscr{L}}{u}\right) = R_C I^+\left(t - \frac{\mathscr{L}}{u}\right) \tag{21}$$

Similarly, if there are only backward-traveling waves existing at the load, then

$$V^-\left(t - \frac{\mathscr{L}}{u}\right) = -R_C I^-\left(t + \frac{\mathscr{L}}{u}\right) \tag{22}$$

Neither (21) nor (22) satisfies (20) unless $R_L = R_C$, in which case we say that the line is *matched* at the load. In this case, there would be no backward-traveling wave at the load since (22) does not satisfy (20) even for $R_L = R_C$. Consequently, in order to satisfy the general terminal condition in (20) we must have forward- and backward-traveling waves in existence at the load.

The discontinuity in the line produced by the load resistor can be seen to result in a wave being reflected in the form of a backward-traveling wave in the same fashion as are the uniform plane waves incident-normal to plane, material boundaries discussed in the preceding chapter. In fact, the analogy is very striking. For the transmission line, we may define a *voltage reflection coefficient* at the load as the ratio of the backward- and forward-traveling voltage waves at $z = \mathscr{L}$; that is,

$$\Gamma_L = \frac{V^-(t + \mathscr{L}/u)}{V^+(t - \mathscr{L}/u)} \tag{23}$$

From (19) we obtain at $z = \mathscr{L}$

$$V(\mathscr{L}, t) = V^+\left(t - \frac{\mathscr{L}}{u}\right) + V^-\left(t + \frac{\mathscr{L}}{u}\right)$$

$$= V^+\left(t - \frac{\mathscr{L}}{u}\right)(1 + \Gamma_L) \tag{24a}$$

$$I(\mathscr{L}, t) = I^+\left(t - \frac{\mathscr{L}}{u}\right) + I^-\left(t + \frac{\mathscr{L}}{u}\right)$$

$$= \frac{V^+(t - \mathscr{L}/u)}{R_C} - \frac{V^-(t + \mathscr{L}/u)}{R_C}$$

$$= \frac{V^+(t - \mathscr{L}/u)}{R_C}(1 - \Gamma_L) \tag{24b}$$

From (17) and (23) we see that a *current reflection coefficient* at the load may be defined as

$$\frac{I^-(t + \mathscr{L}/u)}{I^+(t - \mathscr{L}/u)} = -\Gamma_L \tag{25}$$

Thus, the current reflection coefficient is the negative of the voltage reflection coefficient. Substituting (24) into (20), we obtain

$$R_L = R_C \frac{1 + \Gamma_L}{1 - \Gamma_L} \tag{26}$$

or

$$\Gamma_L = \frac{R_L - R_C}{R_L + R_C} \tag{27}$$

The analogy between the transmission-line case and the case of a uniform plane wave incident-normal to a plane, dielectric boundary is an important one. In the case of an electric field and a magnetic field of a uniform plane wave incident from a region having an intrinsic impedance η_1 normal to the surface of a region having an intrinsic impedance of η_2, we obtain the reflected fields from

$$\Gamma = \frac{E_r}{E_i}$$

$$= \frac{\eta_2 - \eta_1}{\eta_2 + \eta_1}$$

$$= -\frac{H_r}{H_i} \tag{28}$$

Note that the reflection coefficient for the magnetic field is the negative of the reflection coefficient for the electric field. This is necessary so that power flow of the individual waves will be in the proper direction and is analogous to the current reflection coefficient.

The reader should compare the above results for the transmission line to those for the uniform plane wave case outlined in (28). In fact, it is now clear why the forward and backward voltage waves in (19*a*) have the same sign and the forward and backward current waves in (19*b*) have opposite signs. The reason is that V^+ and V^- are both in the direction assumed for the total voltage V, whereas I^+ and I^- must be directed in opposite directions since the total line current I is defined to be in the $+z$ direction.

This reflection of waves by the load discontinuity is illustrated in Fig. 7.9. The reflection process can be viewed as a mirror that produces, as a reflected wave

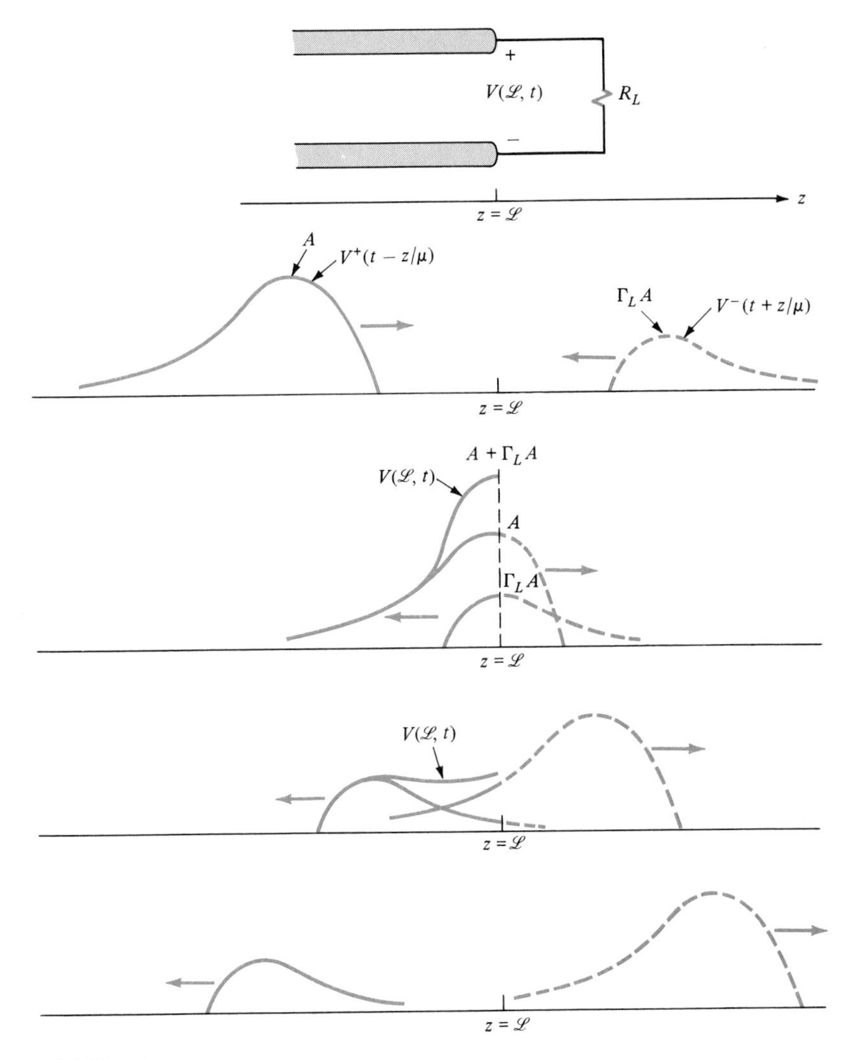

FIGURE 7.9
The process of pulse reflection at a termination.

V^-, a replica of V^+ that is "flipped around," and all points on the V^- waveform are the corresponding points on the V^+ waveform multiplied by Γ_L. Note that the total voltage at the load, $V(\mathscr{L}, t)$, is the sum of the individual waves present at the load at each time, as shown in (24*a*).

Now let us consider the portion of the line at the source, $z = 0$, shown in Fig. 7.10*a*. When we initially connect the source to the line, we may reason that a forward-traveling wave will be propagated down the line. We would not expect

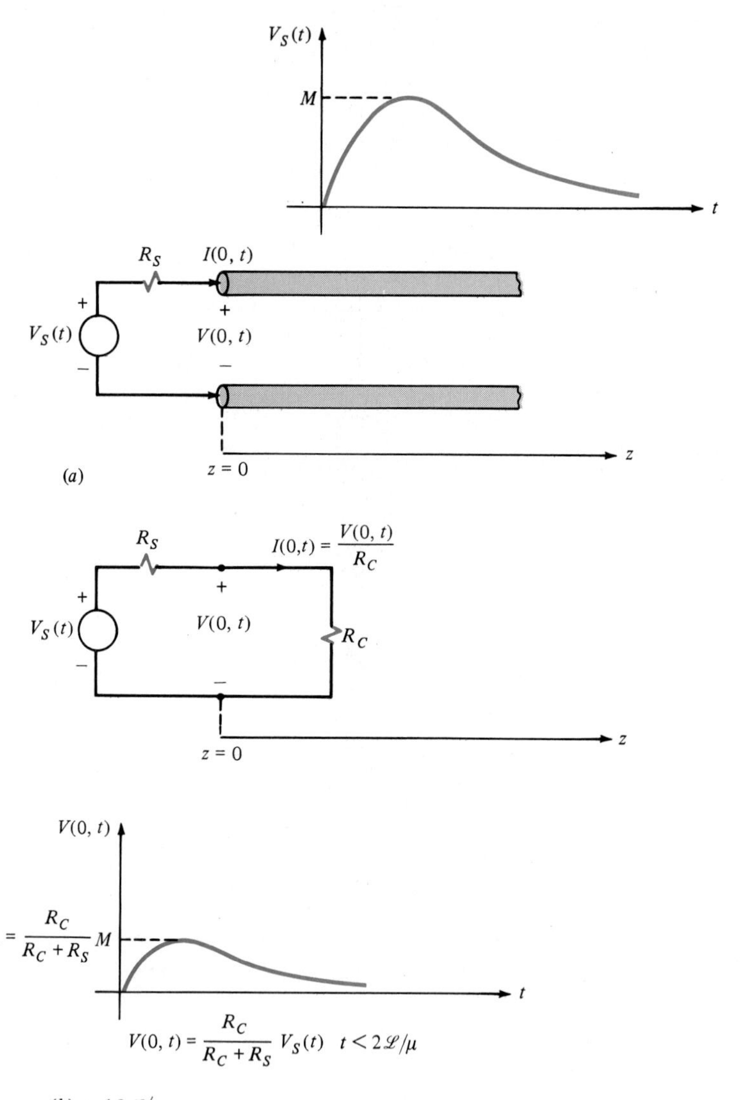

FIGURE 7.10

The equivalent input circuit of a line for $t < 2\mathscr{L}/u$.

a backward-traveling wave to appear on the line until this initial forward-traveling wave has reached the load, a time delay of $T = \mathscr{L}/u$, since the load is presumed to have no source within it to produce a backward-traveling wave and the incident wave will not have arrived to produce a reflected wave. That portion of the incident wave which is reflected at the load will require an additional time T to move from the load back to the source at $z = 0$. Therefore, for $0 \leq t < 2\mathscr{L}/u$, no backward-traveling waves will appear at $z = 0$, and for any time less than $2\mathscr{L}/u$ the total voltage and current at $z = 0$ will consist only of forward-traveling waves, V^+ and I^+; therefore,

$$V(0, t) = V^+\left(t - \frac{0}{u}\right) \tag{29a}$$

$$\begin{aligned} I(0, t) &= I^+\left(t - \frac{0}{u}\right) \\ &= \frac{V^+(t - 0/u)}{R_C} \qquad 0 \leq t < \frac{2\mathscr{L}}{u} \end{aligned} \tag{29b}$$

Since the ratio of the total voltage and current on the line is R_C for $0 \leq t < 2\mathscr{L}/u$ as shown by (29), the line appears to have an input resistance of R_C over this time interval, as shown in Fig. 7.10*b*. Thus, the forward-traveling wave that is initially launched is related to $V_S(t)$ by

$$\begin{aligned} V(0, t) &= V_S(t) - R_S I(0, t) \\ &= V_S(t) - \frac{R_S}{R_C} V(0, t) \end{aligned} \tag{30a}$$

so that

$$V(0, t) = \frac{R_C}{R_C + R_S} V_S(t) \qquad 0 \leq t < \frac{2\mathscr{L}}{u} \tag{30b}$$

and the initially launched wave is of the same shape as $V_S(t)$, but the corresponding points on $V_S(t)$ are reduced in magnitude by the voltage-divider relationship $R_C/(R_C + R_S)$, as shown in Fig. 7.10*b*.

This initially launched wave travels toward the load, requiring a time $T = \mathscr{L}/u$ for the leading edge of the pulse to reach the load. When the pulse reaches the load, a reflected pulse is initiated, as shown in Fig. 7.9. This reflected pulse requires $T = \mathscr{L}/u$ for its leading edge to reach the source. At the source we can, in a similar fashion, obtain a voltage reflection coefficient, which is

$$\Gamma_S = \frac{R_S - R_C}{R_S + R_C} \tag{31}$$

as the ratio of the incoming incident wave (which was the reflected wave at the load) and the reflected portion of this wave (which is sent back toward the load).

A forward-traveling wave is therefore initiated at the source in the same fashion as at the load. This forward-traveling wave has the same shape as the incoming backward-traveling wave (which was the original pulse sent out by the source and reflected at the load), but corresponding points on the incoming wave are reduced by Γ_S.

This process of repeated reflections continues as re-reflections at the source and load. At any time, the total voltage (current) at any point on the line is the sum of the values of all the individual voltage waves (current waves) existing on the line at that point and time, as is shown by (19).

EXAMPLE 7.1 Consider the transmission line shown in Fig. 7.11*a*. At $t = 0$ a 30-V battery with zero source resistance is attached to the line. Sketch the distribution of voltage along the line for several instants of time.

Solution The load and source voltage reflection coefficients are

$$\Gamma_L = \frac{R_L - R_C}{R_L + R_C}$$

$$= \tfrac{1}{3}$$

$$\Gamma_S = \frac{R_S - R_C}{R_S + R_C}$$

$$= -1$$

and the time required to transit the line in one direction is 2 μs. At $t = 0$ a 30-V pulse is sent down the line, and the line voltage is zero prior to the arrival of the pulse and 30 V after the pulse has passed. (See Fig. 7.11*b*.) At $t = 2.5$ μs the pulse has arrived at the load and a backward-traveling pulse of magnitude $30\Gamma_L = 10$ V is sent back toward the source. (See Fig. 7.11*c*.) When this reflected pulse arrives at the source, a pulse of magnitude Γ_S of the incoming pulse, or $\Gamma_S\Gamma_L 30 = -10$ V, is sent back toward the load. (See Fig. 7.11*d*.) This pulse travels to the load, at which time a reflected pulse of Γ_L of this incoming pulse, or $\Gamma_L\Gamma_S\Gamma_L 30 = -3.33$ V, is sent back toward the source. (See Fig. 7.11*d*.) At each point on the line, the total line voltage is the sum of the waves present on the line at that point.

EXAMPLE 7.2 Repeat Example 7.1 where the voltage source is a pulse of 30 V but the duration is 1 μs. The source resistance remains zero.

Solution The distribution of the voltage along the line is shown in Fig. 7.12 at various instants of time. Note that since the voltage source does not remain

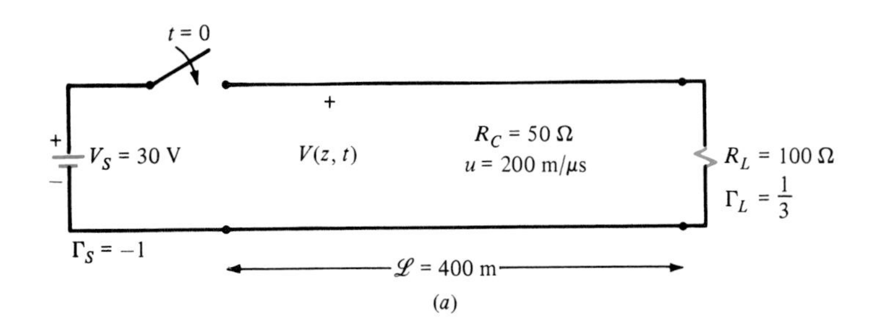

(a)

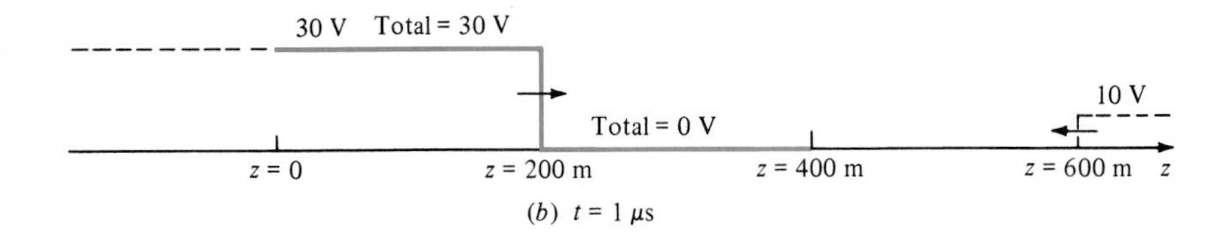

(b) t = 1 μs

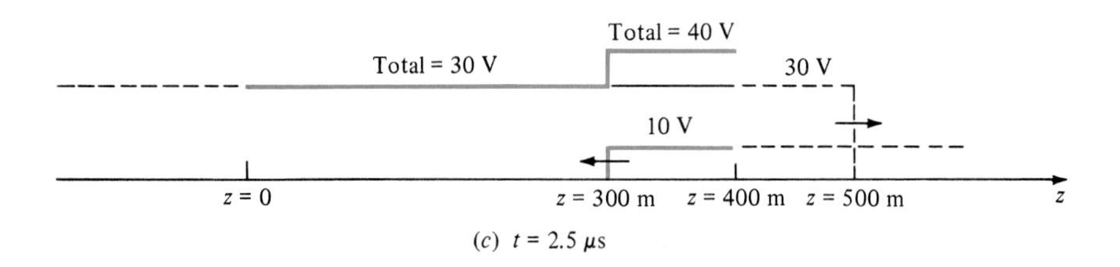

(c) t = 2.5 μs

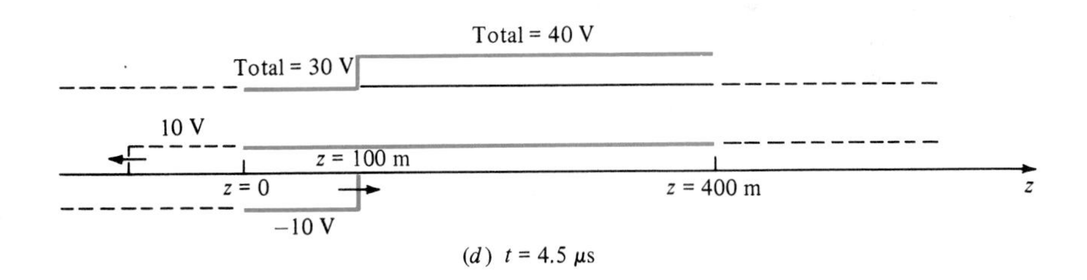

(d) t = 4.5 μs

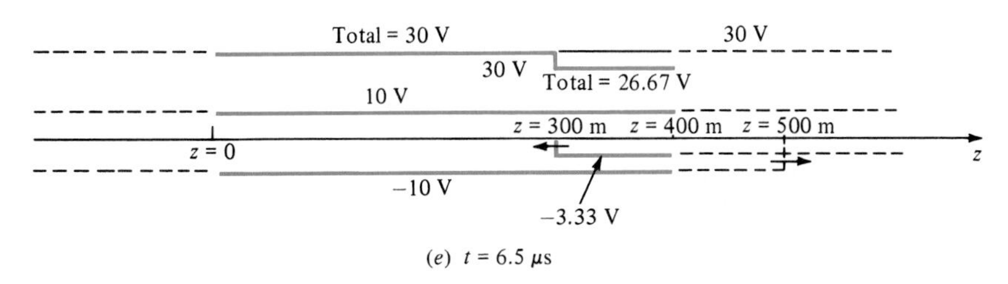

(e) t = 6.5 μs

FIGURE 7.11
Example 7.1.

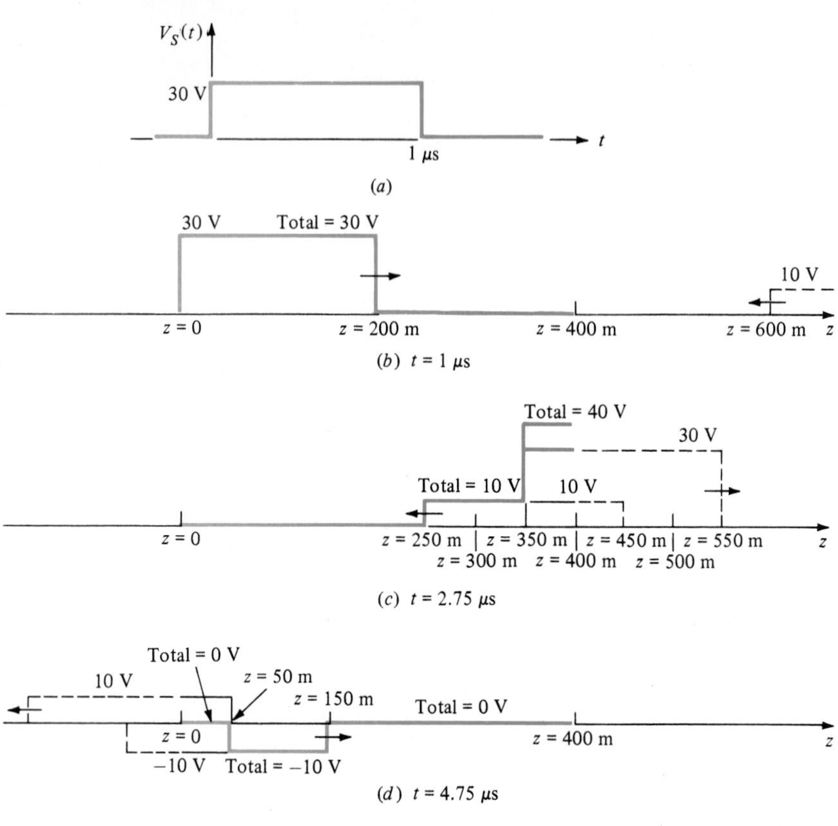

FIGURE 7.12
Example 7.2.

constant for $t > 0$ as in Example 7.1 but returns to zero at $t = 1\ \mu s$, the traveling waves are zero along portions of the line. It is important to observe that when a sketch of the line voltage (or current) as a function of position along the line is obtained, the pulse shape appears to be a "flipped around" replica of $V_S(t)$. (See Fig. 7.9.) In this example, $V_S(t)$ is rectangular and the distinction is not apparent. This observation is obvious when one realizes that the point on the sketch of $V_S(t)$ at $t = 0$ is the first point to proceed down the line: the leading edge of the pulse.

A convenient way of keeping track of these reflections is with a lattice diagram (Fig. 7.13*a*). In the lattice diagram, the horizontal axis is labeled as distance down the line and the vertical axis is labeled as time in increments of the total time required to transit the line in one direction: $\mathscr{L}/u$. Consider $V_S(t)$ as a set of voltage points and apply superposition. Let us examine a point on the initially launched pulse shown in Fig. 7.13*b* having magnitude K at time t'. This point on the pulse travels toward the load and is reflected, resulting in a corresponding point on the reflected waveform of magnitude $K\Gamma_L$. The lattice diagram shows this process in a convenient manner and allows us to obtain the

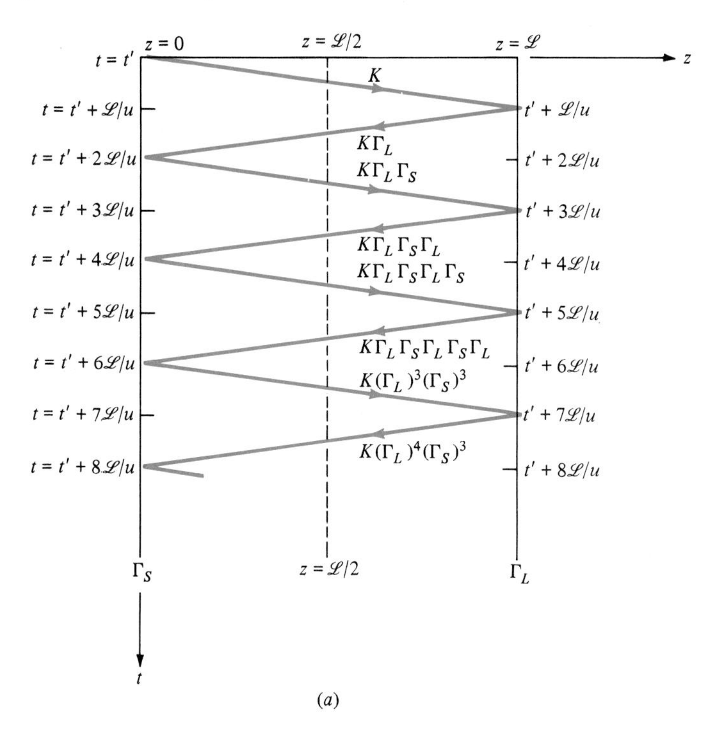

(*a*)

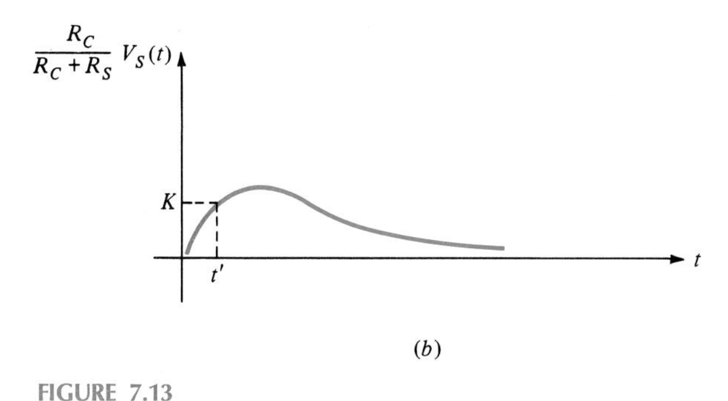

(*b*)

FIGURE 7.13

The lattice diagram.

value of the total line voltage $V(z, t)$ at any point and time on the line. For example, at $t = t' + \mathscr{L}/u$ the total voltage at $z = \mathscr{L}$ is $K + K\Gamma_L = K(1 + \Gamma_L)$. At $z = \mathscr{L}/2$, a point midway down the line, the total line voltage is $K\Gamma_L$ at $t = t' + \frac{3}{2}\mathscr{L}/u$. By following the movement of various points on the initially launched waveform [which differs from $V_S(t)$ only by $R_C/(R_C + R_S)$], we can sketch the total voltage at any point on the line. This can be done for the line current also, but in this case the initially launched wave is

$$I^+\left(t - \frac{0}{u}\right) = \frac{V_S(t)}{R_C + R_S} \tag{32}$$

and we must replace Γ_S and Γ_L in the lattice diagram by the current reflection coefficients $-\Gamma_S$ and $-\Gamma_L$. However, the simplest method of sketching the line voltage and current is to visualize and sketch the individual forward- and backward-traveling waves and to combine all those present at an instant to produce the total line voltage and current distributions at that instant, as was done in the previous examples.

Generally we are interested only in the line voltage and current at the ends of the line. Thus, instead of sketching the voltage or current along the line at fixed instants of time, we are interested in sketching the line voltage or current at individual points on the line ($z = 0$ and $z = \mathscr{L}$) for all time. The following example illustrates this problem.

EXAMPLE 7.3 Consider the 400-m length of coaxial cable shown in Fig. 7.14*a*. The cable is terminated in a short circuit ($R_L = 0$) and is driven by a pulse source having an internal resistance of 150 Ω ($R_S = 150$ Ω). The pulse has a magnitude of 100 V and duration of 6 μs, as shown in Fig. 7.14*b*. Sketch the voltage at the input to the line, $V(0, t)$, for 18 μs.

Solution The coaxial cable has parameters of

$$c = 100 \text{ pF/m}$$

$$l = 0.25\ \mu\text{H/m}$$

The characteristic resistance is

$$R_C = \sqrt{\frac{l}{c}} = 50\ \Omega$$

and the velocity of propagation is

$$u = \frac{1}{\sqrt{lc}} = 200 \times 10^6 \text{ m/s}$$

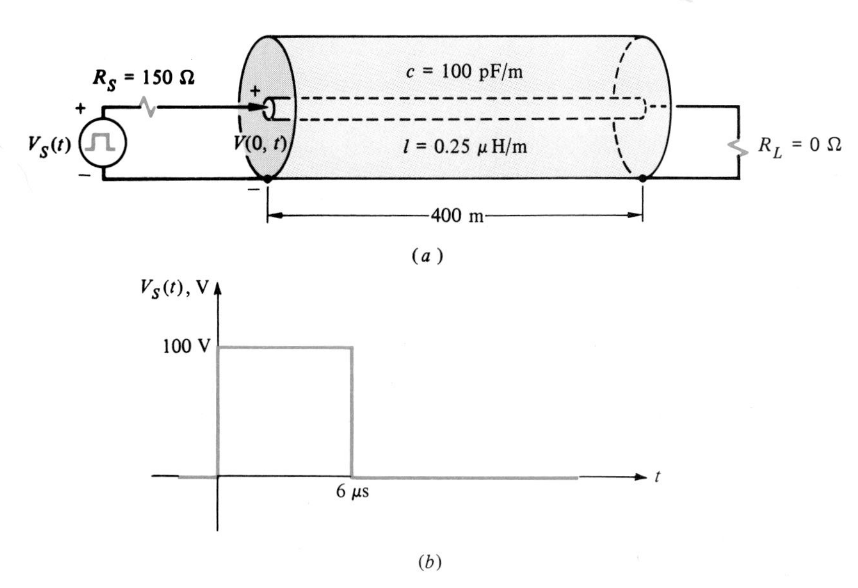

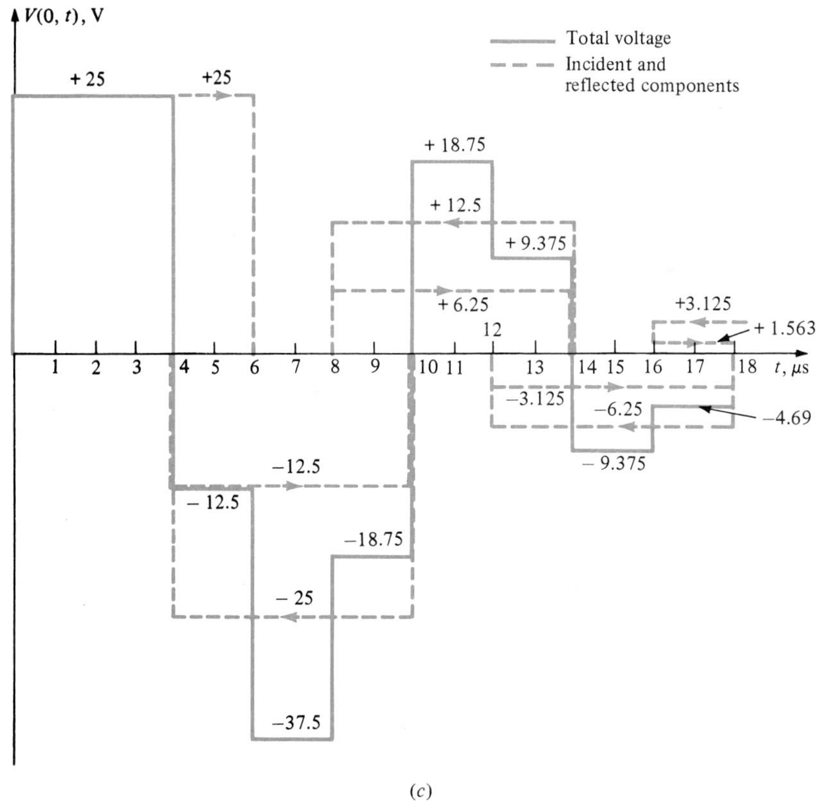

FIGURE 7.14
Example 7.3.

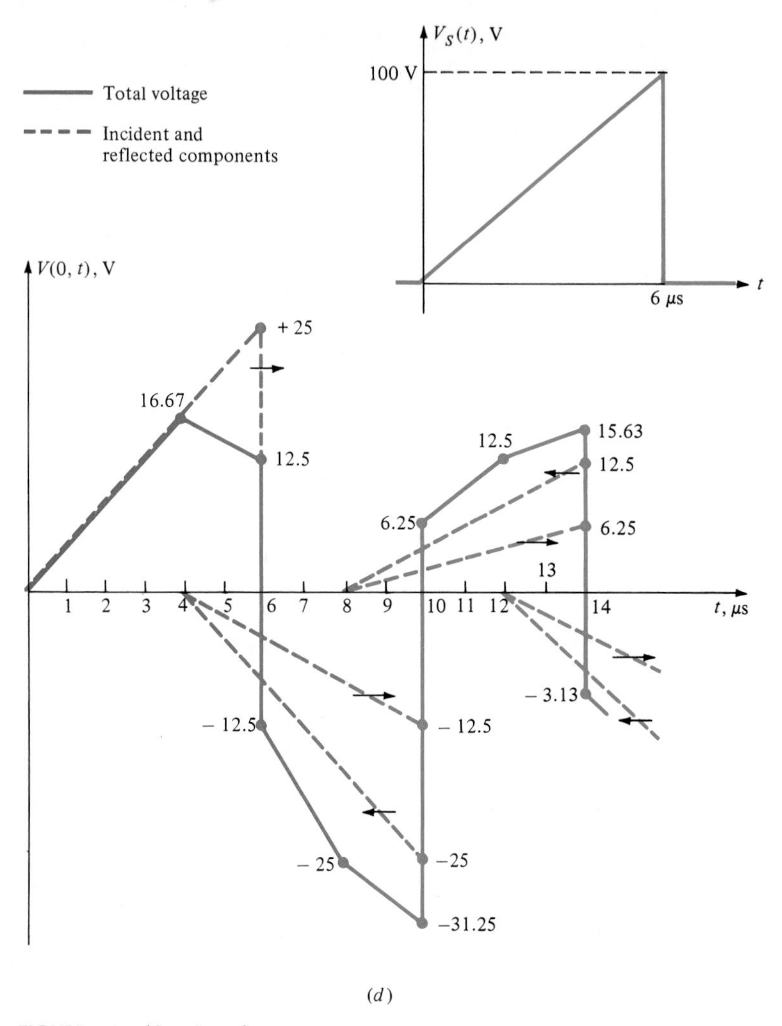

FIGURE 7.14 *(Continued)*

or 200 m/μs. Consequently, the time required for propagation from one end of the cable to the other is 2 μs. The voltage reflection coefficient at the load is

$$\Gamma_L = \frac{R_L - R_C}{R_L + R_C}$$
$$= -1$$

and at the source is

$$\Gamma_S = \frac{R_S - R_C}{R_S + R_C}$$
$$= \tfrac{1}{2}$$

The source initially sees an input resistance to the line of $R_C = 50\ \Omega$. Thus, the initially launched voltage wave is a pulse that has a 6-μs duration and a magnitude of

$$\frac{R_C}{R_C + R_S} V_S = 25\ \text{V}$$

This pulse propagates toward the load, which reflects a pulse with magnitude $\Gamma_L \times 25 = -25$ V; this reflected pulse is re-reflected at the source, producing a pulse traveling back toward the load of $\Gamma_S \times -25 = -12.5$ V; and so forth. The voltage at the source end of the line, $V(0, t)$, is plotted as dashed lines in Fig. 7.14*c* as a function of time, with the components denoted by arrows: $\rightarrow$ denotes the forward-traveling component and $\leftarrow$ denotes the backward-traveling component. The total voltage at the source end is plotted as a solid line and is the combination of all the wave values present at $z = 0$ at any time. The corresponding result for $V_S(t)$ of triangular shape is shown in Fig. 7.14*d*.

It is important to observe that when one is sketching the line voltage *at a particular point on the line as a function of time*, the pulse shape is the same in time orientation as $V_S(t)$. This is in opposition to the case in which the line voltage is sketched *at a particular time as a function of position along the line*, in which case the pulse appears as a flipped-around version of $V_S(t)$. (See Example 7.2 and the accompanying discussion.)

EXAMPLE 7.4 For the transmission-line problem in Example 7.1 shown in Fig. 7.11, sketch the voltage at the load, $V(\mathscr{L}, t)$, and the input current $I(0, t)$ as a function of time for 16 μs.

Solution At $t = 0$ a 30-V pulse is sent out by the source. The leading edge of this pulse arrives at the load at $t = 2\ \mu$s. At this time a pulse of $\Gamma_L 30 = 10$ V is sent back toward the source. This 10-V pulse arrives at the source at $t = 4\ \mu$s and a pulse of $\Gamma_S \Gamma_L 30 = -10$ V is returned to the load. This pulse arrives at the load at $t = 6\ \mu$s and a pulse of $\Gamma_L \Gamma_S \Gamma_L 30 = -3.33$ V is sent back toward the source. The contributions of these waves at $z = \mathscr{L}$ are shown in Fig. 7.15 as dashed lines, and the total voltage is shown as a solid line. Note that the load voltage oscillates about 30 V but asymptotically approaches 30 V: a reasonable result. If we had attached an oscilloscope across the load to display this voltage as a function of time and the time scale were set to 1 ms per division, it would appear that the load voltage immediately assumed a value of 30 V. We would see the picture in Fig. 7.15*b* only if the time scale were sufficiently reduced, say to 1 μs per division.

In order to sketch the load current $I(\mathscr{L}, t)$, we could divide the previously sketched load voltage by R_L. We could also sketch this current directly by using current reflection coefficients, $\Gamma_S = 1$ and $\Gamma_L = -\frac{1}{3}$, and an initial current pulse of 30 $V/R_C = 0.6\ A$. In order to sketch the input current $I(0, t)$, we could similarly directly sketch the forward- and backward-traveling waves as shown in

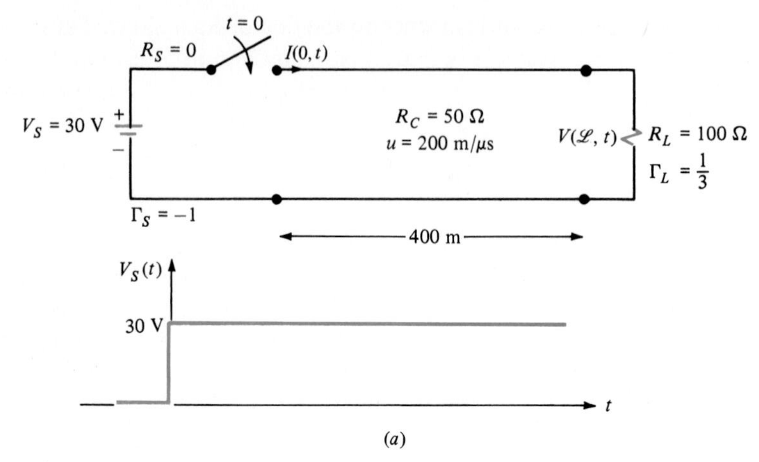

(a)

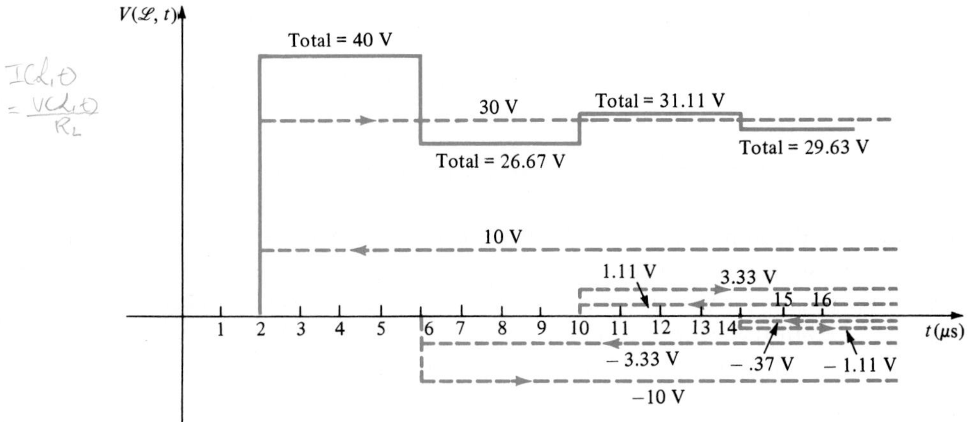

(b)

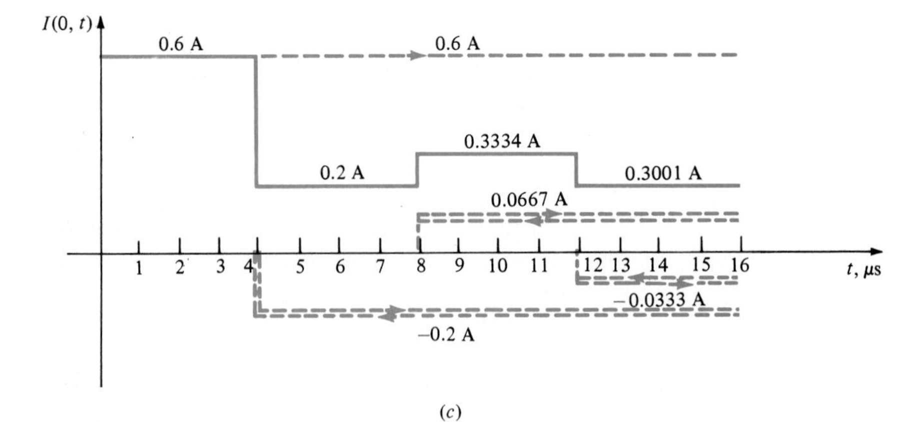

(c)

FIGURE 7.15

Example 7.4.

Fig. 7.15*c* and combine them to obtain the total input current. It would be *absurd* to sketch the input voltage $V(0, t)$ and divide this by R_C since the ratio of *total* voltage and current on the line is *not* R_C except for $t < 2\mathscr{L}/u$. However, we could sketch $I(0, t)$ using a sketch of $V(0, t)$ by realizing that

$$I(0, t) = \frac{V_S(t) - V(0, t)}{R_S}$$

Thus, we could subtract $V(0, t)$ and $V_S(t)$ point by point and divide the result by R_S to obtain $I(0, t)$. But in this example, $R_S = 0$. Thus we have no other recourse but to sketch $I(0, t)$ by sketching the individual forward- and backward-traveling waves directly and combining to obtain the total current. Note that $I(0, t)$ oscillates about the steady-state value of $30V/R_L = 0.3$ A.

7.2 Sinusoidal, Steady-State Excitation of Lossless Lines

We now consider the case in which the source is a sinusoidal waveform; that is,

$$\begin{aligned} V_S(t) &= V_S \cos \omega t \\ &= \text{Re}\,(V_S e^{j\omega t}) \end{aligned} \tag{33}$$

We will assume that the source has been attached to the line for a length of time such that transients have decayed sufficiently, leaving a sinusoidal voltage and current at each point on the line. Thus, we replace the source with its phasor equivalent: $\hat{V}_S = V_S\underline{/0^\circ}$.

We assume that the line voltage and current are in phasor form; that is,

$$V(z, t) = \text{Re}\,[\hat{V}(z)e^{j\omega t}] \tag{34a}$$

$$I(z, t) = \text{Re}\,[\hat{I}(z)e^{j\omega t}] \tag{34b}$$

where $\hat{V}(z)$ and $\hat{I}(z)$ are the phasor line voltage and current, respectively. The transmission-line equations in (9) become, in terms of phasor quantities,

$$\frac{d\hat{V}(z)}{dz} = -j\omega l\hat{I}(z) \tag{35a}$$

$$\frac{d\hat{I}(z)}{dz} = -j\omega c\hat{V}(z) \tag{35b}$$

Differentiating (35*a*) with respect to z and substituting (35*b*) (and vice versa), we obtain

$$\frac{d^2\hat{V}(z)}{dz^2} = -\omega^2 lc\hat{V}(z) \tag{36a}$$

$$\frac{d^2\hat{I}(z)}{dz^2} = -\omega^2 lc\hat{I}(z) \tag{36b}$$

The solutions to (36), which should now be familiar, become

$$\hat{V}(z) = \hat{V}_m^+ e^{-j(\omega/u)z} + \hat{V}_m^- e^{j(\omega/u)z} \tag{37a}$$

$$\hat{I}(z) = \frac{\hat{V}_m^+}{R_C} e^{-j(\omega/u)z} - \frac{\hat{V}_m^-}{R_C} e^{j(\omega/u)z} \tag{37b}$$

where $\hat{V}_m^+$ and $\hat{V}_m^-$ are, in general, complex numbers: $\hat{V}_m^+ = V_m^+ e^{j\theta^+}$, $\hat{V}_m^- = V_m^- e^{j\theta^-}$, and $u = 1/\sqrt{lc}$. The time-domain forms are

$$V(z, t) = V_m^+ \cos(\omega t - \beta z + \theta^+) + V_m^- \cos(\omega t + \beta z + \theta^-) \tag{38a}$$

$$I(z, t) = \frac{V_m^+}{R_C} \cos(\omega t - \beta z + \theta^+) - \frac{V_m^-}{R_C} \cos(\omega t + \beta z + \theta^-) \tag{38b}$$

where the phase constant is

$$\begin{aligned} \beta &= \frac{\omega}{u} \\ &= \omega\sqrt{lc} \qquad \text{rad/m} \end{aligned} \tag{39}$$

Consider the transmission line shown in Fig. 7.16. The line is terminated at $z = \mathscr{L}$ in a complex impedance $\hat{Z}_L$, and the source is a phasor source with a source voltage $\hat{V}_S = V_S \underline{/0^\circ}$ and a complex internal impedance $\hat{Z}_S$. Let us define a complex voltage reflection coefficient at a particular point z on the line as the ratio of the phasor voltages of the backward- and forward-traveling waves:

$$\begin{aligned} \hat{\Gamma}(z) &= \frac{\hat{V}_m^- e^{j\beta z}}{\hat{V}_m^+ e^{-j\beta z}} \\ &= \frac{\hat{V}_m^-}{\hat{V}_m^+} e^{j2\beta z} \end{aligned} \tag{40}$$

In terms of this reflection coefficient, the voltage and current expressions in (37) may be written as

$$\hat{V}(z) = \hat{V}_m^+ e^{-j\beta z}[1 + \hat{\Gamma}(z)] \tag{41a}$$

$$\hat{I}(z) = \frac{\hat{V}_m^+}{R_C} e^{-j\beta z}[1 - \hat{\Gamma}(z)] \tag{41b}$$

We may also define an *input impedance* to the line at any point along the line as the ratio of the total voltage and current on the line, as shown in Fig. 7.16; that is,

$$\hat{Z}_{\text{in}}(z) = \frac{\hat{V}(z)}{\hat{I}(z)} \tag{42}$$

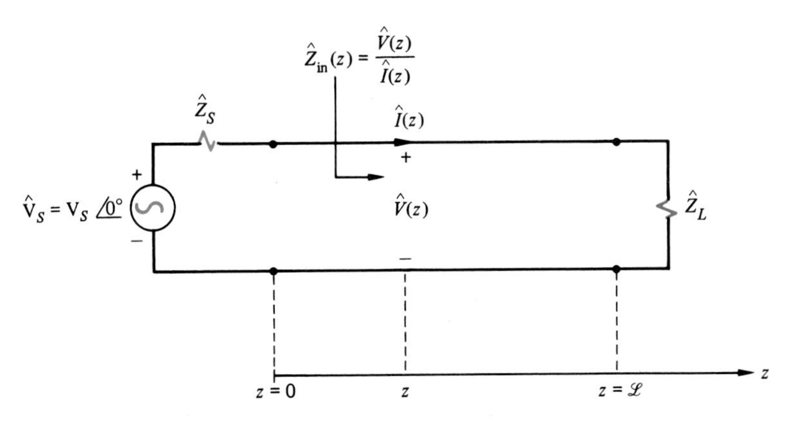

FIGURE 7.16
Illustration of the input impedance at points along a transmission line.

Substituting (41) into (42), we obtain

$$\hat{Z}_{\text{in}}(z) = R_C \frac{1 + \hat{\Gamma}(z)}{1 - \hat{\Gamma}(z)} \tag{43}$$

and find that *the input impedance and reflection coefficient at any point on the line are related.* In particular, at the load $z = \mathscr{L}$ we obtain

$$\begin{aligned} \hat{Z}_{\text{in}}(z = \mathscr{L}) &= \hat{Z}_L \\ &= R_C \frac{1 + \hat{\Gamma}(z = \mathscr{L})}{1 - \hat{\Gamma}(z = \mathscr{L})} \end{aligned} \tag{44}$$

If we define a voltage reflection coefficient at the load from (40) as

$$\hat{\Gamma}(z = \mathscr{L}) = \hat{\Gamma}_L = \frac{\hat{V}_m^-}{\hat{V}_m^+} e^{j2\beta\mathscr{L}} \tag{45}$$

we obtain from (44)

$$\hat{Z}_L = R_C \frac{1 + \hat{\Gamma}_L}{1 - \hat{\Gamma}_L} \tag{46}$$

which may be solved for $\hat{\Gamma}_L$ to yield

$$\hat{\Gamma}_L = \frac{\hat{Z}_L - R_C}{\hat{Z}_L + R_C} \tag{47}$$

a result which again is analogous to the case of normal incidence of uniform plane waves on dielectric boundaries. (In fact, being complex, $\hat{Z}_L$ generally corresponds to $\hat{\eta}_2$, the intrinsic impedance of a lossy region.) Equation (45) may be solved to yield

$$\frac{\hat{V}_m^-}{\hat{V}_m^+} = \hat{\Gamma}_L e^{-j2\beta\mathscr{L}} \tag{48}$$

Substituting (48) into (40) yields

$$\hat{\Gamma}(z) = \hat{\Gamma}_L e^{j2\beta(z-\mathscr{L})} \tag{49}$$

so that *the reflection coefficient at any point on the line, $\hat{\Gamma}(z)$, may be related to the reflection coefficient at the load, $\hat{\Gamma}_L$, which can be directly calculated via* (47). Once $\hat{\Gamma}(z)$ is obtained in this fashion, the input impedance to the line at this point may be found from (43).

EXAMPLE 7.5 A 10-m section of lossless transmission line having $R_C = 50\ \Omega$ and $u = 200$ m/μs is driven by a 26-MHz source having an open-circuit voltage of $\hat{V}_S = 100\underline{/0°}$ V and a source impedance of $\hat{Z}_S = 50 + j0\ \Omega$, as shown in Fig. 7.17*a*. The line is terminated in a load impedance of $\hat{Z}_L = 100 + j50\ \Omega$. Determine the imput impedance to the line and the phasor voltages at the input to the line, $\hat{V}(0)$, and at the load, $\hat{V}(\mathscr{L})$.

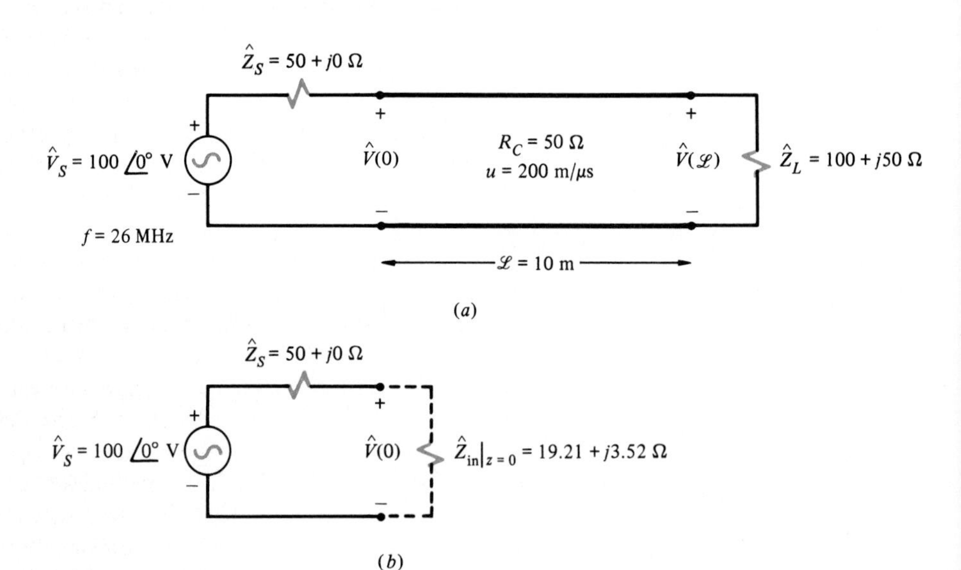

FIGURE 7.17
Example 7.5.

Solution The phase constant is

$$\beta = \frac{\omega}{u}$$

$$= 8.168 \times 10^{-1} \text{ rad/m}$$

and

$$2\beta\mathscr{L} = 16.34 \text{ rad}$$

$$= 936°$$

which is equivalent to $936 - 720 = 216°$. (The line is 1.3λ in electrical length.) The load reflection coefficient is

$$\hat{\Gamma}_L = \frac{\hat{Z}_L - R_C}{\hat{Z}_L + R_C}$$

$$= \frac{50 + j50}{150 + j50}$$

$$= 0.4472\angle 26.57°$$

Thus

$$\hat{\Gamma}(0) = \hat{\Gamma}_L e^{-j2\beta\mathscr{L}}$$

$$= 0.4472\angle 26.57° \times \angle -216°$$

$$= 0.4472\angle -189.4°$$

and

$$\hat{Z}_{\text{in}}\Big|_{z=0} = R_C \frac{1 + \hat{\Gamma}(0)}{1 - \hat{\Gamma}(0)}$$

$$= 19.53\angle 10.39° \ \Omega = 19.21 + j3.52 \quad \Omega$$

The input voltage to the line, $\hat{V}(0)$, can be found by voltage division from Fig. 7.17*b*:

$$\hat{V}(0) = \frac{\hat{Z}_{\text{in}}|_{z=0}}{\hat{Z}_{\text{in}}|_{z=0} + \hat{Z}_S} \hat{V}_S$$

$$= 28.18\angle 7.48° \text{ V}$$

and the time-domain voltage becomes

$$V(0, t) = 28.18 \cos(\omega t - 7.48°) \quad \text{V}$$

The load voltage is slightly more difficult to obtain. From our basic relationship in (41*a*),

$$\hat{V}(z) = \hat{V}_m^+ e^{-j\beta z}[1 + \hat{\Gamma}(z)]$$

we write

$$\hat{V}(0) = \hat{V}_m^+[1 + \hat{\Gamma}(0)]$$

$$\hat{V}(\mathscr{L}) = \hat{V}_m^+ e^{-j\beta\mathscr{L}}(1 + \hat{\Gamma}_L)$$

From the first of these we obtain $\hat{V}_m^+$ as

$$\hat{V}_m^+ = \frac{\hat{V}(0)}{1 + \hat{\Gamma}(0)}$$

$$= 50\underline{/0.033^\circ} \text{ V}$$

Thus

$$\hat{V}(\mathscr{L}) = \hat{V}_m^+ e^{-j\beta\mathscr{L}}(1 + \hat{\Gamma}_L)$$

$$= 70.71\underline{/-99.84^\circ} \text{ V}$$

The time-domain voltage is

$$V(\mathscr{L}, t) = 70.71 \cos(\omega t - 99.84^\circ) \qquad \text{V}$$

A more direct formula for the input impedance to the line may be obtained by substituting (49) and (47) into (43) and expanding:

$$e^{j2\beta(z-\mathscr{L})} = \cos[2\beta(z - \mathscr{L})] + j\sin[2\beta(z - \mathscr{L})] \tag{50}$$

After some manipulation, we obtain

$$\hat{Z}_{\text{in}}(z) = R_C \frac{\hat{Z}_L + jR_C \tan\beta(\mathscr{L} - z)}{R_C + j\hat{Z}_L \tan\beta(\mathscr{L} - z)} \tag{51}$$

The input impedance to the line seen by the source at $z = 0$ becomes

$$\hat{Z}_{\text{in}}\Big|_{z=0} = R_C \frac{\hat{Z}_L + jR_C \tan\beta\mathscr{L}}{R_C + j\hat{Z}_L \tan\beta\mathscr{L}} \tag{52}$$

Several interesting results are obtained from (52). Note that $\beta\mathscr{L}$ can be written as

$$\begin{aligned}\beta\mathscr{L} &= \frac{\omega\mathscr{L}}{u}\\ &= \frac{2\pi\mathscr{L}}{\lambda}\end{aligned} \tag{53}$$

where a wavelength is given by $\lambda = u/f$ in the medium surrounding the conductors.† Suppose the line is terminated in a short circuit ($\hat{Z}_L = 0$). Equation (52) shows that the input impedance to the line is

$$\hat{Z}_{\text{inSC}} = jR_C \tan\frac{2\pi\mathscr{L}}{\lambda} \qquad \hat{Z}_L = 0 \tag{54}$$

The imaginary part of the impedance, $\hat{Z} = jX$, is referred to as the reactance and is plotted in Fig. 7.18*a* for a short-circuit load as a function of the electrical length of the line. Similarly, if the line is terminated in an open circuit ($\hat{Z}_L = \infty$), the input impedance to the line becomes

$$\hat{Z}_{\text{inOC}} = -j\frac{R_C}{\tan(2\pi\mathscr{L}/\lambda)} \qquad \hat{Z}_L = \infty \tag{55}$$

The reactance for an open-circuit load is plotted in Fig. 7.18*b* as a function of electrical line length. By combining (54) and (55), we see that

$$R_C = \sqrt{\hat{Z}_{\text{inSC}}\hat{Z}_{\text{inOC}}} \tag{56}$$

Now suppose that the line length is $\frac{1}{4}\lambda$. For the short-circuit termination in (54), we find that the input impedance appears as an open circuit! A similar result is obtained for a line length $\mathscr{L} = \frac{1}{4}\lambda$ and an open-circuit termination: The line appears as a short circuit! (See also Fig. 7.18.) Similarly, suppose the termination impedance is an inductance, $\hat{Z}_L = j\omega L$. If the line is $\frac{1}{4}\lambda$ in length, the input impedance to the line becomes

$$\begin{aligned}\hat{Z}_{\text{in}}\Big|_{z=0} &= \frac{R_C^2}{\hat{Z}_L}\\ &= -\frac{jR_C^2}{\omega L} \qquad \begin{cases}\mathscr{L} = \frac{1}{4}\lambda\\ \hat{Z}_L = j\omega L\end{cases}\end{aligned} \tag{57}$$

so that the input impedance appears as a capacitive reactance ($C = L/R_C^2$)!

† The velocity of propagation of the line is $u = 1/\sqrt{lc}$. We will show in Sec. 7.4 that $lc = \mu\epsilon$ and thus that the velocity of propagation on the line is identical to that of a uniform plane wave in the medium that surrounds the line.

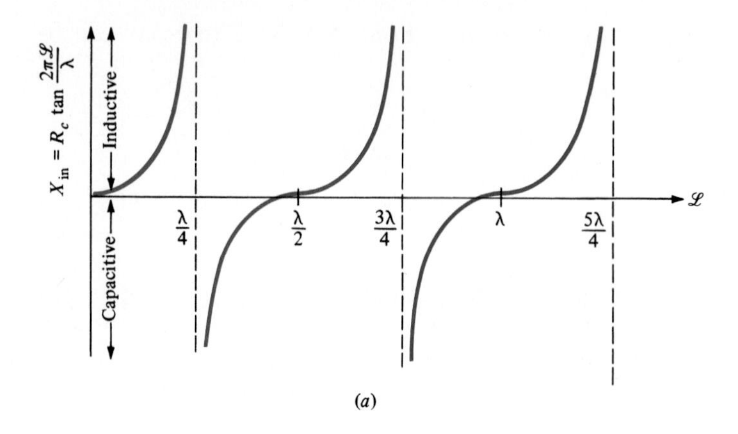

(a)

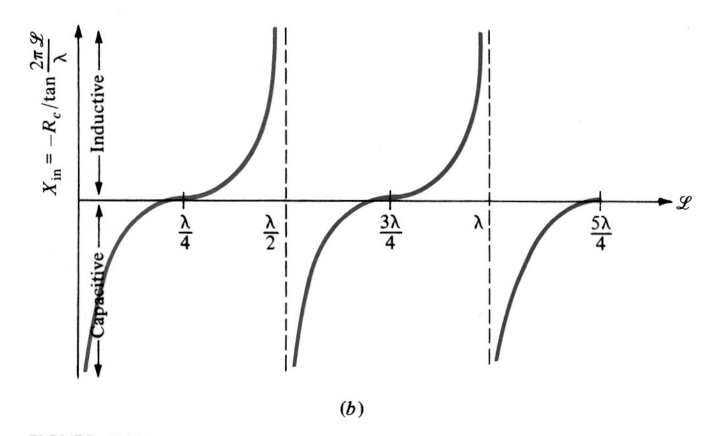

(b)

FIGURE 7.18
Input impedance of a transmission line: *(a)* short-circuit load; *(b)* open-circuit load.

These results for a quarter-wavelength line are also easily seen from the expression for input impedance given in (43). For a line length of $\mathscr{L} = \frac{1}{4}\lambda$,

$$2\beta\mathscr{L} = \frac{4\pi}{\lambda}\mathscr{L} = \pi \tag{58}$$

Thus, from (49)

$$\hat{\Gamma}(0) = \hat{\Gamma}_L \underline{/-180^\circ} = -\hat{\Gamma}_L \qquad \mathscr{L} = \tfrac{1}{4}\lambda \tag{59}$$

and the reflection coefficient at the input to a line that is $\frac{1}{4}\lambda$ long is the negative of the load reflection coefficient. For a short-circuit load, $\Gamma_L = -1$ and $\hat{\Gamma}(0) = 1$. Thus

$$\frac{1 + \hat{\Gamma}(0)}{1 - \hat{\Gamma}(0)} = \infty \qquad \text{and} \qquad \hat{Z}_{\text{inSC}} = \infty \tag{60}$$

Similarly, for an open-circuit load, $\hat{\Gamma}_L = 1$ and $\hat{\Gamma}(0) = -1$. Thus

$$\frac{1 + \hat{\Gamma}(0)}{1 - \hat{\Gamma}(0)} = 0 \qquad \text{and} \qquad \hat{Z}_{\text{inOC}} = 0 \tag{61}$$

Finally, we should observe from (51) that *the input impedance replicates at distances on the line that are separated by multiples of a half-wavelength since*

$$\begin{aligned} \tan \beta(\mathscr{L} - z) &= \tan(\beta\mathscr{L} - \beta z \pm n\pi) \\ &= \tan \beta\left(\mathscr{L} - z \pm \frac{n\pi}{\beta}\right) \\ &= \tan \beta\left(\mathscr{L} - z \pm \frac{n\lambda}{2}\right) \end{aligned} \tag{62}$$

for $n = 1, 2, 3, \ldots$, which implies that

$$\hat{Z}_{\text{in}}(z) = \hat{Z}_{\text{in}}\left(z \pm \frac{n\lambda}{2}\right) \tag{63}$$

7.2.1 Voltage and Current as Functions of Position on the Line

The voltage and current waveforms on the transmission line are the combination of forward- and backward-traveling waves. These combine to yield standing waves on the line as we observed in the previous chapter in the study of uniform, plane waves incident normal to plane boundaries. Again, the analogies between transmission lines and uniform, plane waves incident normal to plane boundaries are striking and should be considered. To show the existence of standing waves on the line, we write the basic voltage and current expressions on the line, given in (41), in terms of the load voltage $\hat{V}(\mathscr{L}) \triangleq \hat{V}_L$ and load current $\hat{I}(\mathscr{L}) \triangleq \hat{I}_L$ by evaluating (41) at $z = \mathscr{L}$:

$$\hat{V}_L = \hat{V}_m^+ e^{-j\beta\mathscr{L}}(1 + \hat{\Gamma}_L) \tag{64a}$$

$$\hat{I}_L = \frac{\hat{V}_m^+}{R_C} e^{-j\beta\mathscr{L}}(1 - \hat{\Gamma}_L) \tag{64b}$$

Solving for $\hat{V}_m^+$ and substituting into (41) and using (49) gives

$$\hat{V}(z) = \frac{e^{j\beta(\mathscr{L} - z)}}{1 + \hat{\Gamma}_L}(1 + \hat{\Gamma}_L e^{-j2\beta(\mathscr{L} - z)})\hat{V}_L \tag{65a}$$

$$\hat{I}(z) = \frac{e^{j\beta(\mathscr{L} - z)}}{1 - \hat{\Gamma}_L}(1 - \hat{\Gamma}_L e^{-j2\beta(\mathscr{L} - z)})\hat{I}_L \tag{65b}$$

Note that these results involve the distance from the load, $\mathscr{L} - z$. Thus, defining the distance from the load as $d = \mathscr{L} - z$, Eqs. (65) become

$$\hat{V}_d = \frac{e^{j\beta d}}{1 + \hat{\Gamma}_L}(1 + \hat{\Gamma}_L e^{-j2\beta d})\hat{V}_L \tag{66a}$$

$$\hat{I}_d = \frac{e^{j\beta d}}{1 - \hat{\Gamma}_L}(1 - \hat{\Gamma}_L e^{-j2\beta d})\hat{I}_L \tag{66b}$$

where $\hat{V}_d$ and $\hat{I}_d$ are the phasor voltage and current, respectively, at a distance d from the load, as shown in Fig. 7.19*a*.

The variations of the magnitudes of these voltages and currents with distance d away from the load are shown in Fig. 7.19 for various choices of load impedance. These variations may be seen by observing the magnitudes of (66):

$$|\hat{V}_d| = \frac{|1 + \hat{\Gamma}_L e^{-j2\beta d}|}{|1 + \hat{\Gamma}_L|}|\hat{V}_L| \tag{67a}$$

$$|\hat{I}_d| = \frac{|1 - \hat{\Gamma}_L e^{-j2\beta d}|}{|1 - \hat{\Gamma}_L|}|\hat{I}_L| \tag{67b}$$

Note that the variations with distance d are contained in the terms $|1 + \hat{\Gamma}_L e^{-j2\beta d}|$ and $|1 - \hat{\Gamma}_L e^{-j2\beta d}|$. These can be viewed as the magnitudes of the sum or difference of the complex numbers $1\underline{/0^\circ}$ and $\hat{\Gamma}_L\underline{/-2\beta d} = \Gamma_L\underline{/\theta_{\Gamma_L} - 2\beta d}$. This is displayed in the so-called crank diagram shown in Fig. 7.20. As d increases, the vector $\hat{\Gamma}_L\underline{/-2\beta d}$ rotates (in the clockwise direction) about the tip of the $1\underline{/0^\circ}$ vector (or complex number). The resultant sum of these two complex numbers is proportional to the magnitude of the line voltage according to (67*a*). The resultant difference of these two complex numbers is proportional to the magnitude of the line current according to (67*b*). Note that $\beta = 2\pi/\lambda$; thus

$$2\beta d = 4\pi\frac{d}{\lambda} \tag{68}$$

and the angle in the crank diagram can be written in terms of wavelengths down the line.

Several cases deserve special consideration:

1 *Short-circuit load*, $R_L = 0$: For this case, $\hat{\Gamma}_L = -1$ and the load voltage $\hat{V}_L$ is zero. The equation for the load current in (67*b*) becomes

$$\begin{aligned}|\hat{I}_d| &= \frac{|1 + e^{-j2\beta d}|}{2}|\hat{I}_L| \\ &= |e^{-j\beta d}|\frac{|e^{j\beta d} + e^{-j\beta d}|}{2}|\hat{I}_L| \\ &= \left|\cos\frac{2\pi d}{\lambda}\right||\hat{I}_L|\end{aligned} \tag{69}$$

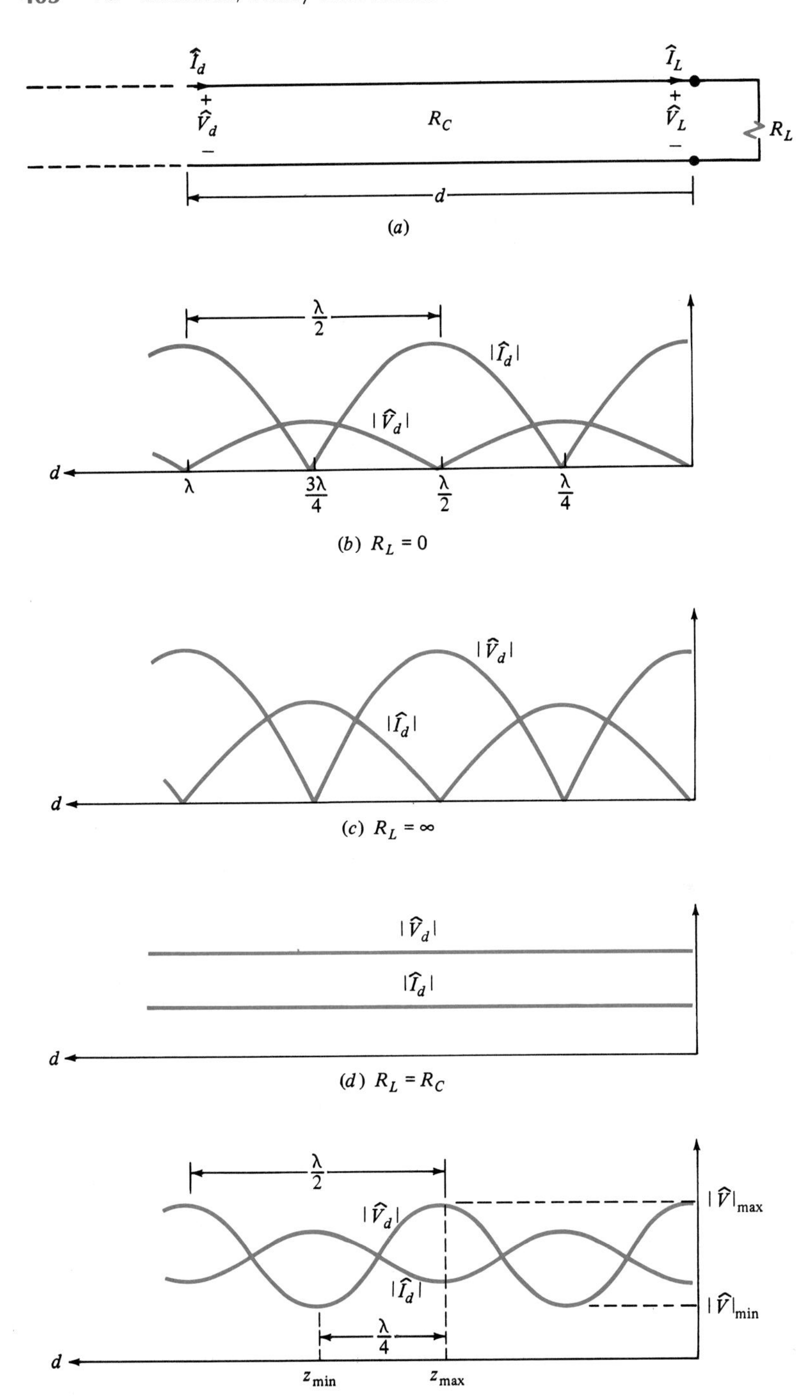

FIGURE 7.19

Line voltage and current as functions of termination impedance.

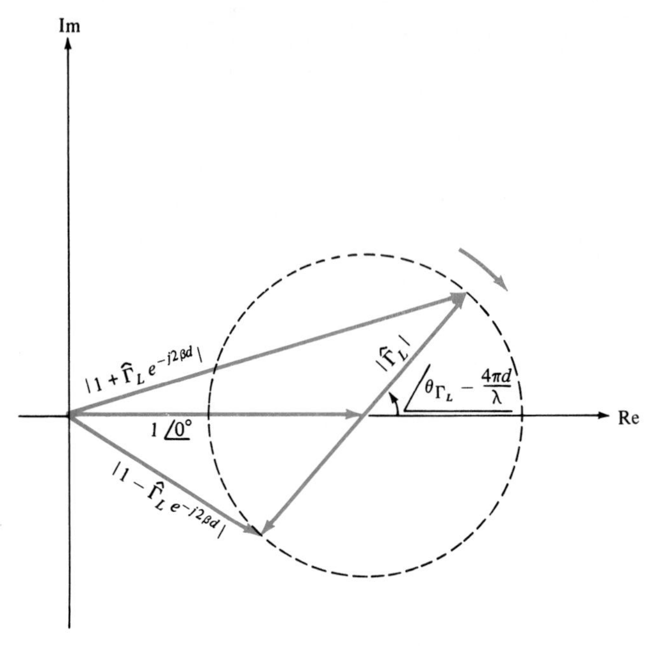

FIGURE 7.20
The crank diagram.

since $\beta d = 2\pi d/\lambda$. This is plotted in Fig. 7.19*b*. Since $\hat{V}_L = 0$, (67*a*) is not an appropriate form for use in plotting the voltage variation. Returning to (41*a*) and substituting (49) gives

$$\hat{V}(z) = \hat{V}_m^+ e^{-j\beta z}(1 + \hat{\Gamma}_L e^{-j2\beta(\mathscr{L} - z)}) \tag{70}$$

Thus

$$|\hat{V}_d| = |\hat{V}_m^+||1 + \hat{\Gamma}_L e^{-j2\beta d}| \tag{71}$$

Substituting $\hat{\Gamma}_L = -1$ gives

$$\begin{aligned} |\hat{V}_d| &= |\hat{V}_m^+||1 - e^{-j2\beta d}| \\ &= |\hat{V}_m^+||e^{-j\beta d}||e^{j\beta d} - e^{-j\beta d}| \\ &= 2|\hat{V}_m^+|\left|\sin\frac{2\pi d}{\lambda}\right| \end{aligned} \tag{72}$$

Thus, the magnitude of the load voltage varies as the sine of the electrical distance from the load. This is also plotted in Fig. 7.19*b*.

Note that a maximum and adjacent minimum are separated by $\lambda/4$. Similarly, adjacent maxima as well as adjacent minima are separated by $\lambda/2$. These results can be clearly seen from the crank diagram in Fig. 7.20. The voltage maxima will occur when the vectors $1\underline{/0^\circ}$ and $\hat{\Gamma}_L\underline{/\theta_{\Gamma_L} - 4\pi d/\lambda}$ are pointing in the same direction. Conversely, voltage minima will occur when these vectors are directed in opposite directions. If a maximum occurs at d_1 and a minimum occurs at d_2, these points must be separated by

$$\frac{4\pi d_1}{\lambda} - \frac{4\pi d_2}{\lambda} = \pi$$

or

$$d_1 - d_2 = \frac{\lambda}{4}$$

Similar results apply to the current maxima and minima. We will find this to be a general result regardless of the load impedance. The crank diagram in Fig. 7.20 makes these results immediately clear.

2 *Open-circuit load*, $R_L = \infty$: For this case, $\hat{\Gamma}_L = +1$ and the load current $\hat{I}_L$ is zero. The equation for the load voltage in (67*a*) becomes

$$|\hat{V}_d| = \left|\cos\frac{4\pi d}{\lambda}\right| |\hat{V}_L| \tag{73}$$

Similarly, (41*b*) yields

$$|\hat{I}_d| = 2\frac{|\hat{V}_m^+|}{R_C}\left|\sin\frac{2\pi d}{\lambda}\right| \tag{74}$$

These results are plotted in Fig. 7.19*c*.

Note again that a maximum and adjacent minimum are separated by $\lambda/4$. Adjacent maxima as well as adjacent minima are again separated by $\lambda/2$. These results are again made clear by the crank diagram in Fig. 7.20.

From the plots in Fig. 7.19, it should be clear why a quarter-wavelength line that is terminated in an open circuit appears as a short circuit. At a quarter-wavelength away from the open circuit, the voltage goes to zero whereas the current is nonzero: a short circuit. Similar remarks apply to the case of a quarter-wavelength line terminated in a short circuit; a quarter-wavelength from the load, the current goes to zero, whereas the voltage is nonzero: equivalent to an open circuit.

3 *General resistive load* R_L: In the case of some general resistive load, $\hat{\Gamma}_L$ is real, so $\theta_{\Gamma_L} = 0$. For this case, it is clear from the crank diagram in

Fig. 7.20 that the voltage and current maxima or minima will occur exactly at the load, since for $\theta_{\Gamma_L} = 0$ the two vectors are aligned for $d = 0$. For $R_L > R_C$, Γ_L will be positive. It is clear from the crank diagram and (67*a* and *b*) that a voltage maximum will occur at the load, whereas a current minimum will occur at the load. This is illustrated in Fig. 7.19*e*. For $R_L < R_C$, Γ_L will be negative and the above properties will be reversed; that is, a current maximum and a voltage minimum will occur at the load. Note that these properties are confirmed for short- and open-circuit loads.

For either $R_L > R_C$ or $R_L < R_C$, the crank diagram shows that a voltage or current maximum and adjacent minimum are again separated by precisely $\lambda/4$. Similarly, the crank diagram shows that adjacent corresponding points on these variations are separated by multiples of $\lambda/2$.

If $R_L = R_C$, the line is said to be *matched*. For this case, $\hat{\Gamma}_L = 0$ and there is no variation in voltage and current along the line. This is shown in Fig. 7.19*d*.

4 *General load* $\hat{Z}_L$: For a general reactive load $\hat{Z}_L$, the above results are seen from the crank diagram to be unchanged, with one exception: For this unrestricted load, $\theta_{\Gamma_L} \neq 0$, and thus the two vectors in the crank diagram will not align at $d = 0$; therefore, the voltage and current maximum and minimum, will *not occur* at the load. Except for this difference, certain general properties hold:

(a) A voltage (and current) maximum and adjacent minimum will be separated by $\lambda/4$.

(b) Corresponding points on the voltage (and current) magnitude waveforms replicate between distances separated by multiples of $\lambda/2$.

(c) The complete (magnitude and phase) line voltage and current replicate between distances separated by multiples of λ.

The crank diagram makes these observations quite clear.

7.2.2. Transmission-Line Matching

Where possible, it is desirable to match transmission lines in order to eliminate reflections. Mismatched lines ($\hat{Z}_L \neq R_C$) give rise to echoes (reflections), as discussed in the previous section. This can be particularly annoying, for example, on telephone circuits. Since it is not always possible to match a line exactly, it is desirable to have a measure of the degree of mismatch. This measure is called the *voltage standing-wave ratio* (*VSWR*) and is defined as the ratio of the magnitude of the maximum voltage on the line to the magnitude of the minimum voltage on the line, or

$$\text{VSWR} = \frac{|\hat{V}|_{\max}}{|\hat{V}|_{\min}} \tag{75}$$

as shown in Fig. 7.19*e*. Note that in the matched case the ratio of voltage maxima on the line to voltage minima is unity, or

$$\text{VSWR} = 1 \qquad \text{for} \qquad \hat{Z}_L = R_C \tag{76a}$$

For $\hat{Z}_L = 0$ or $\hat{Z}_L = \infty$,

$$\text{VSWR} = \infty \qquad \text{for} \qquad \hat{Z}_L = 0 \qquad \text{or} \qquad \hat{Z}_L = \infty \tag{76b}$$

Thus, the VSWR yields a measure of the line mismatch. Note that the VSWR will always be a positive real number and that its value will lie between unity and infinity:

$$1 \le \text{VSWR} < \infty \tag{77}$$

The closer the VSWR is to unity, the better the line match.

Let us now determine a formula for the VSWR in terms of R_C and $\hat{Z}_L$. From the previous results and the crank diagram in Fig. 7.20, we determined that a voltage maximum and an adjacent voltage minimum were separated by $\lambda/4$. From the crank diagram it follows that

$$|\hat{V}_d|_{\max} = 1 + |\hat{\Gamma}| \tag{78a}$$

$$|\hat{V}_d|_{\min} = 1 - |\hat{\Gamma}| \tag{78b}$$

Substituting these into the definition of VSWR given in (75), we obtain

$$\text{VSWR} = \frac{1 + |\hat{\Gamma}|}{1 - |\hat{\Gamma}|} \tag{79}$$

From (49) the reflection coefficient has a constant magnitude at all points along the line. In particular,

$$|\hat{\Gamma}| = |\hat{\Gamma}_L| \tag{80}$$

and (79) becomes

$$\text{VSWR} = \frac{1 + |\hat{\Gamma}_L|}{1 - |\hat{\Gamma}_L|} \tag{81}$$

as is also evident from the crank diagram. Alternatively, if we know the VSWR ratio from measurements, we may compute the magnitude of the reflection coefficients at any point on the line as

$$|\hat{\Gamma}(z)| = |\hat{\Gamma}_L| = \frac{\text{VSWR} - 1}{\text{VSWR} + 1} \tag{82}$$

7.2.3 Power Flow

Now we will consider power flow on the line. From the transverse nature of the fields of the TEM mode, it is clear that power flow occurs in the $\pm z$ direction since the Poynting vector is in the $\pm z$ direction. In terms of the line voltage and current, the average power flow in the $+z$ direction is given by

$$P_{av}(z) = \tfrac{1}{2} \operatorname{Re}[\hat{V}(z)\hat{I}^*(z)] \tag{83}$$

Since

$$\hat{V}(z) = \hat{V}_m^+ e^{-j\beta z}[1 + \hat{\Gamma}(z)] \tag{84a}$$

$$\hat{I}(z) = \frac{\hat{V}_m^+}{R_C} e^{-j\beta z}[1 - \hat{\Gamma}(z)] \tag{84b}$$

(83) becomes

$$P_{av}(z) = \frac{1}{2} \operatorname{Re}\left[\frac{\hat{V}_m^+ \hat{V}_m^{+*}}{R_C}(1 + \hat{\Gamma})(1 - \hat{\Gamma}^*)\right]$$

$$= \frac{1}{2} \frac{|\hat{V}_m^+|^2}{R_C} \operatorname{Re}(1 + \hat{\Gamma} - \hat{\Gamma}^* - \hat{\Gamma}\hat{\Gamma}^*) \tag{85}$$

But note that $\hat{\Gamma} - \hat{\Gamma}^*$ yields a purely imaginary number and that $\hat{\Gamma}\hat{\Gamma}^* = |\hat{\Gamma}|^2$. Thus, (85) becomes

$$P_{av}(z) = \frac{1}{2} \frac{|\hat{V}_m^+|^2}{R_C}(1 - |\hat{\Gamma}|^2) \tag{86}$$

Since $|\hat{\Gamma}(z)| = |\hat{\Gamma}_L|$, (86) becomes

$$P_{av}(z) = \frac{1}{2} \frac{|\hat{V}_m^+|^2}{R_C}(1 - |\hat{\Gamma}_L|^2) \tag{87}$$

The result in (87) could have been derived by adding the average powers of the individual waves: an interesting result quite similar to the case of uniform plane waves in lossless media. The voltage and current in the foward-traveling wave may be denoted as

$$\hat{V}^+(z) = \hat{V}_m^+ e^{-j\beta z} \tag{88a}$$

$$\hat{I}^+(z) = \frac{\hat{V}_m^+}{R_C} e^{-j\beta z} \tag{88b}$$

Similarly,

$$\hat{V}^-(z) = \hat{V}_m^- e^{j\beta z} \tag{89a}$$

$$\hat{I}^-(z) = -\frac{\hat{V}_m^-}{R_C} e^{j\beta z} \tag{89b}$$

denote the voltage and current of the backward-traveling wave. The average powers in the forward- and backward-traveling waves become

$$P_{\text{av}}^+(z) = \frac{1}{2}\frac{|\hat{V}_m^+|^2}{R_C} \tag{90a}$$

$$\begin{aligned} P_{\text{av}}^-(z) &= -\frac{1}{2}\frac{|\hat{V}_m^-|^2}{R_C} \\ &= -\frac{1}{2}\frac{|\hat{V}_m^+|^2}{R_C}|\hat{\Gamma}(z)|^2 \\ &= -\frac{1}{2}\frac{|\hat{V}_m^+|^2}{R_C}|\hat{\Gamma}_L|^2 \end{aligned} \tag{90b}$$

Note that from (87), (90*a*), and (90*b*) we have

$$P_{\text{av}}(z) = P_{\text{av}}^+(z) + P_{\text{av}}^-(z) \tag{90c}$$

The ratio of the reflected and incident powers at any point on the line becomes

$$\begin{aligned} \frac{P_{\text{av, reflected}}}{P_{\text{av, incident}}} &= -\frac{P_{\text{av}}^-(z)}{P_{\text{av}}^+(z)} \\ &= |\hat{\Gamma}_L|^2 \end{aligned} \tag{91}$$

When the line is terminated in either an open circuit ($\hat{\Gamma}_L = +1$) or a short circuit ($\hat{\Gamma}_L = -1$), then, from (87), $P_{\text{av}} = 0$ and all incident power is reflected at the load. Thus, there is no *net* flow of average power in the z direction. This, of course, makes sense because the load (either a short circuit or an open circuit) cannot absorb power. Also, we know that for this case the forward- and backward-traveling waves are equal in magnitude at any point on the line, so that at this point the average power flowing to the right in the foward-traveling wave is equal to the average power flowing to the left in the backward-traveling wave; thus, there is no *net* power flow in the z direction. If the line is matched at the load, i.e. $\hat{Z}_L = R_C$, then $|\hat{\Gamma}_L| = 0$ and (87) becomes

$$\begin{aligned} P_{\text{av}}(z) &= \frac{1}{2}\frac{|\hat{V}_m^+|^2}{R_C} \\ &= P_{\text{av}}^+(z) \qquad \hat{Z}_L = R_C \end{aligned} \tag{92}$$

This result again makes sense because, for this matched case, there is no backward-traveling wave.

If we knew the value of $\hat{V}_m^+$ for a particular load $\hat{Z}_L$ and source $\hat{Z}_S$ and $\hat{V}_S$, then (87) would immediately yield the power flow on the line. Ordinarily we do not find $\hat{V}_m^+$. Of particular interest for these problems, however, is the average power delivered to the load $\hat{Z}_L$. This may be computed in the following manner: First compute the input impedance to the line: $\hat{Z}_{in}|_{z=0}$. Then compute the average power delivered to the line from Fig. 7.21 as

$$\begin{aligned} P_{av,\,to\,line} &= \tfrac{1}{2}\,\mathrm{Re}\,[\hat{V}(0)\hat{I}^*(0)] \\ &= \frac{1}{2}\,\mathrm{Re}\left[\hat{V}(0)\,\frac{\hat{V}^*(0)}{\hat{Z}^*_{in}|_{z=0}}\right] \\ &= \frac{1}{2}\,\frac{|\hat{V}(0)|^2}{|\hat{Z}_{in}|_{z=0}}\cos\underline{/\hat{Z}_{in}|_{z=0}} \end{aligned} \tag{93}$$

But

$$\hat{V}(0) = \frac{\hat{Z}_{in}|_{z=0}}{\hat{Z}_{in}|_{z=0} + \hat{Z}_S}\,\hat{V}_S \tag{94}$$

so that $P_{av,\,to\,line}$ may be computed. Now *if* the line is lossless, as we have been assuming, then the average power delivered to the line is equal to the average power delivered to the load:

$$P_{av,\,to\,load} = P_{av,\,to\,line} \tag{95}$$

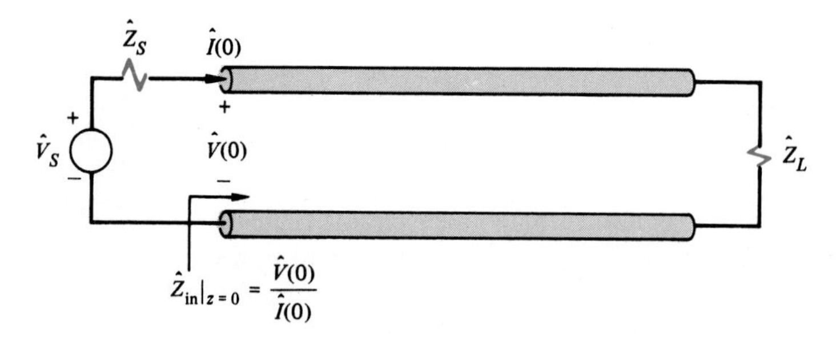

(*a*)

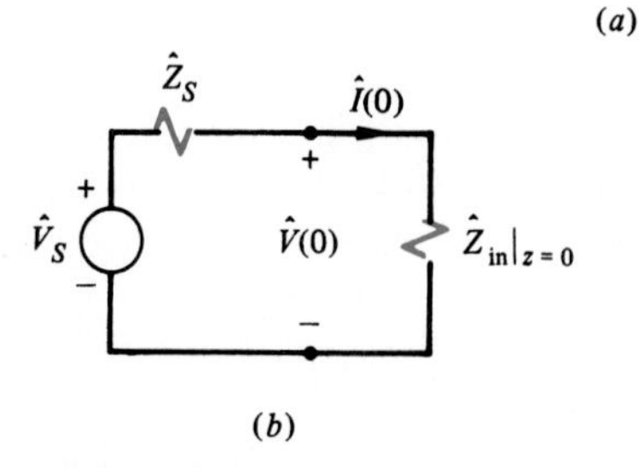

(*b*)

FIGURE 7.21

Input impedance to a transmission line.

EXAMPLE 7.6 Consider the parallel-wire line shown in Fig. 7.22. Compute the time-domain expression for the load voltage and the average power delivered to the load.

Solution The per-unit-length parameters of the line are $c = 200$ pF/m and $l = 0.5$ μH/m, from which we compute

$$R_C = \sqrt{l/c} = 50\ \Omega$$

$$u = \frac{1}{\sqrt{lc}} = 100\ \text{m}/\mu\text{s}$$

The line is 0.3λ in electrical length. The reflection coefficient at the load is

$$\hat{\Gamma}_L = \frac{\hat{Z}_L - R_C}{\hat{Z}_L + R_C} = \frac{100 + j50 - 50}{100 + j50 + 50} = 0.447\underline{/26.6^\circ}$$

and the VSWR is

$$\text{VSWR} = \frac{1 + |\hat{\Gamma}_L|}{1 - |\hat{\Gamma}_L|} = \frac{1 + 0.447}{1 - 0.447} = 2.62$$

First we compute the input impedance to the line. The reflection coefficient at the input to the line is

$$\hat{\Gamma}(z = 0) = \hat{\Gamma}_L e^{j2\beta(0 - \mathscr{L})} = \hat{\Gamma}_L e^{-j2\beta\mathscr{L}}$$

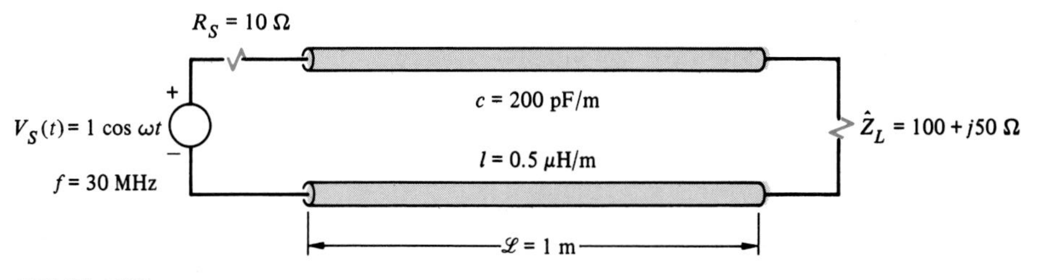

FIGURE 7.22
Example 7.6

and $\beta\mathscr{L} = \omega\mathscr{L}/u = 0.6\pi$ rad $= 108°$, so that

$$\hat{\Gamma}(z = 0) = 0.447\underline{/26.6°}e^{-j216°}$$
$$= 0.447\underline{/-189.4°}$$

Thus

$$\hat{Z}_{\text{in}}\bigg|_{z=0} = R_C \frac{1 + \hat{\Gamma}(z = 0)}{1 - \hat{\Gamma}(z = 0)}$$
$$= 19.54\underline{/10.34°}\ \Omega$$

This result checks with the calculation using (52). The voltage at the input to the line becomes ($\hat{Z}_S = R_S = 10\ \Omega$)

$$\hat{V}(0) = \frac{\hat{Z}_{\text{in}}|_{z=0}}{\hat{Z}_{\text{in}}|_{z=0} + R_S}\hat{V}_S$$
$$= 0.664\underline{/3.5°}\ \text{V}$$

and

$$\hat{I}(0) = \frac{\hat{V}(0)}{\hat{Z}_{\text{in}}|_{z=0}}$$
$$= 34 \times 10^{-3}\underline{/-6.8°}\ \text{A}$$

Thus

$$V(0, t) = 0.664 \cos(60\pi \times 10^6 t + 3.5°) \qquad \text{V}$$

$$I(0, t) = 0.034 \cos(60\pi \times 10^6 t - 6.8°) \qquad \text{A}$$

and

$$P_{\text{av, to line}} = \tfrac{1}{2}\,\text{Re}\,[\hat{V}(0)\hat{I}^*(0)]$$
$$= \frac{1}{2}\frac{|\hat{V}(0)|^2}{|\hat{Z}_{\text{in}}|_{z=0}}\cos\underline{/\hat{Z}_{\text{in}}|_{z=0}}$$
$$= 11.1 \times 10^{-3}\ \text{W}$$

Since the line is lossless,

$$P_{\text{av, to load}} = P_{\text{av, to line}}$$
$$= 11.1 \times 10^{-3}\ \text{W}$$

The load voltage may be computed in the following manner. Since

$$\hat{V}(z) = \hat{V}_m^+ e^{-j\beta z}[1 + \hat{\Gamma}(z)]$$

then
$$\hat{V}(0) = \hat{V}_m^+[1 + \hat{\Gamma}(0)]$$

With the above results we obtain $\hat{V}_m^+$ as

$$\hat{V}_m^+ = \frac{\hat{V}(0)}{1 + \hat{\Gamma}(0)}$$

$$= 1.18\underline{/-3.94^\circ} \text{ V}$$

Thus

$$\hat{V}(\mathscr{L}) = \hat{V}_m^+ e^{-j\beta\mathscr{L}}(1 + \hat{\Gamma}_L)$$

$$= 1.18\underline{/-3.94^\circ} e^{-j108^\circ}(1 + 0.447\underline{/26.6^\circ})$$

$$= 1.668\underline{/-103.8^\circ} \text{ V}$$

In the time domain, the load voltage is

$$V(\mathscr{L}, t) = 1.668 \cos(60\pi \times 10^6 t - 103.8^\circ) \qquad \text{V}$$

As a check on the average power delivered to the load, we may compute, directly,

$$P_{\text{av, to load}} = \tfrac{1}{2} \text{Re}\,[\hat{V}(\mathscr{L})\hat{I}^*(\mathscr{L})]$$

$$= \frac{1}{2}\frac{|\hat{V}(\mathscr{L})|^2}{|\hat{Z}_L|} \cos \underline{/\hat{Z}_L}$$

$$= \frac{1}{2}\frac{(1.668)^2}{111.8} \cos 26.57^\circ$$

$$= 11.1 \times 10^{-3} \text{ W}$$

In computing the above results for sinusoidal line excitation, we require numerous algebraic operations on complex numbers (addition, subtraction, multiplication, division, and conversion from polar to rectangular form and vice versa) in order to arrive at a desired result. The Smith chart, discussed in Appendix C, is an ingenious graphical technique for computing most of the above quantities of interest for sinusoidal line excitation. Use of the Smith chart avoids most of these tedious complex-number operations and at the same time provides considerable insight into the line's behavior which is easily obscured in the above equations.

7.3 Lossy Transmission Lines†

If the line conductors are not perfect conductors, the TEM mode cannot exist since there will be a longitudinal component of the electric field $\mathcal{E}_z$ due to the line currents passing through the imperfect conductors. From a practical standpoint, the line conductors, although not perfect conductors, will usually be sufficiently good conductors that this loss may be included as a reasonable approximation in the transmission-line formulation (which assumes a TEM field structure). Therefore, when we consider lossy line conductors and use the TEM-mode transmission-line formulation, we essentially assume that the field structure is "almost TEM" (sometimes referred to as quasi TEM).

In addition, if the surrounding medium is lossy (either through a conductivity or polarization loss of the medium), this additional loss mechanism may also be accounted for in the transmission-line formulation. We will find in Sec. 7.4 that, as opposed to the case of imperfect conductors, a lossy medium does not in itself preclude the existence of the TEM mode. In other words, a lossy medium and the TEM mode can coexist so long as the medium is homogeneous.

The purpose of this section is to investigate the effects of these losses on the transmission-line formulation and the resulting voltages and currents of the line. In addition, we obtain a derivation of the transmission-line equations from an electromagnetic fields standpoint to show that an equivalent-circuit representation such as in Fig. 7.6 is, in fact, correct.

Consider Fig. 7.23*a*, which shows a Δz section of a two-conductor transmission line. (Parallel wires are shown for illustration.) For the purposes of this discussion, the two wires lie in the xz plane. The conductors and the surrounding medium are considered to be lossy. We construct a flat rectangular surface s in the xz plane between z and Δz and between the surfaces of the two wires. This surface is bounded by the contour c, as shown in Fig. 7.23*a*. Assuming a sinusoidal field variation, the components of the phasor electric fields transverse to the line axis along c are denoted by $\hat{E}_x(x, z)$ and $\hat{E}_x(x, z + \Delta z)$. Note that this transverse field component along c is, in general, a function of x and z. Similarly, the longitudinal component of the electric field along c is denoted by $\hat{E}_z(x_1, z)$ and $\hat{E}_z(x_2, z)$. The component of the transverse magnetic field perpendicular to s is denoted by $\hat{B}_y(x, z)$. Let us apply Faraday's law to this contour:

$$\oint_c \hat{\mathbf{E}} \cdot d\mathbf{l} = -j\omega \int_s \hat{\mathbf{B}} \cdot d\mathbf{s} \tag{96}$$

† For the interested reader, a thorough discussion of the derivation of the transmission-line equations is given in R. B. Adler, L. J. Chu, and R. M. Fano, *Electromagnetic Energy Transmission and Radiation*, Wiley, New York, 1960, chap. 9.

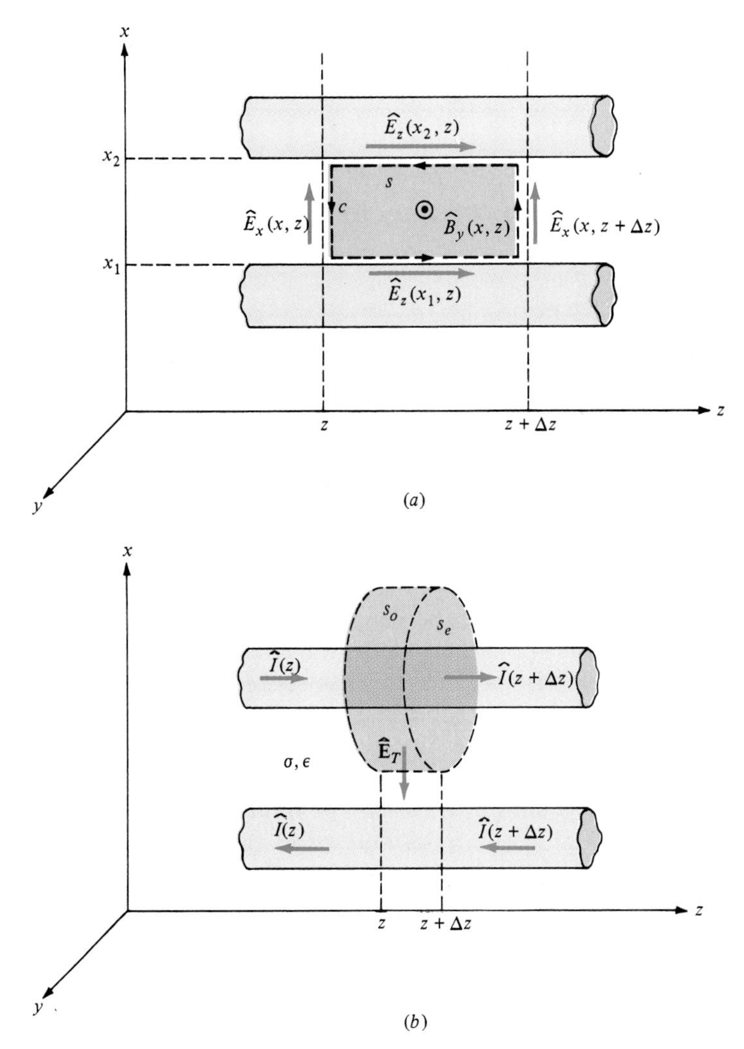

FIGURE 7.23

Contours and surfaces used in the derivation of the transmission-line equations for lossy lines.

We obtain

$$\int_{x_1}^{x_2} [\hat{E}_x(x, z + \Delta z) - \hat{E}_x(x, z)]\, dx + \int_{z}^{z+\Delta z} [\hat{E}_z(x_1, z) - \hat{E}_z(x_2, z)]\, dz$$

$$= -j\omega \int_{z}^{z+\Delta z} \int_{x_1}^{x_2} \hat{B}_y(x, z)\, dx\, dz \qquad (97)$$

Now let us examine the various terms in (97).

If we define the phasor voltage between the wires as

$$\hat{V}(z) = -\int_{x_1}^{x_2} \hat{E}_x(x, z)\, dx \tag{98}$$

then the first term in (97) becomes

$$\hat{V}(z + \Delta z) - \hat{V}(z) = -\int_{x_1}^{x_2} [\hat{E}_x(x, z + \Delta z) - \hat{E}_x(x, z)]\, dx \tag{99}$$

Note that this definition of voltage is, strictly speaking unique [independent of path in a transverse (xy) plane] only if the field structure is TEM. Imperfect conductors invalidate the assumption of a TEM fields structure, in which case the voltage definition in (99) is only an approximation (usually a reasonable one for good conductors).

Suppose a phasor current $\hat{I}(z)$ exists in the upper wire in the positive z direction and returns in the lower wire. For the purposes of discussion, let us assume that the losses in the wires can be lumped as an impedance through which $\hat{I}$ passes. The lossy nature of the conductors will result in a resistance per unit of wire length, r (the net resistance of both wires between z and $z + \Delta z$). In addition, since the wires are not perfect conductors, the current $\hat{I}$ is not confined to the surface of the wires but penetrates into the wires. If we consider a cross section of one wire, we can see that the magnetic flux around some contour internal to the wire links only a portion of the current in the wire. This results in an inductance per unit of wire length, l_i (of both wires), internal to the wires. With these points in mind, the second term in (97) can be written as

$$\int_{z}^{z+\Delta z} [\hat{E}_z(x_1, z) - \hat{E}_z(x_2, z)]\, dz = -(r + j\omega l_i)\, \Delta z\, \hat{I}(z) \tag{100}$$

The term on the right-hand side of (97) is related to the current-produced magnetic flux external to the wires. This flux penetrates the surface s, resulting in an external inductance per unit length given by

$$l_e\, \Delta z = -\frac{\int_{z}^{z+\Delta z} \int_{x_1}^{x_2} \hat{B}_y(x, z)\, dx\, dz}{\hat{I}(z)} \tag{101}$$

[The negative sign is included in (101) to account for the fact that $\hat{I}(z)$ produces a magnetic flux through surface s in the negative y direction.]

Substituting (101), (100), and (99) into (97) results in

$$\hat{V}(z + \Delta z) - \hat{V}(z) = -[(r + j\omega l_i) + j\omega l_e]\, \Delta z\, \hat{I}(z) \tag{102}$$

Dividing both sides of (102) by Δz and taking the limit as $\Delta z \to 0$, we obtain the first transmission-line equation:

$$\frac{d\hat{V}(z)}{dz} = -(r + j\omega l)\hat{I}(z) \tag{103}$$

where we have combined the internal inductance (due to the current internal to the wires) l_i and the external inductance l_e into

$$l = l_i + l_e \tag{104}$$

The second transmission-line equation can be similarly derived by considering the equation of continuity (conservation of charge). Consider the closed surface s shown in Fig. 7.23*b*, which consists of a cylinder of length Δz enclosing the upper wire. The portion of this surface just off the periphery of the cylinder is denoted by s_o, and the portions on the ends of the cylinder are denoted by s_e. The continuity equation applied to this surface is

$$\oint_s \hat{\mathbf{J}} \cdot d\mathbf{s} = -j\omega\hat{q} \tag{105}$$

where $\hat{q}(t) = qe^{j\omega t}$ is the charge enclosed by s. Over the ends of the cylinder,

$$\int_{s_e} \hat{\mathbf{J}} \cdot d\mathbf{s} = \hat{I}(z + \Delta z) - \hat{I}(z) \tag{106}$$

Over the sides of the cylinder, the conductivity of the medium results in a transverse conduction current through s_o (or imaginary displacement current due to polarization loss, which results in the same effect); thus

$$\int_{s_o} \hat{\mathbf{J}} \cdot d\mathbf{s} = g\,\Delta z\,\hat{V}(z) \tag{107}$$

where g is the per-unit-length conductance between the wires produced by the lossy medium. In addition, over the length Δz,

$$\hat{q} = c\,\Delta z\,\hat{V}(z) \tag{108}$$

where c is the per-unit-length capacitance of the structure. Inserting (108), (107), and (106) into (105) yields

$$\hat{I}(z + \Delta z) - \hat{I}(z) + g\,\Delta z\,\hat{V}(z) = -j\omega c\,\Delta z\,\hat{V}(z) \tag{109}$$

Rewriting (109) by dividing by Δz, we obtain, in the limit as $\Delta z \to 0$, the second transmission-line equation:

$$\frac{d\hat{I}(z)}{dz} = -(g + j\omega c)\hat{V}(z) \tag{110}$$

The transmission-line equations in (103) and (110) become

$$\frac{d\hat{V}(z)}{dz} = -\hat{Z}\hat{I}(z) \tag{111a}$$

$$\frac{d\hat{I}(z)}{dz} = -\hat{Y}\hat{V}(z) \tag{111b}$$

where $\hat{Z}$ and $\hat{Y}$ are the per-unit-length impedance and admittance, respectively, given by

$$\hat{Z} = r + j\omega l \tag{112a}$$

$$\hat{Y} = g + j\omega c \tag{112b}$$

The transmission-line equations in (111) can be derived from the per-unit-length circuit shown in Fig. 7.24, which shows that this equivalent circuit is, in fact, a valid representation of the line from an electromagnetic field standpoint. This represents a special case in which circuit theory provides an exact representation of the behavior of electromagnetic fields. The inclusion of imperfect conductors through r and l_i is, however, an approximation.

The results in Sec. 7.2 for sinusoidal excitation of lossless lines can now be extended rather easily to include losses. The solutions to the transmission-line equations for the lossy line given in (111) become

$$\hat{V}(z) = \hat{V}_m^+ e^{-\alpha z} e^{-j\beta z} + \hat{V}_m^- e^{\alpha z} e^{j\beta z} \tag{113a}$$

$$\hat{I}(z) = \frac{\hat{V}_m^+}{\hat{Z}_C} e^{-\alpha z} e^{-j\beta z} - \frac{\hat{V}_m^-}{\hat{Z}_C} e^{\alpha z} e^{j\beta z} \tag{113b}$$

where the propagation constant becomes

$$\begin{aligned} \gamma &= \sqrt{\hat{Z}\hat{Y}} \\ &= \alpha + j\beta \end{aligned} \tag{114}$$

and $\hat{Z}_C$ is the characteristic impedance given by

$$\hat{Z}_C = \sqrt{\frac{\hat{Z}}{\hat{Y}}} \tag{115}$$

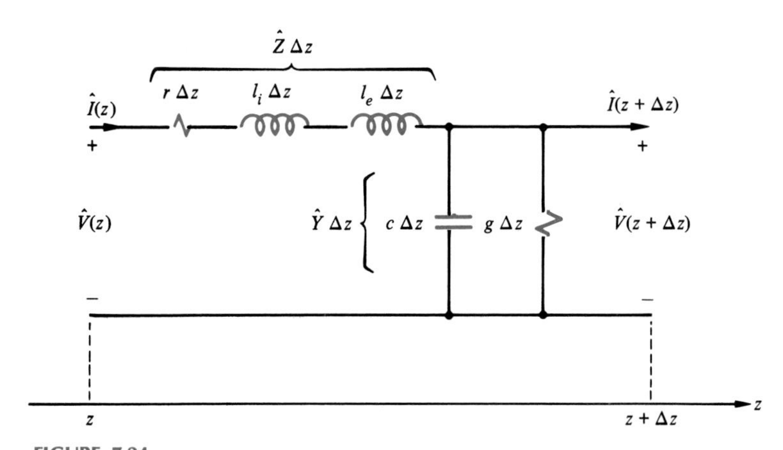

FIGURE 7.24

The per-unit-length model of a lossy transmission line.

Note the similarity of the solution in (113) to that for the lossless line given in (37) and also to the case of uniform plane waves in a lossy medium given in (25) of Chap. 6.

The effects of losses result in three major changes over the lossless case. The first change is that the forward- and backward-traveling waves suffer an attenuation as they move along the line, which is represented by the terms $e^{-\alpha z}$ and $e^{\alpha z}$, respectively, in (113). This is certainly reasonable to expect and is directly analogous to the case of uniform plane waves traveling in a lossy medium; in fact, the results are identical in form. [See (25) in Chap. 6.] The second change is that the forward-traveling voltage and current waves are no longer in phase since $\hat{Z}_C$ is complex, and similarly for the backward-traveling waves. The third change occurs in the velocity of propagation of the voltage and current waves. Note that the velocity of propagation is given by

$$u = \frac{\omega}{\beta} \tag{116}$$

where β is the imaginary part of γ in (114). If a lossless line is considered ($r = l_i = g = 0$), then $\beta = \omega\sqrt{l_e c}$. For lossy lines, β will be larger than for a lossless line, so the velocity of propagation will be less than for a lossless line.

The computational impact of these changes on our previously derived results for lossless lines, however, is not major so long as we are careful to consider certain previously real terms as now being complex. In fact, comparing (37) to (113), we see that the results for the lossy case can be obtained by replacing $e^{\pm j\beta z}$ and R_C in the lossless results by $e^{\pm \gamma z} = e^{\pm \alpha z}e^{\pm j\beta z}$ and $\hat{Z}_C$, respectively. One must be careful, however, to ensure that the results for the lossless case are appropriately modified. For example, sin βz and cos βz for the lossless case become (sinh γz)/j and cosh γz, respectively, for the lossy case, where sinh and cosh are the hyperbolic trigonometric functions. Thus, the expression for input impedance given in (51) becomes, for the lossy case,

$$\hat{Z}_{\text{in}}(z) = \hat{Z}_C \frac{\hat{Z}_L + \hat{Z}_C \tanh \gamma(\mathscr{L} - z)}{\hat{Z}_C + \hat{Z}_L \tanh \gamma(\mathscr{L} - z)} \tag{117}$$

where tanh is the hyperbolic tangent.

Other previously obtained results are similar to those for the lossless case. For example, the reflection coefficient is defined from (113) in a fashion similar to that for the lossless case:

$$\begin{aligned} \hat{\Gamma}(z) &= \frac{\hat{V}_m^- e^{\gamma z}}{\hat{V}_m^+ e^{-\gamma z}} \\ &= \frac{\hat{V}_m^-}{\hat{V}_m^+} e^{2\gamma z} \end{aligned} \tag{118}$$

Thus, we may write (113) as

$$\hat{V}(z) = \hat{V}_m^+ e^{-\gamma z}[1 + \hat{\Gamma}(z)] \tag{119a}$$

$$\hat{I}(z) = \frac{\hat{V}_m^+}{\hat{Z}_C} e^{-\gamma z}[1 - \hat{\Gamma}(z)] \tag{119b}$$

The input impedance to the line is defined as before:

$$\begin{aligned} \hat{Z}_{\text{in}} &= \frac{\hat{V}(z)}{\hat{I}(z)} \\ &= \hat{Z}_C \frac{1 + \hat{\Gamma}(z)}{1 - \hat{\Gamma}(z)} \end{aligned} \tag{120}$$

Thus, the load reflection coefficient may be obtained as

$$\hat{\Gamma}_L = \frac{\hat{Z}_L - \hat{Z}_C}{\hat{Z}_L + \hat{Z}_C} \tag{121}$$

Evaluating (118) at $z = \mathscr{L}$ gives

$$\begin{aligned} \hat{\Gamma}(z) &= \hat{\Gamma}_L e^{2\gamma(z - \mathscr{L})} \\ &= \hat{\Gamma}_L e^{2\alpha(z - \mathscr{L})} e^{j2\beta(z - \mathscr{L})} \end{aligned} \tag{122}$$

Note that (122) illustrates another major difference between the lossy and lossless cases. For the lossless case, the reflection coefficient at two points on the line differed only in phase; that is, the magnitude of the reflection coefficient for a lossless line is the same at all points along the line. For a lossy line, the magnitude of the reflection is a function of position on the line:

$$|\hat{\Gamma}(z)| = |\hat{\Gamma}_L| e^{2\alpha(z-\mathscr{L})} \tag{123}$$

Losses (due to a lossy medium and/or to imperfect conductors) are often specified in terms of the loss per unit length. Consider a lossy line that supports only a forward-traveling wave. This may be a result of the line being matched, $\hat{Z}_L = \hat{Z}_C$, or of the line being infinite in length. The voltage and current are

$$\hat{V}(z) = \hat{V}_m^+ e^{-\alpha z} e^{-j\beta z} \tag{124a}$$

$$\hat{I}(z) = \frac{\hat{V}_m^+}{\hat{Z}_C} e^{-\alpha z} e^{-j\beta z} \tag{124b}$$

The average power flow in the $+z$ direction is

$$\begin{aligned} P_{av}(z) &= \tfrac{1}{2} \operatorname{Re} [\hat{V}(z)\hat{I}^*(z)] \\ &= \frac{|\hat{V}_m^+|^2}{2Z_C} e^{-2\alpha z} \cos \theta_{\hat{Z}_C} \quad \text{W} \end{aligned} \tag{125}$$

The power dissipated in the line losses over a section of line of length d is

$$\begin{aligned} P_{diss} &= P_{av}(z) - P_{av}(z+d) \\ &= P_{av}(z)(1 - e^{-2\alpha d}) \quad \text{W} \end{aligned} \tag{126}$$

The ratio of the two powers is often given in decibels (dB) as

$$10 \log \frac{P_{av}(z)}{P_{av}(z+d)} = 10 \log e^{2\alpha d} \tag{127}$$

Thus, the attenuation constant α can be specified by specifying the line loss per unit length for a matched line.

As was the case for uniform plane wave propagation in a lossy medium, the calculation of the propagation constant γ and the characteristic impedance $\hat{Z}_C$ from the line parameters r, l, g, and c is somewhat involved. At sufficiently high frequencies or for lines with small losses represented by r and g, we may simplify these calculations. In the following we will assume a "low-loss line" such that $r \ll \omega l$ and $g \ll \omega c$; also, we will neglect the conductor internal inductances l_i since usually $l_e > l_i$. For this low-loss line, we may approximate γ as

$$\begin{aligned} \gamma &= \sqrt{(r + j\omega l)(g + j\omega c)} \\ &= j\omega\sqrt{lc}\left(1 + \frac{r}{j\omega l}\right)^{1/2}\left(1 + \frac{g}{j\omega c}\right)^{1/2} \\ &\simeq j\omega\sqrt{lc}\left(1 + \frac{r}{2j\omega l}\right)\left(1 + \frac{g}{2j\omega c}\right) \end{aligned} \tag{128}$$

by use of the binomial expansion. This may be further simplified to

$$\gamma \simeq j\omega\sqrt{lc}\left[1 + \frac{1}{2j\omega}\left(\frac{r}{l} + \frac{g}{c}\right)\right] \tag{129}$$

so that

$$\alpha \simeq \frac{r}{2}\sqrt{\frac{c}{l}} + \frac{g}{2}\sqrt{\frac{l}{c}}$$

$$= \frac{r}{2R_C} + \frac{gR_C}{2} \tag{130a}$$

$$\beta \simeq \omega\sqrt{lc} \tag{130b}$$

(lossless)

where the characteristic resistance for the lossless line is denoted by $R_C = \sqrt{l/c}$. Similarly, the characteristic impedance may be approximated as

$$\hat{Z}_C = \sqrt{\frac{r + j\omega l}{g + j\omega c}}$$

$$= \sqrt{\frac{l}{c}}\left(1 + \frac{r}{j\omega l}\right)^{1/2}\left(1 + \frac{g}{j\omega c}\right)^{-1/2}$$

$$\simeq \sqrt{\frac{l}{c}}\left(1 + \frac{r}{2j\omega l}\right)\left(1 - \frac{g}{2j\omega c}\right)$$

$$\simeq \sqrt{\frac{l}{c}}\left[1 + \frac{1}{2j\omega}\left(\frac{r}{l} - \frac{g}{c}\right)\right]$$

$$\simeq \sqrt{\frac{l}{c}} \qquad \omega > \frac{1}{2}\left(\frac{r}{l} - \frac{g}{c}\right) \tag{131}$$

Thus, the imaginary part of the characteristic impedance may be neglected, and

$$\hat{Z}_C \simeq R_C$$

$$= \sqrt{\frac{l}{c}}$$

$$= ul$$

$$= \frac{1}{uc} \tag{132}$$

for a low-loss line. Similarly, the phase constant can be approximated as that of a corresponding lossless line. Therefore,

$$\beta \simeq \frac{\omega}{u_0}\sqrt{\epsilon_r} \tag{133}$$

where the surrounding (homogeneous) dielectric has $\epsilon = \epsilon_r \epsilon_0$, $\mu \simeq \mu_0$. Lines are therefore typically specified by giving R_C, the type of dielectric (or ϵ_r), and the line loss per unit length (or, equivalently, α).

EXAMPLE 7.7 A typical coaxial cable is RG-58U. The manufacturer lists the nominal characteristic impedance as 53.5 Ω. The attenuation per 100 ft at 100 MHz is listed as 4.5 dB. The interior dielectric is polyethylene, having $\epsilon_r = 2.25$. Determine the per-unit-length inductance, capacitance, and the attenuation constant. Calculate the input impedance to a 1-m section of line at 100 MHz if the line is terminated in a 300-Ω resistive load.

Solution The velocity of propagation is

$$\begin{aligned} u &= \frac{u_0}{\sqrt{\epsilon_r}} \\ &= 2 \times 10^8 \text{ m/s} \end{aligned}$$

Therefore,

$$\begin{aligned} l &= \frac{R_C}{u} \\ &= 0.268\ \mu\text{H/m} \end{aligned}$$

$$\begin{aligned} c &= \frac{1}{uR_C} \\ &= 93.5 \text{ pF/m} \end{aligned}$$

(This calculated value equals the manufacturer's listed nominal value.) From the listed attenuation and (127) we have

$$4.5 = 10 \log e^{2\alpha \times 100}$$

so that

$$\begin{aligned} \alpha &= \frac{1}{200} \ln 10^{4.5/10} \\ &= 5.181 \times 10^{-3} \text{ Np/ft} \\ &= 1.7 \times 10^{-2} \text{ Np/m} \end{aligned}$$

The propagation constant at 100 MHz is therefore

$$\begin{aligned}\gamma &= \alpha + j\beta \\ &= 0.017 + j3.14\end{aligned}$$

For a 300-Ω resistive load, the load reflection coefficient is

$$\begin{aligned}\hat{\Gamma}_L &= \frac{300 - 53.5}{300 + 53.5} \\ &= 0.7\end{aligned}$$

At the input to the 1-m length of line, the reflection coefficient is

$$\begin{aligned}\hat{\Gamma}_{\text{in}} &= \hat{\Gamma}_L e^{-2\alpha} e^{-j2\beta} \\ &= 0.6766\end{aligned}$$

[This is real since $2\beta = 360°$.] Thus, the input impedance is

$$\begin{aligned}\hat{Z}_{\text{in}} &= \hat{Z}_C \frac{1 + \hat{\Gamma}_{\text{in}}}{1 - \hat{\Gamma}_{\text{in}}} \\ &= 277.4\ \Omega\end{aligned}$$

If the line length is changed to 1.2 m, the input impedance becomes $\hat{Z}_{\text{in}} = 71.78\underline{/-66.6°}$.

7.4 The Per-Unit-Length Parameters

As was shown previously, for a lossless line (perfect conductors and lossless surrounding dielectric medium) the per-unit-length external inductance l_e and capacitance c may be computed as though they were static parameters since the field structure for the TEM mode satisfies a static distribution in any transverse (xy) plane. Note that for the lossless case $g = 0$, $r = 0$ and $l_i = 0$, so that $l = l_i + l_e = l_e$. For a uniform line, we may, for example, compute the inductance L and capacitance C for a length d of line by the methods of Chaps. 3 and 4. The per-unit-length parameters become $l_e = L/d$ and $c = C/d$ since all sections of a uniform line are identical in cross section.

These calculations become particularly straightforward for the transmission lines in Fig. 7.1. Consider the calculation of c. We saw in Chap. 3 that as the two wires of the two-wire line in Fig. 7.1a are brought closer together, the charge around the cross-sectional periphery of each wire tends to concentrate on a line between the wire centers (see Fig. 3.22). This is called *proximity effect* and also occurs in the single-wire aboveground case in Fig. 7.1*b*, as the method of images shows (see Chap. 10). Ordinarily, if the ratio of wire separation to wire radius is greater than 10, we can disregard proximity effect and assume a uniform distribution of charge around the wire peripheries. The calculation of c then becomes quite simple, as shown in Example 3.14. Similarly, the calculation of c for the coaxial line in Fig. 7.1*c* is the simplest of all since, because of symmetry, proximity effect is not a factor.

We may also include a lossy medium without invalidating the TEM-mode assumption. This is particularly simple to see if we note that the lossy medium can be characterized by a complex, effective permittivity

$$\hat{\epsilon} = \epsilon\left(1 - j\frac{\sigma}{\omega\epsilon}\right) \tag{134}$$

as shown in Sec. 6.3. The conductivity σ is assumed to include the effects of the ohmic conductivity of the medium as well as polarization loss. The per-unit-length capacitance of the lossless line is a function of ϵ and independent of the permeability of the medium. This can always be written as

$$c = \epsilon K \tag{135}$$

where K is a factor that depends only on the cross-sectional geometry of the line. (See the discussion of static capacitance in Sec. 3.9.)

Now consider a lossy medium. If we substitute the complex, effective permittivity $\hat{\epsilon}$ given in (134) into (135), we obtain

$$\hat{c} = \epsilon\left(1 - j\frac{\sigma}{\omega\epsilon}\right)K \tag{136}$$

Multiplying (136) by $j\omega$ and comparing to (112*b*), we find

$$\begin{aligned} g + j\omega c &= j\omega\hat{c} \\ &= j\omega\epsilon\left(1 - j\frac{\sigma}{\omega\epsilon}\right)K \\ &= \sigma K + j\omega\epsilon K \end{aligned} \tag{137}$$

From this we conclude that

$$g = \sigma K \tag{138}$$

$$c = \epsilon K \tag{139}$$

The important point here is that if we obtain c, then the per-unit-length conductance can be found from

$$g = \frac{\sigma}{\epsilon} c \tag{140}$$

Thus, once we find c, we immediately have the per-unit-length conductance g. (This was also proven in Sec. 3.10.)

This relationship between g and c also has a parallel in the relationship between the per-unit-length external inductance l_e and the per-unit-length capacitance c. In order to show this important relationship, consider Faraday's law and Ampère's law in the homogeneous medium surrounding the (perfect) line conductors:

$$\nabla \times \hat{\mathbf{E}} = -j\omega\mu\hat{\mathbf{H}} \tag{141a}$$

$$\nabla \times \hat{\mathbf{H}} = (\sigma + j\omega\epsilon)\hat{\mathbf{E}} \tag{141b}$$

[We will assume a lossy surrounding medium having conductivity σ and rederive (140).] Expanding (141) for the TEM mode, $\hat{E}_z = \hat{H}_z = 0$, we obtain

$$\frac{\partial \hat{E}_y}{\partial z} = j\omega\mu\hat{H}_x \tag{142a}$$

$$\frac{\partial \hat{E}_x}{\partial z} = -j\omega\mu\hat{H}_y \tag{142b}$$

$$\frac{\partial \hat{E}_y}{\partial x} - \frac{\partial \hat{E}_x}{\partial y} = 0 \tag{142c}$$

$$\frac{\partial \hat{H}_y}{\partial z} = -(\sigma + j\omega\epsilon)\hat{E}_x \tag{142d}$$

$$\frac{\partial \hat{H}_x}{\partial z} = (\sigma + j\omega\epsilon)\hat{E}_y \tag{142e}$$

$$\frac{\partial \hat{H}_y}{\partial x} - \frac{\partial \hat{H}_x}{\partial y} = 0 \tag{142f}$$

Differentiating (142*a*) with respect to z and substituting (142*e*), we obtain

$$\frac{\partial^2 \hat{E}_y}{\partial z^2} - \gamma^2 \hat{E}_y = 0 \tag{143a}$$

and similarly we obtain from the other equations in (142)

$$\frac{\partial^2 \hat{E}_x}{\partial z^2} - \gamma^2 \hat{E}_x = 0 \tag{143b}$$

$$\frac{\partial^2 \hat{H}_x}{\partial z^2} - \gamma^2 \hat{H}_x = 0 \tag{143c}$$

$$\frac{\partial^2 \hat{H}_y}{\partial z^2} - \gamma^2 H_y = 0 \tag{143d}$$

where

$$\gamma^2 = j\omega\mu(\sigma + j\omega\epsilon) \tag{144}$$

and γ is the familiar propagation constant of TEM waves in this lossy medium.

Defining the transverse fields as

$$\hat{\mathbf{E}}_T = \hat{E}_x \mathbf{a}_x + \hat{E}_y \mathbf{a}_y \tag{145a}$$

$$\hat{\mathbf{H}}_T = \hat{H}_x \mathbf{a}_x + \hat{H}_y \mathbf{a}_y \tag{145b}$$

we obtain by substituting (145) into (143)

$$\frac{\partial^2 \hat{\mathbf{E}}_T}{\partial z^2} - \gamma^2 \hat{\mathbf{E}}_T = 0 \tag{146a}$$

$$\frac{\partial^2 \hat{\mathbf{H}}_T}{\partial z^2} - \gamma^2 \hat{\mathbf{H}}_T = 0 \tag{146b}$$

The solutions to (146) are familiar:

$$\hat{\mathbf{E}}_T = \hat{\mathbf{E}}_T^+ e^{-\gamma z} + \hat{\mathbf{E}}_T^- e^{\gamma z} \tag{147a}$$

$$\hat{\mathbf{H}}_T = \hat{\mathbf{H}}_T^+ e^{-\gamma z} + \hat{\mathbf{H}}_T^- e^{\gamma z} \tag{147b}$$

where $\hat{\mathbf{E}}_T^+$, $\hat{\mathbf{E}}_T^-$, $\hat{\mathbf{H}}_T^+$, and $\hat{\mathbf{H}}_T^-$ are, as yet, undetermined vectors with x and y components only. The phasor line voltage and current are defined for this TEM mode via (4) and (6) as

$$\hat{V}(z) = -\int_{c_V} \hat{\mathbf{E}}_T \cdot d\mathbf{l} \tag{148a}$$

$$\hat{I}(z) = \oint_{c_I} \hat{\mathbf{H}}_T \cdot d\mathbf{l} \tag{148b}$$

Clearly, since the contour integrations in (148) are performed in a transverse plane, they are independent of z, and substitution of (147) into (148) yields

$$\hat{V}(z) = \hat{V}_m^+ e^{-\gamma z} + \hat{V}_m^- e^{\gamma z} \tag{149a}$$

$$\hat{I}(z) = \hat{I}_m^+ e^{-\gamma z} + \hat{I}_m^- e^{\gamma z} \tag{149b}$$

where $\hat{V}_m^+$, $\hat{V}_m^-$, $\hat{I}_m^+$, and $\hat{I}_m^-$ are undetermined constants that are the result of the contour integration of $\hat{\mathbf{E}}_T^+$, $\hat{\mathbf{E}}_T^-$, $\hat{\mathbf{H}}_T^+$, and $\hat{\mathbf{H}}_T^-$ in (147) via (148). Comparing (149) to the corresponding expression in (113) derived in the previous section for a lossy line, we identify

$$\begin{aligned}\gamma &= \sqrt{j\omega\mu(\sigma + j\omega\epsilon)} \\ &= \sqrt{j\omega l_e(g + j\omega c)}\end{aligned} \tag{150}$$

From this result we identify the important relationships:

$$l_e g = \mu\sigma \tag{151a}$$

$$l_e c = \mu\epsilon \tag{151b}$$

From these results we obtain our previously derived relationship:

$$\frac{g}{c} = \frac{\sigma}{\epsilon} \tag{152}$$

7.4.1 The External Parameters l_e, c, g

The external per-unit-length parameters, c and l_e, may be computed for static field conditions using the methods of Chap. 3 and Chap. 4, respectively. The per-unit-length conductance, g, may also be computed from static field conditions using the methods of Chap 3. However, only one of these three parameters need be computed as shown previously.

EXAMPLE 7.8 Consider the coaxial transmission line shown in Fig. 7.25*a*. Derive expressions for l_e, c, and g and show that $l_e c = \mu\epsilon$ and that $g = (\sigma/\epsilon)(c)$. The inner wire radius is denoted as r_w, and the inner radius of the shield is denoted as r_s.

Solution Since the TEM-mode field distribution is the same as a static one, we may use the techniques of Chaps. 3 and 4 to compute the per-unit-length quantities l_e, c, and g. First we will compute l_e. Let us suppose that the inner conductor carries a current I that returns on the interior of the outer conductor. By symmetry, these currents will be uniformly distributed around each conductor periphery and the magnetic field intensity will have only a ϕ component, as shown in Fig. 7.25*b*. Applying Ampère's law around a contour of radius r about the inner conductor, we obtain

$$\oint_c \mathbf{H} \cdot d\mathbf{l} = I$$

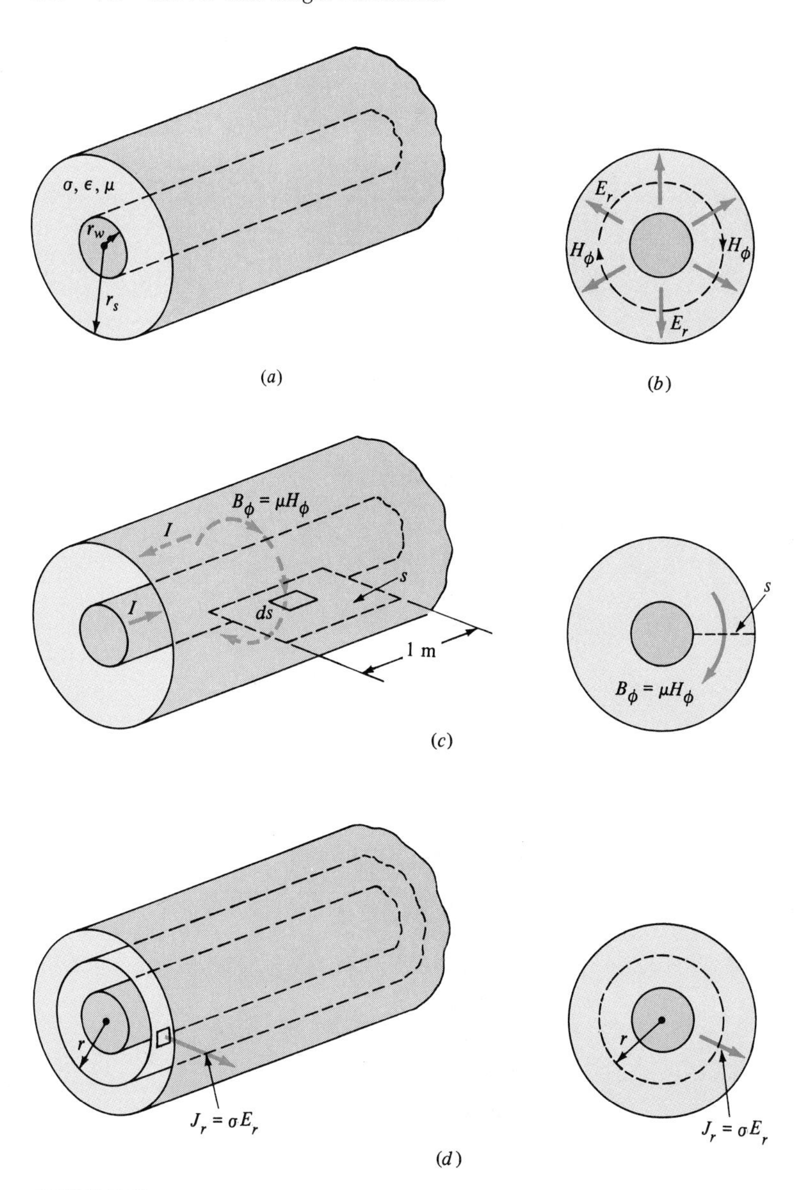

FIGURE 7.25

Example 7.7. Calculation of the per-unit-length parameters for a coaxial cable.

or

$$H_\phi 2\pi r = I$$

so that

$$H_\phi = \frac{I}{2\pi r}$$

The total magnetic flux per unit length linking the current may be obtained by finding the total flux penetrating a flat surface of longitudinal length 1 m between $r = r_w$ and $r = r_s$, as shown in Fig. 7.25*c*; thus, we obtain

$$\begin{aligned}\psi_m &= \int_s \mathbf{B} \cdot d\mathbf{s} \\ &= \int_z^{z+1} \int_{r=r_w}^{r_s} \mu H_\phi \, dr \\ &= \frac{\mu I}{2\pi} \int_{r=r_w}^{r_s} \frac{dr}{r} \\ &= \frac{\mu I}{2\pi} \ln \frac{r_s}{r_w}\end{aligned}$$

From this we obtain

$$\begin{aligned}l_e &= \frac{\psi_m}{I} \\ &= \frac{\mu}{2\pi} \ln \frac{r_s}{r_w} \qquad \text{H/m} \end{aligned} \tag{153}$$

The per-unit-length capacitance can be found by assuming a charge per unit length of q on the inner conductor and $-q$ on the interior of the outer conductor. Once again, because of symmetry we may assume the charge to be uniformly distributed over the periphery of each conductor; the electric field will therefore be in the radial direction, as shown in Fig. 7.25*b*. Constructing a gaussian surface in the form of a cylinder of radius r about the inner conductor and applying Gauss' law, we obtain

$$\oint_s \mathbf{D} \cdot d\mathbf{s} = q$$

or

$$\epsilon E_r 2\pi r = q$$

so that

$$E_r = \frac{q}{2\pi\epsilon r}$$

The voltage between the two conductors is

$$\begin{aligned}V &= -\int_{r=r_s}^{r_w} \mathbf{E} \cdot d\mathbf{l} \\ &= -\int_{r=r_s}^{r_w} \frac{q}{2\pi\epsilon r} \, dr \\ &= \frac{q}{2\pi\epsilon} \ln \frac{r_s}{r_w}\end{aligned}$$

Thus, the per-unit-length capacitance becomes

$$c = \frac{q}{V}$$

$$= \frac{2\pi\epsilon}{\ln (r_s/r_w)} \tag{154}$$

Note that

$$l_e c = \frac{\mu}{2\pi} \ln \frac{r_s}{r_w} \frac{2\pi\epsilon}{\ln (r_s/r_w)}$$

$$= \mu\epsilon$$

as expected.

The per-unit-length conductance g can be found by direct integration. The transverse conduction current is, because of symmetry, also radial and is related to E_r by the conductance of the medium:

$$J_r = \sigma E_r = \frac{\sigma q}{2\pi\epsilon r}$$

The total transverse conduction current (per unit length) can be obtained by integrating J_r over a cylindrical surface of radius r, as shown in Fig. 7.25d:

$$I_T = \int_s J_r \, ds$$

$$= \int_{\phi=0}^{2\pi} \frac{\sigma q}{2\pi\epsilon r} r \, d\phi$$

$$= \frac{\sigma q}{\epsilon}$$

Thus, the per-unit-length conductance is

$$g = \frac{I_T}{V}$$

$$= \frac{\sigma q/\epsilon}{(q/2\pi\epsilon) \ln (r_s/r_w)}$$

$$= \frac{2\pi\sigma}{\ln (r_s/r_w)} \tag{155}$$

Note that $g = (\sigma/\epsilon)c$.

The per-unit-length capacitance for the two-wire line shown in Fig. 7.1*a* is derived in Chap. 10 using image methods. It requires a more complex derivation than does the coaxial cable since, for close proximity of the wires, we cannot assume a uniform distribution of charge around the wire peripheries. For widely separated wires, however, an assumption of uniform charge distribution simplifies the derivation of c considerably, as was shown in Example 3.12. For wires of radius r_w separated a distance D (center to center), the exact result derived in Chap. 10 is

$$c = \frac{\pi\epsilon}{\cosh^{-1}(D/2r_w)} \quad \text{F/m} \tag{156}$$

Since $l_e c = \mu\epsilon$ and $gc = \sigma/\epsilon$, we obtain

$$l_e = \frac{\mu}{\pi}\cosh^{-1}\frac{D}{2r_w} \quad \text{H/m} \tag{157}$$

$$g = \frac{\pi\sigma}{\cosh^{-1}(D/2r_w)} \quad \text{S/m} \tag{158}$$

where $\cosh^{-1} a = \ln(a + \sqrt{a^2 - 1})$.

For the case of one wire of radius r_w at a height H above a ground plane (shown in Fig. 7.1*b*), we may obtain the per-unit-length capacitance from (156) by using the method of images described in Chap. 10. The result is

$$c = \frac{2\pi\epsilon}{\cosh^{-1}(H/r_w)} \quad \text{F/m} \tag{159}$$

and

$$l_e = \frac{\mu}{2\pi}\cosh^{-1}\frac{H}{r_w} \quad \text{H/m} \tag{160}$$

$$g = \frac{2\pi\sigma}{\cosh^{-1}(H/r_w)} \quad \text{S/m} \tag{161}$$

7.4.2 The Conductor Internal Impedance Parameters r and l_i

For imperfect line conductors, we require the per-unit-length parameters of resistance r and internal inductance l_i. The exact calculation of these parameters is somewhat more complicated than of those for the external parameters above. In addition, these parameters depend on the frequency of excitation, whereas the above external parameters l_e, c, and g do not (if we assume that μ, ϵ, and σ, respectively, are frequency-independent). For a detailed discussion of the exact derivation of these parameters, the reader may consult any of numerous

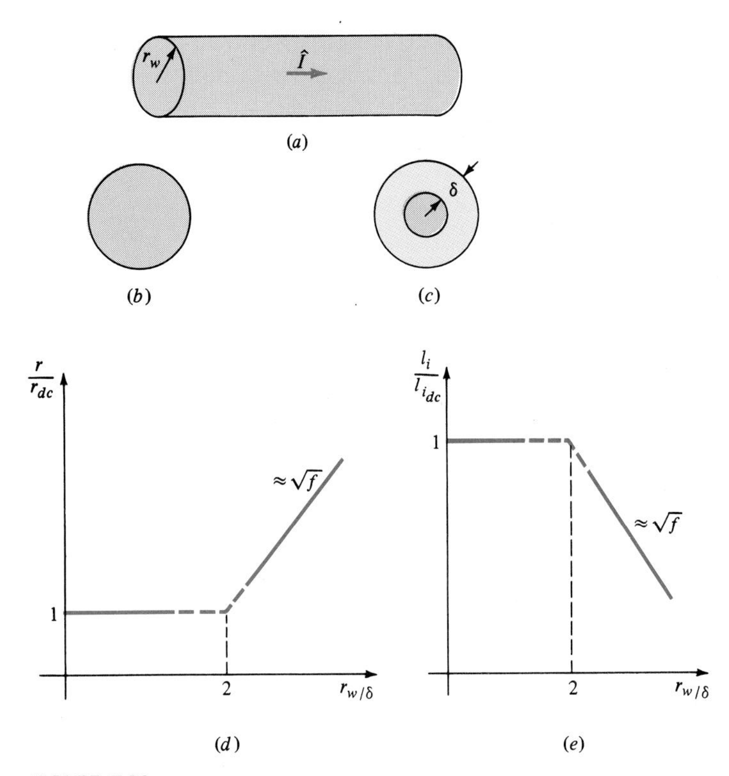

FIGURE 7.26
The per-unit-length resistance and internal inductance of a cylindrical conductor (wire) illustrating skin effect: *(a)* geometry; current distribution over the cross section *(b)* at dc and *(c)* at high frequencies; frequency dependence of *(d)* resistance and *(e)* internal inductance.

textbooks and handbooks on the subject.† Instead of giving exact derivations of r and l_i, we will examine the limiting cases of dc excitation and high-frequency excitation in order to obtain approximate, yet simple, expressions for these quantities.

Consider the wire of radius r_w carrying a current $\hat{I}$ shown in Fig. 7.26*a*. For dc excitation, the current will be uniformly distributed over the wire cross section, as shown in Fig. 7.26*b*. In terms of the conductivity of the wire, σ_w, the dc resistance per unit length is

$$r_{\text{DC}} = \frac{1}{\sigma_w \pi r_w^2} \quad \Omega/\text{m} \tag{162}$$

The per-unit-length internal inductance l_i is obtained in a fashion similar to the calculation for external inductance. However, the flux internal to the wire

† See, for example, C. T. A. Johnk, *Engineering Electromagnetic Fields and Waves*, Wiley, New York, 1975, chap. 9.

links only a part of the wire current. For dc excitation, the current distribution is uniform over the conductor cross section. In a cylinder of radius $r < r_w$, the current enclosed is proportional to the ratio of the areas:

$$I_r = \frac{\pi r^2}{\pi r_w^2} I \tag{163}$$

By symmetry, the magnetic field internal to the wire is circumferentially directed. Using Ampère's law, we obtain the magnetic field along a contour of radius $r < r_w$ as

$$\int_{\phi=0}^{2\pi} H_\phi r \, d\phi = \frac{\pi r^2}{\pi r_w^2} I \tag{164}$$

Thus, the magnetic flux density is

$$\begin{aligned} B_\phi &= \mu_0 H_\phi \\ &= \frac{\mu_0 r}{2\pi r_w^2} I \qquad r \le r_w \end{aligned} \tag{165a}$$

External to the wire, the flux links all the current; thus

$$B_\phi = \frac{\mu_0 I}{2\pi r} \qquad r \ge r_w \tag{165b}$$

Equations (165) are plotted in Fig. 7.27.

Consider an annulus of thickness dr, as shown in Fig. 7.27. The flux penetrating an area along the wire of width dr and length 1 m is obtained from (165a):

$$\begin{aligned} d\psi_m &= B_\phi \, dr \\ &= \frac{\mu_0 I}{2\pi r_w^2} r \, dr \end{aligned} \tag{166}$$

But this flux links only a fraction (r^2/r_w^2) of the total current; therefore, the flux linkages are

$$\begin{aligned} d\Lambda &= d\psi_m \frac{r^2}{r_w^2} \\ &= \frac{\mu_0 I}{2\pi r_w^4} r^3 \, dr \end{aligned} \tag{167}$$

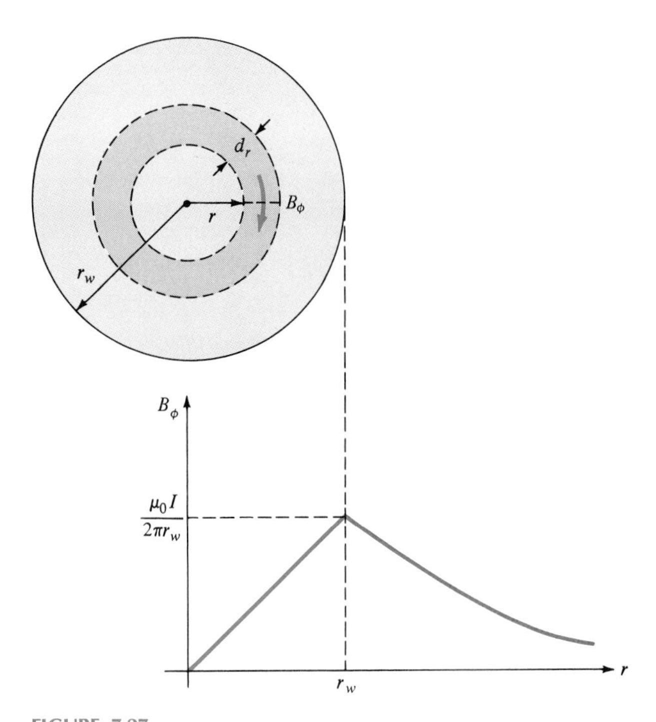

FIGURE 7.27
The interior and exterior magnetic field of a wire at dc.

Integrating from $r = 0$ to $r = r_w$ gives the total internal flux linkages:

$$\Lambda = \int_{r=0}^{r_w} d\Lambda$$

$$= \frac{\mu_0 I}{8\pi} \tag{168}$$

Thus, the per-unit-length internal inductance is

$$l_i = \frac{\Lambda}{I}$$

$$= \frac{\mu_0}{8\pi} = .5 \times 10^{-7} \qquad \text{H/m} \tag{169}$$

For high-frequency excitation, the current density tends to be higher away from the center of the conductor. The majority of the current is concentrated in an annulus at the wire surface of a thickness equal to the skin depth, δ. For a sufficiently high frequency of excitation, the curvature of the wire may be neglected so that the problem is reduced to one of a plane conductor of width $2\pi r_w$ and thickness δ, as shown in Fig. 7.28.

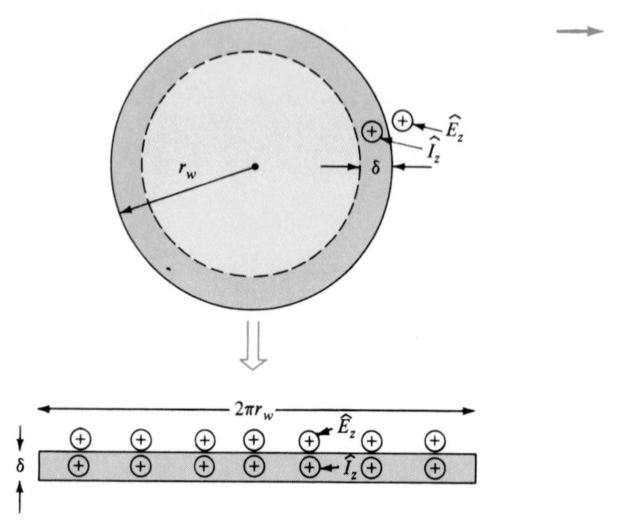

FIGURE 7.28

Determination of the high-frequency resistance of a wire by assuming uniform current distribution within a skin depth of the surface.

The impedance per unit length (in the z direction) of this slab is logically defined as the ratio of the electric field tangent to the surface, $\hat{E}_z$, to the total current through the slab cross section. In order to solve this problem, we examine the case of a semi-infinite plane conductor having a tangential, phasor electric field $\hat{E}_0$ at the surface, as shown in Fig. 7.29. Within the conductor, the displacement current is negligible compared to the conduction current ($\omega\epsilon \ll \sigma_w$); consequently, the fields are governed by the diffusion equation developed in Chap. 6:

$$\frac{d^2\hat{E}_z}{dx^2} = j\omega\mu\sigma_w\hat{E}_z \tag{170}$$

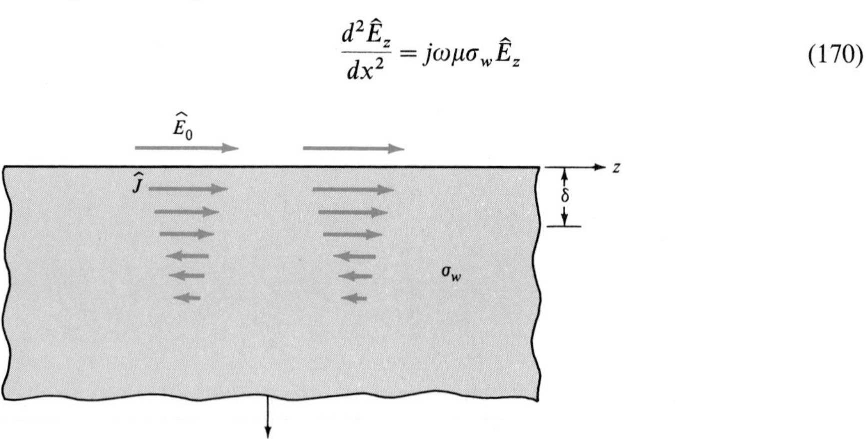

FIGURE 7.29

Calculation of surface impedance of a plane conductor.

Since the current density and the electric field are related by conductivity within the conductor, the diffusion equation also relates the current density:

$$\frac{d^2\hat{J}_z}{dx^2} = j\omega\mu\sigma_w\hat{J}_z \tag{171}$$

The solution to (171) is of the form,

$$\hat{J}_z = C_1 e^{-\gamma x} + C_2 e^{\gamma x} \qquad \text{A/m}^2 \tag{172}$$

where

$$\gamma = \sqrt{j\omega\mu\sigma_w} = \frac{1+j}{\delta} \tag{173}$$

and δ is the skin depth:

$$\delta = \frac{1}{\sqrt{\pi f\mu\sigma_w}} \qquad \text{m} \tag{174}$$

We discard the term $C_2 e^{\gamma x}$ ($C_2 = 0$); otherwise, the current density would increase with x without bound. Thus, the current density is

$$\hat{J}_z = \hat{J}_0 e^{-x/\delta} e^{-j(x/\delta)} \qquad \text{A/m}^2 \tag{175}$$

The total current in the slab (per unit width in the y direction) is

$$\hat{I}_z = \int_0^\infty \hat{J}_z\,dx = \frac{J_0\delta}{1+j} \qquad \text{A/m} \tag{176}$$

At $x = 0$ we find that $\hat{J}_0 = \sigma_w\hat{E}_0$, so that

$$\hat{I}_z = \sigma_w\hat{E}_0\frac{\delta}{l+j} \qquad \text{A/m} \tag{177}$$

Thus, the slab impedance (per unit length and per unit width) is

$$\hat{Z}_s = \frac{\hat{E}_0}{\hat{I}_z} = \frac{1+j}{\sigma_w\delta} \qquad \Omega \tag{178}$$

From this result, we identify the slab resistance and internal inductance (per unit width and length) as

$$R_s = \frac{1}{\sigma_w \delta} \quad \Omega \tag{179a}$$

$$\omega L_s = \frac{1}{\sigma_w \delta} \quad \Omega \tag{179b}$$

Note that all blocks of this material that have an infinite depth and a surface area of 1 m^2 will have this impedance. Thus, Z_s is often referred to as the surface impedance and its units are "ohms per square."

This result can be used to compute the high-frequency internal impedance of the wire on the assumption that the current is concentrated within a thin layer at the surface, as shown in Fig. 7.28. To obtain this internal impedance, divide the above surface impedance by $2\pi r_w$ (since the conductor circumference can be divided into unit widths that are electrically connected in parallel) to yield

$$\begin{aligned} r_{\text{HF}} &= \frac{R_s}{2\pi r_w} \\ &= \frac{1}{2\pi r_w \sigma_w \delta} \quad \Omega/\text{m} \end{aligned} \tag{180a}$$

$$\begin{aligned} \omega l_{i,\,\text{HF}} &= \frac{\omega L_s}{2\pi r_w} \quad \Omega/\text{m} \\ &= \frac{1}{2\pi r_w \sigma_w \delta} \quad \Omega/\text{m} \end{aligned} \tag{180b}$$

This high-frequency conductor resistance r_{HF} can alternatively be found by assuming the current to be uniformly distributed over the surface of an annulus of depth δ from the surface:

$$\begin{aligned} r_{\text{HF}} &\simeq \frac{1}{\sigma_w[\pi r_w^2 - \pi(r_w - \delta)^2]} \\ &\simeq \frac{1}{2\pi r_w \sigma_w \delta} \end{aligned} \tag{181}$$

This can be written in terms of the dc resistance in (162) as

$$r_{\text{HF}} = r_{\text{DC}} \frac{r_w}{2\delta} \tag{182}$$

(127) for a 100-ft length as 3.204 dB/100 ft. Considering the approximations involved, this matches rather well the manufacturer's listed value (measured) of 4.5 dB/100 ft.

7.5 Summary

We considered the primary mode of propagation on two-conductor lines: the TEM mode. For the TEM mode, the electric and magnetic field vectors lie in a plane (xy) perpendicular to the axis (z) of the line. In this sense, these waves are similar to uniform plane waves, although the field vectors of the uniform plane wave are independent of position in this transverse plane (uniform), whereas the transverse fields for the transmission line are not; thus, the waves on transmission lines may be classified only as plane waves. This assumption of the TEM field structure allowed the unique definition of voltage and current in each transverse (xy) plane for nonstatic excitation.

The transmission-line equations were derived in terms of the line voltage and line current and of the per-unit-length inductance l and capacitance c of the line. The solutions to these transmission-line equations were remarkably similar to those of the uniform plane wave: forward- and backward-traveling waves were obtained that travel at a velocity

$$u = \frac{1}{\sqrt{lc}}$$

$$= \frac{1}{\sqrt{\mu\epsilon}}$$

The voltage and current of each wave are related by the characteristic resistance of the line:

$$R_C = \sqrt{\frac{l}{c}}$$

These quantities bear a striking similarity to the corresponding quantities of uniform plane waves. In fact, R_C is analogous to the intrinsic impedance η for uniform plane waves. For a line terminated in a resistance R_L, we obtained a voltage reflection coefficient at the load:

$$\Gamma_L = \frac{R_L - R_C}{R_L + R_C}$$

which was directly analogous to the case of uniform plane waves normally incident on plane, material boundaries. Virtually all of our results for transmission lines have parallels in the uniform plane wave results of the previous chapter, and the reader should try to identify these.

Transients on transmission lines were considered first. We traced the time history of the forward- and backward-traveling voltage and current waves on the line to obtain the time history of the total voltage and current at a particular point on the line.

Sinusoidal, steady-state excitation of the line was considered next. We observed that adjacent corresponding values of the magnitude of the line voltage (line current) are separated by $\lambda/2$. In fact, the input impedance to a section of line replicates for every half-wavelength of distance down the line. A point of voltage (current) maximum on the line and an adjacent point of voltage (current) minimum was separated by $\lambda/4$. Power flow on the line was also considered. The results again bear a striking similarity to the uniform plane wave results of the previous chapter. We found that all of the power in the incident wave is delivered to the load only if the load is matched to the line: $\hat{Z}_L = R_C$.

In the last two sections we considered the extension of these results to lossy lines and the calculation of the per-unit-length parameters of the line. The lossless-line results can be extended quite easily to lossy lines by using a complex propagation constant

$$\gamma = \sqrt{\hat{Z}\hat{Y}}$$

$$= \alpha + j\beta$$

where $\hat{Z}$ and $\hat{Y}$ are the per-unit-length line impedance and admittance, respectively, and a complex, characteristic impedance

$$\hat{Z}_C = \sqrt{\frac{\hat{Z}}{\hat{Y}}}$$

The results again correspond to the results for uniform plane waves in lossy media. A derivation of the transmission-line equations from a fields standpoint was obtained to show that the lumped-circuit derivation was correct. It was found that the line per-unit-length parameters of external inductance l_e, capacitance c, and conductance g, are related as

$$l_e c = \mu\epsilon$$

$$g = \frac{\sigma}{\epsilon} c$$

so that only one of these three parameters need be determined for a particular line. We also considered the calculation of the line internal parameters that are due to imperfect conductors. As opposed to the external parameters, these internal parameters are frequency-dependent due to skin effect and vary as the square root of frequency.

In the next chapter we will find that single conductors are also capable of guiding electromagnetic waves. These hollow single conductors, or waveguides, are typically used to guide waves in the gigahertz frequency range, where the losses of the two-conductor lines considered in this chapter would be prohibitive. On the other hand, for frequencies below the high-megahertz range, the two-conductor transmission lines of this chapter are employed to guide waves since the dimensions of waveguides in this frequency range would be prohibitively large. Thus, each of these guiding structures has a useful and somewhat distinct purpose.

Problems

7-1 Show, by direct substitution, that the expressions for line voltage and current given in (19) satisfy the transmission-line equations given in (9).

7-2 Derive expressions for R_C in terms of

(a) l and u and

(b) c and u.

7-3 For the per-unit-length representations of a lossless transmission line shown in Fig. P7.3, derive the transmission-line equations in the limit as $\Delta z \to 0$. Note that the total per-unit-length inductance and capacitance in each circuit is l and c, respectively.

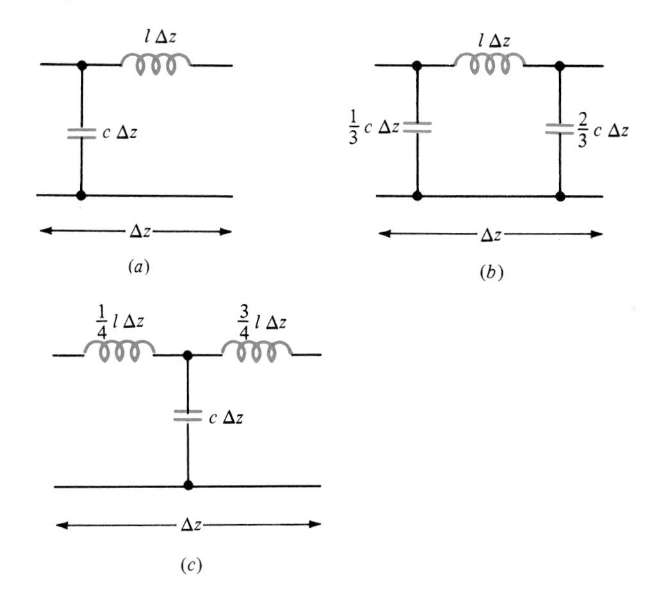

FIGURE P7.3

7-4 Determine the velocity of wave propagation and the characteristic resistance of a lossless coaxial cable if:

(a) $l = 0.25\ \mu\text{H/m}$; $c = 100\ \text{pF/m}$

(b) $c = 50\ \text{pF/m}$; $\epsilon_r = 2.1$ (Teflon)

7-5 Consider a transmission line that has $R_S = 300\ \Omega$, $R_L = 60\ \Omega$, $R_C = 100\ \Omega$, $u = 400\ \text{m}/\mu\text{s}$, $\mathscr{L} = 400\ \text{m}$, and $V_S(t) = 400u(t)\ \text{V}$, where $u(t)$ is the unit step function. Sketch $V(0, t)$, $I(0, t)$, $V(\mathscr{L}, t)$, and $I(\mathscr{L}, t)$, for $0 \le t < 10\ \mu\text{s}$. Do the results converge to expected steady-state values?

7-6 Repeat Prob. 7-5 with $R_L = 0$ (short circuit).

7-7 Repeat Prob. 7-5 with $R_L = \infty$ (open circuit).

7-8 A time-domain reflectometer (TDR) is an instrument used to determine properties of transmission lines. In particular, it can be used to detect the locations of imperfections, such as breaks on the line. The instrument launches a pulse down the line and records the transit time for that pulse to be reflected at some discontinuity and to return to the line input. Suppose a TDR having a source impedance of 50 Ω is attached to a 50-Ω coaxial cable having some unknown length and load resistance. The dielectric of the cable is Teflon ($\epsilon_r = 2.1$). The open-circuit voltage of the TDR is a pulse of duration 10 μs. If the recorded voltage at the input to the line is as shown in Fig. P7.8, determine

(a) the length of the line and
(b) the unknown load resistance.

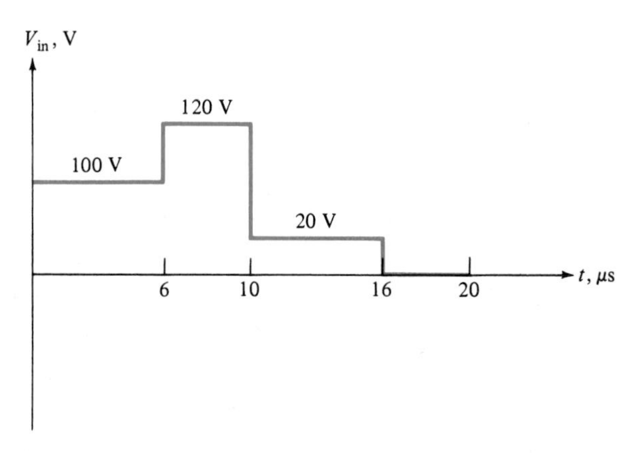

FIGURE P7.8

7-9 A 12-V battery ($R_S = 0$) is attached to an unknown length of transmission line that is terminated in a resistance. If the input current to the line for 6 μs is as shown in Fig. P7.9, determine

(a) the line characteristic resistance and
(b) the unknown load resistance.

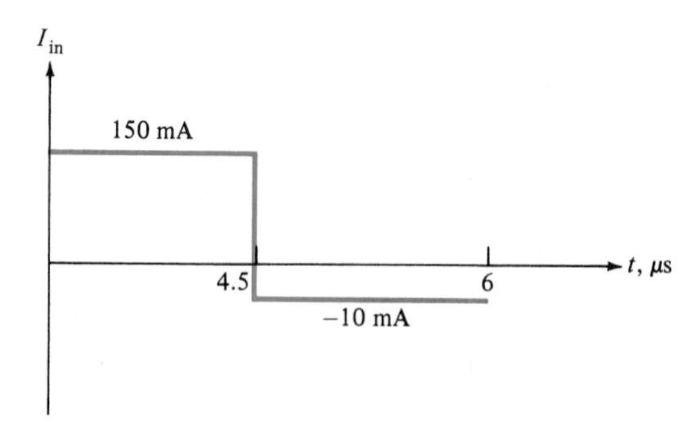

FIGURE P7.9

7-10 Digital data pulses should ideally consist of rectangular pulses. Actual data, however, resemble pulses having a trapezoidal shape with certain rise and fall times. Depending on the ratio of the rise-fall time to the one-way transit time of the transmission line, the received voltage may oscillate about the desired value, possibly causing a digital gate at that end to switch falsely to an undesired state and cause data errors. Matching the line at both ends eliminates this problem, but matching cannot always be accomplished. In order to investigate this problem, consider a line having $R_S = 0$ and $R_L = \infty$. Assume that the source voltage $V_S(t)$ is a ramp waveform given by:

$$
\begin{aligned}
V_S(t) &= 0 && t \leq 0 \\
&= \frac{t}{\tau_r} && t \leq \tau_r \\
&= 1 && t \geq \tau_r
\end{aligned}
$$

where τ_r is the rise time. Sketch the load voltage for line lengths having one-way transit times T such that

(a) $\tau_r = T/10$,
(b) $\tau_r = 2T$,
(c) $\tau_r = 3T$, and
(d) $\tau_r = 4T$.

This example shows that in order to avoid problems resulting from mismatch, one should choose line lengths short enough that $\tau_r \ll T$ for the desired data.

7-11 A transmission line is to be used to transmit digital data, as shown in Fig. P7.11. The data pulses are in the form of trapezoidal pulses with rise-fall times of T, where T is the one-way transit time of the line. Sketch the load and source voltages for $0 \leq t < 8T$ for $R_L = 2\,R_C$ and for $R_L = \frac{1}{2}\,R_C$.

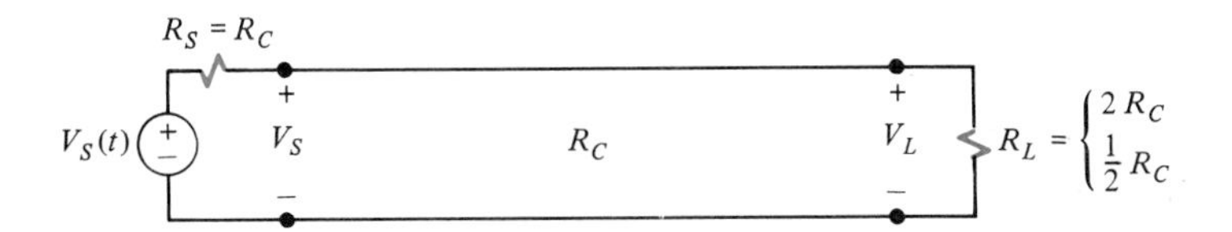

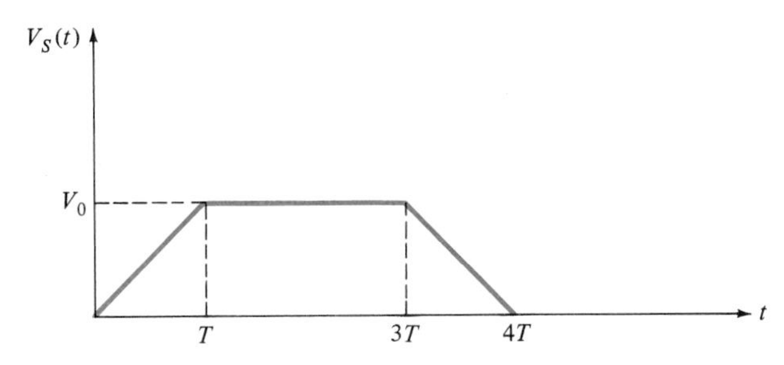

FIGURE P7.11

7-12 Sketch the line voltage as a function of position at $t = T/2, 3T/2, 2T, 3T, 4T$, and $5T$ for the line in Prob. 7-11. Do your sketches verify the plots you obtained in Prob. 7-11?

7-13 A unit step voltage $V_S(t) = u(t)V$ is applied to a line having a one-way delay of T with $R_S = R_C$, $R_L = 3R_C$. Plot the input current to the line as a function of time for $0 \leq t \leq 5T$. Realizing that the line (including the terminations) is a linear system, use this result to plot the load currents for the source voltages shown in Fig. P7.13. Check your results by direct calculation.

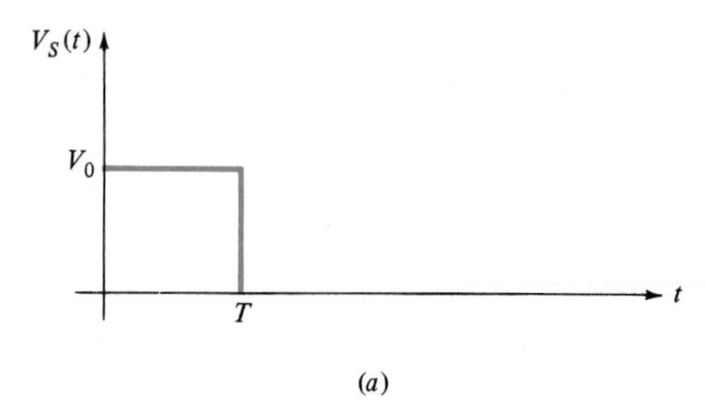

(*a*)

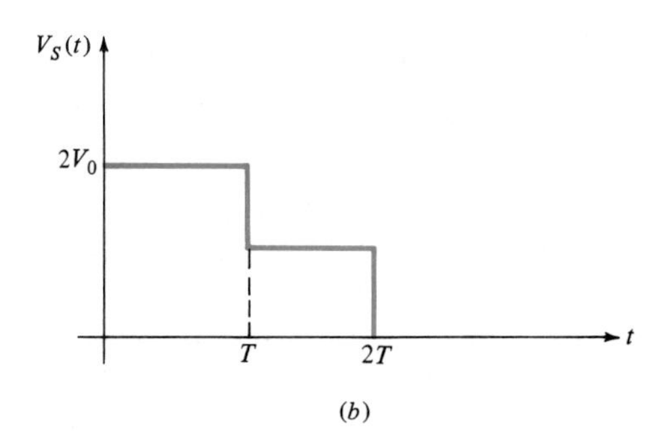

(*b*)

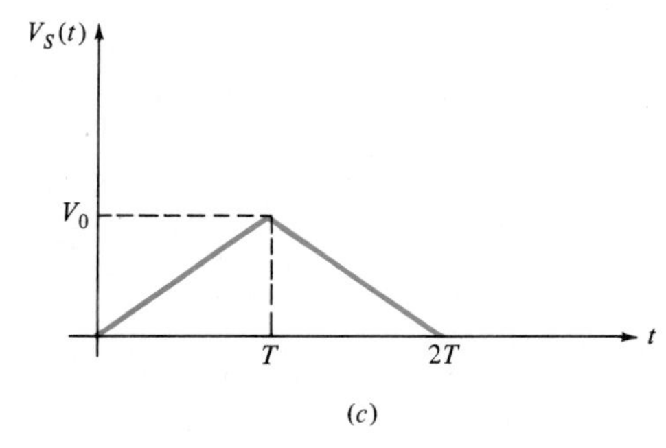

(*c*)

FIGURE P7.13

7-14 A step voltage source having $V_S(t) = 400u(t)$ V and $R_S = 350\ \Omega$ is applied to a length of 50-Ω line that has a short-circuit load. Plot the current in the short-circuit load as a function of the line's one-way delay T for $0 \leq t \leq 12T$. Does the result converge to the expected steady-state value? At what time will the current be within 10 percent of the steady-state value?

7-15 Digital data is being transmitted bidirectionally on a transmission line. Ordinarily the data are transmitted such that there are no conflicts at any point. Model this line as shown in Fig. P7.15 and sketch the voltages at the terminal of each device for $0 \leq t \leq 7T$, where T is the line's one-way delay.

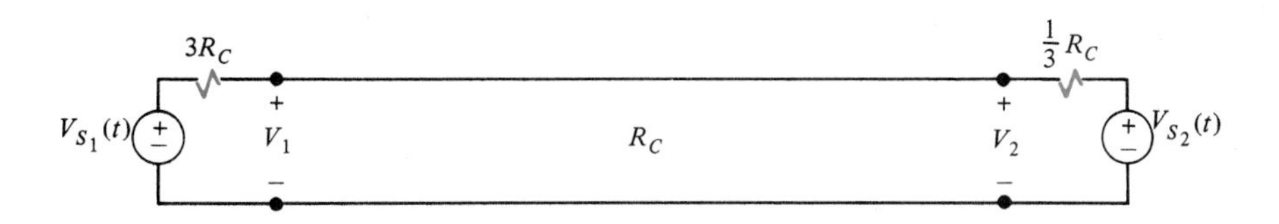

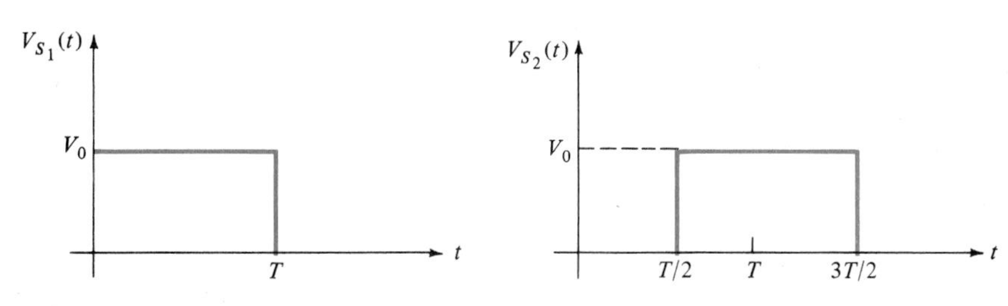

FIGURE P7.15

7-16 A transmission line having $\mathscr{L} = 200$ m, $u = 200$ m/μs, $R_C = 50\ \Omega$, and $R_L = 20\ \Omega$ is driven by a source which has $R_S = 100\ \Omega$ and which has a source voltage that is a rectangular pulse of magnitude 6 V and duration 3 μs. Sketch the input current to the line for a total time of 5 μs.

7-17 Consider a lossless transmission line operated in the sinusoidal steady state. For the following problem specifications, determine **(a)** the line length as a fraction of a wavelength, **(b)** the voltage reflection coefficient at the load and at the input to the line, **(c)** the VSWR, **(d)** the input impedance to the line, **(e)** the time-domain voltage at the line input and at the load, and **(f)** the average power delivered to the load:

(a) $\mathscr{L} = 1$ m; $f = 262.5$ MHz; $R_C = 50\ \Omega$; $\hat{Z}_L = (30 - j200)\ \Omega$; $\hat{Z}_S = (100 + j50)\ \Omega$; $u = 300$ m/μs; $\hat{V}_S = 10\underline{/30^\circ}$ V

(b) $\mathscr{L} = 36$ m; $f = 28$ MHz; $R_C = 150\ \Omega$; $\hat{Z}_S = (500 + j0)\ \Omega$; $\hat{Z}_L = -j30\ \Omega$; $u = 300$ m/μs; $\hat{V}_S = 100\underline{/0^\circ}$ V

(c) $\mathscr{L} = 2$ m; $f = 175$ MHz; $u = 200$ m/μs; $R_C = 100\ \Omega$; $\hat{Z}_S = 50\ \Omega$; $\hat{Z}_L = (200 - j30)\ \Omega$; $\hat{V}_S = 10\underline{/0^\circ}$ V

7-18 Consider a lossless transmission line whose length is $\lambda/4$. Derive relationships between the input and load reflection coefficients and impedances and between the input and load phasor voltages and currents. Simplify your results and write them in terms of R_C, $\hat{Z}_L$, and $\hat{Z}_{in}$.

7-19 Show that the magnitude of the reflection coefficient for a lossless line having a purely reactive load is exactly unity.

7-20 Suppose a $\frac{3}{4}\lambda$ length of line is formed into a closed loop by connecting the appropriate input and output terminals. Determine the impedance between two points, one on one loop and one at the same location but on the other loop.

7-21 Consider a lossless line having a resistive load R_L. Derive an expression for the VSWR in terms of the ratio $k = R_L/R_C$.

7-22 A section of lossless coaxial cable having $R_C = 50\ \Omega$ and $u = 200$ m/μs is terminated in a short circuit and operated at a frequency of 10 MHz. Determine the shortest length of the line such that, at the input terminals, the line appears to be a 100-pF capacitor. Determine the shortest length such that the line appears to be a 1-μH inductor.

7-23 Repeat Prob. 7-22 for an open-circuit load.

7-24 An antenna having an input impedance at 100 MHz of $(72 + j40)\ \Omega$ is connected to a 100-MHz generator via a section of 300-Ω air-filled line of length 1.75 m. If the generator has a source voltage of 10 V and a source impedance of 50 Ω, determine the average power delivered to the antenna.

7-25 A lossless transmission line has a load impedance $\hat{Z}_L = R_C + jX$. In terms of the measured VSWR, determine X. Can X be completely determined?

7-26 A generator with a 50-Ω source resistance is being operated at 100 MHz. A load of $(50 - j100)\ \Omega$ is placed across the terminals of this generator. It is desired to transfer maximum power to this load. If a section of 150-Ω lossless air-filled line is placed in parallel with this load, determine the shortest length of line for which maximum average power is transferred to the load when the line has:

(a) an open-circuit load
(b) a short-circuit load

7-27 A 100-Ω lossless transmission line of unknown length is terminated in an unknown impedance. The input impedance is measured to be $(200 - j50)\ \Omega$. A call is made to an assistant at the other end of the line who is told to remove the load. If the input impedance to the line becomes $-j150\ \Omega$, determine the unknown impedance.

7-28 Two antennas with impedances given by $(100 - j30)\ \Omega$ are connected to a transmitter with three identical lengths of identical lossless transmission line, as shown in Fig. P7.28. Determine the magnitude of the voltage across the terminals of one antenna, $|V_{ant}|$, and the power delivered to that antenna.

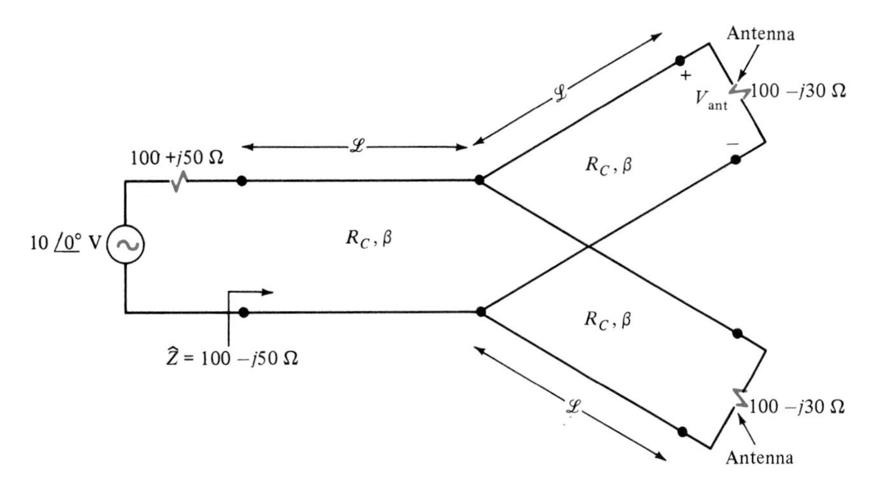

FIGURE P7.28

7-29 Transmission lines whose lengths are much less than a wavelength (i.e., $\mathscr{L} \ll \lambda$) are said to be "electrically short." For a line that is sufficiently short electrically, the distributed-parameter effects are negligible and the line may be modeled as a lumped circuit. Two typical lumped-circuit structures are shown in Fig. P7.29.

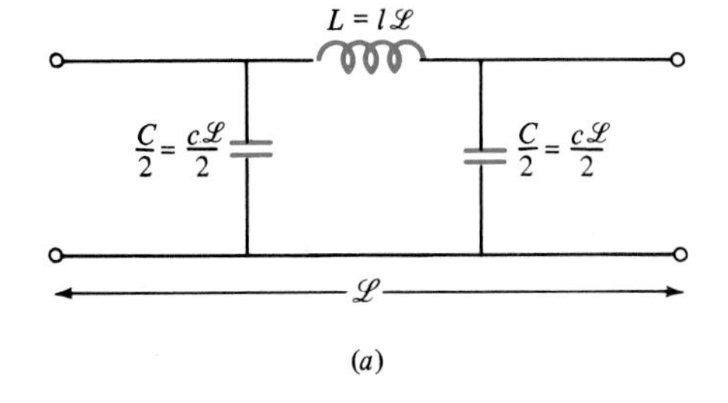

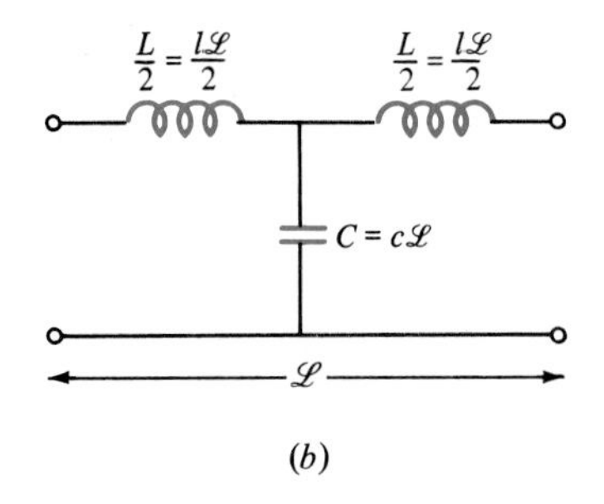

FIGURE P7.29

These structures are chosen to resemble per-unit-length circuits from which the transmission-line equations can be derived. In order to investigate the adequacy of these approximations, compute the normalized input impedance $\hat{Z}_{in}/R_C$ to a line whose length is $\frac{1}{10}\lambda$ for load resistances of

(a) $R_L = 10R_C$
(b) $R_L = R_C$
(c) $R_L = 0.1R_C$

by using each of these structures and compare with the exact result. (Use a standard lumped-circuit program for the approximate calculations. This is the advantage of the lumped-circuit models.) Repeat for a line length of $\frac{1}{100}\lambda$. What can you conclude about the appropriate choice of lumped-circuit model for a particular load? Can you justify why this should be true, based on the structure of the models?

7-30 To investigate electrically short lines further, consider a lossless line that is driven by a 1-V ideal voltage source (zero source impedance) and is terminated in a resistive load. If the effects of the line are negligible, the load voltage should be approximately $1\angle 0^\circ$. In order to investigate this, compute the load voltage for line lengths of $\frac{1}{10}\lambda$ and $\frac{1}{100}\lambda$ for:

(a) $R_L = 10R_C$
(b) $R_L = R_C$
(c) $R_L = \frac{1}{10}R_C$

Does the criterion for a line to be electrically short depend on the termination impedances?

7-31 Suppose a $\frac{3}{8}\lambda$ length of 50-Ω line is driven by a voltage source that has $\hat{V}_S = 100\angle 0$ and $\hat{Z}_S = (50 + j0)\ \Omega$. If the far end of the line is terminated in a short circuit, determine the time-domain current through the short circuit.

7-32 Two antennas having input impedance $\hat{Z}_{\text{ant}} = (73 + j0)\ \Omega$ are fed with lossless lines from a generator, as shown in Fig. P7.32. Determine the average power delivered to each antenna.

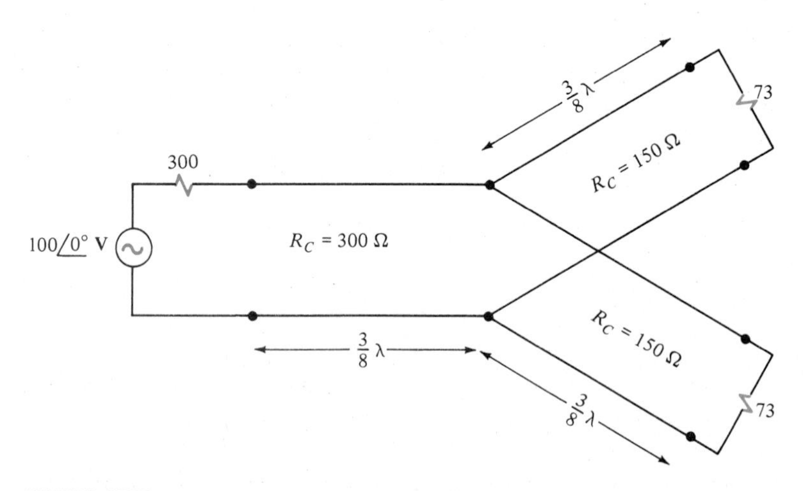

FIGURE P7.32

7-33 Consider a lossy coaxial cable that has $r = 2.25\ \Omega/\text{m}$, $l = 1\ \mu\text{H/m}$, $c = 100\ \text{pF/m}$, and $g \simeq 0$ at 500 MHz. Determine the attenuation constant of the line at

500 MHz. Determine the attenuation per 100 ft of cable length in decibels. Compute the input impedance to a 10-m section of this line at 500 MHz if the load impedance is

(a) an open circuit,
(b) a short circuit, and
(c) $\hat{Z}_L = 10 + j0 \quad \Omega$.

7-34 A lossy transmission line is operated at 100 MHz. Measurements on the line indicate that

$$\hat{Z}_C = 75 + j0 \quad \Omega$$

$$\alpha = 0.02 \text{ Np/m}$$

$$\beta = 3 \text{ rad/m}$$

Determine the per-unit-length resistance, inductance, capacitance, and conductance of the line. If a 7-m length line is terminated in $\hat{Z}_L = (150 + j0)\ \Omega$ and is driven by a source having $\hat{V}_S = 10\underline{/0^\circ}$ V and $\hat{Z}_S = (75 + j0)\ \Omega$, determine the average power delivered to the line and to the load. Repeat these calculations if the line is matched at the load, i.e., $\hat{Z}_L = (75 + j0)\ \Omega$.

7-35 Characteristics of lines are often determined by making short-circuit and open-circuit tests. Consider a lossy line for which the short-circuit input impedance is $235\underline{/30^\circ}\ \Omega$ and the open-circuit input impedance is $120\underline{/-45^\circ}\ \Omega$. Determine $\hat{Z}_C$, α, and β for this cable.

7-36 A length of RG-58U coaxial cable has

$$\hat{Z}_C = 50 + j0 \quad \Omega$$

$$\alpha\mathscr{L} = 2.21 \times 10^{-2} \text{ Np}$$

$$\beta\mathscr{L} = 4.082 \text{ rad/m}$$

If the cable is terminated in a short circuit and driven by a source having $\hat{V}_S = 10\underline{/0^\circ}$ and $\hat{Z}_S = (50 + j0)\ \Omega$, determine the current in the short circuit.

7-37 The characteristic resistance of parallel wire lines, as well as of coaxial cables, typically falls within a rather narrow range of values since l and c depend on the logarithm (natural) of the line dimensions. To illustrate this, compute R_C for a coaxial cable having an inner dielectric of polyethylene ($\epsilon_r = 2.25$) for ratios of shield radius to inner wire radius of $r_s/r_w = 1.5$, 2, 5, and 10. Repeat this for an air-filled, parallel wire line for ratios of wire separation to wire radii of $d/r_w = 2.5$, 5, 10, and 20.

7-38 In many types of coaxial cables, the outer shield is a very thin aluminum foil coated with a Mylar backing instead of a woven braid. A typical cable has the following dimensions:

Shield (aluminum) thickness	0.35 mils
Shield inner radius	32 mils
Inner wire (copper) radius	5 mils
Dielectric (foamed polyethylene)	$\epsilon_r = 1.5$

Calculate the characteristic resistance of this cable. Calculate the loss of a 100-ft length of this cable at 100 MHz.

7-39 Consider a line that is "electrically short" ($\beta\mathscr{L} \ll 1$) and has small losses ($\alpha\mathscr{L} \ll 1$). Show that the input impedance to this line is approximately $(r + j\omega l)\mathscr{L}$ when the load is a short circuit and that the input admittance is approximately $(g + j\omega c)\mathscr{L}$ when the load is an open circuit. Confirm this result for a line having $r = 0.1\ \Omega/\text{m}$, $g = 2.7 \times 10^{-4}$ S/m, $l = 360$ nH/m, $c = 64$ pF/m at 1 MHz, and $\mathscr{L} = 1$ m.

CHAPTER 8

Waveguides and Cavity Resonators

A pair of conductors, such as in a two-wire transmission line considered in the previous chapter, serve to guide energy from one point (a source) to another (a load). There are other types of guiding structures consisting of single, hollow conductors. These structures are referred to as *waveguides*, and a rectangular guide is shown in Fig. 8.1.† Cylindrical waveguides are also used, but only the rectangular guide will be analyzed here. Initially we will assume that the walls of the guide are perfect conductors and that the interior is filled with a lossless ($\sigma = 0$) dielectric (usually air) having material properties ε and μ. Once the analysis of this lossless guide has been completed, we will investigate the effect of imperfect wall conductors and lossy dielectrics. Waveguides are generally used to guide waves at frequencies in the gigahertz range, where the losses of the transmission lines considered in the previous chapter become prohibitive.

Certain other nonmetallic waveguides may also be used to guide waves. One example that is finding increased usage is the fiber-optic cable shown in Fig. 8.2. This type of cable consists of a solid cylinder of inhomogeneous dielectric material such that the relative permittivity of the material is a function of position from the central axis. The varying relative permittivity causes the light

† Strictly speaking, the transmission lines discussed in the preceding chapter are "waveguides" since they certainly guide electromagnetic waves, as we have seen. However, it has become common to refer to them as transmission lines rather than as waveguides.

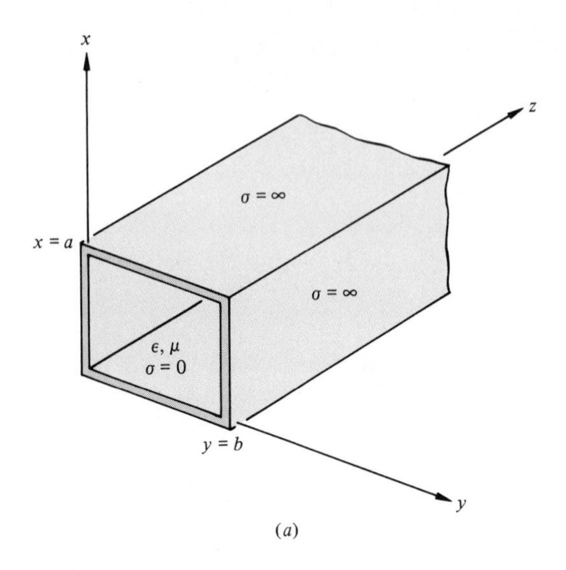

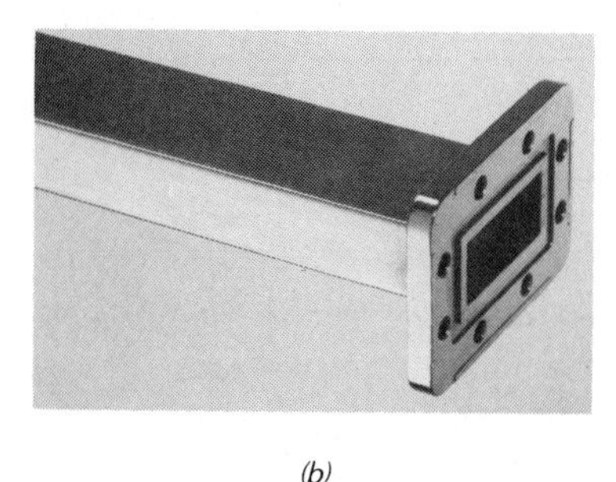

(b)

FIGURE 8.1

Rectangular waveguides; *(a)* parameter definitions; *(b)* a rectangular waveguide used in the X band range (8.2 to 12.4 GHz). *(Andrew Corporation.)*

waves to remain inside the cable, and data or other information is used to modulate the light. Some of the chief advantages of this type of cable over, for example, the conventional wire-type transmission lines considered in Chap. 7 are that (1) there is virtually no interaction or interference coupled between adjacent cables, which is not the case for open-wire lines, and (2) a tremendous weight savings can be realized; one fiber-optic cable can replace conventional cables consisting of several thousand copper-wire pairs. These advantages also apply to the rectangular waveguide shown in Fig. 8.1.

This latter advantage is a result of the very large bandwidths of these guides. Because of these large bandwidths, considerably more information can be carried over these structures than over a two-conductor transmission line. For example, suppose that a signal with frequency components extending from dc to

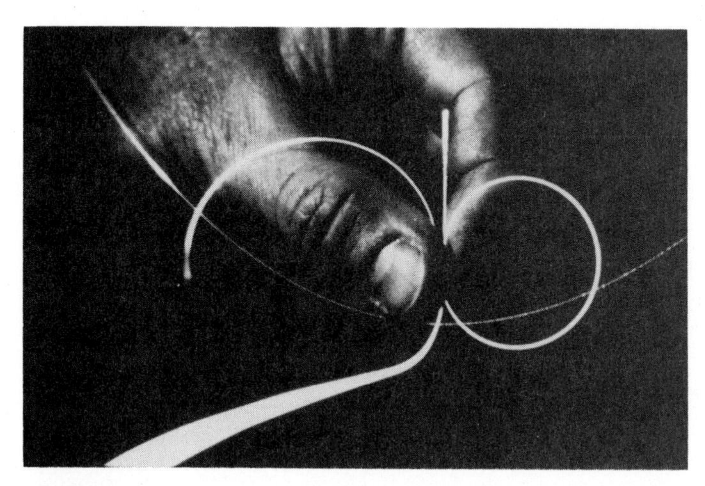

FIGURE 8.2

Loops of a hair-thin glass fiber, illuminated by laser light, represent the transmission medium for lightwave systems. Typically, twelve fibers are embedded between two strips of plastic in a flat ribbon and as many as twelve ribbons are stacked in a cable that can carry more than 40,000 voice channels. *(AT&T Bell Laboratories.)*

1-MHz amplitude modulates a carrier of frequency f_c. For a coaxial-cable transmission line, a carrier frequency of 100 MHz would not incur exhorbitant losses; thus, the frequency span of the AM signal would extend from 99 to 101 MHz, giving a bandwidth $\Delta f/f_c$ of 2 percent. The use of a rectangular waveguide at a carrier frequency of 10 GHz would also not incur prohibitive losses, yet for a similar 2 percent bandwidth we could transmit information having a frequency spread of 100 MHz, some 100 times that of the coaxial line. If an optical guide were used, a carrier frequency of approximately 10^{13} Hz (slightly below the visible region) would be representative; for a 2 percent bandwidth at a carrier frequency of 10^{13} Hz, we would be able to transmit information having a frequency spread of approximately 10^{11} Hz, or 100 GHz. Recently, AT&T Bell Laboratories has announced the ability to transmit digital data at a 4-gigabit rate (4×10^9 bits per second) over a single glass fiber.† The transmission occurred over a 60-mi (100-km) length of fiber. This represented the equivalent of 62,500 two-way voice telephone circuits or 44 broadcast-quality television signals. This illustrates why optical fibers are becoming increasingly attractive for data transmission.

In this chapter we will also investigate the use of rectangular cavities to construct resonant circuits. Construction of resonant circuits using lumped-circuit elements at frequencies above several hundred MHz is not feasible since it is difficult to construct ideal lumped-circuit elements at those frequencies. For example, typical capacitors may appear to be inductors (and vice versa) at these

† *AT&T Record*, March 1985.

frequencies. Furthermore, the impedance of the connection leads of these elements becomes quite large at these frequencies. In addition to the nonideal behavior of lumped-circuit elements, the losses associated with these elements also become large at these frequencies, so that resonant circuits having large quality factors (Q's) are difficult to construct. The cavity resonator provides a means of constructing microwave resonant circuits having large Q's. Cavity resonators may be thought of as rectangular waveguides of finite length having conducting plates at each end, i.e., a rectangular box. This notion will allow us in a simple manner to extend the results obtained for rectangular waveguides to cavity resonators.

8.1 Separability of the Wave Equation: Modes

In order to obtain the structure of the fields in the rectangular guide shown in Fig. 8.1, we must re-examine the wave equations (for sinusoidal field variation) given in Chap. 6, Eqs. (9), in the lossless interior of the guide ($\sigma = 0$):

$$\nabla^2 \hat{\mathbf{E}} = -\omega^2 \mu\epsilon \hat{\mathbf{E}} \tag{1a}$$

$$\nabla^2 \hat{\mathbf{H}} = -\omega^2 \mu\epsilon \hat{\mathbf{H}} \tag{1b}$$

Before proceeding to solve the wave equations, we perform some preliminary calculations. We will be interested in propagation in the z direction: along the longitudinal axis of the guide. Thus, we will assume the forms of all the field components to be

$$\begin{aligned} \hat{E}_x(x, y, z) &= \hat{E}'_x(x, y)e^{-\gamma z} \\ \hat{E}_y(x, y, z) &= \hat{E}'_y(x, y)e^{-\gamma z} \\ &\vdots \\ \hat{H}_z(x, y, z) &= \hat{H}'_z(x, y)e^{-\gamma z} \end{aligned} \tag{2}$$

where the primed components are functions only of x and y. Note that the field vectors given in (2) are in the form of forward-traveling waves (traveling in the $+z$ direction). Backward-traveling waves (traveling in the $-z$ direction) are also of interest; the general form of the field vectors would be

$$\begin{aligned} \hat{E}_x(x, y, z) &= \hat{E}_x^{+\prime}(x, y)e^{-\gamma z} + \hat{E}_x^{-\prime}(x, y)e^{\gamma z} \\ &\vdots \end{aligned}$$

However, to simplify the notion, we will consider only the forward-traveling waves since the results for the backward-traveling waves are quite similar. The constant γ in (2) will be referred to as the *propagation constant.* It is not necessarily the same as γ for uniform plane waves in (10) to (12) of Chap. 6, as we shall see.

Substituting the assumed forms of the field vectors given in (2) into the wave equations given in (1) and matching components gives

$$\begin{aligned}
\frac{\partial^2 \hat{E}_x}{\partial x^2} + \frac{\partial^2 \hat{E}_x}{\partial y^2} + \gamma^2 \hat{E}_x &= -\omega^2 \mu\epsilon \hat{E}_x \\
\frac{\partial^2 \hat{E}_y}{\partial x^2} + \frac{\partial^2 \hat{E}_y}{\partial y^2} + \gamma^2 \hat{E}_y &= -\omega^2 \mu\epsilon \hat{E}_y \\
&\vdots \\
\frac{\partial^2 \hat{H}_z}{\partial x^2} + \frac{\partial^2 \hat{H}_z}{\partial y^2} + \gamma^2 \hat{H}_z &= -\omega^2 \mu\epsilon \hat{H}_z
\end{aligned} \tag{3}$$

This is a result of the fact that $(\partial^2/\partial z^2)e^{-\gamma z} = \gamma^2 e^{-\gamma z}$. Note that this result also applies to the backward-traveling waves (which we do not explicitly consider) since $(\partial^2/\partial z^2)e^{\gamma z} = \gamma^2 e^{\gamma z}$. Combining (3) into vector form shows that the wave equations can be written as

$$\nabla_T^2 \hat{\mathbf{E}} = -(\gamma^2 + \omega^2 \mu\epsilon)\hat{\mathbf{E}} \tag{4a}$$

$$\nabla_T^2 \hat{\mathbf{H}} = -(\gamma^2 + \omega^2 \mu\epsilon)\hat{\mathbf{H}} \tag{4b}$$

where ∇_T is the transverse (to the z axis) operator

$$\nabla_T = \frac{\partial}{\partial x}\mathbf{a}_x + \frac{\partial}{\partial y}\mathbf{a}_y \tag{5}$$

and ∇_T^2 is the transverse laplacian operator

$$\nabla_T^2 = \frac{\partial^2}{\partial x^2} + \frac{\partial^2}{\partial y^2} \tag{6}$$

Now consider Faraday's and Ampère's laws in the interior of the guide ($\sigma = 0$):

$$\nabla \times \hat{\mathbf{E}} = -j\omega\mu\hat{\mathbf{H}} \tag{7a}$$

$$\nabla \times \hat{\mathbf{H}} = j\omega\epsilon\hat{\mathbf{E}} \tag{7b}$$

Substituting the assumed form of the field vectors given in (2) into (7) gives

$$\frac{\partial \hat{E}_z}{\partial y} + \gamma \hat{E}_y = -j\omega\mu \hat{H}_x \tag{8a}$$

$$-\gamma \hat{E}_x - \frac{\partial \hat{E}_z}{\partial x} = -j\omega\mu \hat{H}_y \tag{8b}$$

$$\frac{\partial \hat{E}_y}{\partial x} - \frac{\partial \hat{E}_x}{\partial y} = -j\omega\mu \hat{H}_z \tag{8c}$$

and
$$\frac{\partial \hat{H}_z}{\partial y} + \gamma \hat{H}_y = j\omega\epsilon \hat{E}_x \tag{9a}$$

$$-\gamma \hat{H}_x - \frac{\partial \hat{H}_z}{\partial x} = j\omega\epsilon \hat{E}_y \tag{9b}$$

$$\frac{\partial \hat{H}_y}{\partial x} - \frac{\partial \hat{H}_x}{\partial y} = j\omega\epsilon \hat{E}_z \tag{9c}$$

again because $(\partial/\partial z)e^{-\gamma z} = -\gamma e^{-\gamma z}$. The results in (8) and (9) hold only for forward-traveling waves. For backward-traveling waves, we replace γ in (8) and (9) with $-\gamma$ since $(\partial/\partial z)e^{\gamma z} = \gamma e^{\gamma z}$. Combining (8) and (9) into vector form and using (5) gives

$$\nabla_T \times \hat{\mathbf{E}} - \gamma \mathbf{a}_z \times \hat{\mathbf{E}} = -j\omega\mu \hat{\mathbf{H}} \tag{10a}$$

$$\nabla_T \times \hat{\mathbf{H}} - \gamma \mathbf{a}_z \times \hat{\mathbf{H}} = j\omega\epsilon \hat{\mathbf{E}} \tag{10b}$$

as the reader should verify. [Write $\nabla = \nabla_T + (\partial/\partial z)\mathbf{a}_z$, substitute into (7), and carry out the required operations using $\hat{\mathbf{E}} = \hat{\mathbf{E}}' e^{-\gamma z}$ and $\hat{\mathbf{H}} = \hat{\mathbf{H}}' e^{-\gamma z}$.]

We will now show that, *because of the assumed form of z dependence in (2), we only need to solve for the z components of* $\hat{\mathbf{E}}$ *and* $\hat{\mathbf{H}}$. The other field components, $\hat{E}_x$, $\hat{E}_y$, $\hat{H}_x$, and $\hat{H}_y$, can then be easily computed from a knowlede of $\hat{E}_z$ and $\hat{H}_z$! To show this, we observe that, for example, (8*a*) and (9*b*) can be written as

$$\hat{E}_y = -j\frac{\omega\mu}{\gamma}\hat{H}_x - \frac{1}{\gamma}\frac{\partial \hat{E}_z}{\partial y} \tag{8a}$$

$$\hat{H}_x = -j\frac{\omega\epsilon}{\gamma}\hat{E}_y - \frac{1}{\gamma}\frac{\partial \hat{H}_z}{\partial x} \tag{9b}$$

Combining these gives

$$\hat{E}_y = \frac{1}{\gamma^2 + \omega^2\mu\epsilon}\left(j\omega\mu\frac{\partial \hat{H}_z}{\partial x} - \gamma\frac{\partial \hat{E}_z}{\partial y}\right) \tag{11a}$$

Similarly, by combining the other relations in (8) and (9), we obtain

$$\hat{E}_x = -\frac{1}{\gamma^2 + \omega^2\mu\epsilon}\left(j\omega\mu\frac{\partial \hat{H}_z}{\partial y} + \gamma\frac{\partial \hat{E}_z}{\partial x}\right) \tag{11b}$$

$$\hat{H}_y = -\frac{1}{\gamma^2 + \omega^2\mu\epsilon}\left(j\omega\epsilon\frac{\partial \hat{E}_z}{\partial x} + \gamma\frac{\partial \hat{H}_z}{\partial y}\right) \tag{11c}$$

$$\hat{H}_x = \frac{1}{\gamma^2 + \omega^2\mu\epsilon}\left(j\omega\epsilon\frac{\partial \hat{E}_z}{\partial y} - \gamma\frac{\partial \hat{H}_z}{\partial x}\right) \tag{11d}$$

Therefore, once we obtain the solutions for the z-directed field components $\hat{E}_z$ and $\hat{H}_z$ (and, of course, the propagation constant γ), simple substitution of those components into (11) will give the solutions for all the other field components. This is a general result and depends only on the assumed form given in (2).

Therefore, we need only to solve for the z components of the wave equations within the guide:

$$\frac{\partial^2 \hat{E}_z}{\partial x^2} + \frac{\partial^2 \hat{E}_z}{\partial y^2} = -(\gamma^2 + \omega^2 \mu\epsilon)\hat{E}_z \tag{12a}$$

$$\frac{\partial^2 \hat{H}_z}{\partial x^2} + \frac{\partial^2 \hat{H}_z}{\partial y^2} = -(\gamma^2 + \omega^2 \mu\epsilon)\hat{H}_z \tag{12b}$$

In order to solve these equations, in particular the equation for $\hat{E}_z$ given in (12a), we further assume that the field-vector spatial dependence can be separated as

$$\hat{E}_z(x, y, z) = \hat{X}(x)\hat{Y}(y)e^{-\gamma z} \tag{13}$$

where we identify $\hat{E}'_z(x, y) = \hat{X}(x)\hat{Y}(y)$. In other words, we assume that the phasor electric field vector $\hat{E}_z$ (as well as the magnetic field vector $\hat{H}_z$) can be expressed as the product of three functions, each of which is a function of only one of the coordinate variables. Substituting (13) into (12a) we obtain

$$\frac{d^2\hat{X}(x)}{dx^2}\hat{Y}(y)e^{-\gamma z} + \frac{d^2\hat{Y}(y)}{dy^2}\hat{X}(x)e^{-\gamma z} = -(\gamma^2 + \omega^2\mu\epsilon)\hat{X}(x)\hat{Y}(y)e^{-\gamma z} \tag{14}$$

Dividing both sides of (14) by $\hat{X}\hat{Y}e^{-\gamma z}$, we obtain

$$\frac{1}{\hat{X}(x)}\frac{d^2\hat{X}(x)}{dx^2} + \frac{1}{\hat{Y}(y)}\frac{d^2\hat{Y}(y)}{dy^2} + (\gamma^2 + \omega^2\mu\epsilon) = 0 \tag{15}$$

Since $\hat{X}$ is a function only of x, and $\hat{Y}$ is a function only of y, each term on the left side of (15) must be constant (independent of x, y); otherwise, the equation could not equal zero for all values of x, y. Thus, we denote

$$\frac{1}{\hat{X}(x)}\frac{d^2\hat{X}(x)}{dx^2} = -M^2 \tag{16a}$$

$$\frac{1}{\hat{Y}(y)}\frac{d^2\hat{Y}(y)}{dy^2} = -N^2 \tag{16b}$$

(the sign of these constants and the squaring of them is arbitrarily chosen to simplify the solution) and have

$$-M^2 - N^2 + (\gamma^2 + \omega^2\mu\epsilon) = 0 \tag{17}$$

Once we determine constants M and N, the propagation constant is obtained from (17) as

$$\gamma = \sqrt{M^2 + N^2 - \omega^2 \mu \epsilon} \tag{18}$$

The equations in (16) can be written as

$$\frac{d^2 \hat{X}(x)}{dx^2} + M^2 \hat{X}(x) = 0 \tag{19a}$$

$$\frac{d^2 \hat{Y}(y)}{dy^2} + N^2 \hat{Y}(y) = 0 \tag{19b}$$

The solutions to (19) are of the form

$$\hat{X}(x) = \hat{X}_1 \sin Mx + \hat{X}_2 \cos Mx \tag{20a}$$

$$\hat{Y}(y) = \hat{Y}_1 \sin Ny + \hat{Y}_2 \cos Ny \tag{20b}$$

where $\hat{X}_1$, $\hat{X}_2$, $\hat{Y}_1$, and $\hat{Y}_2$ are, as yet, undetermined constants. The final result for $\hat{E}_z$ becomes [substitute (20) into (13)]

$$\hat{E}_z = (\hat{X}_1 \sin Mx + \hat{X}_2 \cos Mx)(\hat{Y}_1 \sin Ny + \hat{Y}_2 \cos Ny)e^{-\gamma z} \tag{21}$$

and similarly for $\hat{H}_z$.

Note in (11) that there are two possible solutions:

1. $E_z = 0$, $H_z \neq 0$. Because the electric fields for this case are transverse to the guide axis, these field structures are called *transverse electric* (TE) *modes*.
2. $H_z = 0$, $E_z \neq 0$. Similarly, these field structures are called *transverse magnetic* (TM) *modes*.

Note that setting $E_z = H_z = 0$ (the TEM mode) in (11) appears to render all other field components zero. On close inspection, however, we see that if we choose $\gamma = j\omega\sqrt{\mu\epsilon}$, the denominators of these expressions are also zero. Thus, we cannot conclude from these equations that the TEM mode is not a possible mode of propagation. (The fields of a transmission line must also satisfy the wave equations.)

However, it is possible to show that the TEM mode cannot exist within a hollow waveguide. Suppose the TEM mode were to exist. In this case the magnetic field must lie solely in the transverse (xy) plane. These magnetic field lines, however, must form closed paths in this transverse plane since $\nabla \cdot \hat{\mathbf{H}} = 0$. From Ampère's law we observe that the integral of this transverse magnetic field around these closed paths must yield the axial (z-directed) conduction or displacement current. Since $\hat{E}_z = 0$ for the TEM mode, no axial displacement current can exist. Also, since there is no center conductor to carry a conduction current, we conclude that the axial conduction current is zero. Therefore, we conclude that the TEM mode cannot exist in a hollow waveguide. We now examine the TE and TM modes.

8.2 Rectangular Waveguides

The results of the previous section are general and do not depend on the shape or composition of the guiding structure. (Of course, we assumed that the region in which the waves propagate is lossless, $\sigma = 0$; a lossy region could easily be accommodated in the above analysis by replacing ϵ with a complex permittivity $\hat{\epsilon}$, as outlined in Sec. 6.3.) In this section we shall consider the application of those results to the rectangular waveguide shown in Fig. 8.1*a*. To accommodate this specific structure, we shall need to enforce the boundary conditions at the (perfectly conducting) walls of the guide in order to determine the constants $\hat{X}_1$, $\hat{X}_2$, $\hat{Y}_1$, and $\hat{Y}_2$ in the general solution for $\hat{E}_z$ given in (21) in the case of TM modes ($\hat{H}_z = 0$) and a similar form for $\hat{H}_z$ in the case of TE modes ($\hat{E}_z = 0$). The general result will consist of the superposition of these TE and TM solutions, as is shown by (11).

8.2.1 TM Modes

For TM modes, $\hat{H}'_z = \hat{H}_z = 0$, the general solution for $\hat{E}_z$ is given in (21). The boundary conditions are that the tangential electric field must be zero at the walls of the guide; that is,

$$\hat{E}_z = 0 \qquad \text{at } x = 0 \tag{22a}$$

$$\hat{E}_z = 0 \qquad \text{at } x = a \tag{22b}$$

$$\hat{E}_z = 0 \qquad \text{at } y = 0 \tag{22c}$$

$$\hat{E}_z = 0 \qquad \text{at } y = b \tag{22d}$$

Applying (22*a*) and (22*c*) to (21) results in $\hat{X}_2 = \hat{Y}_2 = 0$. The remainder of (21) can be written as

$$\hat{E}_z = \hat{C} \sin Mx \sin Ny \, e^{-\gamma z} \tag{23}$$

where $\hat{C} = \hat{X}_1 \hat{Y}_1$. Applying (22*b*) and (22*d*) to (23) results in

$$\sin Ma = 0 \tag{24a}$$

$$\sin Nb = 0 \tag{24b}$$

This results requires that

$$Ma = m\pi \qquad m = 0, 1, 2, 3, \ldots \tag{25a}$$

$$Nb = n\pi \qquad n = 0, 1, 2, 3, \ldots \tag{25b}$$

or

$$M = \frac{m\pi}{a} \tag{26a}$$

$$N = \frac{n\pi}{b} \tag{26b}$$

The propagation constant in (18) becomes

$$\gamma = \sqrt{\left(\frac{m\pi}{a}\right)^2 + \left(\frac{n\pi}{b}\right)^2 - \omega^2\mu\epsilon} \tag{27}$$

It should be obvious by now that in order for wave propagation to take place, γ must be an imaginary number (or at least have a nonzero imaginary part) since the time-domain forms of the solution will be obtained by multiplying (2) by $e^{j\omega t}$ and taking the real part of the result. Suppose that γ is imaginary:

$$\begin{aligned}\gamma &= j\beta_{mn} \\ &= j\sqrt{\omega^2\mu\epsilon - \left(\frac{m\pi}{a}\right)^2 - \left(\frac{n\pi}{b}\right)^2}\end{aligned} \tag{28}$$

Then $e^{-\gamma z}e^{j\omega t} = e^{j(\omega t - \beta_{mn}z)}$, which is the precise form of the time-domain result needed for propagation. Note that γ will be purely imaginary if

$$\omega^2\mu\epsilon > \left(\frac{m\pi}{a}\right)^2 + \left(\frac{n\pi}{b}\right)^2 \tag{29}$$

On the other hand, suppose that

$$\omega^2\mu\epsilon < \left(\frac{m\pi}{a}\right)^2 + \left(\frac{n\pi}{b}\right)^2 \tag{30}$$

Then

$$\begin{aligned}\gamma &= \alpha_{mn} \\ &= \sqrt{\left(\frac{m\pi}{a}\right)^2 + \left(\frac{n\pi}{b}\right)^2 - \omega^2\mu\epsilon}\end{aligned} \tag{31}$$

and $e^{-\gamma z}e^{j\omega t} = e^{-\alpha_{mn}z}e^{j\omega t}$, which clearly indicates no propagation in the z direction, only attenuation. These nonpropagating modes are said to be *evanescent*. Thus, we require that the guide dimensions a and b and the frequency of excitation satisfy (29) in order for wave propagation to take place along the guide axis.

Clearly, an infinite set of integers m and n may be chosen. Each resulting field structure is designated as the TM_{mn} mode. For each set of integers m and n, there is an associated cutoff frequency for that mode—below which we have attenuation and above which we have propagation. This cutoff frequency becomes, by solving (27) for $\gamma = 0$,

$$f_{c,mn} = \frac{1}{2\pi\sqrt{\mu\epsilon}}\sqrt{\left(\frac{m\pi}{a}\right)^2 + \left(\frac{n\pi}{b}\right)^2} = \frac{u}{2}\sqrt{\frac{m^2}{a^2} + \frac{n^2}{b^2}} \tag{32}$$

where

$$u = \frac{1}{\sqrt{\mu\epsilon}} \tag{33}$$

is the familiar uniform plane wave phase velocity in this medium. These TM modes certainly do not resemble uniform plane waves. Our writing $f_{c,mn}$ in terms of u is therefore simply a condensation of certain more familiar items as an aid to remembering the final result. The phase constant β_{mn} in (28) can be written as

$$\beta_{mn} = 2\pi f\sqrt{\mu\epsilon}\sqrt{1 - \left(\frac{f_{c,mn}}{f}\right)^2} = \beta\sqrt{1 - \left(\frac{f_{c,mn}}{f}\right)^2} \qquad f > f_{c,mn} \tag{34}$$

where

$$\beta = \omega\sqrt{\mu\epsilon} \tag{35}$$

is the uniform plane wave phase constant in this medium.

The form of the other components of the field may be obtained by substituting (23) into (11). The result is

$$\hat{E}_z = \hat{C}\sin Mx \sin Ny\, e^{-j\beta_{mn}z} \tag{36a}$$

$$\hat{E}_x = \frac{-j\beta_{mn}M}{M^2 + N^2}\hat{C}\cos Mx \sin Ny\, e^{-j\beta_{mn}z} \tag{36b}$$

$$\hat{E}_y = \frac{-j\beta_{mn}N}{M^2 + N^2}\hat{C}\sin Mx \cos Ny\, e^{-j\beta_{mn}z} \tag{36c}$$

$$\hat{H}_x = \frac{j\omega\epsilon N}{M^2 + N^2}\hat{C}\sin Mx \cos Ny\, e^{-j\beta_{mn}z} \tag{36d}$$

$$\hat{H}_y = \frac{-j\omega\epsilon M}{M^2 + N^2}\hat{C}\cos Mx \sin Ny\, e^{-j\beta_{mn}z} \tag{36e}$$

$$\hat{H}_z = 0 \tag{36f}$$

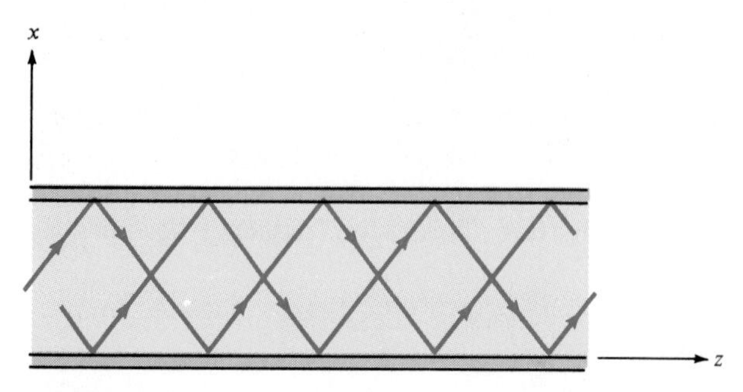

FIGURE 8.3
The decomposition of the wave in a rectangular waveguide into component waves.

where
$$M = \frac{m\pi}{a} \tag{37a}$$

$$N = \frac{n\pi}{b} \tag{37b}$$

Note that the lowest-order TM mode is the TM_{11} mode since setting either m or n equal to zero in (36) renders all field components zero.

The wave propagation in a waveguide can be viewed as the combination of uniform plane waves reflected from the walls of the guide, as illustrated in Fig. 8.3.† The velocity of these uniform plane waves, u, can be determined in the usual fashion as the velocity of the constant phase fronts in each wave, $u = 1/\sqrt{\mu\epsilon}$. However, these waves combine to produce the net z-traveling wave determined above. The velocity of propagation, or more properly the phase velocity, of the resulting constant phase fronts of the z-traveling wave in the guide, however, can be found from our previous results as

$$u_{mn} = \frac{\omega}{\beta_{mn}}$$

$$= \frac{u}{\sqrt{1-(f_{c,mn}/f)^2}} \qquad f > f_{c,mn} \tag{38a}$$

Note that the phase velocity of the resultant wave, u_{mn}, is greater than the speed of light in the material, u. However, u_{mn} is the velocity of the constant phase fronts of the wave, and the energy does not propagate at this speed. The

† See J. D. Kraus and K. R. Carver, *Electromagnetics*, McGraw-Hill, New York, 1973, chap. 13. Also see R. G. Brown, R. A. Sharpe, W. L. Hughes, and R. E. Post, *Lines, Waves, and Antennas*, Ronald Press, New York, 1973, app. E.

process can be visualized as waves approaching the seashore at some angle to the shore. As the wave strikes the shore, the wave crest appears to move along the shore with a velocity greater than that of the wave itself. Wave propagation in the waveguide can be viewed in a similar fashion, as illustrated in Fig. 8.3. In fact, when the individual waves in the guide traveling with velocity u are perpendicular to the guide walls, no net propagation will occur in the z direction; the waves bounce back and forth between the walls. From the seashore example, we observe that the velocity along the shore will be infinite for this situation. Equation (38a) confirms this analogy since for $f = f_{c,mn}$, u_{mn} in (38a) is infinite. Narrowband information-bearing signals propagate along the guide axis at the group velocity (see Example 6.6 and Prob. 8-4):

$$u_{gmm} = \frac{d\omega}{d\beta_{mn}} = u\sqrt{1 - \left(\frac{f_{c,mn}}{f}\right)^2} \tag{38b}$$

and one can show that the phase and group velocities are related by

$$u_{mn} u_{gmn} = u^2 \tag{38c}$$

Note that the group velocity is less than the speed of light in this medium: $u_{gmn} < u$. Thus, the speed of information transmission is less than the speed of light.

Similarly, we may determine a wavelength λ_{mn} of the resultant wave in the guide as the distance required for a phase change of 2π rad in the z direction, or $\lambda_{mn}\beta_{mn} = 2\pi$. Thus, we obtain with (34)

$$\lambda_{mn} = \frac{2\pi}{\beta_{mn}} = \frac{u_{mn}}{f} = \frac{\lambda}{\sqrt{1 - (f_{c,mn}/f)^2}} \qquad f > f_{c,mn} \tag{39}$$

where

$$\lambda = \frac{u}{f} \tag{40}$$

is the uniform plane wave wavelength in this medim.

One final observation may be made concerning analogous quantities in the waveguide problem and those associated with uniform plane waves. If we take the ratios of the transverse electric and magnetic fields given in (36), we obtain

$$\begin{aligned}\frac{\hat{E}_x}{\hat{H}_y} &= \frac{\beta_{mn}}{\omega\epsilon} \\ &= \sqrt{\frac{\mu}{\epsilon}}\sqrt{1-\left(\frac{f_{c,mn}}{f}\right)^2}\end{aligned} \tag{41a}$$

$$\begin{aligned}\frac{\hat{E}_y}{\hat{H}_x} &= -\frac{\beta_{mn}}{\omega\epsilon} \\ &= -\sqrt{\frac{\mu}{\epsilon}}\sqrt{1-\left(\frac{f_{c,mn}}{f}\right)^2}\end{aligned} \tag{41b}$$

From this result, we may define an intrinsic wave impedance of the mode:

$$\eta_{\mathrm{TM},mn} = \eta\sqrt{1-\left(\frac{f_{c,mn}}{f}\right)^2} \quad \Omega \qquad f > f_{c,mn} \tag{42}$$

where

$$\eta = \sqrt{\frac{\mu}{\epsilon}} \tag{43}$$

is the intrinsic impedance of a uniform plane wave in this medium. Thus, the transverse fields are related by

$$\hat{E}_x = \eta_{\mathrm{TM},mn}\hat{H}_y \tag{44a}$$

$$\hat{E}_y = -\eta_{\mathrm{TM},mn}\hat{H}_x \tag{44b}$$

Note that the signs in (44) are consistent with the notion of power flow in the $+z$ direction. (We are considering only forward-propagating waves.) For $f > f_{c,mn}$,

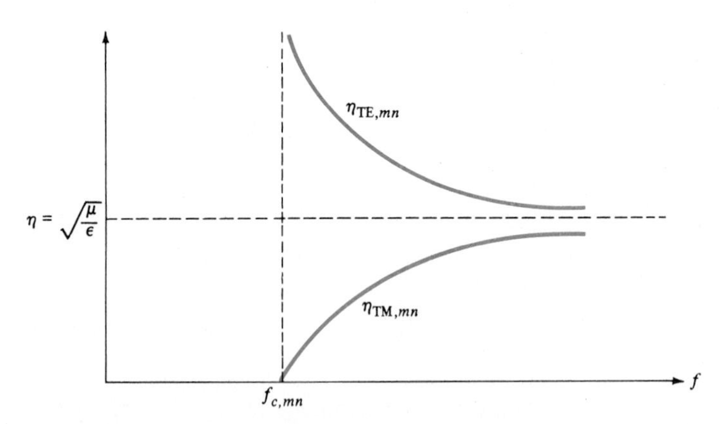

FIGURE 8.4
The wave impedances of the modes as functions of frequency.

the intrinsic wave impedance increases with increasing frequency and asymptotically approaches η, as shown in Fig. 8.4.

The time-domain forms of the fields are obtained by multiplying (36) by $e^{j\omega t}$ and taking the real part of the result:

$$\mathcal{E}_z = C \sin \frac{m\pi x}{a} \sin \frac{n\pi y}{b} \cos\left(\omega t - \frac{2\pi z}{\lambda_{mn}}\right) \tag{45a}$$

$$\mathcal{E}_x = \frac{\beta_{mn}}{h^2} \frac{m\pi}{a} C \cos \frac{m\pi x}{a} \sin \frac{n\pi y}{b} \sin\left(\omega t - \frac{2\pi z}{\lambda_{mn}}\right) \tag{45b}$$

$$\mathcal{E}_y = \frac{\beta_{mn}}{h^2} \frac{n\pi}{b} C \sin \frac{m\pi x}{a} \cos \frac{n\pi y}{b} \sin\left(\omega t - \frac{2\pi z}{\lambda_{mn}}\right) \tag{45c}$$

$$\mathcal{H}_x = -\frac{\omega\epsilon}{h^2} \frac{n\pi}{b} C \sin \frac{m\pi x}{a} \cos \frac{n\pi y}{b} \sin\left(\omega t - \frac{2\pi z}{\lambda_{mn}}\right) \tag{45d}$$

$$\mathcal{H}_y = \frac{\omega\epsilon}{h^2} \frac{m\pi}{a} C \cos \frac{m\pi x}{a} \sin \frac{n\pi y}{b} \sin\left(\omega t - \frac{2\pi z}{\lambda_{mn}}\right) \tag{45e}$$

$$\mathcal{H}_z = 0 \tag{45f}$$

where we have assumed the constant $\hat{C}$ to be real and $h^2 = M^2 + N^2$. The field structure can be plotted in a cross-sectional plane by writing the total transverse electric field from (45b) and (45c):

$$\mathcal{E}_T = \frac{\beta_{mn} C}{h^2}\left(\frac{m\pi}{a} \cos \frac{m\pi x}{a} \sin \frac{n\pi y}{b} \mathbf{a}_x + \frac{n\pi}{b} \sin \frac{m\pi x}{a} \cos \frac{n\pi y}{b} \mathbf{a}_y\right) \sin\left(\omega t - \frac{2\pi z}{\lambda_{mn}}\right) \tag{46}$$

Similarly, the transverse magnetic field becomes, from (45d) and (45e),

$$\mathcal{H}_T = \frac{\omega\epsilon}{h^2} C\left(-\frac{n\pi}{b} \sin \frac{m\pi x}{a} \cos \frac{n\pi y}{b} \mathbf{a}_x + \frac{m\pi}{a} \cos \frac{m\pi x}{a} \sin \frac{n\pi y}{b} \mathbf{a}_y\right) \sin\left(\omega t - \frac{2\pi z}{\lambda_{mn}}\right) \tag{47}$$

For the lowest-order TM_{11} mode, these fields are sketched in Fig. 8.5a for a fixed time ($t = 0$) and at a fixed z ($z = 3\lambda_{11}/4$) such that $\sin(\omega t - 2\pi z/\lambda_{mn}) = 1$. Perhaps the simplest method of sketching these fields is this: First choose

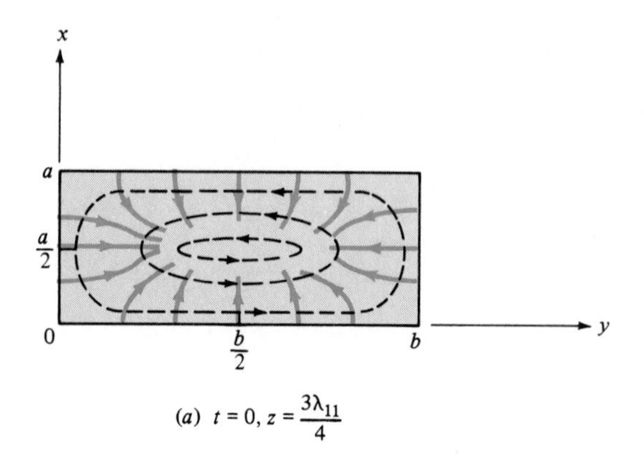

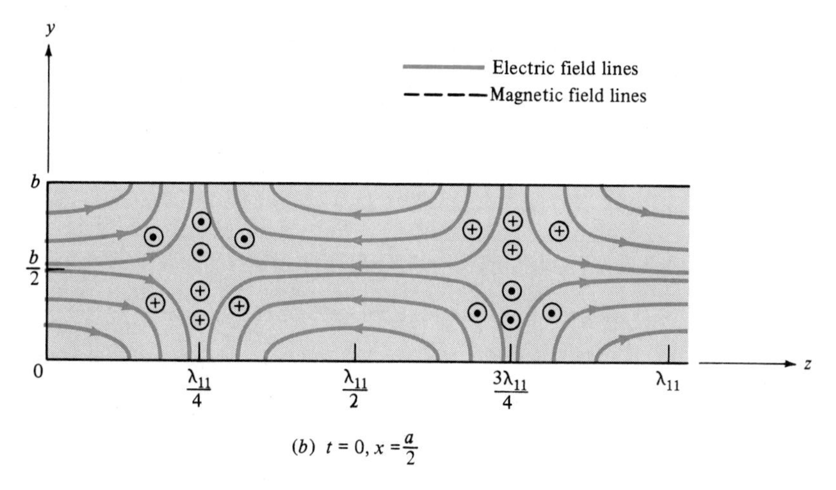

FIGURE 8.5
Field structures of the TM_{11} mode.

a value to t and a value of z such that the t- and z-dependence term, $\sin(\omega t - 2\pi z/\lambda_{mn})$, assumes a value of ± 1; then locate and sketch the orientation of each component—$\mathcal{E}_x$, $\mathcal{E}_y$, $\mathcal{H}_x$, and $\mathcal{H}_y$—at critical points that maximize or minimize the remaining trigonometric terms—$\cos(m\pi x/a)$, $\sin(n\pi y/b)$, $\sin(m\pi x/a)$, and $\cos(n\pi y/b)$—which, for the TM_{11} mode, are $x = 0, a/2, a$, and $y = 0, b/2, b$. The result indicates "streamlines" whereby one can readily sketch a complete field structure.

The fields along the guide axis can be similarly sketched. Typically, these are shown as a cross-sectional sketch in the xz plane for a fixed y and in the yz plane for a fixed x. The values of x or y giving the desired plane are usually chosen to maximize the trigonometric functions of that variable in the equations of (45). For example, we may choose to sketch the fields of the TM_{11} mode in the yz cross section for $x = a/2$. In this cross section, the equations in (45) show that

$\mathcal{E}_x = \mathcal{H}_y = 0$. Thus, only $\mathcal{E}_z$, $\mathcal{E}_y$, and $\mathcal{H}_x$ are to be sketched. For a fixed time, say $t = 0$, these become

$$\mathcal{E}_z\Big|_{\substack{t=0\\x=a/2}} = C \sin \frac{\pi y}{b} \cos \frac{2\pi z}{\lambda_{11}} \tag{48a}$$

$$\mathcal{E}_y\Big|_{\substack{t=0\\x=a/2}} = -\frac{\beta_{11}}{h^2}\frac{\pi}{b} C \cos \frac{\pi y}{b} \sin \frac{2\pi z}{\lambda_{11}} \tag{48b}$$

$$\mathcal{H}_x\Big|_{\substack{t=0\\x=a/2}} = \frac{\omega\epsilon}{h^2}\frac{\pi}{b} C \cos \frac{\pi y}{b} \sin \frac{2\pi z}{\lambda_{11}} \tag{48c}$$

Note that the fields vary in the z direction as a function of electrical distance z/λ_{11}. Thus, the critical points to sketch occur at $z = 0, \lambda_{11}/4, \lambda_{11}/2, 3\lambda_{11}/4$, and λ_{11}. The critical points to sketch in the y direction are $y = 0$, $b/2$, and b. The resulting sketch is shown in Fig. 8.5*b*.

EXAMPLE 8.1 A typical rectangular air-filled waveguide has dimensions $a =$ 0.9 in (2.29 cm) and $b = 0.4$ in (1.02 cm). Determine the cutoff frequencies of the lowest-order mode, TM_{11}, and the TM_{12} and TM_{21} modes. Determine $\eta_{TM,11}$, β_{11}, u_{11}, and λ_{11} for the TM_{11} mode at 18 GHz. Compute the propagation constant for a frequency that is one-half the cutoff frequency of the lowest-order mode.

Solution The cutoff frequencies are

$$f_{c,mn} = \frac{u_0}{2}\sqrt{\left(\frac{m}{a}\right)^2 + \left(\frac{n}{b}\right)^2}$$

$$= 1.5 \times 10^8 \sqrt{\left(\frac{m}{2.29 \times 10^{-2}}\right)^2 + \left(\frac{n}{1.02 \times 10^{-2}}\right)^2}$$

Thus

$$f_{c,11} = 16.16 \text{ GHz}$$

$$f_{c,21} = 19.75 \text{ GHz}$$

$$f_{c,12} = 30.25 \text{ GHz}$$

The factor

$$F_{mn} = \sqrt{1 - \left(\frac{f_{c,mn}}{f}\right)^2}$$

appears in $\eta_{TM,mn}$, β_{mn}, u_{mn}, and λ_{mn}, and for the TM_{11} mode at 18 GHz this becomes

$$F_{11} = 0.44$$

Thus

$$\begin{aligned}\eta_{TM,11} &= \eta_0 F_{11} \\ &= 120\pi F_{11} \\ &= 166.20\ \Omega\end{aligned}$$

$$\begin{aligned}\beta_{11} &= 2\pi f\sqrt{\mu_0\epsilon_0}\, F_{11} \\ &= \beta_0 F_{11} \\ &= 166.20 \text{ rad/m}\end{aligned}$$

$$\begin{aligned}u_{11} &= \frac{u_0}{F_{11}} \\ &= 6.81 \times 10^8 \text{ m/s}\end{aligned}$$

$$\begin{aligned}\lambda_{11} &= \frac{\lambda_0}{F_{11}} \\ &= 3.78 \text{ cm}\end{aligned}$$

The propagation constant of the TM_{11} mode at $\frac{1}{2}f_{c,11} = 8.08$ GHz is

$$\begin{aligned}\gamma_{11} &= \alpha_{11} \\ &= \beta_0\sqrt{\left(\frac{f_{c,11}}{f}\right)^2 - 1} \\ &= \beta_0\sqrt{\left(\frac{16.16 \times 10^9}{8.08 \times 10^9}\right)^2 - 1} \\ &= 293.04\end{aligned}$$

since 8.08 GHz is below cutoff for this mode. Thus, the field vectors in (36) will have z dependence of the form $e^{-\alpha_{11}z}$ instead of $e^{-j\beta_{11}z}$: attenuation of the fields occurs. It is of interest to examine how rapidly these fields attenuate in the z direction. For a distance $z = 1/\alpha_{11}$ they are reduced in amplitude by 37 percent. This occurs for the TM_{11} mode at 8.08 GHz in a distance of 3.41 mm: an extremely rapid attenuation. This illustrates rather dramatically that a mode attenuates very rapidly with distance at frequencies below its cutoff frequency.

8.2.2 TE MODES

The TE modes ($\hat{E}_z = 0$) can be obtained by investigating the solution to (12*b*), which is identical in form to the solution for $\hat{E}_z$ given in (21). Since $\hat{E}_z = 0$, we need to apply the boundary conditions on the tangential electric field at the walls of the guide to $\hat{E}_x$ and $\hat{E}_y$, which are

$$\hat{E}_x = 0 \qquad \text{at } y = 0, y = b \tag{49a}$$

$$\hat{E}_y = 0 \qquad \text{at } x = 0, x = a \tag{49b}$$

The corresponding constraints imposed on $\hat{H}_z$ can be obtained from (11). For example, the conditions in (49) along with $\hat{E}_z = 0$ for this mode, when substituted into (11*a*) and (11*b*), yield

$$\frac{\partial \hat{H}_z}{\partial y} = 0 \qquad \text{at } y = 0 \tag{50a}$$

$$\frac{\partial \hat{H}_z}{\partial y} = 0 \qquad \text{at } y = b \tag{50b}$$

$$\frac{\partial \hat{H}_z}{\partial x} = 0 \qquad \text{at } x = 0 \tag{50c}$$

$$\frac{\partial \hat{H}_z}{\partial x} = 0 \qquad \text{at } x = a \tag{50d}$$

The form of $\hat{H}_z$ is identical to the form of $\hat{E}_z$ given in (21). Differentiating this form with respect to x and with respect to y and applying the conditions in (50) yield

$$\hat{H}_z = \hat{C} \cos Mx \cos Ny\, e^{-j\beta_{mn}z} \tag{51a}$$

Substitution into (11) gives

$$\hat{H}_x = \frac{j\beta_{mn}M}{M^2 + N^2} \hat{C} \sin Mx \cos Ny\, e^{-j\beta_{mn}z} \tag{51b}$$

$$\hat{H}_y = \frac{j\beta_{mn}N}{M^2 + N^2} \hat{C} \cos Mx \sin Ny\, e^{-j\beta_{mn}z} \tag{51c}$$

$$\hat{E}_x = \frac{j\omega\mu N}{M^2 + N^2} \hat{C} \cos Mx \sin Ny\, e^{-j\beta_{mn}z} \tag{51d}$$

$$\hat{E}_y = \frac{-j\omega\mu M}{M^2 + N^2} \hat{C} \sin Mx \cos Ny\, e^{-j\beta_{mn}z} \tag{51e}$$

$$\hat{E}_z = 0 \tag{51f}$$

where M, N, and β_{mn} are the same as for the TM modes. The time-domain forms of these field components are easily obtained by multiplying by $e^{j\omega t}$ and taking the real part of the result (see Prob. 8-12).

The cutoff frequencies, phase velocities, and wavelengths for these TE_{mn} modes are the same as for the TM_{mn} modes. However, the intrinsic wave impedances of the TE_{mn} mode and of the corresponding TM_{mn} mode are different. Taking the ratios of the appropriate transverse field components given in (51), we obtain

$$\frac{\hat{E}_x}{\hat{H}_y} = \frac{\omega\mu}{\beta_{mn}} \tag{52a}$$

$$\frac{\hat{E}_y}{\hat{H}_x} = -\frac{\omega\mu}{\beta_{mn}} \tag{52b}$$

The quantity $\omega\mu/\beta_{mn}$ becomes, by substituting the expression for β_{mn} given in (34), the intrinsic wave impedance of the TE_{mn} mode:

$$\begin{aligned} \eta_{TE,mn} &= \frac{\omega\mu}{\beta_{mn}} \quad \Omega \qquad f > f_{c,mn} \\ &= \frac{\eta}{\sqrt{1-(f_{c,mn}/f)^2}} \end{aligned} \tag{53}$$

where $\eta = \sqrt{\mu/\epsilon}$ is again the intrinsic impedance of a uniform plane wave in this medium. Thus, the transverse field components are related by

$$\hat{E}_x = \eta_{TE,mn}\hat{H}_y \tag{54a}$$

$$\hat{E}_y = -\eta_{TE,mn}\hat{H}_x \tag{54b}$$

and again the signs are consistent with the notion of power flow in the $+z$ direction. Note that for $f > f_{c,mn}$, the intrinsic wave impedance of the TE modes decreases with increasing frequency and asymptotically approaches η, as shown in Fig. 8.4 If we take the product of $\eta_{TE,mn}$ given in (53) and $\eta_{TM,mn}$ given in (42), we find that

$$\begin{aligned} \eta_{TE,mn}\eta_{TM,mn} &= \eta^2 \\ &= \frac{\mu}{\epsilon} \end{aligned} \tag{55}$$

Waveguides are often designed so as to support the lowest-order mode and to exclude the higher-order modes. For example, the lowest-order TE mode is the

TE_{10} mode if we assume a rectangular guide with $b < a$. The field equations for the TE_{10} are obtained from (51) by substituting $m = 1$ and $n = 0$:

$$\hat{H}_z = \hat{C} \cos \frac{\pi x}{a} e^{-j\beta_{10} z} \tag{56a}$$

$$\hat{H}_x = \frac{j\beta_{10} a}{\pi} \hat{C} \sin \frac{\pi x}{a} e^{-j\beta_{10} z} \tag{56b}$$

$$\hat{H}_y = 0 \tag{56c}$$

$$\hat{E}_x = 0 \tag{56d}$$

$$\hat{E}_y = -\frac{j\omega\mu a}{\pi} \hat{C} \sin \frac{\pi x}{a} e^{-j\beta_{10} z} \tag{56e}$$

$$\hat{E}_z = 0 \tag{56f}$$

Recall that the lowest-order TM mode is the TM_{11} mode, since setting $m = 1$ and $n = 0$ (or $n = 1$ and $m = 0$) in (36) results in all field components being zero. The time-domain fields for the TE_{10} mode can be obtained by multiplying (56) by $e^{j\omega t}$ and taking the real part of the result:

$$\mathcal{H}_z = C \cos \frac{\pi x}{a} \cos \left(\omega t - \frac{2\pi z}{\lambda_{10}} \right) \tag{57a}$$

$$\mathcal{H}_x = -\frac{\beta_{10} a}{\pi} C \sin \frac{\pi x}{a} \sin \left(\omega t - \frac{2\pi z}{\lambda_{10}} \right) \tag{57b}$$

$$\mathcal{E}_y = \frac{\omega\mu a}{\pi} C \sin \frac{\pi x}{a} \sin \left(\omega t - \frac{2\pi z}{\lambda_{10}} \right) \tag{57c}$$

where we assume that $\hat{C}$ is real and we have substituted

$$\beta_{10} = \frac{2\pi}{\lambda_{10}} \tag{58}$$

A waveguide can be designed to support only the TE_{10} mode by selecting the guide dimensions such that the operating frequency is greater than the cutoff frequency for this mode given in (32):

$$\begin{aligned} f_{c,10} &= \frac{1}{2\pi\sqrt{\mu\epsilon}} \frac{\pi}{a} \\ &= \frac{1}{2a\sqrt{\mu\epsilon}} \\ &= \frac{u}{2a} \end{aligned} \tag{59}$$

and is less than the cutoff frequency of the next higher mode. Once the guide dimensions have been selected such that it is possible to propagate this mode, one must determine an efficient way of exciting this mode. In order to determine an efficient method, we first must examine the field structure of the TE_{10} mode. The fields are plotted from (57) for $t = 0$ in Fig. 8.6 for various cross sections of the guide. Note in Fig. 8.6*a* that the only nonzero component of the electric field is the y component, and this component of the electric field achieves a maximum at $x = a/2$ and is zero at the upper ($x = a$) and the lower ($x = 0$) walls of the guide (as it must be to satisfy the boundary condition). Also note that $\hat{H}_z$ and $\hat{H}_x$ combine to form closed lines of the resultant magnetic field, as shown in Fig. 8.6*c*.

The usual method of exciting the TE_{10} mode in a rectangular guide is shown in Fig. 8.6*d*. A probe (a coaxial line with the center conductor exposed) is inserted into the wall of the guide; oriented parallel to the y dimension, the probe is located horizontally at $\lambda_{10}/4$ from the shorted end plate of the guide and vertically at $x = a/2$. The probe is thus located where the electric field is a maximum and is oriented parallel to the electric field lines of this mode. This provides an efficient excitation of this TE_{10} mode.

As an illustration of power flow in waveguides, let us compute the average-power Poynting vector for the TE_{10} mode. The average-power vector is

$$\mathbf{S}_{\text{av}} = \tfrac{1}{2}\,\text{Re}\,(\hat{\mathbf{E}} \times \hat{\mathbf{H}}^*) \tag{60}$$

For the TE_{10} mode, (56) yields

$$\begin{aligned}
\hat{\mathbf{E}} \times \hat{\mathbf{H}}^* &= (\hat{E}_y \mathbf{a}_y) \times (\hat{H}_x^* \mathbf{a}_x + \hat{H}_z^* \mathbf{a}_z) \\
&= -\hat{E}_y \hat{H}_x^* \mathbf{a}_z + \hat{E}_y \hat{H}_z^* \mathbf{a}_x \\
&= \frac{\beta_{10}\omega\mu a^2}{\pi^2} |\hat{C}|^2 \sin^2 \frac{\pi x}{a}\,\mathbf{a}_z - \frac{j\omega\mu a}{\pi} |\hat{C}|^2 \sin\frac{\pi x}{a} \cos\frac{\pi x}{a}\,\mathbf{a}_x
\end{aligned} \tag{61}$$

Note that the z component of $\hat{\mathbf{E}} \times \hat{\mathbf{H}}^*$ is purely real, whereas the x component is purely imaginary. Thus

$$\begin{aligned}
\mathbf{S}_{\text{av}} &= -\tfrac{1}{2}\hat{E}_y \hat{H}_x^* \mathbf{a}_z \\
&= \frac{1}{2}\frac{\beta_{10}\omega\mu a^2}{\pi^2} |\hat{C}|^2 \sin^2 \frac{\pi x}{a}\,\mathbf{a}_z \\
&= \frac{1}{2}\frac{|\hat{E}_y|^2}{\eta_{\text{TE},10}}\,\mathbf{a}_z
\end{aligned} \tag{62}$$

and the average-power Poynting vector is, as expected, in the $+z$ direction.

It is a relatively simple matter to show that for any TE or TM mode the average-power density vector in the z direction is given by

$$\begin{aligned}
\mathbf{S}_{\text{av}_z} &= \tfrac{1}{2}\,\text{Re}\,(\hat{E}_x \hat{H}_y^* - \hat{E}_y \hat{H}_x^*)\mathbf{a}_z \\
&= \frac{1}{2}\frac{|\hat{E}_x|^2 + |\hat{E}_y|^2}{\eta_{mn}}\,\mathbf{a}_z
\end{aligned} \tag{63}$$

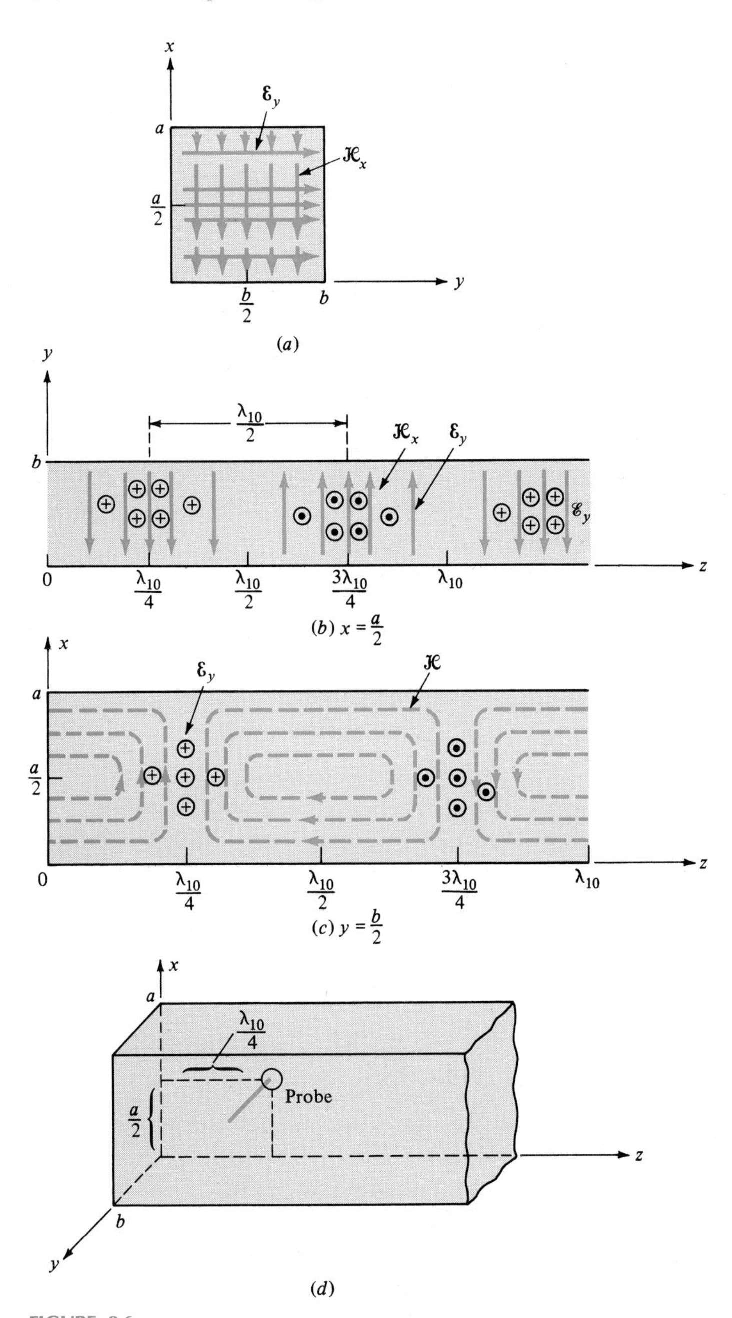

FIGURE 8.6

Field structures of the TE_{10} mode.

where $\eta_{mn} = \eta_{\text{TE},mn}$ for TE modes and $\eta_{mn} = \eta_{\text{TM},mn}$ for TM modes. The total average power transmitted in the $+z$ direction across a cross section of the guide is

$$P_{\text{av}}(z) = \int_{y=0}^{b} \int_{x=0}^{a} \mathbf{S}_{\text{av}} \cdot \mathbf{a}_z \, dx \, dy$$

$$= \frac{1}{2} \int_{y=0}^{b} \int_{x=0}^{a} \frac{|\hat{E}_x|^2 + |\hat{E}_y|^2}{\eta_{mn}} \, dx \, dy \tag{64}$$

EXAMPLE 8.2 Consider the rectangular air-filled waveguide of Example 8.1 with side dimensions $a = 0.9$ in (2.29 cm) and $b = 0.4$ in (1.02 cm). Determine the cutoff frequency of the lowest-order mode. Also, obtain the phase constant, phase velocity, wavelength, and intrinsic impedance of this mode at 7 GHz. If the amplitude of the electric field is 1000 V/m, determine the average power transmitted down the guide in this mode at 7 GHz.

Solution Since $a > b$, the lowest-order mode is the TE_{10} mode with cutoff frequency

$$f_{c,10} = \frac{u_0}{2a}$$

$$= 6.56 \text{ GHz}$$

The common factor

$$F_{mn} = \sqrt{1 - \left(\frac{f_{c,mn}}{f}\right)^2}$$

at 7 GHz becomes $F_{10} = 0.3483$; thus

$$\beta_{10} = \beta_0 F_{10}$$

$$= 51.06 \text{ rad/m}$$

$$u_{10} = \frac{u_0}{F_{10}}$$

$$= 8.613 \times 10^8 \text{ m/s}$$

$$\lambda_{10} = \frac{\lambda_0}{F_{10}}$$

$$= 12.3 \text{ cm}$$

$$\eta_{\text{TE},10} = \frac{\eta_0}{F_{10}}$$

$$= 1082 \ \Omega$$

The average-power density vector in the z direction is given by (62) and $\hat{E}_y$ is given by (56e). Since $\hat{E}_x = \hat{E}_z = 0$ for this mode, the total electric field is $\hat{E}_y$. Given that the amplitude of the electric field is 1000 V/m, from (56e) we have

$$\hat{E}_y = -j1000 \sin \frac{\pi x}{a} e^{-j\beta_{10} z}$$

Thus, the average-power density vector for this mode at 7 GHz is given by

$$\begin{aligned}\mathbf{S}_{\text{av}} &= \frac{1}{2} \frac{|\hat{E}_y|^2}{\eta_{\text{TE},10}} \mathbf{a}_z \\ &= 461.95 \sin^2 \frac{\pi x}{a} \mathbf{a}_z\end{aligned}$$

The total average power transmitted across a cross section of the guide in the z direction is

$$\begin{aligned}P_{\text{av}}(z) &= \int_{x=0}^{a} \int_{y=0}^{b} \mathbf{S}_{\text{av}} \cdot \mathbf{a}_z \, dx \, dy \\ &= 461.95b \int_{x=0}^{a} \sin^2 \frac{\pi x}{a} \, dx \\ &= 230.97ab \\ &= 53.65 \text{ mW}\end{aligned}$$

The fields within the waveguide may be thought of as resulting from currents and charges on the (perfectly conducting) walls of the guide. In order to determine these fields, we utilize the boundary conditions of Chap. 5 at these walls. The discontinuity in the magnetic field intensity vector tangent to the perfectly conducting walls is related to the linear current density on the interior surface of the walls:

$$\begin{aligned}\mathcal{K} &= \mathbf{a}_n \times \mathcal{H} \qquad \text{A/m} \\ &= \mathcal{H}_{\tan}\end{aligned} \tag{65}$$

where $\mathbf{a}_n$ is a unit normal to the wall and pointing into the guide. The quantity $\mathcal{H}_{\tan}$ is the component of $\mathcal{H}$ tangent to the wall. For the TE_{10} mode, for example, we obtain from (57):

Left wall ($y = 0$):

$$\begin{aligned}\mathcal{K} &= -\mathcal{H}_x|_{y=0}\mathbf{a}_z + \mathcal{H}_z|_{y=0}\mathbf{a}_x \\ &= \frac{\beta_{10} a}{\pi} C \sin \frac{\pi x}{a} \sin\left(\omega t - \frac{2\pi z}{\lambda_{10}}\right)\mathbf{a}_z \\ &\quad + C \cos \frac{\pi x}{a} \cos\left(\omega t - \frac{2\pi z}{\lambda_{10}}\right)\mathbf{a}_x \qquad \text{A/m}\end{aligned} \tag{66}$$

Right wall ($y = b$):

$$\mathcal{K} = \mathcal{H}_x|_{y=b}\mathbf{a}_z - \mathcal{H}_z|_{y=b}\mathbf{a}_x$$
$$= -\frac{\beta_{10}a}{\pi} C \sin\frac{\pi x}{a} \sin\left(\omega t - \frac{2\pi z}{\lambda_{10}}\right)\mathbf{a}_z$$
$$- C\cos\frac{\pi x}{a}\cos\left(\omega t - \frac{2\pi z}{\lambda_{10}}\right)\mathbf{a}_x \tag{67}$$

Top wall ($x = a$):

$$\mathcal{K} = \mathcal{H}_z|_{x=a}\mathbf{a}_y$$
$$= -C\cos\left(\omega t - \frac{2\pi z}{\lambda_{10}}\right)\mathbf{a}_y \tag{68}$$

Bottom wall ($x = 0$):

$$\mathcal{K} = -\mathcal{H}_z|_{x=0}\mathbf{a}_y$$
$$= -C\cos\left(\omega t - \frac{2\pi z}{\lambda_{10}}\right)\mathbf{a}_y \tag{69}$$

These are plotted in Fig. 8.7.

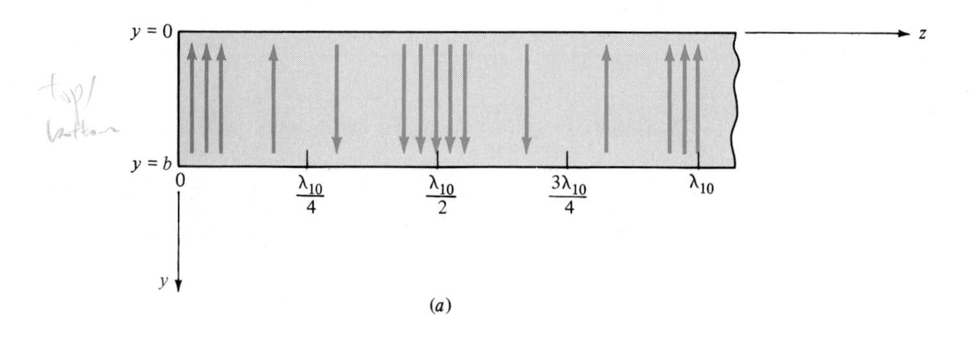

(a)

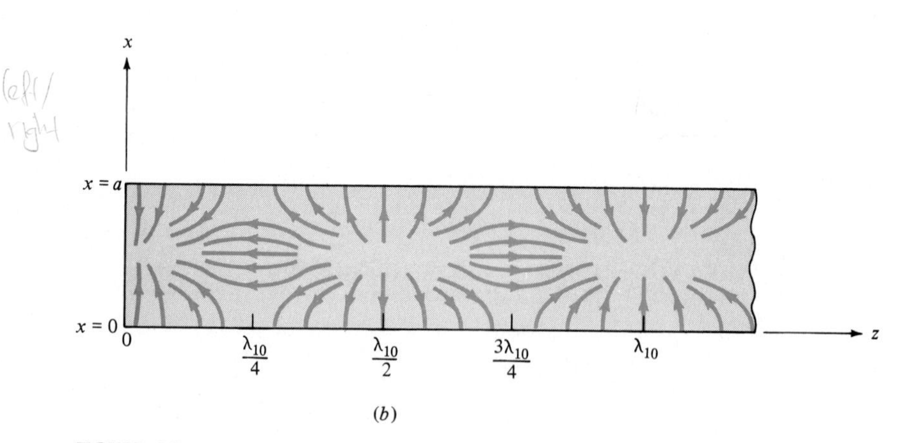

(b)

FIGURE 8.7

Surface currents on the walls of the guide for the TE_{10} mode.

It is frequently necessary to insert probes in a waveguide and to move them along the guide for the purpose of making, say, VSWR measurements (as we did with transmission lines). In order to do this, we must cut a longitudinal slot in one wall of the guide. From the plot in Fig. 8.7*b* we see that at the center of the left (or right) wall the surface currents are longitudinally directed; thus, cutting a slot here would provide minimal perturbation of these surface currents since it would not cut across the surface currents.

The surface charge density is related to the normal component of the electric flux density at the wall via the boundary condition:

$$\begin{aligned} \rho_s &= \mathbf{a}_n \cdot \epsilon \mathcal{E} \\ &= \epsilon \mathcal{E}_{\text{nor}} \end{aligned} \tag{70}$$

where $\mathcal{E}_{\text{nor}}$ is the component of $\mathcal{E}$ normal to the wall. For the TE_{10} mode we have, using $\mathcal{E}_y$ in (57*c*):

Left wall ($y = 0$):

$$\begin{aligned} \rho_s &= \epsilon \mathcal{E}_y|_{y=0} \\ &= \epsilon \frac{\omega \mu a}{\pi} C \sin \frac{\pi x}{a} \sin\left(\omega t - \frac{2\pi z}{\lambda_{10}}\right) \end{aligned} \tag{71}$$

Right wall ($y = b$):

$$\begin{aligned} \rho_s &= -\epsilon \mathcal{E}_y|_{y=b} \\ &= -\epsilon \frac{\omega \mu a}{\pi} C \sin \frac{\pi x}{a} \sin\left(\omega t - \frac{2\pi z}{\lambda_{10}}\right) \end{aligned} \tag{72}$$

Top wall ($x = a$): $\qquad \rho_s = 0 \qquad (73)$

Bottom wall ($x = 0$): $\qquad \rho_s = 0 \qquad (74)$

8.2.3 Waveguide Losses

In the previous sections, in order to simplify the analysis, we assumed that the waveguide walls were perfect conductors; thus, although surface currents existed on these walls, they incurred no losses. In this section we will remove the restriction of perfectly conducting walls and examine the implications. A lossy dielectric that is interior to the guide can easily be considered by replacing the dielectric permittivity ϵ in our previous results with a complex value:

$$\hat{\epsilon} = \epsilon(1 - j \tan \delta) \tag{75}$$

where $\tan \delta$ is the loss tangent of the dielectric.

To account for losses due to imperfectly conducting walls is not as simple. In our previous analyses of wave propagation in a guide having no losses, we noted that the field vectors for each mode vary with distance along the guide as $e^{-\gamma_{mn} z}$. Below the cutoff frequency of the mode the propagation constant is real, $\gamma_{mn} = \alpha_{mn}$, indicating attenuation, whereas above the cutoff frequency the propagation constant is imaginary, $\gamma_{mn} = j\beta_{mn}$, indicating propagation with no attenuation. The effect of losses, whether in the guide walls or the dielectric, is to cause the propagation constant to have a real part at frequencies above cutoff, $\gamma_{mn} = \alpha_{mn} + j\beta_{mn}$. The propagating fields will then suffer attenuation since they vary with distance along the guide as $e^{-\alpha_{mn}z}e^{-j\beta_{mn}z}$. Consequently, the time-average power along the guide will vary as

$$P_{av} = P_0 e^{-2\alpha_{mn}z} \tag{76}$$

where P_0 is the total average power crossing the reference point. Therefore, if we can determine the real part of the propagation constant, α_{mn}, we can determine the power loss in the direction of transmission. Our goal will be to determine this attenuation constant. The method we will use to determine this attenuation constant is to assume that the fields are essentially the same as for the lossless case and to compute the wall losses incurred by these fields. This is a common method of estimating losses and assumes that the losses are small so that the fields are essentially unchanged from the lossless case.

The power loss per unit of distance along the guide is given by

$$\begin{aligned} P_{loss} &= -\frac{\partial P_{av}}{\partial z} \\ &= 2\alpha_{mn} P_{av} \end{aligned} \tag{77}$$

so that

$$\alpha_{mn} = \frac{1}{2}\frac{P_{loss}}{P_{av}} \tag{78}$$

or

$$\alpha_{mn} = \frac{\text{average power lost per unit distance}}{\text{twice the average power transmitted}} \tag{79}$$

The total time-average power transmitted across a cross section of a lossless guide was computed previously. The average-power density along the guide axis can be written as

$$\begin{aligned} \mathbf{S}_{av} \cdot \mathbf{a}_z &= \tfrac{1}{2}\,\text{Re}(\hat{E}_x \hat{H}_y^* - \hat{E}_y \hat{H}_x^*) \\ &= \frac{\eta_{\text{TE, TM}}}{2}(|\hat{H}_y|^2 + |\hat{H}_x|^2) \end{aligned} \tag{80}$$

where $\eta_{\text{TE, TM}}$ is the wave impedance of the mode. Therefore, the total average power transmitted along the guide axis is

$$P_{av} = \frac{\eta_{\text{TE, TM}}}{2} \int_{y=0}^{b} \int_{x=0}^{a} (|\hat{H}_y|^2 + |\hat{H}_x|^2)\, dx\, dy \tag{81}$$

For the important TE_{10} mode, we can determine

$$
\begin{aligned}
P_{\text{av}} &= \frac{\eta |\hat{C}|^2}{2\sqrt{1-(f_{c10}/f)^2}} \int_{y=0}^{b} \int_{x=0}^{a} |\hat{H}_x|^2 \, dx \, dy \\
&= \frac{\eta \beta^2 a^3 b}{4\pi^2} \sqrt{1-(f_{c10}/f)^2} \, |\hat{C}|^2
\end{aligned}
\tag{82}
$$

where $\eta = \sqrt{\mu/\epsilon}$ and $\beta = \omega\sqrt{\mu\epsilon}$. If the guide is air-filled, $\eta = \eta_0$ and $\beta = \beta_0$.

In order to determine the time-average power lost per unit distance, consider Fig. 8.8. As indicated previously, we will assume the fields of the guide to be essentially unaltered in the presence of losses (small losses). If the walls of the guide are not perfect conductors, the components of the magnetic fields tangent to the walls, $\hat{\mathbf{H}}_{\text{tan}}$, will be continuous across the wall. This magnetic field will induce currents and associated electric fields interior to the wall (which are also tangent to the wall's inner surface). These are related to the electric field in the guide walls by the intrinsic impedance of the conducting walls, $\hat{\eta}_c$, as

$$
\hat{E}_{\text{tan}} = \hat{\eta}_c \hat{H}_{\text{tan}} \tag{83}
$$

and are at right angles to each other. The intrinsic impedance of the metal walls was obtained in Sec. 6.3.2 as

$$
\hat{\eta}_c = \sqrt{\frac{\omega\mu_0}{2\sigma_c}}(1+j1) = \frac{1}{\sigma_c \delta}(1+j1) \tag{84}
$$

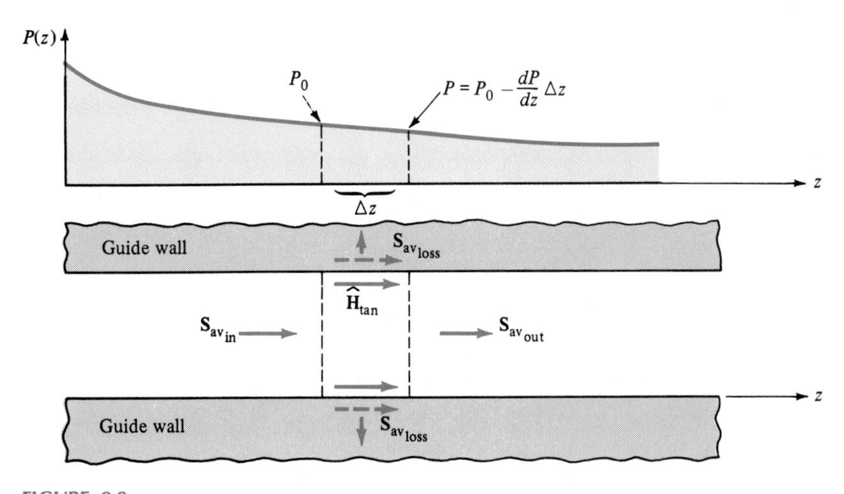

FIGURE 8.8
Illustration of the calculation of guide wall losses.

where $\delta = 1/\sqrt{\pi f \mu_c \sigma_c}$ is the skin depth and $\sigma_c(\mu_c)$ is the conductivity (permeability) of the guide's wall conductor. These fields just interior to the wall result in an average-power density vector directed into the wall:

$$\begin{aligned} S_{\text{av loss}} &= \tfrac{1}{2}|\hat{H}_{\tan}|^2 \operatorname{Re} \hat{\eta}_c \\ &= \frac{1}{2\sigma_c\delta}|\hat{H}_{\tan}|^2 \end{aligned} \tag{85}$$

For a unit distance down the guide, the average-power loss is the integral of (85) over the cross-sectional dimension of the wall. Combining the losses for all four walls yields

$$\begin{aligned} P_{\text{av loss}} = \frac{1}{2\sigma_c\delta}\bigg\{ &\int_{x=0}^{a} (|\hat{H}_x|^2 + |\hat{H}_z|^2)|_{y=0}\, dx \\ &+ \int_{x=0}^{a} (|\hat{H}_x|^2 + |\hat{H}_z|^2)|_{y=b}\, dx \\ &+ \int_{y=0}^{b} (|\hat{H}_y|^2 + |\hat{H}_z|^2)|_{x=0}\, dy \\ &+ \int_{y=0}^{b} (|\hat{H}_y|^2 + |\hat{H}_z|^2)|_{x=a}\, dy \bigg\} \end{aligned} \tag{86}$$

For the TE_{10} mode we obtain

$$\begin{aligned} P_{\text{av loss}} &= \frac{|\hat{C}|^2}{2\sigma_c\delta}\left\{\int_{x=0}^{a}\left(\frac{\beta_{10}^2 a^2}{\pi^2}\sin^2\frac{\pi x}{a} + \cos^2 \pi\frac{x}{a}\right)dx + 2\int_{y=0}^{b} dy\right\} \\ &= \frac{|\hat{C}|^2}{2\sigma_c\delta}\left\{\frac{\beta_{10}^2 a^3}{\pi^2} + a + 2b\right\} \\ &= \frac{|\hat{C}|^2}{\sigma_c\delta}\left[b + \frac{a}{2}\left(\frac{f}{f_{c10}}\right)^2\right] \end{aligned} \tag{87}$$

This latter result is obtained by recognizing that $\beta_{10}^2 = \omega^2\mu\epsilon - (\pi/a)^2$ and $f_{c10} = (1/2\pi\sqrt{\mu\epsilon})\ (\pi/a)$. Substituting (87) and (82) into (79) yields

$$\begin{aligned} \alpha_{10} &= \frac{2\pi^2[b + (a/2)(f/f_{c10})^2]}{\sigma_c\delta\eta\beta^2 a^3 b\sqrt{1 - (f_{c10}/f)^2}} \\ &= \frac{\sqrt{\pi f \mu_c}}{\sqrt{\sigma_c}\eta b}\frac{1 + (2b/a)(f_{c10}/f)^2}{\sqrt{1 - (f_{c10}/f)^2}} \end{aligned} \tag{88}$$

The per-unit-length loss incurred by the imperfect guide walls can be obtained in decibels from (76) as was done for transmission lines:

$$\text{Loss (dB/m)} = 10 \log e^{2\alpha_{mn}}$$
$$= 8.69\alpha_{mn} \tag{89}$$

EXAMPLE 8.3 Sketch the conductor loss versus frequency for the TE_{10} mode in a typical air-filled X-band waveguide considered in Examples 8.1 and 8.2 ($a = 0.9$ in, and $b = 0.4$ in). The conductor walls are brass ($\sigma_c = 1.57 \times 10^7$ S/m, and $\mu_c \simeq \mu_0$).

Solution From the results of the previous examples,

$$f_{c10} = 6.56 \text{ GHz}$$

Thus, (88) yields

$$\alpha_{10} = \frac{4.14 \times 10^{-3}\sqrt{f}[1 + 0.89(6.56/f)^2]}{\sqrt{1 - (6.56/f)^2}}$$

where f is expressed in GHz. The loss in dB/m is sketched in Fig. 8.9. Note that the attenuation shows a minimum at around 15 GHz. This is slightly below the cutoff frequency of the TE_{11} and TM_{11} modes (16.16 GHz) and slightly above the cutoff frequency of the TE_{01} mode (14.76 GHz).

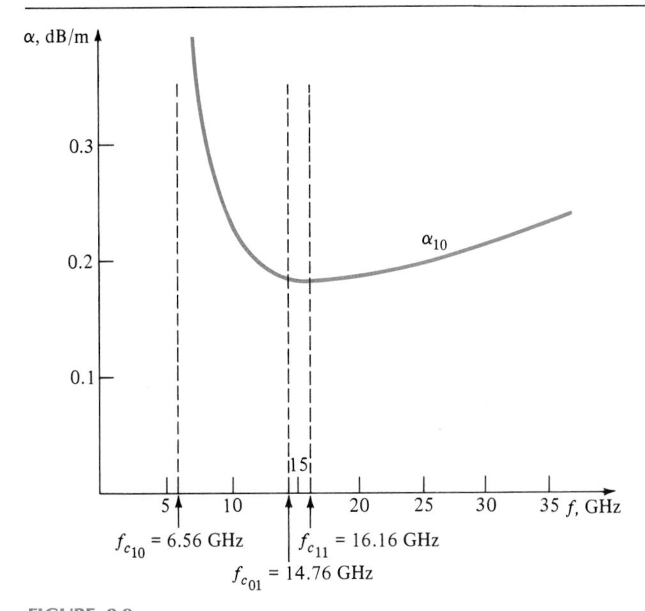

FIGURE 8.9

Example 8.3. Wall loss for the TE_{10} mode.

8.3 Cavity Resonators

As indicated in the introduction to this chapter, it is frequently necessary to construct resonant devices at microwave frequencies. Lumped-circuit elements such as inductors and capacitors are virtually useless due to their nonideal behavior and their large losses at these frequencies. In this section we will examine the construction of devices having microwave resonant frequencies and large Q's (low losses).

Consider the rectangular waveguide. In our previous developments we assumed a z dependence of the form $e^{-\gamma z}$, indicating only forward-traveling waves. Backward-traveling waves having a z dependence $e^{\gamma z}$ would also be possible. The wave equations in (3) would remain unchanged since they relied on the second derivative with respect to z, $\partial^2/\partial z^2$, which yields γ^2 for either case. However, the relations in (8), (9), and (11) would have $-\gamma$ substituted for γ since each appearance of γ results from $\partial/\partial z$.

The possibility of the simultaneous existence of both forward- and backward-traveling waves suggests the possibility of standing waves and resonant behavior. Suppose we insert conducting plates at $z = 0$ and $z = d$, as shown in Fig. 8.10. These conducting plates give rise to reflections of the waveguide fields, resulting in standing waves. In this section we will investigate the resulting fields and the resonant frequencies of this "cavity."

8.3.1 TM Modes

Carrying through the development in Sec. 8.2.1 for the backward-traveling waves gives

$$\hat{E}_z^{\pm} = \hat{C}^{\pm} \sin Mx \sin Ny \, e^{\mp \gamma z} \tag{90}$$

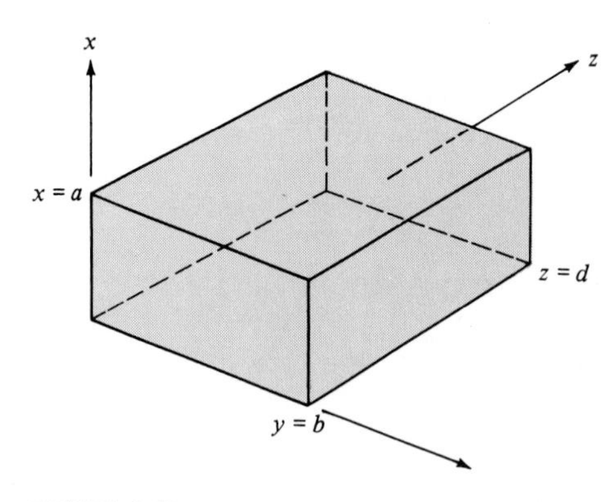

FIGURE 8.10

Parameter definitions for the cavity resonator.

where the upper sign is used for forward-traveling waves and the lower sign is used for backward-traveling waves. The propagation constant remains

$$\gamma^2 = M^2 + N^2 - \omega^2 \mu_0 \epsilon_0 \tag{91}$$

where

$$M = \frac{m\pi}{a} \qquad m = 0, 1, 2, \ldots \tag{92a}$$

$$N = \frac{n\pi}{b} \qquad n = 0, 1, 2, \ldots \tag{92b}$$

The total field is the sum of these forward- and backward-traveling waves:

$$\begin{aligned} \hat{E}_z &= \hat{E}_z^+ + \hat{E}_z^- \\ &= \sin Mx \sin Ny(\hat{C}^+ e^{-\gamma z} + \hat{C}^- e^{\gamma z}) \end{aligned} \tag{93a}$$

Substituting these fields individually into (11) and remembering that γ is to be replaced with $-\gamma$ in these equations for the backward-traveling waves yields, from the results in (36),

$$\hat{E}_x = \frac{M}{M^2 + N^2} \cos Mx \sin Ny(-\gamma\hat{C}^+ e^{-\gamma z} + \gamma\hat{C}^- e^{\gamma z}) \tag{93b}$$

$$\hat{E}_y = \frac{N}{M^2 + N^2} \sin Mx \cos Ny(-\gamma\hat{C}^+ e^{-\gamma z} + \gamma\hat{C}^- e^{\gamma z}) \tag{93c}$$

$$\hat{H}_x = \frac{j\omega\epsilon N}{M^2 + N^2} \sin Mx \cos Ny(\hat{C}^+ e^{-\gamma z} + \hat{C}^- e^{\gamma z}) \tag{93d}$$

$$\hat{H}_y = \frac{-j\omega\epsilon M}{M^2 + N^2} \cos Mx \sin Ny(\hat{C}^+ e^{-\gamma z} + \hat{C}^- e^{\gamma z}) \tag{93e}$$

$$\hat{H}_z = 0 \tag{93f}$$

The boundary conditions imposed by the plates at $z = 0$ and $z = d$ require

$$\left.\begin{aligned} \hat{E}_y &= 0 \\ \hat{E}_x &= 0 \end{aligned}\right\} \quad \text{at } z = 0, z = d \tag{94}$$

From (93*b*) and (93*c*) we see that this requires

$$-\hat{C}^+ e^{-\gamma z} + \hat{C}^- e^{\gamma z} = 0 \tag{95}$$

at $z = 0$ and $z = d$. Thus, $\hat{C}^+ = \hat{C}^- = \hat{C}$ and

$$\gamma = j\frac{p\pi}{d} \triangleq jP \tag{96}$$

where $p = 0, 1, 2, \ldots$ and we define $P = p\pi/d$. Substituting in (93*b*) and (93*c*) gives

$$\hat{E}_x = -\frac{2MP\hat{C}}{M^2 + N^2}\cos Mx \sin Ny \sin Pz \tag{97a}$$

$$\hat{E}_y = -\frac{2NP\hat{C}}{M^2 + N^2}\sin Mx \cos Ny \sin Pz \tag{97b}$$

Using these results in (93*a*), (93*d*), and (93*e*) yields

$$\hat{E}_z = 2\hat{C}\sin Mx \sin Ny \cos Pz \tag{97c}$$

$$\hat{H}_x = \frac{2j\omega\epsilon N\hat{C}}{M^2 + N^2}\sin Mx \cos Ny \cos Pz \tag{97d}$$

$$\hat{H}_y = \frac{-2j\omega\epsilon M\hat{C}}{M^2 + N^2}\cos Mx \sin Ny \cos Pz \tag{97e}$$

$$\hat{H}_z = 0 \tag{97f}$$

The time-domain forms of the fields can easily be obtained by multiplying each phasor by $e^{j\omega t}$ and taking the real part of the result. Since these phasors do not contain isolated, complex exponentials, the time-domain fields will vary with time as $\sin(\omega t)$ or $\cos(\omega t)$. Thus, the fields are oscillatory. (See Prob. 8-27.)

Each set of integers *m*, *n*, and *p* is said to be associated with a TM_{mnp} mode. The resonant frequency of each mode is calculated from (91), substituting (96) for γ:

$$f_{mnp} = \frac{1}{2\pi\sqrt{\mu\epsilon}}\sqrt{\left(\frac{m\pi}{a}\right)^2 + \left(\frac{n\pi}{b}\right)^2 + \left(\frac{p\pi}{d}\right)^2}$$

$$= \frac{u}{2}\sqrt{\left(\frac{m}{a}\right)^2 + \left(\frac{n}{b}\right)^2 + \left(\frac{p}{d}\right)^2} \tag{98}$$

Equations (97) show that neither *m* nor *n* can be zero; otherwise, all field components vanish. However, *p* can be zero. Thus, the lowest order TM mode is the TM_{110} mode.

8.3.2 TE Modes

A similar development can be obtained for the TE modes. The total fields from (51) are

$$\hat{H}_z = \hat{H}_z^+ + \hat{H}_z^-$$
$$= \cos Mx \cos Ny(\hat{C}^+ e^{-\gamma z} + \hat{C}^- e^{\gamma z}) \tag{99a}$$

$$\hat{H}_x = \frac{M}{M^2 + N^2} \sin Mx \cos Ny(\gamma \hat{C}^+ e^{-\gamma z} - \gamma \hat{C}^- e^{\gamma z}) \tag{99b}$$

$$\hat{H}_y = \frac{N}{M^2 + N^2} \cos Mx \sin Ny(\gamma \hat{C}^+ e^{-\gamma z} - \gamma \hat{C}^- e^{\gamma z}) \tag{99c}$$

$$\hat{E}_x = \frac{j\omega\mu N}{M^2 + N^2} \cos Mx \sin Ny(\hat{C}^+ e^{-\gamma z} + \hat{C}^- e^{\gamma z}) \tag{99d}$$

$$\hat{E}_y = \frac{-j\omega\mu M}{M^2 + N^2} \sin Mx \cos Ny(\hat{C}^+ e^{-\gamma z} + \hat{C}^- e^{\gamma z}) \tag{99e}$$

$$\hat{E}_z = 0 \tag{99f}$$

The boundary conditions imposed by the conductors at $z = 0$ and $z = d$ require again that

$$\left.\begin{aligned} \hat{E}_x &= 0 \\ \hat{E}_y &= 0 \end{aligned}\right\} \quad \text{at } z = 0, z = d \tag{100}$$

This requires that $\hat{C}^- = -\hat{C}^+ = \hat{C}$, and again,

$$\gamma = j\frac{p\pi}{d} \triangleq jP \tag{101}$$

This yields

$$\hat{H}_z = j2\hat{C} \cos Mx \cos Ny \sin Pz \tag{102a}$$

$$\hat{H}_x = -j\frac{2MP}{M^2 + N^2} \hat{C} \sin Mx \cos Ny \cos Pz \tag{102b}$$

$$\hat{H}_y = -j\frac{2NP}{M^2 + N^2} \hat{C} \cos Mx \sin Ny \cos Pz \tag{102c}$$

$$\hat{E}_x = -\frac{2\omega\mu N}{M^2 + N^2} \hat{C} \cos Mx \sin Ny \sin Pz \tag{102d}$$

$$\hat{E}_y = \frac{2\omega\mu M}{M^2 + N^2} \hat{C} \sin Mx \cos Ny \sin Pz \tag{102e}$$

$$\hat{E}_z = 0 \tag{102f}$$

The resonant frequency is again given by (98). Equations (102) show that either m or n (but not both) can be zero and that p cannot be zero. Hence, the two lowest order modes are the TE_{011} and TE_{101} modes.

The fields of the TE_{101} mode have the lowest resonant frequency if we assume that $a \geq d > b$. For the TE_{101} mode, the fields become

$$\hat{H}_z = j2\hat{C} \cos \frac{\pi x}{a} \sin \frac{\pi z}{d} \tag{103a}$$

$$\hat{H}_x = -j2 \frac{a}{d} \hat{C} \sin \frac{\pi x}{a} \cos \frac{\pi z}{d} \tag{103b}$$

$$\hat{E}_y = \frac{2\omega\mu a}{\pi} \hat{C} \sin \frac{\pi x}{a} \sin \frac{\pi z}{d} \tag{103c}$$

The electric field is y-directed and has a maximum at $x = a/2$ and $z = d/2$. Thus, in order to excite this mode efficiently, an electric field probe (say a coaxial cable with the center conductor exposed) could be inserted at the center of the side of largest area.

8.3.3 Quality Factor of the Cavity Resonator

The quality factor (or Q) of this resonant structure is defined in a manner similar to that for resonant lumped circuits:

$$\begin{aligned} Q &= 2\pi f \frac{\text{time-average energy stored}}{\text{average-power loss}} \\ &= \omega \frac{W}{P_L} \end{aligned} \tag{104}$$

at this resonant frequency. As in the case of lumped resonant circuits, this gives a measure of the bandwidth of the resonator and indicates the degree of loss in the structure, which tends to broaden the bandwidth. An alternative definition of Q is

$$Q = \frac{f_0}{B} \tag{105}$$

where f_0 is the resonant frequency and B is the bandwidth (or difference in the frequencies) where the oscillation magnitude is reduced from its values at f_0 by 6 dB. This latter result indicates the "sharpness" of this resonance; a larger Q implies a smaller bandwidth.

The time-average stored electric energy is

$$W_e = \frac{\epsilon}{4} \int_v (|\hat{E}_x|^2 + |\hat{E}_y|^2 + |\hat{E}_z|^2)\, dv \tag{106}$$

whereas the time-average stored magnetic energy is

$$W_m = \frac{\mu}{4} \int_v (|\hat{H}_x|^2 + |\hat{H}_y|^2 + |\hat{H}_z|^2)\, dv \tag{107}$$

where the integrals are performed throughout the interior volume of the resonator. These results follow from the observation that the instantaneous stored energy densities are $\frac{1}{2}\epsilon \mathcal{E}^2$ and $\frac{1}{2}\mu \mathcal{H}^2$, respectively, and that the fields vary sinusoidally whereby the time averages give an additional factor of 1/2. The average-power loss is incurred in the walls of the resonator and in the interior dielectric. For a lossless dielectric, the average-power loss is the sum of the average powers dissipated in the six walls. As in the calculation of waveguide wall losses, we will assume small losses, thus the magnetic field tangent to the walls, $\hat{H}_{\tan}$, is the same as in the case of perfectly conducting walls. The average-power density vector interior to and directed into the walls is

$$S_{\text{av loss}} = \tfrac{1}{2} |\hat{H}_{\tan}|^2 \operatorname{Re} \eta_c \tag{108}$$

The power loss is obtained by integrating (108) over the surfaces of the six walls:

$$P_L = \int_s S_{\text{av loss}}\, ds \tag{109}$$

In the case of the TE_{101} mode we obtain

$$\begin{aligned} w_e &= \frac{\epsilon}{4} \int_{x=0}^{a} \int_{y=0}^{b} \int_{z=0}^{d} |\hat{E}_y|^2\, dz\, dy\, dx \\ &= \mu^2 \epsilon a^3 b d f^2 |\hat{C}|^2 \end{aligned} \tag{110}$$

Substituting

$$\begin{aligned} f &= f_{101} \\ &= \frac{1}{2\sqrt{\mu\epsilon}} \sqrt{\frac{1}{a^2} + \frac{1}{d^2}} \end{aligned} \tag{111}$$

into (110) yields

$$w_e = \frac{\mu a b d}{4} \left(1 + \frac{a^2}{d^2}\right) |\hat{C}|^2 \tag{112}$$

Similarly, the time-average stored magnetic energy is

$$\begin{aligned} w_m &= \frac{\mu}{4}\int_{x=0}^{a}\int_{y=0}^{b}\int_{z=0}^{d}(|\hat{H}_x|^2 + |\hat{H}_z|^2)\,dz\,dy\,dx \\ &= \frac{\mu}{4}\left\{\frac{4a^2}{d^2}|\hat{C}|^2\frac{a}{2}b\frac{d}{2} + 4|\hat{C}|^2\frac{a}{2}b\frac{d}{2}\right\} \\ &= \frac{\mu abd}{4}\left(1 + \frac{a^2}{d^2}\right)|\hat{C}|^2 \end{aligned} \tag{113}$$

Thus we see that $w_e = w_m$ as one might expect so that

$$\begin{aligned} w &= w_e + w_m \\ &= 2w_e \\ &= 2w_m \\ &= \frac{\mu abd}{2}\left(1 + \frac{a^2}{d^2}\right)|\hat{C}|^2 \end{aligned} \tag{114}$$

The time-average power loss in the walls is obtained in a fashion similar to the result of Section 8.2.3 as

$$\begin{aligned} P_L &= \frac{1}{2\sigma_c\delta}\left\{2\int_{x=0}^{a}\int_{y=0}^{b}(|\hat{H}_x|^2)|_{z=0,d}\,dx\,dy \right. \\ &\quad + 2\int_{x=0}^{a}\int_{z=0}^{d}(|\hat{H}_x|^2 + |\hat{H}_z|^2)|_{y=0,b}\,dx\,dz \\ &\quad \left. + 2\int_{y=0}^{b}\int_{z=0}^{d}(|\hat{H}_z|^2)|_{x=0,a}\,dy\,dz\right\} \\ &= \frac{|\hat{C}|^2}{\sigma_c\delta d^2}(2a^3b + a^3d + ad^3 + 2bd^3) \end{aligned} \tag{115}$$

Substituting (114) and (115) into (104) gives

$$Q = \sigma_c\delta\,\frac{\pi f_{101}\mu abd(a^2 + d^2)}{2a^3b + a^3d + ad^3 + 2bd^3} \tag{116}$$

EXAMPLE 8.4 Determine the resonant frequency and Q of the TE_{101} mode in an air-filled brass cavity having dimensions $a = 4$ cm, $d = 4$ cm, and $b = 2$ cm.

Solution The resonant frequency is obtained from (98) as

$$\begin{aligned} f_{101} &= 1.5 \times 10^8\sqrt{\frac{1}{(0.04)^2} + \frac{1}{(0.04)^2}} \\ &= 5.3 \text{ GHz} \end{aligned}$$

The Q is calculated from (116) using $\sigma_c = 1.57 \times 10^7$ S/m. The skin depth of brass at the resonant frequency is

$$\delta = \frac{1}{\sqrt{\pi f_{101} \mu \sigma_c}}$$

$$= 1.744 \times 10^{-6}$$

and thus a thin wall could be used. The Q is

$$Q = 5733$$

which is considerably larger than one could obtain with lumped circuits at much lower frequencies.

8.4 Summary

In this chapter we have considered hollow conductors that are used to guide electromagnetic waves in the gigahertz-frequency range: waveguides. We found, as in previous chapters, that in order for wave propagation to take place along the (z) axis of the conductor, the field vectors should exhibit a z dependence of $e^{\pm \gamma z}$. The primary difference between waveguides and the previously considered transmission lines and uniform plane waves lies in the fact that the propagation constant γ is not simply related to the parameters of the medium but depends also on the cross-sectional dimensions of the guide. Each mode of propagation or field structure has a different propagation constant, and each mode has an associated cutoff frequency (which also depends on the guide's cross-sectional dimensions), $f_{c,mn}$. For frequencies of excitation below the cutoff frequency, the mode is simply attenuated. Above this frequency, the mode propagates. This cutoff frequency is determined by examining the propagation constant γ for the mode and finding the frequency at which it changes from purely real to purely imaginary.

The velocity (phase) of propagation, guide wavelength, and intrinsic impedance for each mode are different than for uniform plane waves but are related to them by the factor

$$F_{mn} = \sqrt{1 - \left(\frac{f_{c,mn}}{f}\right)^2}$$

Waveguide losses are examined by modifying the results for the lossless case. The primary modification is that, above the cutoff frequency of each mode, the propagation constant has a real part, whereas it is purely imaginary for the lossless case. The fields vary with z as $e^{-\alpha z} e^{-j\beta z}$. The power flow varies as $e^{-2\alpha z}$.

The portion of the attenuation constant, α, due to dielectric losses can be obtained by substituting the complex permittivity

$$\hat{\epsilon} = \epsilon\left(1 - j\frac{\sigma}{\omega\epsilon}\right)$$

into the lossless result. The portion of the attenuation constant due to losses in the imperfect walls is somewhat more difficult to obtain. We assumed that the imperfect walls modified the fields of the lossless case only slightly. The wall losses were then calculated using the fields obtained for the case of perfectly conducting walls.

Cavity resonators were studied as a means of providing resonant structures at microwave frequencies to be used, for example, to construct microwave oscillators. Placing conducting plates at two ends of a rectangular waveguide results in standing waves in the z direction. This notion allowed calculation of the fields and the resonant frequencies of the resonator. It was shown that cavity resonators having very large Q's at microwave frequencies could be constructed. The use of lumped-circuit elements to construct these devices would not be feasible.

In the next chapter we will examine structures that are used to focus and launch electromagnetic waves. These structures are referred to as antennas.

Problems

8-1 Derive Eqs. (11).

8-2 A typical rectangular waveguide has dimensions of 0.9 in and 0.4 in. If the guide is air-filled and the frequency of operation is 25 GHz, list all possible propagating modes.

8-3 List, in ascending order, the cutoff frequencies of the TE_{01}, TE_{10}, TE_{11}, TM_{11}, TE_{20}, TE_{21}, TE_{12}, TM_{21}, TM_{12}, TE_{22}, and TM_{22} modes in terms of the dominant mode cutoff frequency for rectangular waveguides whose dimensions are

- **(a)** $a = b$,
- **(b)** $a = 2b$,
- **(c)** $a = 2.25b$

8-4 Determine an expression for the group velocity u_{gmn} of narrowband signals propagating within a rectangular waveguide. Show that

$$u_{mn}u_{gmn} = u^2$$

Determine the group velocity of a 12-GHz AM signal propagating in the TM_{11} mode in a square guide whose dimensions are 2 cm × 2 cm.

8-5 A rectangular waveguide is air-filled and operated at 30 GHz. If the cutoff frequency of the TM_{21} mode is 18 GHz, determine the wavelength, phase constant, phase velocity, and intrinsic impedance of this mode. Determine the distance required such that the fields of this mode will be reduced in magnitude by 20 dB (a factor of 10) if the operating frequency is 15 GHz.

8-6 Repeat Prob. 8-5 if the guide is filled with polyethylene ($\epsilon_r = 2.25$).

8-7 Using the expression of $\hat{E}_z$ for the TM modes given in Eq. (23) and the expressions in (11), derive the remaining field-vector expressions in (36*b*) through (36*e*).

8-8 Using the phasor expressions for the TM-mode field vectors given in (36), derive the time-domain expressions for these field vectors given in (45).

8-9 Compare the cutoff frequencies for the TM_{12} mode in air-filled waveguides whose dimensions are:

(a) $a = 0.9$ in; $b = 0.4$ in
(b) $a = 0.4$ in; $b = 0.9$ in
(c) $a = b = 1$ cm
(d) $a = b = 10$ cm

8-10 Sketch the fields of the TM_{21} mode in a cross-sectional view of the guide.

8-11 Derive Eqs. (51) for the phasor fields of the TE modes.

8-12 Determine the time-domain fields for the TE mode from the phasor expressions in (51).

8-13 A rectangular waveguide is to be designed to propagate the dominant TE_{10} mode and to exclude all other modes. The frequency of operation is to be 15 GHz and the guide is air-filled. Determine a set of guide dimensions such that 15 GHz is 25 percent above the TE_{10} cutoff frequency and 25 percent below the cutoff frequency of the next-higher-order mode. Calculate u_{10}, λ_{10}, β_{10}, and η_{TE10} for this guide.

8-14 Determine the dimensions of a square air-filled waveguide having a cutoff frequency for the TE_{10} (TE_{01}) mode of 30 GHz, 3 GHz, 300 MHz, and 30 MHz.

8-15 Determine the cutoff frequencies of the dominant TE_{10} mode for the following air-filled waveguides:

(a) 6.25 in × 3.25 in (L band)
(b) 2.84 in × 1.34 in (S band)
(c) 1.872 in × 0.872 in (C band)
(d) 0.9 in × 0.4 in (X band)
(e) 0.42 in × 0.21 in (K band)
(f) 0.148 in × 0.074 in (V band)

8-16 Determine the average power transmitted down a rectangular air-filled waveguide in the TE_{22} mode at 30 GHz. The components of the electric field are equal in magnitude to 100 V/m.

8-17 Determine an expression for the average power transmitted down a guide for the TM_{11} mode.

8-18 A important use of waveguide theory is in the design of shielded rooms. These rooms are used to isolate sensitive equipment and experiments from the effects of external electromagnetic fields as well as to prevent the radiation from the equipment in the room from reaching the external environment. However, the room cannot be a continuous enclosure since air vents must penetrate the walls so that personnel may operate the test equipment in the room. In order to restrict electromagnetic penetration through these vents while allowing airflow, a "honeycomb" structure is inserted into the vent. This consists of a large number of small rectangular tubes welded together. Each tube may be viewed as a waveguide. The intent is for these tubes to operate as "waveguides below cutoff" in restricting penetration of external (as well as internal) fields. Determine an expression for the attenuation in dB of each tube in terms of the tube length and width. (Assume square tubes.) For widths of 0.5 cm, determine the tube length to give a minimum attenuation of 80 dB.

8-19 Sketch the surface current and charge density on the top wall ($x = a$) of a waveguide for the TM_{11} mode.

8-20 Determine an expression for the ratio f/f_{c10} at which the wall loss of the TE_{10} mode is a minimum. Evaluate this result for dimensions $a = 0.9$ in and $b = 0.4$ in. Repeat for dimensions $a = 0.9$ in and $b = 0.04$ in.

8-21 Determine an expression for the attenuation constant due to wall losses for the TM_{11} mode.

8-22 Evaluate the wall loss for the TE_{10} mode in a copper X-band rectangular waveguide ($a = 0.9$ in, and $b = 0.4$ in) at 7 GHz, 10 GHz, and 12 GHz.

8-23 For an air-filled lossless cavity resonator with dimensions $a = 6$ cm, $b = 5$ cm, and $d = 4$ cm, determine and list, in order of ascending resonant frequencies, the 10 lowest-order modes.

8-24 Shielded rooms, discussed in Prob. 8-18, can be viewed as resonant cavities. Consequently, operation of equipment in that room at a resonant frequency of the cavity should be avoided. A typical shielded room has dimensions of 408 in $\times$ 348 in $\times$ 142 in. Determine all resonant frequencies below 100 MHz.

8-25 Derive Eqs. (97) for the TM modes and Eqs. (102) for the TE modes for rectangular cavities.

8-26 Verify that the TE and TM fields for cavity resonators in (97) and (102) satisfy all boundary conditions.

8-27 Obtain time-domain expressions for the TE and TM fields in a cavity resonator from the phasor expressions in (97) and (102).

8-28 Sketch the fields in a cavity resonator for the TE_{101} mode.

8-29 For a rectangular cavity resonator constructed of perfecty conducting walls, sketch the surface current and charge density on the top wall (at $x = a$) for the TE_{111} model.

8-30 Design a cubic ($a = b = d$) cavity resonator to have a dominant resonant frequency of 7 GHz. Determine the Q of this cavity if the walls are constructed of brass. Repeat this design if the interior is filled with polyethylene ($\epsilon_r = 2.25$).

CHAPTER 9

Antennas

In addition to their primary function of guiding waves from a source to a load, the open-wire transmission lines in Figs. 7.1*a* and *b* and the microstrip line in Fig. 7.2*b* also radiate or transmit energy away from the line; however, they are generally not efficient radiators. Metallic as well as dielectric structures can be designed to launch or radiate waves efficiently into space and to focus or concentrate these waves in a particular direction; these structures are referred to as *antennas*. It can be said that virtually every structure has the inherent capability to radiate electromagnetic energy. The essential question is whether it does so in an efficient and useful manner.

Several examples of common antennas are shown in Fig. 9.1. The blade antenna shown in Fig. 9.1*a* is commonly employed on aircraft to communicate with ground stations, such as the airport tower or air traffic control. The antenna shown in Fig. 9.1*b* is one of a class of log-periodic antennas. These antennas have relatively wide bandwidths; that is, the input impedance and radiation pattern are relatively constant over a fairly wide frequency range. The parabolic antenna in Fig. 9.1*c* and the horn antenna in Fig. 9.1*d* are examples of what may be classified as surface-type antennas, as opposed to the wire-type antennas in Figs. 9.1*a* and *b*. These surface-type antennas provide the ability to focus a transmitted wave strongly, giving these antennas the characteristics of high directivity.

The general principles that we will study in this chapter apply to all antennas. However, upon completing the study of this chapter, the reader should not expect to be able to analyze antenna structures with the same ease as we are now able to analyze most transmission lines and (rectangular) waveguides. Antenna structures are inherently more complicated from an electromagnetic fields standpoint than are transmission lines and waveguides. Our study of antennas in this chapter will only scratch the surface of this topic.

FIGURE 9.1

(a) Monopole (blade) antennas used on aircraft: *(Left)* L band, 950 to 1300 MHz for communication with air traffic control (ATC) and use with distance measuring equipment (DME). *(Center)* UHF, 225 to 400 MHz, communication. *(Right)* VHF, 118 to 136 MHz, communication. *(Douglas Aircraft Company. McDonnell Douglas Corporation, Long Beach, Cal.)*

(b) Log-periodic antenna. (*Scientific Atlanta, Atlanta, Ga.)*

(c) Parabolic antennas. *(Scientific Atlanta, Atlanta, Ga.)*

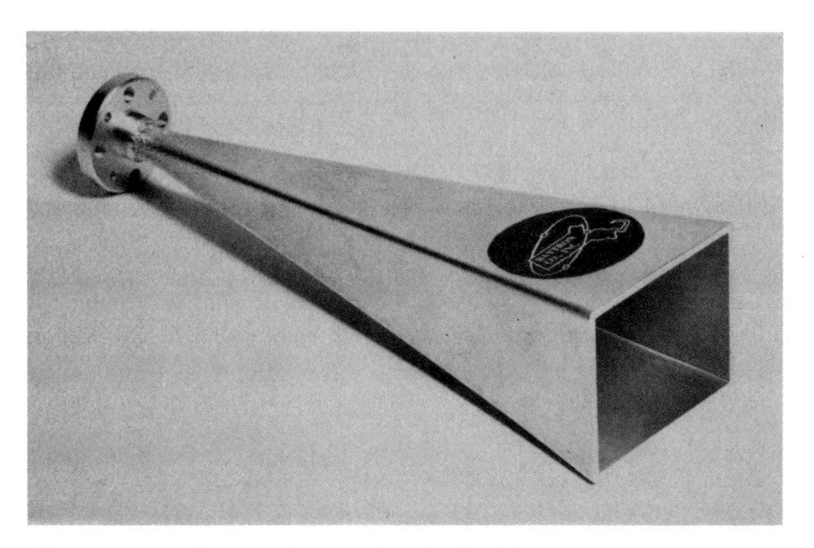

(d) Horn antenna. *(Baytron Company Inc., Medford, Mass.)*

9.1 The Potential Functions

We may determine the forms of the waves radiated from an antenna if we know the current distribution over the surface of the antenna. To show this, we first perform some preliminary vector manipulations. We assume free space and write Maxwell's curl equations for sinusoidal field variation as

$$\nabla \times \hat{\mathbf{E}} = -j\omega\mu_0\hat{\mathbf{H}} \tag{1a}$$

$$\nabla \times \hat{\mathbf{H}} = j\omega\epsilon_0\hat{\mathbf{E}} + \hat{\mathbf{J}}_s \tag{1b}$$

The term $\hat{\mathbf{J}}_s$ in (1*b*) is a phasor current density in the region, and we will treat it as being the source (known) of the fields. We rely on two basic vector identities (Appendix A):

$$\nabla \cdot \nabla \times \mathbf{A} = 0 \tag{2}$$

for any vector field $\mathbf{A}$ and

$$\nabla \times \nabla V = 0 \tag{3}$$

for any scalar field V. Note that since $\nabla \cdot \hat{\mathbf{B}} = 0$, we may write from (2)

$$\hat{\mathbf{B}} = \nabla \times \hat{\mathbf{A}} \tag{4}$$

where $\hat{\mathbf{A}}$ is the magnetic vector potential (considered in Chap. 4). Thus, since $\nabla \cdot \hat{\mathbf{B}} = 0$, we are free to define $\hat{\mathbf{B}}$ as the curl of some auxiliary potential function: the magnetic vector potential $\hat{\mathbf{A}}$. From (1) we write

$$\begin{aligned} \nabla \times \hat{\mathbf{E}} &= -j\omega\hat{\mathbf{B}} \\ &= -j\omega(\nabla \times \hat{\mathbf{A}}) \end{aligned} \tag{5}$$

which becomes

$$\nabla \times (\hat{\mathbf{E}} + j\omega\hat{\mathbf{A}}) = 0 \tag{6}$$

Using the identity in (3), we see that (6) may be written as

$$\hat{\mathbf{E}} + j\omega\hat{\mathbf{A}} = -\nabla\hat{V} \tag{7}$$

where $\hat{V}$ is some scalar potential function. (The negative sign is arbitrarily chosen for later simplification of the results.) Therefore, if we find $\hat{\mathbf{A}}$ and $\hat{V}$, we can find $\hat{\mathbf{B}}$ and $\hat{\mathbf{E}}$ from

$$\hat{\mathbf{B}} = \nabla \times \hat{\mathbf{A}} \tag{8a}$$

$$\hat{\mathbf{E}} = -j\omega\hat{\mathbf{A}} - \nabla\hat{V} \tag{8b}$$

Generally, the simplest technique is to first find $\hat{\mathbf{A}}$ and $\hat{V}$ and then find the field vectors from these. So $\hat{\mathbf{A}}$ and $\hat{V}$ can be viewed as simply intermediate functions on the way toward the determination of the desired field vectors.

From Ampère's law given in (1*b*), we may write

$$\nabla \times (\mu_0 \hat{\mathbf{H}}) = j\omega\mu_0\epsilon_0 \hat{\mathbf{E}} + \mu_0 \hat{\mathbf{J}}_s \tag{9}$$

Substituting (8*a*), we obtain

$$\nabla \times \nabla \times \hat{\mathbf{A}} = j\omega\mu_0\epsilon_0 \hat{\mathbf{E}} + \mu_0 \hat{\mathbf{J}}_s \tag{10}$$

The following vector relationship is becoming commonly used by now and should be familiar:

$$\nabla \times \nabla \times \hat{\mathbf{A}} = \nabla(\nabla \cdot \hat{\mathbf{A}}) - \nabla^2 \hat{\mathbf{A}} \tag{11}$$

where $\nabla^2 \hat{\mathbf{A}}$ is the vector laplacian. Substituting (11) and (8*b*) into (10), we obtain

$$\nabla^2 \hat{\mathbf{A}} + \omega^2 \mu_0 \epsilon_0 \hat{\mathbf{A}} = -\mu_0 \hat{\mathbf{J}}_s + \nabla(\nabla \cdot \hat{\mathbf{A}} + j\omega\epsilon_0\mu_0 \hat{V}) \tag{12}$$

and we have obtained one equation in terms of the potential functions (to be determined), $\hat{\mathbf{A}}$ and $\hat{V}$, and the source current density $\hat{\mathbf{J}}_s$, which is assumed to be known.

At this point we need to invoke some additional information concerning vector fields. It turns out that in order to specify a vector field completely, we must specify not only its curl but also its divergence. This result is known as the *Helmholtz theorem.*† For the magnetic vector potential $\hat{\mathbf{A}}$, we have specified its curl in (4) but have not specified its divergence. Let us choose to specify $\nabla \cdot \hat{\mathbf{A}}$ such that the last term in (12) is zero; that is,

$$\nabla \cdot \hat{\mathbf{A}} = -j\omega\mu_0\epsilon_0 \hat{V} \tag{13}$$

This specification of the divergence of the magnetic vector potential is referred to as the *Lorentz condition.*‡ With the Lorentz condition we have completely specified $\hat{\mathbf{A}}$, and (12) becomes

$$\nabla^2 \hat{\mathbf{A}} + \omega^2 \mu_0 \epsilon_0 \hat{\mathbf{A}} = -\mu_0 \hat{\mathbf{J}}_s \tag{14}$$

Thus we have successfully eliminated $\hat{V}$ from (12) and have the necessary equation relating $\hat{\mathbf{A}}$ to $\hat{\mathbf{J}}_s$ (known). This equation is becoming quite familiar. It is of the form of the wave equation, with the addition of a source term on the right.

---

† For a proof of the Helmholtz theorem, see R. Plonsey and R. E. Collin, *Principles and Applications of Electromagnetic Fields*, McGraw-Hill, New York, 1961, pp. 29–36.

‡ The Lorentz condition in (13) can be shown to be equivalent to the statement of conservation of charge: $\nabla \cdot \hat{\mathbf{J}}_s = -j\omega\hat{\rho}_s$, where $\hat{\rho}_s$ is the volume charge density associated with $\hat{\mathbf{J}}_s$. See R. Plonsey and R. E. Collin, *Principles and Applications of Electromagnetic Fields*, McGraw-Hill, New York, 1961, pp. 322–323.

Expanding the vector laplacian in rectangular coordinates (Appendix A), (14) has three components:

$$\nabla^2 \hat{A}_x + \beta_0^2 \hat{A}_x = -\mu_0 \hat{J}_{s,x} \tag{15a}$$

$$\nabla^2 \hat{A}_y + \beta_0^2 \hat{A}_y = -\mu_0 \hat{J}_{s,y} \tag{15b}$$

$$\nabla^2 \hat{A}_z + \beta_0^2 \hat{A}_z = -\mu_0 \hat{J}_{s,z} \tag{15c}$$

where

$$\beta_0 = \omega\sqrt{\mu_0 \epsilon_0} \tag{16}$$

We can obtain an integral form of the solution to each equation:†

$$\hat{A}_x = \frac{\mu_0}{4\pi} \int_v \frac{\hat{J}_{s,x} e^{-j\beta_0 r}}{r} \, dv \tag{17a}$$

$$\hat{A}_y = \frac{\mu_0}{4\pi} \int_v \frac{\hat{J}_{s,y} e^{-j\beta_0 r}}{r} \, dv \tag{17b}$$

$$\hat{A}_z = \frac{\mu_0}{4\pi} \int_v \frac{\hat{J}_{s,z} e^{-j\beta_0 r}}{r} \, dv \tag{17c}$$

Combining (17) gives the vector expression

$$\hat{\mathbf{A}} = \frac{\mu_0}{4\pi} \int_v \frac{\hat{\mathbf{J}}_s e^{-j\beta_0 r}}{r} \, dv \tag{18}$$

In these equations, v encloses $\hat{\mathbf{J}}_s$ and r is the distance between a differential volume element (the source point) and the point at which we are computing $\hat{\mathbf{A}}$ (the field point), as shown in Fig. 9.2.

In principle, if we know the current distribution over the surface of a radiator or antenna, $\hat{\mathbf{J}}_s$, we may find the components of the magnetic vector potential from (17). From the Lorentz condition in (13) we have

$$\hat{V} = -\frac{\nabla \cdot \hat{\mathbf{A}}}{j\omega\mu_0\epsilon_0} \tag{19}$$

Substituting this into (8*b*) yields

$$\hat{\mathbf{E}} = -j\omega\hat{\mathbf{A}} + \frac{\nabla(\nabla \cdot \hat{\mathbf{A}})}{j\omega\mu_0\epsilon_0} \tag{20}$$

† See C. T. A. Johnk, *Engineering Electromagnetic Fields and Waves*, Wiley, New York, 1975, pp. 596–602.

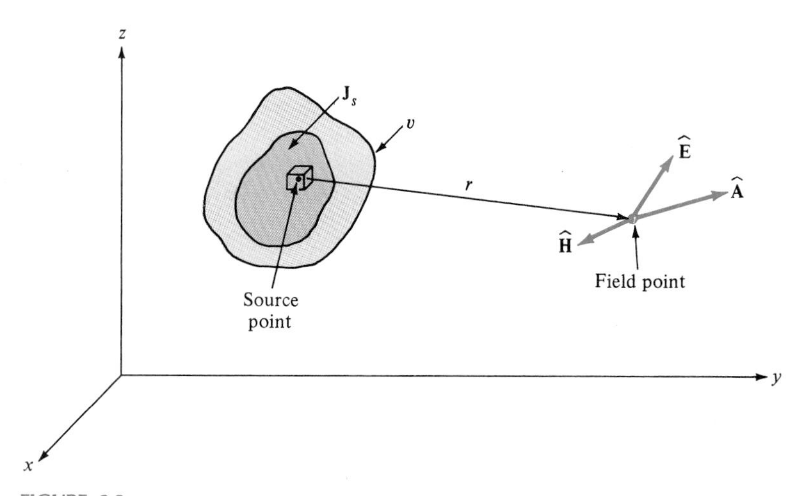

FIGURE 9.2
Computation of the fields due to a given current distribution.

The magnetic field can be found from (8*a*):

$$\hat{\mathbf{H}} = \frac{1}{\mu_0} \nabla \times \hat{\mathbf{A}} \tag{21}$$

Therefore, we may then find the electric and magnetic fields at all points in space resulting from this source current distribution in terms of the magnetic vector potential. Alternatively, the electric field may be obtained from the magnetic field via Ampère's law at points away from the current distribution ($\hat{\mathbf{J}}_s = 0$):

$$\hat{\mathbf{E}} = \frac{1}{j\omega\epsilon_0} \nabla \times \hat{\mathbf{H}} \tag{22}$$

Thus, knowing the source current distribution $\hat{\mathbf{J}}_s$ determines all of the field vectors. Quite obviously, however, we have traded one difficult problem (determining $\hat{\mathbf{E}}$ and $\hat{\mathbf{H}}$) for another difficult problem (determining $\hat{\mathbf{J}}_s$). What, then, is the advantage? Quite often, either by experiment or physical reasoning, we can deduce the form or approximate form of $\hat{\mathbf{J}}_s$ for certain simple structures more easily than we can deduce the form of $\hat{\mathbf{E}}$ and $\hat{\mathbf{H}}$ radiated by these structures. (A good example of this is given in Sec. 9.3.) Also, since $\hat{\mathbf{J}}_s$ is localized to the antenna surface, it appears simpler, theoretically, to characterize $\hat{\mathbf{J}}_s$ than to characterize $\hat{\mathbf{E}}$ and $\hat{\mathbf{H}}$ at every point in space. If we can deduce $\hat{\mathbf{J}}_s$, we can find the resulting fields (or an approximation of them) at every point in space by using the above development.

For complex antenna shapes, however, such as are shown in Fig. 9.1 (and some relatively simple ones too), it is a difficult matter to deduce the form of $\hat{\mathbf{J}}_s$

over the surface of the antenna. For these problems, a more practical, approximate method of solution involves the use of the Friis transmission equation discussed in Sec. 9.7. For wire-type antennas, a very good, but approximate, method makes use of large-scale digital computers. An example of this technique is the method of moments discussed in Chap. 10 for static field problems. The method-of-moments techniques allow an approximate solution for the fields radiated from some relatively complex structures.†

9.2 Elemental Dipole Antennas

Given the complexities of antennas, as was mentioned in the beginning of this chapter, exact solutions for most antenna structures have not been obtained. Solutions generally rely on certain approximations, such as assuming the form of the current distribution over the antenna surface (but even with this assumption, the problem of computing the fields of the antenna remains a difficult one to solve). As a further simplification, the solution is typically restricted to field points at large distances from the antenna (the far field). In this section we will investigate two ideal antennas—the electric and magnetic dipoles. For these antennas, we may obtain exact solutions for the fields. These solutions provide considerable insight into the behavior of other, more realistic antennas for which exact solutions have not been obtained.

9.2.1 The Electric (Hertzian) Dipole

A simple antenna for which we can calculate all the fields in a very straightforward fashion is the electric, or Hertzian, dipole illustrated in Fig. 9.3. The antenna consists of an infinitesimal current element of length dl carrying a phasor current $\hat{I}$ that is assumed to be the same (in magnitude and phase) at all points along the element length. From (17) we see that the magnetic vector potential has only a z component, which can be determined by application of (17c). Assuming the element length to be infinitesimal, (17c) reduces to

$$\hat{A}_z = \frac{\mu_0}{4\pi} \hat{I}\, dl \frac{e^{-j\beta_0 r}}{r} \tag{23}$$

In spherical coordinates (see Appendix A)

$$\hat{A}_r = \hat{A}_z \cos\theta \tag{24a}$$

$$\hat{A}_\theta = -\hat{A}_z \sin\theta \tag{24b}$$

$$\hat{A}_\phi = 0 \tag{24c}$$

† W. L. Stutzman and G. A. Thiele, *Antenna Theory and Design*, Wiley, New York, 1981, chap 7.

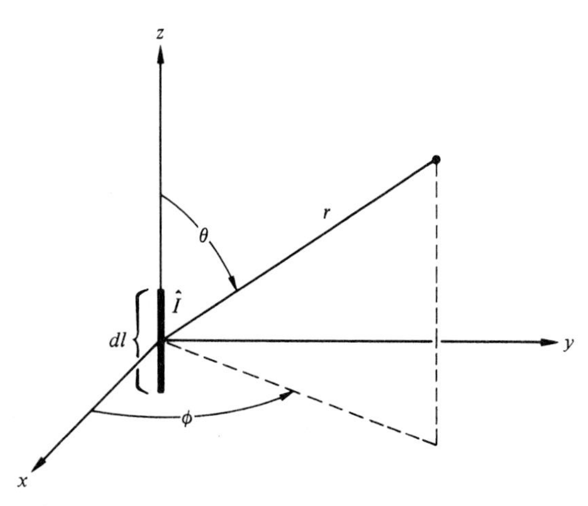

FIGURE 9.3
The elemental electric, or Hertzian, dipole antenna.

Using (21), we can determine the components of the magnetic field as

$$\hat{H}_r = 0 \tag{25a}$$

$$\hat{H}_\theta = 0 \tag{25b}$$

$$\hat{H}_\phi = \frac{\hat{I}\, dl}{4\pi} \beta_0^2 \sin\theta \left(j \frac{1}{\beta_0 r} + \frac{1}{\beta_0^2 r^2} \right) e^{-j\beta_0 r} \tag{25c}$$

Similarly, using (25) and (22), we can obtain the components of the electric field as

$$\hat{E}_r = 2 \frac{\hat{I}\, dl}{4\pi} \eta_0 \beta_0^2 \cos\theta \left(\frac{1}{\beta_0^2 r^2} - j \frac{1}{\beta_0^3 r^3} \right) e^{-j\beta_0 r} \tag{26a}$$

$$\hat{E}_\theta = \frac{\hat{I}\, dl}{4\pi} \eta_0 \beta_0^2 \sin\theta \left(j \frac{1}{\beta_0 r} + \frac{1}{\beta_0^2 r^2} - j \frac{1}{\beta_0^3 r^3} \right) e^{-j\beta_0 r} \tag{26b}$$

$$\hat{E}_\phi = 0 \tag{26c}$$

where

$$\eta_0 = \sqrt{\frac{\mu_0}{\epsilon_0}} \tag{27}$$

is the intrinsic impedance of free space. Note that the fields in (25) and (26) can be viewed as functions of "electrical" distance away from the antenna since each r is multiplied by β_0 and $\beta_0 r = 2\pi(r/\lambda_0)$.

Now let us consider the various components of the field. First note that the magnetic field has only one nonzero component, $\hat{H}_\phi$. Note also that for zero frequency (dc), $\beta_0 = 0$ and (25) reduces to

$$\hat{H}_\phi = \frac{\hat{I}\,dl \sin\theta}{4\pi r^2} \tag{28}$$

which is an expression of the Biot-Savart law for an infinitesimal current element, as discussed in Chap. 4. For nonzero frequency, there are two portions of this magnetic field; one part varies as $1/r$ and the other part varies as $1/r^2$. The amplitudes of the $1/r$ and $1/r^2$ terms are equal at $r = 1/\beta_0 = \lambda_0/2\pi \simeq \lambda_0/6$. The $1/r^2$ term is called the *induction field*, and this term dominates close to the current element. At large distances away from the current element, the $1/r$ term dominates. This $1/r$ term is called the *far-field component* of $\hat{\mathbf{H}}$:

$$\hat{\mathbf{H}}_{\text{farfield}} = \frac{j\beta_0\,\hat{I}\,dl \sin\theta}{4\pi r}\,e^{-j\beta_0 r}\mathbf{a}_\phi \tag{29}$$

Now consider the components of the electric field given in (26). Note that there are only two nonzero components of the field: $\hat{E}_r$ and $\hat{E}_\theta$. The $\hat{E}_\theta$ component contains a $1/r$ term, while the $\hat{E}_r$ component does not. The $1/r$ term will again be called the far-field term:

$$\hat{\mathbf{E}}_{\text{farfield}} = \frac{j\eta_0\beta_0\hat{I}\,dl \sin\theta}{4\pi r}\,e^{-j\beta_0 r}\mathbf{a}_\theta \tag{30}$$

Comparing (30) to (29), we observe that the electric and magnetic far fields are in time phase and orthogonal. Their magnitudes are related by η_0:

$$\frac{|\hat{\mathbf{E}}_{\text{farfield}}|}{|\hat{\mathbf{H}}_{\text{farfield}}|} = \left.\frac{\hat{E}_\theta}{\hat{H}_\phi}\right|_{\text{large } r} = \eta_0 \tag{31}$$

The $1/r^2$ terms in $\hat{E}_r$ and $\hat{E}_\theta$ are again called the induction-field terms. The components $\hat{E}_\theta$ and $\hat{E}_r$ also contain $1/r^3$ terms which dominate at very close distances and which are called the *electrostatic*, or *dipole*, *terms*. The origin of the name for these terms stems from the fact that the electric field of an electrostatic dipole consisting of a pair of point charges $+q$ and $-q$ separated by dl is (Sec. 3.6.2)

$$\mathbf{E} = \frac{q\,dl}{4\pi\epsilon_0 r^3}(2\cos\theta\,\mathbf{a}_r + \sin\theta\,\mathbf{a}_\theta) \tag{32}$$

If we note that $I = dq/dt$, then we may write $\hat{I} = j\omega\hat{q}$, where $\hat{q} = qe^{j\omega t}$. Substituting $\hat{I}$ in (26) and allowing $\omega \to 0$, we obtain a result identical to that of the electrostatic dipole given in (32). The presence of the $1/r^3$ terms in $\hat{E}_\theta$ and $\hat{E}_r$ is reasonable to expect when we realize that a current flowing from one end of the infinitesimal current element to the other in Fig. 9.3 will produce a separation of charge at the two ends of the element: a dipole of charge.

It is also of interest to compute the Poynting vector for the current element. The average-power density vector is

$$\mathbf{S}_{\text{av}} = \tfrac{1}{2}\,\text{Re}\,(\hat{\mathbf{E}} \times \hat{\mathbf{H}}^*) \tag{33}$$

From (25) and (26) we observe that there will be two components of $\hat{\mathbf{E}} \times \hat{\mathbf{H}}^*$:

$$\hat{\mathbf{E}} \times \hat{\mathbf{H}}^* = \hat{E}_\theta \hat{H}_\phi^* \mathbf{a}_r - \hat{E}_r \hat{H}_\phi^* \mathbf{a}_\theta \tag{34}$$

Substituting the relations in (25) and (26) into (34), we find that the θ component in (34) is purely imaginary; therefore, this component yields no contribution to the average-power density vector in (33). We are then left with

$$\begin{aligned}\mathbf{S}_{\text{av}} &= \tfrac{1}{2}\,\text{Re}\,(\hat{E}_\theta \hat{H}_\phi^*)\mathbf{a}_r \\ &= \frac{\omega^2 u_0 |\hat{I}|^2\, dl^2 \sin^2\theta}{32\pi^2 u_0 r^2}\,\mathbf{a}_r \\ &= 15\pi\left(\frac{dl}{\lambda_0}\right)^2 |\hat{I}|^2 \frac{\sin^2\theta}{r^2}\,\mathbf{a}_r \qquad \text{W/m}^2\end{aligned} \tag{35}$$

This shows that average power is flowing away from the current element: our first hint of "radiation." It is instructive to note that this average-power density vector could be computed solely from the far-field expressions given in (29) and (30), as the reader should verify.

A plot of $|\mathbf{S}_{\text{av}}|$ for a fixed distance r about the current element is referred to as the *power pattern* of the antenna. This concept is also applied to other antennas, as will be discussed later in this chapter. However, three-dimensional plots are usually not feasible, so antenna patterns are usually plotted by fixing θ for various values and plotting the variation of ϕ, and vice versa. From (35), however, we see that $|\mathbf{S}_{\text{av}}|$ is independent of ϕ, as is obvious from the physical symmetry in Fig. 9.3, yet that it varies with θ as $\sin^2\theta$, as (35) shows. In this case, we may sketch a simple three-dimensional plot of the power pattern, as shown in Fig. 9.4. In this plot, the distance from the origin to a point on this doughnut-shaped pattern is the relative average power for that particular θ. Note that no power is being radiated in the direction along the dipole length. The power pattern therefore gives a simple visualization of the ability of the antenna to

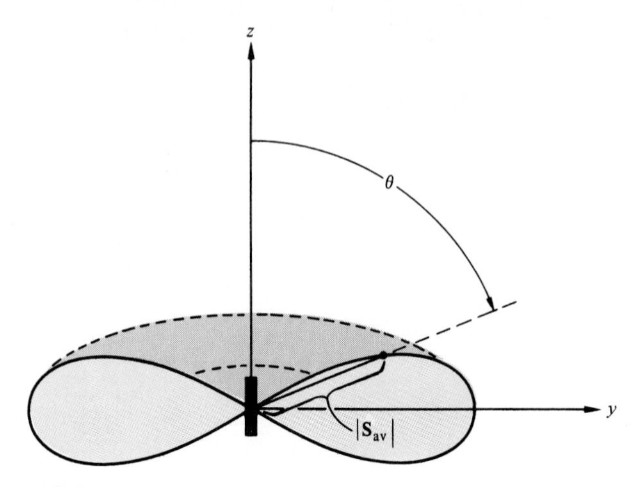

FIGURE 9.4
The radiated power density pattern of an elemental electric (Hertzian) dipole.

direct or focus its radiated power in a particular direction (in this case, broadside to the antenna).

We may obtain the total average power radiated by integrating (35) over some closed surface that encloses the current element. Since $\mathbf{S}_{av}$ is directed solely in the radial direction, a convenient choice of surface would be a sphere of radius r centered on the coordinate system center, as shown in Fig. 9.5. We obtain

$$\begin{aligned} P_{\text{rad}} &= \oint_s \mathbf{S}_{av} \cdot d\mathbf{s} \\ &= \int_{\theta=0}^{\pi} \int_{\phi=0}^{2\pi} \left[15\pi \left(\frac{dl}{\lambda_0} \right)^2 |\hat{I}|^2 \frac{\sin^2 \theta}{r^2} \right] r^2 \sin\theta \, d\phi \, d\theta \\ &= 80\pi^2 \left(\frac{dl}{\lambda_0} \right)^2 \frac{|\hat{I}|^2}{2} \quad \text{W} \end{aligned} \tag{36}$$

Denoting $\hat{I}/\sqrt{2} = \hat{I}_{\text{rms}}$, we can compute a radiation resistance:

$$\begin{aligned} R_{\text{rad}} &= \frac{P_{\text{rad}}}{|\hat{I}_{\text{rms}}|^2} \\ &= 80\pi^2 \left(\frac{dl}{\lambda_0} \right)^2 \quad \Omega \end{aligned} \tag{37}$$

The radiation resistance represents a fictitious resistance which dissipates the same amount of power as that radiated by the Hertzian dipole when both carry the same value of rms current, $\hat{I}_{\text{rms}}$.

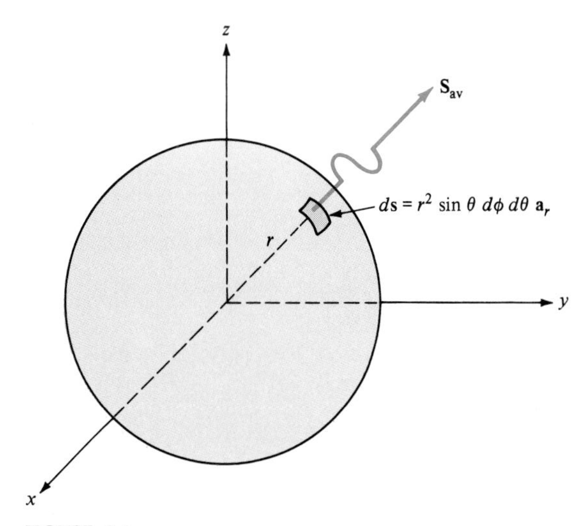

FIGURE 9.5

A closed surface (sphere) for the computation of the total radiated power of a Hertzian dipole.

The Hertzian dipole is a very ineffective radiator. For example, for a length $dl = 1$ cm and a frequency of 300 MHz ($\lambda_0 = 1$ m), the radiation resistance is 79 mΩ. In order to radiate 1 W of power, we require a current of 3.6 A. If the frequency is changed to 3 MHz ($\lambda_0 = 100$ m), the radiation resistance is 7.9 μΩ and the current required to radiate 1 W is 356 A!

It is of considerable importance to note from (29) and (30) that the far fields of this current element can be written as

$$\hat{\mathbf{E}}_{\text{farfield}} = j\hat{E}_0 \frac{e^{-j\beta_0 r}}{r} \mathbf{a}_\theta \tag{38a}$$

$$\begin{aligned}\hat{\mathbf{H}}_{\text{farfield}} &= j\frac{\hat{E}_0}{\eta_0}\frac{e^{-j\beta_0 r}}{r}(\mathbf{a}_r \times \mathbf{a}_\theta) \\ &= \frac{1}{\eta_0}(\mathbf{a}_r \times \hat{\mathbf{E}}_{\text{farfield}})\end{aligned} \tag{38b}$$

where $\hat{E}_0 = (\eta_0 \beta_0 \hat{I} \, dl \sin\theta)/4\pi$. Thus, at a point, these radiated fields resemble uniform plane waves—in that along a line from the current element to the point, $\hat{\mathbf{E}}$ and $\hat{\mathbf{H}}$ are perpendicular to this line, are orthogonal, and are related by η_0. Strictly speaking, these waves are not uniform plane waves but are more properly classified as spherical waves. Thus, these waves resemble uniform plane waves only locally, that is, in the immediate vicinity of the point.

The time-domain forms of the far fields are obtained by multiplying (38) by $e^{j\omega t}$ and taking the real part of the result. In doing so, we obtain (assuming $\hat{I} = I\underline{/0^\circ}$)

$$\mathcal{E}_{\text{farfield}} = \frac{E_0}{r} \cos\left[\omega\left(t - \frac{r}{u_0}\right) + 90^\circ\right]\mathbf{a}_\theta$$

$$= -\frac{E_0}{r} \sin\left[\omega\left(t - \frac{r}{u_0}\right)\right]\mathbf{a}_\theta \tag{39a}$$

$$\mathcal{H}_{\text{farfield}} = \frac{E_0}{\eta_0 r} \cos\left[\omega\left(t - \frac{r}{u_0}\right) + 90^\circ\right](\mathbf{a}_r \times \mathbf{a}_\theta)$$

$$= -\frac{E_0}{\eta_0 r} \sin\left[\omega\left(t - \frac{r}{u_0}\right)\right](\mathbf{a}_r \times \mathbf{a}_\theta) \tag{39b}$$

The term $t - r/u_0$ in both expressions represents a phase delay. In other words, the current waveform $I \cos \omega t$ produces the electric field waveform $\mathcal{E}$, but corresponding points on these waveforms will be delayed in occurrence at distance r by $t_0 = r/u_0$ seconds where $u_0 = 1/\sqrt{\mu_0 \epsilon_0}$ is the velocity of light in the medium (free space in this case). In the time domain, this represents a finite time of propagation, t_0, from the source point to the field point.

Why have we considered this rather ideal "antenna"? The answer lies in the fact that for many other (and structurally much more complicated) antennas, the fields will be of the same general form as in (38) and (39) in the *far field* of the antenna. A definition of far-field conditions for more complicated antenna structures will be given in a subsequent subsection. For the present discussion, it suffices to consider the far-field region as being sufficiently far from the antenna so that the fields resemble (locally) uniform plane waves. For the Hertzian dipole, we might consider the far field to be a radial distance from the antenna such that the $1/r$ terms in the field expressions in (25) and (26) dominate the $1/r^2$ and $1/r^3$ terms. This was found to occur for $r > \lambda_0/2\pi$. A more realistic definition of the near-field–far-field boundary will be given in Sec. 9.7. The criterion will be the distance at which the ratio of $\hat{E}_\theta$ in (26*b*) and $\hat{H}_\phi$ in (25*c*) is approximately equal to η_0 and turns out to be $r \simeq 3\lambda_0$. The electric field and magnetic field in the far fields of many other antennas will be orthogonal, and $\hat{\mathbf{E}}_{\text{farfield}} \times \hat{\mathbf{H}}_{\text{farfield}}$ will yield the direction of power flow: this being the radial direction from the antenna. The far fields fall off in magnitude as $1/r$. In the near field (closer to the antenna than the far field) the field structure is much more complicated than that of the Hertzian dipole. However, the near field is generally not of interest since it does not contribute to net average radiated power, and the receiving and transmitting antennas are usually separated so that they are in the far-field region of each other. Thus, this study of the simple Hertzian dipole has illustrated many of the properties of interest for more complicated antennas.

9.2.2 The Magnetic Dipole (Loop)

A dual to the elemental electric dipole can be found in the elemental magnetic dipole, or current loop, shown in Fig. 9.6. A very small loop of radius b lying in the xy plane carries a phasor current $\hat{I}$. This loop constitutes a vector magnetic-dipole moment, as discussed in Sec. 4.5:

$$\hat{\mathbf{m}} = \hat{I}\pi b^2 \mathbf{a}_z \qquad \mathrm{A \cdot m^2} \tag{40}$$

where πb^2 is the area enclosed by the loop. The magnetic vector potential can be found from the vector expression of (18):

$$\begin{aligned} \hat{\mathbf{A}} &= \frac{\mu_0}{4\pi} \int_v \hat{\mathbf{J}} \frac{e^{-j\beta_0 r'}}{r'} \, dv \\ &= \frac{\mu_0}{4\pi} \int_c \hat{\mathbf{I}} \frac{e^{-j\beta_0 r'}}{r'} \, dl \end{aligned} \tag{41}$$

where the integral is evaluated along the loop (the location of the source current). Evaluation of (41) is somewhat difficult. We make the approximation

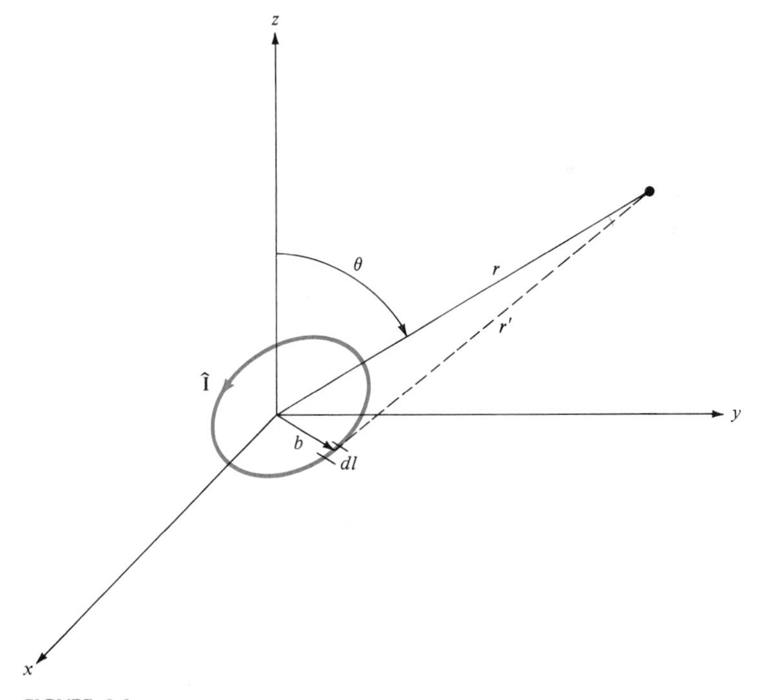

FIGURE 9.6
The elemental magnetic dipole (loop) antenna.

that the loop is "electrically small" in order that the exponential term may be approximated as

$$\begin{aligned} e^{-j\beta_0 r'} &= e^{-j\beta_0 r} e^{-j\beta_0 (r'-r)} \\ &\simeq e^{-j\beta_0 r}[1 - j\beta_0(r' - r)] \end{aligned} \tag{42}$$

The assumption of an electrically small loop is embodied in the expansion of the second exponential since we have assumed that $\beta_0(r' - r) = [2\pi(r' - r)]/\lambda_0 \ll 1$. In other words, we assume that the electrical distances r'/λ_0 and r/λ_0 are approximately the same. Substituting (42) into (41) yields

$$\hat{\mathbf{A}} \simeq \frac{\mu_0}{4\pi} e^{-j\beta_0 r}\left[(1 + j\beta_0 r)\oint_c \frac{\hat{\mathbf{I}}\, dl}{r'} - j\beta_0 \oint_c \hat{\mathbf{I}}\, dl\right] \tag{43}$$

The second integral, $\oint_c \hat{\mathbf{I}}\, dl$, evaluates to zero since the vector current $\hat{\mathbf{I}}$ is assumed to be constant in magnitude and phase around the loop and since its average direction integrated around the loop is zero. In other words,

$$\begin{aligned} \oint_c \hat{\mathbf{I}}\, dl &= \hat{I} b \int_{\phi=0}^{2\pi} \mathbf{a}_\phi \, d\phi \\ &= 0 \end{aligned}$$

where $dl = b\, d\phi$. Therefore, (43) becomes

$$\hat{\mathbf{A}} \simeq \frac{\mu_0(1 + j\beta_0 r)e^{-j\beta_0 r}}{4\pi} \oint_c \frac{\hat{\mathbf{I}}\, dl}{r'} \tag{44}$$

With the exception of the factor $(1 + j\beta_0 r)e^{-j\beta_0 r}$, this integral is the same as for a small loop carrying a dc current (discussed in Chap. 4). Multiplying that result [Eq. (76) of Chap. 4] by $(1 + j\beta_0 r)e^{-j\beta_0 r}$ yields the desired result:

$$\hat{\mathbf{A}} = \frac{\mu_0 \hat{m}}{4\pi r^2}(1 + j\beta_0 r)e^{-j\beta_0 r} \sin\theta\, \mathbf{a}_\phi \tag{45a}$$

where $\hat{m}$ is the magnetic-dipole moment

$$\hat{m} = \hat{I}\pi b^2 \tag{45b}$$

Substituting $\hat{A}$ into (21) and using (22) yields the electric and magnetic fields:

$$\hat{E}_r = 0 \tag{46a}$$

$$\hat{E}_\theta = 0 \tag{46b}$$

$$\hat{E}_\phi = -j\frac{\omega\mu_0\hat{m}\beta_0^2}{4\pi}\sin\theta\left(j\frac{1}{\beta_0 r}+\frac{1}{\beta_0^2 r^2}\right)e^{-j\beta_0 r} \tag{46c}$$

$$\hat{H}_r = 2\frac{j\omega\mu_0\hat{m}\beta_0^2}{4\pi\eta_0}\cos\theta\left(\frac{1}{\beta_0^2 r^2}-j\frac{1}{\beta_0^3 r^3}\right)e^{-j\beta_0 r} \tag{47a}$$

$$\hat{H}_\theta = \frac{j\omega\mu_0\hat{m}\beta_0^2}{4\pi\eta_0}\sin\theta\left(j\frac{1}{\beta_0 r}+\frac{1}{\beta_0^2 r^2}-j\frac{1}{\beta_0^3 r^3}\right)e^{-j\beta_0 r} \tag{47b}$$

$$\hat{H}_\phi = 0 \tag{47c}$$

Comparing (46) and (47) to the field expressions for the electric dipole given in (25) and (26), we see the duality that exists between these two structures. Observations about the electric (magnetic) field of the electric dipole apply to the magnetic (electric) field of the magnetic dipole.

The far field of the magnetic dipole is characterized by the $1/r$ dependent terms:

$$\hat{\mathbf{E}}_{\text{farfield}} = \frac{\omega\mu_0\hat{m}\beta_0}{4\pi r}\sin\theta\, e^{-j\beta_0 r}\mathbf{a}_\phi \tag{48a}$$

$$\begin{aligned}\hat{\mathbf{H}}_{\text{farfield}} &= -\frac{\omega\mu_0\hat{m}\beta_0}{4\pi\eta_0 r}\sin\theta\, e^{-j\beta_0 r}\mathbf{a}_\theta \\ &= \frac{1}{\eta_0}(\mathbf{a}_r \times \hat{\mathbf{E}}_{\text{farfield}})\end{aligned} \tag{48b}$$

As was the case for the electric dipole, the far field of the magnetic dipole is such that the fields (1) decay as $1/r$, (2) lie in a (local) plane perpendicular to the radial direction, and (3) are related by η_0.

Once again, the average-power density vector is radially directed and given by

$$\begin{aligned}\mathbf{S}_{\text{av}} &= \tfrac{1}{2}\,\text{Re}\,(-\hat{E}_\phi\hat{H}_\theta^*\mathbf{a}_r + \hat{E}_\phi\hat{H}_r^*\mathbf{a}_\theta) \\ &= -\tfrac{1}{2}\,\text{Re}\,(\hat{E}_\phi\hat{H}_\theta^*)\mathbf{a}_r \\ &= \frac{1}{2\eta_0}\left(\frac{\omega\mu_0|\hat{m}|\beta_0}{4\pi}\right)^2\frac{\sin^2\theta}{r^2}\mathbf{a}_r \\ &= 1860\left(\frac{A}{\lambda_0^2}\right)^2|\hat{I}|^2\frac{\sin^2\theta}{r^2}\mathbf{a}_r\end{aligned} \tag{49}$$

where $A = \pi b^2$ is the area of the current loop. This result could be obtained strictly from the far fields. The total average power radiated is obtained by integrating (49) over the surface of a sphere, yielding

$$\begin{aligned} P_{\text{rad}} &= \oint_s \mathbf{S}_{\text{av}} \cdot d\mathbf{s} \\ &= 1860\left(\frac{A}{\lambda_0^2}\right)^2 |\hat{I}|^2 \int_{\theta=0}^{\pi} \int_{\phi=0}^{2\pi} \left(\frac{\sin^2\theta}{r^2}\right) r^2 \sin\theta \, d\phi \, d\theta \\ &= 15{,}585|\hat{I}|^2\left(\frac{A}{\lambda_0^2}\right)^2 \end{aligned} \tag{50}$$

Thus, the radiation resistance is

$$\begin{aligned} R_{\text{rad}} &= \frac{P_{\text{av}}}{|\hat{I}_{\text{rms}}|^2} \\ &= 31{,}170\left(\frac{A}{\lambda_0^2}\right)^2 \end{aligned} \tag{51}$$

Like the Hertzian dipole, the magnetic dipole is not an effective radiator. Consider a loop of radius 1 cm. At 300 MHz, the radiation resistance is 3.08 mΩ. In order to radiate 1 W, the loop requires a current of 18 A. At 3 MHz, the radiation resistance is 3.08×10^{-11} Ω and the current required to radiate 1 W is 1.8×10^5 A!

EXAMPLE 9.1 Consider a square loop carrying a current $\hat{I}$, as shown in Fig. 9.7*a*. Assume that the loop is electrically small at the frequency of interest so that the current $\hat{I}$ is the same at all points of the loop (magnitude and phase). Model this loop as four Hertzian dipoles and determine an expression for the far-field electric field at a distance R away from each side in the plane of the loop. Evaluate this expression at a distance of 3 m for a loop having $d = 1$ cm and a 100-MHz current of $\hat{I} = 1\underline{/0^\circ}$ A. Compare this result to that of a magnetic dipole of the same area.

Solution At $y = R$, the far-field electric field is due solely to the sides at $y = d/2$ and $y = -d/2$. The fields due to the sides at $x = d/2$ and $x = -d/2$ cancel because of symmetry and because the currents in these sides go in opposite directions. The fields due to the sides at $y = d/2$ and $y = -d/2$ are in the x direction and oppositely directed. As shown in Fig. 9.7*b*, the resultant electric field is the sum of the far-field expressions for these sides given in (30):

$$\hat{E}_x = \frac{j\eta_0\beta_0\hat{I}d}{4\pi}\left(\frac{e^{-j\beta_0(R-d/2)}}{R-d/2} - \frac{e^{-j\beta_0(R+d/2)}}{R+d/2}\right)$$

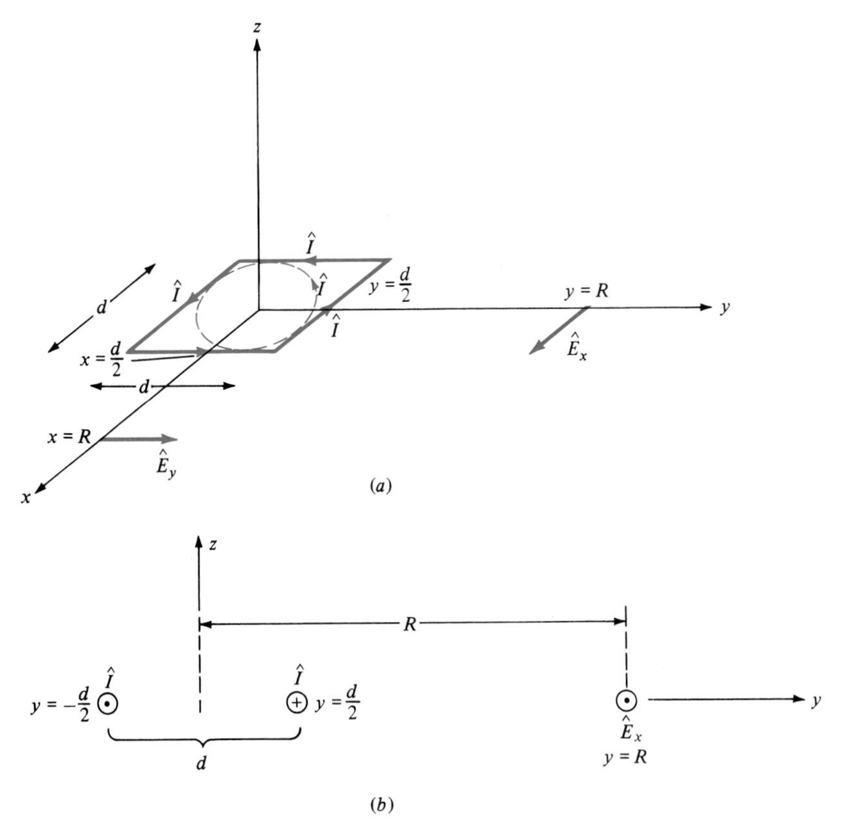

FIGURE 9.7.

Example 9.1. Representing the radiation from a square loop as the superposition of radiated fields of Hertzian dipoles.

The terms $(R - d/2)$ and $(R + d/2)$ in the denominators may be approximated by R for $R \gg d$, leaving

$$\hat{E}_x = \frac{j\eta_0 \beta_0 \hat{I} d}{4\pi R} e^{-j\beta_0 R}(e^{j\beta_0(d/2)} - e^{-j\beta_0(d/2)})$$

$$= -\frac{2\eta_0 \beta_0 \hat{I} d}{4\pi R} e^{-j\beta_0 R} \sin \beta_0 \frac{d}{2}$$

Since the loop is assumed to be electrically small, $\sin \beta_0(d/2) \simeq \beta_0(d/2)$; thus

$$\hat{E}_x \simeq -\frac{\eta_0 \beta_0^2 \hat{I} d^2}{4\pi R} e^{-j\beta_0 R}$$

$$= -\frac{\omega\mu_0 \beta_0 \hat{I} d^2}{4\pi R} e^{-j\beta_0 R}$$

$$= -1184 \frac{\hat{I} d^2}{\lambda_0^2 R} e^{-j\beta_0 R}$$

Similarly, at $x = R$, $\hat{E}_y = -\hat{E}_x$. Comparing this result to the far-field result for a magnetic dipole given in (48a), we see that the results are identical if we identify the dipole moment of this result as

$$\hat{m} = \hat{I}d^2$$

Thus, the far-field result for a small loop is somewhat independent of the shape of the loop and depends only on the area of the loop, d^2.

Evaluating this result for $d^2 = 1\ \text{cm}^2$, $\hat{I} = 1\underline{/0^\circ}$ A, 100 MHz, and $R = 3$ m gives

$$|\hat{E}_x| = 4.39\underline{/0^\circ}\ \text{mV/m}$$

9.2.3 Radiation Patterns of the Elemental Dipoles

From the preceding we have seen that the average power radiated by the dipoles can be computed from the far fields of those dipoles. In addition, antennas used for communication purposes will normally be in the far field of each other. Thus, the far fields of an antenna are of primary importance.

None of the elemental dipoles radiates uniformly in all directions. For example, the far fields of an electric dipole are zero along a line coinciding with the axis of the dipole (the z axis). Similarly, the far fields of a magnetic dipole are zero along a line coinciding with the axis of the loop (the z axis). For both dipoles, the far fields are a maximum in the xy plane (perpendicular to the axis of the electric dipole or in the plane of the loop).

A pictorial representation of the directive properties of an antenna's radiation is referred to as its *radiation pattern*, or *antenna pattern*. This pattern is obtained by plotting the magnitude of the far-field electric field at a fixed distance from the antenna as a function of θ and ϕ. Only the variation of the electric field need be plotted since the magnetic field is related to the electric field by η_0 in the far field. In these plots, the distance from the origin is proportional to the magnitude of the field. The average-power density of the Hertzian dipole was plotted in Fig. 9.4 using this method.

For the Hertzian dipole, the patterns are symmetrical about the z axis, are independent of the ϕ coordinate, and depend on θ as $\sin\theta$, as shown by (30):

$$|E_{\text{farfield}}| \sim \sin\theta$$

The plot of this pattern has the shape of a doughnut, as shown in Fig. 9.8a. Three-dimensional plots of the antenna patterns are not generally needed. Rather, plots of the variation with $\theta(\phi)$ for a fixed $\phi(\theta)$, such as are shown in Figs. 9.8b and c, are more commonly used. Similar results are obtained for the magnetic dipole since these are also functions of $\sin\theta$.

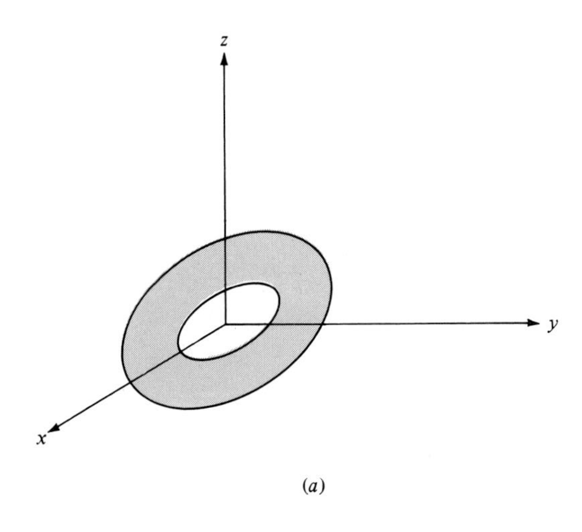

(a)

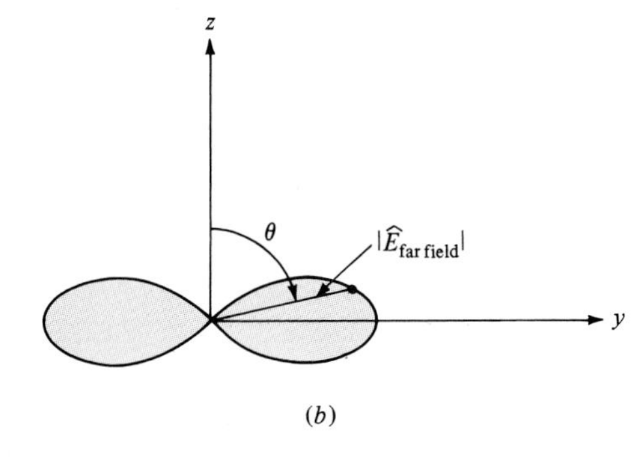

(b)

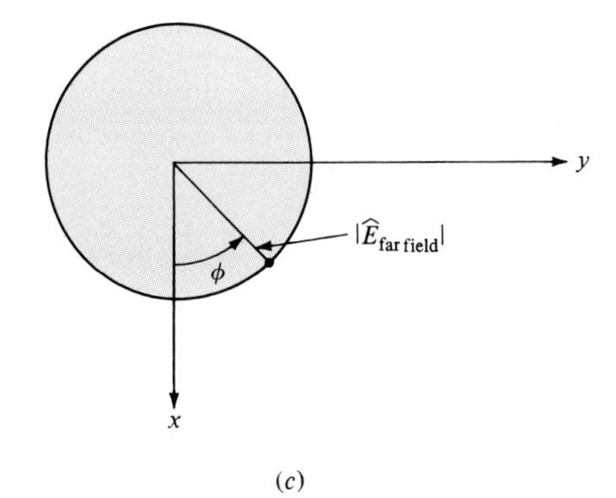

(c)

FIGURE 9.8

Antenna patterns of the Hertzian dipoles. *(a)* Three-dimensional pattern, *(b)* function of θ for $\phi = 90^\circ$, *(c)* function of ϕ for $\theta = 90^\circ$.

9.3 Long-Dipole and Monopole Antennas

The Hertzian dipole considered in Sec. 9.2.1 is an obviously impractical antenna for several reasons. Primarily, the length of the dipole was considered to be infinitesimal in order to simplify the computation of the magnetic vector potential via (18). Also, the current along the Hertzian dipole was assumed to be constant along the dipole; this assumption required the current to be nonzero at the endpoints of the dipole: an unrealistic and, moreover, physically impossible situation since the surrounding medium, free space, is nonconductive. Moreover, the Hertzian dipole is a very ineffective radiator since the radiation resistance is quite small, requiring large currents in order to radiate significant power. The magnetic dipole suffers from similar problems. In this section we will consider two practical and more frequently used antennas: the long-dipole and monopole antennas.

The long-dipole antenna (or, simply, the dipole antenna) consists of a thin wire that is fed or excited via a voltage source inserted at the midpoint, as shown in Fig. 9.9*a*. Each leg is of length $l/2$.

The monopole antenna (Fig. 9.9*b*) consists of a single leg perpendicular to a ground plane. The monopole is fed at its base. For the purposes of analysis, the ground plane is considered to be infinite and perfectly conducting. In practice, this ideal ground plane is approximated. On aircraft, the metallic fuselage simulates the ground plane. For ground-based stations, the earth simulates, to some degree, this ground plane. Since the earth is much less of an approximation to a perfectly conducting plane than is a metal, ground-based stations are usually augmented by a grid of wires lying on the ground to simulate the ground plane.

We found in Sec. 9.1 that if we know the current distribution over the surface of an antenna, we may compute the magnetic vector potential in the space surrounding the antenna via (18), and from this we may compute the fields via (21) and (22). We will assume that the wires of the dipole are thin, so that the only variation of the current on the wire surface is along the wire lengths. Although we still do not know this distribution, we may make a reasonable guess (which turns out to be relatively accurate). In order to justify this guess, let us reconsider the two-wire transmission line having an open-circuit load shown in Fig. 9.10*a*. From the results of Chap. 7 we know that the phasor current $\hat{I}(z)$ will be distributed sinusoidally with position along the wire and that, because of the open-circuit load, the current must go to zero at the endpoints. For example, from the results of Chap. 7 we may write the equation for the phasor line current at a distance d from the open-circuit load as

$$\hat{I}_d = j\frac{1}{R_C}\sin\beta_0 d\hat{V}_L \tag{52}$$

where $\hat{V}_L$ is the load voltage; thus, the current is distributed as shown in Fig. 9.10*a*. If we separate the wires as shown in Fig. 9.10*b*, we might expect this current to remain unchanged so far as the distribution is concerned. Suppose

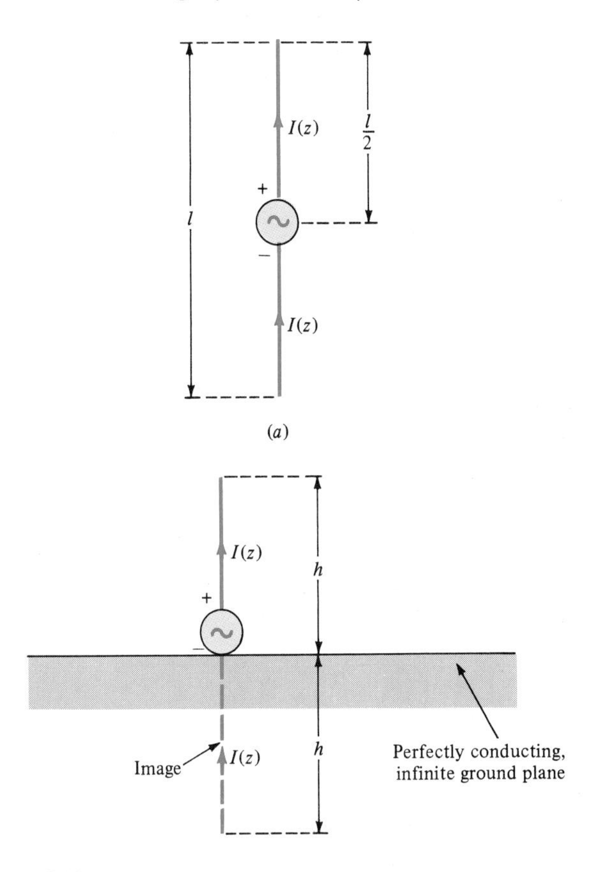

FIGURE 9.9
(a) The dipole antenna, and *(b)* the monopole antenna.

that we further separate the wires, resulting in the dipole antenna shown in Fig. 9.10*c*. It turns out that the current distribution remains very similar to the parallel-wire line: namely, sinusoidal. Experimental measurements bear out this result, and it can also be shown directly.†

Placing the center of the dipole at the origin of a spherical coordinate system as shown in Fig. 9.11*a*, with the dipole directed along the z axis, we may therefore write an expression for the current distribution along the wire as [see (52)]

$$\hat{I}(z) = \hat{I}_m \sin \beta_0\left(\frac{l}{2} - z\right) \qquad 0 < z < \frac{l}{2} \tag{53a}$$

$$\hat{I}(z) = \hat{I}_m \sin \beta_0\left(\frac{l}{2} + z\right) \qquad -\frac{l}{2} < z < 0 \tag{53b}$$

† For a discussion, see C. T. A. Johnk, *Engineering Fields and Waves*, Wiley, New York, 1975, pp. 609–611.

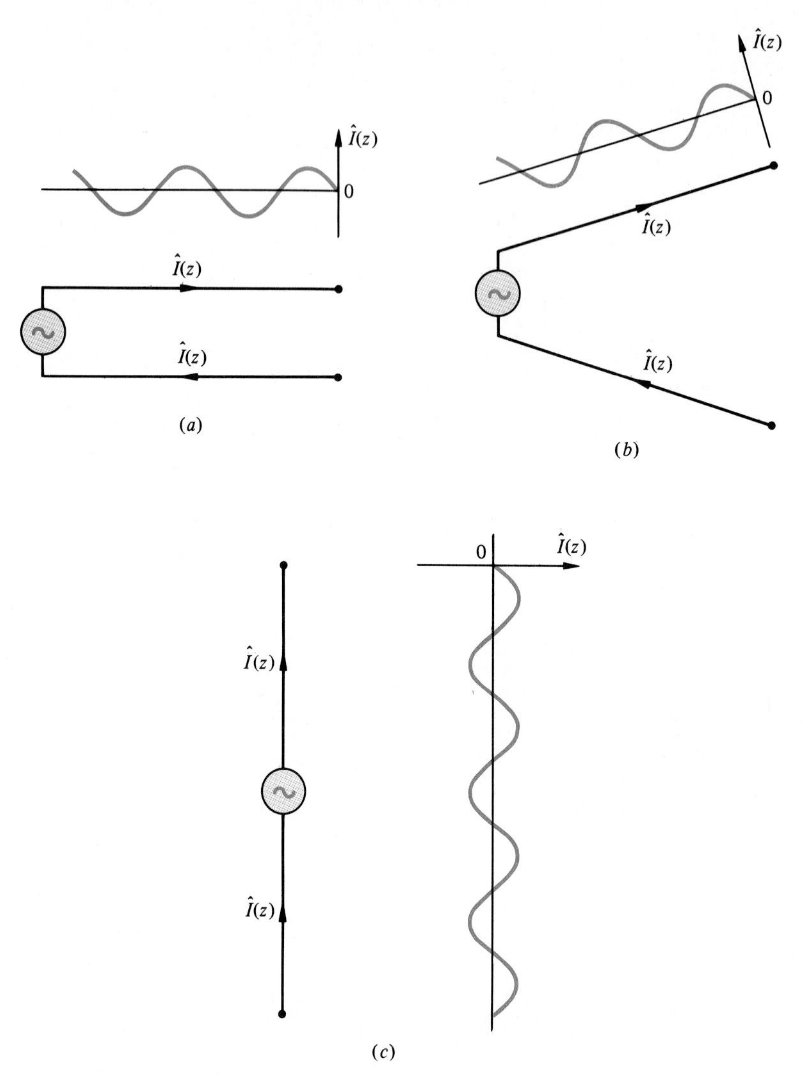

FIGURE 9.10

Current distribution on a long linear dipole. Evolution from an open-circuited transmission line.

where each leg of the dipole is of length $l/2$ and $\hat{I}_m$ is a constant. Note that these expressions are more reasonable for the dipole current distribution than is the constant current assumed for the Hertzian dipole since the currents go to zero at the ends of the wires ($z = \pm l/2$).

Now that we have assumed the current distribution along the dipole, we may compute the magnetic vector potential from (18) and the fields from (21) and (22). A more direct method would be to use the Hertzian dipole results of Sec. 9.2.1 and to consider the fields of the long dipole as being the superposition of

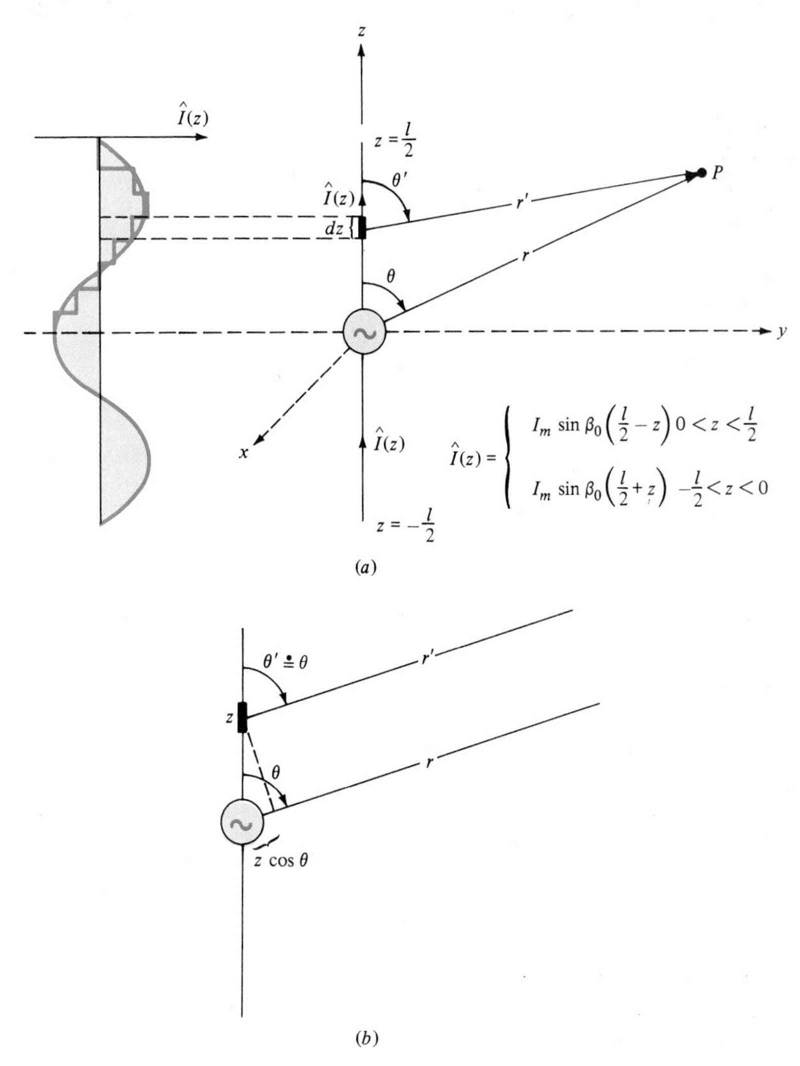

FIGURE 9.11
(a) The long linear dipole; *(b)* far-field approximation.

the fields due to many small Hertzian dipoles of length dz having a current that is constant and equal to the value of $\hat{I}(z)$ at that point along the dipole, as shown in Fig. 9.11*a*. For example, consider the infinitesimal segment dz shown in Fig. 9.11*a*. If the field point of interest, P, is located a distance r away from the center of the dipole and angle θ, and a distance r' and angle θ' from the current element, then the electric field due to this element can be obtained from (26) by substituting r', θ', and dz for r, θ, and dl, respectively, in (26) and also substituting the expression for $\hat{I}$ given in (53). The total electric field is then the superposition (integral) of the contributions from all these infinitesimal elements.

The resulting integral becomes quite complicated. Thus, we shall make another simplifying assumption: Let us consider only the far fields of the dipole. In this case, the far-field electric field in (30) due to this element has only a θ component and a $1/r$ dependence. Thus, we have

$$d\hat{E}_\theta = j\eta_0\beta_0 \frac{\hat{I}(z)\sin\theta'}{4\pi r'} e^{-j\beta_0 r'}\,dz \tag{54}$$

Our ultimate objective is to obtain the fields at P in terms of the radial distance r from the midpoint of the dipole and angle θ. Since we are considering only the far field, the radial distance from the center of the dipole to point P, r, and the radial distance from the current element to point P, r', will be approximately equal ($r \simeq r'$) and angles θ and θ' will be approximately equal ($\theta \simeq \theta'$), as shown in Fig. 9.11*b*.

We may substitute $r' = r$ into the denominator of (54), but we should not substitute this into the $e^{-j\beta_0 r'}$ term for the following reason: This term may be written as

$$e^{-j\beta_0 r'} = \underline{\left/ -\frac{2\pi r'}{\lambda_0} \right.} \tag{55}$$

and its value depends not on the physical distance r' but on the electrical distance r'/λ_0; therefore, even though r' and r may be approximately equal, the term in (55) may depend significantly on the difference in electrical distances. For example, suppose $r = 1000$ m and $r' = 1000.5$ m and the frequency is $f = 3 \times 10^8$ Hz. We have ($\lambda_0 = 1$ m)

$$\begin{aligned} \beta_0 r' &= 2\pi(1000) \\ &= 360{,}000^\circ \end{aligned} \tag{56a}$$

$$\begin{aligned} \beta_0 r &= 2\pi(1000.5) \\ &= 360{,}180^\circ \end{aligned} \tag{56b}$$

Note that the fields at 1000 m away and those only 0.5 m farther away are 180° out of phase! We will see a more striking example of this in the next section. The far fields from two antennas that are widely spaced physically but separated on the order of a wavelength may actually be completely out of phase and add destructively at a point to yield a resulting field of zero.

Thus, it is not a reasonable approximation to substitute r' for r in (55). However, we may still simplify (55) and write the result in terms of r. Consider Fig. 9.11*b*, which shows the two radial distances r and r' as being approximately parallel. Thus, we are assuming that the field point is sufficiently far, physically, from the antenna. From Fig. 9.11*b* we may obtain

$$r' \simeq r - z\cos\theta \tag{57}$$

Substituting (57) into the phase term in (54) and r into the denominator, we obtain

$$d\hat{E}_\theta = j\eta_0\beta_0 \frac{\hat{I}(z)\sin\theta}{4\pi r} e^{-j\beta_0(r - z\cos\theta)}\,dz \tag{58}$$

The total electric field is the sum of these contributions:

$$\hat{E}_\theta = \int_{z=-l/2}^{z=l/2} j\eta_0\beta_0 \frac{\hat{I}(z)\sin\theta}{4\pi r} e^{-j\beta_0 r} e^{j\beta_0 z\cos\theta}\,dz \tag{59}$$

Substituting the expressions for $\hat{I}(z)$ given in (53), we obtain

$$\hat{E}_\theta = \frac{j\eta_0\beta_0\hat{I}_m \sin\theta\, e^{-j\beta_0 r}}{4\pi r}\left[\int_{z=-l/2}^{0} \sin\beta_0\left(\frac{l}{2} + z\right) e^{j\beta_0 z\cos\theta}\,dz + \int_{z=0}^{l/2} \sin\beta_0\left(\frac{l}{2} - z\right) e^{j\beta_0 z\cos\theta}\,dz\right] \tag{60}$$

In the first integral, z is negative; thus, we may change signs on z and change the limits of integration to yield

$$\begin{aligned}\hat{E}_\theta &= \frac{j\eta_0\beta_0\hat{I}_m \sin\theta\, e^{-j\beta_0 r}}{4\pi r}\left[\int_{z=0}^{l/2} \sin\beta_0\left(\frac{l}{2} - z\right) e^{-j\beta_0 z\cos\theta}\,dz + \int_{z=0}^{l/2} \sin\beta_0\left(\frac{l}{2} - z\right) e^{j\beta_0 z\cos\theta}\,dz\right] \\ &= j\frac{\eta_0\beta_0\hat{I}_m \sin\theta\, e^{-j\beta_0 r}}{4\pi r}\int_{z=0}^{l/2} \sin\beta_0\left(\frac{l}{2} - z\right)\left(e^{j\beta_0 z\cos\theta} + e^{-j\beta_0 z\cos\theta}\right) dz\end{aligned} \tag{61}$$

Noting that

$$\cos(\beta_0 z\cos\theta) = \frac{e^{j\beta_0 z\cos\theta} + e^{-j\beta_0 z\cos\theta}}{2} \tag{62}$$

we see that (61) becomes

$$\hat{E}_\theta = j\frac{\eta_0\beta_0\hat{I}_m \sin\theta\, e^{-j\beta_0 r}}{2\pi r}\int_{z=0}^{l/2} \sin\beta_0\left(\frac{l}{2} - z\right)\cos(\beta_0 z\cos\theta)\,dz \tag{63}$$

Integrating (63), we obtain

$$\hat{E}_\theta = j\frac{\eta_0\hat{I}_m e^{-j\beta_0 r}}{2\pi r}\frac{\cos[\beta_0(l/2)\cos\theta] - \cos\beta_0(l/2)}{\sin\theta} \tag{64}$$

The θ variation term in (64) will be denoted by

$$F(\theta) = \frac{\cos[\beta_0(l/2)\cos\theta] - \cos\beta_0(l/2)}{\sin\theta}$$

$$= \frac{\cos(\pi l/\lambda_0 \cos\theta) - \cos\pi l/\lambda_0}{\sin\theta} \tag{65}$$

since $\beta_0 = 2\pi/\lambda_0$. Thus, the electric field in the far field of the dipole is

$$\hat{\mathbf{E}} = \hat{E}_\theta \mathbf{a}_\theta \tag{66a}$$

where

$$\hat{E}_\theta = j\frac{\eta_0 \hat{I}_m e^{-j\beta_0 r}}{2\pi r} F(\theta)$$

$$= j\frac{60\hat{I}_m e^{-j\beta_0 r}}{r} F(\theta) \tag{66b}$$

The magnetic field in the far-field region of a Hertzian dipole is orthogonal to the electric field and related by η_0. [See (31).] If we carry through the above development for the magnetic field, we obtain the obvious result that in the far field of the dipole

$$\hat{\mathbf{H}} = \hat{H}_\phi \mathbf{a}_\phi$$

$$= \frac{\hat{E}_\theta}{\eta_0}\mathbf{a}_\phi \tag{67}$$

where $\hat{E}_\theta$ is given by (66*b*).

The most frequently encountered case is the half-wave dipole, in which the total dipole length is $l = \lambda_0/2$. Substituting into (65), we obtain

$$F(\theta) = \frac{\cos(\pi/2\cos\theta)}{\sin\theta} \qquad \begin{cases} \text{half-wave dipole} \\ l = \dfrac{\lambda_0}{2} \end{cases} \tag{68}$$

The electric field will be a maximum for $\theta = 90°$ (broadside to the antenna). For this case, $F(90°) = 1$ and the maximum electric field for the half-wave dipole becomes $|\hat{E}|_{\max} = 60\hat{I}_m/r$.

It is particularly illustrative to plot the electric field (far-field) pattern about the antenna to show the directive nature of the antenna, as was done for the elemental dipoles. From (66*b*) we need only plot $F(\theta)$ and the pattern is independent of ϕ. The electric field patterns for various electrical lengths of the dipole are shown in Fig. 9.12. The power patterns would be quite similar since $\mathbf{S}_{av} = (\frac{1}{2}|\hat{E}_\theta|^2/\eta_0)\mathbf{a}_r$ in the far field.

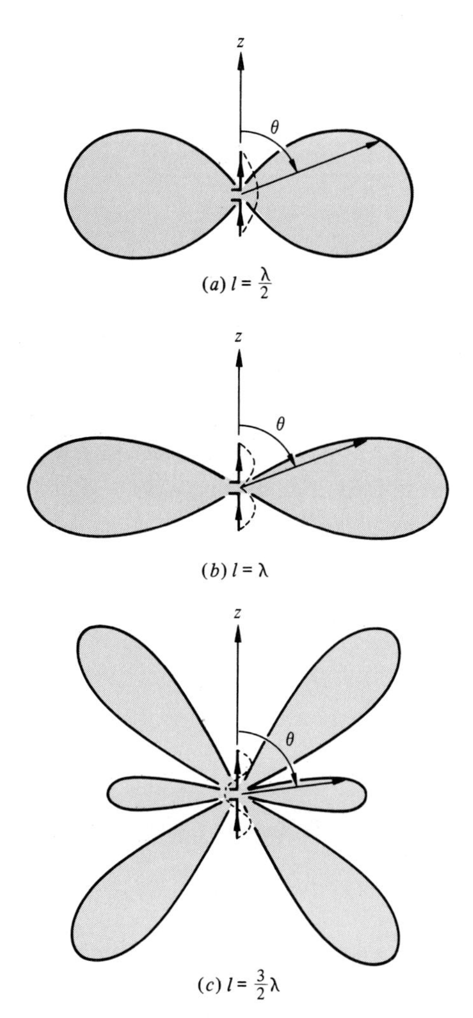

FIGURE 9.12
Patterns and current distributions of dipoles of various lengths.

Now let us compute the average power radiated by the dipole. From (66) and (67) the average-power Poynting vector in the far field is given by

$$\mathbf{S}_{\text{av}} = \tfrac{1}{2}\,\text{Re}\,(\hat{E}_\theta \hat{H}_\phi^*)\mathbf{a}_r$$
$$= \frac{1}{2}\frac{|\hat{E}_\theta|^2}{\eta_0}\mathbf{a}_r \qquad \text{W/m}^2 \tag{69}$$

Substituting (66) into (69) yields

$$\mathbf{S}_{\text{av}} = \eta_0 \frac{|\hat{I}_m|^2}{8\pi^2 r^2} F^2(\theta)\mathbf{a}_r \qquad \text{W/m}^2$$
$$= \frac{4.77|\hat{I}_m|^2}{r^2} F^2(\theta)\mathbf{a}_r \tag{70}$$

The total average power radiated is obtained by integrating (70) over some closed surface that encloses the dipole. The obvious choice for this closed surface is a sphere, and we obtain

$$P_{\text{rad}} = \int_{\theta=0}^{\pi} \int_{\phi=0}^{2\pi} \mathbf{S}_{\text{av}} \cdot \mathbf{a}_r r^2 \sin\theta \, d\theta \, d\phi$$

$$= \frac{\eta_0 |\hat{I}_m|^2}{4\pi} \int_{\theta=0}^{\pi} F^2(\theta) \sin\theta \, d\theta \qquad \text{W}$$

$$= 30|\hat{I}_m|^2 \int_{\theta=0}^{\pi} F^2(\theta) \sin\theta \, d\theta \qquad \text{W} \tag{71}$$

The integral in (71) is not integrable in closed form. For the special case of a half-wave dipole, the integral may be evaluated numerically with a digital computer to yield

$$\int_{\theta=0}^{\pi} F^2(\theta) \sin\theta \, d\theta \simeq 1.2186 \qquad \text{half-wave dipole} \tag{72}$$

Substituting this result into (71), we obtain

$$P_{\text{rad}} = 36.5|\hat{I}_m|^2 \qquad \text{W} \tag{73}$$

Note from (53) that for the half-wave dipole, the current at the input to the antenna ($z = 0$) is

$$\hat{I}_{\text{in}} = \hat{I}_m \sin \beta_0 \frac{l}{2} \qquad \text{half-wave dipole}$$

$$= \hat{I}_m \tag{74}$$

since $\beta_0 = 2\pi/\lambda_0$ and $l = \lambda_0/2$, so $\beta_0 l/2 = \pi/2$. Thus, the amplitude of the current distribution $\hat{I}_m$ is the value of the input current at the terminals of the half-wave dipole. The rms value is $\hat{I}_m/\sqrt{2}$, so (73) becomes

$$P_{\text{rad}} = 73|\hat{I}_{\text{rms}}|^2 \qquad \text{W} \qquad \text{half-wave dipole} \tag{75}$$

Clearly, then, if we know the rms value of the *input current* at the terminals of a half-wave dipole, we may find the total averge power radiated by multiplying the square of the rms current by 73 Ω. This result suggests that we define a radiation resistance of this half-wave dipole as

$$R_{\text{rad}} = 73 \ \Omega \qquad \text{half-wave dipole} \tag{76}$$

Thus
$$P_{\text{rad}} = R_{\text{rad}} \frac{|\hat{I}_m|^2}{2}$$
$$= R_{\text{rad}} |\hat{I}_{\text{rms}}|^2 \qquad \text{W} \tag{77}$$

Now consider the monopole antenna shown in Fig. 9.9*b*. As will be discussed in Chap. 10, for the purposes of analysis we may replace the infinite, perfectly conducting ground plane with the image of the monopole, as indicated in Fig. 9.9*b*. The current distribution along the portion below ground is the image of the current distribution along the original monopole, which means that the monopole and its image are identical to the dipole. Thus, we see that the fields of the monopole are identical to those of the dipole. This observation applies only to the region above the ground plane since there are no fields below the ground plane.

There is one important difference in this analogy between the dipole and the monopole. Although the field patterns are the same, the monopole radiates only half the power of the dipole: the power radiated out of the half-sphere above the ground plane. Thus, the radiation resistance for the monopole is half that of the corresponding dipole. In particular, for a quarter-wave monopole of length $h = \lambda_0/4$ (which corresponds to a half-wave dipole), we have

$$R_{\text{rad}} = 36.5\ \Omega \qquad \text{quarter-wave monopole} \tag{78}$$

Up to this point we have not considered the total input impedance $\hat{Z}_{\text{in}}$ seen at the terminals of a dipole or monopole antenna. The input impedance will, in general, have a real and an imaginary part as

$$\hat{Z}_{\text{in}} = R_{\text{in}} + jX_{\text{in}} \tag{79}$$

and the input current at the antenna terminals is found from the current distribution in (53) at $z = 0$ as

$$\hat{I}_{\text{in}} = \hat{I}_m \sin \frac{\beta_0 l}{2} \tag{80}$$

Note that for a half-wave dipole or a quarter-wave monopole, $l = \lambda_0/2$ and $\hat{I}_{\text{in}} = \hat{I}_m$. This also occurs if the dipole length l is some odd multiple of $\lambda_0/2$ since $\beta_0 l/2$ will be an odd multiple of $\pi/2$ and $|\sin \beta_0 l/2| = 1$ for these cases. Similar results apply to the monopole. The total average power delivered to the antenna is

$$P_{\text{ant}} = \frac{|\hat{I}_{\text{in}}|^2}{2} R_{\text{in}}$$
$$= \frac{|\hat{I}_m|^2}{2} R_{\text{in}} \sin^2 \frac{\beta_0 l}{2} \qquad \text{W} \tag{81}$$

If the antenna is lossless (constructed of perfectly conducting wires), then the total average power delivered to the antenna is simply the total average power radiated: $P_{\text{ant}} = P_{\text{rad}}$. By comparing (81) and (77), we observe that R_{rad} and R_{in} are related by

$$R_{\text{rad}} = R_{\text{in}} \sin^2 \frac{\beta_0 l}{2} \tag{82a}$$

where the radiation resistance is defined from (71) as

$$R_{\text{rad}} = 60 \int_{\theta=0}^{\pi} F^2(\theta) \sin \theta \, d\theta \tag{82b}$$

This is often referred to as "the radiation resistance referred to a point of current maximum" whose value is $\hat{I}_m$. For a half-wave dipole or a quarter-wave monopole, $l = \lambda_0/2$, and thus $R_{\text{in}} = R_{\text{rad}}$ and $\hat{I}_{\text{in}} = \hat{I}_m$. For dipole lengths that are not odd multiples of a half-wavelength, we also have $P_{\text{ant}} = P_{\text{rad}}$ (if the antenna is lossless) but $R_{\text{in}} \neq R_{\text{rad}}$ since $\hat{I}_{\text{in}} \neq \hat{I}_m$. This occurs because the maximum value of the current distribution for these antennas, $\hat{I}_m$, does not occur precisely at $z = 0$ (the antenna input), as shown by (53). Thus, the relationship in (82*a*) can be viewed as giving the radiation resistance referred to the antenna terminals, R_{in}, whereas the radiation resistance defined by $R_{\text{rad}} = 2(P_{\text{rad}}/|\hat{I}_m|^2)$ and by (82*b*) is referred to the *point of current maximum* on the antenna.

There is an additional term in R_{in} if the wires of the antenna are not perfect conductors. One can compute the per-unit-length resistance of these wires using the results for the round wire developed in Sec. 7.4.2 for transmission lines. Once the per-unit-length resistance of the wires, r_{wire}, is computed, the total ohmic power loss can be obtained by adding the power loss in the incremental segments:

$$P_{\text{loss}} = \int_{-l/2}^{l/2} r_{\text{wire}} \frac{|\hat{I}(z)|^2}{2} \, dz \tag{83}$$

Thus, the net ohmic resistance, referred to the input terminals, is

$$\begin{aligned} R_{\text{loss}} &= \frac{P_{\text{loss}}}{|\hat{I}_{\text{in}}|^2/2} \\ &= \frac{r_{\text{wire}} \int_{-l/2}^{l/2} |\hat{I}(z)|^2 \, dz}{|\hat{I}_{\text{in}}|^2} \end{aligned} \tag{84}$$

This expression can be evaluated by substituting the expressions for current distribution given in (53) and the expression for input current given in (80). (See Prob. 9-17.)

Obtaining the imaginary or reactive part of $\hat{Z}_{in}$, X_{in}, is more difficult. For a half-wave dipole, $X_{in} = 42.5\ \Omega$, and for a quarter-wave monopole, $X_{in} = 21.25\ \Omega$.† It is common to construct the dipole such that its length is slightly less than a half-wavelength. For this case, the reactive term X_{in} can be made zero and the real part remains approximately equal to R_{rad}. Similar comments apply to monopoles.

For dipoles that are not a half-wavelength long, and for monopoles not a quarter-wavelength long, X_{in} can be a significant part of the input impedance. Moreover, X_{in} is a function of the wire radius, whereas for half-wave dipoles and quarter-wave monopoles, X_{in} is independent of the wire radius.† In addition, this reactance may be positive or negative, representing an inductive or capacitive component of the input impedance. For a dipole shorter than a half-wavelength, X_{in} is negative, representing a capacitive reactance. For a dipole longer than a half-wavelength, X_{in} is positive, representing an inductive reactance. To radiate the same amount of power, this reactance requires a larger voltage at the input to the antenna than if the reactance were zero. Thus, cancellation networks (loading coils) are required at the input to these antennas to cancel the reactive component of $\hat{Z}_{in}$. Therefore, it is desirable to have the electrical length of a dipole equal to approximately $\lambda_0/2$ (and of a monopole, approximately $\lambda_0/4$) at the operating frequency so that X_{in} is small.

EXAMPLE 9.2 Consider a 1-m-long dipole antenna that is driven by a 150-MHz source having a source resistance of 50 Ω and an open-circuit voltage of 100 V, as shown in Fig. 9.13*a*. Determine the total average power dissipated in the antenna and the total average power radiated. The antenna is constructed of 20-gauge (radius of 4.06×10^{-4} m) copper wires.

Solution At 150 MHz, this 1-m dipole is exactly $\lambda_0/2$ in electrical length so that $R_{rad} = 73\ \Omega$ and $X_{in} = 42.5\ \Omega$. The 20-gauge wires have radii much larger than a skin depth at this frequency ($\delta = 5.4 \times 10^{-6}$ m). Thus, the high-frequency approximation for wire resistance developed in Sec. 7.4.2 may be used to compute

$$\begin{aligned} r_{wire} &= \frac{R_s}{2\pi r_w} \\ &= \frac{1}{2\pi r_w}\sqrt{\frac{\omega\mu}{2\sigma}} \\ &= \frac{1}{2\pi r_w \sigma\delta} \\ &= 1.25\ \Omega/\text{m} \end{aligned}$$

† See E. C. Jordan and K. G. Balmain, *Electromagnetic Waves and Radiating Systems*, Prentice-Hall, Englewood Cliffs, N.J., 1968.

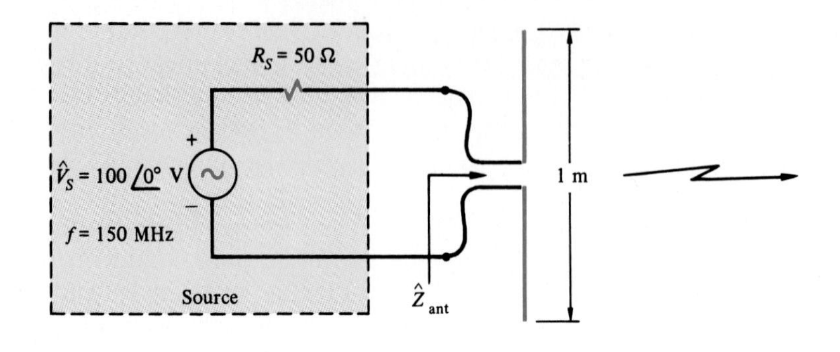

(a)

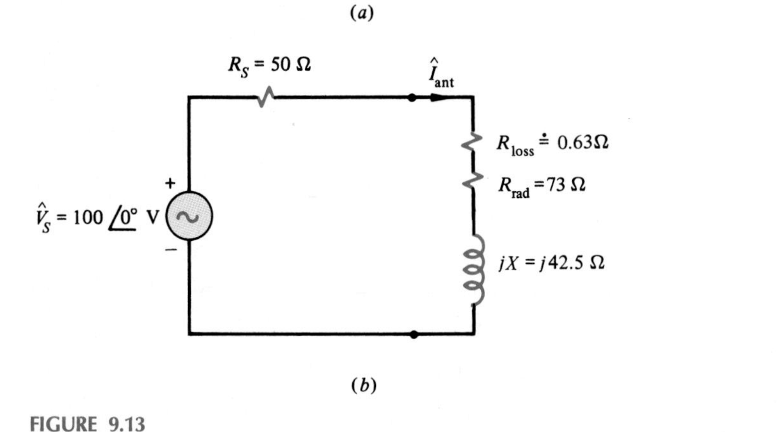

(b)

FIGURE 9.13
Example 9.2. *(a)* Problem definition and *(b)* equivalent circuit representation of the antenna.

The ohmic power loss of the wires is obtained by adding the power loss in the incremental segments via (83):

$$P_{\text{loss}} = \frac{r_{\text{wire}}}{2} \int_{-l/2}^{l/2} |\hat{I}(z)|^2 \, dz$$

Therefore, the ohmic loss is related to the area under the square of the current distribution. For a half-wave dipole we obtain, substituting (53) into (84),

$$P_{\text{loss}} = r_{\text{wire}} \frac{l}{2} \frac{|\hat{I}_m|^2}{2}$$

Therefore, the net ohmic resistance of the dipole is

$$\begin{aligned} R_{\text{loss}} &= r_{\text{wire}} \frac{l}{2} \\ &= \frac{r_{\text{wire}}}{2} \\ &= 0.63\ \Omega \end{aligned}$$

Since the dipole is a half-wave dipole, the total input impedance to the antenna as seen by the source is

$$\hat{Z}_{\text{ant}} = R_{\text{loss}} + R_{\text{rad}} + jX_{\text{in}}$$
$$= 0.63 + 73 + j42.5 \quad \Omega$$

as shown in Fig. 9.13*b*. In order to determine the average power delivered to the antenna, we may compute the antenna input current $\hat{I}_{\text{ant}}$ as

$$\hat{I}_{\text{ant}} = \frac{\hat{V}_S}{R_S + \hat{Z}_{\text{ant}}}$$
$$= \frac{100\angle 0^\circ}{50 + 73.63 + j42.5}$$
$$= 0.765\angle -18.97^\circ \text{ A}$$

The total average power dissipated in antenna losses is

$$P_{\text{loss}} = \tfrac{1}{2}|\hat{I}_{\text{ant}}|^2 R_{\text{loss}}$$
$$= 184 \text{ mW}$$

The total average power radiated is

$$P_{\text{rad}} = \tfrac{1}{2}|\hat{I}_{\text{ant}}|^2 R_{\text{rad}}$$
$$= 21.36 \text{ W}$$

The radiation efficiency is the ratio of the radiated power to the power delivered to the antenna. This becomes

$$e = \frac{R_{\text{rad}}}{R_{\text{rad}} + R_{\text{loss}}}$$
$$= 0.991$$

or 99.1 percent. It should be noted that the power transfer to the antenna can be improved by using a stub tuner (discussed in Appendix C) between the source and the antenna. This is frequently done.

9.4 Antenna Arrays

The patterns (power or electric field) of the Hertzian dipole, the magnetic dipole, the long dipole, and the monopole are omnidirectional in any plane perpendicular to the antenna axis. This characteristic follows from the symmetry of these structures. The reader may have noticed that many standard broadcast antenna stations consist not of one single, vertical antenna above ground (monopole), but of several such antennas. The function of such an "array" is to produce particular directional properties of the radiated field. For example, suppose that we wish to produce a radiation pattern that does not interfere with another broadcasting station, as shown in Fig. 9.14*a*. A single, vertical-wire antenna perpendicular to the ground will, because of symmetry, have an omnidirectional pattern in any plane parallel to the ground. On the other hand, if we utilize two vertical-wire antennas to transmit the signal, as shown in Fig. 9.14*b*, we can adjust the terminal currents and antenna spacing so that the far fields of each antenna will, at various points, add or cancel to produce a directional pattern of the overall array.

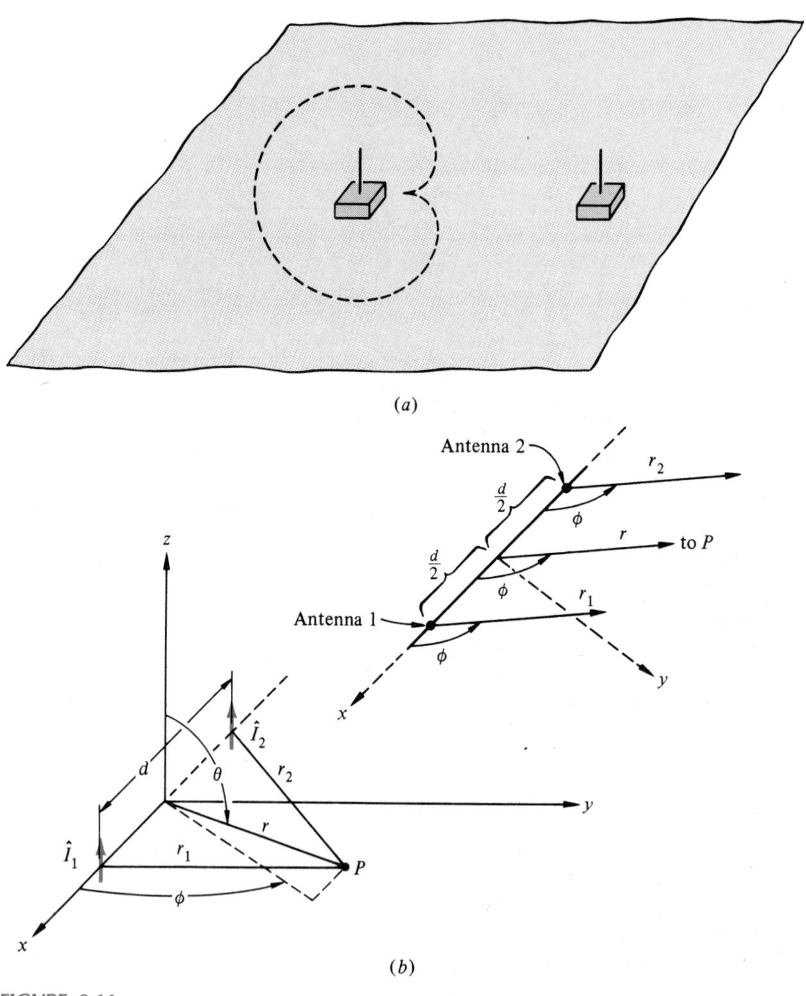

FIGURE 9.14
Antenna arrays.

For example, suppose that the two antennas in Fig. 9.14*b* are half-wave dipoles or quarter-wave monopoles above ground. The terminal phasor currents to the antennas are denoted as $\hat{I}_1$ and $\hat{I}_2$. To determine the proper phase relationship of the currents to produce a desired pattern, we first determine the field at point *P* in Fig. 9.14*b*. We assume that point *P* is in the far field of each antenna. From (66*b*) let us write the electric fields at *P* due to each individual antenna in the form

$$\hat{E}_{\theta 1} = \frac{\hat{K}I\underline{/\alpha}}{r_1} e^{-j\beta_0 r_1} \tag{85a}$$

$$\hat{E}_{\theta 2} = \frac{\hat{K}I\underline{/0}}{r_2} e^{-j\beta_0 r_2} \tag{85b}$$

where $\hat{I}_1 = I\underline{/\alpha}$ and $\hat{I}_2 = \hat{I}\underline{/0}$ and we assume both currents to have the same magnitude. Note that the phase difference between the two currents is α and that the current of antenna 1 leads that of antenna 2 by α. This is usually accomplished by feeding both antennas in parallel with the signal to be transmitted and by placing a phase-shifting network in the line feeding one of the antennas.

Clearly, the forms of the electric field in (85) follow from our previous results; the far fields decay as $1/r$ and have phase dependence $e^{-j\beta_0 r}$, where r is the distance to the field point from the center of the antenna. The constant $\hat{K}$ is simply a grouping of the remaining terms that depend only on θ.

The forms in (85) are also valid, in fact, for dipole (monopole) antennas that are *not* a half-wavelength (quarter-wavelength) long [see (66*b*)]. For such antennas, however, $\hat{I}_1$ and $\hat{I}_2$ are generally not the terminal currents to the antennas. Those terminal currents are nevertheless related by some constant to $\hat{I}_1$ and $\hat{I}_2$ [see (80) and the accompanying discussion], and for identical antennas this constant may be absorbed into the constant $\hat{K}$. The following results apply to these more general cases of any antenna length.

The factor $\hat{K}$ will be a function of θ [see (30) or (66*b*)]. However, we will usually be interested in the variation of the resulting pattern as ϕ is varied for some fixed θ (usually $\theta = \pi/2$ in the xy plane) and a fixed distance r from the center of the array. Thus, for these purposes we may disregard the dependence on θ.

Combining the two contributions given in (85), we obtain the total field at point *P* as

$$\begin{aligned}
\hat{E}_\theta &= \hat{E}_{\theta 1} + \hat{E}_{\theta 2} \\
&= \hat{K}I\left(\frac{e^{-j\beta_0 r_1}}{r_1} e^{j\alpha} + \frac{e^{-j\beta_0 r_2}}{r_2}\right) \\
&= \hat{K}Ie^{j\alpha/2}\left(\frac{e^{-j\beta_0 r_1}e^{j\alpha/2}}{r_1} + \frac{e^{-j\beta_0 r_2}e^{-j\alpha/2}}{r_2}\right)
\end{aligned} \tag{86}$$

To simplify this equation, we assume that P is sufficiently far from the origin that $r_1 \simeq r_2 \simeq r$, where r is the distance from the midpoint of the array to point P. We cannot, however, assume this in the $e^{-j\beta_0 r_1}$ and $e^{-j\beta_0 r_2}$ terms since these terms represent the phase difference in the two fields. As for the long dipole, a small physical difference between r_1 and r_2 can produce a significant difference in these phase terms depending on β_0 (which depends on frequency). For these terms, however, we can simplify the result by again considering the radius vectors along r_1 and r_2 to be approximately parallel to r. In this case, as in the simplification used in the long dipole of the previous section, we may write, as shown in Fig. 9.14b,

$$r_1 \simeq r - \frac{d}{2}\cos\phi \tag{87a}$$

$$r_2 \simeq r + \frac{d}{2}\cos\phi \tag{87b}$$

where d is the separation between the two antennas. Substituting (87) into (86), we obtain

$$\begin{aligned}\hat{E}_\theta &= \frac{\hat{K}I}{r}\, e^{j\alpha/2} e^{-j\beta_0 r}\left(e^{j(\beta_0 (d/2)\cos\phi + \alpha/2)} + e^{-j(\beta_0(d/2)\cos\phi + \alpha/2)}\right) \\ &= \frac{2\hat{K}I}{r}\, e^{j\alpha/2} e^{-j\beta_0 r} \cos\left(\beta_0 \frac{d}{2}\cos\phi + \frac{\alpha}{2}\right)\end{aligned} \tag{88}$$

This may be written as

$$\hat{E}_\theta = 2e^{j\alpha/2}\left[\hat{K}\frac{Ie^{-j\beta_0 r}}{r}\right] F_{\text{array}}(\theta, \phi) \tag{89}$$

where the *array factor* is defined as

$$F_{\text{array}}(\theta, \phi) = \cos\psi \tag{90}$$

and

$$\psi = \beta_0 \frac{d}{2}\cos\phi + \frac{\alpha}{2} \tag{91}$$

The resultant field is the product of the pattern of the individual (identical) elements in brackets and the array factor (which depends only on the antenna spacing and phasing of currents). This is referred to as the principle of *pattern multiplication* and is a general result, as we will see.

Our interest centers on the magnitude of $\hat{E}_\theta$ as a function of ϕ for fixed values of r and θ (usually for $\theta = \pi/2$). Let us write the argument of the array factor as

$$\begin{aligned}\psi &= \tfrac{1}{2}(\beta_0\, d\cos\phi + \alpha) \\ &= \frac{\pi d}{\lambda_0}\cos\phi + \frac{\alpha}{2}\end{aligned} \tag{92}$$

which is a function of electrical distance d/λ_0 from the array. Then the magnitude of (88) may be written as

$$|\hat{E}_\theta| = \frac{M}{r}|\cos\psi| \tag{93}$$

where $M = 2|\hat{K}|I$. The shape of the power pattern is obviously related to the shape of the electric field pattern since in the far field

$$|\hat{E}_\theta| = \eta_0|\hat{H}_\phi| \tag{94}$$

We will, however, be concerned only with the electric field pattern since the location of any minima and maxima in this pattern will correspond to the location of minima and maxima in the power pattern. Note that the pattern will be symmetrical about a line joining the antennas (the x axis in Fig. 9.14*b*). We can adjust not only the phase difference of the antenna currents but also the antenna spacing in order to arrange the pattern shape, as shown by (93) and the following example.

EXAMPLE 9.3 Suppose that the antenna spacing is one-half wavelength, $d = \lambda_0/2$, and that there is no phase difference between the currents, $\alpha = 0$. Sketch the electric field pattern.

Solution For this situation,

$$\psi = \frac{\pi}{2}\cos\phi$$

The array factor becomes

$$\cos\psi = \cos\left(\frac{\pi}{2}\cos\phi\right)$$

For a fixed distance r about the array, the pattern therefore has nulls at values of $\phi = 0°$ and $\phi = 180°$, as shown in Fig. 9.15*a*. The patterns for $d = \lambda_0/2$ and $\alpha = \pi$ and for $d = \lambda_0/4$ and $\alpha = \pi/2$ are shown in Figs. 9.15*b* and *c*, respectively.

The locations of the pattern nulls can be computed in the following manner. The pattern nulls occur at values of ϕ such that $\cos\psi = 0$; that is,

$$\begin{aligned}\psi &= \frac{\pi d}{\lambda_0}\cos\phi + \frac{\alpha}{2} \\ &= \pm\frac{\pi}{2}, \pm\frac{3\pi}{2}, \ldots\end{aligned} \tag{95}$$

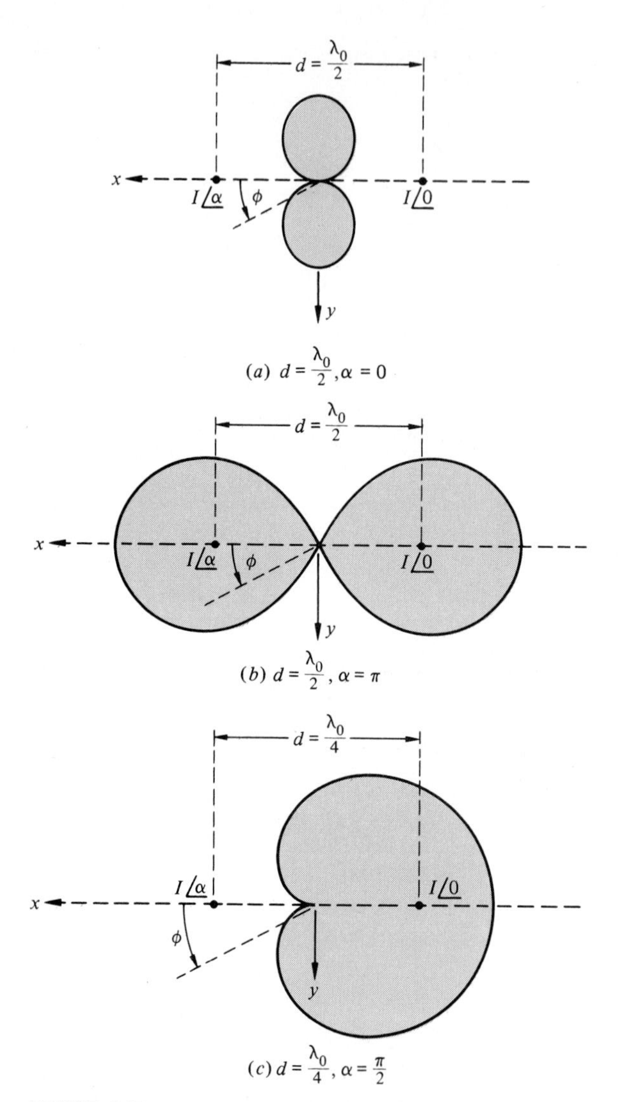

FIGURE 9.15

Example 9.3. Patterns for various spacings and current phase differences.

For $d = \lambda_0/4$ and $\alpha = \pi/2$ in Example 9.3, we require

$$\frac{\pi}{4}\cos\phi + \frac{\pi}{4} = \pm\frac{\pi}{2}, \pm\frac{3\pi}{2}, \ldots$$

or

$$\cos\phi = 1, -3, 5, -7, \ldots$$

The values -3, -7, and 5 are not allowable; thus, the nulls appear only at $\phi = 0°$, as shown in Fig. 9.15*c*.

The maxima and minima in the pattern may be found by differentiating $\cos\psi$ with respect to ϕ and setting the result equal to zero:

$$\frac{d}{d\phi}\cos\left(\frac{\pi d}{\lambda_0}\cos\phi+\frac{\alpha}{2}\right)=0 \tag{96}$$

This results in

$$-\sin\psi\left(-\frac{\pi d}{\lambda_0}\sin\phi\right)=0 \tag{97}$$

or

$$\sin\phi\sin\left(\frac{\pi d}{\lambda_0}\cos\phi+\frac{\alpha}{2}\right)=0 \tag{98}$$

The $\sin\phi=0$ condition results in maxima or minima at $\phi=0$ and $\phi=180°$. This condition is a direct result of the pattern being symmetrical with respect to a line through the two antennas. The other condition yields

$$\frac{\pi d}{\lambda_0}\cos\phi+\frac{\alpha}{2}=0,\ \pm\pi,\ \pm 2\pi,\ldots \tag{99}$$

For $d=\lambda_0/4$ and $\alpha=\pi/2$ in Example 9.3, we obtain

$$\frac{\pi}{4}\cos\phi+\frac{\pi}{4}=0,\ \pm\pi,\ \pm 2\pi$$

or

$$\cos\phi=-1,3,-5,\ldots$$

Obviously, the values $3, -5, \ldots$ are not allowable, and the maxima or minima occur at $\phi=180°$. Since this result may locate maxima as well as minima, one should check each result to determine whether a maximum or a minimum is obtained.

It is important to note that the preceding results can be visualized quite easily. To see this, consider Fig. 9.14*b*. Signals from the two antennas arriving at the distant field point suffer phase shifts of $e^{-j\beta_o r_1}$ and $e^{-j\beta_o r_2}$. There is an additional phase difference between the two received signals due to the difference in phases of the currents. For example, consider Example 9.3 shown in Fig. 9.15. For $d=\lambda_0/2$ and $\alpha=0$, signals arriving at the field point for $\phi=0$ differ in phase by $\beta_0 d=\pi$, because the signal from antenna 2 has to travel an additional distance d over the distance traveled by antenna 1; thus, the signals are 180° out of phase and therefore cancel, creating a null in this direction. For $\phi=90°$, the signals travel the same distance to reach the field point; also, since the antenna currents have the same phase, there is no net phase difference between the two received signals and they add, producing a pattern maximum.

Next consider the problem in Fig. 9.15*b*,where $d = \lambda_0/2$ again but $\alpha = \pi$. For $\phi = 0$, the signal arriving from antenna 2 again has a phase of $-\beta_0 d = -180°$ relative to that of antenna 1 owing to the separation $d = \lambda_0/2$; in addition, the received signal from antenna 2 has a further phase of $-180°$ relative to that from antenna 1 since the current of antenna 2 lags the current of antenna 1 by 180°; thus, the two signals are in phase and add at $\phi = 0°$. At $\phi = 90°$, no distance-induced phase shift is incurred, but the received signals are 180° out of phase and cancel since the antenna currents are 180° out of phase.

The problem in Fig. 9.15*c* is slightly more subtle. For $\phi = 0°$, the signal from antenna 2 suffers a phase shift of $-\beta_0 d = -\pi/2$ relative to that of antenna 1; in addition, the phase of antenna 2 current lags that of antenna 1 current by $-\pi/2$, so that the received signal from antenna 2 has an additional phase of $-\pi/2$ relative to that from antenna 1, resulting in a net phase difference of $-180°$ and therefore cancellation. However, for $\phi = 180°$, the signal arriving from antenna 1 has a phase relative to that of antenna 2 of $-\pi/2$ owing to the distance separation, but since the current of antenna 1 leads that of antenna 2 by $\pi/2$, the arriving signals are in phase and produce a pattern maximum.

Many of the patterns obtained by earlier methods can be verified quite easily for their reasonableness by these methods at $\phi = 0°, 90°, 180°$, and $270°$. These methods can also be used at other points, but the calculation requires one to determine the phase shift caused by the distance difference. This phase shift is, according to (87), $\beta_0(r_1 - r_2) = -2\pi d/\lambda_0 \cos\phi$.

9.4.1 Pattern Multiplication

The resultant pattern for two dipoles (monopoles) given in (89) is the product of the pattern for the individual (identical) elements of the array and the array factor. This result is referred to as *pattern multiplication* and can be extended to larger numbers of array elements, as the following shows.

The patterns of more than two antennas can be obtained by combining the fields from all the antennas in a fashion similar to the case of two antennas above. However, it is often simpler to observe that the resulting pattern can be considered to be a superposition of the individual antenna patterns or groups of patterns in the array. This technique becomes quite simple if the antennas of the array can be grouped into pairs such that the patterns of the individual pairs of antennas are identical in shape and orientation.

As an example, consider Fig. 9.16*a*, which shows an array of four vertical-wire antennas as seen in the $\theta = \pi/2$ plane. The magnitudes of the currents are the same, and the currents of antennas 1 and 3 lead the currents of antennas 2 and 4, respectively, by α degrees. The current of antenna 2 leads the current of antenna 4 by δ degrees. Antennas 1 and 2 are separated by d, as are antennas 3 and 4. Antennas 2 and 3 are separated by $D - d$. If we consider antennas 1 and 2 as a group and antennas 3 and 4 as a group, then these individual groups produce two identical patterns, A and B, separated by D with phase difference δ, as

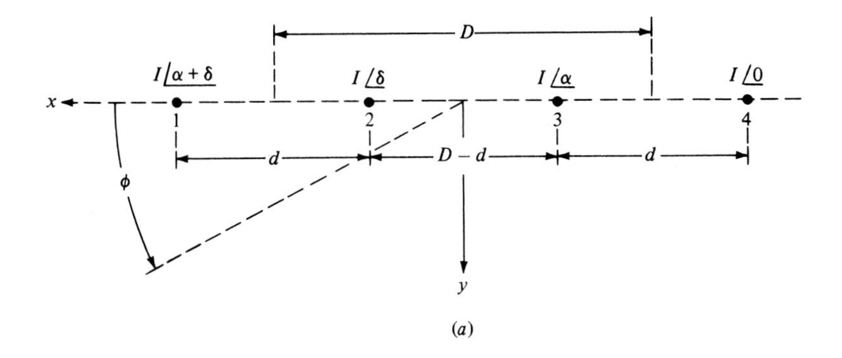

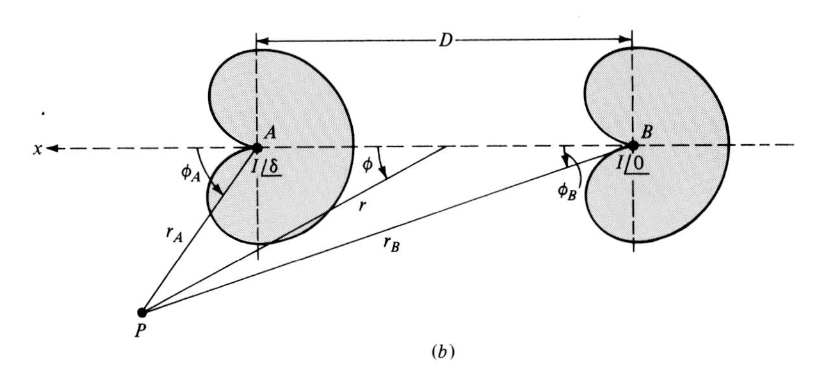

FIGURE 9.16
Pattern mutiplication for a four-element array.

shown in Fig. 9.16*b*. From the result for an individual pair given in (88), we may write the electric field expressions as

$$\hat{E}_{\theta_A} = \frac{2\hat{K}I}{r_A} e^{-j\beta_0 r_A} e^{j\alpha/2} e^{j\delta} \cos\left(\beta_0 \frac{d}{2} \cos\phi_A + \frac{\alpha}{2}\right) \tag{100a}$$

$$\hat{E}_{\theta_B} = \frac{2\hat{K}I}{r_B} e^{-j\beta_0 r_B} e^{j\alpha/2} \cos\left(\beta_0 \frac{d}{2} \cos\phi_B + \frac{\alpha}{2}\right) \tag{100b}$$

Assuming that the field point P is sufficiently far removed from the center of the array, we may assume that $\phi_A \simeq \phi_B \simeq \phi$ in (100). Also, we may assume that $r_A = r_B = r$ in the denominators of (100). For the phase terms $e^{-j\beta_0 r_A}$ and $e^{-j\beta_0 r_B}$, we use an approximation similar to the case of two elements and obtain

$$\begin{aligned}\hat{E}_\theta &= \hat{E}_{\theta_A} + \hat{E}_{\theta_B}\\ &= \frac{4\hat{K}I}{r} e^{-j\beta_0 r} e^{j\alpha/2} e^{j\delta/2} \cos\left(\beta_0 \frac{d}{2}\cos\phi + \frac{\alpha}{2}\right) \cos\left(\beta_0 \frac{D}{2}\cos\phi + \frac{\delta}{2}\right)\end{aligned} \tag{101}$$

Equation (101) shows that the overall pattern of the array may be found as the product of the array factor for the individual pairs:

$$\cos\left(\beta_0 \frac{d}{2} \cos \phi + \frac{\alpha}{2}\right) \tag{102a}$$

and the array factor for the group of pairs:

$$\cos\left(\beta_0 \frac{D}{2} \cos \phi + \frac{\delta}{2}\right) \tag{102b}$$

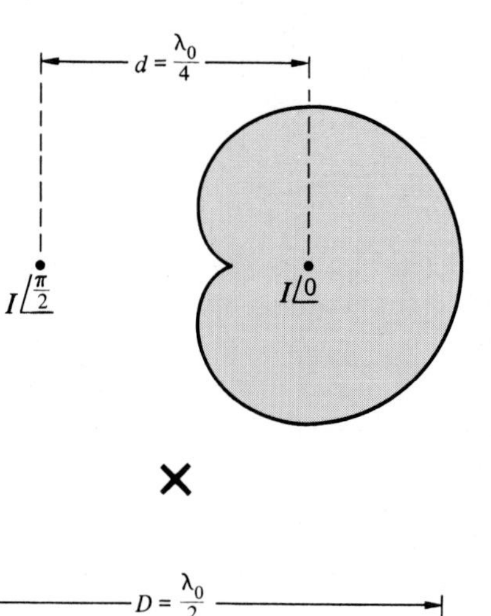

×

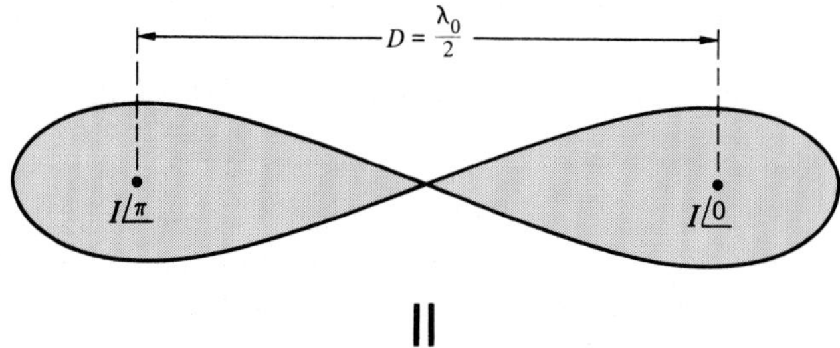

=

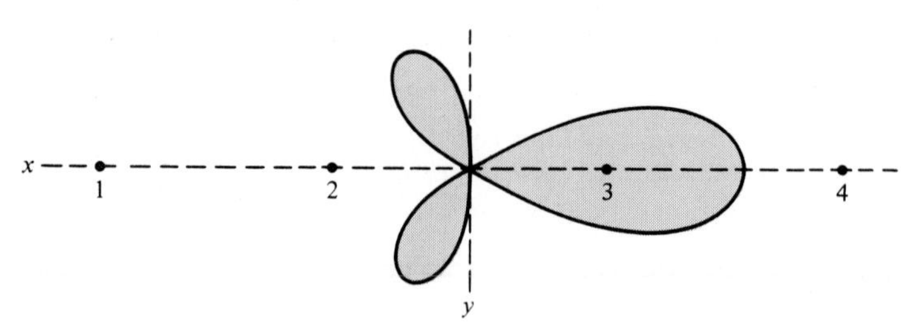

FIGURE 9.17

Example 9.4. Illustration of pattern multiplication.

It should be noted that pattern multiplication is a general result. For example, suppose that an array consists of a group of m identical antennas (or "subarrays"), each having a pattern $F_{\text{ant}}(\theta, \phi)$. If these are replaced by omnidirectional antennas having the same relative separation and current phasing, the pattern of this array is given by $F_{\text{array}}(\theta, \phi)$. Since the far fields of each of the m antennas vary as $e^{-j\beta_0 r}/r$, one can obtain the resulting array pattern as the product $F_{\text{ant}}(\theta, \phi)F_{\text{array}}(\theta, \phi)$.

EXAMPLE 9.4 Consider Fig. 9.16 and suppose that $d = \lambda_0/4$, $\alpha = \pi/2$, $D = \lambda_0/2$, and $\delta = \pi$. Sketch the electric field pattern.

Solution Equation (102*a*) corresponds to the pattern in Fig. 9.15*c*, and (102*b*) corresponds to the pattern in Fig. 9.15*b*. The result is shown in Fig. 9.17.

9.5 Antenna Directivity and Gain

We have seen that a knowledge of the current distribution over the surface of an antenna is, in theory, sufficient to determine all of the fields of that antenna (near as well as far fields). However, for other antennas more complicated in physical structure than the elemental dipoles or the long dipole, determining (or obtaining a reasonable approximation to) the current distribution over the antenna surface is usually a formidable problem in itself. This is particularly true for surface or aperture antennas, such as the horns and parabolics shown in Fig. 9.1. For this reason, the radiation characteristics of an antenna are often characterized in terms of measurable quantities such as directivity and gain. These concepts serve to reduce the characterization of antennas having complicated physical geometries to the measurement of certain properties of the antenna.

The *directive gain* of an antenna $D(\theta, \phi)$ is a measure of the concentration of the radiated power in a particular θ, ϕ direction at a fixed distance r away from the antenna. For the elemental dipoles, the long dipole, and the monopole, we noted that the radiated power is a maximum for $\theta = 90°$ and is zero for $\theta = 0°$ and $\theta = 180°$. To obtain a more quantitative measure of this concentration of radiated power, we will define the radiation intensity $U(\theta, \phi)$.

In Sec. 9.2.1, we found that the radiated average-power density for a simple Hertzian dipole is given by

$$\mathbf{S}_{\text{av}} = \frac{|\mathbf{E}_{\text{far field}}|^2}{2\eta_0}\,\mathbf{a}_r \tag{103}$$

In terms of the items in (38) this becomes

$$\mathbf{S}_{\text{av}} = \frac{E_0^2}{2\eta_0 r^2}\,\mathbf{a}_r \tag{104}$$

For the long dipole and the monopole, as well as for most other more complicated antennas, the field structure at a sufficient distance from the antenna also resembles (locally) a uniform plane wave, and the relationships in (103) and (104) hold for these antennas as well.† (E_0 is, of course, different for other antennas.)

To obtain a power pattern relationship that is independent of distance from the antenna, we multiply (104) by r^2 and define the resulting quantity to be the *radiation intensity* $U(\theta, \phi)$ for the antenna; that is,

$$U(\theta, \phi) = r^2 S_{av} \tag{105}$$

The radiation intensity will be a function only of θ and ϕ and will be independent of distance from the antenna. The total average power radiated will be

$$\begin{aligned} P_{rad} &= \int_{\theta=0}^{\pi} \int_{\phi=0}^{2\pi} S_{av} r^2 \sin\theta \, d\phi \, d\theta \\ &= \int_{\theta=0}^{\pi} \int_{\phi=0}^{2\pi} U(\theta, \phi) \sin\theta \, d\phi \, d\theta \\ &= \oint U(\theta, \phi) \, d\Omega \end{aligned} \tag{106}$$

The quantity $d\Omega = \sin\theta \, d\phi \, d\theta$ is an element of a solid angle Ω, as shown in Fig. 9.18, and the unit of solid angle is the steradian (sr). The units of U are therefore watts per steradian. Note that for $U = 1$, (106) integrates to 4π. The total radiated power is therefore the integral of the radiation intensity over a solid angle of 4π sr. Note also that the average radiation intensity is the total radiated power divided by 4π sr:

$$U_{av} = \frac{P_{rad}}{4\pi} \tag{107}$$

The radiation intensity for more complicated antennas is similarly defined. The directive gain of an antenna in a particular direction, $D(\theta, \phi)$, is the ratio of the radiation intensity in that direction to the average radiation intensity:

$$\begin{aligned} D(\theta, \phi) &= \frac{U(\theta, \phi)}{U_{av}} \\ &= \frac{4\pi U(\theta, \phi)}{P_{rad}} \end{aligned} \tag{108}$$

† See E. C. Jordan and K. G. Balmain, *Electromagnetic Waves and Radiating Systems*, 2d ed., Prentice-Hall, Englewood Cliffs, N.J., 1968, pp. 341–342.

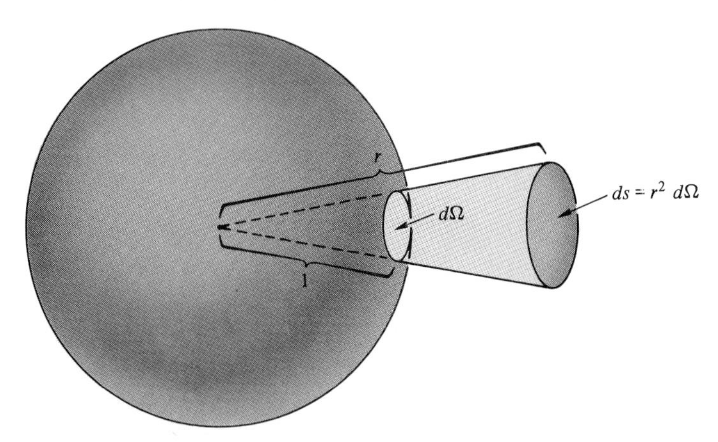

FIGURE 9.18
Surface area of a sphere as a portion of a solid angle.

The directivity of the antenna is the directive gain in the direction that yields a maximum:

$$D_{\max} = \frac{U_{\max}}{U_{\text{av}}} \tag{109}$$

EXAMPLE 9.5 Determine the directive gain and directivity of a Hertzian dipole.

Solution The radiation intensity is found from (35) and (105):

$$\begin{aligned} U(\theta, \phi) &= r^2 S_{\text{av}} \\ &= 15\pi \left(\frac{dl}{\lambda_0}\right)^2 |\hat{I}|^2 \sin^2 \theta \end{aligned}$$

and the radiated power is given in (36):

$$P_{\text{rad}} = 40\pi^2 \left(\frac{dl}{\lambda_0}\right)^2 |\hat{I}|^2$$

Thus, the directive gain is

$$\begin{aligned} D(\theta, \phi) &= \frac{4\pi U(\theta, \phi)}{P_{\text{rad}}} \\ &= 1.5 \sin^2 \theta \end{aligned} \tag{110}$$

The directivity is therefore the directive gain at $\theta = \pi/2$:

$$D_{\max} = 1.5 \tag{111}$$

For the half-wave dipole, we obtain

$$D(\theta, \phi) = \frac{\eta_0}{\pi R_{\text{rad}}} F^2(\theta)$$

$$= 1.64\, F^2(\theta) \tag{112}$$

where $F(\theta)$ is given by (68) and $R_{\text{rad}} = 73\ \Omega$ and

$$D_{\max} = 1.64 \tag{113}$$

which occurs for $\theta = \pi/2$.

The directive gain $D(\theta, \phi)$ of an antenna is simply a function of the shape of the antenna pattern. The *power* gain $G(\theta, \phi)$, on the other hand, takes into account the losses of the antenna. Suppose that a total of P_{app} watts of power is applied to the antenna and that only P_{rad} watts is radiated. The difference is consumed in ohmic losses of the antenna as well as in other inherent losses, such as those in an imperfect ground for monopoles. If we define an efficiency factor e as

$$e = \frac{P_{\text{rad}}}{P_{\text{app}}} \tag{114}$$

then the power gain is related to the directive gain as

$$G(\theta, \phi) = eD(\theta, \phi) \tag{115}$$

where we have defined the power gain as

$$G(\theta, \phi) = \frac{4\pi U(\theta, \phi)}{P_{\text{app}}} \tag{116}$$

For most antennas, the efficiency is nearly 100 percent and thus the power gain and directive gain are nearly equal. Although there are antennas with efficiencies much less than 100 percent, we will assume in the remainder of our discussions that $e = 1$ so that the power gain and directive gain will be the same.†

We also need to discuss the concept of an isotropic point source. An *isotropic point source* is a fictitious lossless antenna that radiates power equally in all directions. Since this "antenna" is lossless, its directive gain and power gain are synonymous. For an isotropic point source radiating or transmitting a total average power of P_T watts, the power density at some distance d away is

$$\mathbf{S}_{\text{av}} = \frac{P_T}{4\pi d^2}\, \mathbf{a}_r \tag{117}$$

---

† Electrically short dipole antennas, for example, may have the ohmic resistance on the order of the radiation resistance, with a resulting efficiency much less than 100 percent.

This relationship is obtained by realizing that since the total transmitted power P_T is being radiated uniformly in the radial direction, the density of power at a distance d away is obtained by dividing P_T by the surface area of a sphere of radius d.

The isotropic point source, although quite idealistic, is useful as a standard or reference antenna to which we refer many of our calculations. For example, since the isotropic point source is lossless, the directive gain and power gain are equal, and both will be designated by G_0. The gain becomes

$$\begin{aligned} G_0(\theta, \phi) &= \frac{4\pi U_0(\theta, \phi)}{P_T} \\ &= 1 \end{aligned} \tag{118}$$

Therefore, the directive gain and power gain of other antennas may be thought of as being determined with respect to an isotropic point source. In certain other cases the gain of an antenna may be referred to the gain of a half-wave dipole. One must be careful to determine the reference antenna. We will assume throughout the remainder of this discussion that all antenna gains are determined with respect to an isotropic point source.

Quite often the gain (directive or power) of an antenna is given in decibels, where

$$G_{\text{dB}} = 10 \log G \tag{119}$$

For example, the Hertzian dipole has a maximum gain of 1.76 dB and the isotropic point source has a maximum gain of 0 dB. The half-wave dipole has a maximum gain of 2.15 dB. Equivalently, we say that the gain of an antenna is the gain over (or with respect to) an isotropic antenna:

$$\begin{aligned} G_{\text{dB}} &= 10 \log \frac{G}{G_0} \\ &= 10 \log G \end{aligned} \tag{120}$$

9.6 Antenna Coupling

Our primary interest, however, is to determine the coupling between two antennas that are separated in space a distance d, as shown in Fig. 9.19*a*. In other words, if we know the relative orientation of the two antennas and the power applied to one of them, we wish to find the power delivered to some load impedance connected to the terminals of the other antenna. The Friis transmission equation, discussed in Sec. 9.7, is a simple, approximate method for calculating this coupling.

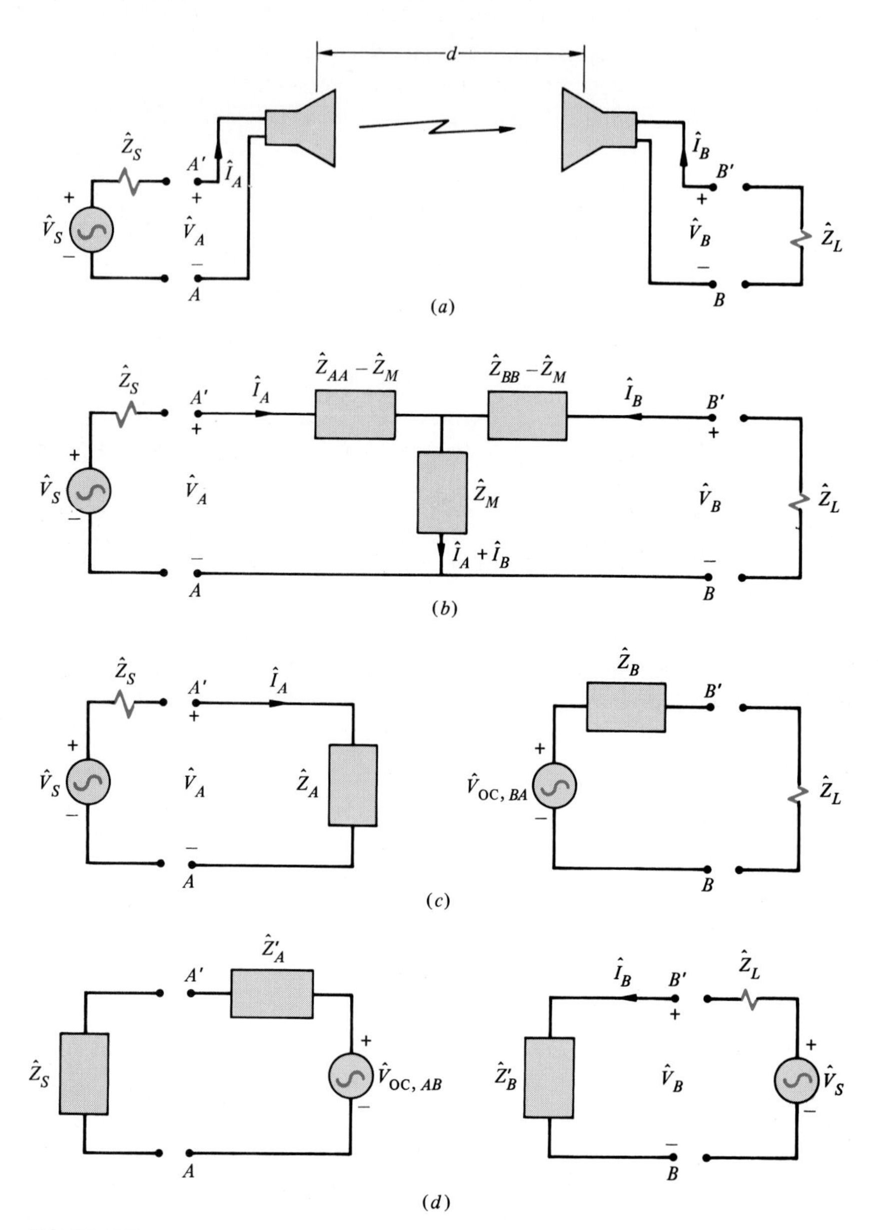

FIGURE 9.19

Equivalent circuits representing the coupling between two antennas.

For the purposes of representing this coupling, we may relate the voltages and currents at the terminals of each antenna as a pair of coupled equations:

$$\hat{V}_A = \hat{Z}_{AA}\hat{I}_A + \hat{Z}_{AB}\hat{I}_B \tag{121a}$$

$$\hat{V}_B = \hat{Z}_{BA}\hat{I}_A + \hat{Z}_{BB}\hat{I}_B \tag{121b}$$

The result in (121) follows from the usual assumption that the intervening medium is linear. If, in addition, the medium is isotropic, then

$$\hat{Z}_{AB} = \hat{Z}_{BA} \tag{122}$$

The usual transmission medium is free space, which is linear and isotropic.

The result in (122) is related to a more general theorem known as the *theorem of reciprocity*.† The reciprocity theorem is applied to antennas as follows. Consider two (possibly different) antennas in a linear, isotropic medium: antennas A and B, shown in Fig. 9.19*a*. If a current $\hat{I}_A$ is applied to the terminals of antenna A and the terminals of antenna B are opened ($\hat{I}_B = 0$), a voltage $\hat{V}_B$ will appear at the terminals of antenna B. If, on the other hand, a current $\hat{I}_B$ is applied to the terminals of antenna B and the terminals of antenna A are opened ($\hat{I}_A = 0$), a voltage $\hat{V}_A$ will appear across the terminals of antenna A. The reciprocity theorem provides the result that the ratios of each driving current to its resulting open-circuit voltages are the same; that is,

$$\left.\frac{\hat{V}_B}{\hat{I}_A}\right|_{\hat{I}_B=0} = \left.\frac{\hat{V}_A}{\hat{I}_B}\right|_{\hat{I}_A=0} \tag{123}$$

From (121) we have

$$\hat{Z}_{AB} = \left.\frac{\hat{V}_A}{\hat{I}_B}\right|_{\hat{I}_A=0} \tag{124a}$$

$$\hat{Z}_{BA} = \left.\frac{\hat{V}_B}{\hat{I}_A}\right|_{\hat{I}_B=0} \tag{124b}$$

and (123) shows that $\hat{Z}_{AB} = \hat{Z}_{BA}$. Thus, (121) becomes

$$\hat{V}_A = \hat{Z}_{AA}\hat{I}_A + \hat{Z}_M\hat{I}_B \tag{125a}$$

$$\hat{V}_B = \hat{Z}_M\hat{I}_A + \hat{Z}_{BB}\hat{I}_B \tag{125b}$$

where $\hat{Z}_{AB} = \hat{Z}_{BA} = \hat{Z}_M$. These equations may be represented as a lumped, two-port equivalent circuit, as shown in Fig. 9.19*b*. The relationships in (125)

---

† See D. T. Paris and F. K. Hurd, *Basic Electromagnetic Theory*, McGraw-Hill, New York, 1969, pp. 502–505.

and the equivalent circuit in Fig. 9.19*b* are an exact representation of the coupling between two antennas; determining the items $\hat{Z}_{AA}$, $\hat{Z}_{BB}$, and $\hat{Z}_M$, however, is essential to the characterization and is usually quite difficult. Nevertheless, several important characteristics of the problem of coupling can be deduced from this representation without the need for determining the values of $\hat{Z}_{AA}$, $\hat{Z}_{BB}$, and $\hat{Z}_M$.

If antenna A is driven by some source with phasor source voltage $\hat{V}_S$ and source impedance $\hat{Z}_S$, and antenna B is terminated in a load impedance $\hat{Z}_L$, we may compute the coupling with the equivalent circuits shown in Fig. 9.19*c*. The input impedance to antenna A, $\hat{Z}_A$, in Fig. 9.19*c* becomes

$$\hat{Z}_A = (\hat{Z}_{AA} - \hat{Z}_M) + \{\hat{Z}_M \| [(\hat{Z}_{BB} - \hat{Z}_M) + \hat{Z}_L]\} \tag{126}$$

where $\|$ means "in parallel." The open-circuit voltage source $\hat{V}_{\text{OC},BA}$ in Fig. 9.19*c* is given by

$$\hat{V}_{\text{OC},BA} = \hat{I}_A \hat{Z}_M \tag{127}$$

and the equivalent source impedance is

$$\hat{Z}_B = (\hat{Z}_{BB} - \hat{Z}_M) + \{\hat{Z}_M \| [(\hat{Z}_{AA} - \hat{Z}_M) + \hat{Z}_S]\} \tag{128}$$

Let us now assume that the load is matched to the receiving antenna; that is,

$$\hat{Z}_L = \hat{Z}_B^* \tag{129}$$

If $\hat{Z}_A$ is written as $R_A + jX_A$, the power delivered by the source to the transmitting antenna (the radiated power for a lossless antenna) becomes

$$P_T = \tfrac{1}{2}|\hat{I}_A|^2 R_A \tag{130}$$

Similarly, if $\hat{Z}_B = R_B + jX_B$, the power delivered to the matched load is

$$P_R = \frac{|\hat{I}_A \hat{Z}_M|^2}{8R_B} \tag{131}$$

Taking the ratio of (130) and (131), we obtain

$$\frac{P_R}{P_T} = \frac{|\hat{Z}_M|^2}{4R_A R_B} \tag{132}$$

Suppose we now place the voltage source $\hat{V}_S$ in the terminal circuit of antenna B but retain the impedances $\hat{Z}_S$ and $\hat{Z}_L$ in their previous locations. The resulting equivalent circuit is shown in Fig. 9.19*d*, where $V_{\text{OC},AB} = \hat{I}_B \hat{Z}_M$. For this situation, we can see that $\hat{Z}'_A = \hat{Z}_A$ and $\hat{Z}'_B = \hat{Z}_B$. Thus, we conclude that *the* *input impedance to an antenna when it is used for transmission is equal to the equivalent source impedance when it is used for reception.* Note that for this statement to be true, we require that the terminal impedances for each antenna remain unchanged. If the antennas are widely separated, $\hat{Z}_M$ will be small in comparison with $\hat{Z}_{AA}$ and $\hat{Z}_{BB}$. For this situation, $\hat{Z}_A \simeq \hat{Z}'_A$ and $\hat{Z}_B \simeq \hat{Z}'_B$ independent of the source and load impedances, $\hat{Z}_S$ and $\hat{Z}_L$ [see (126) and (128)].

With the previous development, we can obtain an additional important result: *The transmission pattern of an antenna is the same as its reception pattern.* To show this, consider Fig. 9.20. Antenna A is held in a fixed position and antenna B is moved about on a sphere of constant radius that is centered at antenna A. Angles θ_B and ϕ_B give the orientation of the axis of antenna B with respect to a radial line between the two antennas. Similarly, angles θ_A and ϕ_A give the orientation of this radial line with respect to the axis of antenna A. Now let θ_B and ϕ_B be fixed, but vary θ_A and ϕ_A by moving antenna B on the surface of the sphere. If antenna B is receiving, the received power at antenna B at various positions on the sphere gives the transmission pattern of antenna A and, according to (132) is related to θ_A and ϕ_A by $\hat{Z}_M(\theta_A, \phi_A)$. Now let antenna B be the transmitter and antenna A be the receiver. Once again, moving antenna B about on the sphere, the received power at antenna A is given by (132), which

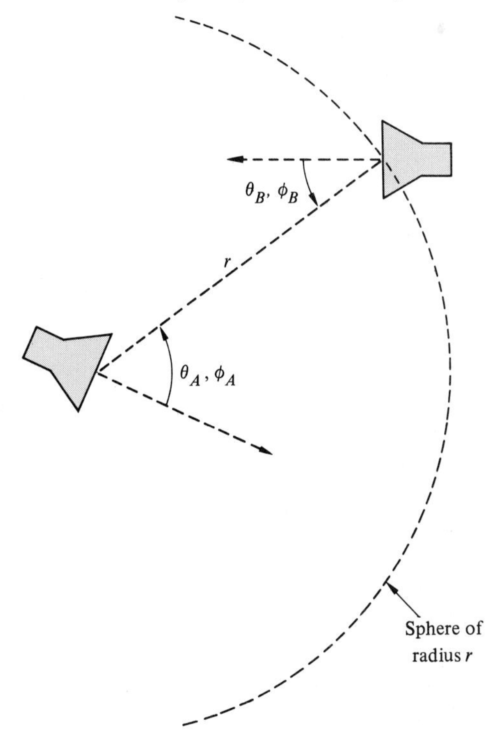

FIGURE 9.20
Demonstration of reciprocity for antenna transmission and reception patterns.

gives the reception pattern of antenna A as a function of θ_A and ϕ_A. This received power is again related to $\hat{Z}_M(\theta_A, \phi_A)$. [Simply interchange R_A and R_B in (132) when the roles of the antennas are reversed.] By reciprocity, $\hat{Z}_M(\theta_A, \phi_A)$ is the same regardless of whether antenna A is transmitting and antenna B is receiving or vice versa, and we obtain the desired result. Thus, the transmission pattern of an antenna may be measured by using it as a receiving antenna—a result that is often useful in pattern measurement.

If the antennas are widely separated, $\hat{Z}_M$ will be quite small in comparison with $\hat{Z}_{AA}$ and $\hat{Z}_{BB}$. For this case, we may approximate (126) and (128) as

$$\hat{Z}_A \simeq \hat{Z}_{AA} \tag{133a}$$

$$\hat{Z}_B \simeq \hat{Z}_{BB} \tag{133b}$$

For this case of widely separated antennas, the input impedance of each antenna is essentially independent of the other antenna and is composed of a real and imaginary part and is of the form

$$\hat{Z}_{\text{in}} = R_{\text{in}} + jX_{\text{in}} \tag{134}$$

The real part, R_{in}, includes a term due to the losses. For a lossless half-wave dipole or quarter-wave monopole, $R_{\text{in}} = R_{\text{rad}}$.

An additional useful concept is that of an antenna's effective aperture. The *effective aperture* of an antenna is related to its ability to extract energy from a passing wave. The effective aperture of an antenna, A_e, is the ratio of the power received (in its load), P_R, to the power density of the incident wave, S_{av}, when the polarization of the incident wave and the polarization of the receiving antenna are matched:

$$A_e = \frac{P_R}{S_{\text{av}}} \quad \text{m}^2 \tag{135}$$

The maximum effective aperture A_{em} is the ratio in (135) when the load impedance is the conjugate of the antenna impedance ($\hat{Z}_L = \hat{Z}_B^*$ in Fig. 9.19c), which means that maximum power transfer to the load takes place. The requirement for matched polarization in (135) will be discussed later. For a linearly polarized incident wave and a receiving antenna that produces linearly polarized waves when it is used for transmission, the requirement for matched polarization essentially means that the antenna is oriented with respect to the incident wave to produce the maximum response; that is, the antenna and the incident electric field vector must be parallel.

EXAMPLE 9.6 Compute the maximum effective aperture of the Hertzian dipole for an incident, linearly polarized, uniform plane wave.

Solution If the dipole is terminated at some gap in its length in an impedance $\hat{Z}_L$, we assume that $\hat{Z}_L = R_{\text{rad}} - jX$, where the input impedance to the dipole is $R_{\text{rad}} + jX$ and the dipole is assumed lossless. Let us suppose that the incident wave is arriving broadside ($\theta = 90°$) to the antenna. We orient the dipole for maximum response; that is, the axis of the dipole is parallel to the electric field of the incident wave $\hat{E}$. The power density in the incident wave is

$$S_{\text{av}} = \frac{1}{2}\frac{|\hat{E}|^2}{\eta_0}$$

and the open-circuit voltage is

$$|\hat{V}_{\text{OC}}| = |\hat{E}|\, dl$$

Since the load is matched for maximum power transfer, the power received is

$$\begin{aligned} P_R &= \frac{|\hat{V}_{\text{OC}}|^2}{8R_{\text{rad}}} \\ &= \frac{|\hat{E}|^2\, dl^2}{8R_{\text{rad}}} \end{aligned}$$

Substituting the value for R_{rad} given in (37), we obtain

$$P_R = \frac{|\hat{E}|^2 \lambda_0^2}{640\pi^2}$$

Thus, the maximum effective aperture is

$$\begin{aligned} A_{em} &= \frac{P_R}{S_{\text{av}}} \qquad \text{m}^2 \\ &= 1.5\,\frac{\lambda_0^2}{4\pi} \end{aligned}$$

The effective aperture is not necessarily related to an antenna's physical "aperture," or area. For the Hertzian dipole, A_{em} depends on frequency through λ_0. Note that

$$A_{em} = \frac{\lambda_0^2}{4\pi} D_{\max}$$

and thus the maximum effective aperture and the directivity of this antenna are related by $\lambda_0^2/4\pi$. We will find this relationship to hold for other antennas as well.

The angle of arrival need not be broadside to the antenna for this relationship to hold. For example, suppose that the incident wave is arriving at an angle θ with respect to the dipole axis but that the electric field is oriented in the θ direction. Thus, this incident field is aligned with the field that would be produced (E_θ) in this direction if the antenna were used for transmission. The open-circuit voltage induced in the antenna is

$$|\hat{V}_{\mathrm{OC}}| = |\hat{E}|\, dl \sin\theta$$

since the component of $\hat{E}$ along the dipole axis is $|\hat{E}| \sin\theta$. Thus, carrying through the above, the maximum effective aperture in this direction is

$$A_{em}(\theta, \phi) = 1.5 \frac{\lambda_0^2}{4\pi} \sin\theta$$

Note that the directivity of this antenna in the direction (θ) of the incoming wave is

$$D(\theta, \phi) = 1.5 \sin\theta$$

Thus

$$A_{em}(\theta, \phi) = \frac{\lambda_0^2}{4\pi} D(\theta, \phi)$$

Reciprocity shows that the reception pattern of an antenna is the same as the transmission pattern of that antenna. We would therefore expect the (directive) gain and effective aperture to be related for any antenna. To show this, we again consider the two antennas shown in Fig. 9.19*a*. Suppose that antenna A is transmitting P_T watts and antenna B is receiving P_R watts in its terminal impedance. The power density at the receiving antenna is the power density of an isotropic radiator radiating P_T watts multiplied by the gain† of the transmitting antenna in the direction of the transmission:

$$S_{\mathrm{av}} = \frac{P_T}{4\pi d^2} G_A \tag{136}$$

Assuming the receiving antenna to be matched to its load and to the polarization of the incident wave, the received power is

$$P_R = S_{\mathrm{av}} A_{emB} \tag{137}$$

† Strictly speaking, since we are discussing the power *radiated* by the antenna, the gain should be directive gain and not power gain. However, the two are simply related by the efficiency of the transmitting antenna.

Substituting (136) into (137), we obtain

$$\frac{P_R}{P_T} = \frac{G_A A_{emB}}{4\pi d^2} \tag{138}$$

Substituting (138) into (132) yields

$$|\hat{Z}_M|^2 = \frac{R_A R_B G_A A_{emB}}{\pi d^2} \tag{139}$$

If the roles of the two antennas are reversed (that is, antenna B is transmitting and antenna A is receiving), we obtain

$$|\hat{Z}_M|^2 = \frac{R_B R_A G_B A_{emA}}{\pi d^2} \tag{140}$$

Comparing (139) and (140), we find that

$$\frac{G_A}{G_B} = \frac{A_{emA}}{A_{emB}} \tag{141}$$

Rewriting (141), we obtain

$$G_A = \frac{G_B}{A_{emB}} A_{emA} \tag{142}$$

and the gain and maximum effective aperture of an antenna are related by some constant. Since antennas A and B are completely general, it suffices to consider one specific antenna for antenna B in order to compute this constant. For the Hertzian dipole,

$$\begin{aligned} \frac{D_{\max}}{A_{em}} &= \frac{1.5}{1.5\lambda_0^2/4\pi} \\ &= \frac{4\pi}{\lambda_0^2} \end{aligned} \tag{143}$$

This result shows that for any antenna,

$$G(\theta, \phi) = \frac{4\pi}{\lambda_0^2} A_{em}(\theta, \phi) \tag{144}$$

The direction for A_{em} (the direction of the incoming incident wave with respect to the receiving antenna) is the direction of G (the gain of the antenna in this direction when it is used for transmission).

There is an important implicit assumption in the result in (144): The polarization of the incoming wave must be matched to the polarization of the receiving antenna. *Polarization* refers to the orientation of the radiated electric field vector at points in the far field and was discussed in Chap. 6, Sec. 6.5. For example, the Hertzian dipole and the long dipole produce far-field-radiated electric fields that are always in the θ direction. The direction of this vector at a point in space is fixed; the amplitude increases and decreases sinusoidally. This type of polarization of $\boldsymbol{\mathcal{E}}$ is referred to as linear polarization; the direction of $\boldsymbol{\mathcal{E}}$ at some point is along a fixed line. For other types of antennas that produce circularly polarized waves, $\boldsymbol{\mathcal{E}}$ rotates so that the tip of $\boldsymbol{\mathcal{E}}$ traces a circle at each point in the far field. The most general form of polarization is elliptical, which encompasses both linear and circular polarizations.

In order that (144) be true, the polarization of the incoming wave must be matched to the polarization of the receiving antenna; that is, the $\boldsymbol{\mathcal{E}}$ vector of the incoming wave must be of the same polarization as the wave that would be transmitted by the receiving antenna in this direction.† If this is not the case, the relationship in (144) becomes‡

$$G(\theta, \phi) = p \frac{4\pi}{\lambda_0^2} A_{em}(\theta, \phi) \tag{145}$$

where p is a polarization mismatch factor which is less than unity, $0 \leq p \leq 1$, and accounts for the degree of polarization mismatch. In this relation we assume that the receiving antenna is also matched to its load. Otherwise, we would need an additional impedance mismatch factor q, $0 \leq q \leq 1$, to multiply this relation to account for this mismatch.

9.7 The Friis Transmission Equation§

We now have the necessary background to derive the Friis transmission equation for the coupling between two antennas. As was pointed out previously, exact calculation of the coupling between two antennas is usually a formidable problem. For this reason, many practical calculations of antenna coupling are carried out, approximately, with the Friis transmission equation, which makes use of the measured gains of the two antennas.

† This definition of matched polarization is somewhat oversimplified. A precise and complete definition that encompasses all antenna and polarization possibilities is quite complicated and lengthy. The interested reader is referred to *Microwave Antenna Measurements*, Scientific-Atlanta, Atlanta, Ga., July 1970, chaps. 2 and 3.

‡ C. T. Tai, "On the Definition of Effective Aperture of Antennas," *IRE Trans. on Antennas and Propagation*, vol. AP-9, March 1961, pp. 224–225.

§ H. T. Friis, "A Note on a Simple Transmission Formula," *Proc. IRE*, vol. 34, May 1946, pp. 254–256.

Consider two antennas in free space (Fig. 9.21). One antenna is transmitting a total of P_T watts of power and the other is receiving P_R watts of power in its terminal impedance. The transmitting antenna has a gain of $G_T(\theta_T, \phi_T)$ and an effective aperture of $A_{eT}(\theta_T, \phi_T)$ in the direction of transmission, θ_T, ϕ_T. The receiving antenna has a gain and effective aperture of $G_R(\theta_R, \phi_R)$ and $A_{eR}(\theta_R, \phi_R)$ in the direction of transmission, θ_R, ϕ_R. From our previous results, the power density at the receiving antenna is [see (136)]

$$S_{\text{av}} = \frac{P_T}{4\pi d^2} G_T(\theta_T, \phi_T) \tag{146}$$

and the received power is

$$P_R = S_{\text{av}} A_{eR}(\theta_R, \phi_R) \tag{147}$$

Substituting (147) into (146), we obtain

$$\frac{P_R}{P_T} = \frac{G_T(\theta_T, \phi_T) A_{eR}(\theta_R, \phi_R)}{4\pi d^2} \tag{148}$$

Replacing the effective aperture of the receiving antenna with its gain via (144) (assuming a matched load and matched polarization so that the effective apertures are the maximum effective apertures):

$$G_R(\theta_R, \phi_R) = \frac{4\pi}{\lambda_0^2} A_{eR}(\theta_R, \phi_R) \tag{149}$$

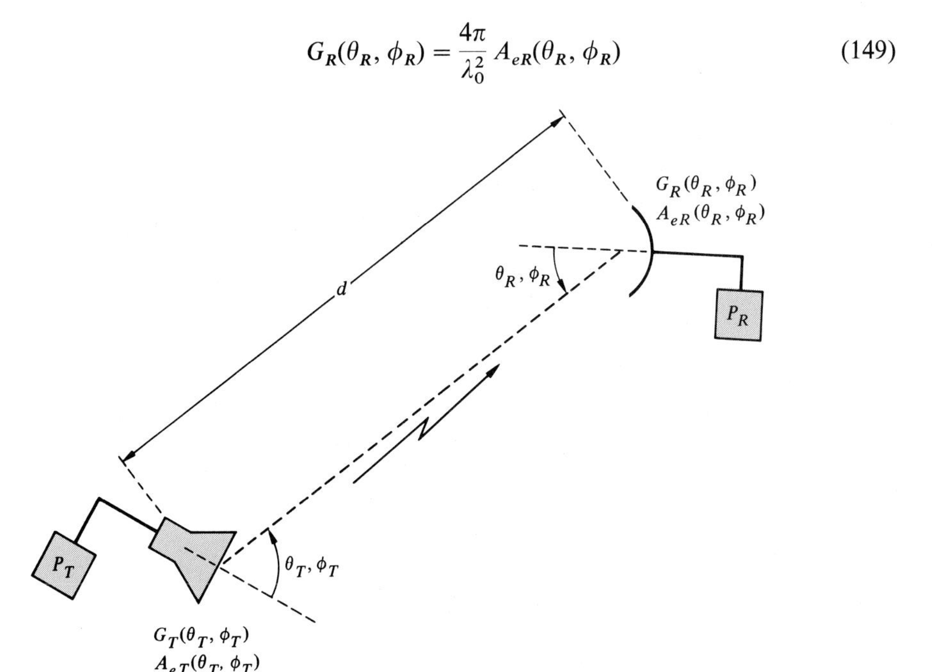

FIGURE 9.21

Configuration for the derivation of the Friis transmission equation.

we obtain the most commonly used version of the Friis transmissison equation:

$$\frac{P_R}{P_T} = G_T(\theta_T, \phi_T)G_R(\theta_R, \phi_R)\left(\frac{\lambda_0}{4\pi d}\right)^2 \tag{150}$$

Note that the coupling between these two antennas decreases with increased separation: a sensible result.

The electric field intensity of the transmitted field at a distance d away from the antenna can also be computed. The power density in the transmitted wave is given by

$$S_{\text{av}} = \frac{1}{2}\frac{|\hat{E}|^2}{\eta_0} \tag{151}$$

and from (146)

$$S_{\text{av}} = \frac{P_T G_T(\theta_T, \phi_T)}{4\pi d^2} \tag{152}$$

Combining (152) and (151), we obtain

$$|\hat{E}| = \frac{\sqrt{60 P_T G_T(\theta_T, \phi_T)}}{d} \tag{153}$$

since $\eta_0 = 120\pi$.

Quite often, in practice, the antenna gains are stated in decibels, as shown in (119). In decibels, the Friis transmission equation given in (150) becomes

$$10 \log \frac{P_R}{P_T} = G_{T,\text{dB}} + G_{R,\text{dB}} - 20 \log f - 20 \log d + 148 \tag{154}$$

One final item should be emphasized. There are a number of assumptions inherent in the Friis transmission equation. In order that the relationship between gain and effective aperture given in (144) be valid, the receiving antenna must be matched to its load impedance and the polarization of the incoming wave; otherwise, the result in (150) will result in an upper limit on the coupling ("worst case"). We also require that the two antennas be in the far field of each other. The far-field criterion is usually taken to be the following. Inherent in the effective-aperture concept is the assumption that the incoming field resembles a uniform plane wave in the vicinity of the receiving antenna. The transmitted wave in the far field of the transmitting antenna resembles a spherical wave from a point source that only locally resembles a uniform plane wave. In order to arrive at a quantitative criterion for the receiving antenna to be in the far field of the transmitting antenna, consider Fig. 9.22*a*. If we denote the maximum

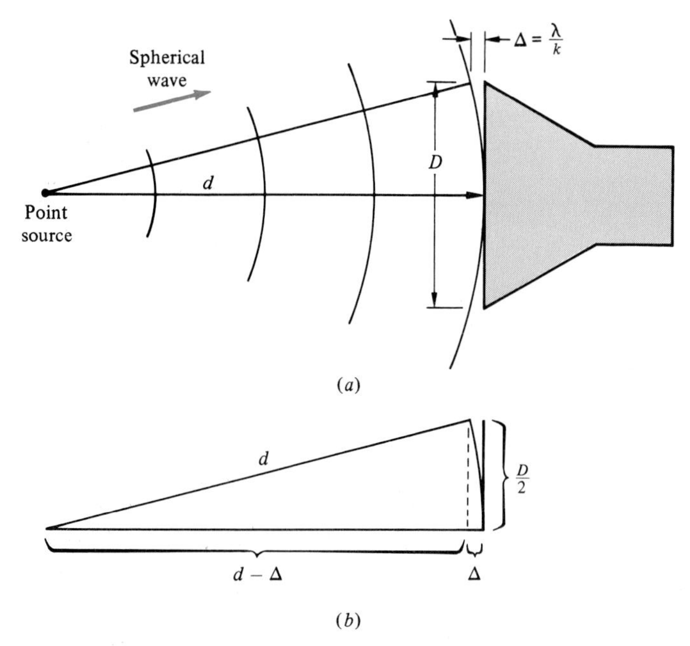

FIGURE 9.22

Derivation of the far-field criterion for surface-type antennas.

dimension of the antenna as D and require that the incident spherical wave differ from a uniform plane wave at the edges of the receiving antenna by some portion of a wavelength, $\Delta = \lambda/k$, we obtain from Fig. 9.22b

$$d^2 = (d - \Delta)^2 + \left(\frac{D}{2}\right)^2$$

$$= d^2 - 2d\Delta + \Delta^2 + \frac{D^2}{4} \tag{155}$$

For $d \gg \Delta$, a reasonable assumption, we obtain from (155)

$$2d\Delta \simeq \frac{D^2}{4} \tag{156}$$

Requiring Δ to be some portion of a wavelength such as $\Delta = \lambda_0/k$, we obtain from (156)

$$d = \frac{kD^2}{8\lambda_0} \tag{157}$$

The usual choice for k is $k = 16$, whereby (157) becomes

$$d_{\text{farfield}} = \frac{2D^2}{\lambda_0} \tag{158}$$

Equation (158) is the widely used far-field criterion for surface-type antennas. Other values for k have been used. For example, if we assume that $\Delta = \lambda_0/32$, then $d_{\text{farfield}} = 4D^2/\lambda_0$. Both criteria are approximations, although (158) has received the most widespread use.

The above result is properly classified as a *phase-curvature* far-field criterion. As has been pointed out previously, the far field should be such that the wave approximates a uniform plane wave in the sense that there be negligible variation in the field over some local plane, the electric and magnetic fields should be orthogonal, and the ratio should approximate the intrinsic impedance of free space, η_0.

For wire-type antennas, a far-field criterion which is different from (158) is generally used. The necessity for this additional criterion stems from the fact that for wire antennas such as monopoles or dipoles of maximum length (dimension) D, points at which $r = 2D^2/\lambda_0$ are often in the near field of these types of antennas. An alternative criterion for the boundary between near and far fields that may be used for wire-type antennas as opposed to surface-type antennas might be the distance at which the ratio of the far-field components, $\hat{E}_\theta$ and $\hat{H}_\phi$, is approximately the intrinsic impedance of free space, $\eta_0 \simeq 377\ \Omega$.

In order to determine this criterion, let us reconsider the fields (total) of the Hertzian dipole given in (25) and (26). The ratio of the far-field components, $\hat{E}_\theta$ in (26*b*) and $\hat{H}_\phi$ in (25*c*) is often referred to as the wave impedance, $\hat{Z}_w$, and is

$$\begin{aligned}\hat{Z}_w &= \frac{\hat{E}_\theta}{\hat{H}_\phi} \\ &= \eta_0 \frac{1 - (2\pi r/\lambda_0)^2 + j(2\pi r/\lambda_0)}{-(2\pi r/\lambda_0)^2 + j(2\pi r/\lambda_0)}\end{aligned} \tag{159}$$

We have seen previously that the $1/r$ dependence terms of the Hertzian dipole fields and the $1/r^2$ dependence terms were equal at $r = \lambda_0/2\pi$, or approximately one-sixth of a wavelength. However, at this distance, evaluation of (159) yields

$$\hat{Z}_w = 0.707\eta_0\underline{/-45^\circ} \qquad r = \frac{\lambda_0}{2\pi} \tag{160}$$

Therefore, this distance is not sufficient for the wave impedance of the far-field components to be η_0, as is required in the far field. Thus, the near-field–far-field transition point should be defined at a farther distance from this antenna. For a distance of $3\lambda_0$ from the Hertzian dipole, we find that

$$\hat{Z}_w = 375.93\underline{/-0.01^\circ} \qquad r = 3\lambda_0 \tag{161}$$

and thus the wave impedance of the far-field components is approximately η_0. For $r = 2\lambda_0$, $\hat{Z}_r = 374.62\underline{/-0.03^\circ}$, and for $r = \lambda_0$, $\hat{Z}_r = 367.68\underline{/-0.23^\circ}$. Thus, any distance between λ_0 and, say, $5\lambda_0$ is sufficient to define the near-field–far-field transition point for a Hertzian dipole from the standpoint of the wave impedance approximating η_0. The most common criterion, however, is $3\lambda_0$.

This far-field criterion of $3\lambda_0$ is also generally applied to other wire-type antennas, such as the long dipole or monopole, with the observation that the fields of these antennas can be considered to be the superposition of the fields of many Hertzian dipoles placed along the antenna. As with surface-type antennas, the near-field–far-field transition distance is not intended to be a precise criterion but is used to give an idea of the *region of transition* from the near field to the far field.

EXAMPLE 9.7 Consider two identical horn antennas separated by a distance of 100 m. Both antennas have directive gains of 15 dB in the direction of transmission and their dimensions are 12 cm by 6 cm. If the transmitting antenna is transmitting 5 W at a frequency of 3 GHz, determine (1) whether the receiving antenna is in the far field of the transmitter, (2) the received power, and (3) the electric field intensity at the receiving antenna.

Solution The far-field distance is on the order of

$$
\begin{aligned}
d_{\text{farfield}} &= \frac{2D^2}{\lambda} \\
&= \frac{2(0.12)^2}{0.1} \\
&= 0.29 \text{ m}
\end{aligned}
$$

The received power is obtained from (150) as

$$
\begin{aligned}
P_R &= P_T G_T G_R \left(\frac{\lambda}{4\pi d}\right)^2 \\
&= 5 \times 31.62 \times 31.62 \times \left(\frac{0.1}{400\pi}\right)^2 \\
&= 31.7 \ \mu\text{W}
\end{aligned}
$$

The magnitude of the electric field in the vicinity of the receiving antenna is obtained from (153) as

$$
\begin{aligned}
|\hat{E}| &= \frac{\sqrt{60 P_T G_T}}{d} \\
&= 0.974 \text{ V/m}
\end{aligned}
$$

9.8 Effect of Ground Reflections on Signal Transmission

In the derivation of the Friis transmission equation in Sec. 9.7, we assumed that the transmitting and receiving antennas were located in an infinite, homogeneous medium. It is perhaps more common to find antennas located above the surface of some conducting (ground) plane, as shown in Fig. 9.23. Perhaps the most common type of ground plane is earth, whose conductivity σ varies depending on the type of soil. The relative permittivity ϵ_r also differs from that of free space, although the variation in this parameter is somewhat less than for the conductivity. For example, dry, sandy soil has $\sigma \simeq 10^{-3}$ S/m and $\epsilon_r \simeq 10$, whereas wet, marshy soil has $\sigma \simeq 10^{-2}$ S/m and $\epsilon_r \simeq 15$.

The mode of propagation of signals over earth depends on the frequency range used. Signals less than a few kilohertz, extremely low frequency (ELF), require very large antennas. These signals propagate by reflection from the ionosphere, which provides a form of waveguide with the earth, and are useful in communicating with submerged submarines since sea water attenuates higher frequencies quite rapidly. Signals in the range up to a few megahertz propagate primarily by a surface wave that travels along the earth's surface; standard AM broadcasting signals propagate primarily via this mode and are strongly influenced by the presence and properties of the earth. Signals from a few megahertz up to around 30 MHz are reflected from the ionosphere and provide communication over very long distances (thousands of miles); amateur shortwave broadcasting is typical of this band.

The frequency band above around 30 MHz will be our concern in this section. In this frequency range the antennas are quite small and may be situated at heights above the ground of several wavelengths. The primary mechanism of signal propagation in this frequency range is illustrated in Fig. 9.23. The received signal consists of a direct wave (as was considered in all previous sections) and a wave reflected at the earth's surface. At frequencies above several gigahertz, the signal transmission (or attenuation) is influenced by scattering of the waves from rain, hills, buildings, etc., but we will omit consideration of these factors to simplify the analysis. The transmitting and receiving antennas are located at heights of h_T and h_R, respectively, above earth. The center-to-center separation is denoted by d, whereas the horizontal separation is denoted by D. The approach we will use is to superimpose the signals of the two waves to give the total received signal. Assuming the antennas to be in the far fields of each other, the received voltage (as well as the incident electric field) at the receiving antenna due to the direct wave will be proportional to

$$\hat{V}_d = \hat{V}_0 F_T(\theta_{Td}) F_R(\theta_{Rd}) \frac{e^{-j\beta_0 d}}{d} \tag{162}$$

where F_T and F_R denote the radiation electric field strength patterns of the transmitting and receiving antennas, respectively [see, for example, (68) for a half-wave dipole], and where θ_{Td} and θ_{Rd} are the angles shown in Fig. 9.23. The reflected wave will contribute

$$\hat{V}_r = \hat{V}_0 F_T(\theta_{Tr}) F_R(\theta_{Rr}) \hat{\Gamma} \frac{e^{-j\beta_0 d_r}}{d_r} \tag{163}$$

where

$$\hat{\Gamma} = \rho \underline{/\phi} \tag{164}$$

is the reflection coefficient at the earth's surface, and d_r is the total path length of this reflected wave. The total received voltage is the sum of (162) and (163):

$$\begin{aligned}
\hat{V} &= \hat{V}_d + \hat{V}_r \\
&= \hat{V}_0 F_T(\theta_{Td}) F_R(\theta_{Rd}) \frac{e^{-j\beta_0 d}}{d} + \hat{V}_0 F_T(\theta_{Tr}) F_R(\theta_{Rr}) \hat{\Gamma} \frac{e^{-j\beta_0 d_r}}{d_r} \\
&= \hat{V}_0 F_T(\theta_{Td}) F_R(\theta_{Rd}) \frac{e^{-j\beta_0 d}}{d} \left[1 + \frac{F_T(\theta_{Tr}) F_R(\theta_{Rr}) \hat{\Gamma}}{F_T(\theta_{Td}) F_R(\theta_{Rd})} \frac{d}{d_r} e^{-j\beta_0 (d_r - d)} \right]
\end{aligned} \tag{165}$$

Thus, the ground reflection modifies the free space propagation by the multiplicative factor

$$\hat{F} = 1 + \frac{F_T(\theta_{Tr}) F_R(\theta_{Rr}) \hat{\Gamma}}{F_T(\theta_{Td}) F_R(\theta_{Rd})} \frac{d}{d_r} e^{-j\beta_0 (d_r - d)} \tag{166}$$

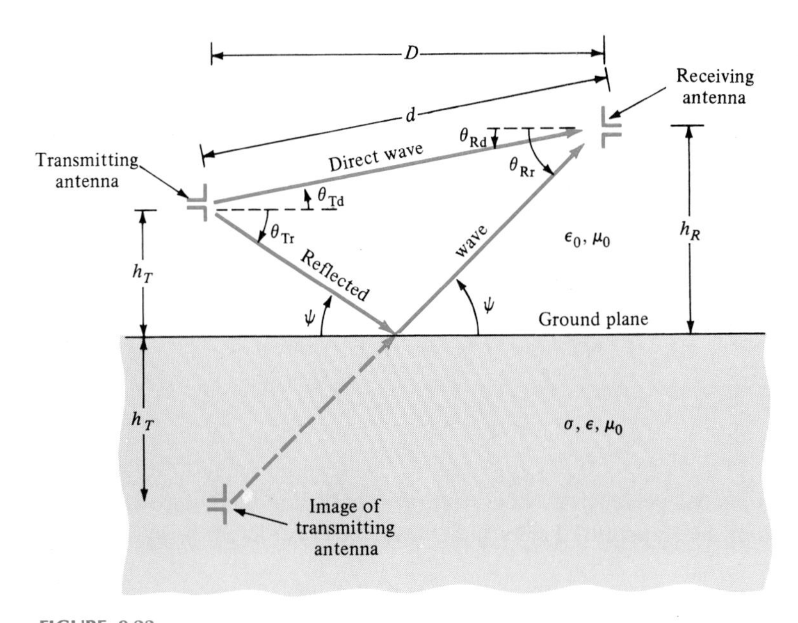

FIGURE 9.23

Illustration of the effect of a ground plane on signal transmission.

Consequently, the Friis transmission equation can be modified to account for ground reflection by multiplying it by the square of the magnitude of $\hat{F}$: $|\hat{F}|^2$. [The Friis transmission equation relates powers, whereas (165) is in terms of voltage or electric fields.] The remainder of this section will be devoted to examining the correction factor $\hat{F}$ in (166).

SIMPLIFICATION OF THE GROUND REFLECTION FACTOR Typical antenna heights and separation distances allow a considerable simplification of (166). Ordinarily, the antenna separation, D, is so large with respect to the antenna heights that we may assume the radiation field strength patterns in the directions of the direct and reflected waves to be the same:

$$F_T(\theta_{Td}) \simeq F_T(\theta_{Tr}) \tag{167a}$$

$$F_R(\theta_{Rd}) \simeq F_R(\theta_{Rr}) \tag{167b}$$

Similarly, we may approximate $d \simeq d_r$ when it does not appear in an exponential phase-shift term. Thus, (166) simplifies to

$$\hat{F} \simeq 1 + \hat{\Gamma} e^{-j\beta_0(d_r - d)} \tag{168}$$

Two problems remain. First we must determine the reflection coefficient $\hat{\Gamma}$. Second, since $d \simeq d_r$, we must be careful in computing the difference $d_r - d$ to avoid round-off error in subtracting these two nearly equal terms. We will investigate the calculation of the reflection coefficient first.

Snell's law of reflection (considered in Chap. 6) can be shown to apply here where the ground plane has finite but nonzero conductivity. Thus, the angles of incidence and reflection must be equal. Instead of defining those angles with respect to a normal to the surface, it is customary in these types of problems to define them with respect to the ground plane, as shown in Fig. 9.23. This angle will be denoted as ψ. The results obtained from Chap. 6 that we shall use here must be modified accordingly. The reflection coefficients obtained in Chap. 6 may be used here to determine the reflection coefficient of the reflected wave. Those results were obtained for the case of both media being lossless ($\sigma_1 = \sigma_2 = 0$). They may easily be modified to fit this situation if we substitute $\epsilon_1 = \epsilon_0$, $\mu_1 = \mu_0$ (medium 1 is free space), and $\mu_2 = \mu_0$:

$$\hat{\epsilon}_2 = \epsilon\left(1 - j\frac{\sigma}{\omega\epsilon}\right) \tag{169}$$

where $\epsilon = \epsilon_0 \epsilon_r$ is the permittivity of earth and σ is its conductivity. Again, two polarizations of the transmitted electric field will be considered: horizontal and vertical.

HORIZONTAL POLARIZATION The horizontal polarization is such that the electric field is parallel to the ground plane. This corresponds to the case of "perpendicular" polarization considered in Chap. 6 (see Fig. 6.21*a*). The reflection coefficient for this case is given in Eq. (188*a*) of Chap. 6. Substituting $\epsilon_1 = \epsilon_0$, $\hat{\epsilon}_2 = \epsilon\sqrt{1 - j(\sigma/\omega\epsilon)}$, $\mu_1 = \mu_0$, and $\mu_2 = \mu_0$ and noting the definition of the angle ψ gives

$$\hat{\Gamma}_h = \frac{\sin\psi - \sqrt{\epsilon_r[1 - j(\sigma/\omega\epsilon)] - \cos^2\psi}}{\sin\psi + \sqrt{\epsilon_r[1 - j(\sigma/\omega\epsilon)] - \cos^2\psi}} \tag{170}$$

For small ψ (grazing incidence), $\hat{\Gamma}_h \simeq -1$. This also applies to the case in which $\sigma/\omega\epsilon$ is very large, which in the limiting case represents a perfectly conducting ground plane.

Curves showing the reflection coefficient in (170) for $\sigma = 10^{-2}$ S/m and $\epsilon_r = 15$ as a function of frequency and angle of incidence are given in Fig. 9.24*a*.

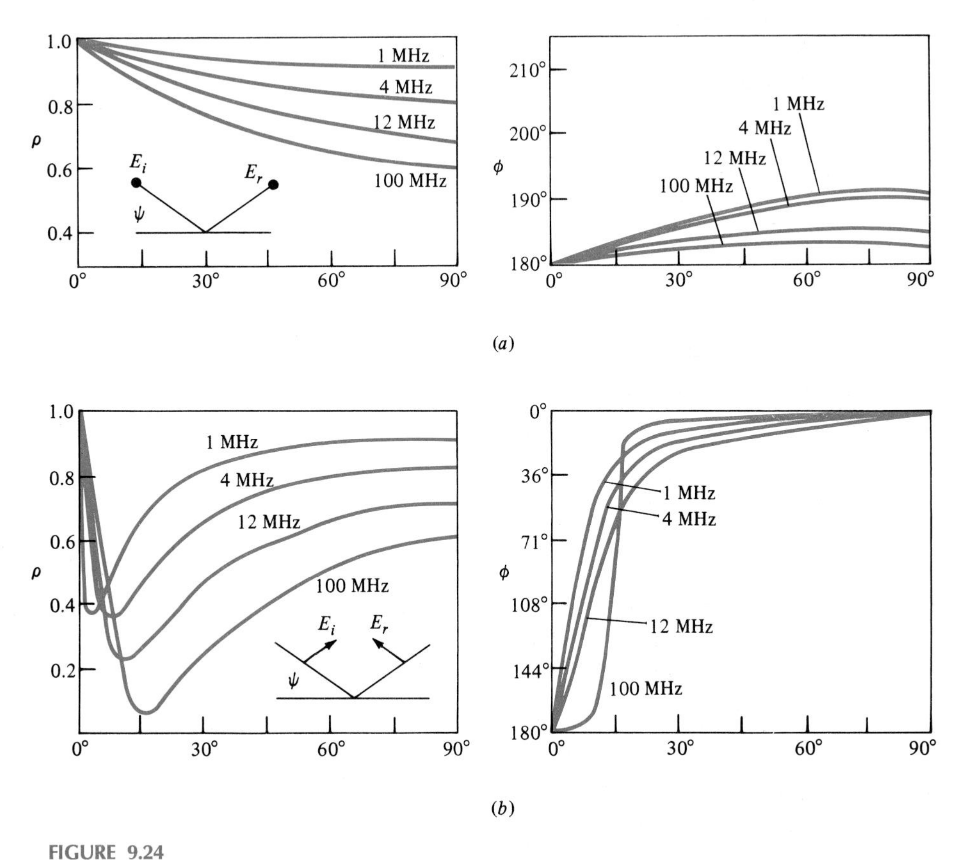

FIGURE 9.24

Reflection coefficients for earth ($\sigma = 10^{-2}$ S/m, $\epsilon_r = 15$) as a function of grazing angle. (R. E. Collin, *Antennas and Radiowave Propagation,* McGraw-Hill, New York, 1985.) *(a)* Horizontal polarization and *(b)* vertical polarization.

The approximation $\hat{\Gamma}_h \simeq -1$ is seen to be reasonable for small angles of incidence.

VERTICAL POLARIZATION Vertical polarization corresponds to the case of "parallel" polarization in Chap. 6. (See Fig. 6.21*b* of Chap. 6.) The reflection coefficient for this case is given in Eq. (194*a*) of Chap. 6. Again substituting $\epsilon_1 = \epsilon_0$, $\hat{\epsilon}_2 = \epsilon\sqrt{1 - j(\sigma/\omega\epsilon)}$, $\mu_1 = \mu_0$, and $\mu_2 = \mu_0$ and noting the definition of the angle ψ in Fig. 9.23 gives

$$\hat{\Gamma}_v = \frac{\epsilon_r[1 - j(\sigma/\omega\epsilon)] \sin\psi - \sqrt{\epsilon_r[1 - j(\sigma/\omega\epsilon)] - \cos^2\psi}}{\epsilon_r[1 - j(\sigma/\omega\epsilon)] \sin\psi + \sqrt{\epsilon_r[1 - j(\sigma/\omega\epsilon)] - \cos^2\psi}} \tag{171}$$

For small ψ (grazing incidence) $\hat{\Gamma}_v \simeq -1$. This is in contrast to the case of a perfectly conducting ground plane, where $\hat{\Gamma}_v = +1$ for all angles of incidence.

Plots of the magnitude and angle of (171) for $\sigma = 10^{-2}$ S/m and $\epsilon_r = 15$ are shown versus frequency and angle of incidence in Fig. 9.24*b*. For vertical polarization, the approximation $\hat{\Gamma}_v \simeq -1$ requires a much smaller angle of incidence than for horizontal polarization. For VHF communication such as TV transmission where frequencies are above 100 MHz, the approximation is reasonable for incidence angles less than, say, 5°.

For the above reasons, we will assume the reflection coefficients for both polarizations to be

$$\hat{\Gamma}_h \simeq \hat{\Gamma}_v \simeq -1 \tag{172}$$

Substituting into (168) yields

$$\hat{F} \simeq 1 - e^{-j\beta_0\Delta} \tag{173}$$

where the path-length difference is denoted by

$$\Delta = d_r - d \tag{174}$$

Rewriting (173) gives

$$\begin{aligned}\hat{F} &= 1 - e^{-j\beta_0\Delta}\\ &= e^{-j\beta_0\Delta/2}(e^{j\beta_0\Delta/2} - e^{-j\beta_0\Delta/2})\\ &= 2je^{-j\beta_0\Delta/2} \sin \beta_0 \frac{\Delta}{2}\end{aligned} \tag{175}$$

Therefore, the magnitude of $\hat{F}$ is

$$|\hat{F}| = 2\left|\sin \beta_0 \frac{\Delta}{2}\right|$$
$$= 2\left|\sin \pi \frac{\Delta}{\lambda_0}\right| \tag{176}$$

where we have substituted $\beta_0 = 2\pi/\lambda_0$.

Note that the correction factor has minimum values of 0 and maximum values of 2. When the correction factor is 0, the direct and reflected waves are 180° out of phase at the receiving antenna and effectively cancel. When the correction factor is 2, the direct and reflected waves are in phase at the receiving antenna and effectively add. From the result in (176) we see that this depends on the difference in path lengths, Δ, as a portion of a wavelength:

$$|\hat{F}| = 0 \qquad \frac{\Delta}{\lambda_0} = 1, 2, 3, \ldots \tag{177a}$$

Similarly

$$|\hat{F}| = 2 \qquad \frac{\Delta}{\lambda_0} = \frac{1}{2}, \frac{3}{2}, \frac{5}{2}, \ldots \tag{177b}$$

Thus, the two waves may add destructively or constructively.

The final investigation concerns determining the path difference Δ. Using the image of the transmitting antenna from Fig 9.23, the path length for the reflected wave can be determined as

$$d_r = \sqrt{D^2 + (h_T + h_R)^2} \tag{178}$$

where D is the horizontal separation between the two antennas:

$$D^2 = d^2 - (h_T - h_R)^2 \tag{179}$$

Substituting (179) into (178) yields

$$d_r = \sqrt{d^2 - (h_T - h_R)^2 + (h_T + h_R)^2}$$
$$= \sqrt{d^2 + 4h_T h_R} \tag{180}$$

By assumption, the antenna separation is large and the antenna heights are small, so that $4h_T h_R \ll d^2$. Therefore, in the computation of Δ we will be subtracting approximately equal numbers, which may result in numerical errors.

To obtain a more accurate value of Δ, we use the binomial expansion

$$(1+x)^n = 1 + \frac{nx}{1!} + \frac{n(n-1)x^2}{2!} + \cdots$$

(which is valid for $|x| < 1$) to expand Δ:

$$\begin{aligned}
\Delta &= d_r - d \\
&= \sqrt{d^2 + 4h_T h_R} - d \\
&= d\left(\sqrt{1 + \frac{4h_T h_R}{d^2}} - 1\right) \\
&\simeq d\left[1 + \frac{1}{2}\left(\frac{4h_T h_R}{d^2}\right) - 1\right] \\
&= \frac{2h_T h_R}{d}
\end{aligned} \tag{181}$$

Thus, the correction factor becomes

$$|\hat{F}| \simeq 2\left|\sin \frac{2\pi h_T h_R}{d\lambda_0}\right| \tag{182}$$

If the argument of the sine term is small, this may be approximated as

$$|\hat{F}| \simeq \frac{4\pi h_T h_R}{d\lambda_0} \tag{183}$$

EXAMPLE 9.8 Consider the problem of communication between an airport traffic control tower and an airplane approaching to land. The tower frequency is 119.1 MHz, and the antenna is vertically polarized and located at a height of 100 ft above the airport surface. An aircraft approaching the airport is at a distance D from the tower. If the height of the aircraft above the ground is 1000 ft, determine the radii of circles centered on the tower at which communication would, ideally, not be possible.

Solution The correction factor will be equal to zero when the path-length difference is a multiple of λ_0. The wavelength of the tower frequency is $\lambda_0 = 2.52$ m $= 8.26$ ft. In terms of the horizontal separation $D \simeq d$,

$$\frac{2h_T h_R}{D} = n\lambda_0 \qquad n = 1, 2, 3, \ldots$$

This gives

$$D = \frac{2h_T h_R}{n\lambda_0}$$

$$= \frac{2{,}419}{n} \quad \text{ft}$$

$$= \frac{0.46}{n} \quad \text{mi}$$

for $n = 1, 2, 3, \ldots$. Therefore, the direct and reflected waves would, theoretically, add destructively to give a transmission null at distance of 0.46 mi. The angle of incidence of the reflected wave can be calculated from Fig. 9.23 as

$$\psi = \tan^{-1} \frac{h_R + h_T}{D} \tag{184}$$

For $D = 0.46$ mi, we calculate that $\psi = 4.7°$. Therefore, for this and greater distances, the above approximations should be reasonably valid. It should be noted that the above nulls in the reception pattern will probably not be observed, given the many factors that do not conform to the ideal assumptions in the derivation; deviation of the earth's surface from a flat plane will no doubt affect the results. However, variations in received signal due to the reflected wave probably will occur.

If the transmitting antenna is a half-wave dipole, a useful result can be obtained. Combining Eqs. (66*b*) and (183), the magnitude of the electric field at the receiving antenna becomes

$$E = \frac{240\pi I h_T h_R}{D^2 \lambda_0} \tag{185}$$

where I is the magnitude of the input current to the dipole. The received electric field is proportional to the heights of the antennas and inversely proportional to the square of the separation distance between them.

9.9 Summary

In this chapter we have studied structures that are used to launch and focus waves. We found that if we know the current distribution over the surface of an antenna, we may, theoretically, obtain the resulting fields radiated by the antenna.

For the Hertzian dipole or infinitesimal current element, we simply postulated this current distribution to be independent of position along the element. We found that the resulting fields exhibit quite different behavior depending on their distance from the current element. For a sufficiently remote distance, in the far field, the fields have characteristics resembling uniform plane waves. In the far field, the electric and magnetic field vectors are orthogonal, related by the intrinsic impedance of free space, and lie (locally) in a plane orthogonal to a radial line from the element. For the infinitesimal current loop (magnetic dipole), similar results are obtained. The far-field vectors of many other more complicated antenna structures are similar to those of the infinitesimal dipoles.

The longer dipole and monopole antennas were considered next. We assumed the current distribution on the dipole to be similar to that of an open-circuited transmission line. The far fields of the long dipole were then computed by considering the dipole current to be a distribution of infinitesimal Hertzian dipoles and superimposing the resulting fields. The far-field structure again was seen to resemble that of a Hertzian dipole. The results for the monopole antenna were obtained from those of the dipole by the method of images. The radiation resistance of each antenna was defined to be the ratio of the total average power radiated by the antenna to the square of the rms current on the antenna. For the special cases of a half-wave dipole and a quarter-wave monopole, the radiation resistance is 73 Ω and 36.5 Ω, respectively, and the input impedances for these antennas are approximately real and equal to the radiation resistance. For other antenna lengths, the reactive component of the input impedance is often dominant.

The dipole and monopole antennas have omnidirectional patterns in a plane perpendicular to the antenna axis. We next considered combining the patterns of an array of two or more dipoles or monopoles to provide certain directional properties of these patterns. We found that by adjusting the separation of these antennas and/or the phase of the currents, the individual patterns could add destructively to produce nulls or minima in the pattern, giving the array certain directional properties.

For more complicated antenna structures, we relied on certain measurable quantities—such as directive gain and power gain—to characterize these antennas. For the calculation of the interaction between two antennas, for example, one to be used as a transmitter and the other to be used as a receiver, we presented a lumped equivalent circuit characterizing this coupling from the standpoint of the antenna terminals. A number of useful general properties were obtained from this characterization. The input impedance to an antenna when it is used for transmission is the same as the equivalent source impedance when it is used for reception. Also, the transmission pattern of an antenna is the same as its reception pattern. For general coupling calculations, we were introduced to the concept of the effective aperture of an antenna as the ratio of the power received at its terminals to the power density of the incident wave. This concept led to the Friis transmission equation, which characterizes the coupling between two antennas in terms of their gains, separation, and frequency of operation. The effect of reflecting surfaces causing additional transmission paths was also considered. These additional transmission paths can give rise to constructive or destructive interference at the receiving antenna.

The analysis of antennas is obviously much more complicated than the analysis of transmission lines or waveguides. The topics considered in this chapter apply generally to all antennas, but the precision of calculated results (such as resulting fields and coupling) is, at this point, much less than we are able to obtain for transmission lines and waveguides.

Problems

9-1 Derive the equations for the fields of a Hertzian dipole given in Eqs. (25) and (26).

9-2 Derive the expression for the average-power density vector for a Hertzian dipole given in (35). Show that this may be calculated solely from the far-field expressions given in (29) and (30).

9-3 Consider a Hertzian dipole of length 1 cm carrying a phasor current of $\hat{I} = 10\underline{/30^\circ}$ A. If the frequency is 100 MHz, determine the electric and magnetic fields at a distance of 10 cm away from the dipole and $\theta = 45^\circ$. Compute the ratios $|\hat{E}_\theta|/|\hat{E}_r|$ and $|\hat{E}_\theta|/|\hat{H}_\phi|$ at this distance. Repeat for distances of 1 m and 10 m and $\theta = 45^\circ$. Determine these distances in wavelengths. Is the result for the 10-m distance expected?

9-4 Compute the radiation resistance of and the total average power radiated by the dipole of Prob. 9-3.

9-5 Suppose that an observer is in the far field of a wire-type antenna such as the Hertzian dipole at a point 100 m away from the antenna. If the observer measures the magnitude of the electric field to be 1 V/m, determine the magnitudes of the electric and magnetic field intensity vectors at a distance of 1000 m along this same radial line. Determine the difference in phase between the field vectors at 100 m and those at 1000 m if the frequency is 500 MHz. (Give this answer as an angle whose magnitude is less than or equal to 360°, and state whether the vectors at 1000 m lead, lag, or are in phase with those at 100 m.) Determine the average power densities at these two points.

9-6 Derive the expressions for the fields of a magnetic dipole given in (46) and (47), using the expression for the magnetic vector potential given in (45).

9-7 Derive the expression for the average-power density vector for a magnetic dipole given in (49). Show that this may be calculated solely from the far-field expressions given in (48).

9-8 Consider a magnetic dipole antenna of radius 1 cm carrying a current of $\hat{I} = 10\underline{/30^\circ}$. If the frequency is 100 MHz, compute the electric and magnetic field intensities at $r = 10$ cm, 1 m, 10 m, and $\theta = 45^\circ$. Determine the ratios $|\hat{H}_\theta|/|\hat{H}_r|$ and $|\hat{E}_\phi|/|\hat{H}_\theta|$ at these distances.

9-9 Compute the radiation resistance of and the total average power radiated by the magnetic dipole of Prob. 9-8.

9-10 Consider a dipole antenna of total length l that is driven by a voltage source at its input. Assume that the antenna is sufficiently short that the current decreases uniformly from its maximum at the center to zero at the endpoints. Derive an expression for the far-field electric field in terms of the input current.

9-11 Determine the magnitudes of the electric and magnetic fields of a half-wave dipole operated at a frequency of 300 MHz at a distance of 100 m in the broadside plane, i.e., $\theta = 90°$. The input current to the terminals is $100\underline{/0°}$ mA. Determine the total average power radiated.

9-12 A lossless quarter-wave monopole antenna is situated above a perfectly conducting ground plane and is driven by a 100-V 300-MHz source that has an internal impedance of 50 Ω. Compute the total average power radiated. Repeat this calculation if a $\frac{1}{5}\lambda$ lossless monopole having an input (20-j50) Ω is substituted. Repeat for a short, $\frac{1}{10}\lambda$ lossless monopole that has an input impedance of (4-j180) Ω.

9-13 A lossless dipole antenna is attached to a source with a length of lossless 50-Ω coaxial cable. The source has an open-circuit voltage of 100 V (rms) and a source impedance of 50 Ω. If the frequency of the source is such that the dipole length is one-half wavelength and the transmission-line length is 1.3 λ, determine the total average power radiated by the antenna and the VSWR on the cable.

9-14 Sketch the current distribution along a dipole for dipole lengths of $\frac{3}{4}\lambda$, λ, $\frac{5}{4}\lambda$, $\frac{3}{2}\lambda$, and 2λ.

9-15 Show that patterns for long dipoles of any length in the ϕ = constant plane always exhibit nulls at $\theta = 0°$ and 180°. Also show that one can sketch the patterns of these antennas from a knowledge of the pattern for $0 \le \theta \le 90°$. Finally show that nulls can exist at angles of θ other than 0° and 180° only if $l = 3\lambda/2, 7\lambda/2, 11\lambda/2, \ldots$.

9-16 Use a digital computer subroutine to integrate (82b) and determine the radiation resistance referred to as a current maximum for dipoles of length $l/\lambda = 1/4, 3/8, 1/2, 5/8, 3/4, 7/8$, and 1. From these results, use (82a) to obtain the radiation resistance referred to the input terminals of the dipole.

9-17 Substitute the relationship for the dipole current given in (53) and the relationship for the input current given in (80) into (84) and determine an expression for the part of the input resistance due to wire loss.

9-18 Repeat Example 9.2 but with the antenna attached to the source via a $\lambda/4$ length of 50-Ω lossless transmission line.

9-19 Two identical monopole antennas are perpendicular to the earth. The antennas are separated by d and fed with currents of equal magnitude, as shown in Fig. 9.14b. Sketch the patterns of the array in a plane parallel to the earth for the following conditions:

a) $d = \lambda_0/2$; $\alpha = 90°$
b) $d = 5\lambda_0/8$; $\alpha = 45°$
c) $d = \lambda_0$; $\alpha = 180°$
d) $d = \lambda_0/4$; $\alpha = 180°$

9-20 Consider a linear array of four vertical monopoles above the earth, as shown in Fig. 9.16. If the separation between adjacent antennas is $\lambda_0/4$, $\alpha = 180°$, and $\delta = 0°$, sketch the electric field pattern at the surface of the earth. Show the locations and values of the resulting maxima and minima in the pattern.

9-21 A standard AM-broadcast-band transmitting station consists of two vertical monopoles above earth. The two antennas are separated by 164 ft, and the transmitting frequency is 1500 kHz. The antennas are fed with signals of equal amplitude and a phase difference of 135°. Sketch the electric field pattern at the surface of the earth. Show the location of all maxima and minima and their relative values.

9-22 Two dipoles are separated by one wavelength. The terminal currents are of equal magnitude but are out of phase by 90°. Sketch the electric field pattern in a plane perpendicular to the dipoles. Show the location of all maxima and minima and their values.

9-23 Four identical monopole antennas have adjacent spacings of $\lambda/2$. The antenna currents are in phase and of equal magnitude. Sketch the resulting electric field pattern along the surface of the earth. Show the location and relative magnitudes of the maxima and minima of the resulting pattern.

9-24 Determine the directive gain and directivity of the elementary magnetic dipole.

9-25 Consider two lossless, widely separated half-wave dipoles. If 10 W is delivered to the transmitting antenna and 1 mW is received in the other antenna in a matched load, determine the mutual impedance Z_m for this transmission path. Determine the received power if the receiving antenna's load, instead of being matched, is changed to $(10 + j0)\ \Omega$.

9-26 Determine the maximum effective aperture of an elemental magnetic dipole and a long dipole. Compare the maximum effective aperture of a half-wave dipole to its physical length.

9-27 An aircraft transmitter is designed to communicate with a ground station. The ground receiver must receive at least 1 μW for proper reception. Assume that both antennas are omnidirectional. After takeoff, the airplane flies over the station at an altitude of 5000 ft. When the airplane is directly over the station, a signal of 500 mW is received by the station. Determine the maximum communication range of the airplane.

9-28 A telemetry transmitter placed on the moon is to transmit data to the earth. The transmitter power is 100 mW and the gain of the transmitting antenna in the direction of transmission is 12 dB. Determine the minimum gain of the receiving antenna in order to receive 1 nW. The distance from the moon to earth is 238,857 mi and the transmitter frequency is 100 MHz.

9-29 A microwave relay link is to be designed. The transmitting and receiving antennas are separated by 30 mi, and the power gain in the direction of transmission for both antennas is 45 dB. If both antennas are lossless and matched and the frequency is 3 GHz, determine the minimum transmitter power if the received power is to be 1 mW.

9-30 An antenna on an aircraft is being used to jam an enemy radar. If the antenna has a gain of 12 dB in the direction of transmission and the transmitted power is 5 kW, determine the electric field intensity in the vicinity of the enemy radar, which is 2 mi away. The frequency of transmission is 7 GHz.

9-31 A lossless half-wave dipole is being driven by a 10-V 50-Ω generator. Determine the electric field intensity in the far field at a distance of 10 km in a plane perpendicular to the antenna. Compute your result by using the Friis transmission equation, and check the result by using Eq. (64).

9-32 A half-wave dipole is oriented perpendicular to the earth which is characterized by $\epsilon_r = 10$ and $\sigma = 10^{-3}$ S/m. The antenna is at a height of 100 ft and the operating frequency is 100 MHz. Determine the change in the total electric field due to the reflected wave at a height of 100 ft and a distance of 1000 ft from the antenna. Determine the exact result, and repeat using the Friis transmission equation with correction factor.

9-33 Two half-wave dipoles are oriented parallel to the earth which is characterized by $\epsilon_r = 10$ and $\sigma = 10^{-2}$ S/m. Both antennas are at a height of 100 ft and are separated by a distance of 300 ft. If one antenna is driven by a 100-MHz 1-V sinusoidal signal, determine the power received at the base of the other antenna into a matched load. Calculate this result directly and by using the Friis transmission formula.

CHAPTER 10

Equations Governing Potential Functions

In Chaps. 3 and 4 we considered the time-stationary, or static, electric and magnetic fields, and in subsequent chapters we considered time-varying electromagnetic fields. We introduced the various laws governing these fields and studied examples showing the determination of fields from these laws. For instance, the application of Coulomb's law yields the static electric field for a given charge distribution, and we can determine the static magnetic field for a given current from Ampère's law.

We determined the fields from a direct application of the various laws. However, all of these problems possessed a rather idealized symmetry to allow a feasible solution. For example, we used Gauss' law in Chap. 3 (Example 3.8) to deduce the electric field about an infinite-length cylinder of charge and assumed the charge to be uniformly distributed around the periphery of the cylinder. For this problem, we obtained the resulting electric field quite easily because of the symmetry of the problem. A direct calculation of the field prior to the discussion of Gauss' law required the evaluation of a rather complicated integral (Example 3.7). The idealized nature of this problem (infinite-length cylinder) and the assumption of a uniform charge distribution were crucial in allowing us to obtain a solution.

The question arises as to how we shall obtain solutions for the field vectors for more complex and nonideal geometries—such as the one shown in Fig. 10.1, in which we may wish to find the resulting potential in the region, given the boundary potential V_0 and the free charge distribution ρ. The techniques outlined in this chapter are intended to be used for these more complicated (but realistic) problems in which a direct application of the techniques of Chaps. 3 and 4 would not be feasible.

In the present chapter we will consider several methods of determining static fields in a given region. Some of these techniques are applicable to time-varying fields, but we will consider only static-field applications. The methods are general in that they are applicable to both electric fields and magnetic fields.

These methods may be divided into two general classes. The first class consists of the solution of the Laplace and Poisson equations derived in Chap. 3. The second class will be referred to as *approximate methods* and will be further subdivided into numerical methods, analog methods, and graphical methods.

The first class of methods—solution of the Laplace and Poisson equations—can be thought of as yielding the exact solution to an approximate problem. The approximation occurs in modeling a physical problem as one which permits a

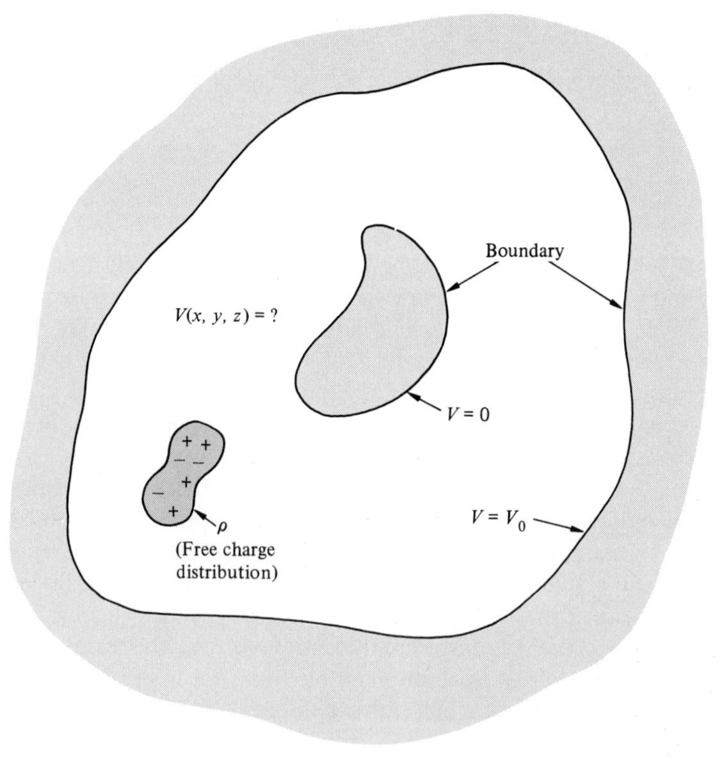

FIGURE 10.1
A general boundary-value problem.

solution of the Laplace and Poisson equations. For example, the physical boundaries of a particular problem may be somewhat irregular (as in Fig. 10.1) and may not fit some coordinate system. We would need to approximate these boundaries so that they fit some desirable geometry that makes a solution of the equations feasible.

The second class of methods may be thought of as yielding approximate solutions to exact problems. Here the approximation comes about in the solution technique. For example, one may approximate a smooth curve as a sequence of small steps. These types of (numerical) approximations simplify the solution of the fundamental equations. Analog methods fall into this class in the sense that a problem analogous to the actual problem will be solved (usually by experimental means). Graphical methods are also approximate solution techniques, with the accuracy of the solution being determined by the accuracy of an associated graphical portrayal of the problem.

10.1 Laplace's Equation

Laplace's equation was derived in Chap. 3 (see Sec. 3.12) for static electric fields. In order to illustrate its use with static magnetic fields, we first derive the equation. We recall from Chap. 3 that the static electric field intensity $\mathbf{E}(x, y, z)$ at a point is related to the scalar potential $V(x, y, z)$ by the gradient relationship

$$\mathbf{E} = -\nabla V \tag{1}$$

We also know from Chap. 3 that in a region having no free charges ($\rho = 0$ in Fig. 10.1) Gauss' law provides

$$\nabla \cdot \mathbf{D} = 0 \tag{2}$$

where $\mathbf{D}$ is the electric flux density vector. For a region that is linear, homogeneous, and isotropic, we may write $\mathbf{D} = \epsilon\mathbf{E}$, and (2) becomes

$$\nabla \cdot \mathbf{E} = 0 \tag{3}$$

The permittivity ϵ may be removed from $\nabla \cdot (\epsilon\mathbf{E})$ since the medium is assumed to be linear, homogeneous, and isotropic. Thus, ϵ is a scalar (isotropic), is independent of $\mathbf{E}$ (linear), and is independent of the coordinate variables x, y, z (homogeneous). Combining (1) and (3) yields

$$\nabla \cdot (-\nabla V) = -\nabla^2 V = 0 \tag{4}$$

or simply

$$\nabla^2 V = 0 \tag{5}$$

where $\nabla^2 \equiv \nabla \cdot \nabla$ and is known as the laplacian. Equation (5) is known as Laplace's equation.

Note that (5) is a partial differential equation; in cartesian coordinates, (5) becomes (see Prob. 10-1)

$$\frac{\partial^2 V}{\partial x^2} + \frac{\partial^2 V}{\partial y^2} + \frac{\partial^2 V}{\partial z^2} = 0 \tag{6}$$

Clearly, (6) implies that in a region free of charges ($\rho = 0$) the potential V everywhere in the region can be found by solving (6) for specified boundary conditions (specified values of V on the boundary of the region, such as are shown in Fig. 10.1). Knowing V, we can obtain **E** from (1).

The essence of this approach—obtaining the fields from a derived partial differential equation—is applicable to a wide range of problems in electromagnetics. Laplace's equation, in particular, is satisfied by the magnetic vector potential **A**. We recall from Chap. 4 that the magnetic flux density **B** is related to the magnetic vector potential **A** by the relationship

$$\mathbf{B} = \nabla \times \mathbf{A} \tag{7}$$

From Ampère's law, in a region free of electric currents we have

$$\nabla \times \mathbf{H} = 0 \tag{8}$$

For a linear, homogeneous, and isotropic region, we have $\mathbf{B} = \mu\mathbf{H}$. Thus, we may rewrite (8) as

$$\nabla \times \mathbf{B} = 0 \tag{9}$$

Equations (7) and (9) may now be combined to obtain

$$\nabla \times \nabla \times \mathbf{A} = 0 \tag{10}$$

Using the vector relationship $\nabla \times \nabla \times \mathbf{A} = \nabla(\nabla \cdot \mathbf{A}) - \nabla^2\mathbf{A}$ (see Appendix A) and recalling from Chap. 4 that $\nabla \cdot \mathbf{A} = 0$, we have

$$\nabla^2\mathbf{A} = 0 \tag{11a}$$

or, in terms of the components of **A**, we may separate (11*a*) into three individual Laplace's equations:

$$\begin{aligned} \nabla^2 A_x &= 0 \\ \nabla^2 A_y &= 0 \\ \nabla^2 A_z &= 0 \end{aligned} \tag{11b}$$

Once we solve (11) for **A**, we may obtain **B** from (7).

Next we observe that $\nabla \cdot \mathbf{B} = 0$, and because $\mathbf{B} = \mu\mathbf{H}$, we also have

$$\nabla \cdot \mathbf{H} = 0 \tag{12}$$

From (8) and (12) we obtain $\nabla \times \nabla \times \mathbf{H} = \nabla(\nabla \cdot \mathbf{H}) - \nabla^2\mathbf{H} = 0$, or

$$\nabla^2\mathbf{H} = 0 \tag{13a}$$

In terms of components of $\mathbf{H}$, (13a) becomes

$$\begin{aligned} \nabla^2 H_x &= 0 \\ \nabla^2 H_y &= 0 \\ \nabla^2 H_z &= 0 \end{aligned} \tag{13b}$$

Similarly, we can show that (see Prob. 10-2)

$$\nabla^2\mathbf{E} = 0 \tag{14a}$$

or

$$\begin{aligned} \nabla^2 E_x &= 0 \\ \nabla^2 E_y &= 0 \\ \nabla^2 E_z &= 0 \end{aligned} \tag{14b}$$

From (5), (11), (13), and (14) we conclude that the potential functions V and $\mathbf{A}$, and magnetic and electric field intensity vectors $\mathbf{H}$ and $\mathbf{E}$, respectively, satisfy Laplace's equation in a source-free region. [To be more precise, we should state that the components of $\mathbf{A}$, $\mathbf{E}$, and $\mathbf{H}$ satisfy Laplace's equations as shown by (11b), (13b), and (14b).]

10.2 Poisson's Equation

Clearly, Laplace's equation must be modified for regions containing sources. For example, in a region having an electric charge density ρ (which we treat as the source of the field), such as in Fig. 10.1, Gauss' law requires that at every point within the region we must have $\nabla \cdot \mathbf{D} = \rho$. Again we assume a linear, homogeneous, and isotropic region, $\mathbf{D} = \epsilon\mathbf{E}$, and we obtain

$$\nabla \cdot \mathbf{E} = \frac{\rho}{\epsilon} \tag{15}$$

Combining (1) and (15) yields

$$\begin{aligned} \nabla \cdot (-\nabla V) &= -\nabla^2 V \\ &= \frac{\rho}{\epsilon} \end{aligned} \tag{16a}$$

or

$$\nabla^2 V = -\frac{\rho}{\epsilon} \tag{16b}$$

This equation is known as Poisson's equation.

Next, considering the magnetic field quantities, from Ampère's law we have

$$\nabla \times \mathbf{H} = \mathbf{J} \tag{17}$$

where **J** is some current density that is treated as the source of **H**. From $\mathbf{B} = \nabla \times \mathbf{A}$ we may write

$$\nabla \times \mathbf{A} = \mu \mathbf{H} \tag{18}$$

Taking the curl of (18) and recalling that $\nabla \cdot \mathbf{A} = 0$, we obtain $\nabla \times \nabla \times \mathbf{A} = \nabla(\nabla \cdot \mathbf{A}) - \nabla^2 \mathbf{A} = -\nabla^2 \mathbf{A}$, or,

$$-\nabla^2 \mathbf{A} = \mu \nabla \times \mathbf{H} \tag{19}$$

Substituting (17) in (19) finally yields

$$\nabla^2 \mathbf{A} = -\mu \mathbf{J} \tag{20a}$$

which is similar to Poisson's equation in terms of the magnetic vector potential. Again, we may separate (20*a*) into its three components as

$$\begin{aligned} \nabla^2 A_x &= -\mu J_x \\ \nabla^2 A_y &= -\mu J_y \\ \nabla^2 A_z &= -\mu J_z \end{aligned} \tag{20b}$$

From (16) and (20) we conclude that in regions containing sources, the potential functions (V and **A**) satisfy Poisson's equation.

The fact that the potential functions satisfy Laplace's equation in a source-free region, and Poisson's equation in a region containing a source, is of considerable help in solving electromagnetic fields within specified regions. We will consider several examples to make this point and will consider solutions for the electrostatic potential function V for illustration. But first we will consider some of the techniques for solving Laplace's and Poisson's equations. Also, we will discuss some of the properties of the solution to these equations.

10.3 Solution to Laplace's Equation

We have already seen in Sec. 10.1 that potential functions and field intensities satisfy Laplace's equation (in a source-free region). Thus, for a given physical problem we should expect a unique (that is, unambiguous) solution to Laplace's equation. Explicitly, this statement is a consequence of the uniqueness theorem, which can be proved as follows.

10.3.1 Uniqueness of Solution

The uniqueness theorem can be stated as follows: Within some closed region, a solution to Laplace's equation given in (5) or Poisson's equation given in (16) for the potential function V is the only solution that satisfies the potential specified over the boundaries of the region. The specifications of the potential over the boundaries of the region are referred to as the boundary conditions for Laplace's (Poisson's) equation, as opposed to the general boundary conditions on the field vectors (such as continuity of tangential electric field). It will become clear in the proof that this theorem also applies to the components of the magnetic vector potential, A_x, A_y, and A_z in (11*b*) and (20*b*), as well as to the components of **H** (13*b*) and **E** (14*b*).

We prove the uniqueness theorem by contradiction; that is, first we assume that there are two possible solutions, V_1 and V_2, to (5) in a given region. From (5) we obtain $\nabla^2 V_1 = \nabla^2 V_2 = 0$. Thus, we may write

$$\nabla^2(V_1 - V_2) = 0 \tag{21}$$

Along the boundary, we also assume that the specified potentials are the same; that is, at all points on the boundary

$$V_1 - V_2 = 0 \tag{22}$$

With the vector identity $\nabla \cdot (\psi \mathbf{F}) = \psi \nabla \cdot \mathbf{F} + \mathbf{F} \cdot (\nabla \psi)$ (see Appendix A), we let the scalar $\psi = V_1 - V_2$ and the vector $\mathbf{F} = \nabla(V_1 - V_2)$, and use the divergence theorem to obtain

$$\int_v \nabla \cdot (\psi \nabla \psi)\, dv = \oint_s (\psi \nabla \psi) \cdot d\mathbf{s} \tag{23}$$

or

$$\int_v \nabla \cdot [(V_1 - V_2)\nabla(V_1 - V_2)]\, dv = \oint_s [(V_1 - V_2)\nabla(V_1 - V_2)] \cdot d\mathbf{s} \tag{24}$$

or

$$\int_v [(V_1 - V_2)\nabla \cdot \nabla(V_1 - V_2)]\, dv + \int_v [\nabla(V_1 - V_2)] \cdot [\nabla(V_1 - V_2)]\, dv$$
$$= \oint_s [(V_1 - V_2)\nabla(V_1 - V_2)] \cdot d\mathbf{s} \tag{25}$$

Finally, we obtain

$$\int_v [(V_1 - V_2)\nabla^2(V_1 - V_2)]\, dv + \int_v |\nabla(V_1 - V_2)|^2\, dv$$

$$= \oint_s [(V_1 - V_2)\nabla(V_1 - V_2)] \cdot d\mathbf{s} \tag{26}$$

The surface **s** is taken to be the interior of the surface bounding the closed region v. In (26) we now use the conditions given by (21) and (22), implying that the first volume integral is zero and the surface integral on the right-hand side of (26) is zero. Consequently, (26) reduces to

$$\int_v |\nabla(V_1 - V_2)|^2\, dv = 0 \tag{27}$$

In (27), $(V_1 - V_2)$ is real, which means that its gradient is real, and the square of a real number is positive. Therefore, the only way (27) can be satisfied is that

$$\nabla(V_1 - V_2) = 0 \tag{28}$$

or

$$V_1 - V_2 = \text{a constant} \tag{29}$$

Equation (29) must hold everywhere, including at the boundary. But (22) requires that at the boundary $V_1 - V_2 = 0$. Therefore, the constant in (29) must be zero, or $V_1 = V_2$ everywhere. Therefore, in a given region Laplace's equation has only one (unique) solution that satisfies the boundary conditions of the region. A similar proof applies to Poisson's equation (see Prob. 10-3).

The boundary conditions that the potential V be specified at every point on the boundary of a region are referred to as the *Dirichlet boundary conditions.* It is also possible (and sometimes advantageous) to specify the rate of change of the potential normal to the boundary at points along the boundary as an alternative boundary condition. This is referred to as the *Neumann boundary condition.* A combination of the two, that is, Dirichlet conditions over a portion of the boundary and Neumann conditions over the remainder, is referred to as *mixed boundary conditions.* The uniqueness property of Laplace's equation was proved for Dirichlet boundary conditions, but it can also be shown to apply to Neumann or mixed boundary conditions in like fashion.† We will deal with Dirichlet conditions exclusively in our future investigations.

---

† See D. T. Paris and F. K. Hurd, *Basic Electromagnetic Theory*, McGraw-Hill, New York, 1969, pp. 148–419.

10.3.2 Superposition of Solutions

Because Laplace's equation is a linear differential equation, the principle of superposition of solutions holds. This principle as applied to Laplace's equation is given by the following. Suppose there are two potential functions V_1 and V_2 that individually satisfy Laplace's equation; that is

$$\nabla^2 V_1 = 0 \tag{30a}$$

$$\nabla^2 V_2 = 0 \tag{30b}$$

By the principle of superposition, the sum of the two functions is also a solution to Laplace's equation; that is,

$$\nabla^2(V_1 + V_2) = 0 \tag{31}$$

which is easily seen if this result is separated as

$$\nabla^2 V_1 + \nabla^2 V_2 = 0 \tag{32}$$

Also, if k is some scalar constant, then kV_1 and kV_2 are also solutions of Laplace's equation since we may write

$$\nabla^2(kV_1) = k\nabla^2 V_1 = 0 \tag{33}$$

The property of superposition also applies to Poisson's equation. For example, suppose in some region that the total charge distribution ρ consists of the sum of two charge distributions, ρ_1 and ρ_2 where $\rho = \rho_1 + \rho_2$. If V_1 is a solution of Poisson's equation for charge distribution ρ_1 alone and if V_2 is a solution of Poisson's equation for ρ_2 alone, that is,

$$\nabla^2 V_1 = -\frac{\rho_1}{\epsilon} \tag{34a}$$

$$\nabla^2 V_2 = -\frac{\rho_2}{\epsilon} \tag{34b}$$

then $V_1 + V_2$ is a solution for the combined charge distribution; that is,

$$\nabla^2(V_1 + V_2) = -\frac{\rho_1 + \rho_2}{\epsilon} \tag{35}$$

since this may be written as

$$\nabla^2 V_1 + \nabla^2 V_2 = -\frac{\rho_1}{\epsilon} - \frac{\rho_2}{\epsilon} \tag{36}$$

This application of superposition to solutions of Laplace's (Poisson's) equation does not allow one to superpose boundary conditions. For example, suppose that the region boundary B is divided into two pieces, B_1 and B_2, so that $B = B_1 + B_2$. If V_1 satisfies Laplace's (Poisson's) equation and the boundary conditions over B_1, and if V_2 satisfies Laplace's (Poisson's) equation and the boundary conditions over B_2, it is obviously not true that $V_1 + V_2$ necessarily satisfies Laplace's (Poisson's) equation *and* the boundary conditions over the entire boundary B.

The concept of linearity and the resulting property of superposition are useful in obtaining a complete solution to Laplace's (or Poisson's) equation that satisfies all boundary conditions. For example, consider a rectangular charge-free region, shown in Fig. 10.2, in which the potential is zero along the sides at $y = 0$ and $y = a$ and $x = 0$. The potential of the side at $x = b$ is specified as V_0. We assume that the potential is independent of z. Solutions of Laplace's equation that satisfy the boundary conditions

$$\begin{aligned} V &= 0 \quad \text{at } y = 0 \\ V &= 0 \quad \text{at } y = a \\ V &= 0 \quad \text{at } x = 0 \end{aligned} \tag{37}$$

(and have no z dependence) are of the form

$$V_n(x, y) = C_n \sinh \frac{n\pi x}{a} \sin \frac{n\pi y}{a} \tag{38}$$

where n is an integer and C_n is independent of x and y (but possibly a function of n). This solution may be verified by direct substitution into Laplace's equation. The solution given in (38) satisfies three of the four boundary conditions. The remaining boundary condition, $V = V_0$ at $x = b$, cannot, however, be satisfied by only one function of the form of (38). However, each function individually

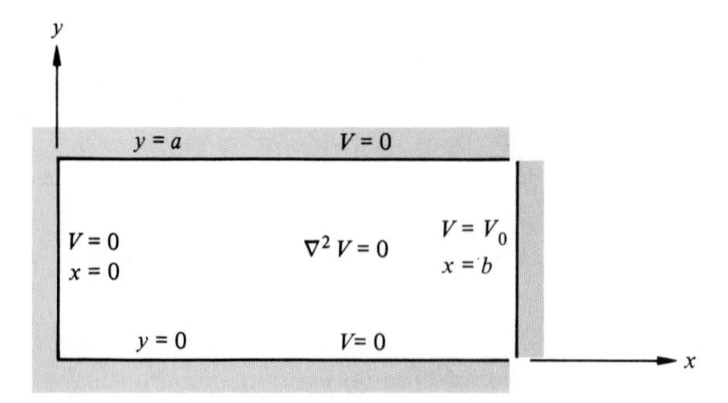

FIGURE 10.2
Illustration of the solution to Laplace's equation for rectangular boundaries.

satisfies Laplace's equation for $n = 1, 2, \ldots$ so that, by superposition, a sum (an infinite sum) of these functions is also a solution to Laplace's equation; that is,

$$V = \sum_{n=1}^{\infty} C_n \sinh \frac{n\pi x}{a} \sin \frac{n\pi y}{a} \tag{39}$$

We will show in Sec. 10.3.4 that the constants C_n can be chosen so that (39) satisfies the last boundary condition at the right boundary ($x = b$). Thus, (39) will satisfy both Laplace's equation in the region and all boundary conditions.

In many problems the potential function is not a function of all three coordinate variables (perhaps due to symmetry). In this regard we say that the field is a one-, two-, or three-dimensional field depending upon whether the field is a function of one, two, or all the coordinate variables. One- and two-dimensional fields provide a considerable simplification in the solution of Laplace's equation. For this reason we will consider one-dimensional fields first before we discuss two- and three-dimensional fields.

10.3.3 Solutions to One-Dimensional Field Problems

A one-dimensional field implies that the field is a function of only one coordinate variable. This class of problems, being the simplest, is considered first. The method of solving such problems will be illustrated by considering examples in rectangular, cylindrical, and spherical coordinates.

EXAMPLE 10.1 Two electrodes are separated by a distance b, and the potentials at the two electrodes are V_1 and V_2 (where $V_1 > V_2$), as shown in Fig. 10.3. Determine the potential between the electrodes, assuming that the potential is invariant with respect to the y and z coordinate variables. Sketch the potential distribution between the electrodes.

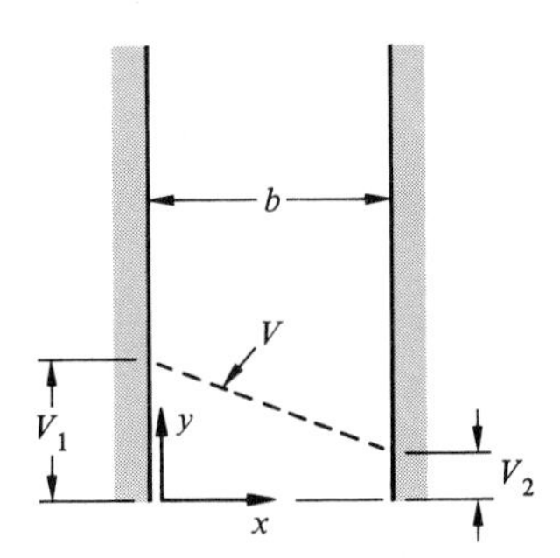

FIGURE 10.3
Example 10.1. Potential distribution between two parallel electrodes.

Solution Because the potential does not vary with respect to the y and z coordinates, we have

$$\frac{\partial^2 V}{\partial y^2} = \frac{\partial^2 V}{\partial z^2} = 0$$

and (6) simply reduces to

$$\frac{d^2 V}{dx^2} = 0 \tag{40}$$

(Note the use of the ordinary derivative rather than of a partial derivative since V is a function of only one variable, x.) Integrating (40) twice, we obtain

$$V = C_1 x + C_2 \tag{41}$$

where C_1 and C_2 are constants. The boundary condition $V = V_1$ at $x = 0$ yields $C_2 = V_1$, and the condition $V = V_2$ at $x = b$ gives $C_1 = (V_2 - V_1)/b$. Substituting these in (41) gives the desired solution

$$V = \frac{1}{b}(V_2 - V_1)x + V_1 \tag{42}$$

which is also sketched in Fig. 10.3.

EXAMPLE 10.2 Find the capacitance of the coaxial cable (per unit of cable length) shown in Fig. 10.4.

Solution In this case, Laplace's equation in cylindrical coordinates reduces to (see Appendix A)

$$\frac{1}{r}\frac{d}{dr}\left(r\frac{dV}{dr}\right) = 0 \tag{43}$$

because the variation of V is with respect to r only. Integrating (43) in two steps yields

$$V = C_1 \ln r + C_2 \tag{44}$$

If we let the conductors be at potentials V_0 and 0 as shown, we have

$$V = \frac{V_0}{\ln (b/a)} \ln \frac{r}{a} \tag{45}$$

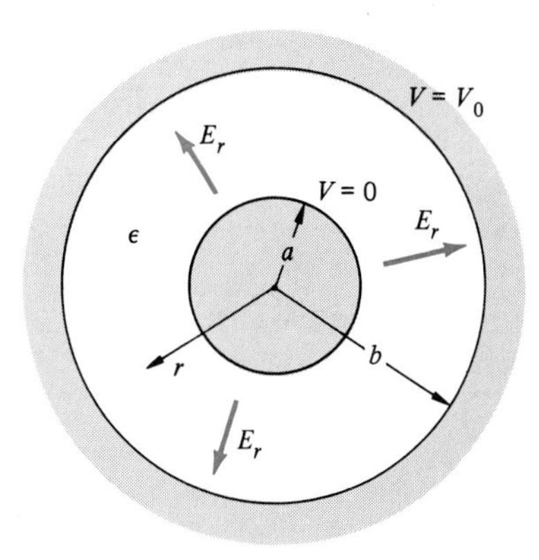

FIGURE 10.4
Example 10.2. Capacitance of a coaxial cable.

Knowing that $\mathbf{E} = -\nabla V$, we obtain the radial component of $\mathbf{E}$, E_r, from (45) as

$$E_r = -\frac{dV}{dr}$$

$$= -\frac{V_0}{\ln(b/a)}\frac{1}{r} \tag{46}$$

The total energy stored in the electric field in the region $a < r < b$ per unit of axial length of the cable is given by (see Sec. 3.8)

$$w_e = \frac{\epsilon}{2}\int_v E_r^2 \, dv$$

$$= \frac{\epsilon}{2}\int_{z=0}^{1}\int_{\phi=0}^{2\pi}\int_{r=a}^{b} rE_r^2 \, dr \, d\phi \, dz \tag{47}$$

Substituting (46) into (47) gives

$$w_e = \frac{2\pi\epsilon}{2}\frac{V_0^2}{[\ln(b/a)]^2}\int_a^b \frac{1}{r}\,dr$$

$$= \frac{\pi\epsilon V_0^2}{\ln(b/a)} \tag{48}$$

In terms of the per-unit-length capacitance c, the same energy can also be expressed as (see Sec. 3.9)

$$w_e = \tfrac{1}{2}cV_0^2 \tag{49}$$

Comparing (48) and (49), we obtain

$$\tfrac{1}{2}cV_0^2 = \frac{\pi\epsilon V_0^2}{\ln(b/a)} \tag{50}$$

and the required per-unit-length capacitance is

$$c = \frac{2\pi\epsilon}{\ln(b/a)} \qquad \text{F/m} \tag{51}$$

Alternatively, we could have obtained this per-unit-length capacitance from the fundamental relationship

$$c = \frac{q}{V} \qquad \text{F/m} \tag{52}$$

where q is the charge per unit length on the outer conductor and V is the potential difference between the outer and inner conductors: $V = V_0$. The charge per unit length q can be obtained by noting that the surface charge on a perfect conductor is numerically equal to the component of the electric flux density vector that terminates normal to the conductor surface. (See Prob. 10-4.)

EXAMPLE 10.3 Between the inner and outer conductors of the spherical shell shown in Fig. 10.5 we have two layers of dielectrics, ϵ_1 and ϵ_2. Find the potential distribution within the region $c < r < a$. The outer conductor is at a potential V_0 and the inner conductor is at zero potential.

Solution Because of symmetry, in spherical coordinates Laplace's equation reduces to (see Appendix A)

$$\frac{d}{dr}\left(r^2 \frac{dV}{dr}\right) = 0$$

which, upon integrating twice, gives

$$V = -\frac{C'}{r} + C''$$

Thus, in regions 1 and 2, respectively, we have

$$V_1 = -\frac{C_1}{r} + C_2 \qquad \text{for } b < r < a \tag{53}$$

$$V_2 = -\frac{C_3}{r} + C_4 \qquad \text{for } c < r < b \tag{54}$$

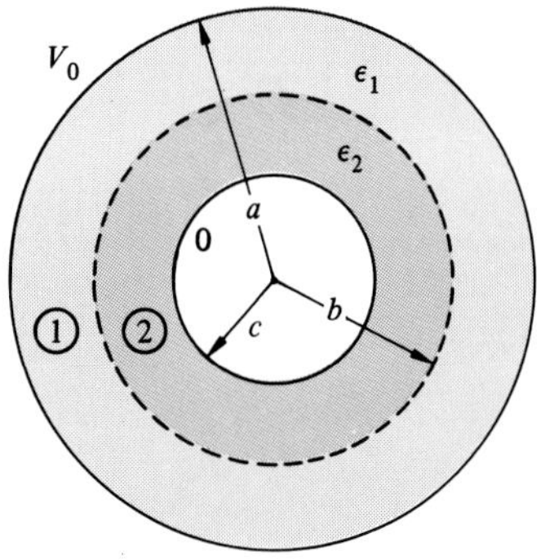

FIGURE 10.5
Example 10.3. Potential distribution of a spherical capacitor.

where the constants C_1, C_2, C_3, and C_4 are to be evaluated from the boundary conditions. At $r = c$, $V_2 = 0$ gives

$$C_4 = \frac{C_3}{c} \tag{55}$$

At $r = a$, $V_1 = V_0$ yields

$$V_0 = -\frac{C_1}{a} + C_2 \tag{56}$$

At $r = b$, we have $V_1 = V_2$, so that

$$-\frac{C_1}{b} + C_2 = -\frac{C_3}{b} + C_4 \tag{57}$$

Also, at the interface the radial components of the electric flux density vector must be continuous as a general boundary condition on the **D** field, or $D_{r1} = D_{r2}$, and we have

$$\epsilon_1 E_{r1} = \epsilon_2 E_{r2}$$

This condition and the relationship $E_r = -\nabla V$ yield, at $r = b$,

$$\epsilon_1 \left.\frac{dV_1}{dr}\right|_{r=b} = \epsilon_2 \left.\frac{dV_2}{dr}\right|_{r=b}$$

$$\epsilon_1 \left.\frac{C_1}{r^2}\right|_{r=b} = \epsilon_2 \left.\frac{C_3}{r^2}\right|_{r=b}$$

and we obtain

$$\epsilon_1 C_1 = \epsilon_2 C_3 \tag{58}$$

We thus have four independent equations involving four unknowns—Eqs. (55), (56), (57), and (58). Solving for unknowns C_1, C_2, C_3, and C_4 and substituting in (53) and (54) finally gives the required potential distribution:

$$V_1 = \left[\frac{1/a - 1/r}{(\epsilon_2/\epsilon_1)(1/c - 1/b) + (1/b - 1/a)} + 1\right] V_0 \qquad b < r < a$$

$$V_2 = \frac{1/c - 1/r}{(1/c - 1/b) + (\epsilon_1/\epsilon_2)(1/b - 1/a)} V_0 \qquad c < r < b$$

10.3.4 Product Solution to Laplace's Equation

The solution to Laplace's equation by direct integration discussed in Sec. 10.3.3 is straightforward for one-dimensional field problems. If the field is two- or three-dimensional, we take the following approach, known as the *product solution* or the *method of separation of variables*.

First we consider Laplace's equation in rectangular coordinates. Thus, we have

$$\frac{\partial^2 V}{\partial x^2} + \frac{\partial^2 V}{\partial y^2} + \frac{\partial^2 V}{\partial z^2} = 0 \tag{59}$$

Let us assume the solution of (59) to be of the form

$$V = X(x)Y(y)Z(z) \tag{60}$$

In (60) notice that the solution is assumed to be a product of three functions: $X(x)$, a function of x alone; $Y(y)$, a function of y only; and $Z(z)$, a function of z only. Substituting (60) into (59) gives

$$\frac{\ddot{X}}{X} + \frac{\ddot{Y}}{Y} + \frac{\ddot{Z}}{Z} = 0 \tag{61}$$

where $X = X(x)$, $Y = Y(y)$, $Z = Z(z)$, $\ddot{X} = \partial^2 X/\partial x^2$, $\ddot{Y} = \partial^2 Y/\partial y^2$, and $\ddot{Z} = \partial^2 Z/\partial z^2$. Notice that (61) must be valid for all values of the variables x, y, and z. In (61) each term involves one variable. For example, the second term is a function of only y and the third term is a function of only z. Thus, the second and third terms do not vary with x, and because the first term is the algebraic sum of the second and third terms $[\ddot{X}/X = -(\ddot{Y}/Y + \ddot{Z}/Z)]$, which do not vary with x, the first term must also not vary with x. Consequently, we have a function of x, $(\ddot{X}/X)$, that does not vary with x. This can be true only if the function is a constant, say k_x^2. Similarly, we conclude that the second and third terms are also constants, say k_y^2 and k_z^2, respectively. We thus have from (61)†

$$k_x^2 + k_y^2 + k_z^2 = 0 \tag{62}$$

The differential equations involving one variable at a time are

$$\ddot{X} - k_x^2 X = 0 \tag{63a}$$

$$\ddot{Y} - k_y^2 Y = 0 \tag{63b}$$

$$\ddot{Z} - k_z^2 Z = 0 \tag{63c}$$

For example, $\ddot{X}/X = k_x^2$ translates to $\ddot{X} = k_x^2 X$, or $\ddot{X} - k_x^2 X = 0$. Solutions to (63*a*) to (63*c*) are of the form

$$X = a_x \cosh k_x x + b_x \sinh k_x x \tag{64a}$$

$$Y = a_y \cosh k_y y + b_y \sinh k_y y \tag{64b}$$

$$Z = a_z \cosh k_z z + b_z \sinh k_z z \tag{64c}$$

From (60) the complete solution to Laplace's equation is the product of the above solutions. Finally, therefore, we obtain

$$V = (a_x \cosh k_x x + b_x \sinh k_x x)(a_y \cosh k_y y + b_y \sinh k_y y)(a_z \cosh k_z z + b_z \sinh k_z z) \tag{65}$$

At this point, three comments should be made. First, this method of separation of variables is applicable to Laplace's equation in cylindrical and spherical coordinates, yielding solutions involving Bessel functions and Legendre polynomials, respectively. We shall consider these later in this section.

Second, from (62) it is clear that at least one of the constants must be negative. We arbitrarily let k_x^2 be negative, and thus (63*a*) becomes

$$\ddot{X} + k_x^2 X = 0 \tag{66}$$

The solution to (66) then becomes

$$X = a'_x \cos k_x x + b'_x \sin k_x x \tag{67}$$

† Compare the similarity of this procedure to the solution for the TE and TM fields in a rectangular waveguide in Chap. 8.

Notice that in (65) the hyperbolic cosine and sine functions ("cosh" and "sinh") in x are changed to circular functions (cosine and sine), and the solution in this case becomes

$$V = (a'_x \cos k_x x + b'_x \sin k_x x)(a_y \cosh k_y y + b_y \sinh k_y y)(a_z \cosh k_z z + b_z \sinh k_z z) \tag{68}$$

Third, in most problems of practical interest we have only two-dimensional fields; that is, the field does not vary with one of the coordinates. In such a case, (59) becomes

$$\frac{\partial^2 V}{\partial x^2} + \frac{\partial^2 V}{\partial y^2} = 0 \tag{69}$$

and the solutions take the form

$$V = (a_x \cosh k_x x + b_x \sinh k_x x)(a'_y \cos k_y y + b'_y \sin k_y y) \tag{70}$$

or

$$V = (a'_x \cos k_x x + b'_x \sin k_x x)(a_y \cosh k_y y + b_y \sinh k_y y) \tag{71}$$

We are now ready to consider examples of applying the method developed in this section.

EXAMPLE 10.4 The potentials on the four sides of a rectangular electrode structure are shown in Fig. 10.2. What is the form of solution to Laplace's equation? Choose the coordinate system as shown.

Solution Notice that the solutions are of the form given in (70) or (71). Let us choose (70). The boundary condition $V = 0$ at $x = 0$ requires that $a_x = 0$. Also, $V = 0$ at $y = 0$ implies that $a'_y = 0$. Consequently, (70) becomes

$$V = C_n \sinh k_x x \sin k_y y$$

where we have combined the constants b_x and b'_y such that $C_n = b_x b'_y$. Furthermore, we observe that from (69) the relationship $k_x^2 + k_y^2 = 0$ implies that $k_x^2 = -k_y^2$ and the magnitudes of k_x and k_y must be the same; that is, $k_x = k_y = k$ and the required form of solution becomes

$$V = C_n \sinh kx \sin ky \tag{72}$$

(This is true, in general, for all two-dimensional problems.) If we had selected (71), the result would have been

$$V = C'_n \sin kx \sinh ky$$

EXAMPLE 10.5 Apply further boundary conditions to evaluate the constants C_n and k in Example 10.4 and thus obtain the complete solution for the potential distribution within the region bounded by the electrodes shown in Fig. 10.2.

Solution From the boundary condition $V = 0$ at $y = a$ for all values of $x(0 < x < b)$ we obtain

$$\sin ka = 0 \qquad \text{or} \qquad k = \frac{n\pi}{a} \tag{73}$$

where n is an integer. Substituting (73) into (72) yields

$$V = C_n \sinh \frac{n\pi x}{a} \sin \frac{n\pi y}{a} \tag{74}$$

The last boundary condition, $V = V_0$ at $x = b$ for all y $(0 < y < a)$, when imposed on (74) cannot be satisfied by (74) alone. But we know that n is an integer. Therefore, we can have an infinite number of solutions, and from Sec. 10.3.2 it follows that the solution has the most general form

$$V = \sum_{n=1}^{\infty} C_n \sin \frac{n\pi y}{a} \sinh \frac{n\pi x}{a} \tag{75}$$

which satisfies the boundary conditions at $y = 0$ and $y = a$ and $x = 0$. To satisfy the condition at $x = b$, we must find C_n. At $x = b$, $V = V_0$ yields, from (75),

$$V_0 = \sum_{n=1}^{\infty} C_n \sinh \frac{n\pi b}{a} \sin \frac{n\pi y}{a}$$

or

$$V_0 = \sum_{n=1}^{\infty} a_n \sin \frac{n\pi y}{a} = f(y) \tag{76}$$

where

$$a_n = C_n \sinh \frac{n\pi b}{a}$$

and (76) is a Fourier sine series. It can be verified (see Prob. 10-5) that the Fourier coefficient a_n is given by

$$a_n = 0 \qquad n \text{ even}$$

$$a_n = \frac{4V_0}{n\pi} \qquad n \text{ odd}$$

Finally, therefore,

$$C_n = \frac{a_n}{\sinh (n\pi b/a)} = \frac{4V_0}{n\pi \sinh (n\pi b/a)} \tag{77}$$

and substituting (77) into (75) gives the final solution as

$$V = \sum_{n\,\mathrm{odd}} \frac{4V_0}{n\pi} \frac{\sinh(n\pi x/a)}{\sinh(n\pi b/a)} \sin\frac{n\pi y}{a} \tag{78}$$

Because (78) denotes the solution to Laplace's equation in rectangular coordinates and in terms of harmonics n, such solutions are also known as *rectangular harmonics.* Problem 10-6 is illustrative of cases in which solutions to Laplace's equation are in the form of the rectangular harmonics.

As we have seen in earlier chapters (Chaps. 3 and 4, for example), a judicious choice of the coordinate system is extremely important in solving a field problem. Thus, the solution to Laplace's equation in rectangular coordinates presented up to this point is appropriate to field problems of rectangular geometry. There exist field problems in which the field distribition is to be obtained within cylindrical structures, such as within cylindrical waveguides, coaxial transmission lines, and electron lenses for electrostatic focusing in cathode-ray tubes. In such cases, it is most appropriate to choose the cylindrical coordinate system to represent the field problem and solve Laplace's equation in cylindrical coordinates. The following development follows that of S. Ramo, J. R. Whinnery, and T. Van Duzer, *Fields and Waves in Communication Electronics*, Wiley, N.Y., 1965, Chap. 3.

In its general form, Laplace's equation in cylindrical coordinates may be written as

$$\frac{1}{r}\frac{\partial}{\partial r}\left(r\frac{\partial V}{\partial r}\right) + \frac{1}{r^2}\frac{\partial^2 V}{\partial \phi^2} + \frac{\partial^2 V}{\partial z^2} = 0 \tag{79}$$

There exists a large class of static field problems in which we have axial symmetry; that is, there are no variations of the field (or potential) with ϕ, and we essentially have a two-dimensional problem. In such cases, (79) reduces to

$$\frac{\partial^2 V}{\partial r^2} + \frac{1}{r}\frac{\partial V}{\partial r} + \frac{\partial^2 V}{\partial z^2} = 0 \tag{80}$$

Proceeding as for the solution in rectangular coordinates, we assume a product solution of the form

$$V(r, z) = R(r)Z(z) \tag{81}$$

which, when substituted into (80), yields

$$\ddot{R}Z + \frac{1}{r}\dot{R}Z + R\ddot{Z} = 0 \tag{82}$$

where $\ddot{R} = d^2R/dr^2$, $\dot{R} = dR/dr$, and $\ddot{Z} = d^2Z/dz^2$. We now separate the variables by dividing (82) by RZ, so that

$$-\frac{\ddot{Z}}{Z} = \frac{\ddot{R}}{R} + \frac{1}{r}\frac{\dot{R}}{R} \tag{83}$$

where $Z = Z(z)$ and $R = R(r)$. Using the same argument as for (61), both sides of (83) must be equal to a constant (known as the separation constant) T^2. Hence, from (83) we obtain

$$\frac{1}{Z}\frac{d^2Z}{dz^2} = T^2 \tag{84}$$

and

$$\frac{1}{R}\frac{d^2R}{dr^2} + \frac{1}{rR}\frac{dR}{dr} = -T^2 \tag{85}$$

Notice that (84) is similar to (63*c*), so we need not pursue it any further. Next, if $T^2 > 0$ in (85), we may rewrite it as

$$\frac{d^2R}{dr^2} + \frac{1}{r}\frac{dR}{dr} + T^2R = 0 \tag{86}$$

which is the *Bessel equation* in its simplest form. To solve (86), we assume a power series solution of the form

$$R(r) = \sum_{i=0}^{\infty} a_i r^i \tag{87}$$

where the a_i are constants. For (87) to be a solution of (86), we substitute (87) into (86) and obtain

$$a_i = a_{2m} = C_1(-1)^m \frac{(T/2)^{2m}}{(m!)^2} \tag{88}$$

where C_1 is an arbitrary constant and m is an integer, $m = 0, 1, 2, 3, \ldots$. Note that $a_1 = a_3 = a_5 = \cdots = 0$ and $a_0 = C_1$. Other selected values are $a_2 = -C_1T^2/4$ and $a_4 = C_1T^4/64$. Hence, (87) and (88) yield

$$\begin{aligned} R(r) &= C_1 \sum_{m=0}^{\infty} \frac{(-1)^m(Tr/2)^{2m}}{(m!)^2} \\ &= C_1\left[1 - \left(\frac{Tr}{2}\right)^2 + \frac{(Tr/2)^4}{(2!)^2}\cdots\right] \\ &= C_1 J_0(Tr) \end{aligned} \tag{89}$$

where $$J_0(Tr) = 1 - \left(\frac{Tr}{2}\right)^2 + \frac{(Tr/2)^4}{(2!)^2} \cdots \tag{90}$$

The symbol J (of an argument) is defined as a Bessel function. In particular, $J_0(Tr)$ is a Bessel function of the first kind and zero order (of the argument Tr).

Because (86) is a second-order differential equation, it must have a second solution with a second arbitrary constant, (89) being the first solution. The mathematics involved in obtaining the second solution is quite cumbersome—so here let us take one form of solution from tables (such as *CRC Mathematical Tables*, 25th ed., CRC Press, Cleveland, Ohio, 1979) as

$$R(r) = C_2 N_0(Tr) \tag{91}$$

where C_2 is an arbitrary constant and $N_0(Tr)$ is known as a Bessel function of the second kind and zero order. Just like $J_0(Tr)$, $N_0(Tr)$ is also a notation for an infinite series; that is,

$$N_0(Tr) = \frac{2}{\pi} \ln\left[\frac{\gamma(Tr)}{2}\right] J_0(Tr) - \frac{2}{\pi} \sum_{m=1}^{\infty} \frac{(-1)^m (Tr/2)^{2m}}{(m!)^2} \left(1 + \frac{1}{2} + \frac{1}{3} + \cdots + \frac{1}{m}\right) \tag{92}$$

where γ is known as Euler's constant and has a value $\gamma = 0.5772\ldots$.

In summary, for $T^2 > 0$, the solution to (80) may be written as

$$V(r, z) = [C_1 J_0(Tr) + C_2 N_0(Tr)][C_3 \sinh(Tz) + C_4 \cosh(Tz)] \tag{93}$$

Before we consider the case when $T^2 < 0$, let us comment on $J_0(Tr)$ and $N_0(Tr)$. These functions, called Bessel functions, are less familar to us than, say, circular ($\sin x$, $\cos x$, etc.) and hyperbolic ($\sinh x$, $\cosh x$, etc.) functions. However, this fact should not cause major concern. We simply observe that $J_0(Tr)$ and $N_0(Tr)$ are merely two of the functions of a whole family of Bessel's functions, which are solutions to the general Bessel equation. [Recall that we termed (86) as only one simple form of the Bessel equation.] Furthermore, the series representing Bessel functions are convergent. And, most important, Bessel functions of various kinds and orders for different arguments have been tabulated.

Returning now to the case $T^2 < 0$, we let $T^2 = -a^2$ or $T = \pm ja$ (where a is real). Then (86) may be written as

$$\frac{d^2R}{dr^2} + \frac{1}{r}\frac{dR}{dr} - a^2 R = 0 \tag{94}$$

In this case, the first solution—from (89) by replacing T with ja—may be written as

$$\begin{aligned} R &= C_5\left[1 + \left(\frac{jar}{2}\right)^2 + \frac{(jar/2)^2}{(2!)^2} + \cdots\right] \\ &= C_5 J_0(jar) \\ &= C_5 I_0(ar) \end{aligned} \tag{95}$$

where I_0 of an argument is a Bessel function of imaginary argument.

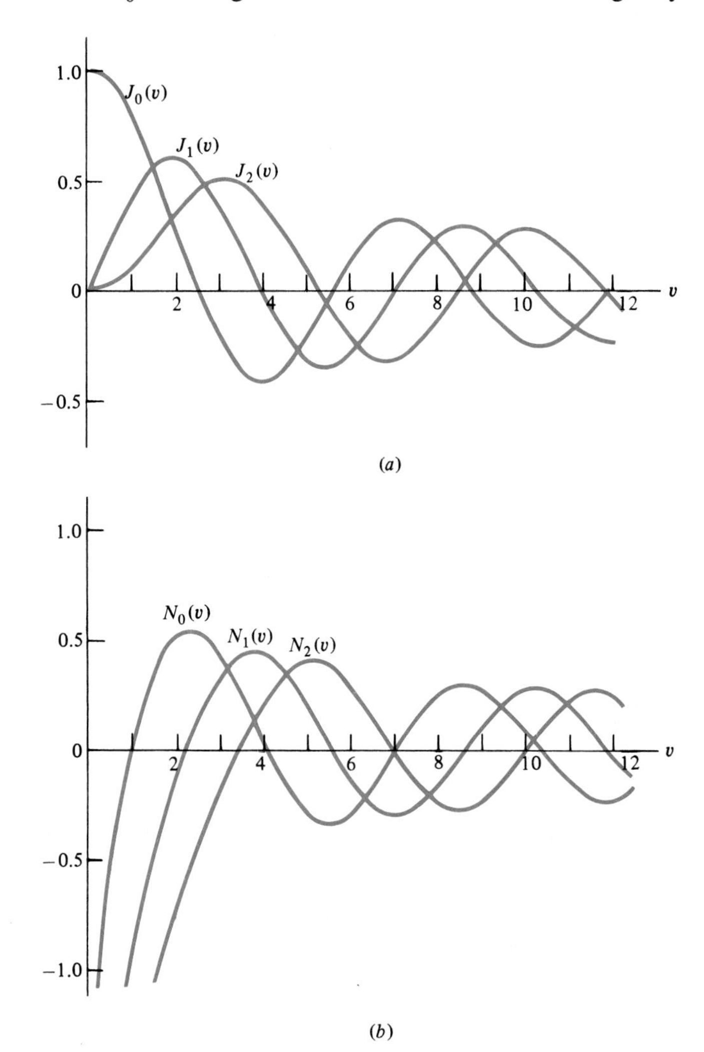

FIGURE 10.6

(a) Bessel functions of the first kind. *(b)* Bessel functions of the second kind.

Proceeding as for $T^2 > 0$, we write the second solution as

$$R = C_6 N_0(jar) = C_6 K_0(ar) \tag{96}$$

Consequently, the complete solution becomes:

$$V(r, z) = [C_5 I_0(ar) + C_6 K_0(ar)][C_7 \sin az + C_8 \cos az] \tag{97}$$

Plots of a number of Bessel functions are shown in Fig. 10.6. Values of Bessel functions of the two kinds, of various orders, and of different arguments may be obtained from these curves. Finally, whereas (93) and (97) give the solutions for the two cases, series solutions are also possible for certain boundary value problems, just as in Example 10.5. The following example illustrates the details.

EXAMPLE 10.6 Find the potential distribution within the cylindrical region shown in Fig. 10.7 subject to the boundary conditions: $V(r, 0) = V(r, l) = 0$, $V(a, z) = V_0$ for $0 < z < l/2$, and $V(a, z) = -V_0$ for $l/2 < z < l$.

Solution Because of the periodicity in the z direction, we choose the general series solution to (80) as

$$V(r, z) = \sum_{n=0}^{\infty} C_n I_0\left(\frac{\pi n r}{l}\right) \sin \frac{\pi n z}{l} \tag{98}$$

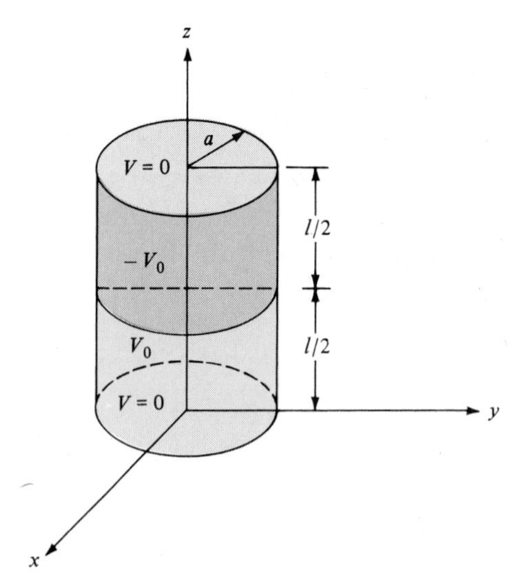

FIGURE 10.7

Example 10.6. Potential distribution within a cylindrical region.

Notice (and verify) that (98) satisfies the boundary conditions at $z = 0$ and $z = l$. To satisfy the remaining two boundary conditions at $r = a$, we set (98) as

$$V(a, z) = \sum_{n=0}^{\infty} C_n I_0\left(\frac{\pi n a}{l}\right) \sin\frac{\pi n z}{l} = f(z) \tag{99}$$

Then, as in Example 10.5, as coefficients for the Fourier sine series we have, from (99),

$$\begin{aligned}
A_n = C_n I_0\left(\frac{\pi n a}{l}\right) &= \frac{2}{l}\int_0^l f(z) \sin\frac{\pi n z}{l}\, dz \\
&= \frac{2}{l}\left(V_0 \int_0^{l/2} \sin\frac{\pi n z}{l}\, dz - V_0 \int_{l/2}^{l} \sin\frac{\pi n z}{l}\, dz\right) \\
&= \frac{2V_0}{\pi n}\left(-\cos\frac{\pi n}{2} + 1 + \cos n\pi - \cos\frac{\pi n}{2}\right) \\
&= 0 \qquad \text{for } n = \text{odd} \\
&= 0 \qquad \text{for } n = 0, 4, 8, 12, \ldots \\
&= \frac{8V_0}{n\pi} \qquad \text{for } n = 2, 6, 10, \ldots
\end{aligned}$$

Hence,

$$V(r, z) = \frac{8V_0}{\pi} \sum_{n=2,6,10,\ldots}^{\infty} \frac{1}{n} \frac{I_0(\pi n r/l)}{I_0(n\pi a/l)} \sin\frac{\pi n z}{l}$$

To complete this section, we finally consider Laplace's equation in spherical coordinates. In a large class of problems, symmetry about the z axis exists. Hence, $\partial V/\partial\phi = 0$ and Laplace's equation in two dimensions becomes

$$\frac{\partial}{\partial r}\left(r^2 \frac{\partial V}{\partial r}\right) + \frac{1}{\sin\theta}\frac{\partial}{\partial\theta}\left(\sin\theta \frac{\partial V}{\partial\theta}\right) = 0 \tag{100}$$

which, when expanded, yields

$$r\frac{\partial^2 V}{\partial r^2} + 2\frac{\partial V}{\partial r} + \frac{1}{r}\frac{\partial^2 V}{\partial\theta^2} + \frac{1}{r\tan\theta}\frac{\partial V}{\partial\theta} = 0 \tag{101}$$

As in the preceding cases for the rectangular and cylindrical coordinates, we assume a product solution

$$V(r, \theta) = R(r)\Theta(\theta) \tag{102}$$

and substitute (102) into (101) to obtain

$$r\ddot{R}\Theta + 2\dot{R}\Theta + \frac{1}{r}R\ddot{\Theta} + \frac{1}{r\tan\theta}R\dot{\Theta} = 0 \tag{103}$$

To separate the variables, we divide (103) by $R\Theta$; consequently,

$$r^2\frac{\ddot{R}}{R} + \frac{2r\dot{R}}{R} = -\frac{\ddot{\Theta}}{\Theta} - \frac{\dot{\Theta}}{\Theta\tan\theta} \tag{104}$$

For a special reason, which will soon be clear, we choose $m(m + 1)$ as the separation constant. Hence, (104) yields

$$r^2\frac{d^2R}{dr^2} + 2r\frac{dR}{dr} - m(m+1)R = 0 \tag{105}$$

and

$$\frac{d^2\Theta}{d\theta^2} + \frac{1}{\tan\theta}\frac{d\Theta}{d\theta} + m(m+1)\Theta = 0 \tag{106}$$

Equation (105) has a solution of the form (the reader should verify this)

$$R(r) = C_1 r^m + C_2 r^{-(m+1)} \tag{107}$$

In order to solve (106), we define a polynomial $P_m(\cos\theta)$ for any integer m by

$$P_m(\cos\theta) = \frac{1}{2^m m!}\left[\frac{d}{d(\cos\theta)}\right]^m (\cos^2\theta - 1)^m \tag{108}$$

and verify that (108) is a solution to (106). Equation (106) is known as *Legendre's equation*, and the polynomials (108), which are the solutions, are called *Legendre polynomials* of order m. Just as Fig. 10.6 shows Bessel functions, forms of Legendre polynomials for $m = 0$, 1, 2, and 3 are shown in Fig. 10.8. Here, the difference between an infinite series (such as a Bessel function) and a polynomial must be noted, since the value of a polynomial can be calculated exactly. The values of a few Legendre polynomials can be obtained from the following set of equations:

$$\begin{aligned} P_0(\cos\theta) &= 1 \\ P_1(\cos\theta) &= \cos\theta \\ P_2(\cos\theta) &= \tfrac{1}{2}(3\cos^2\theta - 1) \\ P_3(\cos\theta) &= \tfrac{1}{2}(5\cos^3\theta - 3\cos\theta) \end{aligned} \tag{109}$$

Of course, Legendre polynomials are available in tables for higher m's.

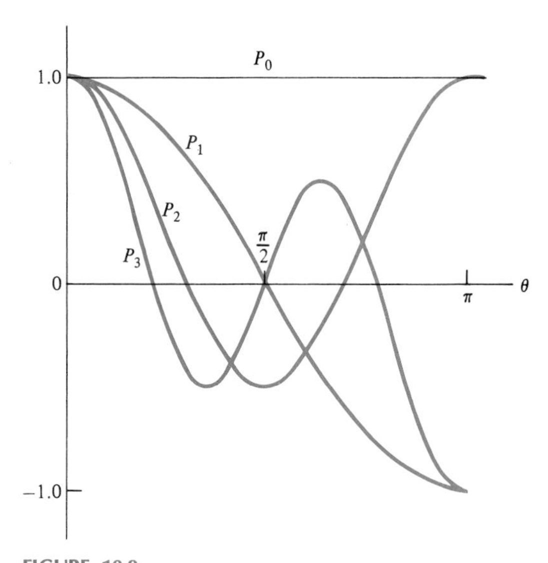

FIGURE 10.8
Legendre polynomials.

Returning to the solution of (106), we have

$$\Theta(\theta) = C_3 P_m(\cos\theta) \tag{110}$$

as one solution. The second solution has a singularity at $\theta = 0$.† Hence, this second solution will never be present in a region that includes the axis of the spherical coordinates; otherwise, the cited reference may be consulted for the procedure to obtain the second solution. For our purposes, the solution

$$V(r, \theta) = (C_4 r^m + C_5 r^{-(m+1)})P_m(\cos\theta) \tag{111}$$

is adequate where C_4 and C_5 are undetermined constants.

10.4 Solution to Poisson's Equation

In principle, the method of solving Laplace's equation is also applicable to Poisson's equation. Therefore, instead of repeating the details, we will consider some examples of solutions of Poisson's equation.

EXAMPLE 10.7 A slab of charge, of charge density ρ, extends to infinity in the y and z directions, as shown in Fig. 10.9. Determine the potential within the slab.

† W. R. Smythe, *Static and Dynamic Electricity*, 2d ed., McGraw-Hill, New York, 1950.

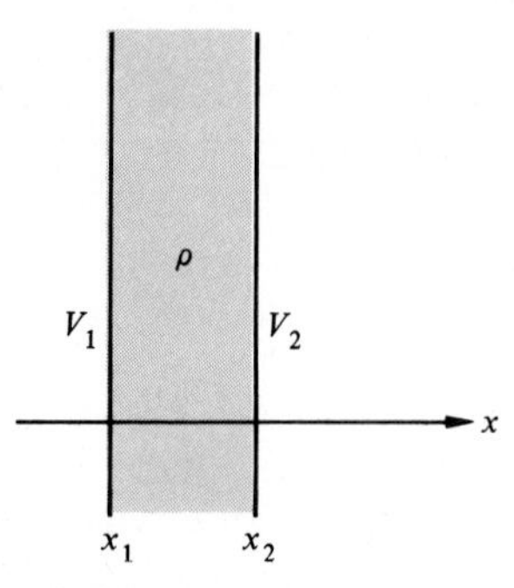

FIGURE 10.9

Example 10.7. Potential distribution with a slab of charge (Poisson's equation).

Solution Given the infinite extent in the y and z directions, the potential within the slab satisfies Poisson's equation in one dimension, or

$$\frac{d^2V}{dx^2} = -\frac{\rho}{\epsilon} \tag{112}$$

Integrating (112) twice, we obtain

$$V = -\frac{\rho x^2}{2\epsilon} + C_1 x + C_2 \tag{113}$$

At $x = x_1$, $V = V_1$, and this becomes

$$V_1 = -\frac{\rho x_1^2}{2\epsilon} + C_1 x_1 + C_2$$

At $x = x_2$, $V = V_2$, and we obtain

$$V_2 = -\frac{\rho x_2^2}{2\epsilon} + C_1 x_2 + C_2$$

Solving for C_1 and C_2, we obtain

$$C_1 = \frac{V_1 - V_2}{x_1 - x_2} + \frac{\rho}{2\epsilon}(x_1 + x_2)$$

$$C_2 = \frac{V_1 x_2 - V_2 x_1}{x_2 - x_1} + \rho\frac{x_1 x_2}{2\epsilon}$$

Substituting these constants into (113) gives the desired solution as

$$V = -\frac{\rho x^2}{2\epsilon} + \left[\frac{V_1 - V_2}{x_1 - x_2} + \frac{\rho}{2\epsilon}(x_1 + x_2)\right]x + \frac{V_1 x_2 - V_2 x_1}{x_2 - x_1} + \rho\frac{x_1 x_2}{2\epsilon}$$

Notice that if we had chosen the origin at x_1, the solution in the previous example would have been a bit simpler. This, of course, does not imply that this choice of origin would have resulted in a different potential (see Prob. 10-7).

10.5 The Method of Images

We have seen in the previous sections that there is one and only one solution to Laplace's equation or Poisson's equation that satisfies certain boundary conditions. The boundary conditions that we have considered for the scalar potential V are the specification of V over the boundaries of the region (Dirichlet conditions).

Consider two regions, region A and region B. We wish to find the scalar potential $V_A(x, y, z)$ in region A that achieves the specified potential V_b on the boundary between the two regions. The method of images is the replacement of region B with an equivalent (or image) charge distribution such that the image charge together with the original charge in region A produce the same boundary conditions, V_b, as before. As far as the potential solution in region A, $V_A(x, y, z)$, is concerned, it is unchanged (as is clear from the uniqueness theorem). The potential solution in region B, $V_B(x, y, z)$, is, of course, changed, but if we are interested only in the solution in region A, this is of no consequence.

A similar analogy for magnetic fields and the magnetic vector potential can be obtained. We will, however, consider only the application to the electric scalar potential solutions. Examples 10.8 and 10.9 will serve to illustrate this concept.

EXAMPLE 10.8 Consider a line charge ρ_l C/m situated at a height h above and parallel to the surface of an infinite ground plane (perfect conductor), as shown in Fig. 10.10*a*. This physical configuration resembles a power transmission line above earth (although the earth is not a perfect conductor). Find the capacitance c per unit of axial length along the line charge between the line and the ground plane.

Solution In order to find c, we must find the potential difference V between the line and the ground plane since the per-unit-length capacitance is

$$c = \frac{\rho_l}{V}$$

Finding the potential difference between the line charge and the ground plane for this structure appears to be somewhat complicated. The method of images, however, allows a considerable simplification. First we observe that the potential is zero along the surface of the ground plane and that the electric field is perpendicular to its surface. If we now replace the ground plane with a line charge $-\rho_l$ at a height h below the surface of the ground plane, as shown in Fig. 10.10*b*, the combination of the two line charges will also obviously produce zero potential at the previous position of the ground-plane surface. Thus, the solution

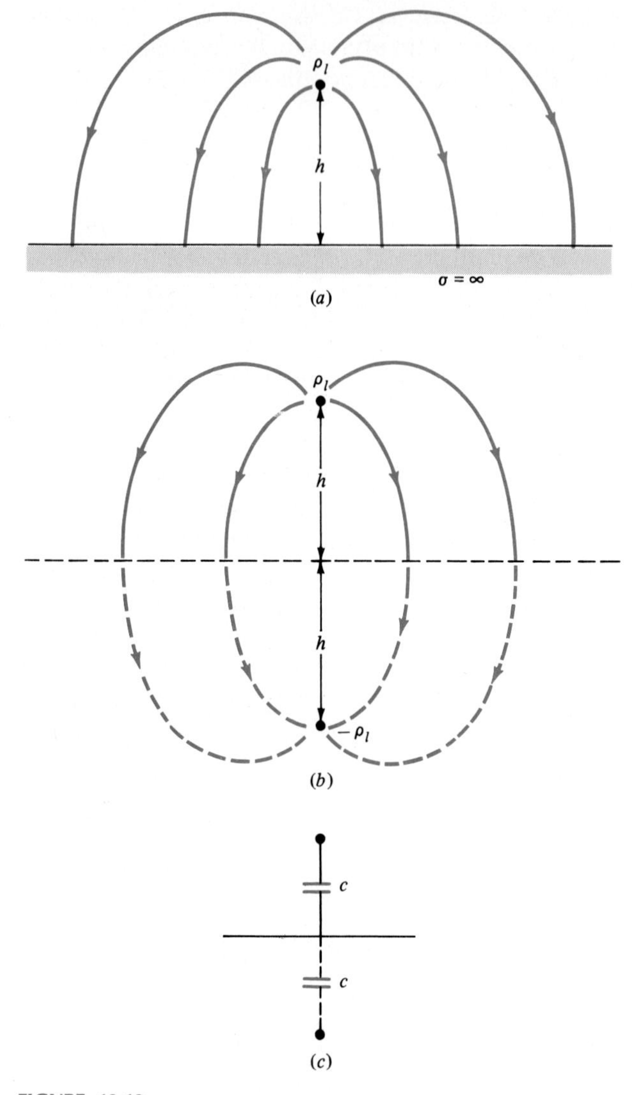

FIGURE 10.10
Example 10.8. The method of images for a line charge above an infinite ground plane.

for the potential in the region above the ground plane in Fig. 10.10*a* is unchanged; the potential solution in the region previously occupied by the ground plane is, of course, changed. Now we have a much simpler problem—that of two line charges separated by a distance $2h$ that bear equal and opposite charge distributions. This problem was solved in Example 3.14. The per-unit-length capacitance of the two-wire structure in Fig. 10.10*b* was found to be

$$c_{\text{two wire}} = \frac{\pi \epsilon_0}{\ln (2h/r_c)}$$

where the line charge has a radius r_c and is located in free space. From Fig. 10.10*c* we see that the per-unit-length capacitance of the single wire above ground is twice the two-wire capacitance, or

$$c = 2c_{\text{two wire}}$$
$$= \frac{2\pi\epsilon_0}{\ln(2h/r_c)}$$

EXAMPLE 10.9 Suppose that the wire in the previous problem has a sufficiently large radius with respect to its height above the ground plane that we cannot assume the charge distribution around its periphery to be uniform (as is implicit in the previous problem). Find the per-unit-length capacitance c.

Solution In this case, even though the charge distribution is nonuniform around the periphery of the wire, we may nevertheless replace the ground plane with a wire at a depth h having an image *distribution* of charge on its periphery, as shown in Fig. 10.11. Note that the charge per unit of line length is a surface

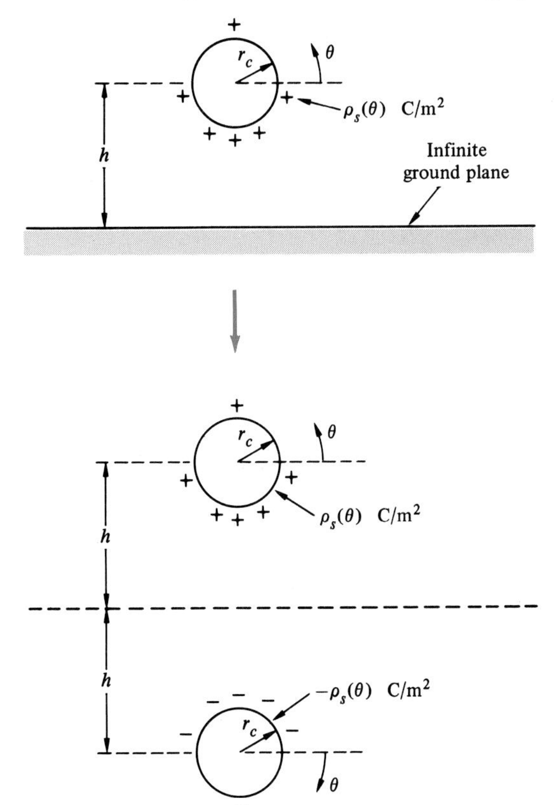

FIGURE 10.11
Example 10.9. The method of images for nonuniform charge distributions on wire peripheries.

charge distribution $\rho_s(\theta)$ C/m^2 as a function of angle θ around the wire periphery. Note that the image distribution is $-\rho_s(\theta)$ and that θ for the image is opposite to that for the original wire. It is clear from symmetry that this image charge distribution will produce the same potential at the original position of the ground-plane surface. (Consider the charge distribution to be made up of a sequence of infinitesimal line charges, each bearing a line charge distribution ρ_l equal to $\rho_s r_c \, d\theta$ evaluated at the particular θ of interest, then apply the concepts of the previous problem by superposing the effects of these individual line charges that compose ρ_s.) Thus, if we can find the per-unit-length capacitance for the two-wire problem, the per-unit-length capacitance of the original wire-above-ground problem is twice this capacitance.

The problem then becomes one of two parallel, infinitely long cylinders of radius r_c that are separated by a distance D, as shown in Fig. 10.12*a*. As is generally the case in the solution of field problems, we will make a guess as to the general form of the solution and verify that it is indeed the solution. Our guess in this problem is related to the observation that the cylinder surfaces form equipotential surfaces. We might suspect that the equipotential surfaces about two line charge distributions would be similar in shape. Thus, we will attempt to determine the separation between two line charges such that a set of their equipotential surfaces coincides with the surfaces of the desired conductors, as shown in Fig. 10.12*b*. Then according to the method of images, we may replace the cylinders with these equivalent line charges and use the resulting fields to compute the capacitance between the cylinders.

First we determine the potential at a point P due to the two line charges. An xy coordinate system is placed midway between the two cylinders and line charges. Clearly, by symmetry the potential will be zero along the y axis ($x = 0$). Thus, we obtain the potential of the point P with respect to this axis. From the results of Example 3.9, the potential of a line charge at a distance R with respect to a distance R_0 is

$$V = -\frac{\rho}{2\pi\epsilon_0}\ln\frac{R}{R_0}$$

where the line charge distribution is ρ C/m. By superposition, the potential at P due to both line charges is

$$\begin{aligned} V_P &= -\frac{\rho}{2\pi\epsilon_0}\ln\frac{R^+}{d/2} + \frac{\rho}{2\pi\epsilon_0}\ln\frac{R^-}{d/2} \\ &= \frac{\rho}{2\pi\epsilon_0}\ln\frac{R^-}{R^+} \end{aligned} \tag{114}$$

where the reference point for both potentials is taken at the origin ($x = 0$, and $y = 0$), although any other point along the y axis could have been chosen since the distances from the line charges to that point are the same and thus cancel out

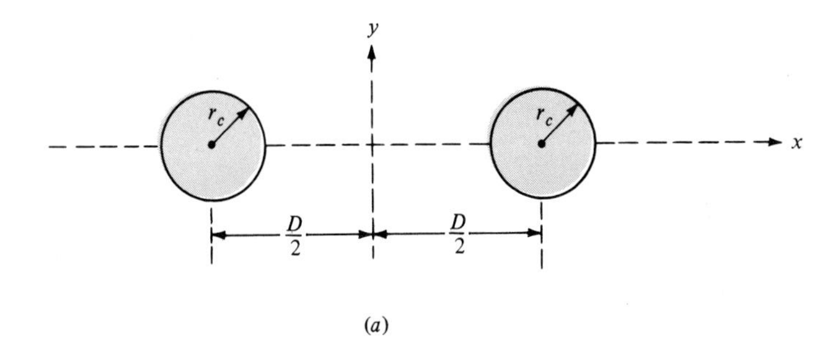

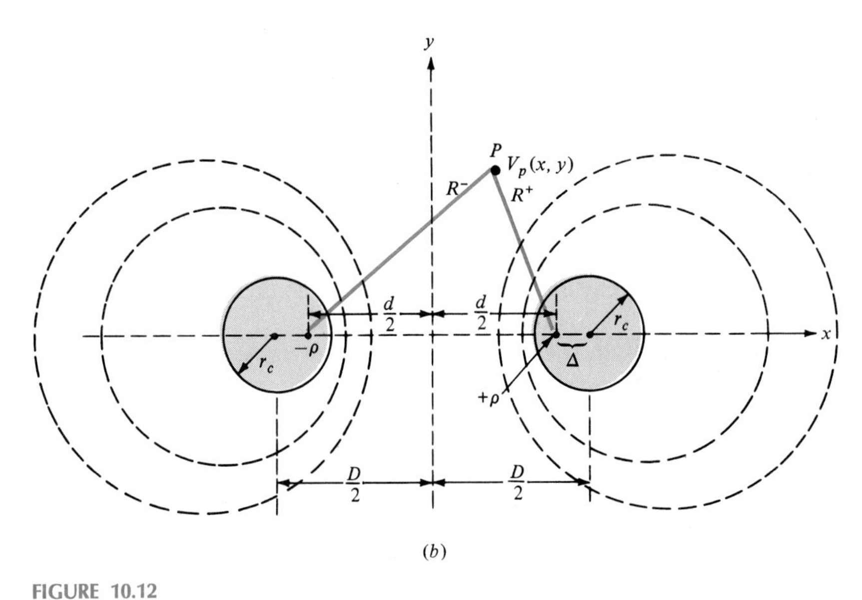

FIGURE 10.12
Example 10.9. Replacement of charge-bearing cylinders (wires) with equivalent line charges.

in the potential expression. From this result we see that points on an equipotential surface are such that the ratio of the distances from each line charge to those points is constant. For an equipotential surface of value V,

$$\frac{R^-}{R^+} = e^{(2\pi\epsilon_0/\rho)V}$$

$$= K \tag{115}$$

where K is a positive constant. The distances R^- and R^+ are given by

$$R^- = \sqrt{(x + d/2)^2 + y^2}$$
$$R^+ = \sqrt{(x - d/2)^2 + y^2}$$

Thus

$$K^2 = \frac{(x + d/2)^2 + y^2}{(x - d/2)^2 + y^2}$$

Writing this result in the form of an equation of a circle centered at $x = x_0$ and $y = 0$ with radius r,

$$(x - x_0)^2 + y^2 = r^2$$

gives

$$x_0 = \frac{d}{2}\frac{K^2 + 1}{K^2 - 1} \tag{116}$$

$$r = \frac{K}{|K^2 - 1|}d \tag{117}$$

Thus, the equipotential surfaces are in the form of cylinders along the x axis located at $x = x_0$ and having radii r. Values of $0 < K < 1$ result from $R^- < R^+$ and yield equipotential surfaces in the left-half plane. Values of $1 < K < \infty$ result from $R^- > R^+$ and yield equipotential surfaces in the right-half plane. As K increases, the circles become smaller and converge on the line charges.

We now wish to obtain the potentials of the equipotential surfaces that coincide with the desired cylindrical conductors. Equating $x_0 = D/2$ and $r = r_c$ in (116) and (117) and eliminating d, we have

$$K^2 - \frac{D}{r_c}K + 1 = 0$$

Solving for K gives

$$K = \frac{D}{2r_c} + \sqrt{\left(\frac{D}{2r_c}\right)^2 - 1}$$

Substituting this into the line charge potential expression in (114) gives the potentials of the equipotential surfaces of the right cylinder as

$$V_0 = \frac{\rho}{2\pi\epsilon_0}\ln\left[\frac{D}{2r_c} + \sqrt{\left(\frac{D}{2r_c}\right)^2 - 1}\right]$$

The potential of the corresponding left cylinder is the negative of this. Thus, the potential difference between the two cylinders is

$$\begin{aligned} V &= 2V_0 \\ &= \frac{\rho}{\pi\epsilon_0}\ln\left[\frac{D}{2r_c} + \sqrt{\left(\frac{D}{2r_c}\right)^2 - 1}\right] \end{aligned}$$

Dividing by the per-unit-length charge on the cylinders, ρ, gives the per-unit-length capacitance between two cylinders of radii r_c that are separated by distance D:

$$c_{\text{two wire}} = \frac{\pi\epsilon_0}{\ln\left[D/2r_c + \sqrt{(D/2r_c)^2 - 1}\right]} \quad \text{F/m}$$

Using the relationship

$$\cosh^{-1} x = \ln\left(x + \sqrt{x^2 - 1}\right)$$

we have

$$c_{\text{two wire}} = \frac{\pi\epsilon_0}{\cosh^{-1}(D/2r_c)} \quad \text{F/m}$$

The capacitance of a single wire of radius r_c at a height $H = D/2$ above an infinite, perfectly conducting plane is, by the method of images,

$$\begin{aligned} c &= 2c_{\text{two wire}} \\ &= \frac{2\pi\epsilon_0}{\cosh^{-1}(H/r_c)} \end{aligned}$$

The method of images can also be used in the case of multiple infinite planes. Consider the line charge shown in Fig. 10.13*a*. Imaging the line charge across the $x = 0$ boundary will preserve the potential along that boundary but not along the $y = 0$ boundary. Similarly, imagining the line charge across the $y = 0$ boundary will preserve the potential along that boundary but not along the $x = 0$ boundary. If we introduce an additional positive line charge, as shown in Fig. 10.13*b*, we observe that the potential will be preserved along both boundaries.

Consider the line charge within a conducting cylinder of radius r_c and infinite length shown in Fig. 10.14*a*. The line charge is displaced from the center of the cylinder a distance Δ. From the results of Example 10.9 we may replace the cylinder with an image line charge $-\rho$ outside the cylinder. In order for us to do so, an equipotential surface due to both line charges must coincide with the original cylinder boundary. This problem corresponds to that of Example 10.9 and Fig. 10.12*b*. Utilizing (116) and (117) of that example where $x_0 = D/2$ and $r = r_c$, we have

$$\begin{aligned} \Delta &= \frac{D}{2} - \frac{d}{2} \\ &= \frac{d}{K^2 - 1} \\ &= \frac{r_c}{K} \end{aligned}$$

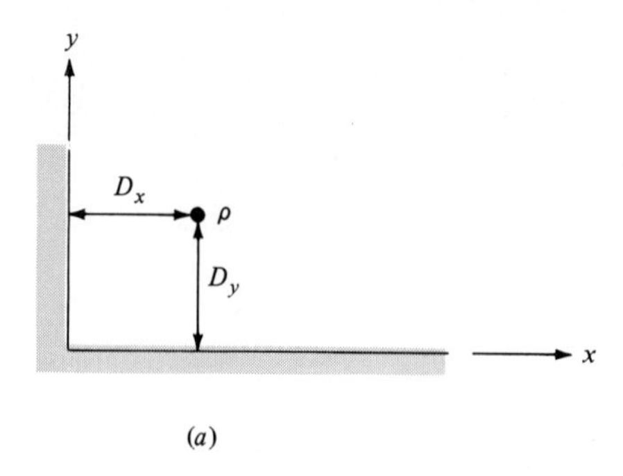

(a)

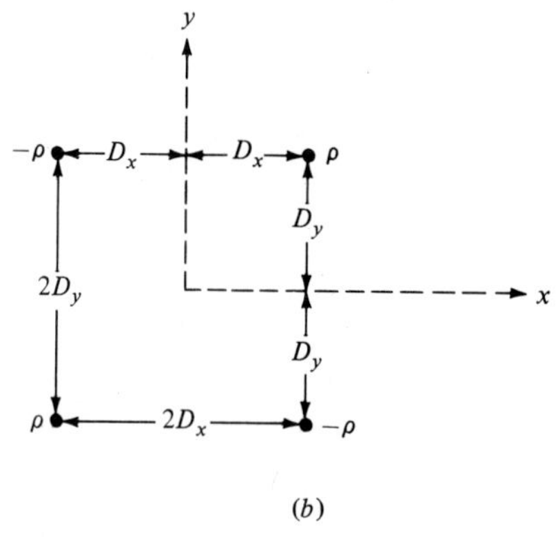

(b)

FIGURE 10.13
Illustration of the replacement of conducting surfaces with image charges.

Comparing Fig. 10.12*b* and Fig. 10.14 we see that

$$\begin{aligned} R &= D - \Delta \\ &= Kr_c \end{aligned}$$

Therefore,

$$\Delta = \frac{r_c^2}{R}$$

Thus, the cylinder may be replaced with a line charge of equal but opposite polarity located at a distance

$$R = \frac{r_c^2}{\Delta}$$

from the axis of the cylinder.

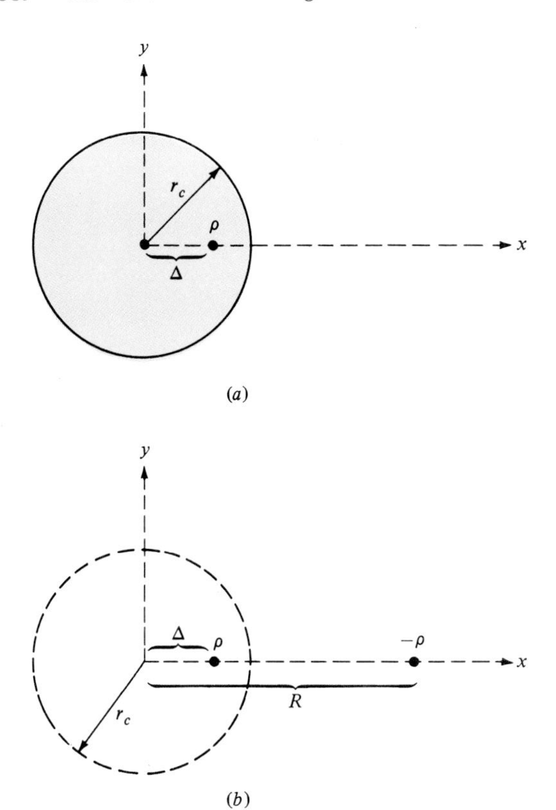

FIGURE 10.14
Illustration of the image of a line charge within a conducting cylinder.

Some imaging problems do not result in a finite number of images. For example, consider the point charge located midway between two parallel, grounded plates of infinite extent shown in Fig. 10.15*a*. Imaging the charge across the upper plate produces charge 2 and zero potential along the position of the upper plate. However, the potential along the position of the lower plate is no longer zero, so we image charge 1 and charge 2 across the lower plate to produce charge 3 and charge 4. Now the potential along the position of the lower plate is zero, but the potential along the position of the upper plate is no longer zero, so we image across this plate to produce charges 5 and 6, and so on. This leads to an infinite set of image charges having alternating signs and separations equal to the plate separation. Although in principle an infinite number of these image charges are required, in practice a finite number consisting of the charges nearest the plates will usually be adequate to approximate the zero potentials at the plate positions.

Up to this point we have concentrated on applying the method of images to static charge distributions, but it can also be applied to currents. Consider the

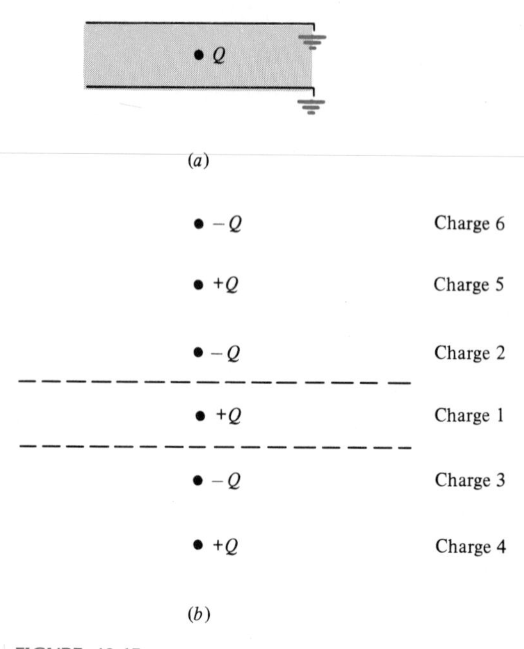

FIGURE 10.15
Illustration of the process of imaging across conducting boundaries which results in an infinite set of image charges.

current I at a height H above and parallel to an infinite, perfectly conducting plane shown in Fig. 10.16*a*. Since current is the movement of charge, we should be able to apply image theory in order to replace the plane with an image current. To do so, we may write

$$I = \rho u$$

where ρ is the charge per unit length along the path and u is the velocity of movement of this charge. The image of the movement of positive charge to the right is the movement of negative charge to the right or, equivalently, the movement of positive charge to the left. Thus, the image current is of the same value as, but opposite in direction to, the original current and is located the same distance below the position of the plane surface.

The image of a current that is directed perpendicularly to the surface is a current of equal value having the same direction as the original current, as shown in Fig. 10.16*b*. In a rough sense we may visualize positive charge accumulating on the top end of the segment and negative charge accumulating on the bottom end. Replacing these "charge accumulations" with their images results in the image current as shown in Fig. 10.16*b*.

The above results can be combined to obtain the image of a current that is neither parallel nor orthogonal to a plane. To do so, we represent the current in

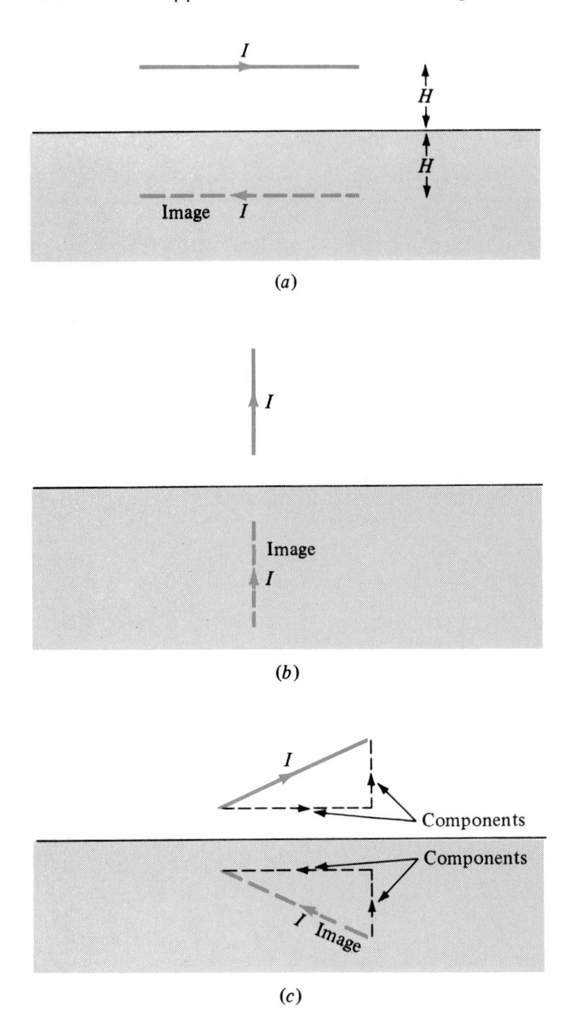

FIGURE 10.16
Illustration of the process of current imaging.

terms of components parallel and orthogonal to the plane, as shown in Fig. 10.16*c*. Replacing the components with their images according to the previous observations yields the correct image current.

10.6 Approximate Methods of Solving Field Problems

In the preceding examples we have generally obtained explicit, analytical expressions (equations) for solving static electromagnetic field problems. Substituting numerical values for the problem parameters into these expressions yields the solution to a particular problem.

An advantage of having analytical expressions is that the general behavior of the solution for different values of the problem parameter may be observed. For example, the capacitance of an air-filled parallel-plate capacitor was obtained in Chap. 3 as $C = \epsilon_0 A/d$, where A is the area of each plate and d is the plate separation. From this result we immediately saw that the capacitance varies inversely as the plate separation.

However, we also observed that analytical expressions are obtainable only in rather idealized cases; for example, in a potential distribution problem, certain symmetries must exist and the boundaries must not be irregular (Fig. 10.17), or in a magnetic circuit problem, saturation must be neglected. In other cases, we make certain simplifying approximations in order to reduce the original problem to one that is amenable to an analytical solution; for example, the capacitance of a parallel-plate capacitor, $C = \epsilon_0 A/d$, is obtained by assuming a uniform distribution of charge over each plate and neglecting fringing of the field at the edges of the plates. (We will examine the adequacy of this assumption in Sec. 10.6.1.)

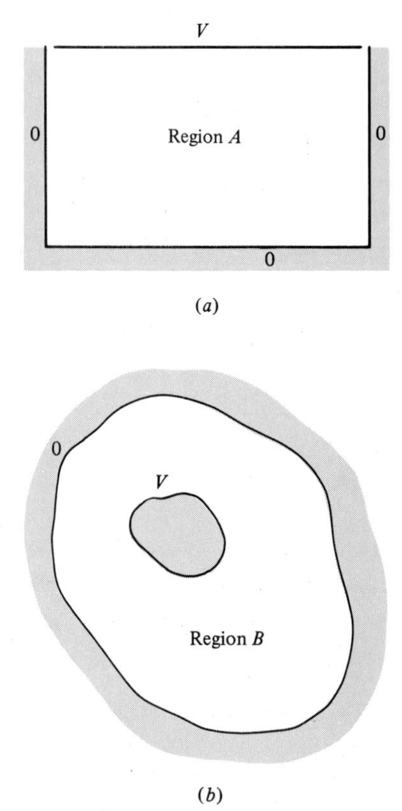

FIGURE 10.17
Analytical solution for potential distribution is feasible in region A *(a)* but not in region B *(b)*.

The ultimate objective in any of these field problems, however, is to obtain a numerical result for a particular problem. For example, knowing a formula for the capacitance of a parallel-plate capacitor provides some insight into its behavior, but the important result is the value of capacitance, e.g., 40×10^{-12} F. The methods for solving field problems discussed in the remainder of this chapter may be classified as approximate methods for obtaining the numerical solution to a problem. Broadly speaking, these techniques fall into one of the three following categories:

1 Numerical methods
2 Analog methods
3 Graphical methods

Before we consider these methods in some detail, it should be reiterated that the analytical methods previously considered yield exact solutions to approximate models, whereas the three techniques listed above result in approximate solutions to exact problems. We will consider these methods briefly; details are available in the references cited at the end of this chapter.

10.6.1 Numerical Methods

We now recall that field phenomena are governed by partial differential equations, such as Laplace's equation, Poisson's equation, and Maxwell's equations. We also realize that the desired information concerning the field is obtained by solving the pertinent differential equation. We have considered explicit analytical solutions earlier, and we now consider obtaining solutions to field problems by numerical methods. We will restrict our discussion to static field problems, although the numerical methods presented here are, in most cases, applicable to time-varying electromagnetic field problems. The references at the end of this chapter give numerous illustrative examples.

There are a number of numerical methods that can be used to solve field problems. For instance, Laplace's equation in a given region can be solved numerically by the finite-element method,[1] the finite-difference method,[2] or the method of moments.[3,4] In the following, the finite-difference method and the method of moments will be applied to static field problems.

Finite-Difference Method In obtaining a solution by the finite-difference method, the partial differential equation (to be solved) is replaced by a system of algebraic equations. The solution to the system of algebraic equations consists of values at discrete points in the region of interest, whereas the solution to the corresponding partial differential equation gives the values everywhere in the region. Because the system of algebraic equations is generally large, the solution is preferably obtained with the aid of a digital computer. Furthermore, to speed up the process of obtaining the solutions, certain special techniques are used (as will be mentioned shortly). Also, the computation is terminated as soon as predetermined accuracy is achieved. Thus, in applying numerical methods we must pay attention to the convergence, error, and accuracy of solutions. Some of these aspects will be considered later in this subsection. But first we observe that a set of finite-difference equations is valid for a given distribution of discrete points in the field. The various standard distributions of these discrete points, called *nodes*, are as follows.

In principle, any form of distribution of nodes may be used, but to simplify the problem, a regular distribution of nodes is almost invariably chosen. Two such regular networks (or *meshes*) showing the nodes are illustrated in Figs. 10.18*a* and *b*; these meshes are the square mesh and the equilateral triangle

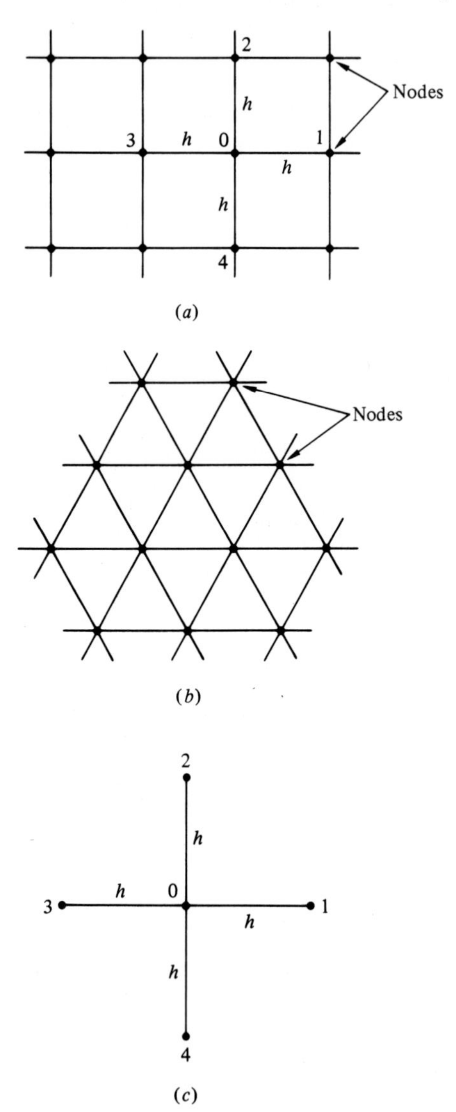

FIGURE 10.18

(a) A square mesh and *(b)* an equilateral triangle mesh, showing regular distribution of nodes; *(c)* a computational molecule of a square mesh or a 4-point star.

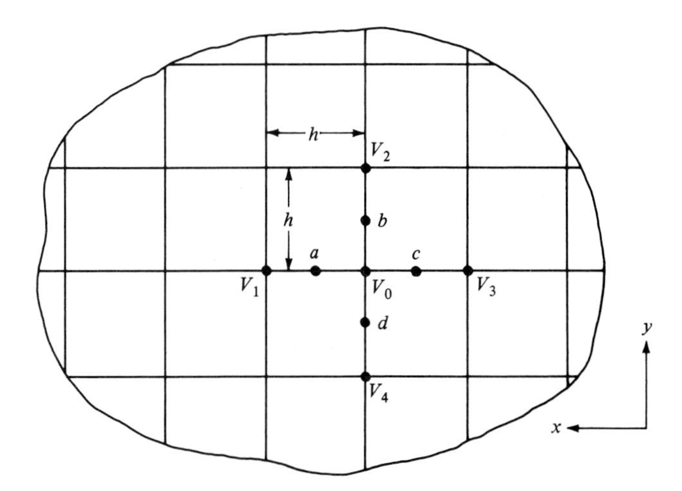

FIGURE 10.19
Illustration of the finite-difference equation method of solving Laplace's equation using a square mesh.

mesh, respectively. The arrangement of nodes in a square mesh, such as nodes 01234 (redrawn in Fig. 10.18*c*), is known as the *computational molecule* or *characteristic star*. It is a symmetrical star since the distances between the outer nodes and the center node, the mesh length, are the same. We choose the mesh length h to be small compared with the dimensions of the field boundaries. For a symmetrical star, we now proceed to obtain the finite-difference equation corresponding to Laplace's equation.

We consider Laplace's equation in two dimensions,

$$\frac{\partial^2 V}{\partial x^2} + \frac{\partial^2 V}{\partial y^2} = 0 \tag{118}$$

where V is the potential in a given region shown in Fig. 10.19. This region is divided into a square mesh of side h. If the unknown V's at the indicated nodes are as shown and if h is sufficiently small, then for a point a midway between the nodes 0 and 1 we have†

$$\left.\frac{\partial V}{\partial x}\right|_a \simeq \frac{1}{h}(V_1 - V_0) \tag{119}$$

And, similarly, for the point c we may write

$$\left.\frac{\partial V}{\partial x}\right|_c \simeq \frac{1}{h}(V_0 - V_3) \tag{120}$$

† More rigorously, (118) can be approximated at node 0 in terms of $V_1, \ldots, V_4$, by expanding these potentials in Taylor's series about V_0 in terms of the partial derivations of V at 0. (See Ref. 2.)

so that at the point 0, from (119) and (120) we obtain

$$\left.\frac{\partial^2 V}{\partial x^2}\right|_0 \simeq \frac{1}{h}\left(\left.\frac{\partial V}{\partial x}\right|_a - \left.\frac{\partial V}{\partial x}\right|_c\right) \simeq \frac{1}{h^2}[(V_1 - V_0) - (V_0 - V_3)] \tag{121}$$

Similarly, if we consider the potentials along the y axis, we have

$$\left.\frac{\partial^2 V}{\partial y^2}\right|_0 \simeq \frac{1}{h^2}[(V_2 - V_0) - (V_0 - V_4)] \tag{122}$$

Equations (121) and (122), when substituted into (118), yield

$$V_0 \simeq \tfrac{1}{4}(V_1 + V_2 + V_3 + V_4) \tag{123}$$

Thus, from (123), V_0 is found from the potential values at the adjacent corners of every square in the mesh. In practice, the computations are carried out in a number of ways, the objectives being that (123) be satisfied at every node and that the computations take a minimum time. Although hand computation is possible for a small number of nodes, it is best to carry out the computations on a digital computer. The procedure for hand computation is illustrated by the following two examples, which show that the finite-difference method involves the repeated modification of the potentials at each node until (123) is satisfied at the nodes within a predetermined degree of accuracy.

As mentioned previously, the iterative method is based on using (123) at every node and on using it repeatedly at the nodes until (123) is satisfied within reasonable accuracy. The iterative procedure for determining the potentials at given nodes is illustrated in the following example.

EXAMPLE 10.10 Compute the potentials at the nodes a, b, c, d, e, and f shown in Fig. 10.20, which also shows the potentials at the boundaries.

Solution The potential function V satisfies Laplace's equation. First we inscribe the square mesh, as shown, and assume that all the specified nodes are at zero potential except the potential at node f. Then we apply (123) at each node in succession. For instance, for the first cycle of computation, $V_f = \frac{1}{4}(30 + 0 + 0 + 0) = 7.5$, $V_a = \frac{1}{4}(10 + 20 + 0 + 0) = 7.5$, and so on. For the next cycle of computation, we use the previously calculated potentials to compute new potentials at the various nodes. Thus, for example, at the end of the second cycle we have $V_f = \frac{1}{4}(5.47 + 12.19 + 30 + 0) = 11.92$, and so on. This process is continued for eight cycles of computations—or until none of the

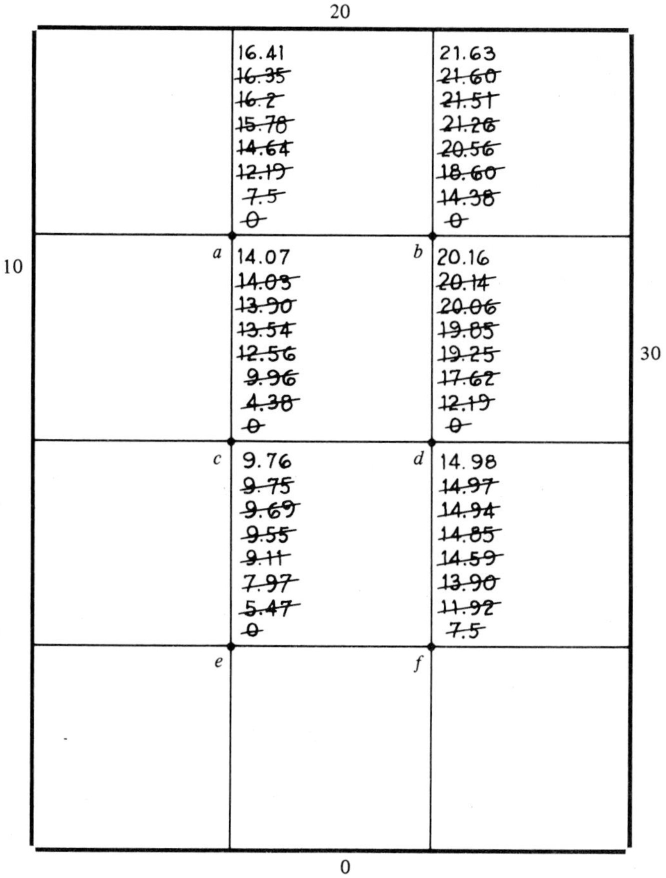

20

	16.41 ~~16.35~~ ~~16.2~~ ~~15.78~~ ~~14.64~~ ~~12.19~~ ~~7.5~~ ~~0~~	21.63 ~~21.60~~ ~~21.51~~ ~~21.28~~ ~~20.56~~ ~~18.60~~ ~~14.38~~ ~~0~~
	a 14.07 ~~14.03~~ ~~13.90~~ ~~13.54~~ ~~12.56~~ ~~9.96~~ ~~4.38~~ ~~0~~	*b* 20.16 ~~20.14~~ ~~20.06~~ ~~19.85~~ ~~19.25~~ ~~17.62~~ ~~12.19~~ ~~0~~
	c 9.76 ~~9.75~~ ~~9.69~~ ~~9.55~~ ~~9.11~~ ~~7.97~~ ~~5.47~~ ~~0~~	*d* 14.98 ~~14.97~~ ~~14.94~~ ~~14.85~~ ~~14.59~~ ~~13.90~~ ~~11.92~~ ~~7.5~~
	e	*f*

10 (left boundary) 30 (right boundary)

0

FIGURE 10.20

Example 10.10 Results of computing potentials at the nodes *a*, *b*, *c*, *d*, *e*, and *f* by iteration.

potentials at the six nodes changes by a predetermined value (say less than 0.5 percent) from its value computed for the preceding cycle (i.e., from its value last crossed out).

In Example 10.10 we obtained the potentials at each node by successive approximation (or repeated modification) of the potentials at each node until (123) was satisfied to a given degree of accuracy. However, we had no measure of the accuracy of the computed potential at a given node during a particular cycle. The degree of accuracy is determined by the residuals at the nodes. Referring to (123), if for a particular node 0 we have

$$V_1 + V_2 + V_3 + V_4 - 4V_0 = R_0 \tag{124}$$

then R_0 is the residual at node 0. Obviously, for the correct values of the potentials $V_0, V_1, \ldots, V_4$, the residual is zero. The relaxation method aims at reducing the residuals to zero, or to a predetermined small value, at all the nodes by suitable variations in V_0's. To achieve an accuracy of 1 to 5 percent, it is generally recommended that the individual residuals eventually be reduced to 0.1 percent of the mean of all the node potentials. This procedure of reducing the residual at one node at a time is known as *point relaxation* and is a very convenient method for hand computation. The method is illustrated in the following example.

EXAMPLE 10.11 Solve the problem of Example 10.10 by the point-relaxation method.

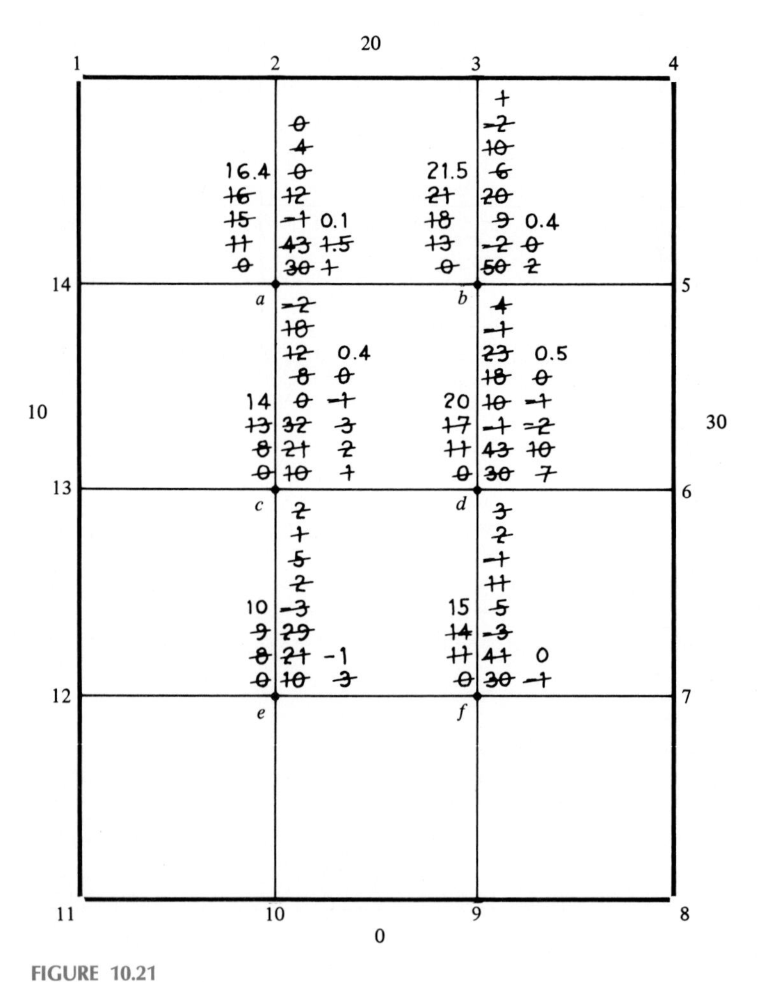

FIGURE 10.21
Example 10.11. Results of computing potentials at the nodes a, b, c, d, e, and f by relaxation.

Solution As in the last example, first we lay out the square mesh as shown in Fig. 10.21 and assume that all the specified nodes are at zero potential. Conventionally, we record the potentials at the top left of each node and write the values of the residuals at the top right of the node. While going through each cycle of computation, we write down the new values of potentials and residuals and cross out the old ones, as shown in Fig. 10.21. Notice that we have attempted to reduce the largest residual occurring at each stage approximately to zero. For the first cycle, the node b has the largest residual, 50. The mean of the surrounding node potentials V_a, V_d, V_5, and V_3 is $50/4 \simeq 13$. Therefore, the potential at node b is changed to 13; from (124) the resulting residual at this node becomes 2, and this changes the residuals at nodes a, b, and d. Next we observe that the largest residual of 43 occurs at node a. The mean of the surrounding node potentials is 11, so we change the potential of node a to 11; this change brings about a change in the residuals at nodes a, b, and c. We look for the node with the largest residual again and adjust the potential at that node to make the residual close to zero. This process is continued until the residuals at all the nodes are reduced to a predetermined minimum.

The last two examples illustrate the two common methods of hand computation of potentials at specified nodes. Clearly, hand computation is cumbersome and becomes very time-consuming if a large number of nodes is involved. For such cases, solutions are generally obtained by a digital computer. Reconsidering (123), we may write down the following set of equations for the nodes a, b, c, d, e, and f of Fig. 10.21. Notice that the nodes at the boundaries have been designated as nodes 1 through 14. Thus, we have

$$\begin{aligned}
-4V_a + V_b + V_c &= -V_2 - V_{14} \\
V_a - 4V_b + V_d &= -V_3 - V_5 \\
V_a - 4V_c + V_d + V_e &= -V_{13} \\
V_b + V_c - 4V_d + V_f &= -V_6 \\
V_c - 4V_e + V_f &= -V_{10} - V_{12} \\
V_d + V_e - 4V_f &= -V_7 - V_9
\end{aligned} \tag{125}$$

In matrix notation, we may rewrite (125) as

$$\underbrace{\begin{bmatrix} -4 & 1 & 1 & 0 & 0 & 0 \\ 1 & -4 & 0 & 1 & 0 & 0 \\ 1 & 0 & -4 & 1 & 1 & 0 \\ 0 & 1 & 1 & -4 & 0 & 1 \\ 0 & 0 & 1 & 0 & -4 & 1 \\ 0 & 0 & 0 & 1 & 1 & -4 \end{bmatrix}}_{[\mathbf{A}]} \underbrace{\begin{bmatrix} V_a \\ V_b \\ V_c \\ V_d \\ V_e \\ V_f \end{bmatrix}}_{[\mathbf{V}]} = \underbrace{\begin{bmatrix} -V_2 - V_{14} \\ -V_3 - V_5 \\ -V_{13} \\ -V_6 \\ -V_{10} - V_{12} \\ -V_7 - V_9 \end{bmatrix}}_{[\mathbf{B}]} \tag{126}$$

or

$$[\mathbf{A}][\mathbf{V}] = [\mathbf{B}] \tag{127}$$

Premultiplying both sides of (127) by $[\mathbf{A}]^{-1}$ yields the desired solution:

$$[\mathbf{V}] = [\mathbf{A}]^{-1}[\mathbf{B}] \tag{128}$$

Various standard numerical methods exist to find $[\mathbf{A}]^{-1}$. Notice that $[\mathbf{A}]$ is a *band matrix* as marked in (126), and for such a case special numerical techniques may be used to obtain a rapid convergence to the final solution. Various methods suitable for computer solutions are given in Ref. 2, and we will not consider these any further.

We close this section by briefly considering the errors affecting the accuracy of numerical solutions. The three common errors are: (1) *round-off error*, (2) *truncation error*, and (3) *computational error*. Round-off error occurs because we can use only a limited number of significant figures in the computations. The truncation error is introduced by expressing a partial differential equation as a difference equation, where certain higher-order terms are neglected (as in the Taylor series expansion of the partial derivatives in Laplace's equation). The computational error is caused by limiting the number of iterations to a finite number in obtaining the solution via iterative methods.

EXAMPLE 10.12 Obtain the solution to the boundary-value problem of Example 10.10.

Solution In this case, the equation to be solved is the matrix equation (127), in which the elements of $[\mathbf{B}]$ are $(-30, -50, -10, -30, -10, -30)$. The system of equations in (127) may be solved by using a computer, which gives these results:

	Equation (128)	*Iterative method (Fig. 10.20)*
V_a	16.44	16.41
V_b	21.66	21.63
V_c	14.10	14.07
V_d	20.19	20.16
V_e	9.77	9.76
V_f	14.99	14.98

Method-of-Moments Techniques In the previous section we discussed a method for the approximate solution of Laplace's equation in which the partial differential equation was approximated as a difference equation. In this section we will discuss an approximate method for solving differential as well as integral equations. This method is generally referred to as the *method of moments.*[3,4]

The method consists of the following general technique. Suppose we wish to find the distribution of charge in a region, $\rho_v(x, y, z)$, given the potential at various points in the region. These quantities are related by (see Chap. 3)

$$V(x, y, z) = \frac{1}{4\pi\epsilon_0} \int_v \frac{\rho_v(x', y', z')}{R}\, dv \tag{129}$$

where R is the distance between a differential volume element of the charge distribution and the point $[x, y, z]$ at which the potential is known. Thus, we are interested in solving an integral for the integrand. Equation (129) is the integral form of Poisson's equation. We could presume a *form* for ρ_v such as

$$\begin{aligned}\rho_v(x, y, z) &= \alpha_1\rho_1(x, y, z) + \alpha_2\rho_2(x, y, z) + \cdots + \alpha_N\rho_N(x, y, z)\\ &= \sum_{i=1}^{N} \alpha_i\rho_i(x, y, z)\end{aligned} \tag{130}$$

where ρ_i are some prechosen functions and α_i are constants to be determined. This procedure is similar to finding the solution of a differential equation by assuming the solution to be in the form of a power series.

Substituting (130) into (129), we obtain

$$\begin{aligned}V_j &= V(x_j, y_j, z_j)\\ &= \sum_{i=1}^{N} K_{ji}\alpha_i\end{aligned} \tag{131a}$$

where

$$K_{ji} = \frac{1}{4\pi\epsilon_0} \int_v \frac{\rho_i}{R_{ij}}\, dv \tag{131b}$$

where V_j is the (known) potential at some point $[x_j, y_j, z_j]$ in the region. Each K_{ji} can be found since ρ_i is known (assumed). A set of N equations in terms of the N unknowns $\alpha_1, \alpha_2, \ldots, \alpha_N$ can be obtained by evaluating (131) at N points in the region. The result is

$$\begin{aligned}V_1 &= K_{11}\alpha_1 + K_{12}\alpha_2 + \cdots + K_{1N}\alpha_N\\ V_2 &= K_{21}\alpha_1 + K_{22}\alpha_2 + \cdots + K_{2N}\alpha_N\\ &\vdots\\ V_N &= K_{N1}\alpha_1 + K_{N2}\alpha_2 + \cdots + K_{NN}\alpha_N\end{aligned} \tag{132}$$

These equations can be solved for $\alpha_1, \alpha_2, \ldots, \alpha_N$ by a variety of standard techniques using a digital computer, and the charge distribution ρ_v can be found from (130).

To illustrate this method, let us suppose that we wish to find the capacitance of a parallel-plate capacitor. In calculating the result $C = \epsilon_0 A/d$ in Chap. 3, we assumed that the total charge Q on each plate was uniformly distributed over that plate, and we neglected the fringing of the field at the edges of the plates. Let us now calculate this capacitance without making these two approximations. Suppose we prescribe the potential of one plate (the absolute potential) as $+V$ volts and the potential of the other plate as $-V$ volts. The potential difference between the two plates is $+V - (-V) = 2V$. If we could determine the surface charge distribution ρ_s over one of the plates, we could determine the total charge on this plate as

$$Q = \int_s \rho_s \, ds \tag{133}$$

where s is the surface of the plate. If both plates are of the same area, then due to the symmetry of the assigned plate potentials, the total charge (and charge distribution) on one plate will be the negative of the total charge (and charge distribution) on the other plate; consequently,

$$C = \frac{Q}{2V} \tag{134}$$

The problem is therefore to determine the charge distribution on the capacitor plates, *given* the potential of each plate.

We will assume a form of the charge distribution on each plate. Two examples are shown in Fig. 10.22. In Fig. 10.22*a*, each plate is divided into n surface elements Δs_i, and the charge distribution is assumed to be constant over each surface element; consequently, in (130) $\rho_i(x, y, z) = 1$. The values of α_i, the charge distribution coefficients on each subarea, are to be determined. The total charge on each surface element is $\alpha_i \, \Delta s_i$ and the total charge on the plate is

$$Q \simeq \sum_{i=1}^{N} \alpha_i \, \Delta s_i$$

A pointwise form of charge distribution is shown in Fig. 10.22*b* in which the charge on each surface element, $\alpha_i \Delta s_i$, is considered to be a point charge concentrated at the center of each surface element.

Now consider the parallel-plate capacitor in Fig. 10.23*a* in which each plate is divided into subareas. The potential at the center of one subarea, V_j, is a result of the charge distribution on this subarea as well as on all the other subareas. This may be written as

$$V_j = K_{j1}\alpha_1 + \cdots + K_{jn}\alpha_n + \cdots + K_{jk}\alpha_k + \cdots + K_{j(2n)}\alpha_{2n} \tag{135}$$

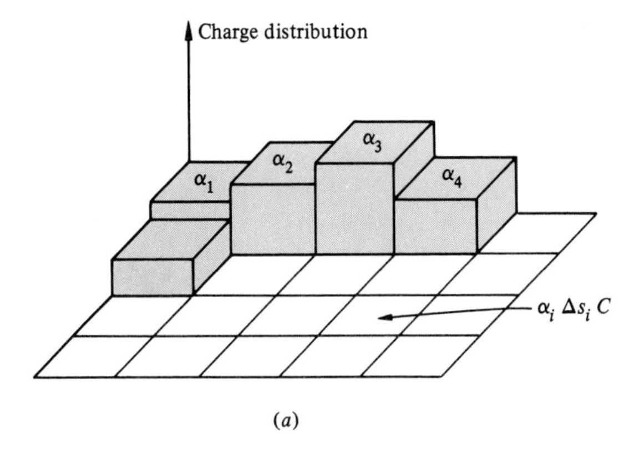

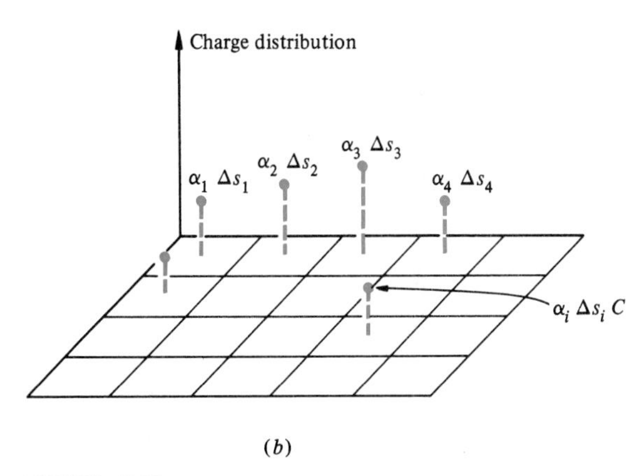

FIGURE 10.22

Two forms of assumed distribution of charge over the surface of a rectangular plate: *(a)* piecewise-constant charge distribution; *(b)* pointwise charge distribution.

Note that $V_1 = \cdots = V_n = +V$ and that $V_{n+1} = \cdots = V_{2n} = -V$. Writing equations of the form of (135) for all the other subareas, we obtain a set of $2n$ equations in terms of the $2n$ unknowns, α_i. These can be written in matrix form as

$$\begin{bmatrix} K_{11} & K_{12} & \cdots & K_{1(2n)} \\ K_{21} & K_{22} & \cdots & K_{2(2n)} \\ \vdots & \vdots & & \vdots \\ K_{n1} & K_{n2} & \cdots & K_{n(2n)} \\ K_{(n+1)1} & K_{(n+1)2} & \cdots & K_{(n+1)(2n)} \\ \vdots & \vdots & & \vdots \\ K_{(2n)1} & K_{(2n)2} & \cdots & K_{(2n)(2n)} \end{bmatrix} \begin{bmatrix} \alpha \\ \alpha_2 \\ \vdots \\ \alpha_n \\ \alpha_{n+1} \\ \vdots \\ \alpha_{2n} \end{bmatrix} = \begin{bmatrix} +V \\ +V \\ \vdots \\ +V \\ -V \\ \vdots \\ -V \end{bmatrix} = \begin{bmatrix} +1 \\ +1 \\ \vdots \\ +1 \\ -1 \\ \vdots \\ -1 \end{bmatrix} \quad \mathrm{V} \tag{136}$$

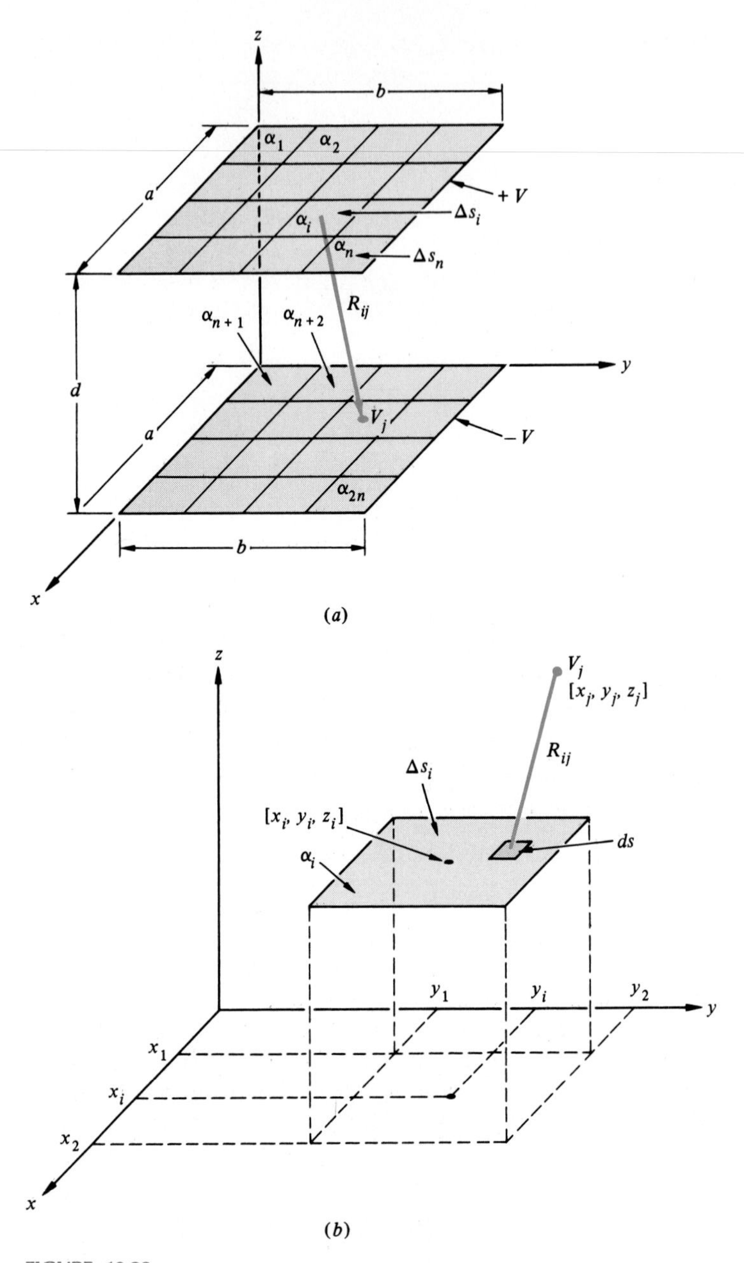

FIGURE 10.23

Computation of the contribution to the potential on a subsection due to charge on another subsection.

Solving this set of simultaneous equations for each α_i, we obtain the total charge on each plate as

$$Q = \sum_{i=1}^{n} \alpha_i \Delta s_i \tag{137}$$

from which we obtain the capacitance via (134).

The essential problem is to determine each K_{ji} in (135). The basic problem is illustrated in Fig. 10.23*b*. The potential V_j at a point $[x_j, y_j, z_j]$ due to a constant charge distribution α_i over a surface is

$$\begin{aligned} V_j &= K_{ji} \\ &= \frac{1}{4\pi\epsilon_0} \int_{y=y_1}^{y_2} \int_{x=x_1}^{x_2} \frac{dx_i\, dy_i}{R_{ij}} \end{aligned} \tag{138}$$

where

$$R_{ij}^2 = (x_j - x_i)^2 + (y_j - y_i)^2 + (z_j - z_i)^2 \tag{139}$$

To simplify the evaluation of this integral, let us assume that the capacitor plates are square (i.e., $a = b$) and that the subareas are square (i.e., $x_2 - x_1 = y_2 - y_1 = \Delta w$). With these assumptions, the integral in (138) for $i = j$ (i.e., the contribution to the potential V_j due to the charge distribution on its own subarea, α_j) becomes[3]

$$\begin{aligned} K_{jj} &= \frac{\Delta w}{\pi\epsilon_0} \ln\left(1 + \sqrt{2}\right) \\ &= 0.8814 \frac{\Delta w}{\pi\epsilon_0} \end{aligned} \tag{140}$$

For $i \neq j$ (that is, for the contribution to the potential V_j due to the charge distribution on some other subarea, α_i), the integral in (138) can also be evaluated in closed form, but the result is complicated. For K_{ji} with $i \neq j$, we will consider α_i to be a pointwise charge distribution; thus, (138) becomes

$$\begin{aligned} K_{ji} &\simeq \frac{\Delta s_i}{4\pi\epsilon_0 R_{ij}} \\ &= \frac{\Delta w^2}{4\pi\epsilon_0 R_{ij}} \end{aligned} \tag{141}$$

With these results in hand, we can evaluate (136) and obtain an approximation to the capacitance via (134) and (137). As the number of subareas is increased, we should obtain a more accurate result for the capacitance.

EXAMPLE 10.13 Determine the capacitance of a parallel-plate capacitor whose plates are square, 1 m on a side, and separated by 1 m.

Solution As a first approximation, we choose subareas, as shown in Fig. 10.24*a*. The entries in (136) become

$$\begin{bmatrix} 31.73 \times 10^9 & 9 \times 10^9 \\ 9 \times 10^9 & 31.73 \times 10^9 \end{bmatrix} \begin{bmatrix} \alpha_1 \\ \alpha_2 \end{bmatrix} = \begin{bmatrix} +1 \\ -1 \end{bmatrix}$$

which can be solved for $\alpha_1 = 43.99 \times 10^{-12}$ C/m^2 and $\alpha_2 = -\alpha_1$. The total charge on each plate is $Q = \alpha_1 \times 1$ m$^2 C$. The approximation to the capacitance

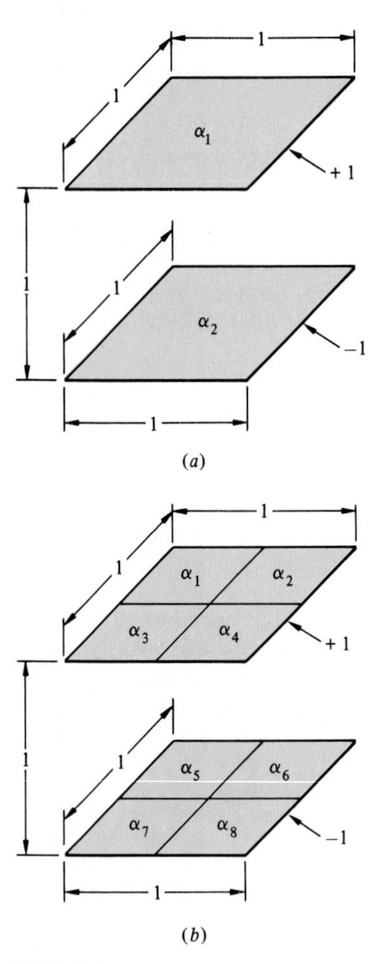

FIGURE 10.24
Example 10.13. Computation of the capacitance between two square plates by the method of moments. *(a)* One subsection per plate and *(b)* four subsections per plate.

becomes $C = Q/2V = 21.997 \times 10^{-12}$ F. This is compared to the approximate value of $C = \epsilon_0 A/d = \epsilon_0 = 8.84 \times 10^{-12}$ F.

As a second approximation, we divide each plate into four subareas, as shown in Fig. 10.24*b*. For these subareas, $\Delta w = \frac{1}{2}$ and the matrix in (136) is 8×8. Equation (136) was solved for this case with a digital computer. The results for the charge coefficients are $\alpha_1 = \alpha_2 = \alpha_3 = \alpha_4 = 48.83 \times 10^{-12}$ and $\alpha_5 = \alpha_6 = \alpha_7 = \alpha_8 = -48.83 \times 10^{-12}$. The total charge on the upper plate is $Q = 48.83 \times 10^{-12} C$. The capacitance becomes $C = Q/2 = 24.41 \times 10^{-12}$ F, as compared with 21.997×10^{-12} F for one subarea on each plate.

The results converge to a more accurate value of capacitance as the number of subareas on each plate is increased.

Note that for this example, the approximate values for the capacitance, 22 pF for $n = 1$ and 24 pF for $n = 4$ (pF denotes picofarad, or 10^{-12} F), were on the order of 2.5 and 2.8 times the value of $C = \epsilon_0 A/d = 8.84$ pF, which was determined by neglecting fringing of the field at the plate edges. It has been indicated that neglecting the fringing of the field is a reasonable assumption so long as the plate separation is sufficiently smaller than the plate area. As an illustration of this, Fig. 10.25 shows the results computed by the method of moments for a square-plate capacitor with plate width w and a plate separation of d, using 36 subareas on each plate.[3] The ratio of the exact capacitance to the

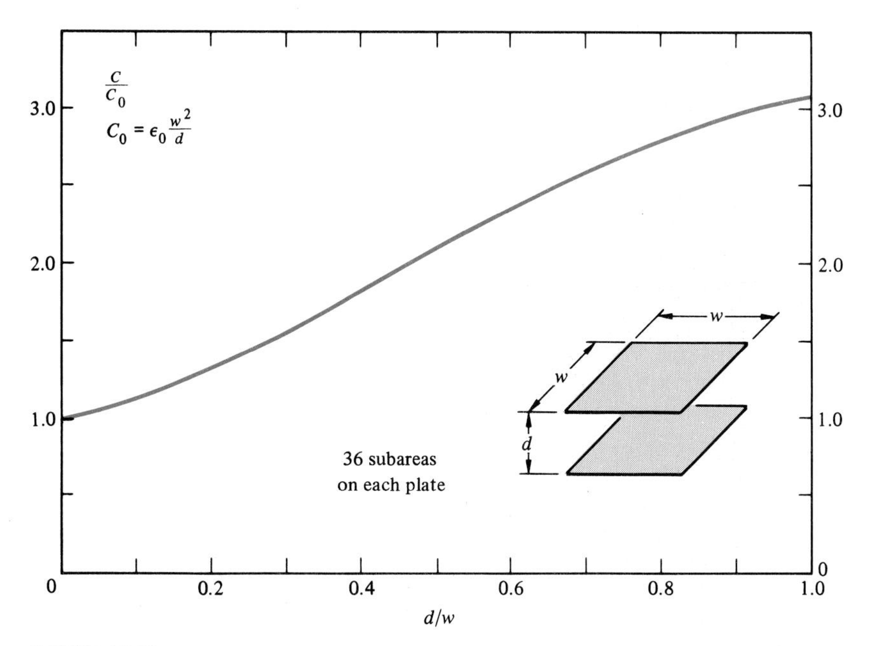

FIGURE 10.25
Values of the capacitance of a parallel-plate capacitor as a function of the ratio of separation d to plate width w computed by the method of moments. See Ref. 3.

capacitance computed by neglecting fringing, $C_0 = \epsilon_0 w^2/d$, is plotted versus the ratio of d/w. Note that for the problem in Example 10.13, $C/C_0 \simeq 3.1$.

Needless to say, the analytical solution of the parallel-plate capacitor problem without neglecting fringing would be virtually impossible. The method of moments permits us to obtain the solution (or an approximation to the solution) where an analytical solution would not be forthcoming. Numerous digital computer programs are available for this calculation.[4] In using these programs, we need as input data only the width of the plates, the plate separation, and the desired number of subareas for each plate. Symmetry can also be used to reduce the required number of simultaneous equations.[4] For example, in the previous problem for four subareas on each plate, we may solve four equations in four unknowns instead of eight equations in eight unknowns since the charge coefficients on the lower plate are (by symmetry) the negative of the corresponding ones on the upper plate: for example, $\alpha_5 = -\alpha_1$.

10.6.2 Analog Methods

Analog and experimental methods offer a quick way of determining field quantities in a given region. We recall from earlier considerations that problems relating to field quantities which satisfy Laplace's equation, Poisson's equation, or the diffusion equation can be formulated as boundary-value problems. For instance, the problem of Example 10.10 requires the solution to Laplace's equation, in two dimensions, for the potentials specified at the boundaries. We also recall the similarity between magnetic circuits and electric circuits (Chap. 4) whereby magnetic circuit problems could be solved by using appropriate electric circuit analogs. In these cases, the problem was generally solved analytically. However, in developing analogs for field problems we resort to experimental methods in obtaining the solutions.

Before we consider the analog method in detail, let us refer to Table 10.1, which summarizes the analogous quantities in dc electric circuits, magnetostatics, and electrostatics. The similarities are drawn on the basis that conductivity, permeability, and permittivity are considered to be analogous quantities, with the corresponding equations being identical in form.

To appreciate the usefulness of the analog method, refer to Fig. 10.26, in which a sheet made of a high-resistivity material is shown to occupy a region between two electrodes which are at two different potentials, V_1 and V_2. Such an arrangement is analogous to a region of a dielectric material separating the two electrodes. The sheet may also be considered to be analogous to an airgap region between two highly permeable magnetized surfaces corresponding to the electrodes. Once the analogy is established, direct measurements of electric circuit quantities (such as voltage and resistance) yield the corresponding magnetostatic or electrostatic field quantities. Although other analogs exist, we first consider the conductive-sheet analog.

Conductive-Sheet Analog The basic requirement for the conductive sheet to be useful for a field analog is that the sheet should be homogeneous and

TABLE 10.1

Analogy between DC Electric Circuits, Magnetostatic, and Electrostatic Quantities

DC Electric Circuit	*Magnetostatic*	*Electrostatic*
Conductivity	Permeability	Permittivity
σ (S/m)	μ (H/m)	ϵ (F/m)
Conductance	Permeance	Capacitance
$G = \sigma A/l$ (S)	$\mathcal{P} = \mu A/l$ (H)	$C = \epsilon A/d$ (F)
Electromotive force	Magnetomotive force	Potential difference
V (V)	$\mathfrak{F}$(At)	V (V)
Current	Magnetic flux	Electric flux
$I = VG$ (A)	$\psi_m = \mathfrak{F}\,\mathcal{P}$(W)	$\psi_e = VC$ (C)
Voltage gradient	Magnetic field intensity	Electric field intensity
E (V/m)	H (A/m)	E (V/m)
Current density	Magnetic flux density	Electric flux density
$J = \sigma E$ (A/m^2)	$B = \mu H$ (W/m^2)	$D = \epsilon E$ (C/m^2)

isotropic and should have a high resistivity. Teledeltos paper is probably the best among the commercially available conductive sheets. Typically, Teledeltos paper has a resistivity of the order of 2000 Ω per square and is available in rolls 50 ft long and 22 in wide.

To solve a boundary-value problem, such as Laplace's equation for given boundary conditions, the Teledeltos paper is laid on a flat insulating surface,

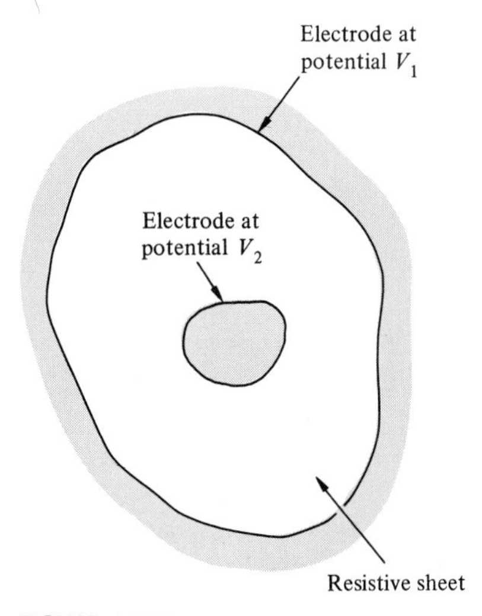

FIGURE 10.26

A conductive sheet analog for determining the potential distribution between two electrodes.

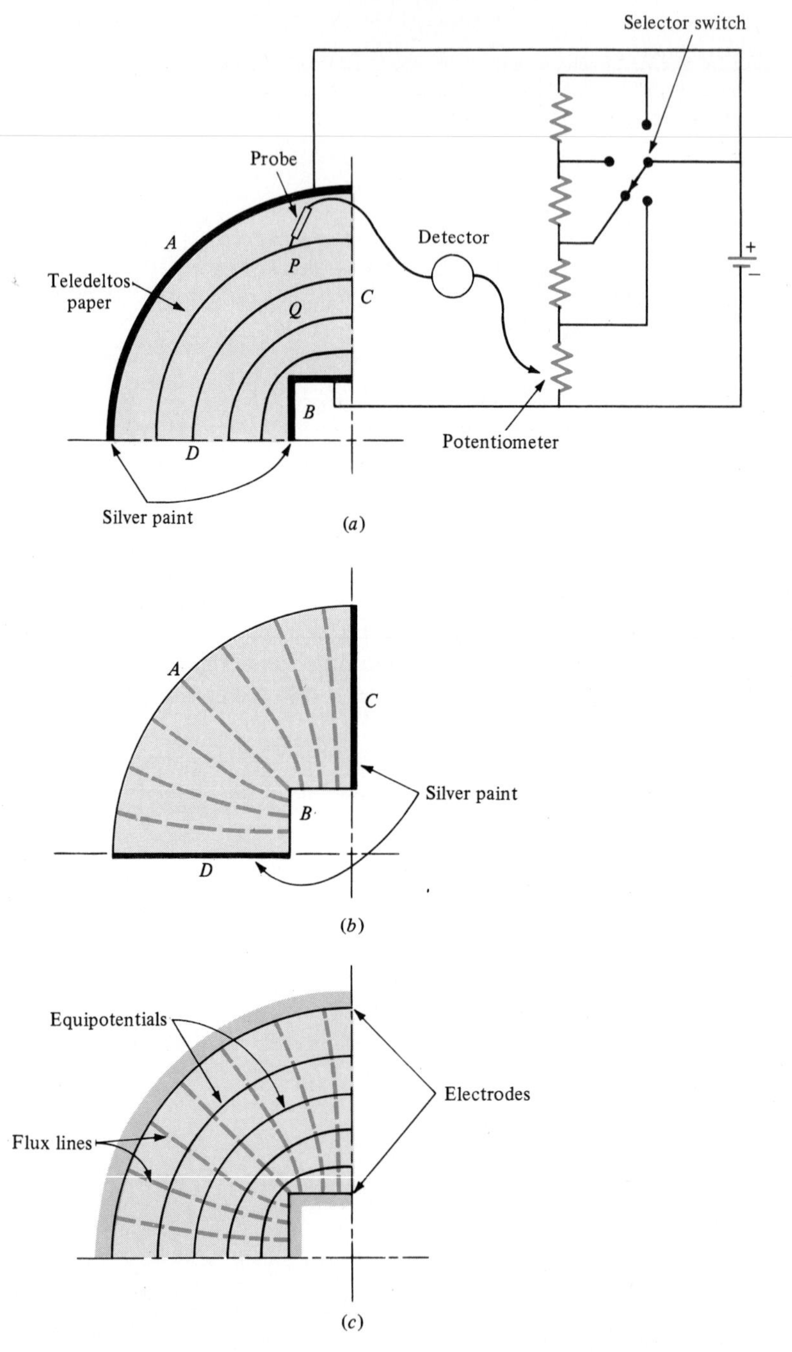

FIGURE 10.27

Example 10.14. *(a)* Experimental arrangement for obtaining equipotentials with Teledeltos paper. *(b)* Equipotentials (representing flux lines) obtained by interchanging conducting and nonconducting boundaries. *(c)* Equipotentials and flux lines.

such as wood. The equipotential boundaries are simulated by painting the pertinent portions with silver paint (Fig. 10.27). Electrodes are then connected to the painted boundaries; the electrodes serve as leads for connection to the measuring equipment for measuring the potentials. Thus, equipotentials can be located and traced (by a graphite pencil) in a given region. The flux lines can also be traced experimentally by interchanging the conducting and nonconducting boundaries. The details are illustrated in the following example.

EXAMPLE 10.14 A portion of a capacitor is shown in Fig. 10.27. Sketch the equipotentials and the electric flux lines using the conductive-sheet analog.

Solution First of all, the electrodes A and B are simulated by silver paint, as shown in Fig. 10.27*a*. The connections are then made as shown. We place the probe at A, set the potentiometer to obtain balance (zero deflection of the detector), and call this setting of the potentiometer 100 percent. Next we reset the potentiometer for balance when the probe is at B, and we term this setting 0 percent. For these settings the selector switch is chosen to remain at an appropriate position.

We then set the potentiometer at 80 percent and move the probe on the Teledeltos paper until balance is obtained. With this setting of the potentiometer, we obtain several points along the curve P that give zero deflection. The curve P is then an equipotential. Next we change the potentiometer setting to 60 percent and repeat the above procedure to obtain the curve Q. Other equipotential curves are traced in a similar manner by setting the potentiometer at 40 percent and at 20 percent.

Having traced the equipotentials, shown in Fig. 10.27*a*, the flux lines can be obtained by plotting the orthogonal equipotentials. In this case, the conducting and nonconducting boundaries are interchanged; that is, we neatly cut out the silver paint at A and B and paint the boundaries C and D. Then we repeat the procedure used to obtain Fig. 10.27*a* and obtain the equipotentials shown in Fig. 10.27*b*. These equipotentials correspond to the flux lines. Finally, Fig. 10.27*c* shows the composite plot of the equipotentials and the flux lines. The usefulness of such plots will be discussed later.

An alternative to the conductive-sheet analog is the electrolytic-tank analog, which is somewhat cumbersome to set up but which can handle three-dimensional field problems as well as problems involving composite media. For details, the reader may refer to Ref. 5.

Impedance-Network Analogs The conductive-sheet (or the electrolytic-tank) analog is a direct analog for solving Laplace's equation. In principle, the analog simulates the partial differential equation exactly; errors are introduced by the experimental nature of the method. We now recall that Laplace's equation can also be solved by expressing it as a *difference equation*; however, the finite difference method does not yield an exact solution. Similarly, an analog simulating the difference equations also yields approximations, and in many applications such approximations are quite acceptable.

Difference equations can be simulated by impedance-network analogs. These impedance-network analogs consisting only of passive lumped circuit elements such as the resistor (R), inductor (L), and capacitor (C) have a wide range of applications. Table 10.2 lists some of the field equations that can be simulated by impedance networks. In the following, however, we will consider in detail only a simple application.

Considering the difference-equation representation of Poisson's equation, we have (see "Finite-Difference Method" in Sec. 10.6.1) for a four-point star (Fig. 10.18*c*)

$$\frac{1}{h^2}(V_1 + V_2 + V_3 + V_4 - 4V_0) = -\frac{\rho}{\epsilon} \tag{142}$$

The resistive-network analog corresponding to the four-point star, Fig. 10.28*a*, is shown in Fig. 10.28*b*; here, for the sake of simplicity, we choose all resistances to be equal. Thus, Kirchhoff's law at node 0 gives I_0, the current out of the node, in terms of the node potentials $V_0, V_1, \ldots, V_4$, or

$$\frac{V_1 - V_0}{R} + \frac{V_2 - V_0}{R} + \frac{V_3 - V_0}{R} + \frac{V_4 - V_0}{R} = I_0 \tag{143}$$

TABLE 10.2

Certain Applications of Impedance-Network Analogs

Equation	*Network*	*Applications*
Laplace's $\nabla^2 V = 0$	R	Magnetic and electric potentials
Poisson's $\nabla^2 V = -\rho/\epsilon$	R	Potentials in regions with charges: e.g., carrier distribution in transistor; space charge in a region
Diffusion $\nabla^2\psi = \partial\psi/\partial t$	RC, LC, or R	Transient current diffusion in a solid conductor
Wave $\nabla^2\psi = \dfrac{\partial^2\psi}{\partial t^2}$	LC	Transmission lines
Maxwell's $\nabla \times \mathcal{E} = -\dfrac{\partial \mathcal{B}}{\partial t}$ $\nabla \times \mathcal{H} = \mathcal{J} + \dfrac{\partial \mathcal{D}}{\partial t}$	RLC	Fields in waveguides and cavity resonators

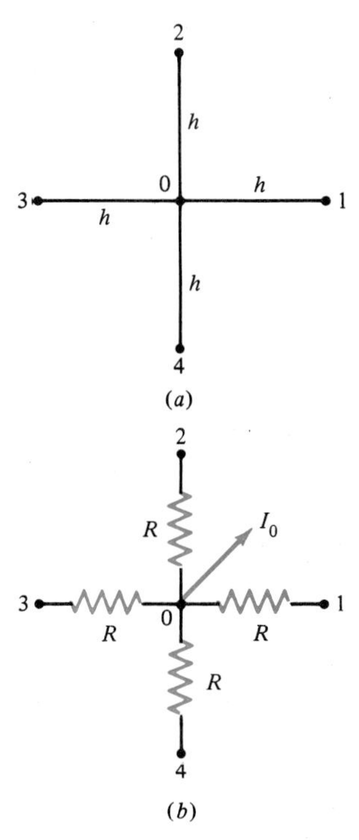

FIGURE 10.28
(a) A four-point finite-difference star; *(b)* resistive-network analog.

which may also be expressed as

$$V_1 + V_2 + V_3 + V_4 - 4V_0 = I_0 R \tag{144}$$

If the network voltage V_N is made proportional to the field potential V—that is, $V_N = kV$ (where k is the proportionality constant)—then comparing (142) and (144) shows that the current out of node 0 must be

$$I_0 = -\frac{kh^2}{R}\frac{\rho}{\epsilon} \tag{145}$$

for a complete analogy. Notice that for rectangular coordinates all resistances are the same, but for other coordinate systems the values of the resistances (in the resistive-network analog) depend on position.

An experimental setup for carrying out the measurements is shown in Fig. 10.29. The current I_0, given by (145), is set by the resistance R_1. In practice, it is convenient to measure I_0 by measuring the potential difference V_s across a standard resistor R_s such that $I_0 = V_s/R_s$. This is accomplished by the potentiometer bridge B_2 and the detector D_2. After the adjustment of the current at one node, the voltage corresponding to ρ/ϵ is measured by using B_1 and D_1. This

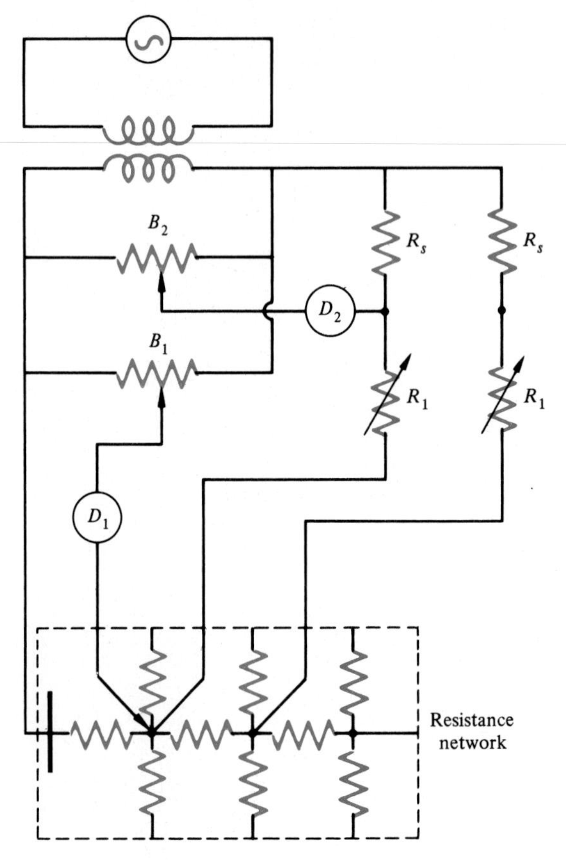

FIGURE 10.29
Circuit for current setting and measurements on a resistance-network analog.

process is repeated at all the nodes to obtain the corresponding potentials. The sources of error in such analogs are the resistor tolerances and mesh sizes.

10.6.3 Graphical Methods

In Secs. 10.6.1 and 10.6.2 we have considered the numerical and analog methods of solving certain field problems. In many two-dimensional field problems involving complex geometrical boundaries, approximate field distributions in regions of interest can be estimated by graphical field-mapping techniques. The field plot may then be used in the proper context, such as to obtain the conductance, capacitance, or permeance of a given region. Information regarding field intensities and regions of saturation can also be obtained from the field plots.

We recall from earlier considerations that static field distributions are governed by Laplace's or Poisson's equation—the former holds for a source-free region, whereas the latter applies in the presence of sources. The fact that analytic functions of a complex variable must satisfy the Cauchy-Riemann equations, and thus are solutions to Laplace's equation, provides us with the basis of graphical field mapping. Thus, if we have a complex variable

$$w(x, y) = u(x, y) + jv(x, y) \tag{146}$$

then the Cauchy-Riemann equations require that

$$\frac{\partial u}{\partial x} = \frac{\partial v}{\partial y} \tag{147}$$

and

$$\frac{\partial u}{\partial y} = -\frac{\partial v}{\partial x} \tag{148}$$

From (147) and (148) it can readily be verified that the functions u and v satisfy Laplace's equation. [Take the partial derivative of (147) with respect to x and the partial derivative of (148) with respect to y and add the two results.] Furthermore, u and v are called *conjugate functions*, and the two families of curves—$u(x, y)$ = a constant and $v(x, y)$ = a constant—intersect at right angles and may be chosen to correspond, respectively, to the magnetic (electric) flux lines and magnetic (electric) equipotentials of a laplacian field. In air, for instance, the magnetic flux lines and the equipotentials intersect orthogonally, (as shown in Fig. 10.30) and form curvilinear squares.

Suppose that Fig. 10.30 represents a region of space in which the magnetic flux lines and equipotentials are drawn. Then the magnetic flux $\Delta\psi_m$ passing through a tube of length l, width w, and thickness unity is given by

$$\Delta\psi_m = \frac{\text{mmf}}{\text{reluctance}} = \Delta u \frac{\mu l}{w} \tag{149}$$

or

$$\frac{\Delta\psi_m}{\mu \Delta u} = \frac{l}{w} = \text{a constant} \tag{150}$$

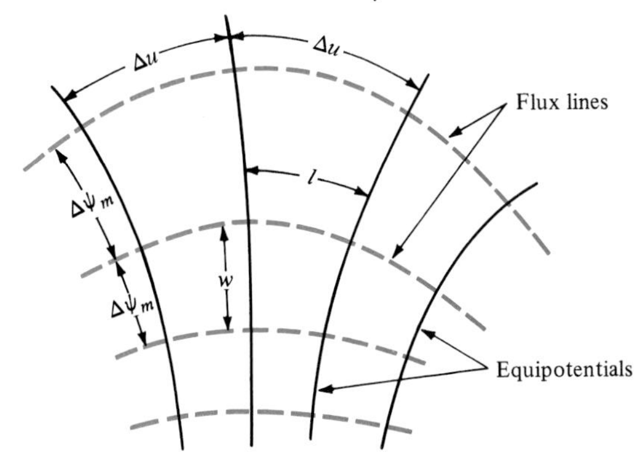

FIGURE 10.30

Orthogonal intersections of magnetic flux lines and equipotentials forming curvilinear squares.

which must be satisfied by the field plot. If there are n squares between two points on an equipotential, the total flux ψ_m crossing the equipotential between these points is

$$\psi_m = n\,\Delta\psi_m \tag{151}$$

And if there are m squares between any two flux lines, then the permeance $\mathcal{P}$ of a field of a region having n squares along an equipotential and m squares along a flux line is, from (150) and (151),

$$\mathcal{P} = \frac{\psi_m}{u} = \mu\frac{n}{m} \qquad \text{H/m} \tag{152}$$

By analogy, we may write for conductance g and capacitance c for a unit-thick field

$$g = \sigma\frac{n}{m} \qquad \text{S/m} \tag{153}$$

and

$$c = \epsilon\frac{n}{m} \qquad \text{F/m} \tag{154}$$

where n and m denote, respectively, the number of flux tubes and the number of potential divisions.

We will now see the applications of the graphical method by the following examples.

EXAMPLE 10.15 The electromagnetic device shown in Fig. 10.31*a* carries a 200-turn coil on the stator pole. Assuming the stator and rotor to be made of infinitely permeable material, calculate the inductance of the coil if the rotor axis

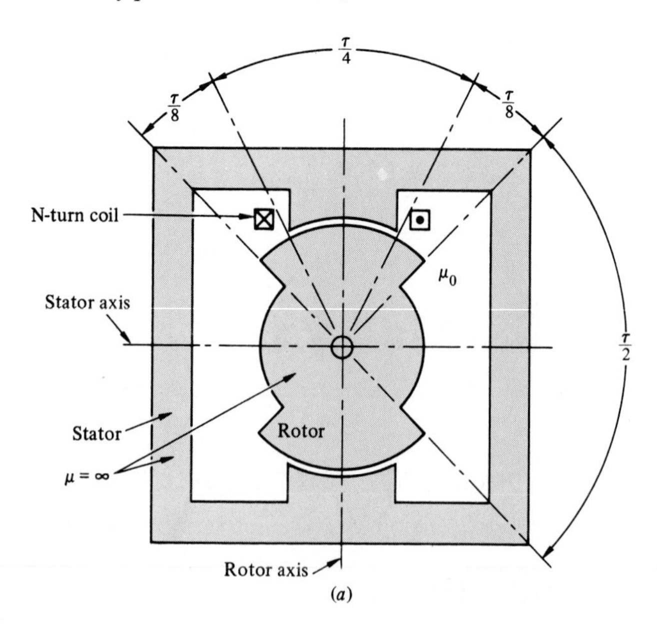

(*a*)

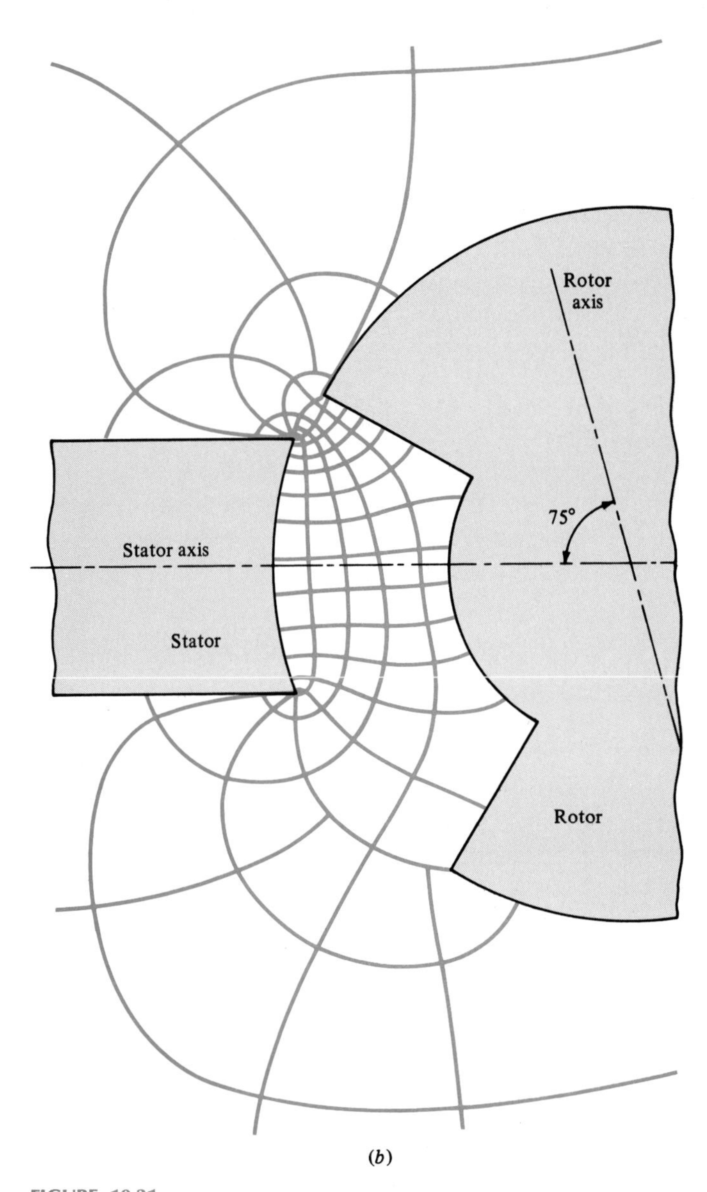

FIGURE 10.31

Example 10.14. *(a)* An electromagnetic device; *(b)* A field map for the given rotor position.

is displaced from the stator axis by 75° and the thickness of the core (into the paper) is 5 cm.

Solution The field plot for the given rotor position by one-half of the device is shown in Fig. 10.31*b*, from which $n = 10$ and $m = 18$. Substituting these into (152) yields

$$\mathcal{P} = \mu_0 \times \tfrac{10}{18} \qquad \text{H/m}$$

and $L = N^2\mathcal{P} \times$ core thickness where N is the number of turns of the coil. Thus, we finally obtain

$$L = (200)^2 \times 4\pi \times 10^{-7} \times \tfrac{10}{18} \times 5 \times 10^{-2} \qquad \text{H}$$

$$= 1.4 \text{ mH}$$

EXAMPLE 10.16 A two-conductor transmission line is shown in Fig. 10.32. Sketch equipotentials and field lines for the region between the conductors, and determine the capacitance per unit length from the plot.

Solution The plot is shown in Fig. 10.32. Because of symmetry, we consider only one quadrant, and we have

$$c = 4c_{\text{quadrant}}$$

and from (154)

$$c = 4\epsilon_0 \frac{n}{m}$$

From Fig. 10.32, $n = 9$ and $m = 4$. For free space, $\epsilon_0 \simeq 10^{-9}/36\pi$. Substituting these values above, we obtain

$$c = \frac{4 \times 10^{-9} \times 9}{36\pi \times 4}$$

$$= 8 \times 10^{-11} \text{ F/m}$$

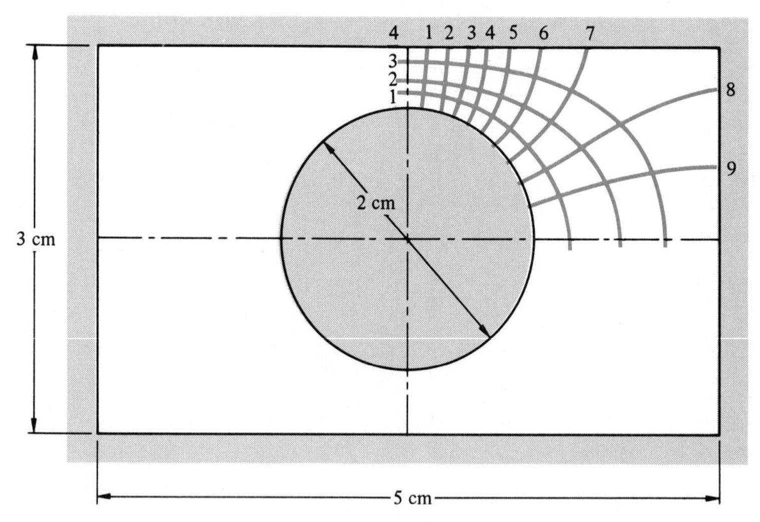

FIGURE 10.32

Example 10.16. Determination of capacitance from a field plot.

10.7 Summary

In this chapter we considered diverse approximate methods for solving field problems that could not be solved analytically. In addition, we derived Laplace's and Poisson's equations for the potential functions introduced in earlier chapters. Using the electrostatic scalar potential for illustration, we found explicit solutions to Laplace's equation for regions having regular boundaries, such as rectangular boxes. We also considered cylindrical and spherical boundaries whose solutions involve Bessel functions and Legendre polynomials, respectively. For irregular boundaries, we found out that explicit solutions were not forthcoming, and for such cases the approximate methods were presented: namely, numerical (or digital) methods, analog methods, and graphical methods. Several illustrative examples were used to show the applications of various methods. We saw that exact solutions are feasible only for approximate (or idealized) models of field problems, whereas approximate methods are applicable to exact models. The choice of the method of solution clearly depends upon the problem at hand and the desired accuracy.

Problems

10-1 Show that Eq. (6) follows from (5) in a cartesian coordinate system.

10-2 Following the procedure of Sec. 10.1, show that in a source-free region the electric field intensity satisfies Laplace's equation.

10-3 Prove that the solution to Poisson's equation for given boundary conditions is unique.

10-4 A perfect conductor carries a surface charge density ρ_s. Determine the normal components of the electric flux density at the conductor surface.

10-5 In Eq. (76) a potential is expressed as an infinite series having only sine terms and unknown coefficients. Using the method of evaluating constants in a Fourier series (from orthogonality properties), determine these constants, and hence show that the results given in the text following (76) are correct.

10-6 For each of the three electrostatic two-dimensional field problems (in rectangular coordinates) shown in Fig. P10.6, write the product form of solution that may be utilized in constructing a series solution to the problem.

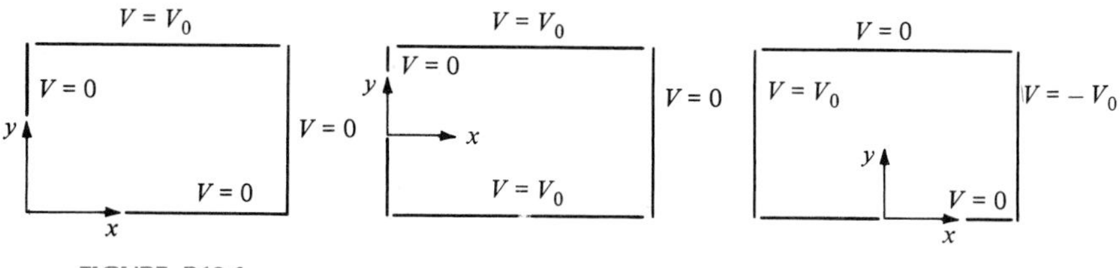

FIGURE P10.6

10-7 For the problem discussed in Example 10.7, choose the origin at x_1. Find the potential distribution within the charge slab, the boundary conditions remaining the same as in Fig. 10.9.

10-8 Two parallel-plate conductors are separated by two dielectric slabs, ϵ_1 and ϵ_2, as shown in Fig. P10.8. Neglecting fringing, obtain the potential distribution between the conductors if the top conductor is at a potential V_0 and the bottom one is grounded.

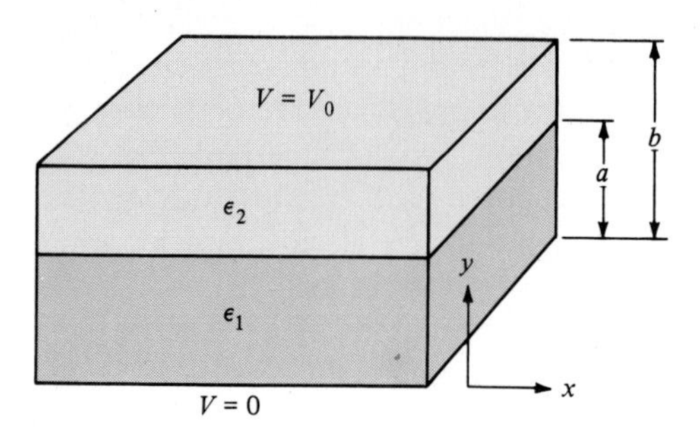

FIGURE P10.8

10-9 Solve Laplace's equation to obtain the potential distribution between the electrodes of the geometries shown in Fig. P10.9.

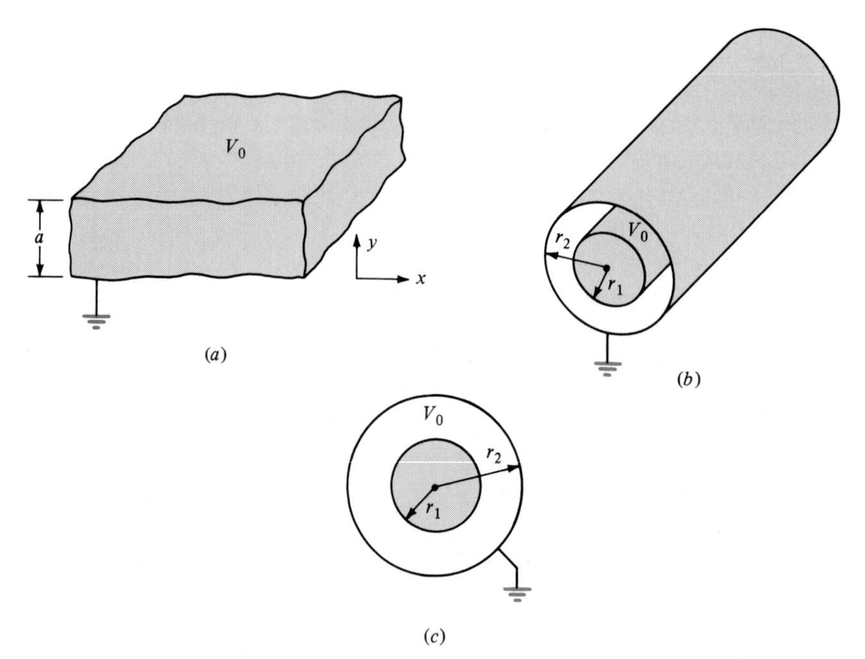

FIGURE P10.9
(a) Parallel plates, *(b)* coaxial cylinders, and *(c)* concentric spheres.

10-10 Find the potential everywhere within the region bounded as shown in Fig. P10.10 and having the boundary conditions as specified.

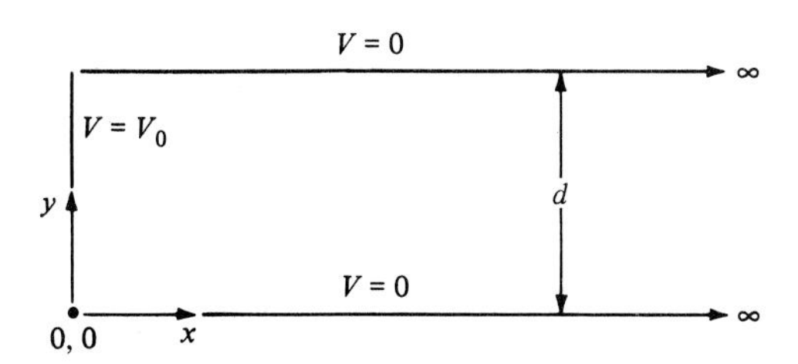

FIGURE P10.10

10-11 For each of the three two-dimensional electrostatic field problems in cylindrical coordinates shown in Fig. P10.11, write the product form of solution that may be utilized in constructing a series solution to the problem.

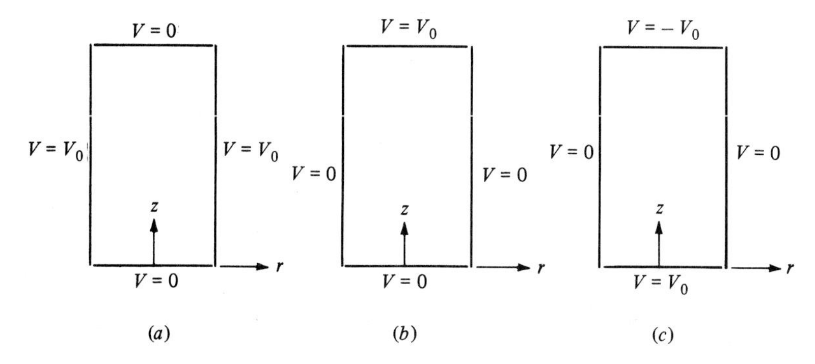

FIGURE P10.11

10-12 The lid of a hollow cylinder of radius a and length l is maintained at potential $J_o(r/a)$. The bottom and the wall of the cylinder are at zero potential. What is the potential within the cylinder?

10-13 For the cylindrical structure shown in Fig. P10-11a, obtain the series for the potential within the cylinder if its radius is a and its length is l.

10-14 A solid sphere of radius a and permeability μ is immersed in a uniform magnetic field $\mathbf{H} = \mathbf{a}_z H_o$, existing in free space. The magnetic field intensity $\mathbf{H}$ is related to the magnetic scalar potential ψ_m by $\mathbf{H} = -\nabla\psi_m$, where ψ_m satisfies Laplace's equation.

(a) Write the form of solution for ψ_m outside the sphere.

(b) Using the given boundary conditions, obtain explicit expressions for ψ_m for $r > a$ and for $r < a$.

(c) Determine the magnetic field intensity and the magnetic flux density within the sphere.

(d) If $\mu \gg \mu_o$, show that, within the sphere, $\mathbf{B} = 3B_o\mathbf{a}_z$ where $\mathbf{B}_o = \mu\mathbf{H}_o$.

10-15 [illegible]e charge ρ_l C/m is located midway between a wedge-shaped corner of perfect [illegible]ductors, as shown in Fig. P10.15. Determine the image charge distribution that may be used to replace the perfect conductors.

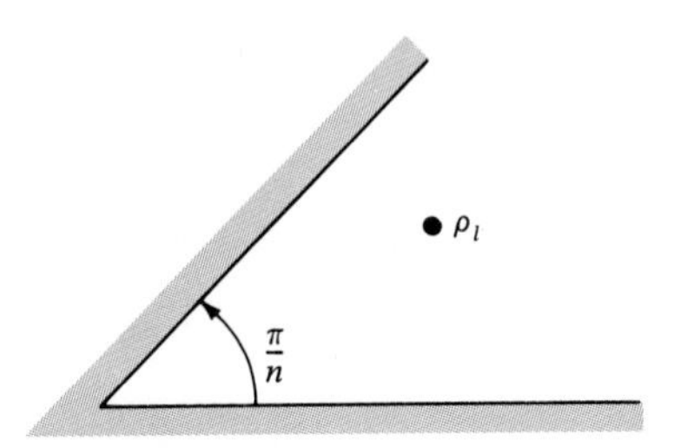

FIGURE P10.15

10-16 For a point charge Q at a distance r away from the center of a conducting sphere of radius R, determine the value and location of the image charge.

10-17 Discuss the applicability of the image concept for the case of a straight conductor, carrying a steady current I, that is placed at a height h above an infinitely permeable surface.

10-18 A two-dimensional potential distribution problem is illustrated in Fig. P10.18. Obtain the potentials at the specified nodes a, b, c, d, e, f, g, and h by using an iterative procedure.

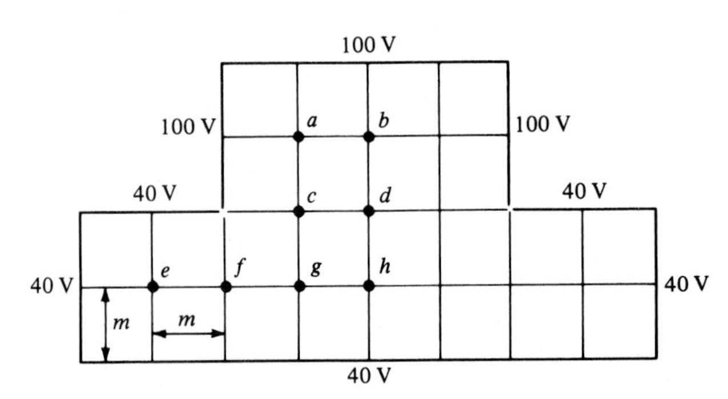

FIGURE P10.18

10-19 Apply the point-relaxation method to solve the problem of Fig. P10.18.

10-20 For Prob. 10-18, express the equations governing the potentials at the specified nodes in matrix form. Solve the problem on a digital computer and compare the results with those of Probs. 10-18 and 10-19.

10-21 Solve numerically for the potentials at the nodes shown in Figs. P10.21a to c, which also show the potentials at the boundaries. Assume a two-dimensional field.

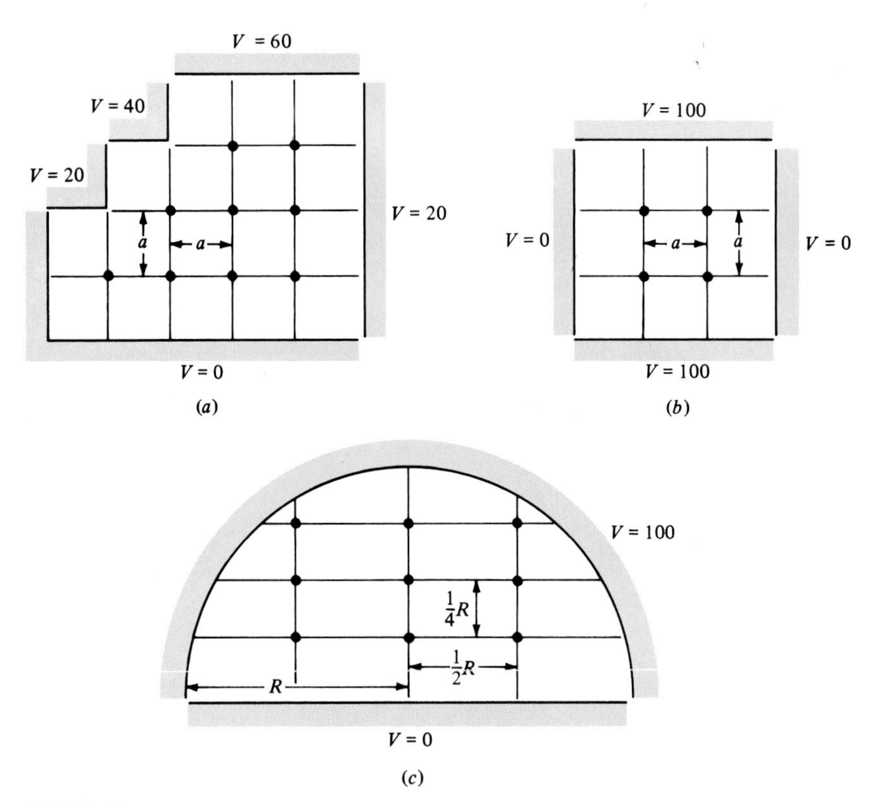

FIGURE P10.21

10-22 **A strip transmission line consists of two infinitely long conducting strips of width W and separation H, as shown in Fig. P10.22. Adapt the method-of-moments**

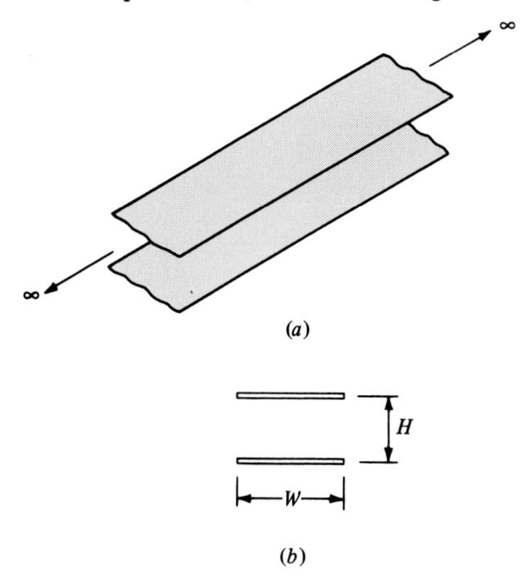

FIGURE P10.22
A strip transmission line.

technique described in Sec. 10.6.1 to this problem to obtain the capacitance per unit of axial length. Assume that the surrounding medium is free space. Write a digital computer program to implement your method, and compute this per-unit-length capacitance for W/H ratios of 10, 5, and 1. Increase the subsection widths in each case to determine convergence. (See A. T. Adams,[4] pp. 179–180 and App. E.)

10-23 Sketch the equipotentials and flux lines for the configuration shown in Fig. P10.21*c*.

10-24 If Fig. P10.21*c* shows the cross section of a half of a cylinder that is filled with a dielectric ϵ, determine the capacitance per unit length of the geometry.

10-25 Three identical cylindrical conductors of infinite length are symmetrically arranged as shown in Fig. P10.25. A potential difference of V_0 volts is applied between two of the conductors, and the third conductor is grounded.

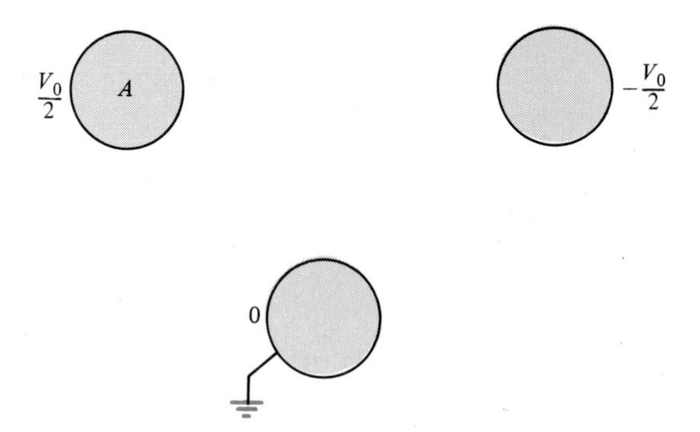

FIGURE P10.25

(a) Sketch the field, showing the equipotentials and the electric field.
(b) Over what portion of cylinder A is the charge density the greatest?

References

1. A. Konrad and P. Silvester, "Scalar Finite-Element Program Package for Two-Dimensional Field Problems," *IEEE Trans. on Microwave Theory and Techniques*, vol. MTT-19, 1971, pp. 952–954.
2. K. J. Binns and P. J. Lawrenson, *Analysis and Computation of Electric and Magnetic Field Problems*, Pergamon Press, New York, 1963.
3. R. F. Harrington, *Field Computation by Moment Methods*, Macmillan, New York, 1968.
4. A. T. Adams, *Electromagnetics for Engineers*, Ronald Press, New York, 1971.
5. D. Vitkovitch (ed.), *Field Analysis: Experimental and Computational Methods*, Van Nostrand, New York, 1966.
6. G. Liebmann, "The Solution of Waveguide and Cavity-Resonator Problems with Resistance-Network Analogue," *Proc. IEE*, vol. 99, no 4, 1952, pp. 260–272.

APPENDIX A

Vector Identities And Vector Operations

A.1 Vector Identities

Several of the most important identities are given below. These may be proved by direct verification for general vector fields $\mathbf{A}(x, y, z)$, $\mathbf{B}(x, y, z)$, and $\mathbf{C}(x, y, z)$ and scalar fields $\psi(x, y, z)$ and $\Phi(x, y, z)$ expressed in a rectangular coordinate system.

$$\nabla(\Phi + \psi) = \nabla\Phi + \nabla\psi$$

$$\nabla\cdot(\mathbf{A} + \mathbf{B}) = \nabla\cdot\mathbf{A} + \nabla\cdot\mathbf{B}$$

$$\nabla\times(\mathbf{A} + \mathbf{B}) = \nabla\times\mathbf{A} + \nabla\times\mathbf{B}$$

$$\nabla(\Phi\psi) = \Phi\nabla\psi + \psi\nabla\Phi$$

$$\nabla\cdot(\psi\mathbf{A}) = \mathbf{A}\cdot\nabla\psi + \psi\nabla\cdot\mathbf{A}$$

$$\nabla\cdot(\mathbf{A}\times\mathbf{B}) = \mathbf{B}\cdot(\nabla\times\mathbf{A}) - \mathbf{A}\cdot(\nabla\times\mathbf{B})$$

$$\nabla\times(\Phi\mathbf{A}) = \nabla\Phi\times\mathbf{A} + \Phi\nabla\times\mathbf{A}$$

$$\nabla\times(\mathbf{A}\times\mathbf{B}) = \mathbf{A}\nabla\cdot\mathbf{B} - \mathbf{B}\nabla\cdot\mathbf{A} + (\mathbf{B}\cdot\nabla)\mathbf{A} - (\mathbf{A}\cdot\nabla)\mathbf{B}$$

$$\nabla\cdot\nabla\Phi = \nabla^2\Phi$$

$$\nabla\cdot\nabla\times\mathbf{A} = 0$$

$$\nabla \times \nabla \Phi = 0$$

$$\nabla \times \nabla \times \mathbf{A} = \nabla(\nabla \cdot \mathbf{A}) - \nabla^2 A$$

$$\nabla(\mathbf{A} \cdot \mathbf{B}) = (\mathbf{A} \cdot \nabla)\mathbf{B} + (\mathbf{B} \cdot \nabla)\mathbf{A} + \mathbf{A} \times (\nabla \times \mathbf{B}) + \mathbf{B} \times (\nabla \times \mathbf{A})$$

$$\mathbf{A} \cdot (\mathbf{B} \times \mathbf{C}) = \mathbf{B} \cdot (\mathbf{C} \times \mathbf{A}) = \mathbf{C} \cdot (\mathbf{A} \times \mathbf{B})$$

$$\mathbf{A} \times (\mathbf{B} \times \mathbf{C}) = \mathbf{B}(\mathbf{A} \cdot \mathbf{C}) - \mathbf{C}(\mathbf{A} \cdot \mathbf{B})$$

Stokes' theorem

$$\int_s (\nabla \times \mathbf{A}) \cdot d\mathbf{s} = \oint_c \mathbf{A} \cdot d\mathbf{l} \qquad \text{open surface } s \text{ bounded by contour } c$$

Divergence theorem

$$\int_v (\nabla \cdot \mathbf{A})\, dv = \oint_s \mathbf{A} \cdot d\mathbf{s} \qquad \text{volume } v \text{ bounded by closed surface } s$$

A.2 Vector Operations

The vector operations of gradient, divergence, and curl are given below for the rectangular, cylindrical, and spherical coordinate systems. These relationships were derived for a rectangular coordinate system in Chap. 2. They may be similarly derived for the cylindrical and spherical systems, although the derivations are more involved than for the rectangular system.

Rectangular coordinates

$$\nabla \Phi = \mathbf{a}_x \frac{\partial \Phi}{\partial x} + \mathbf{a}_y \frac{\partial \Phi}{\partial y} + \mathbf{a}_z \frac{\partial \Phi}{\partial z}$$

$$\nabla \cdot \mathbf{A} = \frac{\partial A_x}{\partial x} + \frac{\partial A_y}{\partial y} + \frac{\partial A_z}{\partial z}$$

$$\nabla \times \mathbf{A} = \mathbf{a}_x \left(\frac{\partial A_z}{\partial y} - \frac{\partial A_y}{\partial z} \right) + \mathbf{a}_y \left(\frac{\partial A_x}{\partial z} - \frac{\partial A_z}{\partial x} \right) + \mathbf{a}_z \left(\frac{\partial A_y}{\partial x} - \frac{\partial A_x}{\partial y} \right)$$

$$\nabla^2 \Phi = \frac{\partial^2 \Phi}{\partial x^2} + \frac{\partial^2 \Phi}{\partial y^2} + \frac{\partial^2 \Phi}{\partial z^2}$$

Cylindrical coordinates

$$\nabla\Phi = \mathbf{a}_r \frac{\partial \Phi}{\partial r} + \mathbf{a}_\phi \frac{1}{r}\frac{\partial \Phi}{\partial \phi} + \mathbf{a}_z \frac{\partial \Phi}{\partial z}$$

$$\nabla \cdot \mathbf{A} = \frac{1}{r}\frac{\partial}{\partial r}(rA_r) + \frac{1}{r}\frac{\partial A_\phi}{\partial \phi} + \frac{\partial A_z}{\partial z}$$

$$\nabla \times \mathbf{A} = \mathbf{a}_r \left(\frac{1}{r}\frac{\partial A_z}{\partial \phi} - \frac{\partial A_\phi}{\partial z}\right) + \mathbf{a}_\phi \left(\frac{\partial A_r}{\partial z} - \frac{\partial A_z}{\partial r}\right) + \mathbf{a}_z \left[\frac{1}{r}\frac{\partial (rA_\phi)}{\partial r} - \frac{1}{r}\frac{\partial A_r}{\partial \phi}\right]$$

$$\nabla^2\Phi = \frac{1}{r}\frac{\partial}{\partial r}\left(r\frac{\partial \Phi}{\partial r}\right) + \frac{1}{r^2}\frac{\partial^2 \Phi}{\partial \phi^2} + \frac{\partial^2 \Phi}{\partial z^2}$$

Spherical coordinates

$$\nabla\Phi = \mathbf{a}_r \frac{\partial \Phi}{\partial r} + \mathbf{a}_\theta \frac{1}{r}\frac{\partial \Phi}{\partial \theta} + \mathbf{a}_\phi \frac{1}{r \sin\theta}\frac{\partial \Phi}{\partial \phi}$$

$$\nabla \cdot \mathbf{A} = \frac{1}{r^2}\frac{\partial}{\partial r}(r^2 A_r) + \frac{1}{r\sin\theta}\frac{\partial}{\partial \theta}(\sin\theta A_\theta) + \frac{1}{r\sin\theta}\frac{\partial A_\phi}{\partial \phi}$$

$$\nabla \times \mathbf{A} = \mathbf{a}_r \left\{\frac{1}{r\sin\theta}\left[\frac{\partial}{\partial \theta}(A_\phi \sin\theta) - \frac{\partial A_\theta}{\partial \phi}\right]\right\}$$

$$+ \mathbf{a}_\theta \left[\frac{1}{r\sin\theta}\frac{\partial A_r}{\partial \phi} - \frac{1}{r}\frac{\partial}{\partial r}(rA_\phi)\right]$$

$$+ \mathbf{a}_\phi \left\{\frac{1}{r}\left[\frac{\partial}{\partial r}(rA_\theta) - \frac{\partial A_r}{\partial \theta}\right]\right\}$$

$$\nabla^2\Phi = \frac{1}{r^2}\frac{\partial}{\partial r}\left(r^2\frac{\partial \Phi}{\partial r}\right) + \frac{1}{r^2\sin\theta}\frac{\partial}{\partial \theta}\left(\sin\theta\frac{\partial \Phi}{\partial \theta}\right) + \frac{1}{r^2\sin^2\theta}\frac{\partial^2 \Phi}{\partial \phi^2}$$

A.3 Conversion between Coordinate Systems

In this section we derive the conversion relationships between the corresponding components of a vector that is expressed in two different coordinate systems. First consider a vector **A** expressed in a rectangular coordinate system as

$$\mathbf{A}(x, y, z) = A_x(x, y, z)\mathbf{a}_x + A_y(x, y, z)\mathbf{a}_y + A_z(x, y, z)\mathbf{a}_z \qquad \text{(A-1)}$$

and in a cylindrical coordinate system as

$$\mathbf{A}(r, \phi, z) = A_r(r, \phi, z)\mathbf{a}_r + A_\phi(r, \phi, z)\mathbf{a}_\phi + A_z(r, \phi, z)\mathbf{a}_z \qquad \text{(A-2)}$$

We wish to find the conversion relationships between A_x, A_y, A_z and A_r, A_ϕ, A_z. The relationships between the coordinate variables were derived in Chap. 2 as

$$\begin{aligned} x &= r\cos\phi & r &= \sqrt{x^2+y^2} \\ y &= r\sin\phi & \tan\phi &= \frac{y}{x} \\ z &= z & z &= z \end{aligned} \tag{A-3}$$

In order to derive the conversion relationships between A_x, A_y, A_z and A_r, A_ϕ, A_z, consider Fig. A.1, which shows a view of the cylindrical coordinate system in the xy plane of the rectangular system. From Fig. A.1 we find that

$$\begin{aligned} A_r &= A_x\cos\phi + A_y\sin\phi \\ A_\phi &= -A_x\sin\phi + A_y\cos\phi \\ A_z &= A_z \end{aligned} \tag{A-4}$$

Similarly, we obtain

$$\begin{aligned} A_x &= A_r\frac{x}{\sqrt{x^2+y^2}} - A_\phi\frac{y}{\sqrt{x^2+y^2}} \\ A_y &= A_r\frac{y}{\sqrt{x^2+y^2}} + A_\phi\frac{x}{\sqrt{x^2+y^2}} \\ A_z &= A_z \end{aligned} \tag{A-5}$$

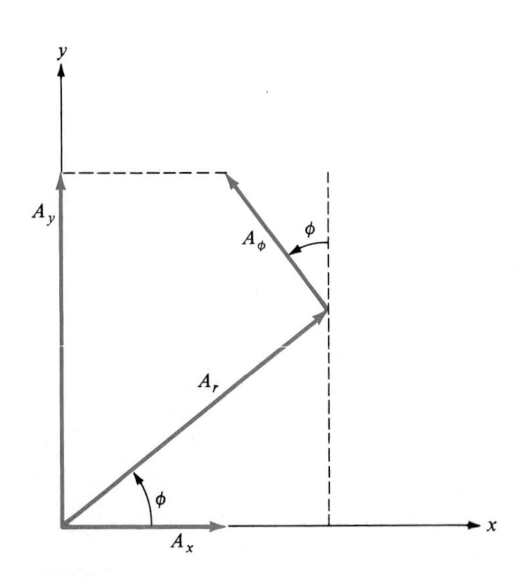

FIGURE A.1
Illustration of the conversion between rectangular and cylindrical coordinates.

Alternatively, we may obtain these components by taking the following dot products:

$$\begin{aligned} A_r &= \mathbf{A} \cdot \mathbf{a}_r = A_x \mathbf{a}_x \cdot \mathbf{a}_r + A_y \mathbf{a}_y \cdot \mathbf{a}_r + A_z \mathbf{a}_z \cdot \mathbf{a}_r \\ A_\phi &= \mathbf{A} \cdot \mathbf{a}_\phi = A_x \mathbf{a}_x \cdot \mathbf{a}_\phi + A_y \mathbf{a}_y \cdot \mathbf{a}_\phi + A_z \mathbf{a}_z \cdot \mathbf{a}_\phi \\ A_z &= \mathbf{A} \cdot \mathbf{a}_z \end{aligned} \tag{A-6}$$

Noting that

$$\begin{aligned} \mathbf{a}_x \cdot \mathbf{a}_r &= \cos\phi \\ \mathbf{a}_y \cdot \mathbf{a}_r &= \sin\phi \\ \mathbf{a}_x \cdot \mathbf{a}_\phi &= -\sin\phi \\ \mathbf{a}_y \cdot \mathbf{a}_\phi &= \cos\phi \\ \mathbf{a}_z \cdot \mathbf{a}_r &= 0 \\ \mathbf{a}_z \cdot \mathbf{a}_\phi &= 0 \end{aligned} \tag{A-7}$$

we obtain (A-4).

The relationship between the coordinate variables of a spherical coordinate system (r, θ, ϕ) and of a rectangular coordinate system (x, y, z) can similarly be shown to be

$$\begin{aligned} x &= r \sin\theta \cos\phi \qquad & r &= \sqrt{x^2 + y^2 + z^2} \\ y &= r \sin\theta \sin\phi \qquad & \theta &= \cos^{-1} \frac{z}{\sqrt{x^2 + y^2 + z^2}} \\ z &= r \cos\theta \qquad & \phi &= \tan^{-1} \frac{y}{x} \end{aligned} \tag{A-8}$$

The conversion between the vector components can also similarly be shown to be

$$\begin{aligned} A_r &= A_x \cos\phi \sin\theta + A_y \sin\phi \sin\theta + A_z \cos\theta \\ A_\theta &= A_x \cos\phi \cos\theta + A_y \sin\phi \cos\theta - A_z \sin\theta \\ A_\phi &= -A_x \sin\phi + A_y \cos\phi \end{aligned} \tag{A-9}$$

by computing

$$\begin{aligned} A_r &= \mathbf{A} \cdot \mathbf{a}_r = A_x \mathbf{a}_x \cdot \mathbf{a}_r + A_y \mathbf{a}_y \cdot \mathbf{a}_r + A_z \mathbf{a}_z \cdot \mathbf{a}_r \\ A_\theta &= \mathbf{A} \cdot \mathbf{a}_\theta = A_x \mathbf{a}_x \cdot \mathbf{a}_\theta + A_y \mathbf{a}_y \cdot \mathbf{a}_\theta + A_z \mathbf{a}_z \cdot \mathbf{a}_\theta \\ A_\phi &= \mathbf{A} \cdot \mathbf{a}_\phi = A_x \mathbf{a}_x \cdot \mathbf{a}_\phi + A_y \mathbf{a}_y \cdot \mathbf{a}_\phi + A_z \mathbf{a}_z \cdot \mathbf{a}_\phi \end{aligned} \tag{A-10}$$

where we may use

$$
\begin{aligned}
\mathbf{a}_r \cdot \mathbf{a}_z &= \cos\theta \\
\mathbf{a}_\theta \cdot \mathbf{a}_z &= -\sin\theta \\
\mathbf{a}_\phi \cdot \mathbf{a}_z &= 0 \\
\mathbf{a}_r \cdot \mathbf{a}_x &= \sin\theta\cos\phi \\
\mathbf{a}_\theta \cdot \mathbf{a}_x &= \cos\theta\cos\phi \\
\mathbf{a}_\phi \cdot \mathbf{a}_x &= -\sin\phi \\
\mathbf{a}_r \cdot \mathbf{a}_y &= \sin\theta\sin\phi \\
\mathbf{a}_\theta \cdot \mathbf{a}_y &= \cos\theta\cos\phi \\
\mathbf{a}_\phi \cdot \mathbf{a}_y &= \cos\phi
\end{aligned}
\tag{A-11}
$$

The conversion from spherical coordinates to rectangular coordinates is

$$
\begin{aligned}
A_x &= A_r \frac{x}{\sqrt{x^2+y^2+z^2}} + A_\theta \frac{xz}{\sqrt{(x^2+y^2)(x^2+y^2+z^2)}} \\
&\quad - A_\phi \frac{y}{\sqrt{x^2+y^2}} \\
A_y &= A_r \frac{y}{\sqrt{x^2+y^2+z^2}} + A_\theta \frac{yz}{\sqrt{(x^2+y^2)(x^2+y^2+z^2)}} \\
&\quad + A_\phi \frac{x}{\sqrt{x^2+y^2}} \\
A_z &= A_r \frac{z}{\sqrt{x^2+y^2+z^2}} - A_\theta \frac{\sqrt{x^2+y^2}}{\sqrt{x^2+y^2+z^2}}
\end{aligned}
\tag{A-12}
$$

EXAMPLE A.1 Convert the vector

$$\mathbf{A} = x\mathbf{a}_x + z\mathbf{a}_y + \mathbf{a}_z$$

to a cylindrical coordinate system.

Solution First change the variables via (A-3):

$$\mathbf{A} = r\cos\phi\,\mathbf{a}_x + z\mathbf{a}_y + \mathbf{a}_z$$

and then convert the components via (A-4):

$$\mathbf{A} = (r\cos^2\phi + z\sin\phi)\mathbf{a}_r + (-r\cos\phi\sin\phi + z\cos\phi)\mathbf{a}_\phi + \mathbf{a}_z$$

APPENDIX B

Faraday's Law for Moving Contours

In Chap. 5 we considered Faraday's law for stationary contours. In this appendix we will derive several alternative forms of Faraday's law valid for moving contours. These formulations are particularly important in the analysis of rotating electric motors and generators as well as of other energy conversion devices.

Faraday's law for stationary contours is given in Eq. (5) of Chap. 5 as

$$\oint_c \boldsymbol{\mathcal{E}} \cdot d\mathbf{l} = -\frac{d}{dt}\int_s \boldsymbol{\mathcal{B}} \cdot d\mathbf{s} \qquad \text{(B-1}a\text{)}$$

where the emf developed along the contour c is

$$\text{emf} = \oint_c \boldsymbol{\mathcal{E}} \cdot d\mathbf{l} \qquad \text{(B-1}b\text{)}$$

Let us consider a fundamental problem: determining the force exerted on a point charge. Suppose that a point charge q is moving with velocity u. The coulomb or electric force exerted on the point charge by some electric field $\boldsymbol{\mathcal{E}}$ is given by

$$\mathbf{F}_e = q\boldsymbol{\mathcal{E}} \qquad \text{N} \qquad \text{(B-2)}$$

and the magnetic force exerted on the point charge by some magnetic field $\mathcal{B}$ is given by

$$\mathbf{F}_m = q\mathbf{u} \times \mathcal{B} \qquad \text{N} \tag{B-3}$$

where $\mathbf{u}$ is the velocity vector of the point charge. The velocity of the point charge is with respect to some reference frame in which $\mathcal{E}$ and $\mathcal{B}$ are defined. The combined (or resultant) force exerted on a point charge in the presence of both fields is the sum of these two effects and is given by the Lorentz force equation:

$$\begin{aligned} \mathbf{F} &= \mathbf{F}_e + \mathbf{F}_m \qquad \text{N} \\ &= q(\mathcal{E} + \mathbf{u} \times \mathcal{B}) \end{aligned} \tag{B-4}$$

In Chap. 3 the electric field resulting from a stationary charge distribution was conveniently defined as the force per unit charge exerted on some small, positive test charge q:

$$\mathbf{E} = \lim_{q \to 0} \frac{\mathbf{F}_e}{q} \tag{B-5}$$

This definition of the electric field and the resulting definition of potential difference (or voltage) as the work per unit charge required to move the charge through the field were consistent and meaningful for stationary charge distributions. Suppose, however, that the test charge is moving in the presence of an electric field and a magnetic field. The force exerted on this charge is given by the Lorentz force equation in (B-4). Note that this force is no longer directly proportional to the electric field. However, if we define a quantity

$$\mathcal{E}' = \mathcal{E} + \mathbf{u} \times \mathcal{B} \tag{B-6}$$

as an effective electric field, then from the Lorentz force equation

$$\mathbf{F} = q\mathcal{E}' \tag{B-7}$$

and the force exerted on the moving test charge is directly proportional to this effective electric field in a manner similar to that for the case of a stationary test charge.

This effective electric field $\mathcal{E}'$ can also be thought of as the electric field in the moving frame of reference of the point charge. An observer moving with the charge may interpret the force given in (B-7) as being due to a static electric field since the charge appears stationary to this moving observer. In terms of this effective electric field, Faraday's law may be stated for moving contours as

$$\oint_c \;' \cdot d\mathbf{l} = -\frac{d}{dt}\int_s \; \cdot d\mathbf{s} \tag{B-8a}$$

and the emf developed along the moving contour is

$$\text{emf} = \oint_c {}' \cdot d\mathbf{l} \tag{B-8b}$$

Once again, the orientation of contour c and related surface s are given by the right-hand rule.

In the example of the wire loop considered in Chap. 5, Sec. 5.1.1 contour c and surface s are stationary with respect to $\mathcal{B}$, in which case $\mathcal{E}' = \mathcal{E}$ and (B-8) reduces to (B-1). Now let us consider motion of the contour. Consider a conducting bar moving with velocity $\mathbf{u}$ with respect to magnetic field $\mathcal{B}$, as shown in Fig. B.1. The movement of the bar and consequently the free charges (electrons) within the bar induces a force $\mathbf{F}_m = q\mathbf{u} \times \mathcal{B}$ on these charges. The charges move to the ends of the bar until this magnetic force $\mathbf{F}_m$ is balanced by the coulomb force $\mathbf{F}_e$ produced by the charge separation. Equilibrium is established almost instantaneously; the time required to reach equilibrium is related to the relaxation-time constant of the conductor. Once equilibrium is established, the magnetic force $\mathbf{F}_m$ is equal in magnitude but opposite in direction to the coulomb force $\mathbf{F}_e$ produced by the charge separation, and the emf $\oint_c \mathcal{E}' \cdot d\mathbf{l}$ appears between the ends of the bar as a resultant separation of charge. Note that $\mathcal{E}' = (\mathbf{F}_m + \mathbf{F}_e)/q$ is zero along the conductor and thus that $\int \mathcal{E}' \cdot d\mathbf{l}$ along the conductor is zero. For a perfect conductor, this charge separation occurs instantaneously; thus, $\mathcal{E}' = 0$ along the bar instantaneously.

Taking the time derivative in (B-8) under the integral, we obtain

$$\oint_c \mathcal{E}' \cdot d\mathbf{l} = -\int_s \frac{\partial}{\partial t}(\mathcal{B} \cdot d\mathbf{s})$$

$$= -\int_s \left(\frac{\partial \mathcal{B}}{\partial t} \cdot d\mathbf{s} + \mathcal{B} \cdot \frac{\partial d\mathbf{s}}{\partial t} \right) \tag{B-9}$$

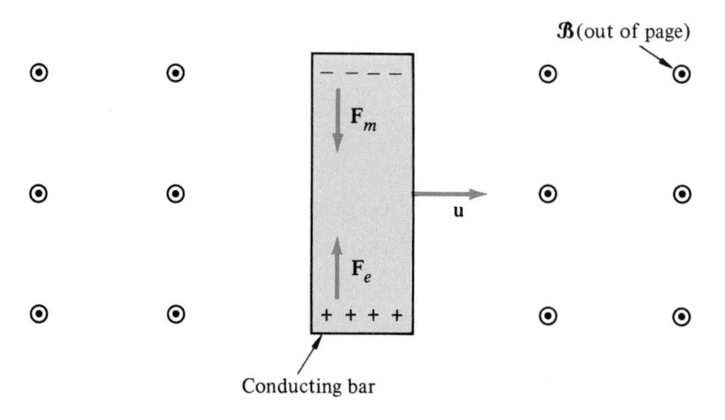

FIGURE B.1
Illustration of the force on changes in a moving, conducting bar immersed in a magnetic field.

We will now show that (B-9) has an equivalent form:

$$\oint_c \mathcal{E}' \cdot d\mathbf{l} = -\int_s \frac{\partial \mathcal{B}}{\partial t} \cdot d\mathbf{s} + \oint_c (\mathbf{u} \times \mathcal{B}) \cdot d\mathbf{l} \tag{B-10}$$

and we will show the equivalence of the following forms of Faraday's law:

$$\text{(i)} \quad \oint_c \mathcal{E}' \cdot d\mathbf{l} = -\frac{d}{dt}\int_s \mathcal{B} \cdot d\mathbf{s} \tag{B-11a}$$

$$\text{(ii)} \quad \oint_c \mathcal{E}' \cdot d\mathbf{l} = -\int_s \frac{\partial \mathcal{B}}{\partial t} \cdot d\mathbf{s} + \oint_c (\mathbf{u} \times \mathcal{B}) \cdot d\mathbf{l} \tag{B-11b}$$

$$\text{(iii)} \quad \oint_c \mathcal{E} \cdot d\mathbf{l} = -\int_s \frac{\partial \mathcal{B}}{\partial t} \cdot d\mathbf{s} \tag{B-11c}$$

$$\text{(iv)} \quad \nabla \times \mathcal{E} = -\frac{\partial \mathcal{B}}{\partial t} \tag{B-11d}$$

$$\text{(v)} \quad \nabla \times \mathcal{E}' = -\frac{\partial \mathcal{B}}{\partial t} + \nabla \times (\mathbf{u} \times \mathcal{B}) \tag{B-11e}$$

We first prove the equivalence of (i) and (ii). Thus, we need to prove the equivalence of the right-hand sides of (i) and (ii):

$$\frac{d}{dt}\int_s \mathcal{B} \cdot d\mathbf{s} \stackrel{?}{=} \int_s \frac{\partial \mathcal{B}}{\partial t} \cdot d\mathbf{s} - \oint_c (\mathbf{u} \times \mathcal{B}) \cdot d\mathbf{l} \tag{B-12}$$

Note that the left-hand side of (B-12) is the time rate of change of the magnetic flux

$$\psi_m = \int_s \mathcal{B} \cdot d\mathbf{s} \tag{B-13}$$

penetrating s under those conditions in which we consider (1) the possibility of movement of surface s and related contour c and (2) a time variation of the magnetic field $\mathcal{B}$. Consider Fig B.2a, which shows contour c at time t and its displaced location at time $t_1 = t + \Delta t$, where Δt is some suitably small time increment. The surface s_1 is the surface s at its location for t_1 (and is of the same size and shape as surface s). The time rate of change of the magnetic flux penetrating s is given by

$$\begin{aligned} \frac{d\psi_m}{dt} &= \frac{d}{dt}\int_s \mathcal{B} \cdot d\mathbf{s} \\ &= \lim_{\Delta t \to 0} \frac{1}{\Delta t}\left(\int_{s_1} \mathcal{B}_1 \cdot d\mathbf{s} - \int_s \mathcal{B} \cdot d\mathbf{s}\right) \end{aligned} \tag{B-14}$$

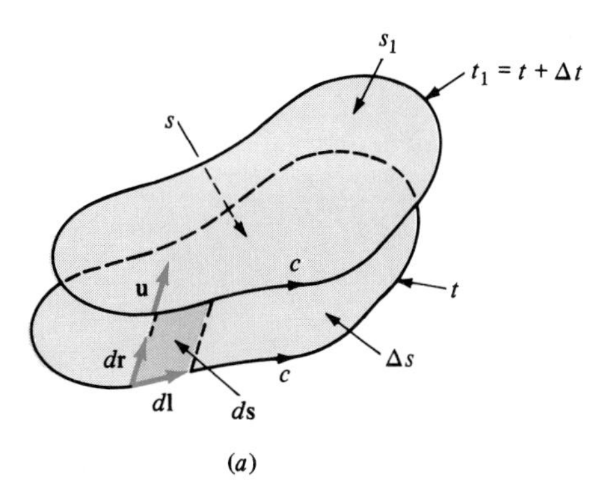

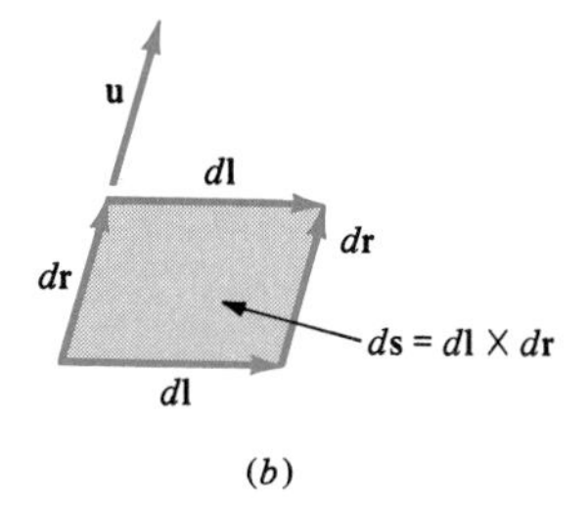

FIGURE B.2
Parameter definitions for a moving contour.

where $\mathcal{B}_1$ is the magnetic flux density vector at time $t_1 = t + \Delta t$. We first note that since

$$\oint_s \mathcal{B} \cdot d\mathbf{s} = 0 \tag{B-15}$$

for any closed surface s, we may write

$$\int_{s1} \mathcal{B}_1 \cdot d\mathbf{s} + \int_{\Delta s} \mathcal{B}_1 \cdot d\mathbf{s} = \int_s \mathcal{B}_1 \cdot d\mathbf{s} \tag{B-16}$$

where Δs is the surface swept out by the movement of contour c as shown in Fig. B.2a. Consider a portion of c, $d\mathbf{l}$, which is moving with velocity $\mathbf{u}$. This portion of the contour moves a distance of $d\mathbf{r} = \mathbf{u}\, dt$, so

$$\mathbf{u} = \frac{d\mathbf{r}}{dt} \tag{B-17}$$

as shown in Fig. B.2a. The portion of Δs produced by this movement of the contour is

$$\begin{aligned} d\mathbf{s} &= d\mathbf{l} \times d\mathbf{r} \\ &= d\mathbf{l} \times \mathbf{u}\, \Delta t \end{aligned} \tag{B-18}$$

Thus, the portion of the flux through side Δs is given by

$$\int_{\Delta s} \mathcal{B}_1 \cdot d\mathbf{s} = \int_{\Delta s} \mathcal{B}_1 \cdot (d\mathbf{l} \times \mathbf{u}) \, \Delta t \tag{B-19}$$

From Appendix A we have the vector identity

$$\begin{aligned} \mathbf{A} \cdot (\mathbf{B} \times \mathbf{C}) &= \mathbf{B} \cdot (\mathbf{C} \times \mathbf{A}) \\ &= \mathbf{C} \cdot (\mathbf{A} \times \mathbf{B}) \end{aligned} \tag{B-20}$$

so that

$$\begin{aligned} \mathcal{B}_1 \cdot (d\mathbf{l} \times \mathbf{u}) &= d\mathbf{l} \cdot (\mathbf{u} \times \mathcal{B}_1) \\ &= (\mathbf{u} \times \mathcal{B}_1) \cdot d\mathbf{l} \end{aligned} \tag{B-21}$$

Substituting (B-21) into (B-19), we obtain

$$\int_{\Delta s} \mathcal{B}_1 \cdot d\mathbf{s} = \Delta t \oint_c (\mathbf{u} \times \mathcal{B}_1) \cdot d\mathbf{l} \tag{B-22}$$

Now the magnetic flux at time t_1, $\mathcal{B}_1$, can be approximated for sufficiently small Δt by

$$\mathcal{B}_1 \simeq \mathcal{B} + \frac{\partial \mathcal{B}}{\partial t} \Delta t \tag{B-23}$$

Substituting (B-23) into (B-22), we obtain

$$\int_{\Delta s} \mathcal{B}_1 \cdot d\mathbf{s} \simeq \Delta t \oint_c (\mathbf{u} \times \mathcal{B}) \cdot d\mathbf{l} + (\Delta t)^2 \oint_c \left(\mathbf{u} \times \frac{\partial \mathcal{B}}{\partial t} \right) \cdot d\mathbf{l} \tag{B-24}$$

for sufficiently small Δt. Now we substitute (B-24) and (B-23) into (B-16) to obtain

$$\begin{aligned} \int_{s_1} \mathcal{B}_1 \cdot d\mathbf{s} \simeq \int_s \mathcal{B} \cdot d\mathbf{s} + \Delta t \int_s \frac{\partial \mathcal{B}}{\partial t} \cdot d\mathbf{s} - \Delta t \oint_c (\mathbf{u} \times \mathcal{B}) \cdot d\mathbf{l} \\ - (\Delta t)^2 \oint_c \left(\mathbf{u} \times \frac{\partial \mathcal{B}}{\partial t} \right) \cdot d\mathbf{l} \end{aligned} \tag{B-25}$$

Substituting (B-25) into (B-14), we obtain

$$\begin{aligned} \frac{d}{dt} \int_s \mathcal{B} \cdot d\mathbf{s} &= \lim_{\Delta t \to 0} \frac{1}{\Delta t} \left(\int_{s_1} \mathcal{B}_1 \cdot d\mathbf{s} - \int_s \mathcal{B} \cdot d\mathbf{s} \right) \\ &= \int_s \frac{\partial \mathcal{B}}{\partial t} \cdot d\mathbf{s} - \oint_c (\mathbf{u} \times \mathcal{B}) \cdot d\mathbf{l} \end{aligned} \tag{B-26}$$

and we have proved the equivalence of (i) and (ii).

The equivalence of (iii) can be shown by substituting the relationships between $\mathcal{E}'$, $\mathcal{E}$, and $\mathcal{B}$ given in (B-6) into (ii). Form (iv) can be obtained by applying Stokes' theorem to (iii). Similarly, (v) can be obtained by applying Stokes' theorem to (ii).

Thus, Faraday's law for moving contours can be written as

$$\oint_c \mathcal{E}' \cdot d\mathbf{l} = -\int_s \frac{\partial \mathcal{B}}{\partial t} \cdot ds + \oint_c (\mathbf{u} \times \mathcal{B}) \cdot d\mathbf{l} \tag{B-27}$$

The first term on the right-hand side of (B-27) gives the portion of the induced emf $\oint_c \mathcal{E}' \cdot d\mathbf{l}$ due to the time rate of change of $\mathcal{B}$ penetrating s. The second term gives the portion of the induced emf that results from the movement of contour c with respect to $\mathcal{B}$, where $\mathbf{u}$ is the velocity of the portion of c, $d\mathbf{l}$, with respect to $\mathcal{B}$. The first term on the right-hand side of (B-27) is referred to as *transformer emf* since the time-changing magnetic flux in the stationary windings of a transformer produces the same effect. The second term is referred to as *motional emf*, such as that produced in rotating electric machines. Substituting the relationship between $\mathcal{E}$ and $\mathcal{E}'$ given in (B-6) into (B-27), we obtain

$$\oint_c \mathcal{E} \cdot d\mathbf{l} = -\int_s \frac{\partial \mathcal{B}}{\partial t} \cdot ds \tag{B-28}$$

Note that for stationary contours c, $\mathbf{u} = 0$, $\mathcal{E}' = \mathcal{E}$, and (B-28) and (B-1) are identical; thus, one may differentiate $\mathcal{B}$ with respect to time first and then integrate the result over s, as in (B-28), or integrate $\mathcal{B}$ over s first and then differentiate the result with respect to time, as in (B-1). If the contour is not stationary, the order of differentiation and integration of $\mathcal{B}$ is important.

The point form of Faraday's law can be obtained by applying Stokes' theorem to (B-28), resulting in

$$\nabla \times \mathcal{E} = -\frac{\partial \mathcal{B}}{\partial t} \tag{B-29}$$

EXAMPLE B.1 As an example of the application of Faraday's law, consider Fig. B.3. A pair of thin, perfectly conducting wires forms a set of rails along which another thin, perfectly conducting wire of length L is moving with velocity u. A magnetic field $\mathcal{B}$ is perpendicular to this loop, and we wish to find the emf generated across a small gap in the loop. We will consider two cases: one with constant $\mathcal{B}$ and the other with time-varying $\mathcal{B}$. In all cases, since $\mathcal{E}' = 0$ along the wires, the total emf appears across the gap and is equal to $-d\psi_m/dt$, where ψ_m is the net magnetic flux penetrating the loop [as in (B-13)].

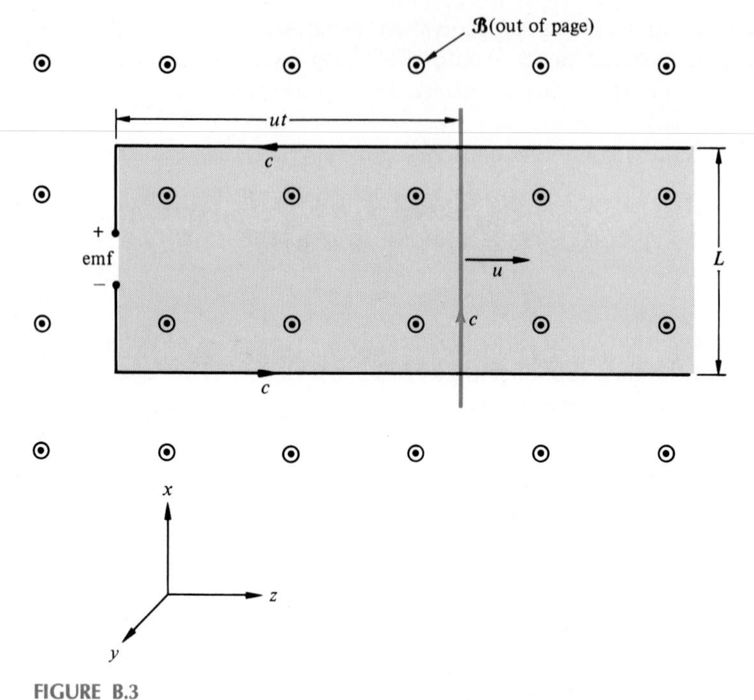

FIGURE B.3

Example B.1. The application of Faraday's law to moving contours.

Solution

Case 1 Suppose $\mathcal{B}$ is constant and given by

$$\mathbf{B} = B_0 \mathbf{a}_y$$

The net flux penetrating the loop becomes

$$\psi_m = B_0 Lut$$

and the emf (with the polarity shown) becomes

$$\begin{aligned} \text{emf} &= -\frac{d\psi_m}{dt} \\ &= -B_0 Lu \end{aligned}$$

Note that the polarity of the induced emf is actually opposite to that shown in Fig. B.3. This is easily seen by considering the direction of the magnetic force $\mathbf{F}_m = q\mathbf{u} \times \mathbf{B}$ exerted on the changes of the bar and their resulting direction of movement. This result can be obtained in an alternative fashion by using the form of Faraday's law given by (B-27). Since $\partial\mathcal{B}/\partial t = 0$, (B-27) becomes

$$\text{emf} = \oint_c (\mathbf{u} \times \mathcal{B}) \cdot d\mathbf{l}$$

$$= \oint_c (-uB_0\mathbf{a}_x) \cdot d\mathbf{l}$$

$$= \int_0^L -uB_0\, dx$$

$$= -B_0 Lu$$

Case 2 Suppose $\mathcal{B}$ is time-varying and given by

$$\mathcal{B} = B_0 \cos \omega t\, \mathbf{a}_y$$

The net flux penetrating the loop becomes

$$\psi_m = B_0 Lut \cos \omega t$$

and the emf appearing at the terminals becomes

$$\text{emf} = -\frac{d\psi_m}{dt}$$

$$= -B_0 Lu(\cos \omega t - \omega t \sin \omega t)$$

Using the form of Faraday's law given in (B-27), we also obtain

$$\text{emf} = -\int_s \frac{\partial \mathcal{B}}{\partial t} \cdot d\mathbf{s} + \oint_c (\mathbf{u} \times \mathcal{B}) \cdot d\mathbf{l}$$

$$= -\int_s (-\omega B_0 \sin \omega t\, \mathbf{a}_y) \cdot (dx\, dz\, \mathbf{a}_y)$$

$$+ \oint_c (-uB_0 \cos \omega t\, \mathbf{a}_x) \cdot d\mathbf{l}$$

$$= \omega B_0 \sin \omega t \int_{z=0}^{ut} \int_{x=0}^{L} dx\, dz - uB_0 \cos \omega t \int_{x=0}^{L} dx$$

$$= \omega B_0\, Lut \sin \omega t - uB_0 L \cos \omega t$$

EXAMPLE B.2 Consider a loop of perfectly conducting wire with side dimensions of L and W rotating with radian frequency ω in a magnetic field, as shown in Fig. B.4. If the magnetic field is constant and directed in the y direction, whereby

$$\mathbf{B} = B_y \mathbf{a}_y$$

with the plane of the loop rotating about the z axis, determine the emf induced in a small gap in the loop.

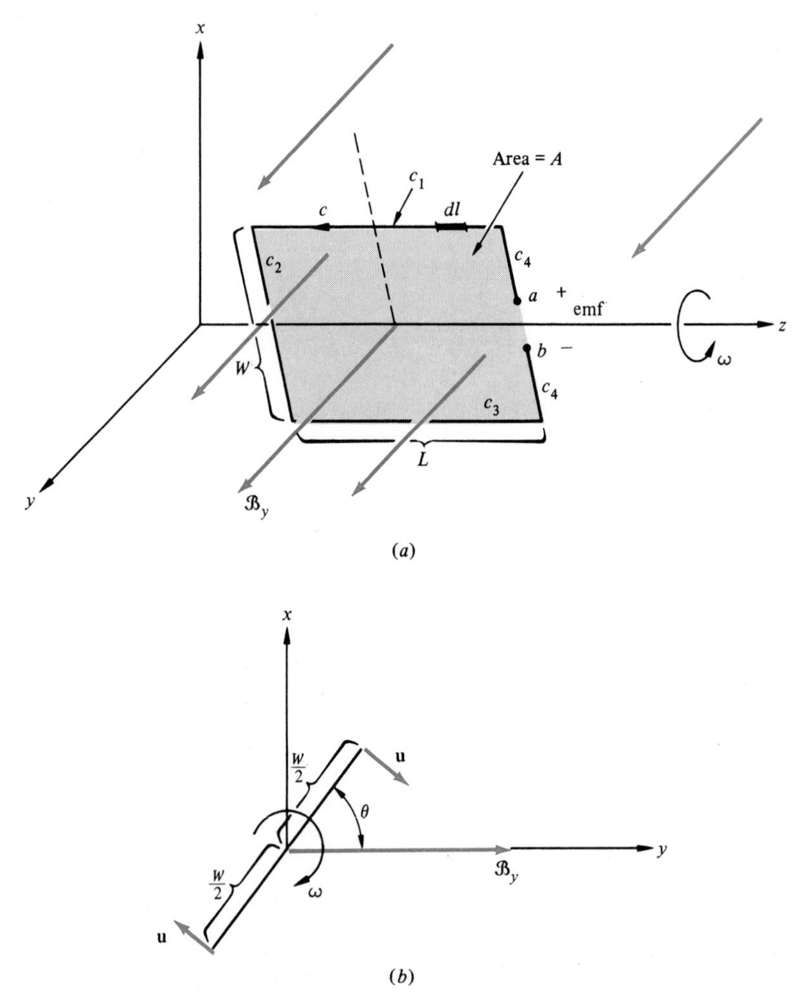

FIGURE B.4
Example B.2. Faraday's law applied to a rotating loop in a magnetic field: an elementary electric generator.

Solution If we denote the instantaneous angle between the plane of the loop and the magnetic field as θ, then the flux passing through the loop with area $A = LW$ is given by

$$\psi_m = B_y A \sin \theta$$

Let us now suppose that at $t = 0$ and $\theta = 0°$. Then $\theta = \omega t$ and

$$\psi_m = B_y A \sin \omega t$$

The induced emf is found from Faraday's law in (B-8) as

$$\begin{aligned} \text{emf} &= \oint_c \boldsymbol{\mathcal{E}}' \cdot d\mathbf{l} \\ &= -\frac{d\psi_m}{dt} \end{aligned}$$

where the direction of the contour c is shown in Fig. B.4. This becomes

$$\underset{\text{wire}}{\int_a^b} \boldsymbol{\mathcal{E}}' \cdot d\mathbf{l} + \underset{\text{gap}}{\int_b^a} \boldsymbol{\mathcal{E}}' \cdot d\mathbf{l} = -\omega B_y A \cos \omega t$$

Since

$$\underset{\text{wire}}{\int_a^b} \boldsymbol{\mathcal{E}}' \cdot d\mathbf{l} = 0$$

we obtain the emf induced across the gap as

$$\begin{aligned} \text{emf} &= \underset{\text{gap}}{\int_b^a} \boldsymbol{\mathcal{E}}' \cdot d\mathbf{l} \\ &= -\omega B_y A \cos \omega t \end{aligned}$$

as shown in Fig. B.4*a*. Similarly, using (B-27) we obtain

$$\begin{aligned} \text{emf} &= -\int_s \cancelto{0}{\frac{\partial \boldsymbol{\mathcal{B}}}{\partial t}} \cdot d\mathbf{s} + \oint_c (\mathbf{u} \times \boldsymbol{\mathcal{B}}) \cdot d\mathbf{l} \\ &= \oint_c (\mathbf{u} \times \boldsymbol{\mathcal{B}}) \cdot d\mathbf{l} \end{aligned}$$

Along the portions of the contour c_1 and c_3 of length L, we obtain

$$\int_{c_1+c_3} (\mathbf{u} \times \mathcal{B}) \cdot d\mathbf{l} = -2uLB_y \cos \omega t$$

since $\mathbf{u} \times \mathcal{B} = -uB_y \cos \omega t \ \mathbf{a}_z$ along contour c_1 and $\mathbf{u} \times \mathcal{B} = uB_y \cos \omega t \ \mathbf{a}_z$ along contour c_3. Along the portions of the contour c_2 and c_4 of length W, we obtain

$$\int_{c_2+c_4} (\mathbf{u} \times \mathcal{B}) \cdot d\mathbf{l} = 0$$

since $\mathbf{u} \times \mathcal{B}$ is perpendicular to these portions of c. Thus, the induced emf is given by

$$\text{emf} = -2uLB_y \cos \omega t$$

In order to show that this result is the same as that derived from an application of (B-8), we observe from Fig. B.4*b* that the sides of length L move an incremental arc length of

$$dr = r \, d\theta$$

where $r = W/2$ and $\theta = \omega t$. Therefore,

$$dr = \frac{W}{2} \omega \, dt$$

and the instantaneous velocity of these sides is

$$\begin{aligned} u &= \frac{dr}{dt} \\ &= \frac{W}{2} \omega \end{aligned}$$

Thus, we obtain

$$\begin{aligned} \text{emf} &= -2uLB_y \cos \omega t \\ &= -2\left(\frac{W}{2} \omega\right) LB_y \cos \omega t \\ &= -\omega AB_y \cos \omega t \end{aligned}$$

as obtained from a direct application of the form of Faraday's law given in (B-8).

Example B.2 illustrates the method by which an alternating voltage is produced in a rotating electric generator: Several loops of wire are rotated in a uniform, dc magnetic field, and the resulting ac voltage (or emf such as $\omega AB_y \cos \omega t$ obtained in Example B.2) is connected through slip rings to external terminals, as shown in Fig. B.5*a*. The terminal voltage waveform is shown in Fig. B.5*b*, and an equivalent circuit is given in Fig. B.5*c*; the resistance R_a and inductance L_a of the loop (armature) are included. If a load such as a resistor is attached to the two terminals, a current will pass through the loop and supply energy to the load. This current will in turn produce a magnetic field that tends to oppose the original magnetic field—and thus the loop rotation—so that energy must be expended in rotating the loop by the prime mover. Steam turbines, gasoline engines, etc., may serve as the prime movers, and conversion of energy from one form (mechanical) to another (electrical) is achieved.

Example B.2 shows a method for producing a sinusoidal (ac) voltage. A dc (or, more properly, a unidirectional) voltage may be produced by using, instead of slip rings, a split-ring commutator, as shown in Fig. B.6*a*. The commutator consists of a segmented copper ring with carbon brushes that make contact with the external terminals labeled *a* and *b*. A dc current I_f passes through the field winding, which produces a dc magnetic field in the space between the poles labeled N and S. If we assume this field to be uniform and neglect fringing, then the results of Example B.2 show that an emf (or voltage) $\omega AB \cos \omega t$ is produced in the rotating loop. The brushes are arranged such that the voltage produced at the terminals remains of one polarity, resulting in a pulsating dc voltage as shown in Fig. B.6*b*. An equivalent circuit is shown in Fig. B6*c*.

This pulsating voltage may be smoothed by employing several loops whose planes are displaced from each other by a uniform angle; each loop is attached to its own set of commutator segments. For example, if we arrange another loop perpendicular to the original loop, the resulting voltage shown in Fig. B.6*d* is the sum of the voltages of each loop which is displaced in phase by 90°. Three loops whose planes are separated by 60° would result in the sum of three sinusoidal voltages displaced in phase by 60°, and thus a further smoothing of this pulsating dc voltage would take place. Practical dc generators make use of many such loops, and the resulting terminal voltage is almost indistinguishable from the true dc voltage produced by a battery. For the practical dc generator, the open-circuit terminal voltage is again proportional to the radian frequency of rotation, ω, and the magnetic flux B; i.e., $V_g = k\omega B$, where k is some constant.

A dc electric motor can also be obtained simply by driving the terminals of the dc generator with a voltage source consisting of a battery (or dc generator) E_S having source resistance R_S, as shown in Fig. B.7. If the loop is rotating with angular frequency ω, a back emf $V_g = k\omega B$ is produced, as described previously. If the applied voltage E_S is larger than this back emf V_g, a current I_a will pass through the loop (the armature), which produces a magnetic field in opposition to the field thus producing the net torque of an electric motor.

An interesting discussion of the applications (and pitfalls) of Faraday's law is found in L. V. Bewley, *Flux Linkages and Electromagnetic Induction*, Dover, New York, 1964.

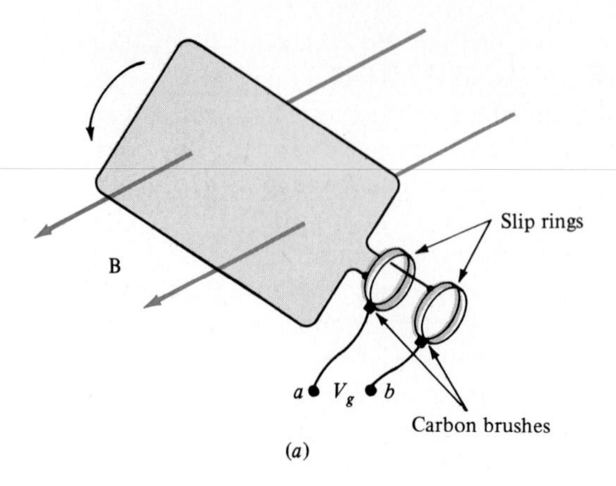

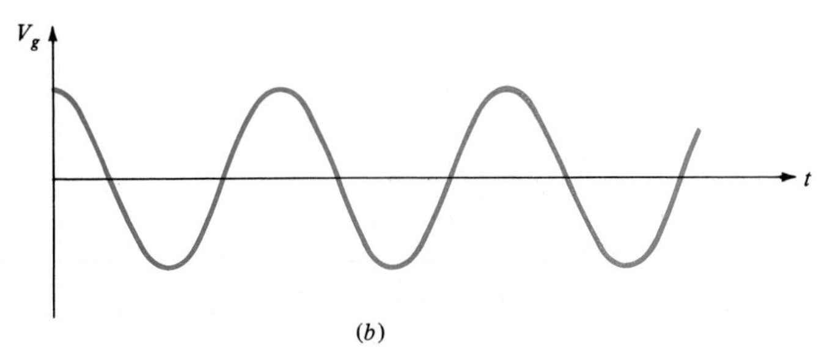

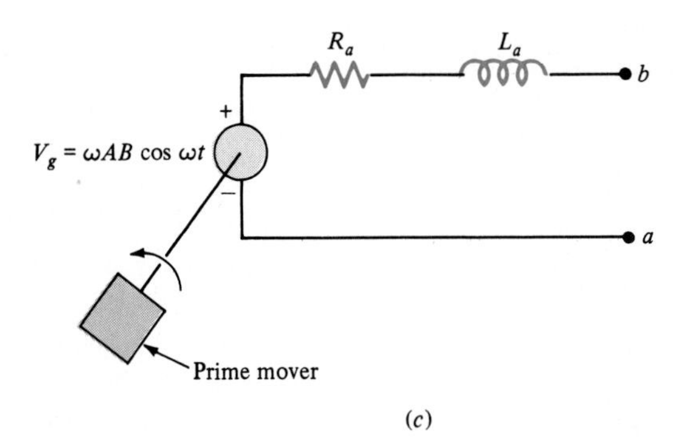

FIGURE B.5

Illustration of an ac generator.

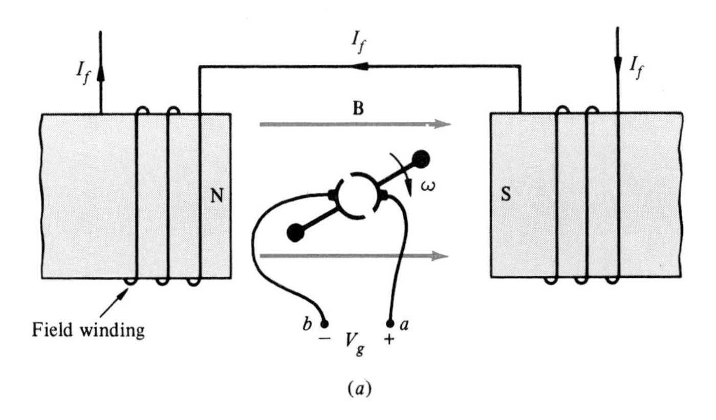

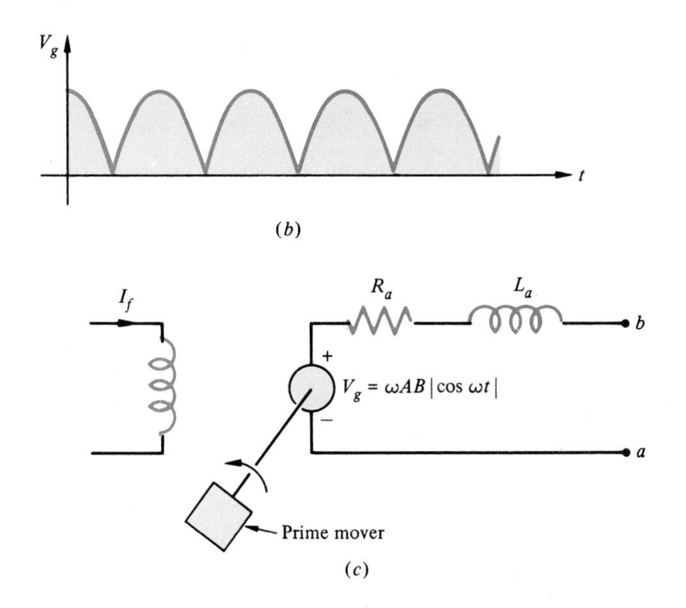

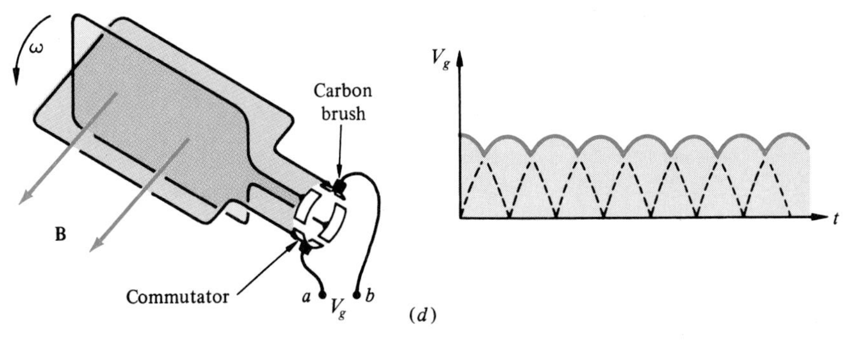

FIGURE B.6

Illustration of a dc generator.

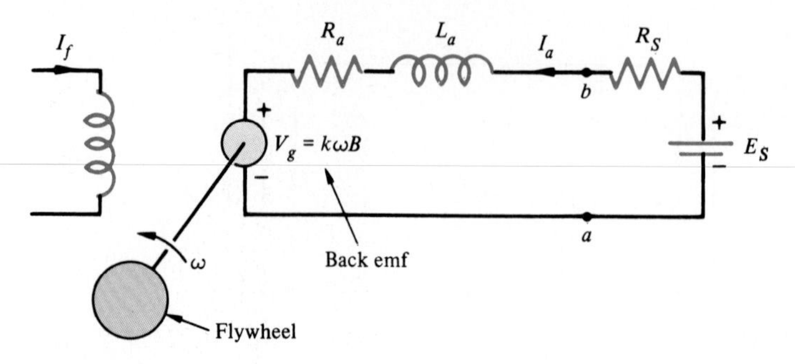

FIGURE B.7
Illustration of a dc motor.

Problems

B-1 A conducting disk rotates at a constant speed ω_m in a uniform magnetic field B directed axially, as shown in Fig. PB.1. The inner and outer radii are r_i and r_o, respectively. Determine the induced voltage available at the terminals a and a'. This device is known as the Faraday disk generator and is the primitive form of the homopolar generator.

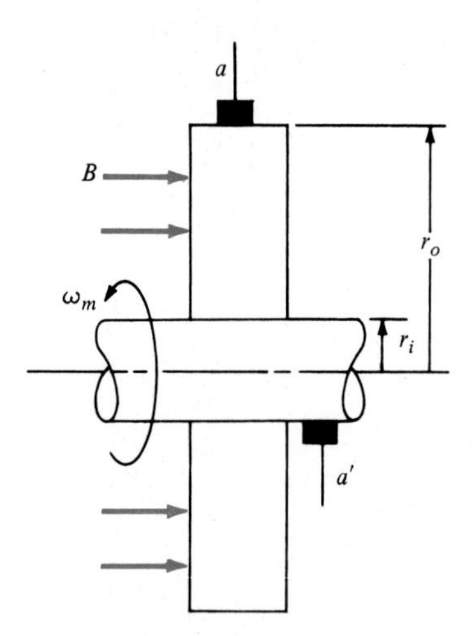

FIGURE PB.1

B-2 The disk of Fig. PB.1 is fed with a current I at the terminal a, and the magnetic field B remains the same as in Prob. B-1. Evaluate the torque developed by the machine.

B-3 A dc generator having a liquid-metal armature is proposed in Fig. PB.3. Determine the number of turns, N_a, such that the terminal voltage is independent

FIGURE PB.3

of the load current (under steady-state). The channel is c m long and the fluid flows at u m/s out of the paper.

B-4 A closed contour is stationary in space and is located in a time-varying field B. Obtain an expression for the induced emf in terms of the magnetic vector potential.

B-5 A rectangular loop is placed in the magnetic field of a very long straight conductor carrying a current I, as shown in Fig. PB.5, which also shows the various dimensions. If the loop moves in the x direction at a velocity u, determine the induced voltage in the loop at some instant in time.

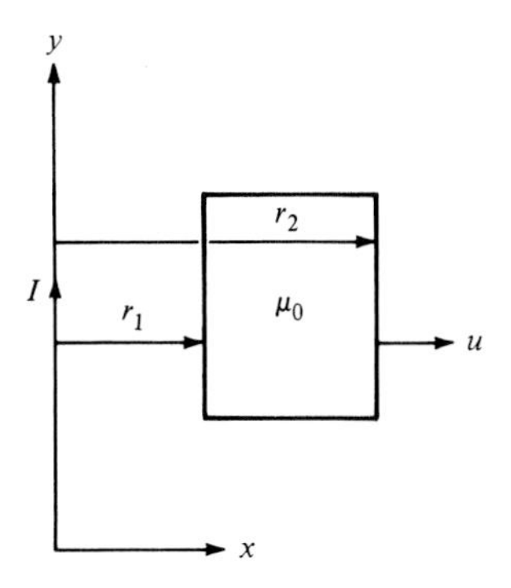

FIGURE PB.5

B-6 A circular loop of a conductor is located in a uniform B field, as shown in Fig. PB.6. The loop has a radius R and rotates at a constant angular velocity ω_m about the axis ab, as shown. If the axis of rotation is perpendicular to the B field, determine the emf induced in the loop.

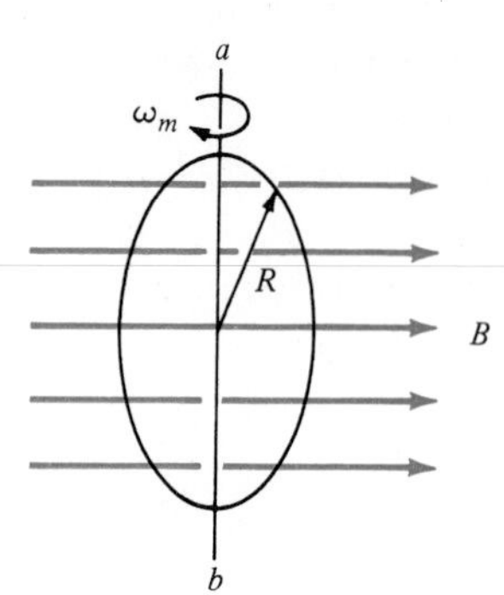

FIGURE PB.6

B-7 A thin conducting sheet of conductivity σ is located in the xz plane in a magnetic field, the y component of which is given by

$$B_y = B_m \cos \beta x \cosh \beta y \cos \omega t$$

Assuming the electric field to exist only in the z direction, determine

(a) the induced electric field and
(b) the current density in the sheet.

B-8 If the sheet of Prob. B-7 is allowed to move with a velocity $\mathbf{u} = u_x \mathbf{a}_x$, what is the induced electric field in the z direction?

B-9 A magnetic field given by

$$\mathbf{B} = \mathbf{a}_y B_y = B_m \sin(\omega t - \beta x)\mathbf{a}_y$$

represents a traveling field. Assume that the electric field is z-directed only.

(a) What is the velocity of the traveling field?
(b) Determine the electric field induced in a conductive sheet that is
  (1) held stationary in the above field;
  (2) moving with a velocity u_x in the x direction in the given field.

B-10 A rectangular channel of the geometry shown in Fig. PB.10 is located in a dc magnetic field $\mathbf{B} = B\mathbf{a}_y$. A conducting fluid of conductivity σ flows through the

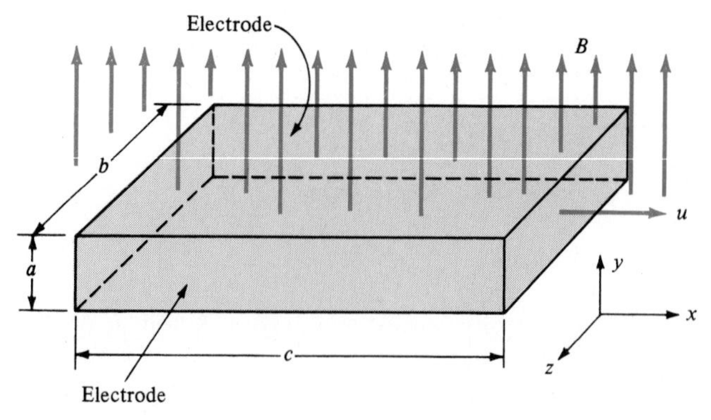

FIGURE PB.10

channel with a velocity $\mathbf{u} = u\mathbf{a}_x$. Such a configuration forms the basis of hydromagnetic generators and liquid-metal pumps.

(a) Determine the voltage available at the open-circuited terminals marked "Electrode."

(b) A resistive load R is now connected to the electrodes such that the fluid now carries a current I. Evaluate the following quantities:

(1) generator output

(2) ohmic loss in the fluid

(3) pressure drop across the channel

APPENDIX C

The Smith Chart†

In the sinusoidal, steady-state analysis of transmission lines (considered in Chap. 7), we require numerous algebraic operations with complex numbers (addition, subtraction, multiplication, and division) in order to obtain certain quantities of interest. The *Smith chart* is an ingenious graphical technique that avoids virtually all of these tedious complex-number operations. This reason alone should be sufficient motivation for its study. In addition to this advantage, the Smith chart also provides numerous other important results not readily apparent from the equations of the transmission line.

An example is the determination of the input impedance to a transmission line, given its electrical length and its load impedance. For example, consider the uniform, lossless transmission line with characteristic resistance R_C and phase constant $\beta = 2\pi/\lambda$ shown in Fig. C.1. As discussed in Sec. 7.2, the input impedance to the line at any point along the line, $\hat{Z}_{\text{in}}(z)$, may be determined in the following manner. The voltage reflection coefficient $\hat{\Gamma}(z)$ at any point z on the line and the input impedance to the line at the same point, $\hat{Z}_{\text{in}}(z)$, are related by

$$\begin{aligned}\hat{Z}_{\text{in}}(z) &= \frac{\hat{V}(z)}{\hat{I}(z)} \\ &= R_C \frac{1 + \hat{\Gamma}(z)}{1 - \hat{\Gamma}(z)} \qquad \text{(C-1)}\end{aligned}$$

and the voltage reflection coefficient at any two points on the line, z_1 and z_2, are related by

$$\hat{\Gamma}(z_1) = \hat{\Gamma}(z_2)e^{j2\beta(z_1 - z_2)} \qquad \text{(C-2)}$$

† P. H. Smith, "Transmission Line Calculator," *Electronics*, January 1939; and P. H. Smith, "An Improved Transmission Line Calculator," *Electronics*, January 1944.

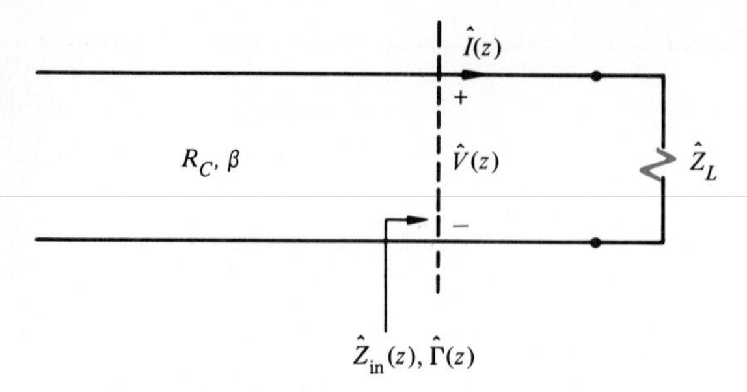

FIGURE C.1
Input impedance to a transmission line.

where the phase constant β is given by

$$\beta = \frac{2\pi}{\lambda} = \frac{\omega}{u} \tag{C-3}$$

where u is the velocity of propagation of the wave on the line in the surrounding (homogeneous) medium characterized by μ and ϵ ($u = 1/\sqrt{\mu\epsilon}$), ω is the radian frequency of excitation, and λ is a wavelength in the surrounding medium ($\lambda = u/f$).

The key to understanding the Smith chart (which is crucial to its proper and effective use) lies in the fact that the Smith chart relates, graphically, the input impedance at some point on the line and the voltage reflection coefficient *at that point* given by (C-1) and (C-2). Consider the equation for input impedance given in (C-1). We first define the *normalized input impedance* as

$$\hat{z}_{\text{in}} = \frac{\hat{Z}_{\text{in}}(z)}{R_C} = \frac{1 + \hat{\Gamma}(z)}{1 - \hat{\Gamma}(z)} = r + jx \tag{C-4}$$

where r and x are the real and imaginary parts, respectively, of $\hat{z}_{\text{in}}$. Similarly, the reflection coefficient at this point may be written in terms of its real and imaginary parts, p and q, respectively, as

$$\hat{\Gamma}(z) = |\hat{\Gamma}(z)|\underline{/\theta_{\hat{\Gamma}}} = p + jq \tag{C-5}$$

Substituting (C-5) into the expression for (normalized) input impedance in (C-4), we obtain

$$\hat{z}_{\text{in}} = r + jx = \frac{1 + p + jq}{1 - p - jq} \tag{C-6}$$

If we equate the real and imaginary parts of this expression, we may obtain two equations:

$$\left(p - \frac{r}{r+1}\right)^2 + q^2 = \frac{1}{(r+1)^2} \qquad \text{circles of constant } r \tag{C-7a}$$

$$(p-1)^2 + \left(q - \frac{1}{x}\right)^2 = \frac{1}{x^2} \qquad \text{circles of constant } x \tag{C-7b}$$

Equations (C-7) are equations of circles. In particular, (C-7*a*) is the equation of a circle of radius $1/(r+1)$ and centered at $p = r/(r+1)$ and $q = 0$. Equation (C-7*b*) is the equation of a circle of radius $1/x$ and centered at $p = 1$ and $q = 1/x$.

The Smith chart is a graphical portrayal of the two equations in (C-7). For example, the voltage reflection coefficient $\hat{\Gamma} = p + jq$ may be plotted in the pq plane. Equation (C-7*a*) gives the relationship between the real part of $\hat{z}_{in}$, r, and p and q, as shown in Fig. C.2*a*. Equation (C-7*b*) gives the relationship between

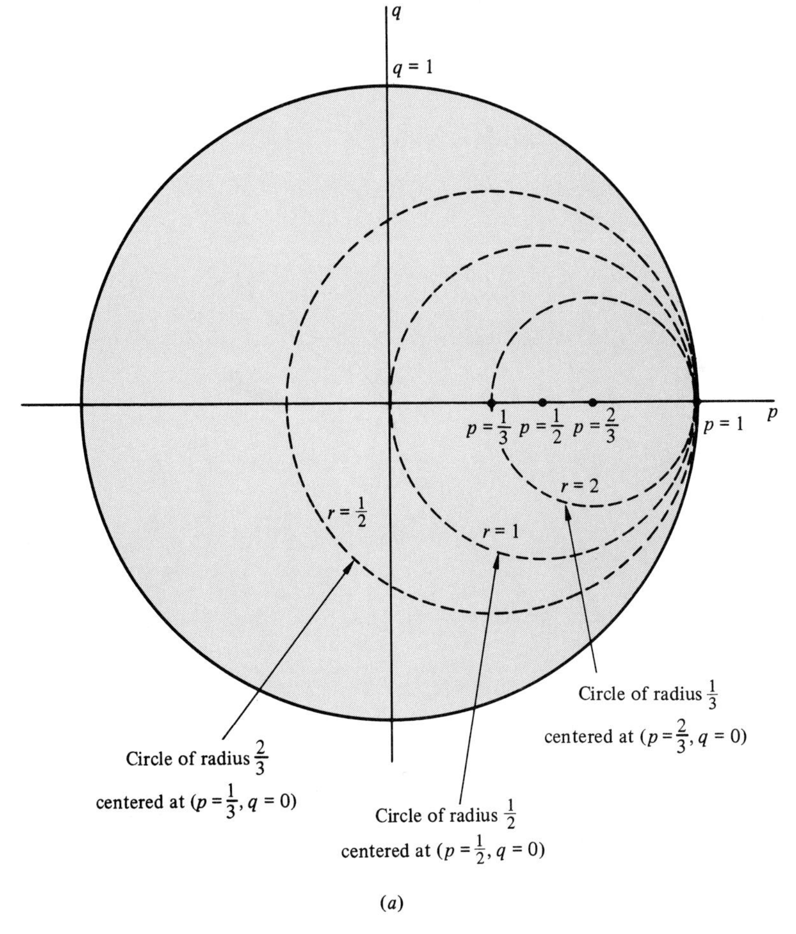

FIGURE C.2

Graphical portrayal of *(a)* equation (C-7*a*).

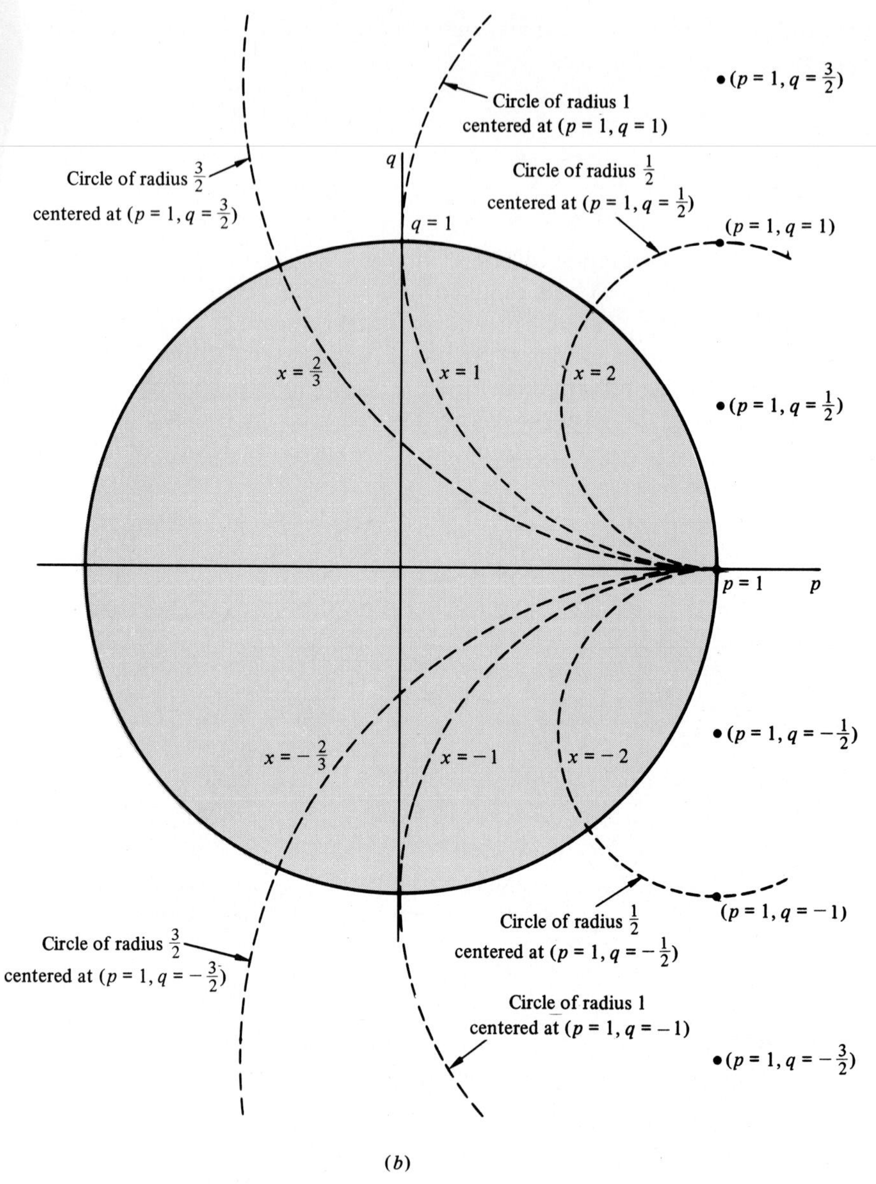

(*b*)

FIGURE C.2 (*continued*)
Graphical portrayal of (*b*) equation (C-7*b*).

the imaginary part of $\hat{z}_{\text{in}}$, x, and p and q, as shown in Fig. C.2*b*. If we superimpose the graphs of the two equations in (C-7), shown in Fig. C.2, we obtain the Smith chart shown in Fig. C.3. The important result from the foregoing is the fact that the voltage reflection coefficient *and* the normalized input impedance *at the same point on the line* are related by the Smith chart, as shown in Fig. C.4. On the outer rim of the chart is a scale labeled "Angle of Reflection Coefficient in Degrees" from which $\theta_{\hat{\Gamma}}$ can be read. At the bottom of the chart are several scales, one of which is labeled "Reflection, Coeff., Vol." A compass can be used to measure the length of the $\hat{\Gamma}$ vector, $|\hat{\Gamma}|$, using this scale. The remaining scales will be explained in subsequent discussions. The key point is that the Smith chart is a graphical relationship between (C-4) for the normalized input impedance and (C-5) for the reflection coefficient *at that point on the line.*

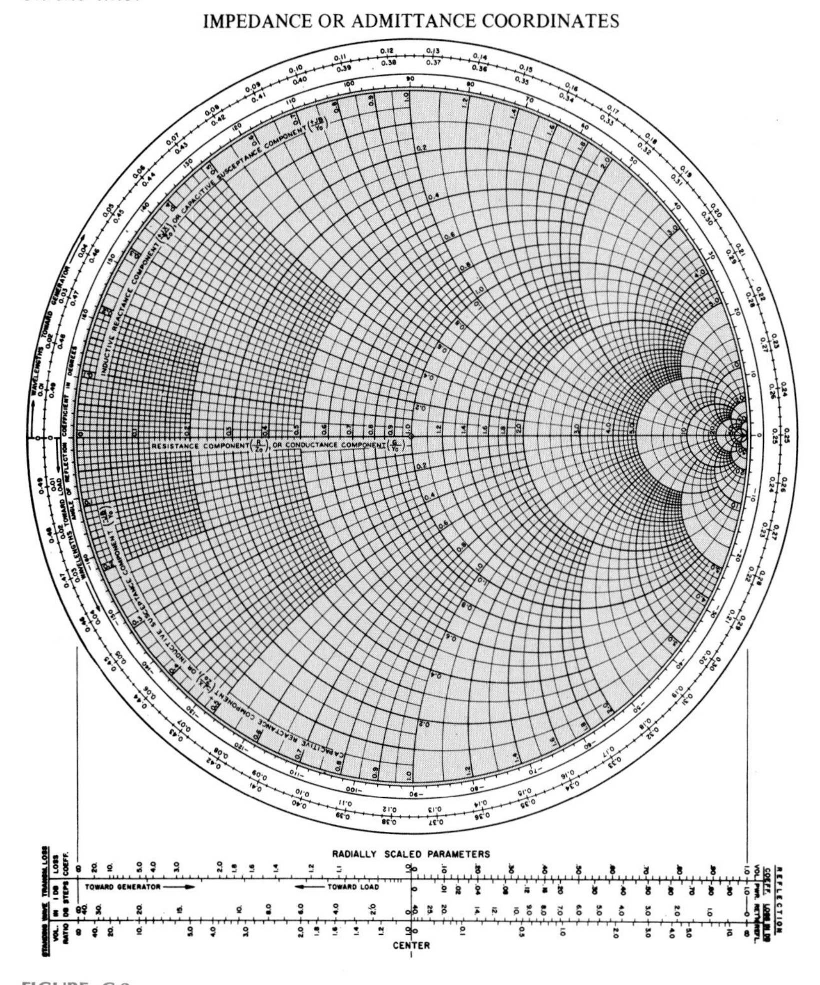

FIGURE C.3

The Smith chart.

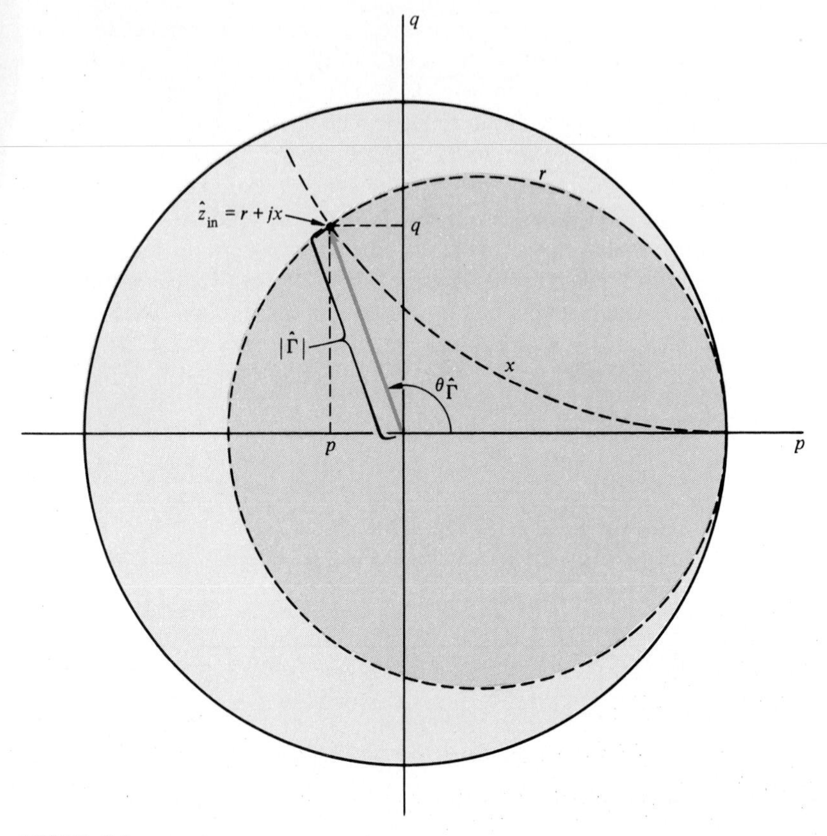

FIGURE C.4

Illustration of the relationship between reflection coefficient and normalized input impedance of a transmission line on a Smith chart.

To illustrate some applications, let us suppose that we wish to find the input impedance to a uniform, lossless line of length $\mathscr{L}$, given R_C, β, and $\hat{Z}_L$. At the load, (C-4) becomes

$$\hat{z}_L = \frac{\hat{Z}_L}{R_C}$$

$$= \frac{1 + \hat{\Gamma}_L}{1 - \hat{\Gamma}_L} \tag{C-8}$$

which is plotted in Fig. C.5. From the location of $\hat{z}_L$ we immediately have $\hat{\Gamma}_L = |\hat{\Gamma}_L| \underline{/\theta_{\hat{\Gamma}_L}}$. The reflection coefficient at a distance $\mathscr{L}$ away from the load is given by (C-2), with $z_1 = 0$ and $z_2 = \mathscr{L}$:

$$\hat{\Gamma} = \hat{\Gamma}_L e^{-j2\beta\mathscr{L}} \tag{C-9}$$

The reflection coefficient may be obtained from $\hat{\Gamma}_L$ by subtracting $2\beta\mathscr{L}$ from $\theta_{\hat{\Gamma}_L}$, as shown in Fig. C.5. From this we immediately find the normalized input impedance $\hat{z}_{\text{in}}$.

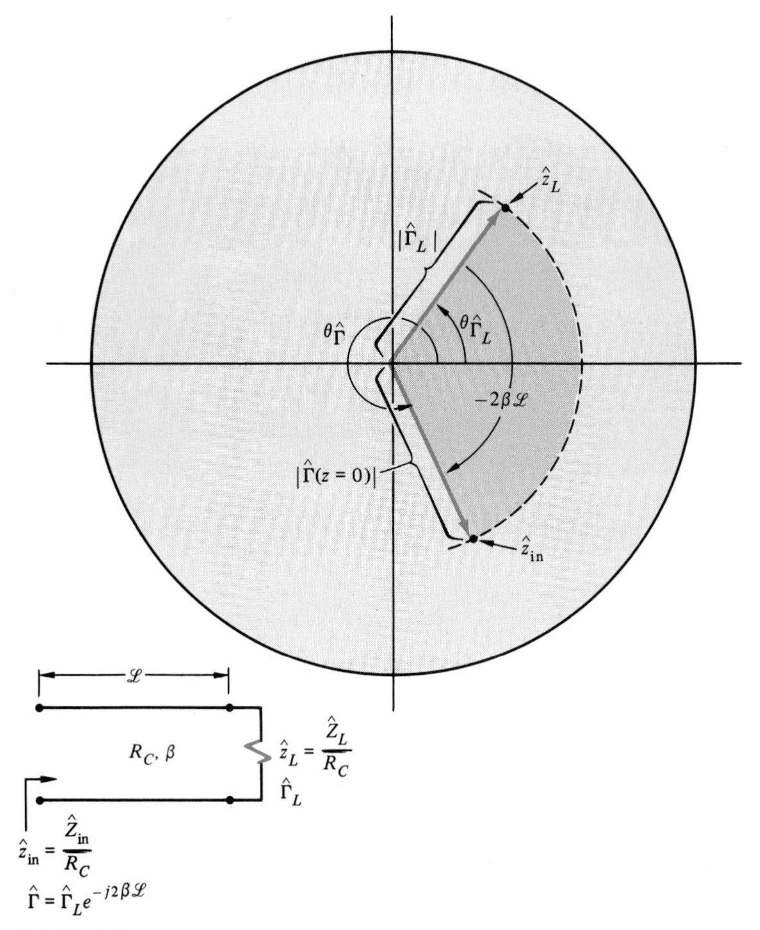

FIGURE C.5
Determining the input impedance at a point on a transmission line using the Smith chart.

The angle $2\beta\mathscr{L}$ can be read directly on the previously mentioned reflection-coefficient-angle scale on the outer rim of the chart. Alternatively, we may perform this rotation in terms of the line length as a portion of a wavelength instead of $2\beta\mathscr{L}$. If the homogeneous medium surrounding the line is characterized by μ and ϵ, then the velocity of propagation of the TEM waves on the line is

$$u = \frac{1}{\sqrt{\mu\epsilon}} \quad \text{m/s} \tag{C-10}$$

A wavelength is then defined as

$$\lambda = \frac{u}{f} \tag{C-11}$$

where f is the frequency of operation. The length of the line, $\mathscr{L}$, is $\mathscr{L}/\lambda$ portions of a wavelength. One of the outer scales of the Smith chart is labeled wavelengths "Toward Generator." This scale may be used when transferring the normalized load impedance $\hat{z}_L$ down the line (toward decreasing z) and will be referred to as the TG (toward generator) scale. Similarly, an additional chart is labeled wavelengths "Toward Load" and will be referred to as the TL (toward load) scale. This scale may be used when moving from some point on the line toward the load (in the direction of increasing z). Two examples will serve to illustrate this basic use of the Smith chart.

EXAMPLE C.1 Consider a transmission line of length 10 m that is immersed in polyethylene ($\epsilon_r = 2.25$, and $\mu_r = 1$) and operated at a frequency of 34 MHz. If the load impedance is $\hat{Z}_L = (50 + j100)\ \Omega$ and the characteristic resistance is $R_C = 50\ \Omega$, determine the input impedance to the line.

Solution The velocity of propagation is

$$u = \frac{1}{\sqrt{\mu\epsilon}}$$

$$= \frac{3 \times 10^8}{\sqrt{2.25}} = 2 \times 10^8 \text{ m/s}$$

and a wavelength at the operating frequency is

$$\lambda = \frac{u}{f}$$

$$= \frac{2 \times 10^8}{34 \times 10^6} = 5.882 \text{ m}$$

Thus, the length of the line in terms of wavelengths is

$$\mathscr{L} = 10 \text{ m} \times \frac{1}{5.882 \text{ m}/\lambda} = 1.70\lambda$$

The load impedance at the operating frequency is $\hat{Z}_L = (50 + j100)\ \Omega$ and the characteristic resistance is $R_C = 50\ \Omega$. Thus, the normalized load impedance is

$$\hat{z}_L = \frac{\hat{Z}_L}{R_C}$$

$$= 1 + j2$$

which is plotted on the Smith chart in Fig. C.6. Rotating clockwise on the TG scale, 1.70λ results in the normalized input impedance shown in Fig. C.6 of

$$\hat{z}_{\text{in}} = 0.29 - j0.82$$

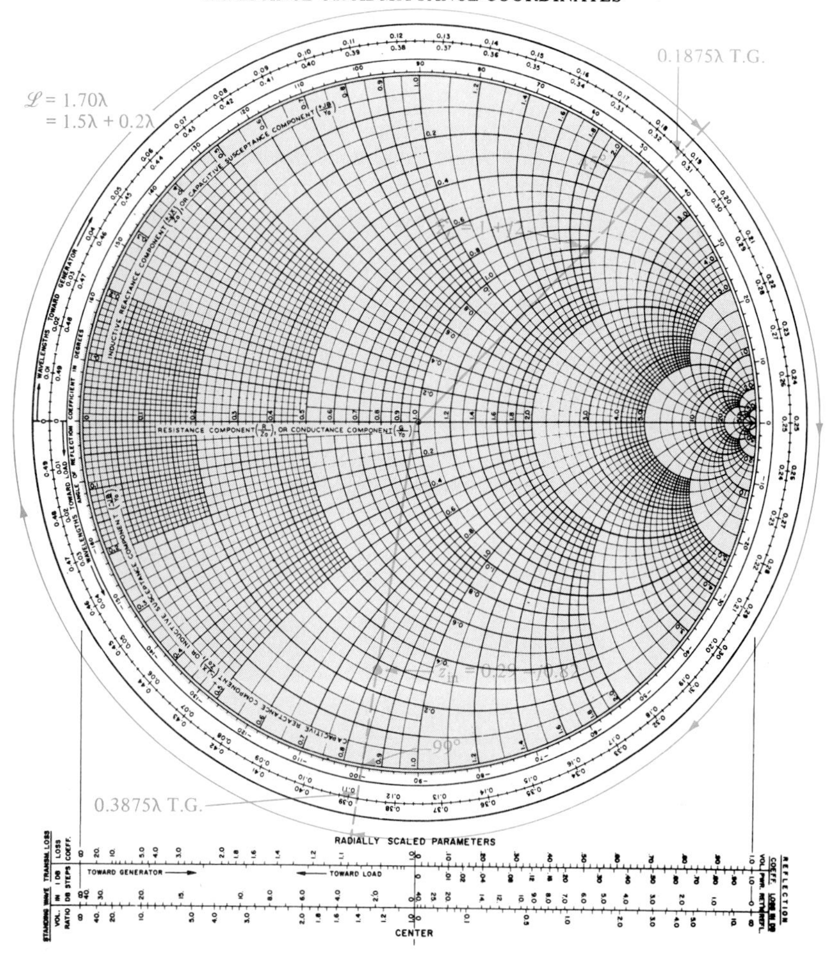

FIGURE C.6

Example C.1. Determining input impedance given the load impedance.

The input impedance becomes

$$\hat{Z}_{\text{in}} = \hat{z}_{\text{in}} R_C$$
$$= 14.5 - j41 \qquad \Omega$$

The exact value calculated from the results of Sec. 7.2 is

$$\hat{Z}_{\text{in}} = 14.52 - j40.52 \qquad \Omega$$

Notice that in moving the full electrical length of the line, 1.7λ, we complete three full revolutions around the chart (1.5λ) plus an additional 0.2λ, ($1.5\lambda + 0.2\lambda = 1.7\lambda$). This again shows, as was pointed out in Sec. 7.2, that the input impedance to the line replicates at points that are separated by multiples of a half-wavelength. In terms of the length of the line in degrees, we have

$$2\beta\mathscr{L} = 1224.07^\circ$$

Rotation around the chart by this angle (in the clockwise direction) three full revolutions, or 1080°, leaves 144.07°. Subtracted from the angle $\hat{\Gamma}_L$, 45°, this results in the angle of $\hat{\Gamma}$ of −99.07°, as shown in Fig. C.6. Also, we may obtain from the Smith chart

$$\hat{\Gamma}_L = 0.71\underline{/45^\circ}$$

and the exact value is

$$\hat{\Gamma}_L = 0.707\underline{/45^\circ}$$

Similarly, from the Smith chart we obtain

$$\hat{\Gamma}|_{z=0} = 0.71\underline{/-99^\circ}$$

whereas the exact value is

$$\begin{aligned}\hat{\Gamma}|_{z=0} &= 0.707\underline{/-1179.07^\circ} \\ &= 0.707\underline{/-99.07^\circ}\end{aligned}$$

EXAMPLE C.2 Suppose the input impedance to a line is $\hat{Z}_{\text{in}} = (30 - j40)\ \Omega$ and the load impedance is $\hat{Z}_L = (20 + j40)\ \Omega$. If the line has a characteristic resistance of $R_C = 100\ \Omega$ and velocity of propagation of $u = 250$ m/μs and is operated at a frequency of 30 MHz, determine the length of the line.

Solution The normalized impedances are

$$\begin{aligned}\hat{z}_{\text{in}} &= 0.3 - j0.4 \\ \hat{z}_L &= 0.2 + j0.4\end{aligned}$$

which are plotted in Fig. C.7. On the TG scale, we move from $\hat{z}_L$ (0.062λ) to $\hat{z}_{\text{in}}$ (0.435λ) for a total line length of

$$\mathscr{L} = 0.435\lambda - 0.062\lambda = 0.373\lambda$$

Note that this length of line is unique only within multiples of a half-wavelength. Thus

$$\mathscr{L} = 0.373\lambda \pm \frac{n\lambda}{2} \qquad n = 1, 2, 3, \ldots$$

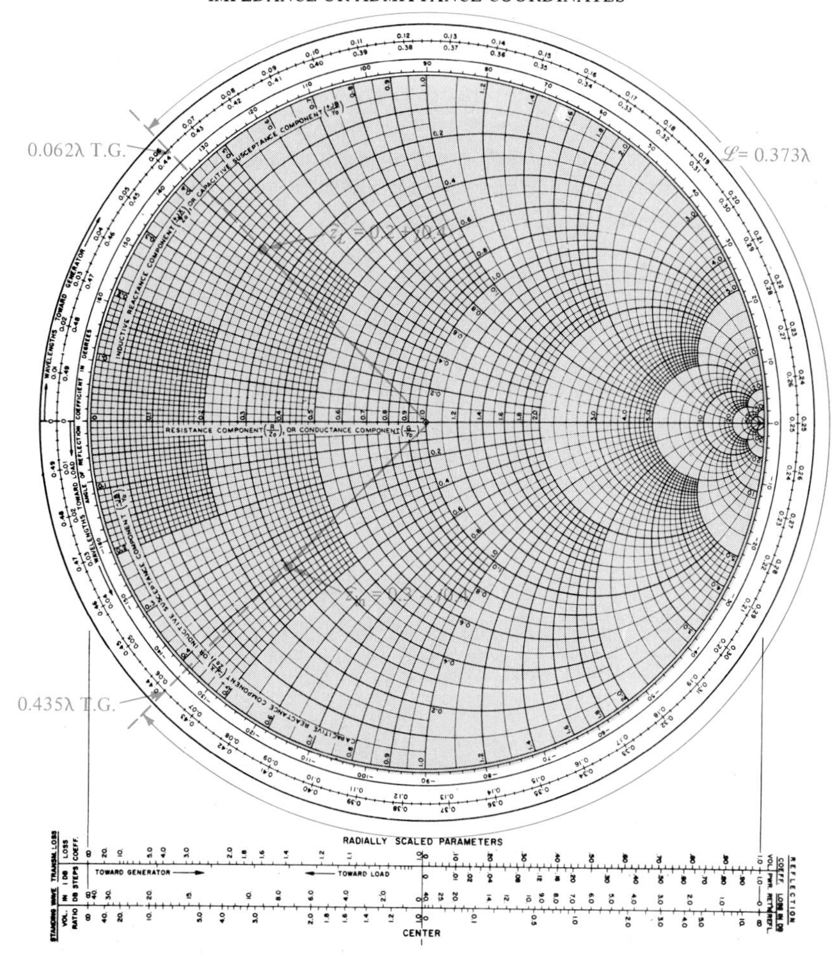

FIGURE C.7
Example C.2. Determining line length given the input and load impedances.

The shortest length of line is, however, 0.373λ. The physical length is determined from

$$\lambda = \frac{u}{f}$$

$$= \frac{250 \times 10^6}{30 \times 10^6} = 8.33 \text{ m}$$

Therefore,

$$\mathscr{L} = 0.373\lambda$$

$$= 0.373\lambda \times 8.33 \text{ m}/\lambda = 3.11 \text{ m}$$

C.1 Additional Results

Several additional and interesting results may be obtained from the Smith chart. We may prove the following results (obtained in Sec. 7.2). At a voltage maximum (minimum) on the line, $\hat{z}_{\text{in}}$ is real, i.e., purely resistive, and $\hat{z}_{\text{in}} =$ VSWR (1/VSWR). This results from some rather simple observations concerning the equation for the input impedance given in (C-1). In terms of the normalized input impedance, (C-4) becomes

$$\hat{z}_{\text{in}} = \frac{1 + \hat{\Gamma}}{1 - \hat{\Gamma}} \tag{C-12}$$

However, the equation for the phasor line voltage at this point becomes, from Sec. 7.2,

$$\hat{V}(z) = \hat{V}_m^+ e^{-j\beta z}(1 + \hat{\Gamma}) \tag{C-13}$$

We showed in Sec. 7.2 that at a voltage maximum, (C-13) becomes (see Fig. 7.20)

$$|\hat{V}(z)|_{\text{max}} = |\hat{V}_m^+|(1 + |\hat{\Gamma}|) \tag{C-14}$$

Thus, at a voltage maximum,

$$1 + \hat{\Gamma} = 1 + |\hat{\Gamma}| \tag{C-15}$$

Similarly, at this voltage maximum

$$1 - \hat{\Gamma} = 1 - |\hat{\Gamma}| \tag{C-16}$$

and (C-12) becomes, at this point,

$$\hat{z}_{\text{in}} = \frac{1 + |\hat{\Gamma}|}{1 - |\hat{\Gamma}|} \tag{C-17}$$

However, from (C-2) we see that

$$|\hat{\Gamma}| = |\hat{\Gamma}_L| \tag{C-18}$$

that is, the *magnitude* of the voltage reflection coefficient is the same at all points along the line: a well-established fact by now. Thus, (C-17) becomes

$$\begin{aligned} \hat{z}_{\text{in}} &= \frac{1 + |\hat{\Gamma}_L|}{1 - |\hat{\Gamma}_L|} \\ &= r_{\text{max}} \\ &= \text{VSWR} \end{aligned} \tag{C-19}$$

Similarly, at a voltage minimum on the line (see Fig. 7.20 again),

$$\begin{aligned} 1 + \hat{\Gamma} &= 1 - |\hat{\Gamma}| \\ &= 1 - |\hat{\Gamma}_L| \end{aligned} \tag{C-20}$$

and

$$\begin{aligned} 1 - \hat{\Gamma} &= 1 + |\hat{\Gamma}| \\ &= 1 + |\hat{\Gamma}_L| \end{aligned} \tag{C-21}$$

so that at a voltage minimum,

$$\begin{aligned} \hat{z}_{\text{in}} &= \frac{1 - |\hat{\Gamma}_L|}{1 + |\hat{\Gamma}_L|} \\ &= r_{\text{min}} \\ &= \frac{1}{\text{VSWR}} \end{aligned} \tag{C-22}$$

Thus, in order to find the VSWR on a particular transmission line, we plot the normalized load impedance $\hat{z}_L$, which determines $|\hat{\Gamma}_L|$. Rotating this vector until it crosses the horiziontal axis of the Smith chart results in an input impedance to the line that is purely real. Two possible values separated by $\lambda/2$ may be located; the larger value is the VSWR, and the smaller is 1/VSWR. The VSWR may also be found directly by using the scale at the bottom of the chart labeled "Standing-Wave Vol. Ratio." If $|\hat{\Gamma}|$ is transferred from the chart to this scale, the VSWR may be read directly from the scale.

EXAMPLE C.3 Consider a transmission line terminated in an impedance of $\hat{Z}_L = (150 - j200)\ \Omega$. If the line has a characteristic resistance of $R_C = 100\ \Omega$, find the VSWR and the shortest distance from the load at which the input impedance to the line is purely resistive.

Solution These results are summarized in Fig. C.8. The normalized load impedance is $\hat{z}_L = 1.5 - j2$, which is plotted in Fig. C.8. The result is

$$\begin{aligned} \frac{1}{\text{VSWR}} &= r_{\text{min}} \\ &= 0.22 \end{aligned}$$

or

$$\text{VSWR} = 4.55$$

Note that on the TG scale, the input impedance to the linear appears purely resistive at 0.5λ TG. Thus, the shortest distance from the load at which this occurs is $0.5\lambda - 0.302\lambda = 0.198\lambda$. At this point, the normalized input impedance is

$$\hat{z}_{\text{in}} = 0.22$$

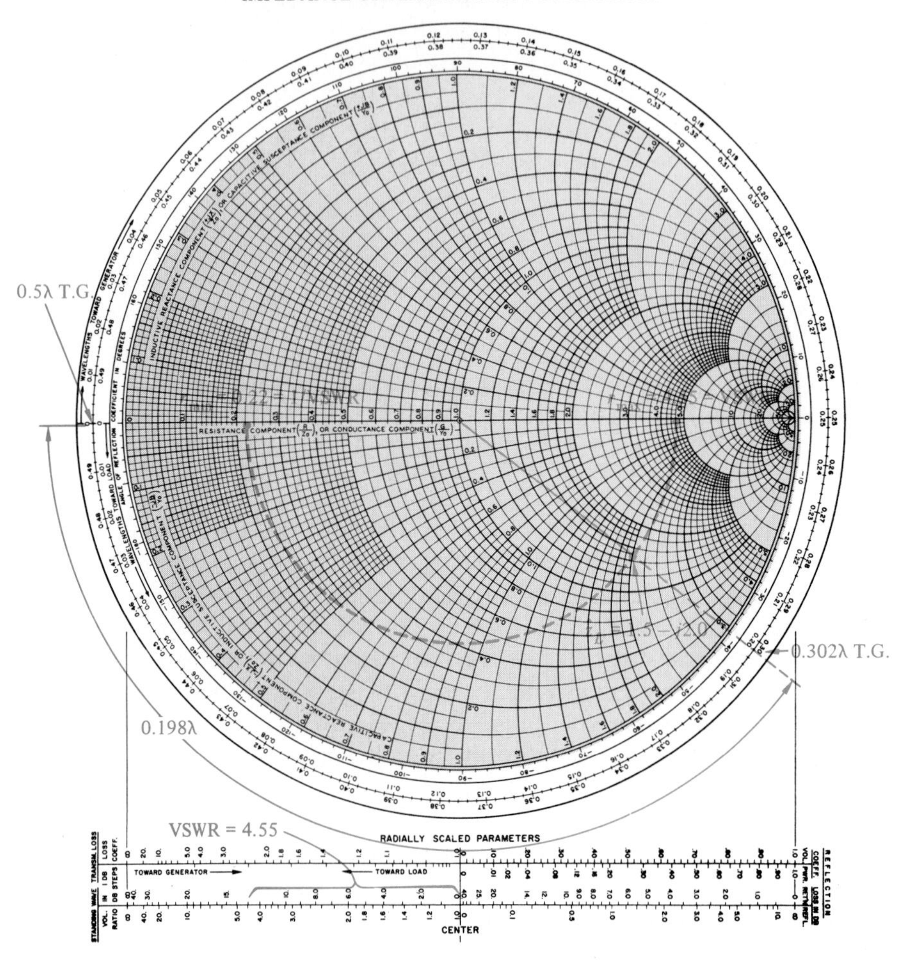

FIGURE C.8

Example C.3. Determining VSWR with a Smith chart.

so that

$$\hat{Z}_{in} = \hat{z}_{in} R_C$$

$$= 22\ \Omega$$

A distance of $\lambda/4$ further results in an input impedance of

$$\hat{z}_{in} = r_{max}$$

$$= 4.55$$

or

$$\hat{Z}_{in} = 455\ \Omega$$

An additional interesting result is the use of the Smith chart to invert a complex number. The use of the Smith chart for this purpose results, again, from the expression for the input impedance given in (C-1). At a distance $\mathscr{L}$ from the load, (C-1) with (C-2) becomes, in terms of the normalized input impedance,

$$\hat{z}_{\text{in}} = \frac{1 + \hat{\Gamma}_L e^{-j2\beta\mathscr{L}}}{1 - \hat{\Gamma}_L e^{-j2\beta\mathscr{L}}} \tag{C-23}$$

Suppose we move down the line an additional distance of precisely $\lambda/4$. In this case, (C-23) becomes

$$\begin{aligned}\hat{z}_{\text{in}} &= \frac{1 + \hat{\Gamma}_L e^{-j2\beta\mathscr{L}} e^{-j2(2\pi/\lambda)(\lambda/4)}}{1 - \hat{\Gamma}_L e^{-j2\beta\mathscr{L}} e^{-j2(2\pi/\lambda)(\lambda/4)}} \\ &= \frac{1 - \hat{\Gamma}_L e^{-j2\beta\mathscr{L}}}{1 + \hat{\Gamma}_L e^{-j2\beta\mathscr{L}}}\end{aligned} \tag{C-24}$$

since $\beta = 2\pi/\lambda$ and

$$\begin{aligned}e^{-j2\beta(\lambda/4)} &= e^{-j\pi} \\ &= -1\end{aligned} \tag{C-25}$$

Note that (C-24) is precisely the inverse of (C-23). This provides a convenient means of transforming an impedance to its corresponding admittance. If we wish to invert an impedance (or complex number) $\hat{Z}$, we may arbitrarily choose a characteristic resistance R_C with which to normalize $\hat{Z}$ so that $\hat{z} = \hat{Z}/R_C$ lies at a position on the Smith chart at which we may accurately plot the point. Rotate $\hat{z}$ one-half revolution around the chart ($\lambda/4$) and read the value of $\hat{y} = 1/\hat{z}$. The inverse of $\hat{Z}$, $\hat{Y}$, is given by

$$\hat{Y} = \frac{\hat{y}}{R_C} \tag{C-26}$$

EXAMPLE C.4 Given an impedance $\hat{Z} = (100 + j150)\ \Omega$, find the inverse or admittance $\hat{Y} = 1/\hat{Z}$.

Solution First choose an R_C to place $\hat{z}$ in a region of the chart so that it may be accurately plotted. For example, a choice of $R_C = 100$ seems reasonable, resulting in $\hat{z} = 1 + j1.5$. The results are shown in Fig. C.9. The normalized admittance is $\hat{y} = 0.3 - j0.47$, or $\hat{Y} = \hat{y}/R_C = (0.3 - j0.47) \times 10^{-2}$ S. The exact value is $(0.308 - j0.462) \times 10^{-2}$ S.

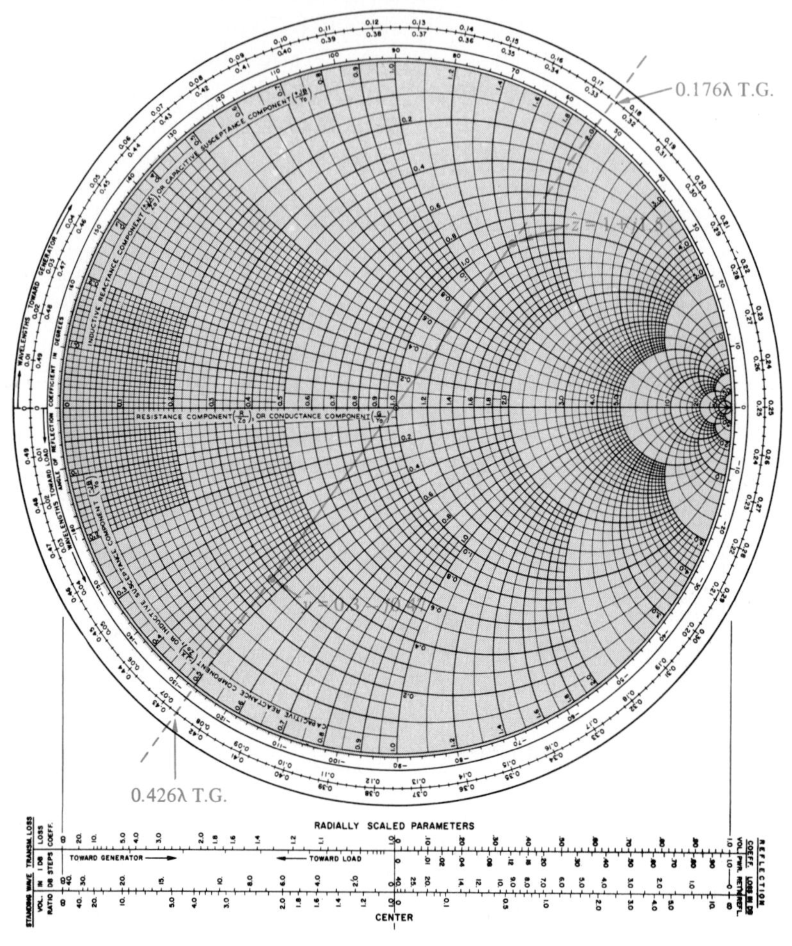

FIGURE C.9

Example C.4. Determining admittance from impedance with a Smith chart.

The above result also shows that input impedances separated by $\lambda/4$ on a line are related by

$$Z_{\text{in},1} = R_C^2/Z_{\text{in},2}$$

C.2 High-Frequency Measurement of Impedance

Suppose we wish to measure the input impedance to some device at a certain frequency. At first this may seem to be a rather straightforward measurement. However, the unknown impedance must inevitably be connected to the measurement device (impedance bridge, network analyzer, etc.) with a length of

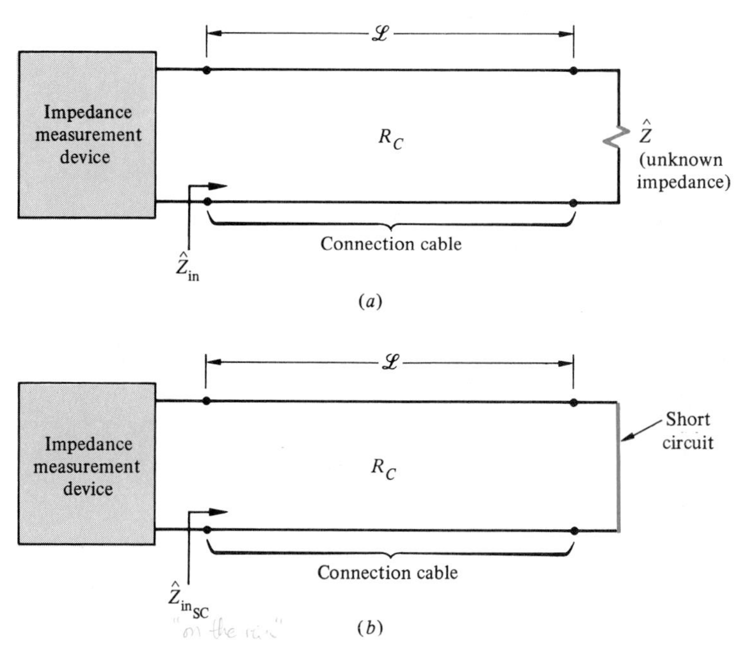

FIGURE C.10
Illustration of the effect of a connection cable on the high-frequency measurement of impedance.

transmission line or connection cable. The actual impedance being measured is not the unknown impedance—it is the impedance of the connection cable that is terminated in the unknown impedance. If the connection cable is a significant portion of a wavelength at the measurement frequency (e.g., longer than, say, $\frac{1}{20}\lambda$), then the impedance being measured is not the desired impedance. We must have a way of "calibrating out" the connection cable so that we measure only the desired unknown impedance. A method of doing this is summarized below.

Consider Fig. C.10. With the connection cable attached to the measurement device, we replace the unknown impedance terminating this cable with a short circuit, as shown in Fig. C.10*b*. With the short circuit attached, suppose the measurement device reads $\hat{Z}_{in,SC}$. Normalize this to R_C and plot the result, $\hat{z}_{in,SC}$, on the Smith chart. If we move toward the short (TL) until we locate an impedance of $0 + j0$, we will have determined the electrical length of the connection cable: this being the distance on the TL scale required to move from $\hat{z}_{in,SC}$ to $0 + j0$. Once this has been done, we may now determine the true value of the unknown impedance by reconnecting this impedance as shown in Fig. C.10*a* and determining $\hat{Z}_{in}$ read by the measurement device. Plot the normalized value, $\hat{z}_{in} = \hat{Z}_{in}/R_C$, on the Smith chart and move on the TL scale the electrical length of the line determined by the short-circuit test.

EXAMPLE C.5 Suppose the short-circuit test in Fig. C.10*b* yields $\hat{Z}_{in,SC} = j100\,\Omega$. (Note that the input impedance to a lossless line terminated in a short circuit or open circuit can have no real part, as was determinated in Sec. 7.2.)

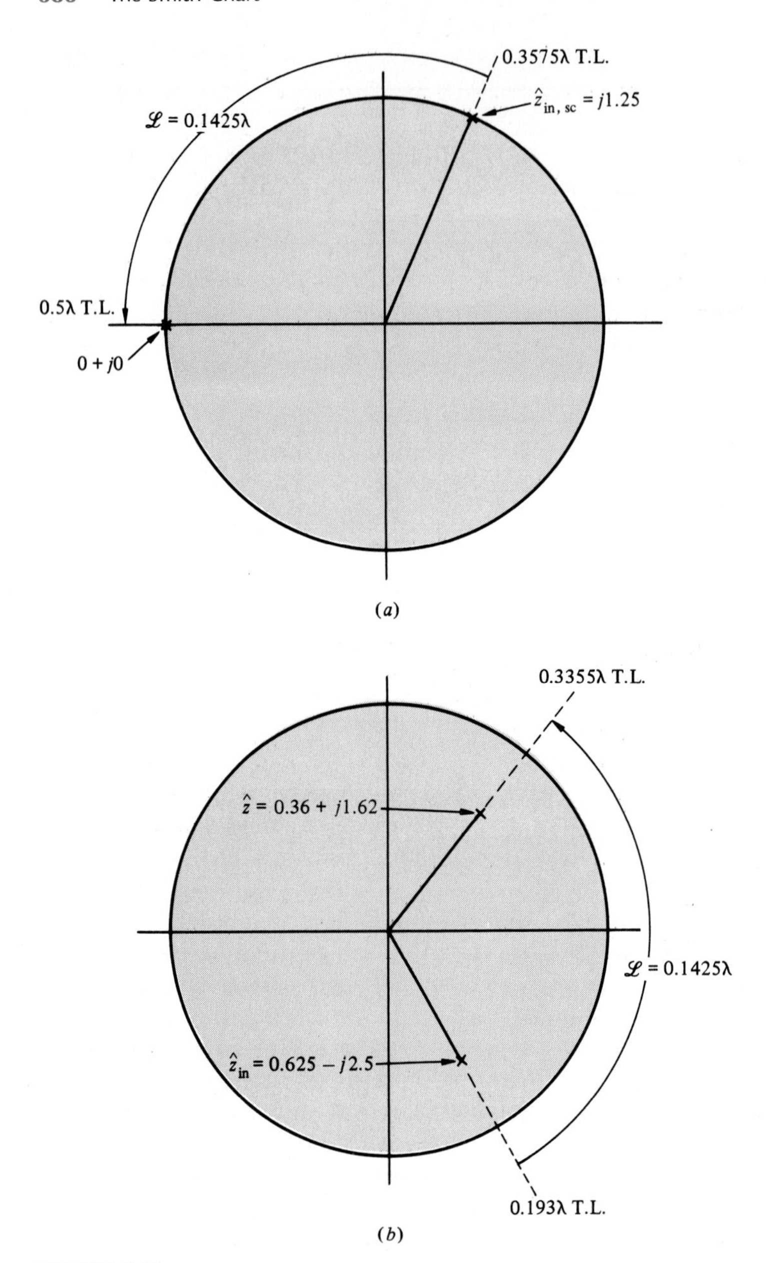

FIGURE C.11

Example C.5. High-frequency measurement of impedance.

Find the unknown impedance $\hat{Z}$ if the characteristic resistance of the line is $R_C = 80\ \Omega$ and the impedance measured with $\hat{Z}$ attached is $(50 - j200)\ \Omega$.

Solution The normalized input impedance is $\hat{z}_{\text{in,SC}} = j100/80 = j1.25$. Locating this on the Smith chart, we determine the electrical length of the line to be $0.5\lambda - 0.3575\lambda = 0.1425\lambda$, as shown in Fig. C.11*a*. If we now reconnect the unknown impedance to the cable, the measurement equipment reads $\hat{Z}_{\text{in}} = (50 - j200)\ \Omega$. Normalizing, we obtain $\hat{z}_{\text{in}} = 0.625 - j2.5$, which is plotted in Fig. C.11*b*. Rotating counterclockwise (toward the load) the electrical length of the line, we arrive at $0.193\lambda + 0.1425\lambda = 0.3355\lambda$ TL and obtain

$$\hat{z} = 0.36 + j1.62$$

The unknown impedance then becomes

$$\begin{aligned}\hat{Z} &= \hat{z}R_C \\ &= 28.8 + j129.6 \quad \Omega\end{aligned}$$

and the effect of the connection cable has been "calibrated out."

C.3 Use of Slotted-Line Data

Accurate measurement of impedances at high frequencies, such as above 100 MHz, can be a difficult task. In Sec. C.2 we investigated a measurement method that utilized some impedance-measuring device, such as a bridge or network analyzer. An alternative method is the use of the slotted line, as shown in Fig. C.12*a*. The slotted line consists of a length of coaxial line with a longitudinal slot in the solid outer cylinder. A movable carriage containing a small probe that extends a short distance into the interior of the cable is mounted on the slot. A scale is attached to this arrangement to permit an accurate determination of the longitudinal position of the probe, and a detector is connected to the probe. The detector readings are directly related to the line voltage at corresponding points along the line so that a determination of the VSWR and position of voltage maxima and minima can be made.

The key to understanding the use of a slotted line is the important fact, developed in Sec. C.1 and Chap. 7, that at a point of voltage maximum (minimum) on the line, the normalized input impedance to the line at that point will be purely resistive and a maximum (minimum) and will be equal to the VSWR (1/VSWR). Consider Fig. C.12*b*. We may measure the VSWR by moving the probe along the slotted line and taking the ratio of the largest to smallest readings. At points labeled A, $\hat{z}_{\text{in}} = 1/\text{VSWR}$; at points B, $\hat{z}_{\text{in}} = \text{VSWR}$. Thus, we know and can plot on the Smith chart the value of the normalized input impedance at each of these points. In order to find the normalized value of load

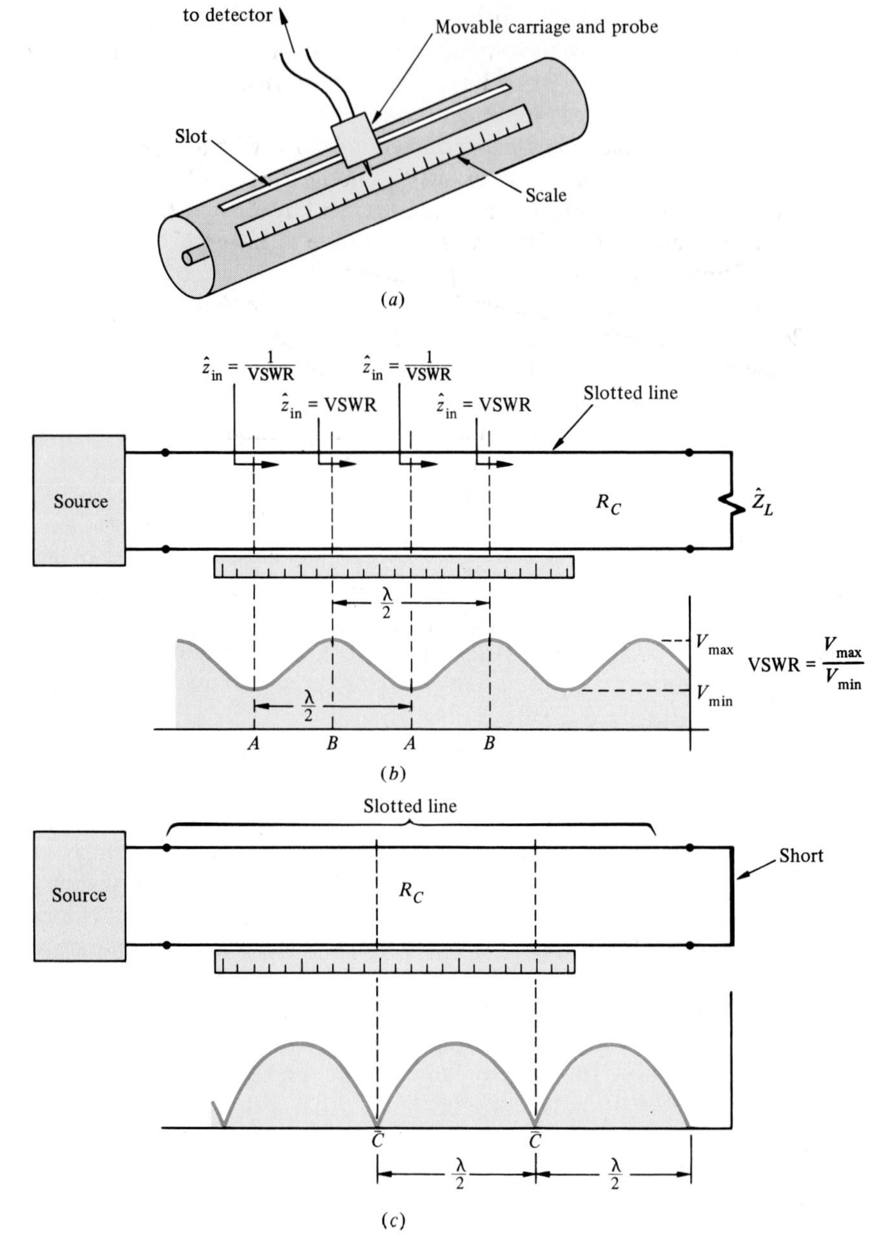

FIGURE C.12

Measurement of impedance using a slotted line.

impedance $\hat{z}_L$, we need only know the electrical distance from any of these points to the load. If we know the exact distance from any of these points to the load, we can simply rotate the plotted point $\hat{z}_{in}$ this electrical distance on the TL scale to find $\hat{z}_L$.

The distance to the load is found by replacing the load with a short circuit, as shown in Fig. C.12*c*. Actually, the exact distance to the load is not needed because of the following important result: We found in Chap. 7 that the input impedance to a section of lossless transmission line replicates for every half-wavelength of line; that is, the input impedances to the line at points separated by multiples of a half-wavelength are identical. From the short-circuit test in Fig. C.12*c*, we see that, effectively, the load can be thought of as being positioned at any of the points labeled *C*. Locating the position of any of these effective positions on the scale, we may now move $\hat{z}_{in}$ determined at *A* or *B* on the Smith chart toward the nearest position *C*. If we move from position *A* or *B* to the right to arrive at *C*, we move on the TL scale of the Smith chart. If we move from *A* or *B* to the left to arrive at *C*, we move on the TG scale of the Smith chart.

With the load replaced by a short circuit, as in Fig. C.12*c*, we may determine the length along the scale corresponding to $\lambda/2$ by moving the probe between the voltage nulls and may thus determine a wavelength. (Maxima would also suffice but nulls are more easily located than maxima since the variation of voltage with distance is sharper around a null than around a maximum, as evidenced in Fig. C.12*c*.) Slotted lines are typically air-filled (which simplifies movement of the probe); thus, the frequency, which we do not accurately know, may also be determined by dividing the velocity of propagation, $u_0 \simeq 3 \times 10^8$ m/s, by the wavelength previously determined by the short-circuit test.

EXAMPLE C.6 Suppose that, in the short-circuit test, points *C*, the locations of the voltage nulls, are at 10 cm, 30 cm, and 50 cm on the scale, as shown in Fig. C.13*a*, and that when the short circuit is replaced by the unknown load impedance, the minimum voltage points are found to be at 25 cm, 45 cm, and 65 cm and the maximum voltage points at 15 cm, 35 cm, and 55 cm. The VSWR is determined to be 4. If the characteristic resistance of the slotted line is 50 Ω, determine the value of the unknown impedance and the frequency of operation.

Solution From the short-circuit test we find that

$$\frac{\lambda}{2} = 20 \text{ cm} \qquad \text{or} \qquad \lambda = 40 \text{ cm}$$

The frequency of operation is found from

$$f = \frac{u_0}{\lambda}$$

$$= \frac{3 \times 10^8}{40 \times 10^{-2}} = 750 \text{ MHz}$$

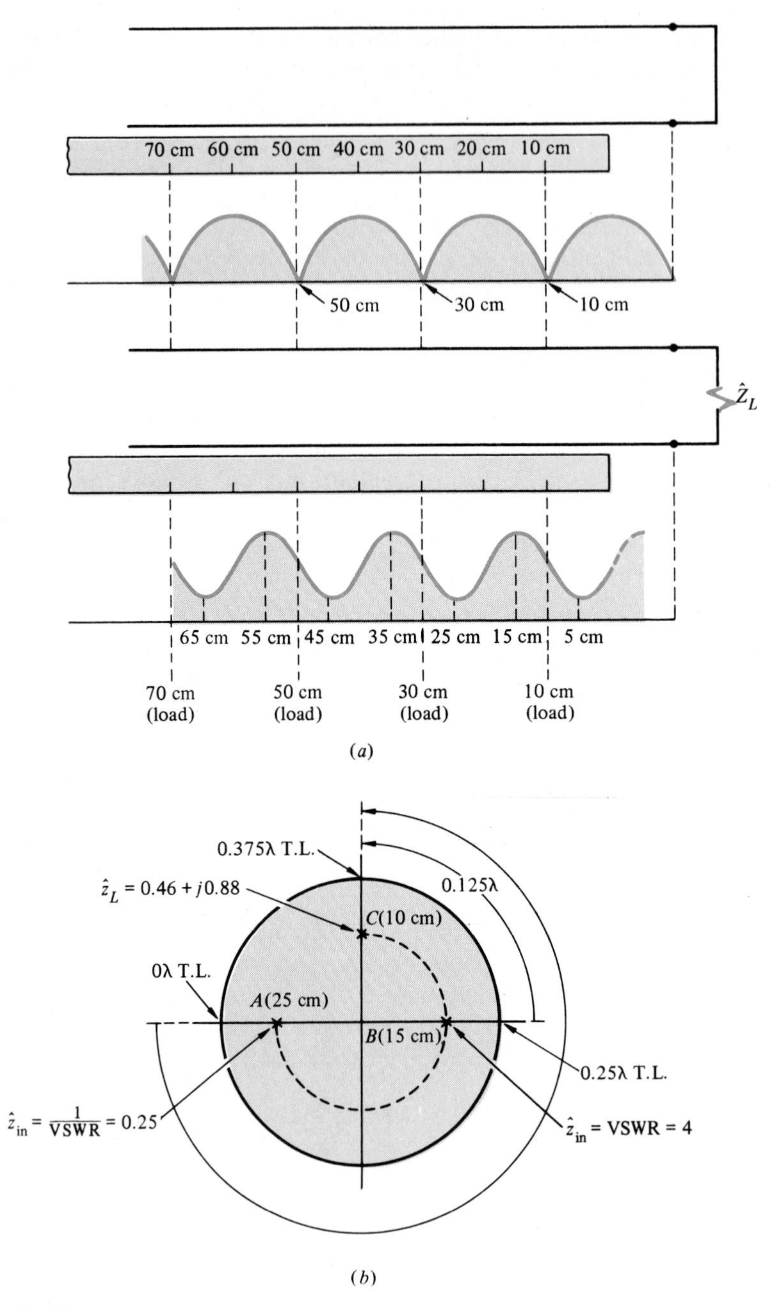

FIGURE C.13
Example C.6. Use of slotted-line data.

At a point of voltage maximum (15 cm) we move to the right to arrive at 10 cm, the nearest effective position of the load. This requires a movement of the Smith chart a distance of $\frac{5}{40}\lambda = 0.125\lambda$ on the TL scale, as shown in Fig. C.13*b*. This results in

$$\hat{z}_L = 0.46 + j0.88$$

and

$$\begin{aligned}\hat{Z}_L &= \hat{z}_L R_C \\ &= (0.46 + j0.88) \times 50 \\ &= 23 + j44 \quad \Omega\end{aligned}$$

We might also move from a point of voltage minimum at A to C. At 25 cm, $\hat{z}_{\text{in}} = 1/\text{VSWR} = 0.25$, which is also plotted on the Smith chart in Fig. C.13*b*. We now move from A (25 cm) to C (10 cm) an electrical distance of

$$\frac{25 - 10}{40}\lambda = \frac{15}{40}\lambda = 0.375\lambda$$

as shown in Fig. C.13*b* and obtain the same answer for $\hat{Z}_L$ as before.

C.4 Transmission-Line Matching

It was clear from the results of Sec. 7.2 that, for a lossless transmission line connecting a source ($\hat{V}_S, \hat{Z}_S$) and a load ($\hat{Z}_L$), maximum power would be transferred from the source to the load only if the input impedance to the line, $\hat{Z}_{\text{in}}$, was the complex conjugate of the source impedance: that is, $\hat{Z}_{\text{in}} = \hat{Z}_S^*$. Now let us consider a special case for the source impedance. Suppose that the source impedance is real and equal to the characteristic resistance of the line, R_C, as shown in Fig. C.14*a*. (The majority of the commercially available high-frequency signal generators have source impedances of 50 Ω, and since most coaxial cables are 50-Ω cables, this assumption of $\hat{Z}_S = R_C$ is quite realistic.) If the load were matched to the line ($\hat{Z}_L = R_C$), then the input impedance would equal R_C and maximum power transfer to the line (and to the load if the line is lossless) would take place. The stub-matching problem that we will consider in this section is basically the determination of some auxiliary network which, when attached to the line, results in $\hat{Z}_{\text{in}} = R_C$ when the line is not matched ($\hat{Z}_L \neq R_C$).

C.4.1 Single-Stub Tuners

The basic idea behind all stub tuners can be illustrated by referring to Fig. C.14*b*. A point along the line can be found such that the input admittance $\hat{Y}_{\text{in}}$ has a real part equal to $1/R_C$ and some imaginary part X; that is,

$$\hat{Y}_{\text{in}} = \frac{1}{R_C} + jX \tag{C-27}$$

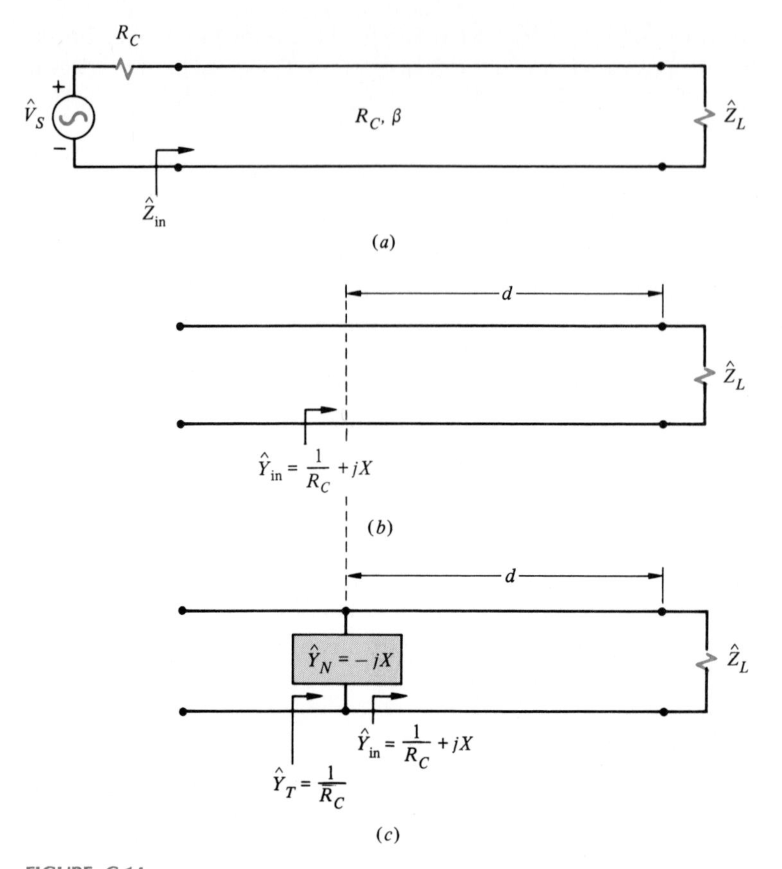

FIGURE C.14
Matching a transmission line using an auxiliary network.

This fact can be easily seen if we note that the admittance in (C-27) normalized to R_C is

$$\begin{aligned} \hat{y}_{\text{in}} &= \hat{Y}_{\text{in}} R_C \\ &= 1 + jXR_C \end{aligned} \tag{C-28}$$

Basically, then, we are stating that if we plot the normalized load admittance

$$\begin{aligned} \hat{y}_L &= \hat{Y}_L R_C \\ &= \frac{R_C}{\hat{Z}_L} \\ &= \frac{1}{\hat{z}_L} \end{aligned} \tag{C-29}$$

on a Smith chart, we may always find a distance down the line such that $\hat{y}_L$—when rotated around the Smith chart by this distance—intersects the unity real-part circle.† Obviously, this may always be done. We do not know (or care) what value of the imaginary part, XR_C in (C-28), will result.

Now, if we move down the line a distance d such that $\hat{Y}_{\text{in}}$ is given by (C-27), we simply need to place a network that has an admittance of $\hat{Y}_N = -jX$ across (in parallel with) the line at this point to cancel the reactive portion of $\hat{Y}_{\text{in}}$, as shown in Fig. C.14*c*. The combined input admittance of the line of length d and the parallel network is

$$\begin{aligned}\hat{Y}_T &= \hat{Y}_{\text{in}} + \hat{Y}_N \\ &= \left(\frac{1}{R_C} + jX\right) - jX \\ &= \frac{1}{R_C}\end{aligned} \tag{C-30}$$

and at all points to the left of this point of attachment the line appears matched.

This is the basic problem we will consider. The next question we need to answer is "What is the structure of the matching network?" Clearly this network must provide an admittance that is purely reactive (no real part), which may be either capacitive or inductive since X can be positive or negative. An obvious and simple way of doing this would be to parallel the line with either a capacitor or an inductor having the appropriate value of admittance. The difficulty with this idea is that these lumped elements do not behave as simple inductors or capacitors at frequencies above the high-megahertz range. Below this frequency range, however, they will usually be satisfactory for the matching network. In the higher-frequency ranges it is more desirable to use a section of transmission line called a *stub*, or *stub tuner* for this matching network. A section of line having R_C, β, and length l with a short-circuit load will also satisfy the requirements, since from Chap. 7 the input admittance is

$$\hat{Y}_N = -j\frac{1}{R_C \tan \beta l} \qquad \text{short-circuit load} \tag{C-31}$$

---

† The normalized input admittance is

$$\hat{y}_{\text{in}} = \frac{1}{\hat{z}_{\text{in}}} = R_C \hat{Y}_{\text{in}} = \frac{1 - \hat{\Gamma}(z)}{1 + \hat{\Gamma}(z)}$$

Comparing this to (C-4) and (C-2) by which the Smith chart was constructed, we see that we can use the chart to deal with admittances rather than impedances if we add 180° to $\hat{\Gamma}(z)$. This amounts to rotating the chart by 180°. Actually, such rotation is not necessary since we are dealing with relative differences in distances.

Similarly, the input admittance of a line having R_C, β, and length l with an open-circuit load has an input admittance of

$$\hat{Y}_N = j\frac{\tan \beta l}{R_C} \quad \text{open-circuit load} \tag{C-32}$$

Although a lumped inductor or a lumped capacitor will suffice for the matching network at the lower frequencies of operation, we will consider only the use of short-circuited or open-circuited stubs.

The basic problem is then illustrated in Fig. C.15*a* for a short-circuited stub. The stub line is chosen to be of the same type as the line to be matched, although we could have selected a different type of line for the stub having a different R_C and β. The various admittances are shown as normalized to the characteristic resistance of the appropriate line:

$$\hat{y} = \hat{Y}R_C \tag{C-33}$$

We first plot the normalized load impedance $\hat{z}_L = \hat{Z}_L/R_C$ on the Smith chart and rotate 180° to find the normalized load admittance $\hat{y}_L = \hat{Y}_L R_C$, as shown in Fig. C.15*b* (step 1). Then we move distance d_1 (toward the generator) until we intersect the unity real-part circle (step 2). At this point the normalized input impedance to the line is $\hat{y}_{\text{in}} = 1 + jx$. In other words, the actual admittance is given by (C-27). We could also have rotated further a total distance d_2 and found an input admittance of $\hat{y}_{\text{in}} = 1 - jx$, which is also shown in Fig. C.15*b*.

Now we need to find that length of a short-circuited stub which will cancel the reactive part of this input admittance. If we place the stub at distance d_1 from the load, the stub must have a normalized input admittance of $\hat{y}_N = -jx$; denote this length of stub as l_1. If we place the stub at distance d_2 from the load, the stub must have a normalized input admittance of $\hat{y}_N = +jx$; denote this length of stub as l_2. Determining these stub lengths is now a simple matter. Plot the stub load admittance (normalized), which is $\infty + j\infty$, as shown in Fig. C.15*b*. We move toward the generator along the stub until we find this reactance (on the outer circle of the chart) (step 3). For d_1 we need $\hat{y}_N = -jx$, or a stub length l_1. If we had chosen d_2, we would need $\hat{y}_N = +jx$, or a stub length l_2. Since the stub is assumed lossless, maximum power will be transferred from the source having $\hat{Z}_S = R_C$ to the load $\hat{Z}_L$.

Open-circuited stubs could have been used instead, but they are more difficult to construct and are not ordinarily used. The only difference that open-circuited stubs would make in the analysis is that since the admittance of an open circuit is $0 + j0$, we would start at the left side of the chart for the rotation in step 3 instead of at the right side (as in Fig. C.15*b*). The distances of stub placement, d_1 and d_2, would remain unchanged.

EXAMPLE C.7 Consider a transmission line having $R_C = 50\ \Omega$ and $\hat{Z}_L = (25 - j50)\ \Omega$. Determine two locations and lengths of short-circuited stubs to match this line.

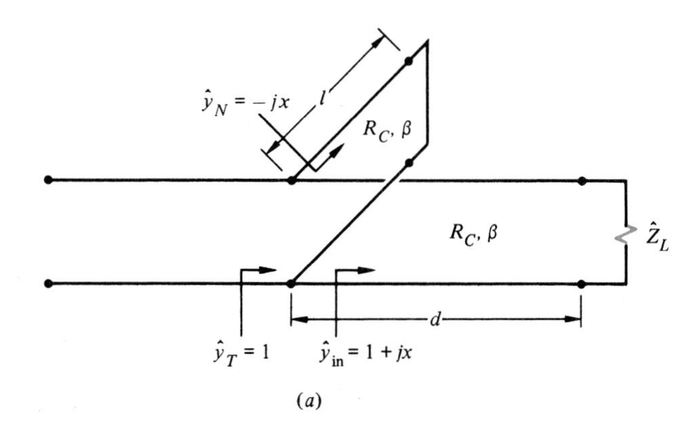

(a)

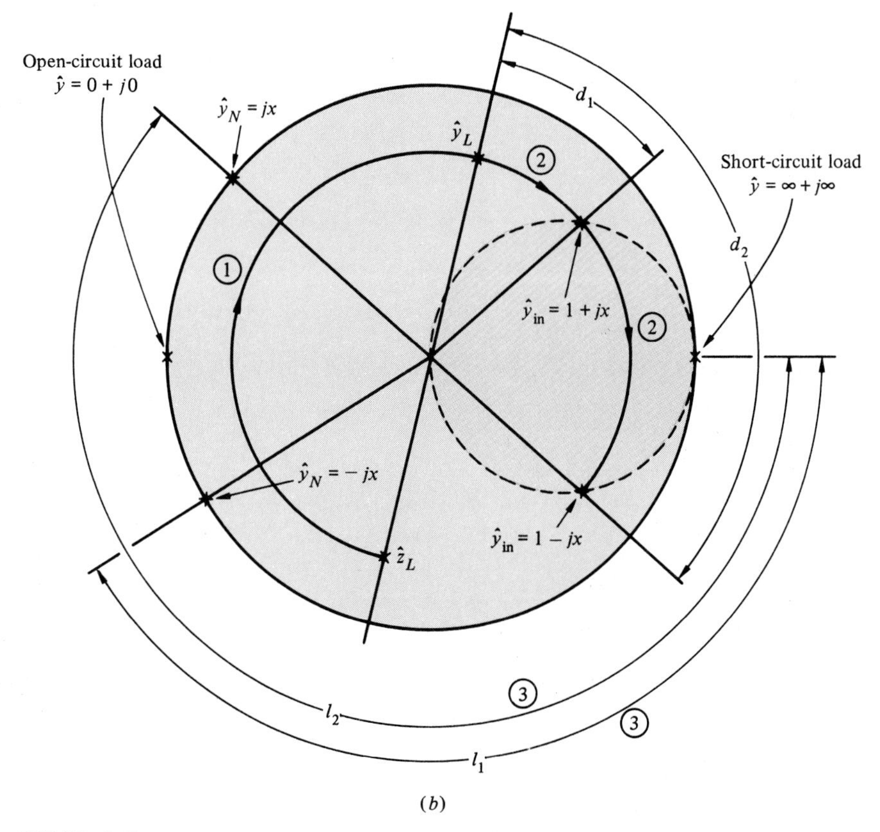

(b)

FIGURE C.15

Illustration of the use of parallel, short-circuit stub tuners for matching.

Solution The normalized impedance is

$$\hat{z}_L = 0.5 - j1.0$$

and is plotted in Fig C.16*a*. The normalized load admittance is

$$\hat{y}_L = 0.4 + j0.8$$

and occurs at 0.115λ on the TG scale. Rotating to the unity real-part circle produces

$$\hat{y}_{\text{in}} = 1 + j1.6$$

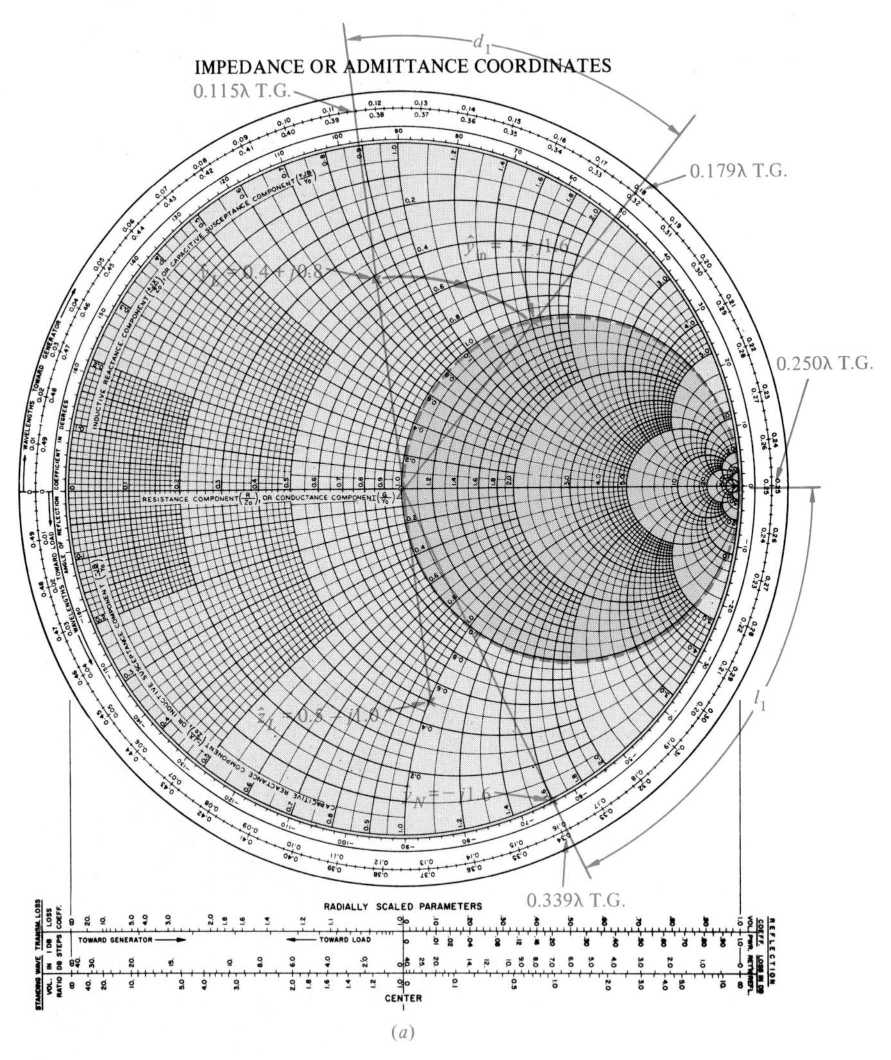

(*a*)

FIGURE C.16

Example C.7. Stub tuner matching.

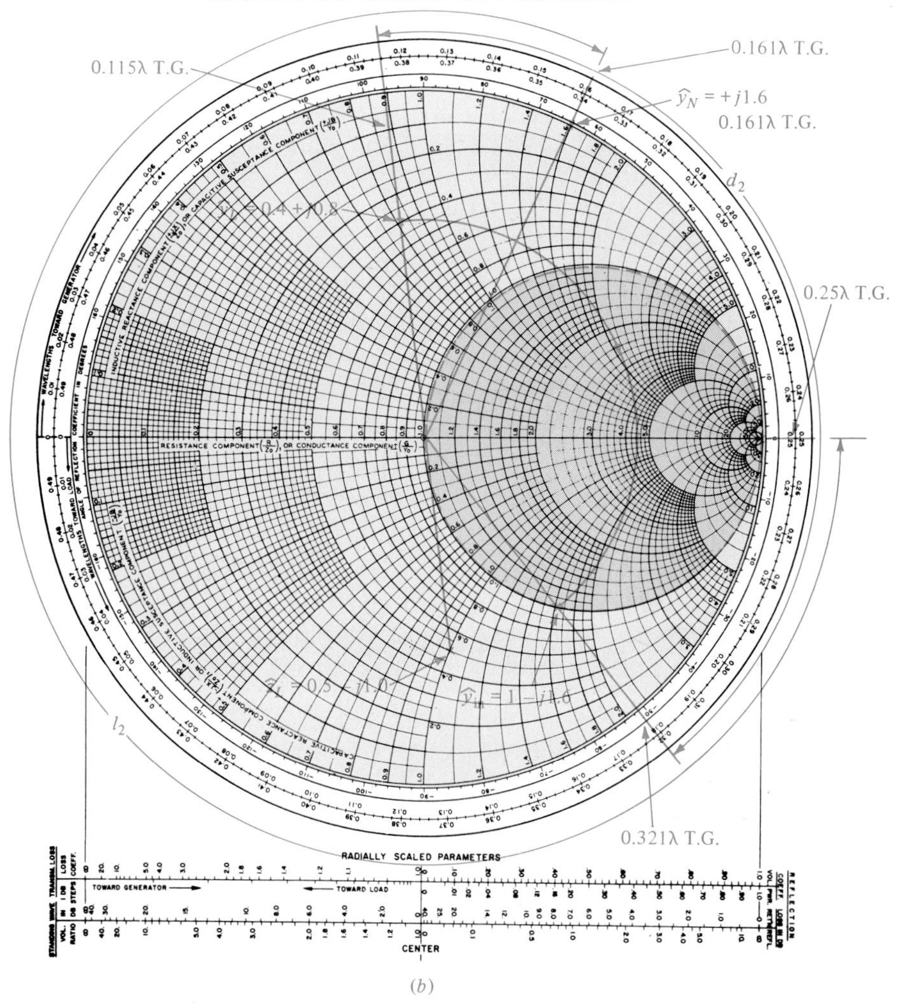

(*b*)

located at 0.179λ on the TG scale. Thus

$$d_1 = 0.179\lambda - 0.115\lambda = 0.064\lambda$$

The length of one short-circuited stub is

$$l_1 = 0.339\lambda - 0.250\lambda = 0.089\lambda$$

The other possible solution is illustrated in Fig. C.16*b* and is

$$d_2 = 0.321\lambda - 0.115\lambda = 0.206\lambda$$

$$l_2 = 0.161\lambda + 0.25\lambda = 0.411\lambda$$

If we had chosen to use open-circuited stubs, the distances of stub placement from the load, d_1 and d_2, would remain unchanged. However, the lengths of the stubs would be different. For stub placement distance $d_1 = 0.064\lambda$, we need to obtain a stub input admittance of $-j1.6$. From Fig. C.16*a*, if we start at an admittance of the open circuit of $0 + j0$ and move toward the generator to obtain $y_N = -j1.6$, we simply add 0.25λ to the short-circuited stub length:

$$d_1 = 0.064\lambda$$

$$l_1 = 0.089\lambda + 0.25\lambda = 0.339\lambda$$

Similarly, for the second solution for stub placement, $d_2 = 0.206\lambda$, we require an open-circuited stub length yielding an input admittance of $j1.6$. From Fig. C.16*b* the stub length is 0.25λ less than the corresponding short-circuited stub length:

$$d_2 = 0.206\lambda$$

$$l_2 = 0.411\lambda - 0.25\lambda = 0.161\lambda$$

Single-stub tuners are easily constructed for coaxial cables, as shown in Fig. C.17. A section of cable is soldered to the original line and the inner conductors are connected together. A shorting plug or plunger may be moved inside the stub section so that the length of the shorted stub may be varied.

C.4.2 Double-Stub Tuners

Single-stub tuners can theoretically be used to match any load impedance, but the practical difficulty with their use is that changing the load impedances requires changing the position of the stub. For the case of the coaxial cable shown in Fig. C.17, this is obviously very inconvenient. Double-stub tuners considered in this section use two short-circuited stubs located at fixed positions on the line, as shown in Fig. C.18. One stub is located at the load, and the other is located a distance d from the load. As before, we will assume that the lines used to construct the stubs are of the same type as the original line (although the

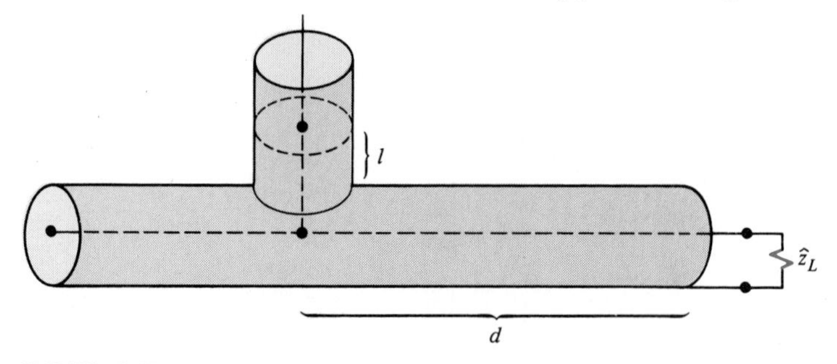

FIGURE C.17

Illustration of a stub tuner for a coaxial cable.

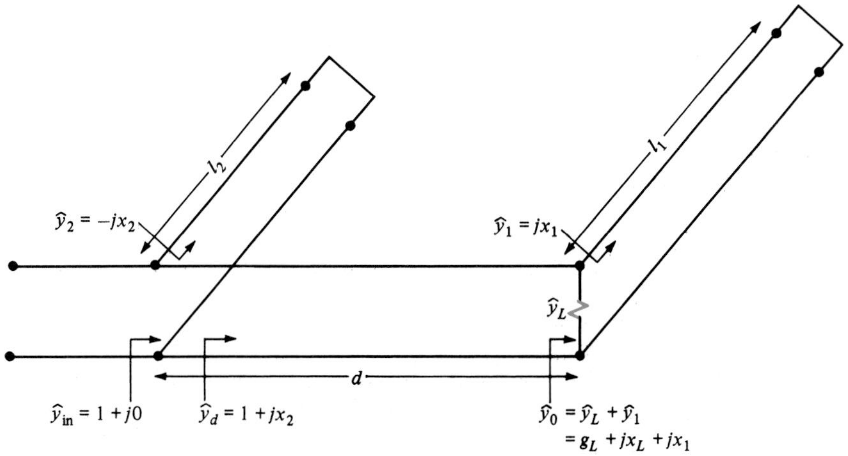

FIGURE C.18
Matching with a double-stub tuner.

following procedure can be adapted to situations where the lines are not all the same type).

The key to understanding the double-stub tuner lies in determining the desired result and "working backward" to uncover the procedure. For example, in Fig. C.18 we need an input admittance to the left of the second stub of

$$\hat{y}_{\text{in}} = 1 + j0 \tag{C-34}$$

If the input admittance of the second stub is

$$\hat{y}_2 = -jx_2 \tag{C-35}$$

then the input admittance to the right of this second stub is

$$\begin{aligned}\hat{y}_d &= 1 + jx_2 \\ &= \frac{1 - \hat{\Gamma}_d}{1 + \hat{\Gamma}_d}\end{aligned} \tag{C-36}$$

If the load admittance is

$$\hat{y}_L = g_L + jx_L \tag{C-37}$$

and the admittance of the first stub is

$$\hat{y}_1 = jx_1 \tag{C-38}$$

then the admittance to the left of the load is

$$
\begin{aligned}
\hat{y}_0 &= \hat{y}_L + \hat{y}_1 \\
&= g_L + j(x_L + x_1) \\
&= \frac{1 - \hat{\Gamma}_0}{1 + \hat{\Gamma}_0}
\end{aligned} \tag{C-39}
$$

The reflection coefficient immediately to the left of the load, $\hat{\Gamma}_0$ in (C-39), and the reflection coefficient immediately to the right of the second stub position, $\hat{\Gamma}_d$ in (C-36), are related by (C-2):

$$\hat{\Gamma}_d = \hat{\Gamma}_0 e^{-j2\beta d} \tag{C-40a}$$

or

$$\hat{\Gamma}_0 = \hat{\Gamma}_d e^{j2\beta d} \tag{C-40b}$$

The locus of $\hat{y}_d$ in (C-36) is the unity real-part matching circle. However, because of the relationship in (C-40), *the locus of $\hat{y}_0$ is the unity real-part matching circle rotated toward the load an angle of $2\beta d$.*

Therefore, we perform the following steps in the reverse order of the above:

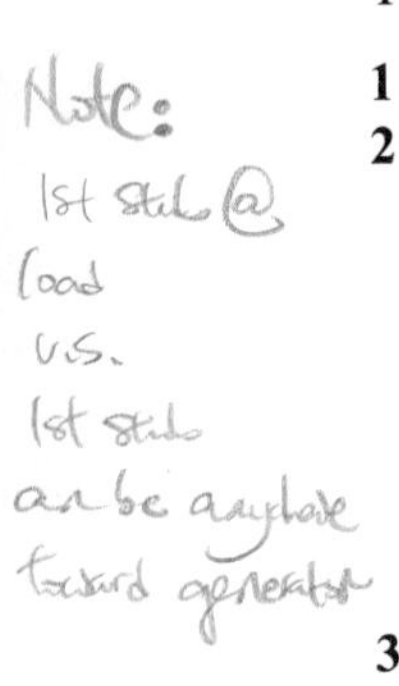

1. Select a distance d at which to place the second stub.
2. Draw the unity real-part matching circle rotated a distance of $2\beta d$ or d in wavelengths toward the load, as shown in Fig. C.19. This construction is quite simple if we realize that the unity real-part matching circle has a radius of 0.5 centered on the $r = 3$, $x = 0$ point. (See Fig. C.2.) Therefore, once we draw the radial line through the unity real-part matching circle and rotate it toward the load $2\beta d$, we may use a compass to bisect the line, and the displaced unity real-part circle can be drawn.
3. Plot the normalized load admittance $\hat{y}_L = g_L + jx_L$.
4. Rotate this point on a constant g_L circle until it intersects the rotated unity real-part $\hat{y}_0$ circle. This step essentially determines the sum $x_L + x_1$.
5. Read the admittance

$$\hat{y}_0 = g_L + jx_L + jx_1$$

6. Determine the required length of the first stub to yield this value of

$$\hat{y}_1 = jx_1$$

7. Rotate this $\hat{y}_0$ point on the constant-radius circle a distance $2\beta d$ toward the generator until it intersects the unity real-part matching ($\hat{y}_d$) circle.

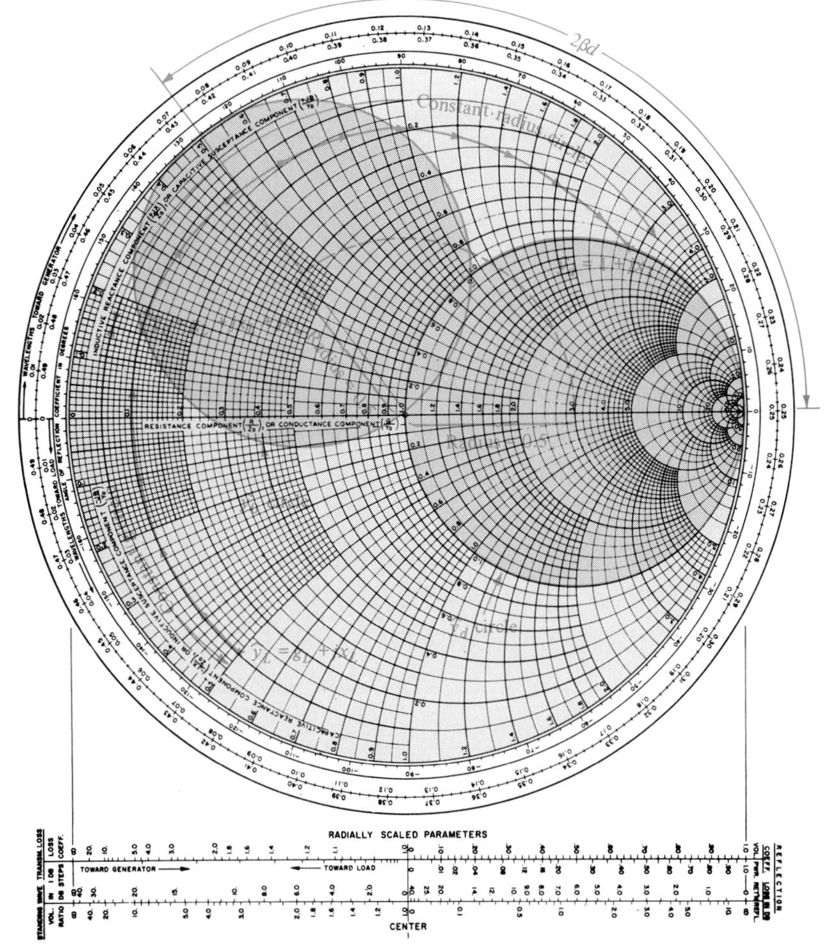

FIGURE C.19

Illustration of the use of a Smith chart in the design of a double-stub tuner.

8 Read

$$\hat{y}_d = 1 + jx_2$$

9 Determine the length of the second stub to produce an input admittance of

$$\hat{y}_2 = -jx_2$$

Note that two intersections with the $\hat{y}_0$ circle are possible.

EXAMPLE C.8 Design a double-stub tuner to match a load of $\hat{Z}_L = (25 + j40)\ \Omega$ to a 50-Ω line. Position the second stub $\frac{1}{8}\lambda$ from the load.

Solution Plot the normalized load impedance

$$\hat{z}_L = 0.5 + j0.8$$

and rotate 180° to obtain

$$\hat{y}_L = 0.562 - j0.899$$

as shown in Fig. C.20. Rotate the unity matching ($\hat{y}_d$) circle $\lambda/8$ toward the load to obtain the $\hat{y}_0$ circle. Rotate $\hat{y}_L$ to intersect this circle, giving

$$\hat{y}_0 = 0.562 + j0.12$$

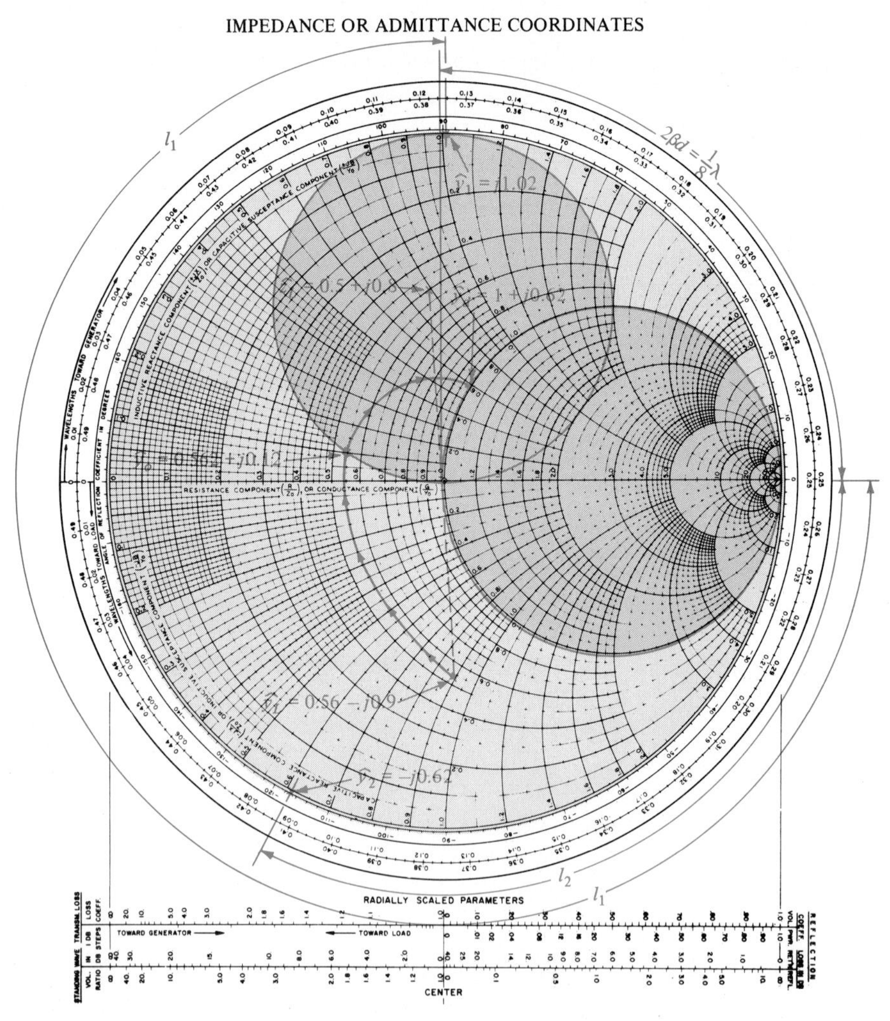

FIGURE C.20
Example C.8. Double-stub tuner matching.

Since $x_L = -j0.9$, the admittance of the first stub must be

$$\begin{aligned}\hat{y}_1 &= jx_1 \\ &= j0.12 - (-j0.9) \\ &= j1.02\end{aligned}$$

This gives a required length of the first stub of

$$\begin{aligned}l_1 &= 0.25\lambda + 0.126\lambda \\ &= 0.376\lambda\end{aligned}$$

Rotating the $\hat{y}_0$ point on a constant-radius circle until it intersects the $\hat{y}_d$ circle yields

$$\hat{y}_d = 1 + j0.62$$

Thus, the admittance of the second stub is

$$\hat{y}_2 = -j0.62$$

The length of the second stub is

$$\begin{aligned}l_2 &= 0.413\lambda - 0.25\lambda \\ &= 0.163\lambda\end{aligned}$$

A second solution gives

$$\begin{aligned}\hat{y}_0 &= 0.562 + j1.87 \\ \hat{y}_d &= 1 - j2.6 \\ \hat{y}_1 &= j2.77 \\ \hat{y}_2 &= j2.6 \\ l_1 &= 0.194\lambda + 0.25\lambda \\ &= 0.444\lambda \\ l_2 &= 0.191\lambda + 0.25\lambda \\ &= 0.441\lambda\end{aligned}$$

Not all load impedances can be matched for a specific choice of d. In Example C.8, normalized load admittances having a real part greater than 2 will not have intersections with the $\hat{y}_0$ circle. However, this problem can be remedied by choosing d such that an intersection with the $\hat{y}_0$ circle is possible.

C.4.3 Quarter-Wave Transformers

An alternative method of matching is achieved by inserting a quarter-wave length of transmission line having a characteristic resistance R_c' in front of the load, as shown in Fig. C.21. The input impedance to a quarter-wavelength section of line having a characteristic impedance R_c' is given by (C-1):

$$\hat{Z}_{\text{in}} = R_c' \frac{1 - \hat{\Gamma}_L}{1 + \hat{\Gamma}_L} = \frac{R_c'^2}{\hat{Z}_L} \tag{C-41}$$

where we have used (C-2) to obtain the reflection coefficient at the input and $2\beta d = \pi$ for $d = \lambda/4$. Thus, in order to match $\hat{Z}_L$ to a line having a characteristic impedance R_c, we may use a line having

$$R_c' = \sqrt{R_c \hat{Z}_L} \tag{C-42}$$

As an example, to match a 300-Ω load to a 75-Ω line requires a $\lambda/4$ section of line having $R_c' = 150\ \Omega$. Note that the VSWR on the line that is matched is unity, whereas it is not unity on the matching section. For this example, VSWR $= 2$ on the matching section.

This technique is obviously feasible only for resistive loads. Also, we must calculate the length of this line by using the wavelength λ' for this line, which is the same wavelength as that of the line to be matched only if both lines have the same dielectric material. This method also holds for only one frequency, as do the stub tuners of the previous sections.

C.4.4 Broadband Matching and Pads

Broadband matching (matching over a wide frequency range) is a more difficult problem. Typically, resistive pads (illustrated in Fig. C.22) are used to

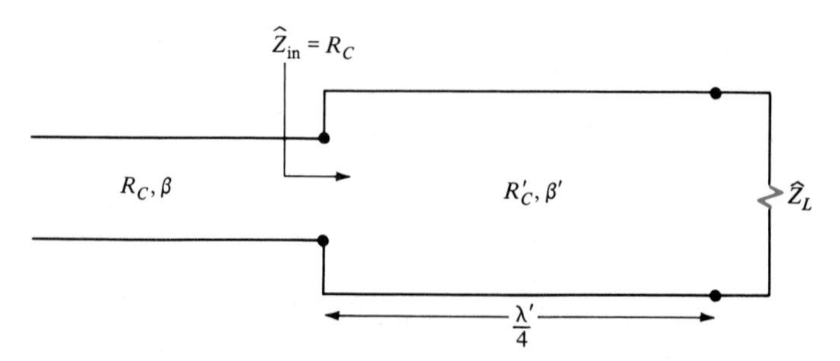

FIGURE C.21
Illustration of a quarter-wave transformer.

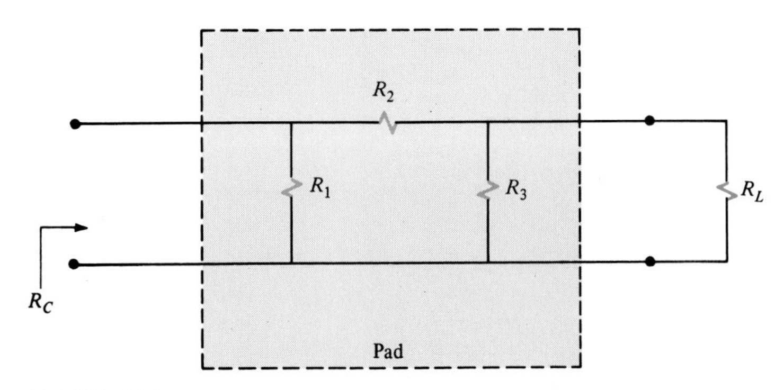

FIGURE C.22
A broadband matching network or pad.

match loads over a wide frequency range. Being resistive, these pads give insertion loss not present with the previous methods and are quite expensive. For example, a typical precision 50-Ω 6-dB pad (6-dB insertion loss), which may be used from dc to 1 GHz, costs over \$100 (U.S.). The resistor values used to produce a 6-dB 50-Ω pad are shown in Fig. C.23*a*. A commercially available, precision 50-Ω 10-dB pad is shown in Fig. C.23*b*. The insertion loss is defined as the ratio of the power delivered to the load with and without the pad:

$$\text{Insertion loss (dB)} = 10 \log \frac{P_{L,\,\text{without pad}}}{P_{L,\,\text{with pad}}} \tag{C-43}$$

A rudimentary idea of how the pad may accomplish matching for a wide range of load impedances is this: If we choose R_3 in Fig. C.22 to be much less than the value of all possible values of load impedance used and choose R_2 to be much greater than R_3, then the input impedance seen looking into the pad is

$$\begin{aligned} R_{\text{in}} &= R_1 \| (R_2 + R_3 \| R_L) \\ &\simeq R_1 \| R_2 \end{aligned} \tag{C-44}$$

Of course, the smaller R_3, the wider the range of load resistances that may be accommodated. The larger R_2, the less important R_3 becomes. However, both options result in increased insertion loss. Typically, the larger the insertion loss that one can tolerate, the better the matching ability of the pad.

The structures shown in Figs. C.22 and C.23 are referred to as "pi" structures, with reference to their similarity to π. It is also possible to choose other structures. A dual to the pi structure is the "tee" structure, with reference to its similarity to the letter T. The resistors R_1 and R_3 (usually with $R_1 = R_3$) form the top arms of the T, and R_2 is connected to the lower conductor between R_1 and R_3.

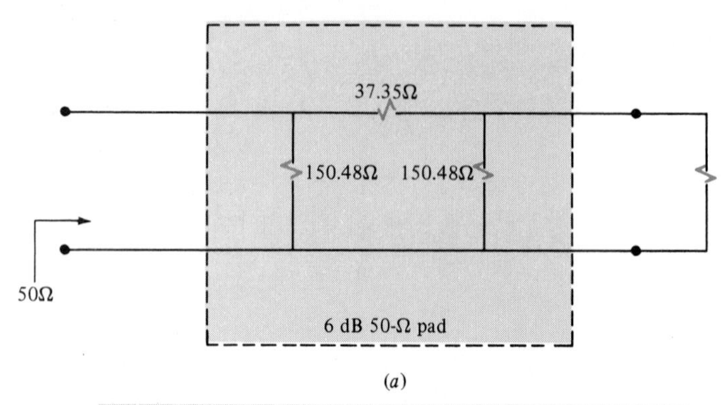

(*a*)

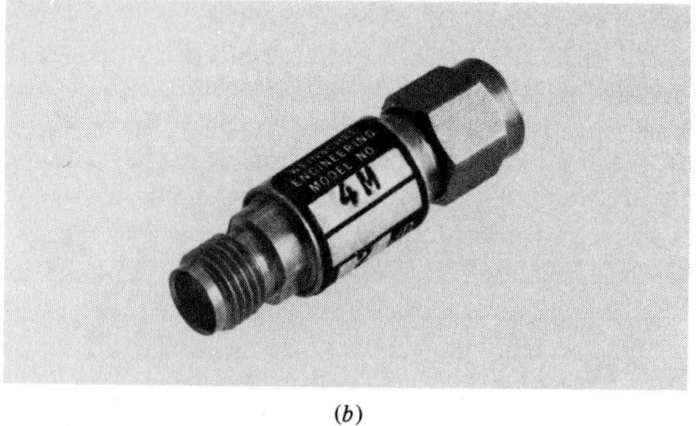

(*b*)

FIGURE C.23

A 6-db, 50-Ω pad. *(a)* Element values, *(b)* physical configuration. (Courtesy of Weinschel Engineering, Gaithersburg, MD.)

EXAMPLE C.9 Design a 6-dB 75-Ω pad.

Solution We may choose the pi structure in Fig. C.22. For the input impedance to be the characteristic resistance of the line to be matched, R_C, we require

$$\frac{1}{R_C} = \frac{1}{R_1} + \frac{1}{R_2 + R_3 \| R_L} \tag{C-45a}$$

Also,

$$\text{IL} = 20 \log \frac{1}{\left[\dfrac{R_3 \| R_L}{R_2 + R_3 \| R_L}\right]} \tag{C-45b}$$

where IL denotes the insertion loss (in dB). Rewriting (C-45*b*) gives

$$\frac{R_2 + R_3 \| R_L}{R_3 \| R_L} = 10^{\text{IL}/20} \tag{C-46}$$

Substituting (C-45a) gives

$$\frac{1}{R_3 \| R_L} = \left(\frac{1}{R_C} - \frac{1}{R_1}\right) 10^{\mathrm{IL}/20} \tag{C-47}$$

To simplify the solution, we will choose $R_3 = R_1$. Substituting gives

$$\frac{R_1 + R_L}{R_1 R_L} = \left(\frac{1}{R_C} - \frac{1}{R_1}\right) 10^{\mathrm{IL}/20} \tag{C-48}$$

Solving for R_1 gives

$$R_1 = \frac{R_L(1 + X)}{(R_L/R_C)X - 1} \tag{C-49}$$

where

$$X = 10^{\mathrm{IL}/20} \tag{C-50}$$

Substituting IL = 6 dB, $R_L = R_C$, and $R_C = 75\ \Omega$ gives

$$R_1 = 225.71\ \Omega$$
$$= R_3$$

The value of R_2 can be obtained from (C-46) as

$$R_2 = (R_3 \| R_L)(X - 1)$$
$$= 56.03\ \Omega$$

If this pad is used to match a load that is not 75 Ω (for which the pad was designed), the VSWR on the 75-Ω line will not be unity. However, the pad still provides some matching. For example, the input impedance to the pad for a short-circuit load will be 44.89 Ω, whereas for an open-circuit load it will be 125.32 Ω. For either case, the VSWR on the 75-Ω line will be 1.67, which is quite good. For more realistic loads, the VSWR will be closer to unity.

C.5 Applications of Smith Charts to Uniform Plane Wave Propagation

Throughout our study of transmission lines in Chap. 7, we noted numerous direct analogies to the problem of uniform plane waves having normal incidence on plane boundaries (which were considered in Chap. 6). These analogies allow important simplifications of the computational details involved in uniform plane wave problems via the use of the Smith chart. Using the Smith chart for problems of this type also makes certain results immediately transparent.

Consider the case of a plane boundary separating two lossless regions, as shown in Fig. C.24. Forward- and backward-traveling uniform plane waves are

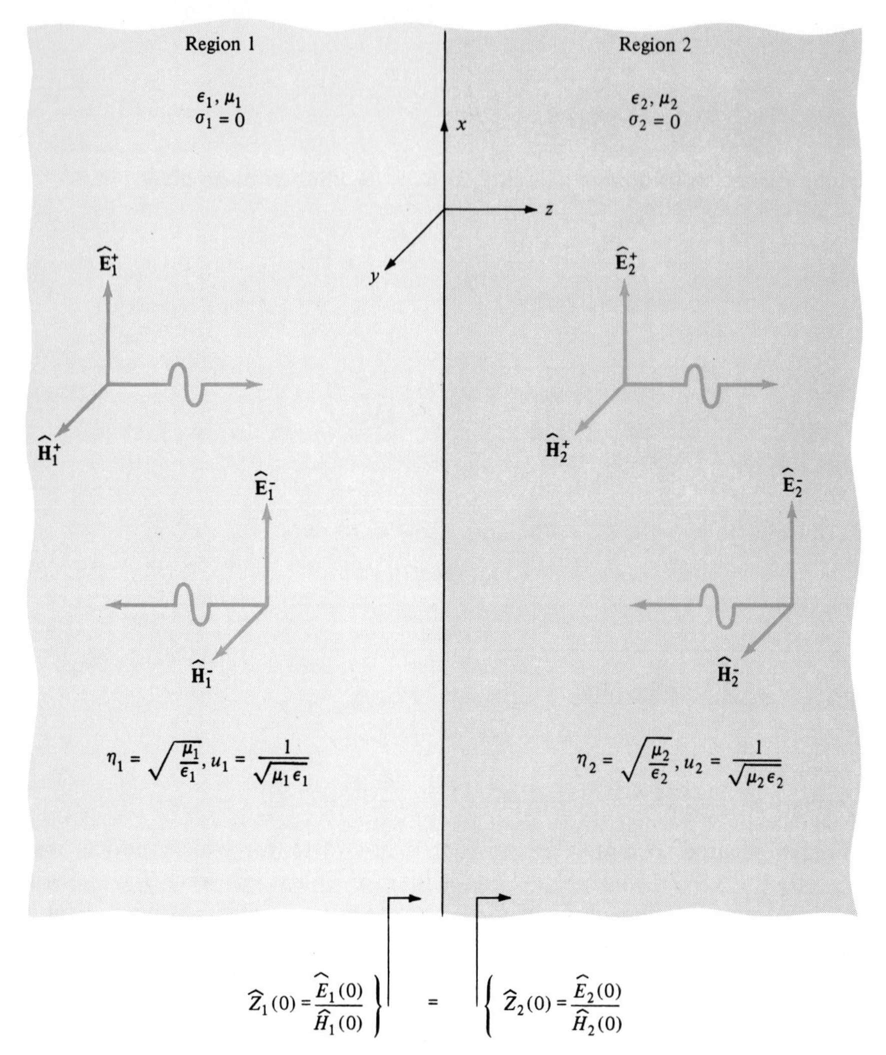

FIGURE C.24
Illustration of the use of wave impedances in problems involving uniform plane waves.

assumed to exist in each region, and these are propagating normal to the interface. The total fields in region 1 are

$$\hat{\mathbf{E}}_1(z) = \hat{\mathbf{E}}_1^+ + \hat{\mathbf{E}}_1^-$$

$$= (\hat{E}_1^+ e^{-j\beta_1 z} + \hat{E}_1^- e^{j\beta_1 z})\mathbf{a}_x \tag{C-51a}$$

$$\hat{\mathbf{H}}_1(z) = \hat{\mathbf{H}}_1^+ + \hat{\mathbf{H}}_1^-$$

$$= \left(\frac{\hat{E}_1^+}{\eta_1} e^{-j\beta_1 z} - \frac{\hat{E}_1^-}{\eta_1} e^{j\beta_1 z}\right)\mathbf{a}_y \tag{C-51b}$$

Similarly, in region 2 the total fields are

$$\hat{\mathbf{E}}_2(z) = \hat{\mathbf{E}}_2^+ + \hat{\mathbf{E}}_2^-$$
$$= (\hat{E}_2^+ e^{-j\beta_2 z} + \hat{E}_2^- e^{j\beta_2 z})\mathbf{a}_x \tag{C-52a}$$
$$\hat{\mathbf{H}}_2(z) = \left(\frac{\hat{E}_2^+}{\eta_2} e^{-j\beta_2 z} - \frac{\hat{E}_2^-}{\eta_2} e^{j\beta_2 z}\right)\mathbf{a}_y \tag{C-52b}$$

where the usual quantities are defined for each region:

$$\eta_1 = \sqrt{\frac{\mu_1}{\epsilon_1}}$$
$$= \sqrt{\frac{\mu_{r1}}{\epsilon_{r1}}}\,\eta_0 \tag{C-53a}$$
$$\beta_1 = \omega\sqrt{\mu_1\epsilon_1}$$
$$= \frac{\omega}{u_0}\sqrt{\mu_{r1}\epsilon_{r1}} \tag{C-53b}$$
$$\eta_2 = \sqrt{\frac{\mu_2}{\epsilon_2}}$$
$$= \sqrt{\frac{\mu_{r2}}{\epsilon_{r2}}}\,\eta_0 \tag{C-53c}$$
$$\beta_2 = \omega\sqrt{\mu_2\epsilon_2}$$
$$= \frac{\omega}{u_0}\sqrt{\mu_{r2}\epsilon_{r2}} \tag{C-53d}$$

We may define reflection coefficients in each region as the ratio of the backward- and forward-traveling waves:

$$\hat{\Gamma}_1(z) = \frac{\hat{E}_1^- e^{j\beta_1 z}}{\hat{E}_1^+ e^{-j\beta_1 z}}$$
$$= \frac{\hat{E}_1^-}{\hat{E}_1^+} e^{j2\beta_1 z} \tag{C-54a}$$
$$\hat{\Gamma}_2(z) = \frac{\hat{E}_2^- e^{j\beta_2 z}}{\hat{E}_2^+ e^{-j\beta_2 z}}$$
$$= \frac{\hat{E}_2^-}{\hat{E}_2^+} e^{j2\beta_2 z} \tag{C-54b}$$

In a manner similar to the case of transmission lines, we may define *wave impedances* as the ratio of the *total* electric and magnetic fields:

$$\hat{Z}_1(z) = \frac{\hat{E}_1(z)}{\hat{H}_1(z)} \tag{C-55a}$$

$$\hat{Z}_2(z) = \frac{\hat{E}_2(z)}{\hat{H}_2(z)} \tag{C-55b}$$

According to the general boundary conditions, the total electric and magnetic fields must be continuous across the boundary; thus

$$\hat{E}_1(0) = \hat{E}_2(0) \tag{C-56a}$$

$$\hat{H}_1(0) = \hat{H}_2(0) \tag{C-56b}$$

Therefore, the wave impedances must also be continuous across the boundary:

$$\hat{Z}_1(0) = \hat{Z}_2(0) \tag{C-57}$$

Using either (C-51), (C-54*a*), and (C-55*a*) or (C-52), (C-54*b*), and (C-55*b*), it is a simple matter to show that the reflection coefficient at some point in a medium can be found from the wave impedance at that point and from the intrinsic impedance of the medium in a manner directly analogous to the case of transmission lines:

$$\hat{\Gamma}_1(z) = \frac{\hat{Z}_1(z) - \eta_1}{\hat{Z}_1(z) + \eta_1} \tag{C-58a}$$

$$\hat{\Gamma}_2(z) = \frac{\hat{Z}_2(z) - \eta_2}{\hat{Z}_2(z) + \eta_2} \tag{C-58b}$$

For example, writing (C-51) in terms of the reflection coefficient in (C-54*a*) yields

$$\hat{E}_1(z) = \hat{E}_1^+ e^{-j\beta_1 z}[1 + \hat{\Gamma}_1(z)] \tag{C-59a}$$

$$\hat{H}_1(z) = \frac{\hat{E}_1^+}{\eta_1} e^{-j\beta_1 z}[1 - \hat{\Gamma}_1(z)] \tag{C-59b}$$

Taking the ratio of these gives the wave impedance in region 1:

$$\hat{Z}_1(z) = \eta_1 \frac{1 + \hat{\Gamma}_1(z)}{1 - \hat{\Gamma}_1(z)} \tag{C-60}$$

from which we obtain (C-58*a*). Similarly, (C-52) may be written as

$$\hat{E}_2(z) = \hat{E}_2^+ e^{-j\beta_2 z}[1 + \hat{\Gamma}_2(z)] \tag{C-61a}$$

$$\hat{H}_2(z) = \frac{\hat{E}_2^+}{\eta_2} e^{-j\beta_2 z}[1 - \hat{\Gamma}_2(z)] \tag{C-61b}$$

Taking the ratio gives the wave impedance in region 2:

$$\hat{Z}_2(z) = \eta_2 \frac{1 + \hat{\Gamma}_2(z)}{1 - \hat{\Gamma}_2(z)} \tag{C-62}$$

The wave impedances at two points in each region are related by (C-54):

$$\hat{\Gamma}_1(z) = \hat{\Gamma}_1(0)e^{j2\beta_1 z} \tag{C-63a}$$

$$\hat{\Gamma}_2(z) = \hat{\Gamma}_2(0)e^{j2\beta_2 z} \tag{C-63b}$$

Once the reflection coefficient is found, we may obtain the wave impedance via (C-60) or (C-62). The important point to note here is that *these are precisely the same relationships used to construct the Smith chart.* Thus, the Smith chart can be used to relate the reflection coefficients and consequently the wave impedances at two points *in the same medium.*

Although the reflection coefficients are not continuous across a boundary, the wave impedances are continuous. Thus, we may use this fact to reflect wave impedances through multiple-interface boundaries. For example, consider the case of a plane boundary separating two lossless regions, as shown in Fig. C.25*a*. An incident uniform plane wave and the resulting reflected uniform plane wave are normal to the boundary in region 1, and a transmitted wave is produced in region 2. The analogy to a lossless transmission line is shown in Fig. C.25*b*. The total phasor electric and phasor magnetic fields in region 1 are

$$\begin{aligned}\hat{E}_1(z) &= \hat{E}_i + \hat{E}_r \\ &= \hat{E}_1^+ e^{-j\beta_1 z} + \hat{E}_1^- e^{j\beta_1 z}\end{aligned} \tag{C-64a}$$

$$\begin{aligned}\hat{H}_1(z) &= \hat{H}_i + \hat{H}_r \\ &= \frac{\hat{E}_1^+}{\eta_1} e^{-j\beta_1 z} - \frac{\hat{E}_1^-}{\eta_1} e^{j\beta_1 z}\end{aligned} \tag{C-64b}$$

where η_1 and β_1 are defined for region 1:

$$\begin{aligned}\eta_1 &= \sqrt{\frac{\mu_1}{\epsilon_1}} \\ &= \sqrt{\frac{\mu_{r1}}{\epsilon_{r1}}}\,\eta_0\end{aligned} \tag{C-65a}$$

$$\begin{aligned}\beta_1 &= \omega\sqrt{\mu_1\epsilon_1} \\ &= \frac{\omega}{u_0}\sqrt{\mu_{r1}\epsilon_{r1}}\end{aligned} \tag{C-65b}$$

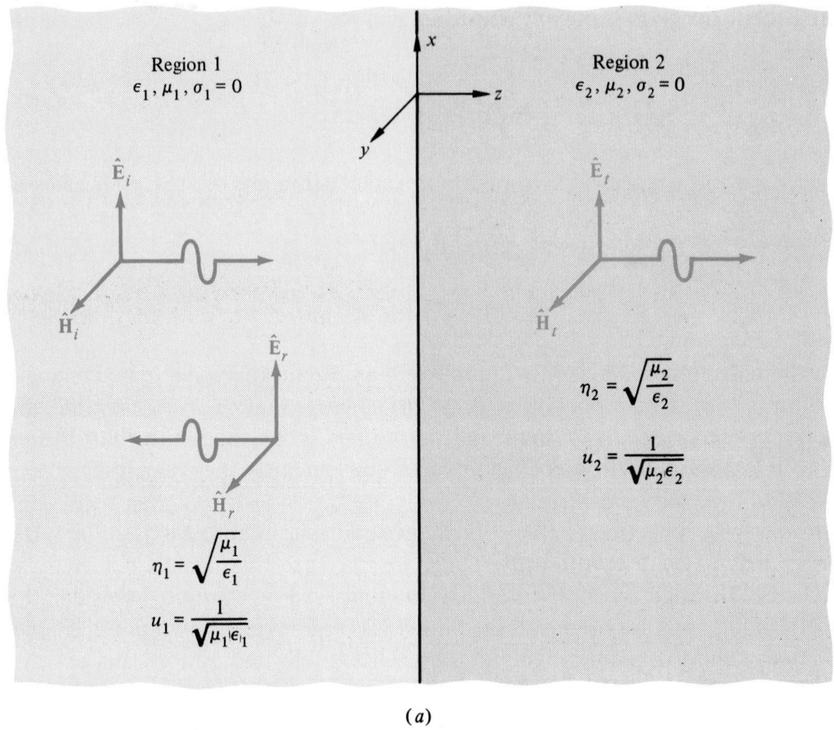

(*a*)

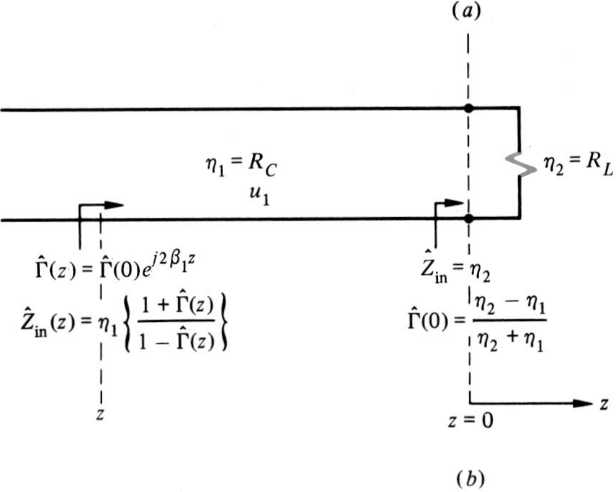

(*b*)

FIGURE C.25
Illustration of the analogy between uniform plane waves incident on a boundary and a terminated transmission line.

Since only a forward-traveling (transmitted) wave is present in region 2, the wave impedance at all points in that region is simply the intrinsic impedance of that medium:

$$\hat{Z}_2(z) = \frac{\hat{E}_t}{\hat{H}_t} = \eta_2 \tag{C-66}$$

By continuity of the wave impedances,

$$\hat{Z}_1(0) = \hat{Z}_2(0) = \eta_2 \tag{C-67}$$

The reflection coefficient can then be obtained in region 1 at the interface as

$$\hat{\Gamma}_1(0) = \frac{\hat{Z}_1(0) - \eta_1}{\hat{Z}_1(0) + \eta_1} = \frac{\eta_2 - \eta_1}{\eta_2 + \eta_1} \tag{C-68}$$

in a fashion analogous to that for a terminated transmission line. The Smith chart can then be used to give the wave impedance at any other point in region 1.

Note that the above results are directly analogous to the results for transmission lines:

$$\begin{aligned} \eta_2 &\to R_L \\ \eta_1 &\to R_C \\ \hat{E}(z) &\to \hat{V}(z) \\ \hat{H}(z) &\to \hat{I}(z) \\ \hat{Z}(z) &\to \hat{Z}_{\text{in}}(z) \end{aligned} \tag{C-69}$$

where R_L is the load resistance on the transmission line, R_C is the characteristic resistance of the line, $\hat{V}(z)$ and $\hat{I}(z)$ are the phasor line voltage and current respectively, and $\hat{Z}_{\text{in}}(z)$ is the input impedance to the line at z. The standing wave ratio is similarly defined as the ratio of the maximum value of the total electric field to its minimum value.

EXAMPLE C.10 We want to construct a radome to house a radar antenna that operates at 15 GHz. We will assume that the radius of the radome is sufficiently large that its interior surface is in the far field of the radar antenna and thus the waves are (locally) uniform plane waves at the radome surface. If the radome is to be constructed of a lossless dielectric with $\epsilon_r = 2.5$ and $\mu_r = 1$, determine the required thickness so that the radome appears transparent to the radar signal: that is, so that no reflections occur at the interior surface of the radome.

Solution The problem is shown in Fig. C.26. We wish to find the thickness of the radome, d, such that the wave impedance in the region containing the radar transmitter is η_0. This may also be thought of as the "input impedance" at the interface by analogy to transmission-line problems. The intrinsic impedance of the radome is

$$\eta = \sqrt{\frac{\mu_r}{\epsilon_r}}\,\eta_0$$
$$= 0.632\eta_0$$

The intrinsic impedance of the free-space region external to the radome, η_0, can be considered to be the "load" for the "transmission line" consisting of the radome section. The normalized load impedance for the radome region is

$$\hat{z}_L = \frac{\eta_0}{\eta}$$
$$= \sqrt{2.5} = 1.58$$

which is located on the Smith chart at 0.25λ TG. It is clear that in order that the input impedance to the radome surface be η_0, we must move "toward the

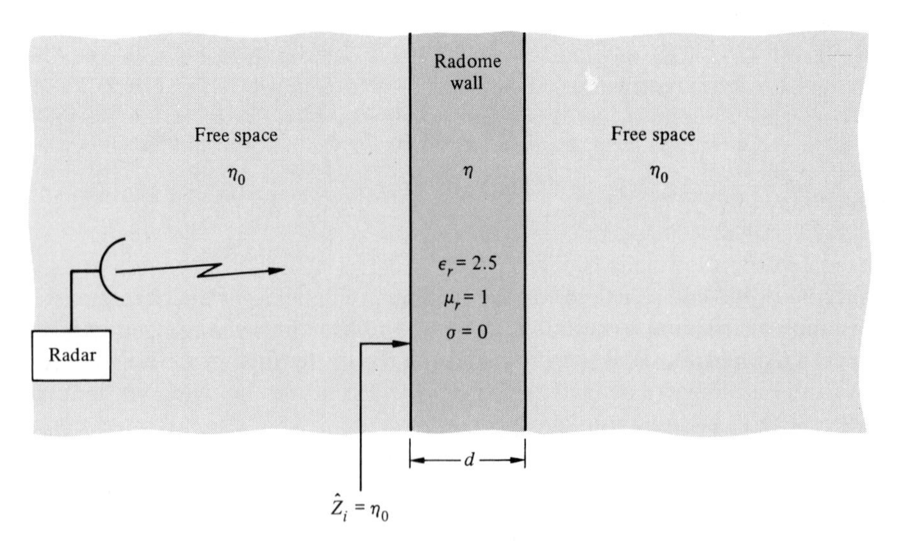

FIGURE C.26
Example C.10. Design of a radome to minimize reflections.

generator" such that $\hat{z}_{in} = \hat{z}_L$. Clearly this requires one full rotation around the chart, or $d = 0.5\lambda$. The radome thickness should thus be $\lambda/2$, as should have been obvious, since line impedances replicate for line lengths that are a multiple of a half-wavelength. To determine the required physical thickness of the dielectric, we must compute the wavelength in the dielectric:

$$\lambda = \frac{u}{f}$$

But $u = u_0/\sqrt{\epsilon_r \mu_r} = 1.897 \times 10^8$, so

$$\lambda = 1.265 \text{ cm}$$

Thus

$$d = 0.5\lambda = 6.325 \text{ mm}$$

The result that the required thickness be precisely $\lambda/2$ for Example C.10 should have been obvious in the first place. The utility of these results, however, is considerable when multiple dielectric regions are considered. Determining the required thickness of a region to achieve some general input impedance is, in general, quite difficult from the input impedance expression in (C-60). With the Smith chart, the problem is almost trivial.

Consider the radome problem in Example C.10 shown in Fig. C.27. For this problem, let us determine all the fields without requiring matching at the first interface. Incident and reflected waves are present to the left of the dielectric, whereas only a transmitted wave is present to the right of the dielectric. Within the dielectric, forward- and backward-traveling waves are present. The problem consists of determining the reflected and transmitted fields, given the incident field. It would be possible to solve this problem without using the Smith chart, but the computational details would be more involved.

The total electric fields in each region are

$$\hat{\mathbf{E}}_1(z) = \hat{E}_i e^{-j\beta_o z}[1 + \hat{\Gamma}_1(z)]\mathbf{a}_x \tag{C-70a}$$

$$\hat{\mathbf{E}}_2(z) = \hat{E}^+ e^{-j\beta z}[1 + \hat{\Gamma}_2(z)]\mathbf{a}_x \tag{C-70b}$$

$$\hat{\mathbf{E}}_3(z) = \hat{E}_t e^{-j\beta_o z}\mathbf{a}_x \tag{C-70c}$$

If we can determine the reflection coefficients at each interface, $\hat{\Gamma}_1(0)$ and $\Gamma_2(d)$, we may obtain the fields in the following manner. First, the total electric field in region 1 at the interface is $\hat{E}_i[1 + \hat{\Gamma}_1(0)]$. Transferring this across the boundary by continuity of the total tangential electric field gives

$$\hat{E}_i[1 + \hat{\Gamma}_1(0)] = \hat{E}^+[1 + \hat{\Gamma}_2(0)] \tag{C-71}$$

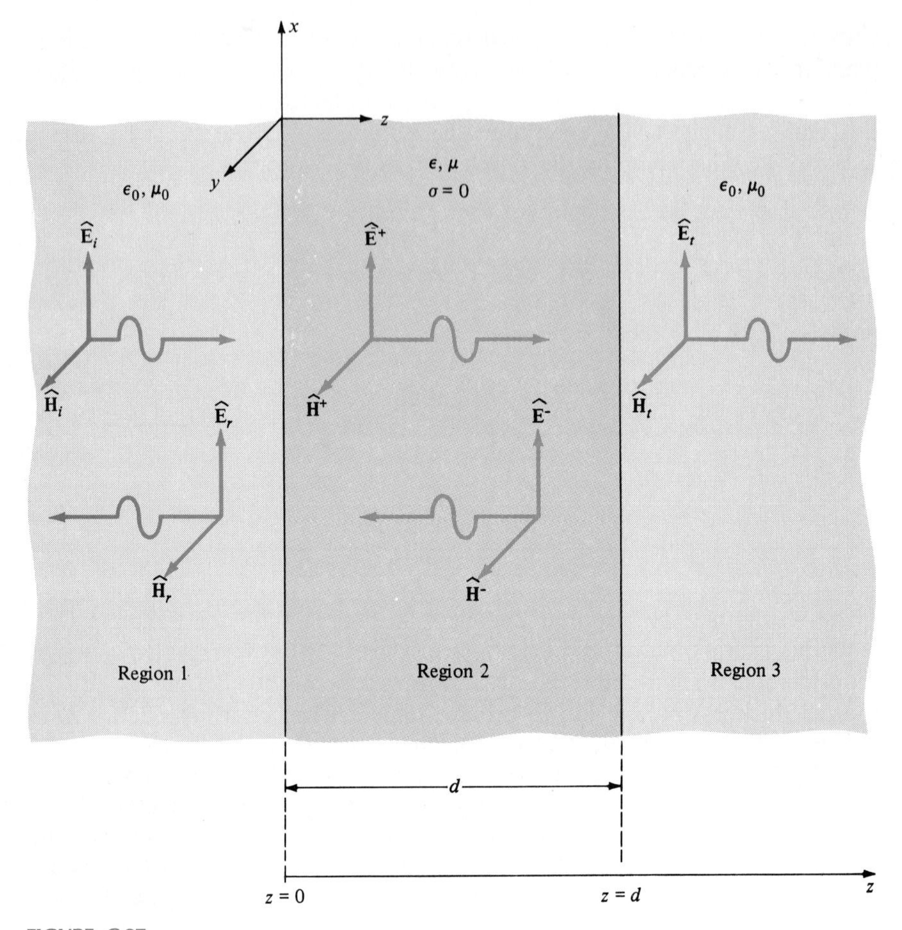

FIGURE C.27
Illustration of the use of the Smith chart in multiregion problems.

from which we obtain

$$\hat{E}^+ = \frac{[1 + \hat{\Gamma}_1(0)]}{[1 + \hat{\Gamma}_2(0)]}\hat{E}_i \tag{C-72}$$

The total electric field in the second region at the second interface is

$$\begin{aligned}\hat{E}_2(d) &= \hat{E}^+ e^{-j\beta d}[1 + \hat{\Gamma}_2(d)] \\ &= \frac{1 + \hat{\Gamma}_2(d)}{1 + \hat{\Gamma}_2(0)}[1 + \hat{\Gamma}_1(0)]e^{-j\beta d}\hat{E}_i \end{aligned} \tag{C-73}$$

by using (C-72). By continuity of the tangential electric field at the second interface, we obtain

$$\hat{E}_t = \hat{E}_2(d) \tag{C-74}$$

Thus, we only need to determine the reflection coefficients on either side of each interface. These reflection coefficients may be found by using the Smith chart.

EXAMPLE C.11 Consider a radome constructed of Teflon ($\epsilon_r = 2.1$, and $\mu_r = 1$) and an incident 10-GHz radar signal of amplitude 1000 V/m. If the radome thickness is 7 mm, determine the percentage of incident power transmitted through the radome.

Solution First we determine the thickness of the Teflon in wavelengths. At 10 GHz,

$$\begin{aligned}\lambda_{\text{Teflon}} &= \frac{u}{f} \\ &= \frac{3 \times 10^8}{\sqrt{2.1} \times 10^{10}} \\ &= 20.7 \text{ mm}\end{aligned}$$

so that 7 mm is 0.338λ. The wave impedance in the third region is simply η_0, as is the wave impedance in Teflon at the second interface. Therefore, the intrinsic impedance of free space normalized so that of Teflon gives the normalized wave impedance in Teflon at the second interface:

$$\begin{aligned}\hat{z}_2(d) &= \frac{\eta_0}{\eta_{\text{Teflon}}} \\ &= \sqrt{2.1} \\ &= 1.45\end{aligned}$$

Plotting this on the Smith chart and rotating 0.338λ on the TG scale gives the normalized wave impedance in the Teflon to the right of the first interface:

$$\hat{z}_2(0) = 0.81 + j0.27$$

The unnormalized wave impedance is

$$\begin{aligned}\hat{Z}_2(0) &= \hat{z}_2(0)\eta_{\text{Teflon}} \\ &= 211 + j70.2 \quad \Omega\end{aligned}$$

Thus, the wave impedance in the first region at the first interface is

$$\begin{aligned}\hat{Z}_1(0) &= \hat{Z}_2(0) \\ &= 211 + j70.2\end{aligned}$$

Therefore,

$$\hat{\Gamma}_1(0) = \frac{\hat{Z}_1(0) - \eta_0}{\hat{Z}_1(0) + \eta_0}$$

$$= 0.304\underline{/150^\circ}$$

The reflected electric field is therefore

$$\hat{E}_r = 304\underline{/150^\circ} \text{ V/m}$$

The average power in the incident wave is proportional to $|\hat{E}_i|^2$, whereas the average power in the reflected wave is proportional to $|\hat{E}_r|^2$. Thus, the average power in the transmitted wave is proportional to $|\hat{E}_i|^2 - |\hat{E}_r|^2$ because the Teflon is lossless. Therefore, the ratio of the transmitted to the incident power is

$$\frac{P_t}{P_i} = 1 - |\Gamma_1(0)|^2$$

$$= 90.7\,\%$$

In order to check this, we will compute the transmitted power directly. The transmitted electric field is

$$\hat{E}_t = \frac{[1 + \hat{\Gamma}_2(d)][1 + \hat{\Gamma}_1(0)]}{[1 + \hat{\Gamma}_2(0)]} e^{-j\beta d}\hat{E}_i$$

We have already computed $\hat{\Gamma}_1(0)$. In order to compute $\hat{\Gamma}_2(0)$, we observe that

$$\hat{z}_2(0) = \frac{1 + \hat{\Gamma}_2(0)}{1 - \hat{\Gamma}_2(0)}$$

and that $\hat{z}_2(0) = 0.81 + j0.27$. Solving for $\hat{\Gamma}_2(0)$ yields

$$\hat{\Gamma}_2(0) = \frac{\hat{z}_2(0) - 1}{\hat{z}_2(0) + 1}$$

$$= 0.18\underline{/117^\circ}$$

Similarly,

$$\hat{\Gamma}_2(d) = \hat{\Gamma}_2(0)e^{j2\beta_2 d}$$

$$= 0.18\underline{/0^\circ}$$

since $2\beta_2 d = 243^\circ$. Substituting gives

$$\hat{E}_t = 0.952\underline{/-120^\circ}\, \hat{E}_i$$

Therefore, the transmitted power (through a 1-m² area of each surface) is

$$P_t = \frac{1}{2}\frac{|\hat{E}_t|^2}{\eta_0}$$

and

$$P_i = \frac{1}{2}\frac{|\hat{E}_i|^2}{\eta_0}$$

so that

$$\frac{P_t}{P_i} = 90.7\%$$

as before.

C.6 Use of the Smith Chart for Lossy Lines

The voltage and current relationships for a lossy line are given from the results of Sec. 7.3 by

$$\hat{V}(z) = \hat{V}_m^+ e^{-\alpha z} e^{-j\beta z}[1 + \hat{\Gamma}(z)] \tag{C-75a}$$

$$\hat{I}(z) = \frac{\hat{V}_m^+}{\hat{Z}_C} e^{-\alpha z} e^{-j\beta z}[1 - \hat{\Gamma}(z)] \tag{C-75b}$$

where the reflection coefficient is given by

$$\hat{\Gamma}(z) = \hat{\Gamma}_L e^{2\alpha(z-\mathscr{L})} e^{j2\beta(z-\mathscr{L})} \tag{C-76}$$

Thus, the input impedance to a line of length d is given by

$$\hat{Z}_{\text{in}} = \hat{Z}_C \frac{1 + \hat{\Gamma}_L e^{-2\alpha d} e^{-j2\beta d}}{1 - \hat{\Gamma}_L e^{-2\alpha d} e^{-j2\beta d}} \tag{C-77}$$

Normalizing this result gives

$$\hat{z}_{\text{in}} = \frac{\hat{Z}_{\text{in}}}{\hat{Z}_C}$$

$$= \frac{1 + (\hat{\Gamma}_L e^{-2\alpha d}) e^{-2\alpha d}}{1 - (\hat{\Gamma}_L e^{-2\alpha d}) e^{-2\alpha d}} \tag{C-78}$$

Comparing this to the case of a lossless line we see that the only change in our procedure is to alter the *length* of the reflection coefficient vector by the factor $e^{-2\alpha d}$:

$$|\hat{\Gamma}_d| = |\hat{\Gamma}_L| e^{-2\alpha d} \tag{C-79}$$

Therefore we plot the normalized load impedance, $\hat{z}_L$, and rotate by an angle $\underline{/-2\beta d}$. As we do so, we "spiral inward" by the factor $e^{-2\alpha d}$ to obtain the normalized input impedance. If we find the normalized load impedance from the normalized input impedance we "spiral outward" by the factor $e^{2\alpha d}$.

EXAMPLE C.12 Consider an RG-58U coaxial cable which has $\hat{Z}_C = 53.5\,\Omega$ (approximately real) and an attenuation of 4.5 dB per 100 feet at 100 MHz. The interior dielectric is polyethylene ($\epsilon_r = 2.25$). If a 100 foot section is driven by a 100 MHz source and terminated in an impedance of $\hat{Z}_L = 100 + j150\,\Omega$, determine the input impedance.

Solution The attenuation constant is

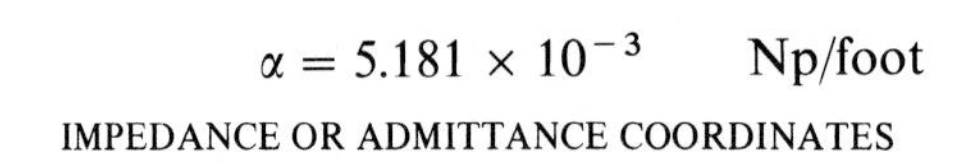

$$\alpha = 5.181 \times 10^{-3} \qquad \text{Np/foot}$$

IMPEDANCE OR ADMITTANCE COORDINATES

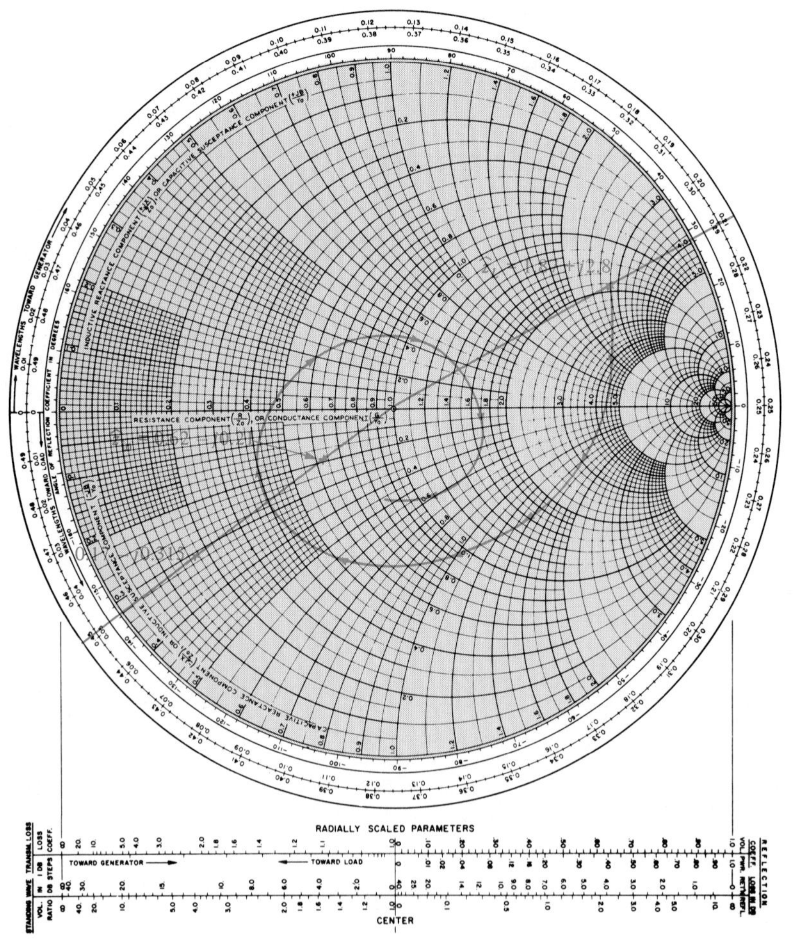

FIGURE C.28

Example C.12. Use of the Smith chart for lossy transmission lines.

Thus

$$e^{-2\alpha d} = 0.355$$

The velocity of propagation is, assuming low loss,

$$u \simeq \frac{u_o}{\sqrt{\epsilon_r}}$$
$$= 2 \times 10^8 \text{ m/s}$$

Thus, the line length (100 ft = 30.48 m) in wavelengths at 100 MHz is

$$\mathscr{L} = 15.24\lambda$$

or equivalently 0.24λ. The normalized load impedance is

$$\hat{z}_L = 1.87 + j2.804$$

located at 0.21λ on the TG scale, as shown in Fig. C.28. Rotating clockwise 0.24λ (or 15.24λ) to 0.45λ on the TG scale results in a value of $0.17 - j0.315$. Scaling this distance from the origin by the factor $e^{-2\alpha d} = 0.355$ results in

$$\hat{z}_{\text{in}} = 0.62 - j0.21$$

Thus, the input impedance is

$$\hat{Z}_{\text{in}} = 33.17 - j11.235$$
$$= 35.021\underline{/-18.7^\circ}$$

The exact value calculated with the results of Sec. 7.3 is

$$\hat{Z}_{\text{in}} = 35.21\underline{/-17.984^\circ}$$
$$= 33.49 - j10.87$$

Problems

C-1 Determine the electrical length of the following transmission lines (i.e., determine the actual length as a fraction of a wavelength):

(a) $f = 3$ MHz; $\mathscr{L} = 130$ m; $\mu_r = 1$; $\epsilon_r = 1$
(b) $f = 2.5$ GHz; $\mathscr{L} = 5$ cm; $\mu_r = 1$; $\epsilon_r = 2.5$
(c) $f = 480$ MHz; $\mathscr{L} = 0.5$ cm; $l = 5$ μH/m; $c = 20$ pF/m

C-2 Determine the electrical length of the lines in Prob. C-1, $\beta\mathscr{L}$, in degrees.

C-3 Determine the input impedance, VSWR, and voltage reflection coefficient at the load for the following transmission lines:

(a) $\hat{Z}_L = (25 - j50)\ \Omega$; $R_C = 50\ \Omega$; $\mathscr{L} = 0.4\lambda$

(b) $\hat{Z}_L = (100 + j50)\ \Omega$; $R_C = 100\ \Omega$; $\mathscr{L} = 1.3\lambda$
(c) $\hat{Z}_L = (100 - j80)\ \Omega$; $R_C = 75\ \Omega$; $\mathscr{L} = 0.6\lambda$
(d) $\hat{Z}_L = -j375\ \Omega$; $R_C = 300\ \Omega$; $\mathscr{L} = 0.8\lambda$

C-4 Determine the load impedance, VSWR, and load reflection coefficient for the following transmission lines:

(a) $\hat{Z}_{\text{in}} = (30 - j100)\ \Omega$; $R_C = 50\ \Omega$; $\mathscr{L} = 0.4\lambda$
(b) $\hat{Z}_{\text{in}} = 50\ \Omega$; $R_C = 75\ \Omega$; $\mathscr{L} = 1.3\lambda$
(c) $\hat{Z}_{\text{in}} = (150 + j230)\ \Omega$; $R_C = 100\ \Omega$; $\mathscr{L} = 0.6\lambda$
(d) $\hat{Z}_{\text{in}} = j250\ \Omega$; $R_C = 100\ \Omega$; $\mathscr{L} = 0.8\lambda$

C-5 Determine the shortest lengths of the following transmission lines, the VSWR, and $\hat{\Gamma}_L$:

(a) $\hat{Z}_{\text{in}} = -j20\ \Omega$; $\hat{Z}_L = j50\ \Omega$; $R_C = 100\ \Omega$
(b) $\hat{Z}_{\text{in}} = (50 - j200)\ \Omega$; $\hat{Z}_L = (60 - j45)\ \Omega$; $R_C = 100\ \Omega$
(c) $\hat{Z}_{\text{in}} = (30 + j50)\ \Omega$; $\hat{Z}_L = (200 + j200)\ \Omega$; $R_C = 100\ \Omega$
(d) $\hat{Z}_{\text{in}} = (135 + j0)\ \Omega$; $\hat{Z}_L = (60 - j37.5)\ \Omega$; $R_C = 75\ \Omega$

C-6 Using the Smith chart, show the following properties of lossless transmission lines:

(a) $|\hat{\Gamma}(z)| \leq 1$
(b) The input impedance replicates for distances separated by a multiple of a half-wavelength.
(c) The VSWR for a line having a purely reactive load is infinite, and the magnitude of the reflection coefficient is unity.
(d) The input impedance at any point on a line having a purely reactive load cannot have a real part.
(e) Adjacent points on a line where the input impedance is purely resistive are separated by a quarter-wavelength.
(f) The input impedance to a quarter-wavelength transmission line is $R_C^2/\hat{Z}_L$.

C-7 Determine the inverse of the following complex numbers:

(a) $\hat{C} = 100 + j30$
(b) $\hat{C} = 25 \times 10^{-3} - j50 \times 10^{-3}$
(c) $\hat{C} = 50 - j100$
(d) $\hat{C} = 5 - j10$

Check your results by direct computation.

C-8 A lossless 100-Ω transmission line is connected between a 50-Ω source and a load of $\hat{Z}_L = (100 + j70)\ \Omega$. Determine a length of this line such that maximum power is delivered from the source to the load. Can you always choose such a length of line for any load impedance? Can you always choose a length of line such that the input impedance has no imaginary part and how will this affect maximum power transfer?

C-9 We want to determine the value of an unknown impedance, $\hat{Z}_L$, attached to a length of transmission line having a characteristic resistance of 100 Ω. Removing the load yields an input impedance of $-j80\ \Omega$. With the unknown impedance attached, the input impedance is $(30 + j40)\ \Omega$. Determine the unknown impedance.

C-10 In a slotted-line experiment, with the load replaced with a short circuit, the voltage minima are located at 15 cm, 25 cm, 35 cm, etc. With the load attached, the voltage minima are located at 8 cm, 18 cm, 28 cm, etc., and the VSWR is 8. If the slotted line is air-filled and has a characteristic resistance of 100 Ω, determine the frequency of operation and the value of the load impedance.

C-11 Repeat Prob. C-10, except that instead of replacing the load with a short circuit, replace it with an open circuit. All other specifications are unchanged.

C-12 A lossless air-filled slotted line with $R_C = 100\ \Omega$ has an unknown load, $\hat{Z}_L$, that produces a VSWR of 4. When the load is removed, the voltage minima shift 2 cm closer to the load. If the frequency of operation is 1 GHz, determine the load impedance.

C-13 An air-filled 100-Ω parallel-wire line is driven by a 300-MHz source and terminated in a mismatched load of $(40 - j20)\ \Omega$. We want to transfer maximum power from the source to the load. In order to do this, we will place either an inductor or a capacitor across the line at a distance d from the load. Determine the shortest distance d at which to place this element; also determine the type of element and its value.

C-14 Repeat Prob. C-13 if the load impedance is changed to $(25 - j25)\ \Omega$ and the line characteristic resistance is changed to 50 Ω.

C-15 A section of 300-Ω lossless parallel-wire line that is air-insulated is used to connect a half-wave antenna to a source. The input impedance to the antenna is $(73 + j42.5)\ \Omega$. The open-circuit voltage of the source is 100 V at 375 MHz, and the internal source impedance is 300 Ω. Determine the length (shortest) and location from the antenna at which a 300-Ω parallel-wire short-circuited stub tuner is to be placed such that maximum power is transferred to the antenna. Determine the power delivered to the antenna and the VSWR on the portion of the line between the tuner and the antenna.

C-16 Repeat Prob. C-15 using an open-circuited stub tuner.

C-17 A somewhat less practical tuner than the parallel short-circuited stub tuner is the series short-circuited (or open-circuited) stub tuner, in which the stub is inserted in series with one wire of a two-wire line instead of in parallel with the line. Design a series short-circuited stub tuner for the specifications of Prob. C-15.

C-18 Redesign Prob. C-17 for a series open-circuited stub tuner.

C-19 A 100-Ω line is terminated in an impedance of $(100 + j100)\ \Omega$. Determine the shortest distance d from the load at which point the line voltage has the smallest magnitude.

C-20 A section of 100-Ω transmission line is used to connect a load of $(250 - j100)\ \Omega$ to a source that has a source resistance of 100 Ω. Determine the smallest distance from the load at which point a parallel short-circuited stub tuner may be placed to achieve maximum power transfer. Determine the required length of this parallel short-circuited stub tuner if the line used to construct it has a characteristic resistance of:

(a) $R_C = 100\ \Omega$
(b) $R_C = 50\ \Omega$

C-21 Design a double-stub tuner for use on a 300-Ω line having a load impedance of $(100 + j100)\ \Omega$. Place the second stub at a distance of $\lambda/8$ from the load.

C-22 Design a double-stub tuner to match a load of $(25 + j25)\ \Omega$ on a 100-Ω line. Choose a distance at which to place the second stub to be some integer multiple of $\lambda/16$.

C-23 Determine the range of load admittances that cannot be matched with a double-stub tuner when the second stub is placed at a distance of:

(a) $\lambda/16$
(b) $\lambda/8$
(c) $3\lambda/16$
(d) $\lambda/4$
(e) $3\lambda/8$

C-24 A half-wave dipole antenna having a measured input impedance of 73 Ω (approximately real) at 150 MHz is to be connected to a 150-MHz source with a section of 300-Ω parallel-wire line. A section of parallel-wire air-insulated line having a characteristic resistance R_C is to be inserted between the input terminals of the antenna and the 300-Ω line. Determine R_C and the required length of this line to obtain a match. If this matching line is to be constructed using 20-gauge wire that has a radius of 16 mils, determine the required conductor separation of this matching line.

C-25 Design a 10-dB 300-Ω pad. Use a pi structure. Determine the VSWR on a 300-Ω line if this pad is placed at the end of the line and is unterminated.

C-26 Repeat Prob. C-25 using a T structure.

C-27 An antenna having an input impedance of $(73 + j42.5)\ \Omega$ is to be connected to a source. The connection line consists of a 50-Ω line of length 0.2λ in series with a 300-Ω line of 0.6λ, with the 300-Ω section connected to the antenna. Determine the impedance seen by the source.

C-28 Two antennas having identical input impedances of $(73 + j42.5)\ \Omega$ are connected to a source as shown in Fig. PC-28. Determine the input impedance $\hat{Z}_{in}$.

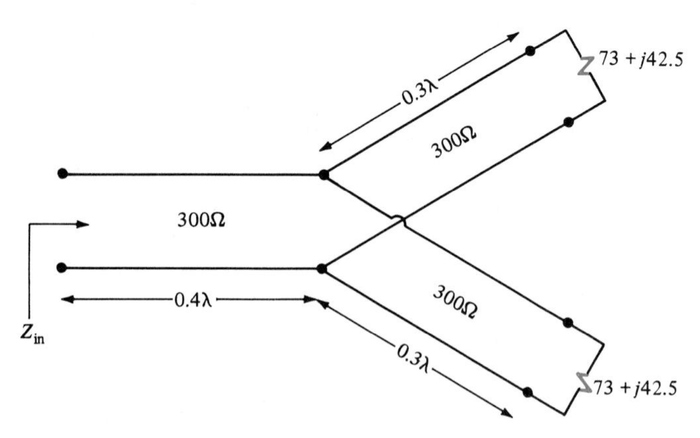

FIGURE PC.28

C-29 Consider the multiple-interface problem shown in Fig. C.27, in which region 2 is composed of a lossless material having $\epsilon_r = 9$ and $\mu_r = 4$. If the incident wave has an electric field intensity of 100 V/m and a frequency of 100 MHz and if region 2 is 20 cm thick, write complete time-domain expressions for the incident, reflected, and transmitted field vectors.

C-30 A low-loss coaxial cable has the following specifications at 400 MHz:

$$\hat{Z}_C \simeq 75 + j0 \quad \Omega$$

$$\alpha = 0.05 \text{ Np/m}$$

$$u = 0.667u_o$$

Determine the input impedance to a 11.175-m length of this cable at 400 MHz if the line is terminated in:

(a) a short circuit
(b) an open circuit
(c) a 300-Ω resistor

Compute the exact results using the results of Chap. 7.

C-31 A shielded room has walls of sheet steel ($\sigma = 0.58 \times 10^7$ S/m, and $\mu_r = 1000$). It is intended to protect sensitive electronic equipment from interfering outside sources. If a 30-MHz uniform plane wave is incident normal to one wall, determine the electric fields reflected at the wall surface and transmitted through the wall as a percentage of the incident electric field if the wall thickness is one skin depth.

ANSWERS

Chapter 2

2-1 $|\mathbf{T}| = 911.23$ lb; $|\mathbf{L}| = 3367.4$ lb.

2-2 (*a*) $5\mathbf{a}_x + 2\mathbf{a}_y - 5\mathbf{a}_z$; (*b*) 7.35; (*c*) $0.68\mathbf{a}_x + 0.27\mathbf{a}_y - 0.68\mathbf{a}_z$.

2-3 (*a*) 3.74; (*b*) 1.47.

2-4 (*a*) $3\mathbf{a}_x + 4\mathbf{a}_y - 3\mathbf{a}_z$.
(*b*) $-2\mathbf{a}_x + 2\mathbf{a}_y - 3\mathbf{a}_z$.
(*c*) $-\mathbf{a}_x + 8\mathbf{a}_y - 9\mathbf{a}_z$.
(*d*) 3.74.
(*e*) $0.41\mathbf{a}_x + 0.41\mathbf{a}_y - 0.82\mathbf{a}_z$.
(*f*) 7.
(*g*) 7.
(*h*) $-\mathbf{a}_x - 7\mathbf{a}_y - 4\mathbf{a}_z$.
(*i*) $\mathbf{a}_x + 7\mathbf{a}_y + 4\mathbf{a}_z$.
(*j*) -19.

2-5 (*a*) 0.8; (*b*) 109.1°; (*d*) $-0.115\mathbf{a}_x - 0.808\mathbf{a}_y - 0.577\mathbf{a}_z$.

2-6 (*a*) $-2\mathbf{a}_x + 5\mathbf{a}_y + \mathbf{a}_z$; (*b*) 4; (*c*) $4\mathbf{a}_x + 4\mathbf{a}_y - 4\mathbf{a}_z$;
(*d*) 40.9°; (*e*) 2; (*f*) $-2\mathbf{a}_x$.

2-7 (*b*) 10.8; (*d*) $-0.19\mathbf{a}_r + 0.9\mathbf{a}_\phi - 0.38\mathbf{a}_z$; (*f*) 57.56°;
(*g*) $-0.65\mathbf{a}_r + 0.18\mathbf{a}_\phi + 0.74\mathbf{a}_z$.

2-8 (*b*) -1.29.
(*c*) $-\dfrac{\pi^2}{6}\mathbf{a}_r - \dfrac{\pi}{2}\mathbf{a}_\theta - \dfrac{5\pi}{3}\mathbf{a}_\phi$.
(*d*) $0.69\mathbf{a}_r - 0.72\mathbf{a}_\theta$.
(*e*) 4.04.
(*f*) 102.74°.
(*g*) $-0.288\mathbf{a}_r - 0.275\mathbf{a}_\theta - 0.918\mathbf{a}_\phi$.

2-9 1.

2-10 $\alpha = -2, \beta = -2\pi$.
2-11 $\mathbf{a}_n = -0.82\mathbf{a}_x - 0.41\mathbf{a}_y + 0.41\mathbf{a}_z$.
2-14 1.57 m^3.
2-15 5.21 m^2.
2-17 $63.43°$.
2-18 (*b*) $0.45\mathbf{a}_x + 0.89\mathbf{a}_y$; (*c*) 22.36.
2-19 $3/\sqrt{6}$.
2-20 (*a*) 16 J; (*b*) -1 J; (*c*) 4 J.
2-21 (*a*) $-7/6$ J; (*b*) $-3/2$ J.
2-22 1.
2-23 0.
2-24 (*a*) -2; (*b*) -2.
2-26 (*a*) 1.
2-28 $3\pi/2$.
2-33 $[2/(1 + r)^2]\mathbf{a}_\phi = 2\mathbf{a}_\phi$, at $r = 0$; and $\frac{1}{2}\mathbf{a}_\phi$, at $r = 1$.
2-34 (*a*) 0; (*b*) -1.
2-35 π.
2-38 1/2.
2-39 (*a*) $-1/6$; (*b*) $-1/6$.
2-40 $\pi/2$.
2-41 0.

CHAPTER 3

3-1 $\pi/2$ C.
3-2 13 C.
3-3 $1.718\mathbf{a}_x + 0.859\mathbf{a}_y + 2.577\mathbf{a}_z$ N.
3-4 (*a*) 0; (*b*) $127.28\mathbf{a}_z$; (*c*) $114.15(\mathbf{a}_x + \mathbf{a}_y)$ N.
3-5 1.0 cm.
3-6 1.0×10^{-11} N.
3-7 $2.38\mathbf{a}_z$.
3-9 $(\rho_l/2\epsilon_0 R)\mathbf{a}_r$.
3-10 $(\rho_s/2\epsilon_0)\mathbf{a}_z$.
3-11 0; $(\rho_s/\epsilon_0)\mathbf{a}_y$.
3-12 0.
3-13 8310 V/m.
3-14 $a\rho_l/[2\pi\epsilon b(a - b)$.
3-15 $4R^3\rho_v/3r^3\epsilon$; $4r\rho_v/3\epsilon$.
3-17 $(\rho_s/\pi\epsilon_0 W) \tan^{-1}(W/2D)$.
3-18 2 μC.
3-19 $ak \cos ar/r^2$.
3-20 $2\pi ka^2$ C.
3-21 54633 V/m, at $z = 10$ m.
3-22 (*a*) $Q/6$ C; (*b*) 50 μC; (*c*) $50(\cos\theta_1 - \cos\theta_2)$ C.

3-24 (*a*) $\rho_s R^2 \mathbf{a}_r / \epsilon r^2$; (*c*) $\rho_s R^2/\epsilon r$; $\rho_s R/\epsilon$.
3-25 Q; 0.
3-28 $V_0 - \dfrac{1}{4}\dfrac{\rho}{\epsilon} r^2$.
3-29 $\dfrac{\rho_s}{2\epsilon}(\sqrt{a^2+h^2} - h)$ V.
3-30 -98875 V.
3-31 4500 V.
3-32 34 kV.
3-33 $2\pi\epsilon$.
3-35 $4\pi\epsilon ab/(b-a)$ F.
3-36 $-kR^2/2\epsilon_0$ V.
3-37 (*b*) 48 V; (*c*) $253.5\epsilon_0$ J/m^3
3-38 $\epsilon_0 \epsilon A/(\epsilon_0 + 2\epsilon)$ F.
3-39 3.8 μF; 2.5×10^5 V/m, at $r = 0.01$.
3-40 692.5 kV/m, at $r = 4$ cm.
519.4 kV/m, at $r = 2$ cm.
3-42 -14 V.
3-43 1.88×10^{-16} m.
3-44 7.84×10^{-19} C.
3-45 (*a*) 38.5 J.
(*b*) 0.16×10^{-9} F.
(*c*) 693 kV.
3-47 $\frac{8}{9}$ kV; $\frac{1}{9}$ kV.
3-48 $(\epsilon l/\alpha) \ln (r_2/r_1)$ F.
3-49 $(2\pi a^2 R^5)/15\epsilon_0$ J.
3-50 $-\frac{1}{2}\epsilon_0 A\alpha^3 e^{-\alpha r}$ C/m^3.
3-51 4.24×10^6 S/m.
3-52 $2\pi\sigma/\ln(b/a)$ S.
3-53 $4\pi\epsilon_0 a$ F
3-54 $Q^2 x/2\epsilon A$ N
3-55 271.48 kV

Chapter 4

4-1 $2.92 \times 10^{-6}\,(0.69\mathbf{a}_x - 0.226\mathbf{a}_y + 0.69\mathbf{a}_z)$ T.
4-2 $\dfrac{Ia^2}{2(h^2+a^2)^{3/2}}\mathbf{a}_z$ A/m.
4-3 $\mu_0 I/4R$ T.
4-4 $\dfrac{\pi}{120}\rho_s NR\,\mathbf{a}_z$ A/m.
4-5 $\dfrac{3\mu_0 I a^2 yz}{4(z^2+a^2)^{5/2}}$ T.

4-7 $\dfrac{I}{\pi a^2}\,\mathbf{a}_z$ A/m².

4-8 $\dfrac{\mu_0 I l}{2\pi} \ln \dfrac{r^2}{r_1}$ Wb.

4-9 (*b*) 3.9 percent; (*c*) 2.0 percent.

4-11 0.045×10^{-3} T.

4-13 12.6 mWb.

4-14 $\frac{1}{2} J r\,\mathbf{a}_\phi$ A/m.

4-16 (*a*) $\dfrac{\mu_0}{4} J(r_2^2 - r_1^2)$ A.

(*b*) $\dfrac{\mu_0}{4} b(b - 2x)$ A.

(*c*) $\dfrac{1}{2} \mu_0 J b$ T.

4-19 $\theta_2 = 2.86°$.

4-20 1.828 mWb.

4-21 10.76 A.

4-22 (*a*) 0.042 J; (*b*) 0.286 J.

4-23 5.66 mH.

4-25 948.7 A.

4-26 (*c*) 1.03 mH; 1.05 mH.

4-29 100 turns.

4-30 0.8 T.

4-32 $\psi_m^2(b - g)$ N.

4-33 $\dfrac{1}{4} k_1 I_m^2 \sin 2\delta$ Nm.

4-34 (*b*) 600 N.

Chapter 5

5-1 1.01 mV.

5-2 75.4 mV.

5-3 5.584 V.

5-4 −0.628 V.

5-6 $-100\,BL \cos t$ V.

5-7 $2 \times 10^{-4} \sin \omega t$ A.

5-8 $V_1 \neq V_2$.

5-9 (*a*) −0.1/3 V; (*b*) 0.2/3 V; (*c*) 0.1/3 V; (*d*) 0.2/3 V.

5-10 60 MHz.

5-11 1.39 mA.

5-13 $\alpha^2 + \beta^2 = \mu_0 \epsilon_0 \omega^2$.

5-14 $\omega = \beta/\sqrt{\mu_0 \epsilon_0}$ r/s.

5-17 1 N/m.

5-18 $-Blu$.

5-19 $\alpha \dfrac{\epsilon_1}{\epsilon_2} \mathbf{a}_x + \beta \mathbf{a}_y + \delta \mathbf{a}_z$ V/m.

5-20 $\alpha \mathbf{a}_x + \dfrac{\mu_2}{\mu_1} \beta \mathbf{a}_y + \dfrac{\mu_2}{\mu_1} \delta \mathbf{a}_z$ T.

5-21 $\alpha \mathbf{a}_x + \dfrac{\epsilon_2}{\epsilon_1} \beta \mathbf{a}_y + \dfrac{\epsilon_2}{\epsilon_1} \delta \mathbf{a}_z$ C/m^2.

5-22 $\dfrac{\mu_1}{\mu_2} \alpha \mathbf{a}_x + \beta \mathbf{a}_y + \delta \mathbf{a}_z$ A/m.

5-23 $\alpha_e = -\rho_s/\epsilon_0,\ \beta_e = 0,\ \gamma_e = 0.$
$\alpha_h = 0,\ \beta_h = \mu_0 \mathcal{K}_y,\ \gamma_h = \mu_0 \mathcal{K}_z.$

5-24 $\mathcal{B}_2 = \mu_0(\mathbf{a}_x - 0.6\mathbf{a}_y + 0.8\mathbf{a}_z).$

5-26 $\dfrac{10}{9}$ W.

5-27 $\dfrac{10}{9}$ W.

5-28 1.74×10^{-4} W.

5-33 (*a*) 36 Hz; (*b*) $4 \times 10^{-6} \cos 72\pi t$ A/m^2.

5-34 $\mathcal{E}_z = \mathcal{E}_{zm} (\cosh \gamma y)/(\cosh \gamma h).$

5-36 0.5 mH.

Chapter 6

6-4 $\mathcal{E} = 100 \cos(1.6\pi \times 10^9\, t - 25.13x)\mathbf{a}_z$ V/m.
$\mathcal{H} = -0.398 \cos(1.6\pi \times 10^9\, t - 25.13x)\mathbf{a}_x$ A/m

6-5 $\hat{\eta} = 195.12 + j152.34.$

6-7 (*a*) 1.89×10^8 m/s; (*b*) 3.77 cm; (*c*) 166.6 rad/m;
(*d*) $\mathcal{H} = 0.0422 \cos(10\pi \times 10^9 t - 166.67z)\mathbf{a}_y$ A/m.

6-8 62.83; 0.31; $251.33\ \underline{/45^\circ}$.

6-11 8.89 MHz.

6-13 $\epsilon_r = 4.0.$

6-14 (*a*) 73.3 m; (*d*) 2.3 m; (*f*) 23.2 cm.

6-15 1.1 W, at 7 GHz.

6-18 $\hat{\mathbf{E}} = 2Ee^{-j\beta z}\mathbf{a}_x.$

6-19 (*a*) Circular, left hand; (*b*) circular, right hand;
(*c*) elliptical, right hand.

6-20 5×10^5 m/s; 10^6 m/s; anomalous dispersion.

6-21 $1.5 \times 10^8\, n$ Hz.

6-22 $P_{\text{av}} = 51.97$ W.

6-23 $P_{\text{av}} = 59.8\ \mu$W.

6-24 $P_{\text{av}} = 904.78$ W.

6-25 $P_{\text{av}} = 0.1267\ \mu$W.

6-27 24

6-28 35.8 percent power lost.
6-30 0.9216.
6-31 28.3 ft.
6-32 0.1125 μm.
6-33 122 dB.
6-34 3 m.
6-35 150 MHz.
6-36 281 MHz.

Chapter 7

7-4 (*a*) 2×10^8 m/s; 50 Ω; (*b*) 2.0702 m/s; 96.61 Ω.
7-8 (*a*) 621.059 m; (*b*) 75 Ω.
7-9 (*a*) 80 Ω; (*b*) 262.7 Ω.
7-14 15 T.
7-17 (*a*) $0.933 \underline{/-297.5^\circ}$; (*b*) 28.85; (*c*) $6.42 + j82.1$; (*d*) $4.85 \underline{/64.39^\circ}$; (*f*) 11.164 mW.
7-20 $jR_c/2$.
7-22 2.86 m
7-23 7.14 m.
7-24 76.68 mW.
7-26 (*a*) 1.0817 m; (*b*) 0.3317 m.
7-27 $(130 + j85)$ Ω.
7-28 3.69 V.
7-31 $2 \cos(\omega t - 135^\circ)$A.
7-32 1.04 W.
7-33 (*a*) $892.56 \underline{/0^\circ}$ Ω; (*b*) $11.204 \underline{/0^\circ}$ Ω; (*c*) $20.969 \underline{/0^\circ}$ Ω.
7-34 1.5 Ω/m; 2.667×10^{-4} S/m; 358 nH/m; 63.7 pF/m; 156.1 mW; 28 mW.
7-35 $\alpha = 0.487$ Np/m; $\beta = 59.68$ r/m.
7-36 $1.16 \cos(\omega t + 373.37^\circ)$ A.
7-38 90.9 Ω; 10.6 dB/100 ft.

Chapter 8

8-2 TE_{01}, TE_{10}, TE_{11}, TM_{11}, TE_{20}, TE_{21}, TM_{21}, TE_{30}, TE_{31}, TM_{31}.
8-4 1.403×10^8 m/s.
8-5 5.525 mm.
8-6 4.604 mm.
8-9 (*a*) 6.562 GHz; (*b*) 19.75 GHz.
8-13 5×10^8 m/s; 3.33 cm; 60π rad/m; 628.3 Ω.
8-14 5 mm, at 30 GHz.
8-15 (*a*) 945 MHz; (*f*) 39.9 GHz.

8-16 4.595 mW.
8-18 1.466 cm.
8-19 15.4 GHz; 13.47 GHz.
8-22 0.253 dB/m; 0.1084 dB/m; 0.098 dB/m.
8-30 3.03 cm, 6654; 2.02 cm, 4437.

Chapter 9

9-3 $\dfrac{|\hat{E}_\theta|}{|\hat{E}_r|} = 0.479; \dfrac{|\hat{E}_\theta|}{|\hat{H}_\phi|} = 1724.6$, at $r = 10$ cm.
9-4 $8.773 \times 10^{-3}\ \Omega$; 438.65 mW.
9-5 0°; 1.33 mW/m^2; 13.3 μW/m^2.
9-8 0.8579; 460.23 at $r = 1$ m.
9-9 $3.8 \times 10^{-5}\ \Omega$; 1.9 mW.
9-11 60 mV/m; 159.15 μA/m; 182.5 mW.
9-12 23 W for $\lambda/4$ monopole.
9-13 43.1 W; 2.18.
9-18 19.17 W.
9-24 1.5, at $\theta = 90°$.
9-25 0.336 mW.
9-27 670 m.
9-28 92.14 dB.
9-29 36.81 W.
9-30 0.68 V/m.
9-31 0.461 mV/m.
9-32 $1.186 \angle{-4.306°}$.
9-33 6.6×10^{-8} W.

Chapter 10

10-4 $D_n = \rho_s \quad \text{C/m}^2$.
10-6 $C_1 \sin kx \sinh ky$.
$C_1 \sin kx \cosh ky$.
$C_1 \sinh kx \cos ky$.
10-10 $\displaystyle\sum_{n,\text{odd}} \frac{4V_0}{n\pi} e^{-(n\pi x/a)} \sin \frac{n\pi y}{a}$
10-11 (a) $C_1 I_0(kr) \sin(kz)$; (c) $C_1 J_0(kr) \sinh(kz)$
10-12 $\dfrac{C}{\sinh(l/a)} J_0\left(\dfrac{r}{a}\right) \sinh\left(\dfrac{z}{a}\right)$.
10-18 94, 92, 84, 80, 45.4, 61.8, 61.7 and 61 V on nodes *a*, *b*, *c*, *d*, *e*, *f*, *g* and *h*, respectively.

Appendix B

B-1 $\frac{1}{2} B\omega_m(r_0^2 - r_i^2)$ V.

B-2 $\frac{1}{2} BI\,(r_0^2 - r_i^2)$ Nm.

B-3 $1/\mu\sigma uc$.

B-6 $\pi\omega_m R^2 B \sin \omega_m t$ V.

Appendix C

C-1 (*a*) 1.3.

C-2 (*b*) 237.17°.

C-3 (*a*) $(182.5 - j70)\ \Omega$; 4.3; $0.62 \angle{-83^\circ}$.

C-4 (*a*) $(7.25 - j27.25)\ \Omega$; 9; $0.8 \angle{-122^\circ}$.

C-5 (*a*) $0.396\ \lambda$; ∞; $1 \angle{127^\circ}$.

C-8 $0.152\ \lambda$.

C-9 $(32 - j49)\ \Omega$.

C-10 $(36 - j132)\ \Omega$.

C-12 $(115 + j160)\ \Omega$.

C-13 $C = 5.305$ pF.

C-14 $L = 26.53$ nH.

C-15 $d = 32.24$ cm; $l = 7.36$ cm; $P_{\text{load}} = 4.167$ W; VSWR = 4.2.

C-17 $d = 23.84$; $l = 12.72$ cm.

C-19 $0.338\ \lambda$.

C-20 (*a*) $0.114\ \lambda$; (*b*) $0.167\ \lambda$.

C-21 $l_1 = 0.49\ \lambda$; $l_2 = 0.409\ \lambda$.

C-23 (*b*) Cannot match $\hat{y}_L$ with $g_L \geq 2$.

C-24 $R_C = 148\ \Omega$; $\mathscr{L} = 75$ cm.

C-25 1.22.

C-27 $(4.5 - j25)\ \Omega$.

C-28 $(690 - j450)\ \Omega$.

INDEX